Gamma-Ray Burst Science in 2030

Gamma-Ray Burst Science in 2030

Editors

Elena Moretti
Francesco Longo

MDPI • Basel • Beijing • Wuhan • Barcelona • Belgrade • Manchester • Tokyo • Cluj • Tianjin

MDPI

Editors

Elena Moretti
Institut de Fısica d'Altes Energies
The Barcelona Institute of Science and Technology
Bellaterra (Barcelona)
Spain

Francesco Longo
Department of Physics
University of Trieste
INFN, sezione di Trieste
Trieste
Italy

Editorial Office
MDPI
St. Alban-Anlage 66
4052 Basel, Switzerland

This is a reprint of articles from the Special Issue published online in the open access journal *Galaxies* (ISSN 2075-4434) (available at: www.mdpi.com/journal/galaxies/special_issues/Gamma-Ray_Burst).

For citation purposes, cite each article independently as indicated on the article page online and as indicated below:

LastName, A.A.; LastName, B.B.; LastName, C.C. Article Title. *Journal Name* **Year**, *Volume Number*, Page Range.

ISBN 978-3-0365-6899-7 (Hbk)
ISBN 978-3-0365-6898-0 (PDF)

Contents

About the Editors

Elena Moretti

Dr. Elena Moretti sudied particle physics at the University of Padova and successively obtained her Ph.D. in Physics at the University of Trieste with a thesis on prompt high-energy gamma emissions from GRBs. Her research followed this same line in the following years when she moved first to the Oskar Klein Centre in Stockholm and then to the Max Planck institute in Munich. Her interest in high-energy emissions from GRBs expanded to the search of very high-energy (VHE) emissions. Working with the observations of the Fermi-LAT and Fermi-GBM data, she started searching for a GRB emission in the VHE observations obtained with the MAGIC telescopes, optimising the tools and procedures for possible detection. In January 2019, Dr. Moretti was one of the first people to directly witness the detection of a GRB in the VHE band (>100GeV). This unprecedented result achieved with the MAGIC telescopes opened a new phase of study for these spectacular and powerful events.

Francesco Longo

Prof. Francesco Longo is an Associate professor in Experimental Physics at the Department of Physics of the University of Trieste. He is currently the coordinator of the Physics PhD program at the University of Trieste. He is a research associate at the Trieste section of the National Institute of Nuclear Physics (INFN). His scientific interests span from high-energy astrophysical phenomena in the solar system to Gamma Ray Bursts. He graduated in 1998 with a thesis on the possible correlation of GRB with Gravitational wave events and obtained his PhD degree in 2002 with a thesis on the search for Quantum Gravity using GRB data. He is currently involved in space and ground-based gamma ray experiments.

Preface to "Gamma-Ray Burst Science in 2030"

The study of Gamma Ray Bursts (GRBs) has been a very active field since their discovery, by the Vela satellites, about fifty years ago.

Recent years have witnessed exciting breakthroughs that further stimulated the interest of the scientific community in the GRB field. The recent identification of a spectral component beyond synchrotron emission in the afterglow phase increased the debate about particle acceleration and emission processes at place in a GRB. The ubiquity of the extra spectral component and the environmental conditions under which it is produced are still under debate. Moreover, whether this component is also present in the earlier prompt phase or not is another open question.

The recent advancements in the GRB field were achieved thanks to the multi-wavelength and multi-messenger approach. Being able to simultaneously collect data on a burst with different instruments is the key for a deep and complete understanding of the phenomena under study. Indeed, the multi-messenger and multi-wavelength observation of the neutron-star-neutron-star merging event GW 170817 (detected both electromagnetically and via gravitational waves) has already led to a step forward in our comprehension of GRB jet dynamics and morphology.

These observations of the environment, the emission processes and the jet dynamics are key tools to unveil the nature of progenitors of GRB of short and long durations and the connection of GRB to other astrophysical questions, such as the origin of the heavy elements or the high redshift universe stellar population.

As these discoveries increased the curiosity of the community, the next generation of instruments will be ready in the coming years, providing the tools for continuing explorations in this field. Future missions will provide a more complete picture of the phenomena by measuring the polarization of the bulk of prompt emissions and by detecting high-redshift GRBs.

The purpose of this reprint is to give an overview of the main challenges and problems that the GRB field is currently confronted with to propose some possible solutions and to explore the tools that the community will have in their hands in the near future. We are convinced that this Special Issue will become a reference that pictures the current state of the art of the GRB field and envisions it in a close future.

Elena Moretti and Francesco Longo
Editors

galaxies

MDPI

Review

The Structure of Gamma Ray Burst Jets

Om Sharan Salafia [1,2,3,*,†] and Giancarlo Ghirlanda [2,3,†]

1 Dipartimento di Fisica "G. Occhialini", Università Degli Studi di Milano-Bicocca, Piazza Della Scienza 3, I-20126 Milano, MI, Italy
2 INFN-Sezione di Milano-Bicocca, Piazza Della Scienza 3, I-20126 Milano, MI, Italy
3 INAF-Osservatorio Astronomico di Brera, Via E. Bianchi 46, I-23807 Merate, LC, Italy
* Correspondence: om.salafia@inaf.it
† These authors contributed equally to this work.

Abstract: Due to relativistic bulk motion, the structure and orientation of gamma-ray burst (GRB) jets have a fundamental role in determining how they appear. The recent discovery of the GW170817 binary neutron star merger and the associated GRB boosted the interest in the modeling and search for signatures of the presence of a (possibly quasi-universal) jet structure in long and short GRBs. In this review, following a pedagogical approach, we summarize the history of GRB jet structure research over the last two decades, from the inception of the idea of a universal jet structure to the current understanding of the complex processes that shape the structure, which involves the central engine that powers the jet and the interaction of the latter with the progenitor vestige. We put some emphasis on the observable imprints of jet structure on prompt and afterglow emission and on the luminosity function, favoring intuitive reasoning over technical explanations.

Keywords: gamma-ray burst: general; relativistic processes; magnetohydrodynamics

Citation: Salafia, O.S.; Ghirlanda, G. The Structure of Gamma Ray Burst Jets. *Galaxies* **2022**, *10*, 93. https://doi.org/10.3390/galaxies10050093

Academic Editors: Elena Moretti, Francesco Longo and Yosuke Mizuno

Received: 21 June 2022
Accepted: 23 August 2022
Published: 30 August 2022

1. Introduction

Jets, in the form of collimated outflows of plasma possibly endowed with magnetic fields, are ubiquitous in astrophysics. They typically extend over orders of magnitude in distance from their birthplace (from parsec scales in protostars to >kpc scales in galaxies hosting supermassive black holes), in redshift (with jet signatures being detected in association with the most distant galaxies known so far up to $z \sim 9$) and in luminosity, reaching the largest values in gamma-ray bursts (GRBs).

GRBs are luminous, extra-galactic transients powered by compact objects (black holes–BH, neutron stars–NS) produced by the core-collapse of a massive star or by the merger of a compact object binary (most likely NS-NS or NS-BH). In the most widely accepted scenario, the 'central engine' (that is, the system consisting of the compact object and possibly a surrounding accretion disk) launches a bipolar relativistic collimated outflow. Bulk energy dissipation in such an outflow produces a bright, highly variable, non-thermal 'prompt' emission in the X-ray/γ-ray band. The outflow deceleration by the external circum-burst medium produces the long-lasting multi-wavelength 'afterglow' emission extending from the γ-rays through the optical to the radio band.

The presence of *relativisic* outflows in GRBs is supported by some theoretical arguments and a few compelling observational pieces of evidence. The very fast prompt emission light curve variability requires the source to be very compact, but the observation of non-thermal prompt emission spectra extending above MeV photon energies indicates that the source is optically thin to pair production by photon–photon annihilation. This apparent contradiction can hardly be reconciled without invoking highly relativistic expansion, which eases the constraints by both decreasing the comoving photon energy by a Γ factor (the bulk expansion Lorentz factor) and increasing the source size limit imposed by variability by a Γ^2 factor (e.g., [1–5]). Even more directly, the apparent size increase of $\sim$0.3 pc in

$\sim$50 rest-frame days as measured for the first time in the nearby GRB 030329 [6] suggested an apparently superluminal expansion speed, indicating relativistic bulk motion [7]. *Collimation* of GRB outflows is required to reduce the otherwise huge γ-ray isotropic equivalent. This nomenclature refers to the energy and/or luminosity of a GRB computed assuming isotropic emission. Because of relativistic 'beaming' (i.e., aberration) of radiation, the vast majority of the observable photon flux comes from emitting regions moving within a tiny $1/\Gamma$ angle around the line of sight, making an isotropic outflow (and hence isotropic emission) essentially indistinguishable from one that expands radially within a $\theta_j \gtrsim 1/\Gamma$ collimation angle [8]. Energy reaching $E_{\gamma,\rm iso} \sim 10^{54-55}$ erg (e.g., GRB 990123 [9] and GRB 130427A [10]), which would require the mind-boggling conversion of 1–5 $M_\odot$ rest mass energy into γ-rays with 100% efficiency without invoking collimation. If the outflow is collimated within an angle θ_j, such an energy budget is reduced by a 'beaming' factor $f_b = (1 - \cos\theta_j) \sim \theta_j^2/2 \approx 0.004(\theta_j/5^\circ)^2$. The collimation angle θ_j is typically estimated from a steepening of the afterglow light curve around a few days after the initial gamma-ray burst, interpreted as the signature of the presence of a jet [8,11–15], see also Section 4.2. Such a feature, often referred to as a 'jet break', arises as the relativistic beaming angle $1/\Gamma$ (which increases during the afterglow phase due to the deceleration of the blastwave, i.e., the expanding shock produced as the jet expands within the external medium) becomes comparable to θ_j [8], allowing the observer to 'see' the jet borders. Typical collimation angles estimated from the observation of jet breaks range Opening angles as small as $\theta_j < 1^\circ$ have been reported in some studies, e.g., [16,17]. We caution that, while opening angles as small as these are not impossible in principle, these estimates are based on assumptions on the interstellar medium density and prompt emission efficiency, and they rely on the interpretation of an observed steepening in the afterglow light curve as a jet break. For the latter interpretation to hold, the steepening must be achromatic, i.e., it must show up independently of the observing band, but it is often impossible to verify it due to the absence of multi-wavelength observations at the relevant time. For these reasons, such estimates must be taken with a grain of salt. from $\theta_j \sim 4^\circ$ in 'long' GRBs [18] to $\theta_j \sim 16^\circ$ in 'short' GRBs [19].

For simplicity, the jets of GRBs have been long modeled as a conical outflow with a constant energy per unit solid angle $\mathrm{d}E/\mathrm{d}\Omega(\theta)$ and bulk Lorentz factor $\Gamma(\theta)$ within its aperture $\theta \leq \theta_j$ (here, θ is the angle from the jet symmetry axis). This basic model is typically referred to as the 'uniform', 'homogeneous' or 'top-hat' jet structure model. If the jet is observed within θ_j, the steepening in the afterglow light curve is used to infer the jet opening angle θ_j from where the true burst energy can be derived $E_\gamma \sim E_{\gamma,\rm iso}\theta_j^2/2$. It was found by [18,20,21] that E_γ is narrowly distributed around 10^{51} erg, suggesting a standard energy reservoir in GRBs. Within a 'top-hat' jet model, this implies that $E_{\gamma,\rm iso}$ scales as θ^{-2}. On the other hand, some authors [22–24] soon realized that the same observations could be explained assuming GRB jets possess a universal *structure* $\mathrm{d}E/\mathrm{d}\Omega \propto \theta^{-2}$ (see Section 2.1 for a precise definition).

The evolving interest in the structure of GRB jets can be seen in Figure 1, where we have collected from the NASA ADS all the papers mentioning "gamma-ray burst" and "structured jet" or "jet structure" in their abstracts. The red line is the cumulative distribution of the grey histogram and shows two clear "steps": an initial growing interest in structured jets corresponding to the 2000–2006 period and a recent "explosion" of interest prompted by the discovery and interpretation of the gamma-ray burst GRB 170817A associated to the first binary neutron merger gravitational wave event [25–27].

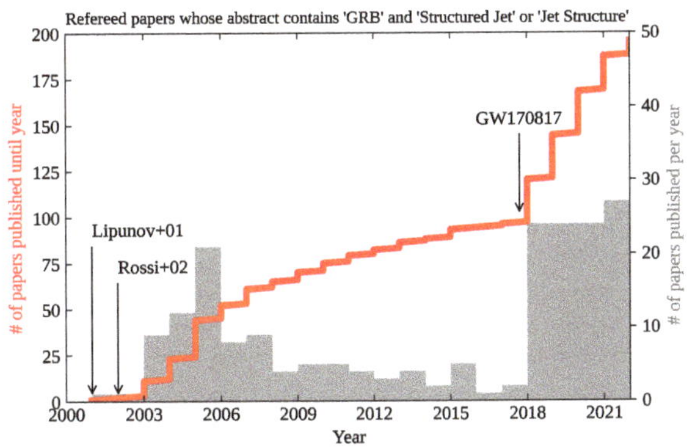

Figure 1. Timeline of scientific papers about GRB structured jets. The red solid line shows the cumulative number (shown on the left-hand axis) of refereed papers that contain "gamma-ray burst" and "structured jet" or "jet structure" in their abstract, according to the NASA ADS [28]. The grey histogram (number shown on the right-hand axis) shows the corresponding number of papers published per year. The dates of the two seminal papers [23,24] and of the GW170817 discovery [25–27] are annotated. We note that alternative nomenclatures with respect to the 'structured jet'/'jet structure' used here exist, hence the actual number of papers on the subject could be higher.

The initial interest in structured jets in the early 2000s was in part driven by attempts at explaining the diversity of GRB energetics within a unifying scenario where all jets share a universal structure. Two analytical functions were explored initially to describe the jet structure: a power law jet with $\mathrm{d}E/\mathrm{d}\Omega \propto \theta^{-2}$, as suggested by the E_γ clustering described above and supported by early analytical studies [29] and numerical simulations [30] of the jet emerging from its progenitor star envelope (see Section 2.2); a Gaussian jet with $\mathrm{d}E/\mathrm{d}\Omega \propto \exp(-(\theta/\theta_\mathrm{c})^2/2)$ [22,24,31,32], where most of the jet energy is contained within two times the 'core' opening angle θ_c, which is a more realistic representation of a nearly sharp-edged jet. Less continuous structures, such as one composed of two nested uniform jets (a narrow, fast and energetic jet surrounded by a wider, slower and weaker layer. Notably, a two-component jet structure has also been proposed to interpret jets in radio galaxies [33]) were considered, motivated by the possibility to explain the optical afterglow bumps observed in a few GRBs [20,34].

In the same period, many attempts were made at identifying, in the observational data then available, distinctive features of a structured jet. Modeling of the afterglow light curve of GRB030329 [35] suggested a structured jet as a viable interpretation of the low-frequency data, although alternative interpretations were not excluded. The sharpness of the light curve change across the jet break time, which in the structured jet scenario provides a measure of the viewing angle θ_v [24], depends on the jet structure and on the viewing angle, with sharper breaks corresponding to larger θ_v [36]. However, the jet lateral expansion also affects the shape of the light curve around the jet break time [37]. Attempts at testing the universal jet structure model [38–40] were mainly limited by the few events with measured redshifts and jet breaks [41]. Linear polarization measurements of the afterglow emission were also considered as diagnostics for the jet structure [42,43], despite the polarization depends also on the configuration of the magnetic field in the emission region [36,44,45]. Considerably different rates of GRB afterglows without a corresponding prompt emission detection (so-called orphan afterglows) are predicted in the case of a structured jet with respect to the conical uniform scenario [46–53].

Owing to the difficulties in identifying distinctive signatures in the available data of the structured jet scenario (see [54]), the community started to lose interest in it during the 2006–2017 period. The discovery of GRB 170817A [26,55,56] associated with the GW170817 gravitational wave source [25] suddenly changed everything (see also Section 5): after more than six months of monitoring of the puzzling non-thermal afterglow of GRB 170817A, a structured jet appeared as the only scenario able to provide a self-consistent interpretation of the shallow evolution of the afterglow light curves [57–65] and of the proper motion [66] and small size [67] of the very long baseline interferometry (VLBI) images of the source (see [68] for a review of the multi-messenger aspects of GW170817, and [69] for a more general review of electromagnetic counterparts of compact binary mergers).

Why is the structure important? The structure determines the properties of the emission for different observers, therefore determining in part the distribution of observable properties and the detectability of these sources. The jet structure carries information about the processes that shape it (the jet-launching mechanism and the interaction between the jet and the ambient medium surrounding the central engine) and is therefore an indirect probe of otherwise unobservable phenomena. Several works developing the concept of the jet structure, its origin, and how it determines the observed properties of GRBs appeared in the literature in the last five years. The presence of a jet with some structure appears unavoidable, considering the phases following the formation of the central engine and, therefore, a growing part of the community is starting to systematically consider GRB observations under this more realistic assumption when interpreting both their prompt and afterglow emission components. However, often the available observations are insufficient to allow for distinguishing between a structured jet from a less realistic assumption of a uniform jet. Most likely, the combination of several observables and the further development of numerical simulations will lead to constraining the structure of GRB jets in the near future.

The scope of this review is that of introducing the general definition of jet structure (Section 2.1) and present a very intuitive description of how the jet acquires its angular structure (Section 2.2). A very simplified overview of the mechanisms responsible for the jet launch (Section 2.3) and for its propagation up to where it can freely expand (Section 2.4) is provided. The possible signatures of the presence of a structured jet on the observed properties of the prompt GRB emission are presented in Section 3. The afterglow emission from a structured jet considering different possible structures and its dependence on the key jet structure parameters is summarized in Section 4. Finally, in Section 5, we briefly review the observations of the non-thermal electromagnetic counterparts of GW170817 and their interpretation in the structured jet scenario.

2. Origin of the Jet Structure

The mechanisms that shape the structure of a gamma-ray burst relativistic jet are complex and not entirely understood. In this section, we will summarize some general ideas about these processes in a pedagogical manner.

2.1. General Definition of Jet Structure

At a fixed time t, assuming axisymmetry, radial expansion and a relativistic equation of state $p = e_{\rm int}/3$ (where p is the pressure and $e_{\rm int}$ is the internal energy density, both measured in the comoving frame of the fluid) for simplicity, the jet structure can be described by four functions of the radial coordinate r and polar angle θ (measured from the jet axis), namely the modulus of the four-velocity $\Gamma\beta = u(r,\theta,t)$ (where Γ is the bulk Lorentz factor and $\beta = (1-\Gamma^{-2})^{1/2}$), the comoving rest-mass density $\rho'(r,\theta,t)$, the dimensionless enthalpy $h(r,\theta,t) = 1 + 4e_{\rm int}/3\rho' c^2$ and the magnetization $B^2/4\pi\rho' c^2 = \sigma(r,\theta,t)$ (where B is the comoving magnetic field strength, assumed transverse with respect to the expansion). Often, it is possible to limit the discussion to cold ($h \sim 1$) and highly relativistic ($u \sim \Gamma$) parts of the fluid, in which case the rest-mass density, magnetization and Lorentz factor $\Gamma(r,\theta,t)$ are sufficient. If the radial structure is unimportant and the focus is on kinetic energy, then the description of the jet structure can be accomplished by two angular functions: the kinetic energy per unit solid angle. If the time t at which the expression is valuated is such that the outflow is still in an acceleration phase, the appropriate Lorentz factor here is the "terminal" one, i.e., the one that can be estimated assuming the available internal (and/or magnetic) energy will be eventually converted to kinetic energy.

$$\frac{\mathrm{d}E}{\mathrm{d}\Omega}(\theta,t) = \int_0^\infty (\Gamma(r,\theta,t)-1)\Gamma(r,\theta,t)\rho'(r,\theta,t)c^2r^2\mathrm{d}r \tag{1}$$

and the average Lorentz factor

$$\Gamma(\theta,t) = \left(\frac{\mathrm{d}E}{\mathrm{d}\Omega}\right)^{-1}\int_0^\infty (\Gamma(r,\theta,t)-1)\Gamma^2(r,\theta,t)\rho'(r,\theta,t)c^2r^2\mathrm{d}r, \tag{2}$$

The latter description, applied to the "coasting" phase (see below), is the most widely adopted one, as it is sufficient for a basic description of the link between the prompt and afterglow emission observables and the jet structure in many contexts. The purpose of the above definitions is to clarify the connection between the three-dimensional physical properties of the outflow and the functions that are customarily used to describe its angular structure. A variety of similar, but not identical, definitions can be found in the literature: the essence of the arguments presented here does not depend on the precise definition.

2.2. Stages in the Life of a Relativistic Jet

In order to understand the structure of gamma-ray burst relativistic jets, we need to have a global view of how a jet is formed and how it evolves throughout its life. For that purpose, let us briefly summarize the main stages of the jet evolution with reference to the scheme shown in Figure 2:

- **Birth–jet launch:** the jet is launched by the central engine. Different mechanisms have been considered to power the jet, depending on the nature of the central engine (e.g., an accreting BH [70] or a magnetar [71]). Further details are provided in Section 2.3;
- **Infancy–jet head formation:** the jet material expands within the low-density funnel where it formed until it collides with the dense ambient that surrounds the central engine (the progenitor star envelope or the merger ejecta), which we term 'the progenitor vestige'. A forward-reverse shock structure forms—the jet 'head' [72]—where the jet momentum flux is counterbalanced by the ram pressure of the vestige material (as seen in the headrest frame—see Section 2.4);
- **Childhood–jet propagation through the progenitor vestige:** the jet head, which is sustained by fresh jet material flowing across the reverse shock, propagates through the progenitor vestige [30,73–75]. Due to the absence of lateral confinement, as soon as the head has slowed down enough as to become causally connected in the transverse direction, shocked material (both from the vestige and from the jet) is cast aside to form a hot, over-pressured cocoon that shrouds the jet and slowly expands laterally. The cocoon pressure is typically sufficient [75] to balance the lateral momentum flux of the jet material that flows from the central engine, leading to the formation of an oblique shock—the 're-collimation' or 're-confinement' shock [76]–where the lateral component of the jet momentum is dissipated, turning the flow from radial into cylindrical. The jet is therefore collimated by its own cocoon (Section 2.4);
- **Adolescence–breakout:** the jet head reaches the steep density gradient that marks the outer edge of the progenitor vestige. The head forward shock thus accelerates [77,78], the reverse shock disappears, and the jet and cocoon material starts flowing freely out of the open channel: this process is broadly referred to as the 'jet breakout'. During this process, the forward shock transitions from an optically thick region (where photon pressure dominates and the shock is therefore radiation-mediated) to an optically thin region: during this transition, photons from the hot downstream are released producing the 'shock breakout' emission [79], which represents the first observable electromagnetic emission in the jet's life. Childhood and adolescence are further described in Section 2.4;

- **Adulthood–free expansion**: the flow of fresh jet material from the central engine stops or diminishes significantly, setting a finite radial extent of the resulting outflow, which is now better described as an inhomogeneous shell [80] that expands radially away from the progenitor at relativistic speed. After the jet breakout, the vast majority of the shell is still optically thick to Compton scattering [3,5,81,82] (both off electrons associated with baryons in the outflow and potentially off pairs that can form within the outflow as a consequence of energy dissipation events) for another few orders of magnitude in radius. Initially, in this expansion phase, radial density gradients remain frozen ('coasting phase', [80,83]) until radial pressure waves have the time to cross the outflow, leading to a radial spreading phase. During the free expansion phase, radial inhomogeneities in the bulk Lorentz factor can lead to the development of internal shocks [84], which have long been considered one of the main candidate mechanisms for the dissipation and subsequent radiation of the outflow's energy. Internal-shock-induced turbulence [85] has also been proposed as a possible triggering mechanism for magnetic reconnection (see also, e.g., [86,87]), which represents the other leading scenario for the dissipation of the outflow's energy in this phase;
- **Seniority–external shock:** the shell expands into the external low-density medium that surrounds the progenitor, which can be just the interstellar medium (ISM) or a stellar bubble inflated by the progenitor's stellar wind [88]. As soon as the shell has swept a sufficient amount of external medium, corresponding to a rest mass energy equal to the shell's kinetic energy divided by the square of its bulk Lorentz factor [80,89], the expansion starts to be affected: a forward-reverse shock structure forms, with the reverse shock quickly crossing the entire shell [90,91], initiating the deceleration of the latter and the transfer of its energy to the forward-shocked external medium. Soon after the start of the deceleration, the forward shock settles into a self-similar expansion phase [83,92], erasing any memory of the details of the shell radial structure. The angular structure remains unaffected as most of the shocked shell is out of causal contact in the transverse direction;
- **Senility–lateral spreading and transition to non-relativistic expansion:** as soon as the transverse sound crossing time scale becomes shorter than the dynamical expansion time scale (or, in other terms, the angular size of causally connected regions starts to exceed the reciprocal of the local bulk Lorentz factor), pressure waves start to level out angular inhomogeneities, initiating a lateral expansion phase [11,32,43,89,93–95] which increases the shock working surface, therefore increasing the shock deceleration rate. The shock soon transitions to a non-relativistic expansion phase, slowly converging towards the Sedov–Taylor spherical blastwave behavior.

The above brief account should help in making clear that the jet structure evolves throughout the life of a GRB jet. Three crucial phases can be identified in such an evolution: (1) the jet *launch* where the initial energy breakdown (internal, kinetic, magnetic), its angular profile and time dependence (all of which could contribute in principle to determining the properties of the jet at later stages) is set by the central engine and its evolution; (2) the *interaction* of the jet with the progenitor vestige, comprising the *propagation* of the jet head, the formation of the cocoon and the jet *breakout*; (3) the *expansion* of the jet in the external medium, and the subsequent *deceleration*. The first phase sets the initial conditions of the problem and thus determines, along with the properties of the progenitor, the structure at later stages. On the other hand, it is possible that memory of the details of the initial conditions is lost along the evolution: for example, as the jet transitions from adulthood to seniority (i.e., when the external shock enters the self-similar blastwave phase) the radial structure of the jet is erased (over a time scale that depends on the presence or absence of a low-velocity tail).

Given the time-dependent nature of the system, no univocal definition of jet structure exists: the most useful definition must be determined depending on the particular application and on the observational or theoretical aspects under consideration. Yet, in modeling the prompt and afterglow emission, often the most relevant structure is that corresponding to the jet's adulthood, i.e., during the free expansion following the breakout from the progenitor vestige. This is the phase during which the prompt emission is believed to be produced, and the structure during this phase also constitutes the initial condition for producing the afterglow emission.

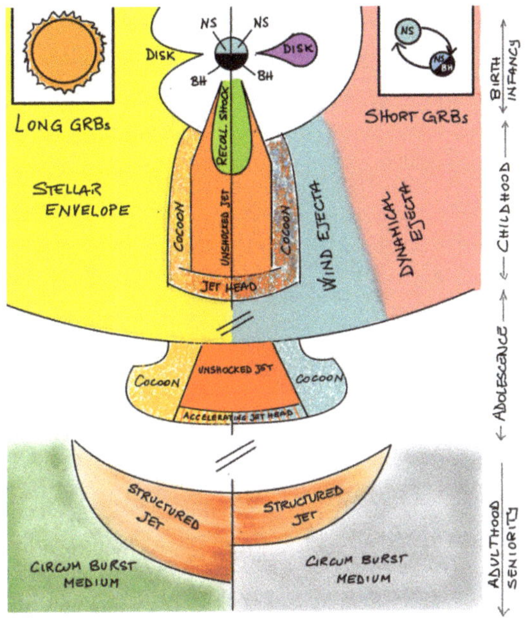

Figure 2. Artist's impression of the different phases in the early evolution of a GRB jet (long and short GRBs schematically represented on the left-hand and right-hand side, respectively). The putative progenitors are depicted in the stamps on the top corners. The formation of a compact central object (BH or NS) accompanied by an accretion disk powers a relativistic jet, determining the properties at its base ("birth"). The jet expands within the progenitor vestige (stellar envelope—left—or merger ejecta and disk winds—right) and starts interacting with it ("infancy"). The cocoon formed by ambient and jet-shocked material produces an inward pressure that collimates the jet as it propagates ("childhood"). As the jet breaks out ("adolescence") its head accelerates and the cocoon blows out. The subsequent phases ("adulthood and seniority", described in the text) are responsible for the prompt and afterglow emissions. See Section 2.2 for the description of the phases through which a jet evolves.

2.3. *Models of Jet-Launching Central Engines*

The leading jet-launching mechanism, by analogy with other relativistic jets such as those observed in active galactic nuclei (AGN, [96]) and microquasars [97,98], is the Blandford–Znajek (BZ) mechanism [70]. This is a process by which the rotational energy of a spinning black hole (BH) is extracted in presence of a large-scale, poloidal magnetic field threading the horizon. The magnetic field is sustained by an accretion disk, and the mechanism requires the formation of a "force-free" [99] magnetosphere close to the poles of the BH, which is only possible when the magnetic field energy density in the polar region exceeds the rest-mass energy density. In other words, a low-density funnel must be present along the BH rotation axis, and this is the region where the jet forms. The jet luminosity produced by the process follows [70]

$$L_{\rm BZ} \propto B^2 M_{\rm BH}^2 a^2, \tag{3}$$

with higher-order (a^4) corrections when a approaches unity [100,101]. Here, B is the strength of the radial component of the magnetic field at the horizon, $M_{\rm BH}$ is the black hole's gravitational mass and a is its dimensionless spin parameter. The normalization constant depends on the magnetic field geometry and on the accretion disk properties [100]. Typical expected values in the GRB context are $L_{\rm BZ} \sim 10^{49} - 10^{51}$ erg/s, which match the loose observational luminosity constraints set by the observed gamma-ray energies and durations, and by the collimation angles inferred from afterglow observations (see Section 3.1). Jets launched by this process are expected to start off as magnetically-dominated outflows (i.e., $\sigma \gg 1$ at the jet base).

Alternative jet-launching scenarios include energy deposition by neutrino-antineutrino ($\nu\bar{\nu}$) pair annihilation in the funnel above the BH [102–107], and a proto-magnetar central engine [71,108–110]. The $\nu\bar{\nu}$ luminosity for the former mechanism could be provided by the hot, inner parts of the accretion disk [111], and a jet-powered by such a process would be dominated by internal energy ($\eta/\rho' c^2 \gg 1$) at its base, which would then be converted to kinetic energy by hydrodynamic acceleration. We note, though, that the $\nu\bar{\nu}$ mechanism seems unable to explain the large luminosities [112] and energies [107] of some GRBs, and global simulations of jet launching in the aftermath of a binary neutron star merger seem to indicate that a jet powered by such mechanism would be unable to break out of the dense ejecta cloud that surrounds the merger remnant [113].

2.4. Models of Jet Propagation through the Progenitor Vestige

The jet launched by the central engine must initially propagate through the dense surrounding region constituted by the progenitor vestige, that is, the stellar envelope in the case of a collapsar or the ejecta cloud in the case of a compact binary merger. The details of the propagation and its final outcome depend on the properties of the jet at its base, its duration, and the properties of the vestige. The main features of the jet evolution during this phase can be understood based on a relatively simple hydrodynamical model ([30,76,114–117], see, e.g., [117–119] for the extension to the highly magnetized jet case, which presents some quantitative differences, despite the general picture remaining similar), whose main features are the following (see Figure 3 for a sketch): the jet is represented by an outflow with luminosity $L_{\rm j}$, expanding radially at a speed $\beta_{\rm j,0}$ within an aperture angle $\theta_{\rm j,0}$ at its base. The outflow interacts with the vestige at a height z above the central engine, where a forward-reverse shock structure forms, called the jet head. The jet head advancement speed $\beta_{\rm h}$ through the vestige, whose density at a height z is $\rho_{\rm v}$ (and which can be expanding outwards at a speed $\beta_{\rm v}$), is set by the balance between the jet momentum flux that crosses the reverse shock and the ram pressure of the vestige material as seen from the forward shock downstream frame, namely (e.g., [74,75,120,121], neglecting the vestige pressure).

$$\Gamma_{\rm j}^2\Gamma_{\rm h}^2(\beta_{\rm j}-\beta_{\rm h})^2\rho_{\rm j}' h_{\rm j} c^2 = \Gamma_{\rm h}^2\Gamma_{\rm v}^2(\beta_{\rm h}-\beta_{\rm v})^2\rho_{\rm v}c^2, \tag{4}$$

where $\Gamma_{\rm x} = (1-\beta_{\rm x}^2)^{-1/2}$, $\beta_{\rm j}$ is the dimensionless jet speed just before crossing the reverse shock, $\pi\theta_{\rm j}^2 z^2$ is the reverse shock working surface, and we used $\rho_{\rm j}' h_{\rm j} = L_{\rm j}/\pi\theta_{\rm j}^2 z^2 \beta_{\rm j}\Gamma_{\rm j}^2 c^3$. This can be solved for $\beta_{\rm h}$, which gives

$$\beta_{\rm h} = \frac{\beta_{\rm j} + \tilde{L}^{-1/2}\beta_{\rm v}}{1+\tilde{L}^{-1/2}}, \tag{5}$$

where the dimensionless quantity $\tilde{L} = \Gamma_{\rm j}^2\rho_{\rm j}' h_j/\Gamma_{\rm v}^2\rho_{\rm v} = L_{\rm j}/\pi\theta_{\rm j}^2 z^2\beta_{\rm j}\Gamma_{\rm v}^2\rho_{\rm v}c^3$ is what sets the overall properties of the jet advancement [117]. When the head advancement is sub-relativistic, then $\beta_{\rm h} \sim \beta_{\rm v} + \tilde{L}^{1/2}\beta_{\rm j}$; when it is relativistic, then $\Gamma_{\rm h} \sim \tilde{L}^{1/4}/\sqrt{2}$.

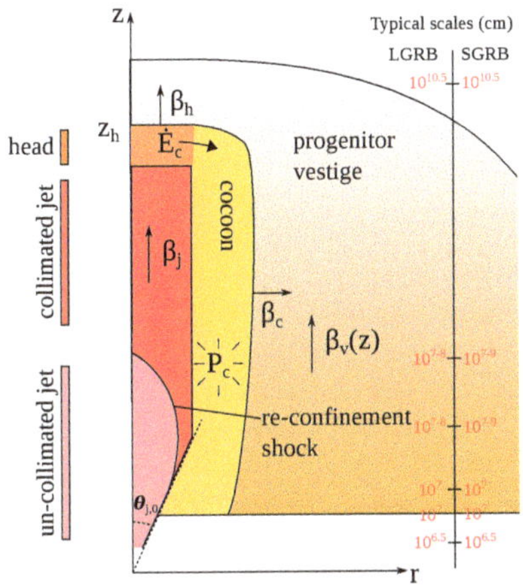

Figure 3. Sketch of the main elements in a basic hydrodynamical model of the jet propagation through the progenitor vestige. Adapted from [114].

If the head advancement speed β_h is sufficiently low, $\Gamma_\mathrm{h}\beta_\mathrm{h} \lesssim 1/\sqrt{3}\theta_\mathrm{j}$ [114], the head is causally connected by sound waves in the transverse direction: the lack of lateral confinement in the head then causes the shocked material (both that of the jet and that of the vestige) to flow laterally and form an over-pressured cocoon that shrouds the jet. The cocoon slowly expands laterally within the vestige at a speed proportional to the square root of the ratio of its pressure to the average vestige density [75,120], but this is typically not sufficient to prevent a pressure build-up as it is filled with an increasing amount of shocked material flowing from the head. When the cocoon pressure becomes comparable to the transverse momentum flux of the jet at its base, a "re-confinement shock" forms [76] where such transverse momentum is dissipated, collimating the jet into a cylindrical flow. The condition for such a self-collimation can be written approximately as $\tilde{L} \lesssim \theta_{\mathrm{j},0}^{-4/3}$ [75]. The self-collimation reduces the reverse shock working surface, therefore favoring the jet head propagation. The increase in the head speed, on the other hand, reduces the energy flow to the cocoon, therefore affecting its ability to effectively collimate the jet. In self-collimated jets, the final jet opening angle at breakout is thus set by these competing effects.

In presence of a homologous expansion of the vestige (as expected in the case of compact binary merger ejecta), another self-regulation effect arises [122]: if the jet head stalls (i.e., $\beta_\mathrm{h} \sim \beta_\mathrm{v}$) and the jet is self-collimated (so that the head working surface is constant—but this is ensured by the fact that $\beta_\mathrm{h} \sim \beta_\mathrm{v}$ implies $\tilde{L} \ll 1$ and hence the self-collimation condition $\tilde{L} \lesssim \theta_{\mathrm{j},0}^{-4/3}$ is certainly satisfied), the expansion has the effect of easing the head propagation because it reduces $\tilde{L}$. If the jet is launched shortly after the onset of the vestige homologous expansion (so that the jet duration is much longer than such delay), the result is that the jet's ability to break out depends solely on the ratio of the jet energy to that of the expanding vestige [122], regardless of the jet duration.

The importance of the jet propagation phase, from the observational point of view, stems from the fact that the jet structure after breakout carries the imprint of the jet-vestige interaction. If the rearrangement of the jet after the breakout (and before the prompt emission is produced) does not erase such memory, it is possible in principle to extract information about the progenitor (and possibly also about the central engine) from the jet structure at a stage when observable emission can be produced (e.g., [74]). The extent to which the resulting structure is determined by the central engine, the jet-vestige interaction, or both, is still a matter of debate. For what concerns binary neutron star mergers, the recent three-dimensional, special-relativistic hydrodynamical numerical simulations by [123] seem to suggest that turbulence at the jet-cocoon interface (see also [124]) tend to erase the details of the injected jet structure (i.e., the angular structure as initially set by the central engine), which could lend support to the hypothesis that "adult" jets from binary neutron star mergers share a quasi-universal structure (e.g., [24,125]). Yet, the development of such turbulence seems strongly suppressed in magnetohydrodynamic simulations with a significantly magnetized injected jet [126] and, moreover, in simulations that use a different jet injection technique (e.g., [127]) come to the opposite conclusion.

2.5. Jet and Cocoon Breakout

When the jet head forward shock reaches the steep density gradient that characterizes the outer edge of the progenitor vestige, it starts accelerating in much the same way as a supernova shock does as it approaches the outer edge of the stellar envelope [77,128–131]. The main differences between the jet forward shock breakout and a supernova shock breakout are that the former is relativistic and highly anisotropic, both features having a strong impact on the resulting dynamics and emission [78,132–136]. As explained in Section 2.2, the head forward shock separates the shocked vestige material from the unshocked one (see Figure 2). Below the inner part of the forward shock, within an angle $\theta_{\rm j,bo}$ (the *jet angle at breakout*), is the head reverse shock: since the jet material crossing the latter is (by definition) faster than the head, its ram pressure ensures that the reverse shock keeps up with the accelerating forward shock. As a result, the head material remains dense and optically thick to Compton scattering during the breakout, and therefore its internal energy contributes to the expansion rather than being radiated. At angles larger than $\theta_{\rm j,bo}$, on the other hand, the forward shock breakout is accompanied by the expansion of the underlying shocked material (which is part of the upper cocoon): the forward shock transitions from radiation-mediated to collisionless, liberating photons in what is sometimes called the "cocoon shock breakout emission" [79,137,138] (with a typical temperature of several tens of keV, [132]), and the underlying material becomes gradually transparent, giving rise to a "cooling" emission (typically peaking in the UV, [132]). The expansion of the cocoon after the shock breakout is sometimes called the cocoon "blowout" [74,114]. The entire breakout process is followed by a rearrangement of the jet and cocoon material into an inhomogeneous shell, which is what is most often referred to as the structured jet. In the following section, we focus on the structural properties of the latter, as found from numerical simulations.

2.6. Expected General Features of the Jet Structure in GRBs

Despite the complexity of the involved processes, the modeling and understanding of the birth and evolution of relativistic jets have been addressed since the late seventies, initially prompted by observations of radio galaxies [72]. A big deal of the understanding of jet launching and its evolution has been obtained through numerical simulations of the central engine (typically within general-relativistic magnetohydrodynamics, GRMHD) and of the jet propagation (usually within special-relativistic hydrodynamics–RHD–or magnetohydrodynamics–RMHD–framework, see [139] for a recent review). In the GRB context, a widely adopted approach (due in part to computational limitations) to the simulation of the propagation and breakout phase has been that of "injecting" a jet with properties based on an educated guess into a model of the progenitor vestige (in two dimensions, e.g., [30,140], and three dimensions, e.g., [116,123,141,142], but see e.g., [113,119,143]). While this helps in limiting the needed computational resources by leaving out the central engine region from the computational domain, it prevents a direct connection between the properties and evolution of the central engine and those of the jet at larger scales: in particular, this approach does not allow for a self-consistent description of: (1) the central engine variability (due, e.g., to the stochastic fluctuations in the accretion rate due to a turbulent disk); (2) the evolution of the jet luminosity (linked to that of the accretion rate and/or of the central compact object); and possibly (3) orientation; the (4) injected jet structure (angular distributions of kinetic luminosity and magnetization); and (5) the effects of the central engine on the vestige (e.g., gravity and the accretion disk winds). On top of this, the idealized nature of the progenitor models employed in most of these studies can affect the results by introducing exact symmetries that are not present in nature. The steady advancement of computational methods and resources has led recently to many important works that investigated some of these limitations (e.g., [144–152]). These include three-dimensional GRMHD simulations that self-consistently cover jet launch (usually within the Blandford–Znajek paradigm), propagation and breakout [150–152], even though these still feature idealized initial conditions and do not include a treatment of neutrinos, whose contribution to cooling, transport of momentum and energy can have prominent effects on the central regions. Yet, these simulations currently constitute some of the most detailed investigations that can shed light on the GRB jet structure. Figure 4 shows the jet structures obtained from GRMHD simulations in the collapsar case (top panels-[151]) and in the case of binary neutron star mergers (bottom panels-[152]). In qualitative agreement with previous works, these investigations find a jet angular structure after breakout that broadly features a narrow *core* (with an opening angle of few degrees) with approximately uniform Lorentz factor and energy density (mostly containing jet material that crossed the collimation shock, but did not reach the head before breakout), surrounded by a wider structure (sometimes called the jet *wings*, and typically composed of an inner part of mixed jet and cocoon material—where the amount of mixing depends on the jet magnetization, e.g., [142]—surrounded by a wider, cocoon-dominated part) where both the average Lorentz factor and the energy density fall off relatively quickly with the angle (typically as steep power laws $\propto \theta^a$ with $a \lesssim -3$, or as Gaussians). A minority of the simulations find a "hollow" jet core with a lower energy density and Lorentz factor along the axis with respect to that at the core edge [150], but it is unclear whether this is a genuine result or a numerical artifact [153].

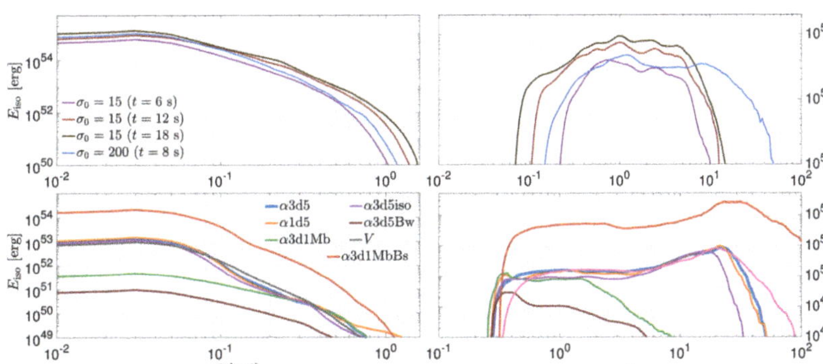

Figure 4. Jet structures resulting from GRMHD simulations representing collapsars (upper panels, adapted from [151]) and binary neutron star merger remnants (lower panel, adapted from [152]). Left-hand panels show the isotropic-equivalent energy $E_{\rm iso} = 4\pi \mathrm{d}E/\mathrm{d}\Omega$ from a late snapshot of the simulations (significantly later than the jet breakout), while right-hand panels show the distribution of energy in the four-velocity modulus $u = \Gamma\beta$ space. Different colors refer to different initial conditions, as detailed in [151,152].

3. Prompt Emission from a Structured Jet

The origin of the prompt emission of GRBs is not well understood yet. Generally speaking, the emission is thought to be powered by some mechanism that dissipates the dominant form of energy in the jet (either kinetic or magnetic), transforming it into internal energy. In most scenarios, part of the latter takes the form of relativistic particles with a non-thermal energy distribution, which produces the observed emission by means of some radiative process at the photosphere or beyond [154]. Different dissipation mechanisms operating within the relativistic outflow and different radiation mechanisms (typically synchrotron or inverse Compton) have been proposed, but no compelling evidence has been found yet in support of any of the envisaged scenarios. Still, the jet structure might have a leading role in explaining at least some of the prompt emission features.

3.1. Observed Temporal and Spectral Poperties of GRB Prompt Emission

The GRB observed duration distribution is bimodal, which leads to the division into short and long *duration* events (with an observer frame separation at $\sim$2 s). The average spectral properties of events in the two duration classes display some differences, with short GRBs featuring on average harder spectra with respect to long ones [155], the difference residing primarily in the low-energy part of the spectrum [156,157]. Short events are detected at a lower rate with respect to long events (with an instrument-dependent short/long ratio of 1:3 for *CGRO*/BATSE, 1:5 for *Fermi*/GBM and of 1:9 for *Swift*/BAT). However, several different instrumental and cosmological effects shape (and hence bias) these observed properties [53,158,159].

The prompt emission has no apparent periodicity [160] and features a power density spectrum consistent [161,162] with turbulent dissipation processes. Different methods were employed to measure the minimum *variability* timescale of prompt emission light curves, resulting in different distributions [163,164]. Ref. [164] reports similar rest-frame distributions for the minimum variability timescale in long and short GRBs, centered around 0.5 s and extending down to 10 ms in $\sim$10% of the events. Another interesting feature of the prompt emission is the presence of *spectral lags*. These consist in the fact that pulses observed in the lower energy bands of the gamma-ray instruments are seen to lag behind the corresponding pulses in the higher energy bands [165]. This feature seems to be more commonly present in long GRBs, while short events typically have lags consistent with zero [166].

Most attempts at identifying the fundamental building blocks of GRB prompt emission light curves adopted parametric functions to represent pulses (e.g., [165,167]). With the caveat that these methods are applied to large samples of light curves of GRBs with unknown redshift, the results show apparent (observer frame) differences between short and long GRBs [168].

On the longer timescales, periods of activity can at times be separated by quiescent phases. In a sizable fraction ($\sim$ 15%) of long GRBs, a long quiescence phase (reaching, in some cases, $>$100 s) separates *precursor* activity from the main emission episode [169]. Precursors have also been identified in short GRBs [170]. Long apparent quiescences also separate the main event from late time pulses, or *flares*, often observed in the X-ray band by *Swift*/XRT. X-ray flares share some common properties with the prompt emission [171] and are thus often interpreted as linked to late-time central engine activity. No significant differences in the average spectral properties of precursors and main emission episodes have been identified [172], while flares appear clearly softer.

The prompt emission of GRBs is characterized by a *non-thermal* spectrum. The presence of thermal-like emission has been identified in a few cases either during the initial phases of the burst [173] or along its full duration [174], with no evidence of such emission component in short bursts [29]. A combination of multiple emission components (e.g., the sum of a power law with a high energy cutoff and a black body) was also adopted to interpret observed spectra [175,176]. Observationally, the spectral energy distribution of GRB prompt emission typically peaks at 0.1–1 MeV and the low (resp. high) energy spectrum, below (resp. above) the peak, is consistent with a power-law with photon index ~ -1 (resp. -2.5).

3.2. On the Synchrotron Origin of GRB Prompt Emission

The interpretation of the prompt emission spectrum as synchrotron from shock-accelerated electrons faced the contradicting evidence of the observed GRB spectra being harder, below the SED peak than the expected synchrotron spectral shape. Given the physical conditions of the emission region, in particular, a large magnetic field $B \sim 10^{4-6}$ Gauss, shock accelerated electrons should cool rapidly [177] producing a spectrum with photon index -1.5 below the synchrotron characteristic SED peak. One possible solution to this issue [178–180] considered that electrons do not cool efficiently (so-called marginally fast cooling scenario) so that the separation between the characteristic synchrotron frequency (identified as the peak of the νF_ν spectrum) and the cooling frequency is relatively small. As such, the hardest synchrotron spectral power law (i.e., the single electron spectrum with photon index $-2/3$) would become visible in the observer energy range. The tension with observations would be solved by admitting that the fitted empirical Band function captures an average spectral index between the two characteristic ones (i.e., $-2/3$ and $-3/2$ below and above the cooling break, respectively) [181]. This interpretation was recently proved valid by the discovery [182–185] in long *Swift* and *Fermi* GRBs of a spectral break distributed in the 1–100 keV range. Overall, these studies find that in a sizable fraction of bright GRBs (possibly limited by current detectors' performances - see [181]) there is a break, located at energies a factor $\sim$10 below the characteristic SED peak. Remarkably, the power-law indices below and above the break are consistent with the single electron synchrotron photon spectral index ($-2/3$) and cooling synchrotron photon index ($-3/2$), respectively. Further support for the synchrotron origin of the prompt emission was obtained by fitting a synchrotron model directly to the data [186,187].

While these results, after three decades of debate, represent a step forward to unveiling the synchrotron nature of the prompt emission, they present further challenges [188]. If the break is interpreted as the cooling synchrotron frequency, it implies a small magnetic field ($B \sim 10$ G) in the emission region. If the latter is located relatively close to the central engine (as suggested by the observed small variability timescales) the Synchrotron Self Compton (SSC) emission would become relevant though its signature has not been clearly observed at GeV energies by *Fermi*/LAT. Possible solutions, consider emission in a downstream decaying magnetic field (e.g., [189]) or proton-synchrotron emission [188] (but see [190]).

3.3. Correlations between Spectral Peak Frequency and Energetics

Other key features of the prompt emission, common to both long and short GRBs, are the observed correlations between the rest frame SED peak energy ($E_{\rm peak}$) and the burst energy or luminosity, considering isotropic emission ($E_{\rm iso}$-[191] and $L_{\rm iso}$ [192]) or accounting for the GRB jet aperture angle (i.e., $E_\gamma = E_{\rm iso}(1-\cos\theta_{\rm j})$, see [21]). Short and long GRBs, owing to their different duration, follow two nearly parallel correlations in the $E_{\rm peak} - E_{\rm iso}$ plane while they show a similar $E_{\rm peak} - L_{\rm iso}$ correlation [193,194]. While these correlations may be subject to instrumental selection effects, their physical nature is corroborated by the existence, within individual GRBs, of similar relations between the same observables as a function of time along the prompt emission duration [157,195].

3.4. Impact of Jet Structure on the Prompt Emission Observables

One big difficulty in interpreting the prompt emission properties of GRBs in terms of the jet structure comes from the fact that the prompt emission mechanism is not well understood: for this reason, a typical approach is to assume that the prompt emission simply transforms some fraction of the kinetic (or magnetic) energy into radiation. Given a prompt emission efficiency $\eta_\gamma(\theta)$ (which represents the fraction of the available kinetic/magnetic energy at θ that is radiated in gamma-rays), the dependence of the prompt emission properties ($E_{\rm peak}$, $E_{\rm iso}$, $L_{\rm iso}$) on the viewing angle $\theta_{\rm v}$ (the *apparent structure* in the language of [125,196]) is set by the energy and Lorentz factor angular profiles, Equations (1) and (2). In particular, the bolometric isotropic-equivalent energy can be computed as [125,197,198]

$$E_{\rm iso}(\theta_{\rm v}) = \int_0^{2\pi} {\rm d}\phi \int_0^{\pi/2} \sin\theta\,{\rm d}\theta\, \frac{\delta^3(\theta,\phi,\theta_{\rm v})}{\Gamma(\theta)} \eta_\gamma(\theta) \frac{{\rm d}E}{{\rm d}\Omega}(\theta), \tag{6}$$

where $\delta = \Gamma^{-1}(1-\beta\cos\alpha)^{-1}$ is the Doppler factor, with α being the angle between the line of sight and the radial expansion direction, which can be expressed as [125]

$$\cos\alpha = \cos\theta\cos\theta_{\rm v} + \sin\theta\sin\phi\sin\theta_{\rm v}. \tag{7}$$

Figure 5 shows $E_{\rm iso}(\theta_{\rm v})$ corresponding to a few different $\mathrm{d}E/\mathrm{d}\Omega$ and Γ profiles, assuming a constant η_γ at all angles. A general feature that is demonstrated in the figure is that at some viewing angles (typically close to the jet axis) the emission is dominated by material moving along the line of sight, resulting in $E_{\rm iso}(\theta_{\rm v}) \sim 4\pi\mathrm{d}E/\mathrm{d}\Omega(\theta = \theta_{\rm v})$. Far off-axis, on the other hand, the flux received by the observer is spread over a larger portion of the jet, corresponding to regions with the most favorable combination of a large intrinsic luminosity and a sufficiently low Lorentz factor as to avoid a too severe de-beaming of radiation away from the line of sight. The steeper the Lorentz factor decay as a function of θ, the shallower the $E_{\rm iso}$ decay at large $\theta_{\rm v}$, as the de-beaming is less severe in broader and more energetic portions of the jet.

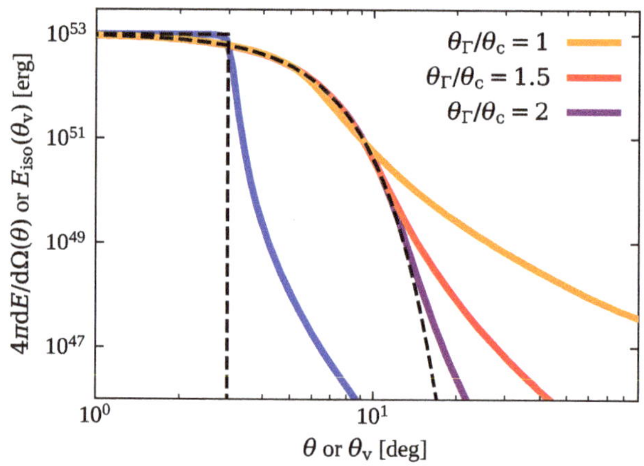

Figure 5. Apparent and intrinsic structure for a uniform and a Gaussian structured jet. Black dashed lines represent $4\pi\mathrm{d}E/\mathrm{d}\Omega$ at an angle $\theta = \theta_{\rm v}$ from the jet axis. Colored solid lines show $E_{\rm iso}(\theta_{\rm v})$ for four different models: a uniform jet model with $\mathrm{d}E/\mathrm{d}\Omega = 10^{53}/4\pi$ erg and $\Gamma = 300$ within an angle $\theta_{\rm j} = 3^\circ$ (blue) and three Gaussian models with $\mathrm{d}E/\mathrm{d}\Omega = (10^{53}/4\pi)\exp[-(\theta/\theta_{\rm c})^2/2]$ and $\Gamma\beta = 300\exp[-(\theta/\theta_\Gamma)^2/2]$, with $\theta_{\rm c} = 3^\circ$ and three different values of θ_Γ (orange, red and purple, with the corresponding $\theta_\Gamma/\theta_{\rm c}$ ratios given in the legend). Adapted from [125].

In order to compute $E_{\rm peak}(\theta_{\rm v})$, one needs to make a further assumption about the spectral shape $S'_\nu(\nu', \theta)$ of the radiation as measured in the jet comoving frame (e.g., [197,199]): the simplest assumption, often adopted in the literature, is that of a fixed spectrum at all angles, which yields $E_{\rm peak}(\theta_{\rm v}) \propto \Gamma(\theta = \theta_{\rm v})$ for viewing angles at which the emission is dominated by material on the line of sight, and a shallower decrease at larger viewing angles (see, e.g., Figure 2 in [199]). In long GRBs, this simple assumption (along with a Gaussian ansatz for the jet structure) is sufficient [125] to reproduce the observed $E_{\rm iso} - E_{\rm peak}$ correlation [191], while in short GRBs it leads to a clear tension with the observed $E_{\rm peak}$ of GRB170817A [197]. A more 'physical' approach to the modeling of the expected peak spectral energy would require one to assume a particular prompt emission scenario, such as, e.g., that of internal shocks, and explicitly compute the typical comoving photon energy at each angle for a given jet structure [200].

The computation of $L_{\rm iso}(\theta_{\rm v})$ also requires additional assumptions [144,145]. The duration of the prompt emission is usually assumed to be linked to that of the central engine activity. This is based on the idea that the prompt emission is composed of short pulses, each corresponding to a dissipation event in the jet. The events happen around a typical radius R, where the jet material travels with a typical bulk Lorentz factor Γ, and their intrinsic duration is very short, so that the observed duration is set by the (Doppler-contracted) angular time scale $t_{\rm ang} \sim R/\Gamma^2 c$ (i.e., the maximal arrival time difference between photons emitted within an angle $1/\Gamma$ with respect to the line of sight), which is assumed to be much shorter than the central engine activity duration T_{CE}. As a result, the observed duration is $T_{GRB} \sim T_{CE}$ and the average isotropic-equivalent luminosity is simply $L_{iso} \sim E_{iso}/T_{CE}$. If the spread in R and Γ is not too large, then this holds true also for off-axis observers, as long as the single-pulse duration remains much shorter than T_{CE} [201], and in such situation Equation (6) divided by T_{CE} can be used to compute the average $L_{iso}(\theta_v)$ [114,197,199,202,203]. Pulse overlap can result in a shallower dependence on the viewing angle [201], but the dependence will eventually steepen at sufficiently large angles such that the duration of single pulses will become comparable or longer to T_{CE}.

Useful discussions on the transformation between on- and off-axis isotropic-equivalent energies, durations, luminosities and peak photon energies, along with useful analytical approximations, can be found in [204,205].

3.5. The GRB Luminosity Function and the Jet Structure

A key property of the GRB population is the luminosity function (LF). The LF can be defined as the probability density of the isotropic-equivalent luminosity at a particular redshift $P(L_{iso}\,|\,z)$ or, equivalently, as the comoving rate density in a differential luminosity bin, $\mathrm{d}R_z/\mathrm{d}L_{iso} = R_z P(L_{iso}\,|\,z)$, where R_z is the event rate density (GRBs per comoving Gpc^3 yr) at redshift z. If the population does not feature a luminosity evolution with redshift, then the local LF $\mathrm{d}R_0/\mathrm{d}L_{iso} = R_0 P(L_{iso})$ is sufficient to describe the luminosity distribution in the population. In the context of a structured jet, the luminosity of each event depends both on the intrinsic properties of the underlying jet, and on its viewing angle. Hence, the luminosity function is shaped at least in part by viewing angle effects [24]. If the jet structure is universal (i.e., all jets share the same properties) then the LF is entirely determined by viewing angle effects: for a population with isotropic orientations and a monotonic viewing-angle-dependent luminosity $L_{iso}(\theta_v)$ (common to all events), the LF can be obtained [24,206] by application of the chain rule, namely

$$\frac{\mathrm{d}R_0}{\mathrm{d}L_{iso}} = \frac{\mathrm{d}R_0}{\mathrm{d}\theta_v}\frac{\mathrm{d}\theta_v}{\mathrm{d}L_{iso}} = R_0\left(\frac{\partial}{\partial\theta_v}L_{iso}(\theta_v)\right)^{-1}\sin\theta_v\bigg|_{\theta_v=f^{-1}(L_{iso})}, \tag{8}$$

where f^{-1} is the inverse of $L_{iso}(\theta_v)$, i.e., $f^{-1}(L_{iso}(\theta_v)) = \theta_v$. In the simple case of a power law dependence of the luminosity on the viewing angle, $L_{iso}(\theta_v) \propto \theta_v^{-a}$, and using the small-angle approximation $\sin\theta_v \sim \theta_v$, one obtains [206] $\mathrm{d}R_0/\mathrm{d}L_{iso} \propto L^{-1-2/a}$. The fact that this asymptotes to L^{-1} when $a \to \infty$ shows that, due to viewing angle effects, the LF cannot in principle be shallower than L^{-1} due to the contribution of off-axis jets, no matter how suppressed their emission is at large viewing angles. In practice, the spectrum of a far off-axis jet would be much softer than a typical GRB, the duration much longer, and the light curve smooth [207]. These features would lead to a different classification than a GRB. As an additional note, the $L^{-1-2/a}$ behavior breaks down when the small angle approximation $\sin\theta_v \sim \theta_v$ becomes invalid. The assessment of the contribution of far off-axis jets to the GRB LF thus requires additional care.

In practice, even in the case in which the progenitor parameter space for which a relativistic jet can be launched is very narrow (as suggested, for example, by the relatively small spread in the peak luminosities of supernovae associated with long GRBs, e.g., [208,209]), some spread in the jet properties within the population is unavoidable. For that reason, even in the wildest unification fantasy the best that can be expected is a *quasi-universal* jet structure with parameters that are spread around a "typical" value. The LF in a quasi-universal structured jet scenario can then be seen as a convolution of probability distributions,

$$P(L_{iso}) = \int_0^{\pi/2} P(L_{iso}\,|\,\theta_v)P(\theta_v)\,\mathrm{d}\theta_v, \tag{9}$$

where $P(\theta_v) = \sin\theta_v$ and $P(L_{iso} \mid \theta_v)$ is the probability distribution of the isotropic-equivalent luminosity at a given viewing angle, which is in turn induced by the probability distributions of the jet structure parameters (Salafia et al. in preparation).

The black symbols with error bars in the top panel of Figure 6 show the LF of long GRBs as obtained by combining samples of high luminosity (HL) GRBs with measured z and estimated isotropic equivalent luminosities $L_{iso} \geq 10^{50}$ erg/s [210]. The plot also shows the extension of the LF to intermediate luminosities (IL) $10^{48} \leq L_{iso}/(\mathrm{erg\,s^{-1}}) < 10^{50}$ where, due to selection effects, only lower limits on the intrinsic rate density can be placed [206], and to the low luminosity range (LL – $L_{iso} \lesssim 10^{47}$ erg s^{-1}) dominated by few events detected in the very local Universe [206]. In the bottom panel of Figure 6, the green and purple lines and bands show the inverse cumulative LFs of short GRBs obtained by two different studies [211,212] based on the observed properties of the *Swift* and *Fermi* samples.

The LF of both long and short GRBs extends over eight orders of magnitudes in luminosity and presents a steep decay $L^{-\alpha}$ with $\alpha \gtrsim 3$ at high luminosities ($L > L_{break} \approx 10^{52}$ erg/s). In the structured jet framework, the most luminous events are likely those observed within the core opening angle. As discussed above, the intermediate and faint end of the LF (for $L < L_{break}$) is at least in part shaped by the jet structure [206]. In long GRBs, the binned LF can be reproduced by a quasi-universal structured jet model [206] with a power law dependence of the luminosity on the viewing angle $L_{iso}(\theta_v) = L_c \min(1, (\theta_v/\theta_c)^{-a})$ with $a \gtrsim 6$ and θ_c and L_c narrowly distributed around 5° and 3×10^{52} erg/s, respectively, (cyan solid line in the top panel of Figure 6). In short GRBs, due to the scarcer data and the absence of an agreement on the general features of the LF, the situation is more unclear. Yet, an attempt at deriving a quasi-universal jet structure model from modeling the jet propagation through, and breakout from, binary neutron star merger ejecta [114] does produce an LF (blue solid line in Figure 6) with a steep decay above a break at $L_{break} \sim 3 \times 10^{52}$ erg/s and a relatively less steep distribution at lower luminosities, which gets shallower as it extends to the LL range where the observation of GRB170817A [27] places some constraints [67]. A shallower (e.g., a power law $L_v \propto \theta_v^{-2}$) structure would result in a steeper LF that would overproduce the LL long GRBs rate density (blue symbol in Figure 6, top panel) or, in short GRBs, the binary neutron star merger rate (pink shaded region in Figure 6, bottom panel).

3.6. *Jet Structure and the* $E_{peak} - E_{iso}$ *Correlation*

A fraction of the GRBs (around one-third) that trigger the Burst Alert Telescope (BAT) onboard *Swift* end up with a measurement of their redshifts [213]. This allows, in most cases. The estimate of E_{iso} and L_{iso} require to measure the SED peak which is often possible thanks to the detection of the burst also by the *Fermi* satellite which provides a broad band (10keV-40MeV) energy spectral coverage., to estimate rest frame properties such as E_{peak}, E_{iso} and L_{iso}. Figure 7 shows the GRBs from [214] on the (E_{iso}, E_{peak}) plane (black symbols), demonstrating the apparent correlation between these two quantities [191]. Such a correlation is naturally expected in a quasi-universal structured jet scenario, given the common dependence of E_{iso} and E_{peak} on the viewing angle (e.g., [215]). Assuming Gaussian profiles $\eta_\gamma \mathrm{d}E/\mathrm{d}\Omega = \epsilon_0 \exp(-\theta^2/\theta_c^2)$ and $\Gamma(\theta) = 1 + (\Gamma_c - 1)\exp(-\theta^2/\theta_c^2)$, an angle-independent comoving peak SED photon energy $E'_{peak} = 1\,\mathrm{keV}$ [216], and a quasi-universal structured jet scenario in which the structure parameters in the population are narrowly distributed around typical values $\langle\epsilon_0\rangle = 3 \times 10^{53}$, $\langle\Gamma_0\rangle = 800$, $\langle\theta_c\rangle = 3^\circ$, the authors of [125] could reproduce both the LF of long GRBs and the observed $E_{peak} - E_{iso}$ correlation, as shown by the model distribution represented in Figure 7. The horizontal dispersion in the figure corresponds to just considering a 0.5 dex log-normal dispersion of the core energy density ϵ_0. As shown by the color-coded viewing angle, within this interpretation the known long GRBs are observed within $\theta_v \lesssim 3\theta_c \sim 9^\circ$, which is consistent with the constraints derived by [198]. In the bottom left corner, corresponding to $E_{iso} < 10^{51}$ erg and $E_{peak} < 30$ keV, should reside jets observed at larger viewing angles. If these were detected with next-generation instruments with wide-field, highly sensitive hard X-ray monitors, they could probe the expected bending of the $E_{peak} - E_{iso}$ correlation induced by large viewing angles and in turn help constraining the quasi-universal jet structure scenario.

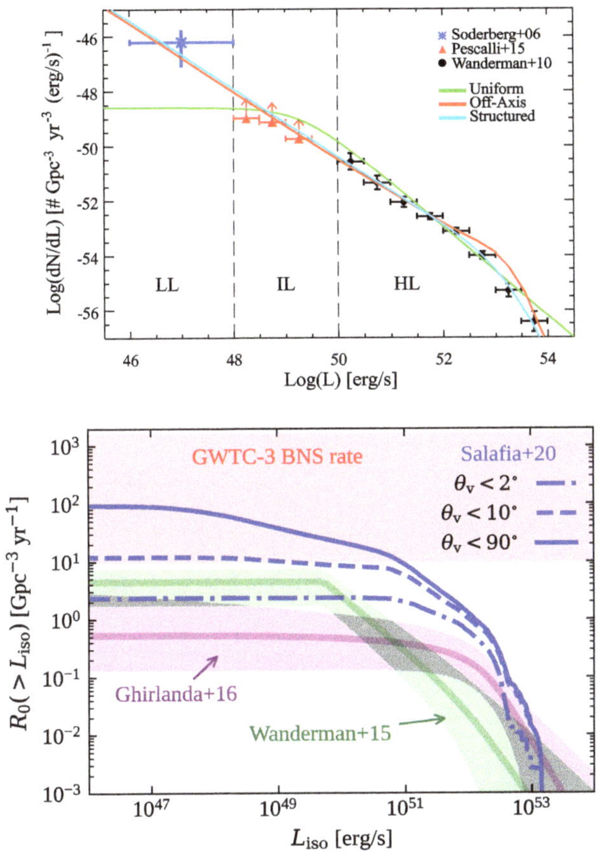

Figure 6. Top: luminosity function of Long GRBs as obtained by [210] (black symbols) extended to low luminosities by [206] (red and blue symbols). Models considering a uniform jet (only seen on-axis – green – or isotropically oriented – red) or a structured jet with a steep power law profile (cyan) are shown. The separation in low, intermediate, and high luminosity (LL, IL, HL) GRBs is indicated by the dashed vertical lines. **Bottom**: models of the (inverse cumulative) SGRB luminosity function. Models fitted to observed properties of short GRBs (detected at cosmological distances) are shown by the green [211] and purple [212] thick transparent lines and bands (medians and 90% credible regions). The luminosity function obtained by [114] by computing the jet structure from a semi-analytical calculation of the jet propagation and breakout is shown by the blue lines (contributions by jets observed in different intervals of viewing angle are shown), arbitrarily normalized to a local rate density $R_0 = 100\,\mathrm{Gpc}^{-3}\,\mathrm{yr}^{-1}$. The local BNS merger rate density constraint from [217], i.e., $10 \leq R_{0,\mathrm{BNS}}/\mathrm{Gpc}^3\,\mathrm{yr} \leq 1700$, is shown by the pink shaded region.

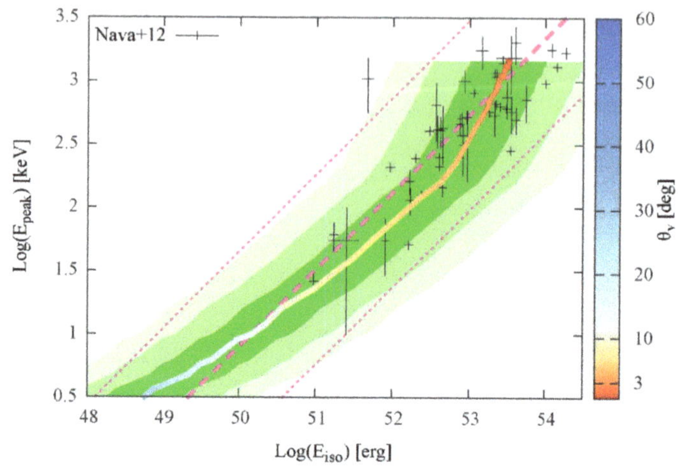

Figure 7. Rest frame peak energy $E_{\rm peak}$ versus isotropic equivalent energy $E_{\rm iso}$ of long GRBs. The data points (cross symbols) are a flux-limited sample of bright *Swift* bursts. The dashed (dotted) line shows the correlation regression line (and its 3σ scatter). The color-coded solid line shows the values of $E_{\rm peak}$ and $E_{\rm iso}$ assuming a structured Gaussian jet seen under progressively larger viewing angles-vertical color-code bar). The green shadows, representing the 1, 2, 3σ confidence levels around the color-coded line, are obtained considering a dispersion of the core energy of 0.5 dex around a nominal value of 3×10^{53} erg. Figure reproduced from [125].

3.7. Jet Structure and 'Late-Prompt' Emission

The observed X-ray emission of GRBs extending after the prompt phase up to a few hours is often characterized by a steep decay [218] of the flux transitioning to a shallow (so-called *plateau*) phase [219]. On top of this, X-ray flares are often observed [220–222]. Intriguingly, these features of the early X-ray emission can be explained within a structured jet scenario: the steep-plateau shape is what an observer nearly aligned with the jet axis would see, while an off-axis observer should see a more uniform power law decay [207,223]. In this interpretation, the X-ray light curve up to the end of the plateau phase would have an internal origin, being produced by prompt emission photons reaching the observer from increasingly high-latitude parts of the jet (hence the delayed arrival time, [224]). Flares have been explained, in the context of a structured jet, as late-time internal dissipation episodes whose brightness and spectral hardness are reduced by the debeaming effects for an observer with a viewing angle far from the jet axis [225].

4. Afterglow Emission from a Structured Jet

As explained in Section 2.2, the expansion of the jet material in the "circum-burst medium" (the relic stellar wind from the progenitor massive star in long GRBs. In most long GRBs, the afterglow seems to be best modeled assuming a homogeneous interstellar medium (ISM), which poses a challenge to the massive star progenitor scenario for long GRBs [88], or a tenuous interstellar medium in short GRBs) leads to the formation of a shock, which is widely believed to be the main source of the broad-band GRB afterglow emission [89,226–228]. Hereon, we focus on the case of a homogeneous ISM composed of hydrogen, with number density n, and discuss the impact of jet structure on the expected afterglow emission.

4.1. Jet Structure and the Early Afterglow

Initially, as the expanding jet material is highly relativistic, there is no causal contact between regions at angular distances $\gtrsim 1/\Gamma$ and hence the shock dynamics only depends on local quantities. Calling $E(\theta) = 4\pi {\rm d}E/{\rm d}\Omega(\theta)$, as soon as the isotropic-equivalent ISM mass is swept by the jet, $M(\theta) = nm_\mathrm{p}4\pi R(\theta)^3/3$ (where m_p is the proton mass and R is the outer radius of the jet or, more precisely, that of the forward shock) becomes comparable to $E(\theta)/\Gamma^2(\theta)c^2$, a reverse shock starts to propagate backwards (as seen in the rest-frame of the contact discontinuity that separates the shocked ISM and jet materials) through the jet, initiating the deceleration of the latter. Assuming a jet radial width $\Delta_0 \sim cT_\mathrm{CE}$ after breakout, the initial deceleration phase proceeds differently depending on the local Sedov length $l_\mathrm{S}(\theta) = (3E(\theta)/4\pi nm_\mathrm{p}c^2)^{1/3}$ and bulk Lorentz factor $\Gamma(\theta)$. Assuming free expansion, the "sound-crossing" radius at which radial sound waves (assuming a relativistic sound speed $c_\mathrm{s} = c/\sqrt{3}$) cross the shell is $R_\mathrm{s}(\theta) \sim \sqrt{3}\Gamma(\theta)^2\Delta_0$. The "deceleration" radius at which the interaction with the ISM becomes relevant is $R_\mathrm{d}(\theta) \sim l_\mathrm{S}(\theta)\Gamma^{-2/3}(\theta)$. If $R_\mathrm{s}(\theta) > R_\mathrm{d}(\theta)$, the portion of the jet is said to be in the "thick shell" regime, where the deceleration starts before radial pressure waves can smooth out radial inhomogeneities and cause any significant radial spreading [80,90]: in this case, the reverse shock is relativistic (i.e., the velocity of the unshocked jet is relativistic as seen from the contact discontinuity that separates the shocked ISM and shocked jet material) and crosses the whole jet shell at a radius $R_\mathrm{cross}(\theta) \sim l_\mathrm{S}(\theta)^{3/4}\Delta_0^{1/4}$. Conversely, if $R_\mathrm{s}(\theta) < R_\mathrm{d}(\theta)$ the jet portion is in the "thin shell" regime, where it reaches the deceleration radius after undergoing a significant radial spread, which washes out radial inhomogeneities and leads to an effective jet radial with $\Delta(\theta) \sim R/\Gamma(\theta)^2$. In this case, the reverse shock remains Newtonian and crosses the shell at $R_\mathrm{cross}(\theta) \sim R_\mathrm{d}(\theta)$. Interestingly, as shown in Figure 8, assuming $E(0) = 10^{54}$ erg and $\Gamma(0) = 1000$ (the dependence on these values is weak) and adopting a Gaussian profile $\propto \exp(-(\theta/\theta_\mathrm{c})^2/2)$ for both quantities, for most short GRBs (with $T_\mathrm{CE} \lesssim 2$ s and $n \lesssim 1\,\mathrm{cm}^{-3}$) the deceleration is entirely in the thin shell regime (see also [229]), while for long GRB jets (typically with $T_\mathrm{CE} \sim 30$ s and $n \sim 1\,\mathrm{cm}^{-3}$) it proceeds in the thick shell regime within an inner region $\theta < \theta_\mathrm{thick}$ which corresponds typically to the jet core, $\theta_\mathrm{thick}/\theta_\mathrm{c} \sim 1$. More generally, within the above Gaussian structured jet assumption, the existence of a transition angle θ_thick corresponds to the condition $\Gamma(0) > (l_\mathrm{S}(0)/\sqrt{3}cT_\mathrm{CE})^{3/8} \sim 430\,E_{54}^{1/8}n_0^{-1/8}T_\mathrm{CE,1}^{-3/8}$, in which case $\theta_\mathrm{thick} = (2\theta_\mathrm{c}/\sqrt{7})[8\ln\Gamma(0) - 3\ln(l_\mathrm{S}(0)/\sqrt{3}cT_\mathrm{CE})]^{1/2}$.

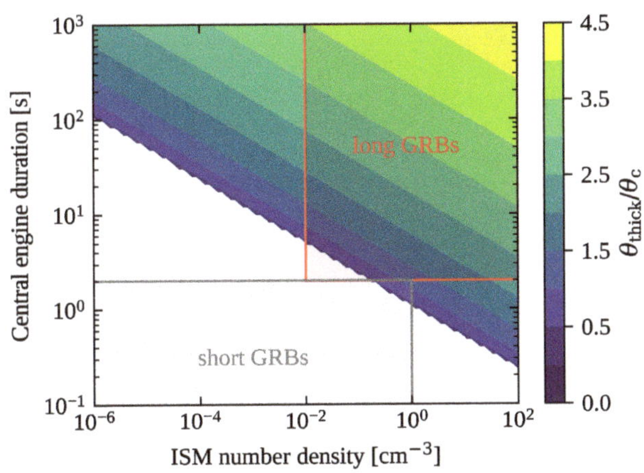

Figure 8. Angle θ_thick within which the deceleration takes place in the thick shell regime, in units of the core angle θ_c, in a structured jet with Gaussian energy $\mathrm{d}E/\mathrm{d}\Omega(\theta)$ and bulk Lorentz factor $\Gamma(\theta)$ profiles $\propto \exp(-(\theta/\theta_\mathrm{c})^2/2)$, assuming $4\pi\mathrm{d}E/\mathrm{d}\Omega(0) = 10^{54}$ erg and $\Gamma(0) = 1000$, as a function of the ISM number density n and central engine duration T_CE. Boxes show the regions of the plane where most long (red box) and short (grey box) GRBs are expected to lie.

During this phase, diffusive shock acceleration of electrons [230–233] can take place at both the forward and reverse shocks, leading to synchrotron [234] and possibly inverse Compton radiation. For viewing angles close to the jet axis, where the emission is dominated by material moving close to the line of sight, the reverse shock emission is expected to peak at the observer time $t_{\rm pk,RS} \sim (1+z)R_{\rm cross}(\theta_{\rm v})/\Gamma(\theta_{\rm v})^2 c$ that corresponds to the shock crossing time. In the thin shell regime, this matches the peak time of the forward shock emission. The light curve of the reverse shock emission in the Optical (where the peak of the synchrotron spectrum at $t_{\rm pk,RS}$ is expected to lie for "standard" parameters, [234]) and X-rays are expected to display a rapid rise and decay before and after the peak, therefore appearing as a flare. In the radio, the expected decay is slower (as the synchrotron peak moves rapidly to lower frequencies after the peak), with possible late-time bumps [235], even though this critically depends on how rapidly the shock-generated magnetic field decays after the reverse shock has disappeared [236]. The emission as seen by a far off-axis observer may be instead dominated by material moving at a different angle, or more generally consist of a comparable amount of radiation from a broader portion of the jet, resulting in a delayed and smoother light curve (see [229] for examples in short GRBs).

In parts of the jet where the deceleration proceeds in the thick shell regime, the radial structure plays a role in the reverse-forward shock dynamics, and thus in shaping the early afterglow emission. This is particularly relevant in far off-axis parts of the jet that contain blown-out cocoon material (which is expected to feature a broad velocity profile, e.g., [79,116,138]) and/or if the central engine does not turn off abruptly at $T_{\rm CE}$, but rather decays slowly, resulting in a jet with a low-velocity tail that contains a non-negligible amount of energy. In that case, the reverse shock can be long-lived, with slower material gradually catching up, with modified dynamics of both the reverse and forward shocks: this is often called a *refreshed* shock [237] scenario. This remains currently one of the leading explanations for the X-ray *plateaux* (see Section 3.7). The latter phenomenon is therefore mostly explained as resulting from either a radial or an angular jet structure [238]: the degeneracy between these two options when trying to explain "non-standard" decays in afterglow light curves has been addressed in [239].

4.2. Jet Structure and the Late Afterglow

As soon as the reverse shock has disappeared, the shocked jet shell transfers most of its energy to the shocked ISM [83,91] and the system turns into a *blastwave*, that is, a forward shock whose memory of the details of the original explosion is lost. In this phase, its dynamics depend solely on the (angle-dependent) Sedov length $l_{\rm S}(\theta) \propto (E(\theta)/n)^{1/3}$ and can be described by the evolution of the Lorentz factor of the shocked ISM immediately behind the forward shock, as a function of the forward shock radius (and the angle, given the anisotropy) $\Gamma(R,\theta)$. As long as the blastwave is relativistic, $\Gamma(R,\theta) \gg 1$, its radial structure and evolution are well-described by the self-similar Blandford–McKee solution [92], stably against perturbations [83], and hence $\Gamma(R,\theta) = (R/l_{\rm S}(\theta))^{-3/2}$ [24,92,240]. Lateral energy transfer (which typically proceeds from the jet axis outwards) starts being relevant as soon as transverse causal contact is established, which happens at each angle θ when $[\Gamma(R,\theta)]^{-1} \gtrsim \theta$. From that radius on, lateral pressure waves will start transferring shock energy laterally, gradually smearing out the angular energy profile $E(\theta)$ [31,32,43,241]. As soon as the entire shocked region reaches transverse causal contact, it expands laterally [11,93,95,242], gradually reaching isotropy. At late times, when the whole shock becomes non-relativistic and the expansion quasi-spherical, the shock structure and evolution is well-described [93,94,243] by the Sedov-Taylor self-similar solution [244–246]. This qualitative description should make clear that most of the memory of the angular structure is wiped out during the lateral spreading phase.

The shock dynamics and emission in the structured jet scenario in this phase have been addressed by several studies, e.g., [24,31,32,37,43,57,64,199,241,247–251]. At any time in the evolution, the emission is typically modeled within a parametrized relativistic leptonic diffusive shock acceleration and synchrotron/inverse-Compton emission model [227,228].

Shock-accelerated electrons are assumed to be injected in the shock downstream with a power-law distribution of Lorentz factors, $dn_e/d\gamma \propto \gamma^{-p}$ (where n_e is the comoving electron number density in the immediate downstream, and we focus on the case where $p > 2$), above a minimum Lorentz factor γ_m. Their number is assumed to be a fraction. Most often, the fraction is set to $\chi_e = 1$, despite the theoretical expectation being $\chi_e \sim$ few $\times 10^{-2}$ [233,252]. Yet, in some GRB afterglows with broad-band, high-cadence datasets, $\chi_e < 1$ has been shown to provide a substantially better fit to the data, e.g., [236,253]. χ_e of the total shocked ISM electrons, and their energy density is assumed to be a fraction ϵ_e of the internal energy in the shock downstream (which in a strong shock depends only on the shock velocity/Lorentz factor and on the adiabatic index, being set by shock-jump conditions [254]): the minimum Lorentz factor γ_m is entirely determined once the shock Lorentz factor Γ and the χ_e, ϵ_e and p parameters are given. A random magnetic field generated by small-scale turbulence behind the shock [255] is assumed to hold a fraction ϵ_B of the energy density. The electron population in the shock downstream evolves as fresh electrons are injected, and as older electrons cool down due to synchrotron, inverse-Compton and adiabatic losses: as a result, the electron distribution in phase space takes approximately the form [228]

$$\frac{dn_e}{d\gamma} \propto \begin{cases} \gamma^{-q} & \gamma_p \leq \gamma < \gamma_0 \\ \gamma^{-p-1} & \gamma \geq \gamma_0 \end{cases}, \tag{10}$$

where $\gamma_p = \min(\gamma_m, \gamma_c)$, $\gamma_0 = \max(\gamma_m, \gamma_c)$,

$$q = \begin{cases} p & \gamma_m \leq \gamma_c \\ 2 & \gamma_m > \gamma_c \end{cases}, \tag{11}$$

and γ_c is the Lorentz factor above which electrons lose most of their energy through synchrotron, inverse-Compton and adiabatic losses in a dynamical time $t'_{\rm dyn} \sim R/\Gamma c$. This is typically given by [227,228]

$$\gamma_c = \frac{6\pi m_e c^2 \Gamma}{\sigma_T B^2 R(1+Y)}, \tag{12}$$

where $Y = u_{\rm rad}/u_B$ is the ratio of radiation energy density. This includes the synchrotron radiation produced by the electrons–which emit and cool both by synchrotron and by inverse-Compton scattering of their own synchrotron photons, i.e., synchrotron-self-Compton [256,257]–and possibly an external radiation field, which can be upscattered by relativistic electrons in the shock downstream giving rise to an additional emission component [258–260]. to magnetic energy density as measured in the shock downstream comoving frame. More complicated electron energy distributions arise when Klein-Nishina effects are important and most electrons cool rapidly, e.g., [178,180].

The above simple model of electron acceleration and cooling thus depends on the "microphysical" parameters p, χ_e, ϵ_e and ϵ_B, and on the shock Lorentz factor Γ. Once the shock dynamics $\Gamma(R, \theta)$ is determined, the luminosity emitted towards an observer at a viewing angle θ_v from electrons in the entire shock can thus be computed for a fixed set of microphysical parameters by integrating the radiative transfer equation after computing the synchrotron (and possibly inverse-Compton) emissivity (and absorption coefficient) over the shock downstream, the shock radial structure is given by the appropriate self-similar solution. In order for the resulting luminosity to be appropriately compared to the observed flux, the integration must be performed over the appropriate equal-arrival-time volume or, in other words, at the appropriate 'retarded' time to account for the different photon paths that lead to the same arrival time to the observer [7,67,261–263]. If the shock is approximated as infinitely thin, this reduces to equal-arrival-time surfaces [43,199,228,264].

Even when limiting the discussion to an isotropic explosion and to synchrotron emission only, the above model leads to rather complex light curves [263]. At any fixed time, the SED is composed of various smoothly-connected power law segments, corresponding to different spectral regimes: the main critical frequencies are the synchrotron frequencies $\nu_{\rm syn}(\gamma) = \delta\gamma^2 eB/2\pi m_e c$ (where δ is the Doppler factor related to bulk motion) corresponding to the electron distribution break Lorentz factors $\gamma_{\rm m}$ and $\gamma_{\rm c}$ (typically referred to as $\nu_{\rm m} = \nu_{\rm syn}(\gamma_{\rm m})$ and $\nu_{\rm c} = \nu_{\rm syn}(\gamma_{\rm c})$), and frequency $\nu_{\rm a}$ below which synchrotron self-absorption becomes important (i.e., the synchrotron self-absorption optical depth $\tau_{\rm ssa}(\nu_{\rm a}) = 1$). The comoving specific synchrotron emissivity at a comoving frequency ν' can be approximated by a series of power-law segments, namely

$$j'_{\nu'} = j'_{\nu',\rm max}\begin{cases} (\nu'/\nu'_{\rm p})^{1/3} & \nu' \leq \nu'_{\rm p} \\ (\nu'/\nu'_{\rm p})^{-(q-1)/2} & \nu'_{\rm p} < \nu' \leq \nu'_0 \\ (\nu'_0/\nu'_{\rm p})^{-(q-1)/2}(\nu'/\nu'_0)^{-p/2} & \nu' > \nu'_0 \end{cases}, \tag{13}$$

where $\nu_0 = \nu_{\rm syn}(\gamma_0)$, $\nu_{\rm p} = \nu_{\rm syn}(\gamma_{\rm p})$, and q, γ_0 and $\gamma_{\rm p}$ have the same meaning as before. The maximum specific emissivity $j'_{\nu',\rm max}$ depends on the microphysical parameters, ISM density and shock Lorentz factor [199,227]. When considering optically thin portions of the shock, the intensity received by the observer is just $I_\nu = \delta^3 j'_{\nu'}(\nu/\delta)\Delta R'$, where $\Delta R'$ is the shock effective thickness, typically of order $\Delta R' \sim R/\Gamma^2$. Hence, if the emission is dominated by a small, optically thin portion of the shock, the observed flux density has the same shape as the emissivity. Synchrotron self-absorption [265] suppresses the emission when $\tau_{\rm ssa} > 1$, introducing low-frequency power law segments with $I_\nu \propto \nu^\alpha$ with $\alpha = 2$ if $\nu < \nu_{\rm p}$ and $\alpha = 5/2$ otherwise. If $\gamma_{\rm c} < \gamma_{\rm m}$, an additional power law segment with $\alpha = 11/8$ emerges below $\nu_{\rm a}$ due to the inhomogeneous cooling stage of electrons behind the shock [266]. Useful figures summarizing all the possible synchrotron spectral regimes in GRB afterglows can be found in [263]. Reference [178] treats additional cases where the electron distribution (and hence the synchrotron spectrum) is modified by a non-monotonic dependence of the cooling rate on the electron Lorentz factor γ due to Klein–Nishina effects.

For a given jet structure, observed off the jet core ($\theta_{obs} > \theta_c$), the emission is dominated, at different times, by different portions of the emitting surface [249,251]. Figure 9 shows that as time increases, the flux seen by the observer is dominated by emissions produced progressively close to the jet axis as a consequence of the competition between the decrease in the shock velocity (implying the increase in the beaming angle and the decrease in the emissivity) and the increasing shock initial energy towards the jet axis with respect to the observer line of sight (center of the coordinate system in Figure 9). This effect has important consequences on the observed afterglow light curves for different jet structures and viewing angles. Figure 10 shows some examples of mono-chromatic afterglow light curves at three characteristic frequencies corresponding to the radio, optical and X-ray bands (from top to bottom). Three different jet structures are considered here: uniform (solid lines), Gaussian (dashed line) power-law with both the energy and the Lorentz factor decreasing as θ^{-3} (dotted line). In all three cases, a jet core opening angle of 3 degrees is considered. As long as the observer's line of sight is within the beaming cone of the jet core, $\theta_{\rm v} - \theta_{\rm c} < 1/\Gamma(R,0)$, the light curves corresponding to different structures are almost indistinguishable. This applies to the entire light curve as long as $\theta_{\rm v} < \theta_{\rm c}$. If the viewing angle is larger, differences among the three structures are apparent mainly in the rising phases of the light curves, with the two non-uniform structures presenting similar shallow rising phases preceding the peak, after which all structures join into the same, core-dominated decay. The enormous difference in the light curves at intermediate viewing angles $\theta_{\rm c} < \theta_{\rm v} \lesssim \text{few} \times \theta_{\rm c}$ stems from the fact that, in the Gaussian and power-law structure cases, the emission is initially dominated by material moving along the line of sight, which is absent in the uniform jet case.

The fact that the emission is dominated by material progressively closer to the jet axis impacts also the apparent displacement of the source centroid as seen in Very Long Baseline Interferometry (VLBI) imaging [67,248,267–270]. The surface brightness of the shock, as seen by a distant observer, corresponds to the intensity I_ν described above. Its distribution $I_\nu(\theta_x, \theta_y)$ on the plane of the sky (where θ_x and θ_y represent two suitably chosen angular coordinates on the relevant sky patch) is most commonly referred to as the 'image' of the source. See the next section for some example surface brightness distributions, which can be measured through VLBI imaging within the limited resolution that can be reached with current facilities. The image centroid (i.e., the mean of the distribution) lies on the projection of the jet axis on the (θ_x, θ_y) plane because the jet axisymmetry induces a reflection symmetry in the image. Such symmetry can also be exploited to speed up the computation of I_ν for the calculation of light curves [199]. The displacement of the centroid before the light curve peak (after which the emission becomes core-dominated) is directly related to the shape of the jet structure.

The fact that the evolution of the pre-peak light curves and in principle also that of the image centroid position contain some information on the jet structure suggests that it is possible to reconstruct (at least partially) the latter information from the observations. This is typically achieved by fitting an analytical structured jet afterglow model to the observed light curves (and centroid displacement), in order to recover the parameters that describe the structure, e.g., [63,67,248,269,271,272]. As demonstrated by [273], the light-curve-based reconstruction can be performed more explicitly by integrating a differential equation derived from standard afterglow theory. Unfortunately [273,274], the accuracy and cadence required for a detailed reconstruction are highly demanding, and global degeneracies remain unbroken unless the evolution of the emission in multiple spectral regimes is observed see also [251].

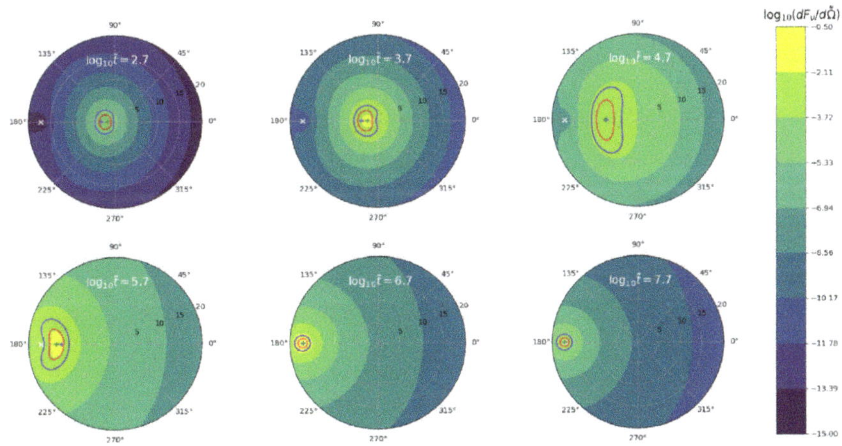

Figure 9. Angular map showing the intensity distribution per unit solid angle (color coded) of emission from the afterglow forward shock, centered on the line of sight. The position of the jet axis is marked by the white cross symbol. The green cross and red (blue) contours show the peak of the intensity and the region containing 50 (80%) of the total flux. A power-law structured jet with core values $\Gamma_c = 1000$ and $\theta_c = 0.03$ rad is considered. The power-law slopes are $a = 4$ and $b = 2$ for the energy and Γ structure, the observer viewing angle is 10 times the core opening angle and the external medium density is constant. The maps correspond to different observing times (as measured in the rest frame of the central engine), from 500 s (**top left**) to 580 days (**bottom right**). Reproduced from [251].

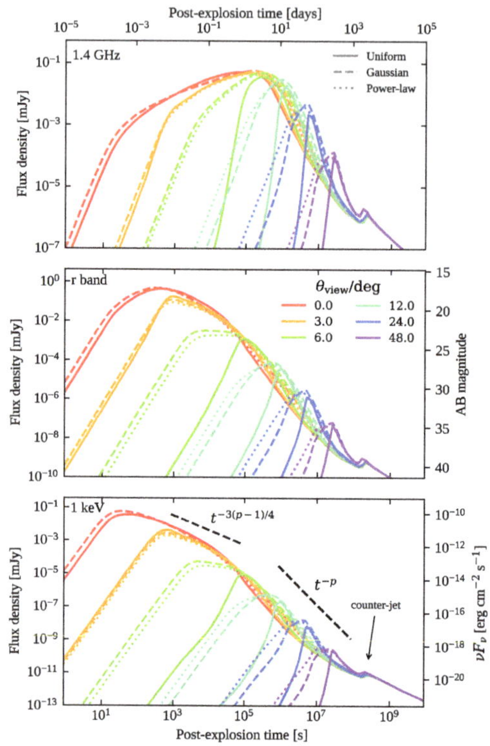

Figure 10. Example synthetic afterglow light curves for jets at $z = 1$ with differing structures, seen under various viewing angles (color-coded as reported in the legend), computed within the "standard afterglow" model described in Section 4.2 assuming $\epsilon_e = 0.1$, $\epsilon_B = 10^{-3}$, $p = 2.2$, and a homogeneous external medium with number density $n = 0.1\,\mathrm{cm}^{-3}$. Each structure can be defined as $\mathrm{d}E/\mathrm{d}\Omega = (E_c/4\pi)f(\theta/\theta_c)$ and $\Gamma = 1 + (\Gamma_c - 1)f(\theta/\theta_c)$: (i) the Uniform structure (solid lines) has $f(x) = \Theta(1 - x)$, where $\Theta(x)$ is the Heaviside step function; (ii) Gaussian jet (dashed lines), with $f(x) = \exp(-x^2/2)$; (iii) Power-law jet, with $f(x) = (1 + x^3)^{-1}$. For all jets, $E_c = 10^{53}$ erg, $\Gamma_c = 300$ and $\theta_c = 3°$. Each panel shows light curves computed at a different observing frequency: 1.4 GHz (top panel), r-band (i.e., 4.6×10^{14} Hz, central panel) and 1 keV (i.e., 2.4×10^{17} Hz, bottom panel). In the bottom panel, we show for comparison the slopes expected for the X-ray (mostly also valid for the Optical) post-peak (for viewing angles close to the jet axis) and post-jet-break light curves. The late-time peak produced by the emission of the counter-jet is also annotated.

5. Gw170817 and GRB170817A: An Observational Test-Bed for Off-Axis Structured Jet Theory

As noted in the introduction, the discovery of GW170817 and its electromagnetic counterparts [25,26] marks a discontinuity in the evolution of the interest of the astrophysical community in structured jets. The reason is that a structured jet observed off-axis provides the most satisfactory and self-consistent explanation for the behavior of the associated short gamma-ray burst GRB170817A [55,56] and of the non-thermal emission observed at the source position starting on the second week after the gravitational wave event in radio [63,66,67,275–280], X-rays [58,60,281–286] and optical around the peak (with the Hubble Space Telescope [287,288] and with the Large Binocular Telescope [67]). Striking evidence in favor of such a scenario came from the latter component: its initial light curve evolution, with an unprecedented shallow increase in flux as $\sim t^{0.8}$ over three months (see Figure 11), sparked a debate within the community about its interpretation. The two main competing scenarios attributed the emission to a mildly relativistic shock propagating into the interstellar medium. In one scenario, the shock was produced by an off-axis structured jet, e.g., [58,62] that successfully broke out of the merger ejecta. In the other, it was due to a quasi-spherical outflow with a velocity profile, with most energy in the slower ejecta (models of the radio surface brightness in the two scenarios are shown in Figure 12).

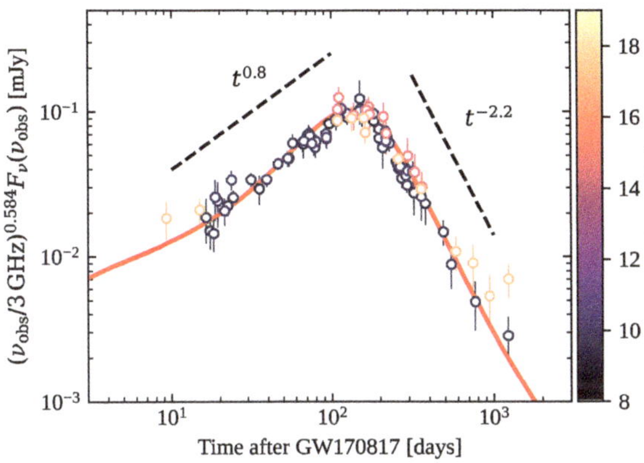

Figure 11. Observations and model of the afterglow of GRB170817A. Colored circles with error bars show flux densities measured in radio, X-ray and optical observations (data from [289]) at the position of GW170817, rescaled to a common frequency of 3 GHz assuming a power-law spectrum $F_\nu \propto \nu^{-0.584}$ [60]. The data points are color-coded according to the color bar on the right in order to show the original observing frequency. The red solid line is the prediction of an off-axis structured jet model with a power-law profile of both the energy and the Lorentz factor, with the same form and similar parameters as that in [67], namely $E(\theta) = E(0)/(1 + (\theta/\theta_c)^{s_E})$ and $\Gamma(\theta) = 1 + (\Gamma(0) - 1)/(1 + (\theta/\theta_c)^{s_\Gamma})$ with $\theta_c = 3.2°$, $E(0) = 4 \times 10^{52}$ erg, $s_E = 4.5$, $\Gamma(0) = 1000$, $s_\Gamma = 3.3$, $n = 10^{-3}\,cm^{-3}$, $\epsilon_e = 0.1$, $\epsilon_B = 10^{-4}$, $p = 2.168$ and $\theta_v = 19°$.

The latter outflow could have been either the result of the jet being present, but 'choked' [279,290] (i.e., the central engine turned off before the head was able to break out), or could have arisen from the rapid conversion of magnetic energy into kinetic energy soon after the merger, e.g., [291,292]. Unfortunately, Nakar and Piran [239] showed that it was impossible to tell apart the two scenarios solely from the light curve evolution before the peak, because a shallow power-law increase in the radio and X-ray flux density could be produced by ejecta with an appropriately chosen angular profile, or velocity profile (or an infinite family of combinations of the two), and in neither case the required parameters were unrealistic. The solution to the riddle was eventually provided by high-resolution VLBI observations [66,67] at 75, 207 and 230 days after the merger, which revealed an apparently superluminal motion of the radio source centroid and a very small projected size of the image [67] see Figure 13. Only the off-axis structured jet scenario has been demonstrated to provide a complete, self-consistent explanation of the light curves and centroid motion to date.

The off-axis viewing angle is most likely in the 15–25° range, see [66,293], which is in good agreement with the binary orbital plane inclination derived from the gravitational wave analysis [25], has been also identified as the culprit of the extremely low luminosity of the GRB170817A prompt emission approximately $L_{iso} \sim 10^{47}$ erg/s, see [26,55,56] when compared to the other known short GRBs with a redshift measurement, since the latter is instead observed within $\sim 2\theta_c$ [198,205]. However, the simple interpretation of GRB170817A as being regular short GRBs with a suppressed flux due to relativistic beaming e.g., [26,294] is not viable. Compactness limits [204,205] indicate that the GRB170817A emission was produced by material moving at a Lorentz factor $\Gamma \gtrsim$ few, but seen under a viewing angle

$\theta_e \lesssim 5°$. Given the viewing angle $\theta_v \gtrsim 15°$ and the opening angle $\theta_c \lesssim 5°$, e.g., [293], this is not compatible with emission originating at the border of the jet core, for which $\theta_e = (\theta_v - \theta_c) > 5°$. The mechanism that produced the observed emission could still have been a similar one as that behind the known short GRB population, but operating well outside the jet core, e.g., [197], or a different mechanism, such as the cocoon shock breakout, e.g., [79,239,290] see Section 2.5.

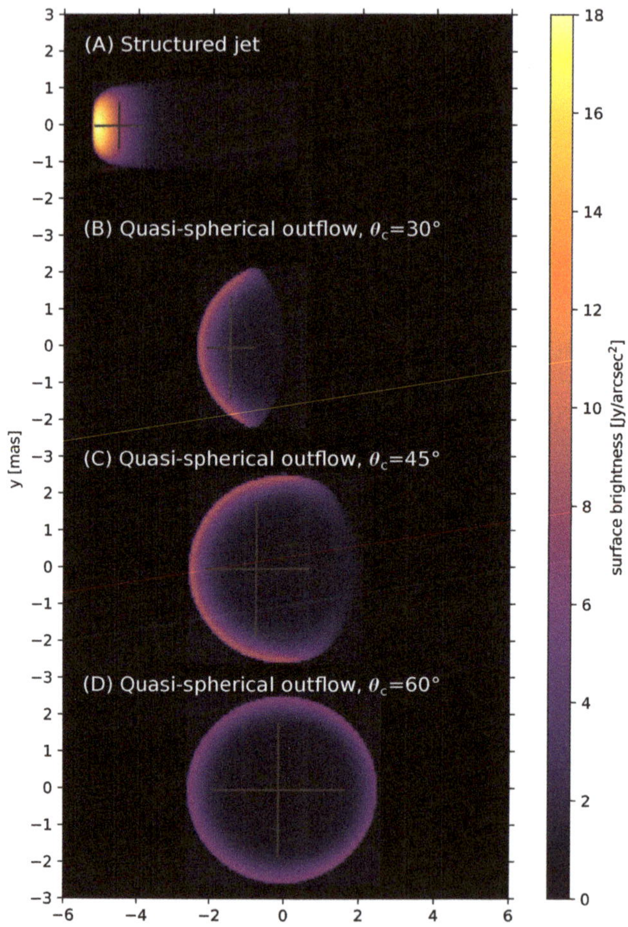

Figure 12. Model images of an off-axis structured jet (A) and of three quasi-spherical explosions (B, C and D) with similar parameters, but differing opening angles θ_c. The angle between the line of sight and the shock symmetry axis in all cases is $\theta_v = 15°$. The images show the surface brightness at 5 GHz, 207 days after the merger, as seen from a distance of 41 Mpc. In each panel, the merger is located at coordinates $(0, 0)$, while the grey cross shows the image centroid and the full width at half the maximum of the image in two perpendicular directions. All models are compatible with the observations of the GW170817 non-thermal multi-wavelength counterpart (shown in Figure 11) before the peak. Adapted from [67].

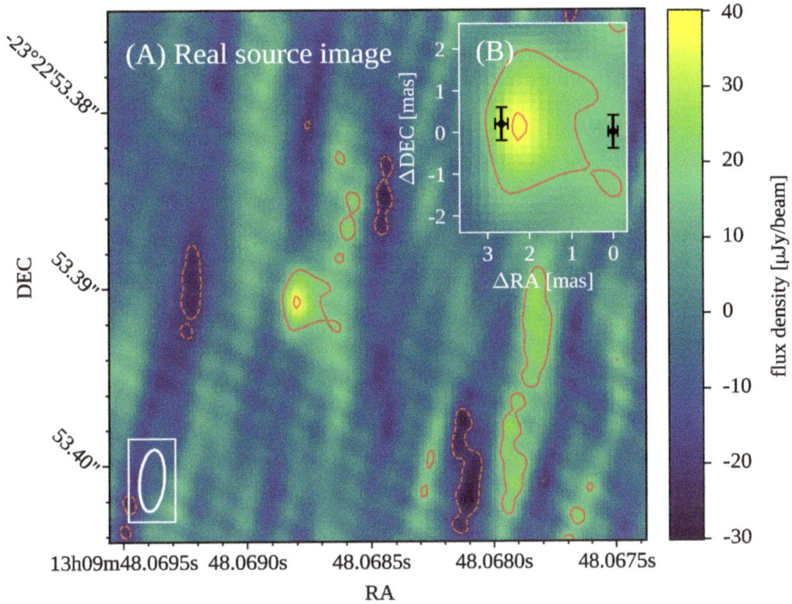

Figure 13. Global VLBI image of the GRB170817A afterglow 207 days after the merger. The main panel (**A**) shows the cleaned surface brightness map (color-coded according to the color bar on the right) of a small region around the position of GW170817, observed 207 days after the merger. Red contours are lines of constant surface brightness corresponding to −20 (dashed), 20 and 40 (solid) μJy/beam. The root-mean-square of the image noise is 8 μJy/beam. The ellipse in the lower left corner of the plot encompasses the full width at half the maximum of the synthesized beam (i.e., the resolution element). The inset, panel (**B**), shows a zoom of the region close to the peak of the surface brightness distribution, with black error bars marking the best fit and one-sigma errors on the centroid position of the source at 75 and 230 days, as measured by [66], with the axes showing the displacement with respect to the position at 75 d. Reproduced from [67].

6. Concluding Remarks

In this review, we attempted at summarizing the historical development of the current ideas on the processes that shape the structure of relativistic jets in gamma-ray bursts and the main observational consequences of that structure. We put some emphasis on the qualitative physical description of the structure formation and on the imprint of the structure on the main observables that characterize the prompt and afterglow emission of gamma-ray bursts, with the aim of providing first guidance (with minimal use of technicalities and a pedagogical approach) to those who encounter these topics for the first time. Inevitably, the review only covers a fraction of the relevant literature, only scratches the surface of most arguments, and represents only a partial account of the huge scientific effort performed by the community in the last few decades on this topic. For example, we entirely neglected the important topic of the polarization of the emission from a structured jet: fortunately, this is brilliantly covered by another review article [45] in this same Special Issue.

Most of the topics presented here are fields of active research and are evolving at a fast pace, especially in the latest years after GW170817. The upcoming science runs of the ground-based gravitational wave detector network, currently comprising the Laser Interferometer Gravitational-wave Observatory (LIGO, [295]), Advanced Virgo [296] and KAGRA [297], will likely soon yield at least one new binary neutron star and/or black hole-neutron star merger event (e.g., [298–307]), hopefully with an associated jet: this will provide new unique insights on the structure of short GRB jets, on the properties of off-axis jet emission in general, and on the incidence of jets, which can be used to constrain the progenitor population (e.g., [308]). With new 'golden' events such as GW170817, direct information on the jet structure can be extracted, e.g., through the methodologies discussed at the end of Section 4.2, provided that a detailed, high-cadence, multi-wavelength dataset will be collected. The availability of VLBI observations will greatly enhance the chances to break the inherent degeneracies in the afterglow modeling and hence in the jet structure. Moreover, new events observed at differing viewing angles will be the perfect route to test the quasi-universal structured jet hypothesis. We look forward to learning much more about the structure of gamma-ray burst jets from these and other observations, and from theoretical advances, in the close future.

Funding: This research was funded by the Italian Ministry of University and Research through the PRIN grant 20179ZF5KS and by the Italian National Institute for Astrophysics (INAF) through the PRIN grant "Towards the SKA and CTA era: discovery, localisation, and physics of transient sources" (1.05.01.88.06).

Acknowledgments: We thank G. Ghisellini for years of fruitful discussions. We thank Yuri Sato for useful comments. We thank the anonymous referees for their comments, which helped in improving the quality and completeness of this review.

Conflicts of Interest: The authors declare no conflict of interest.

Abbreviations

The following abbreviations are used in this manuscript:

GRB	Gamma Ray Burst
NS	Neutron Star
BH	Black Hole
AGN	Active Galactic Nucleus
CGRO	Compton Gamma Ray Observatory
BATSE	Burst And Transient Source Experiment
GBM	Gamma-ray Burst Monitor
BAT	Burst Alert Telescope
XRT	X-ray Telescope
SSC	Synchrotron Self Compton
LAT	Large Area Telescope

References

1. Cavallo, G.; Rees, M.J. A qualitative study of cosmic fireballs and gamma-ray bursts. *MNRAS* **1978**, *183*, 359–365. [CrossRef]
2. Paczynski, B. Gamma-ray bursters at cosmological distances. *Astrophys. J.* **1986**, *308*, L43–L46. [CrossRef]
3. Goodman, J. Are gamma-ray bursts optically thick? *Astrophys. J.* **1986**, *308*, L47. [CrossRef]
4. Krolik, J.H.; Pier, E.A. Relativistic motion in gamma-ray bursts. *Astrophys. J.* **1991**, *373*, 277–284. [CrossRef]
5. Lithwick, Y.; Sari, R. Lower limits on Lorentz factors in gamma-ray bursts. *Astrophys. J.* **2001**, *555*, 540–545.
6. Taylor, G.B.; Frail, D.A.; Berger, E.; Kulkarni, S.R. The Angular Size and Proper Motion of the Afterglow of GRB 030329. *Astrophys. J.* **2004**, *69*, L1.
7. Rees, M.J. Appearance of Relativistically Expanding Radio Sources. *Nature* **1966**, *211*, 468–470. [CrossRef]
8. Rhoads, J.E. How to Tell a Jet from a Balloon: A Proposed Test for Beaming in Gamma-Ray Bursts. *Astrophys. J.* **1997**, *487*, L1.
9. Briggs, M.S.; Band, D.L.; Kippen, R.M.; Preece, R.D.; Kouveliotou, C.; van Paradijs, J.; Share, G.H.; Murphy, R.J.; Matz, S.M.; Connors, A.; et al. Observations of GRB 990123 by theCompton gamma ray observatory. *Astrophys. J.* **1999**, *524*, 82–91.
10. Preece, R.; Michael Burgess, J.; von Kienlin, A.; Bhat, P.N.; Briggs, M.S.; Byrne, D.; Chaplin, V.; Cleveland, W.; Collazzi, A.C.; Connaughton, V.; et al. The First Pulse of the Extremely Bright GRB 130427A: A Test Lab for Synchrotron Shocks. *Science* **2013**, *343*, 51–54.
11. Rhoads, J.E. The Dynamics and Light Curves of Beamed Gamma-Ray Burst Afterglows. *Astrophys. J.* **1999**, *525*, 737–749. [CrossRef]
12. Fruchter, A.S.; Thorsett, S.E.; Metzger, M.R.; Sahu, K.C.; Petro, L.; Livio, M.; Ferguson, H.; Pian, E.; Hogg, D.W.; Galama, T.; et al. Hubble Space Telescope and Palomar Imaging of GRB 990123: Implications for the Nature of Gamma-Ray Bursts and Their Hosts. *Astrophys. J.* **1999**, *519*, L13.

13. Kulkarni, S.R.; Djorgovski, S.G.; Odewahn, S.C.; Bloom, J.S.; Gal, R.R.; Koresko, C.D.; Harrison, F.A.; Lubin, L.M.; Armus, L.; Sari, R.; et al. The afterglow, redshift and extreme energetics of the γ-ray burst of 23 January 1999. *Nature* **1999**, *398*, 389–394. [CrossRef]
14. Stanek, K.Z.; Garnavich, P.M.; Kaluzny, J.; Pych, W.; Thompson, I. BVRI Observations of the Optical Afterglow of GRB 990510*. *Astrophys. J.* **1999**, *522*, L39. [CrossRef]
15. Harrison, F.A.; Bloom, J.S.; Frail, D.A.; Sari, R.; Kulkarni, S.R.; Djorgovski, S.G.; Axelrod, T.; Mould, J.; Schmidt, B.P.; Wieringa, M.H.; et al. Optical and Radio Observations of the Afterglow from GRB 990510: Evidence for a Jet. *Astrophys. J.* **1999**, *523*, L121. [CrossRef]
16. Chandra, P.; Frail, D.A. A Radio-selected Sample of Gamma-Ray Burst Afterglows. *Astrophys. J.* **2012**, *746*, 156. [CrossRef]
17. Lu, R.J.; Wei, J.J.; Qin, S.F.; Liang, E.W. Selection Effects on the Observed Redshift Dependence of Gamma-Ray Burst Jet Opening Angles. *Astrophys. J.* **2012**, *745*, 168. [CrossRef]
18. Frail, D.A.; Kulkarni, S.R.; Sari, R.; Djorgovski, S.G.; Bloom, J.S.; Galama, T.J.; Reichart, D.E.; Berger, E.; Harrison, F.A.; Price, P.A.; et al. Beaming in Gamma-Ray Bursts: Evidence for a Standard Energy Reservoir. *Astrophys. J.* **2001**, *562*, L55. [CrossRef]
19. Fong, W.; Berger, E.; Margutti, R.; Zauderer, B.A. A decade of short-duration gamma-ray burst broadband afterglows: Energetics, circumburst densities, and jet opening angles. *Astrophys. J.* **2015**, *815*, 102.
20. Berger, E.; Kulkarni, S.R.; Frail, D.A. A Standard Kinetic Energy Reservoir in Gamma-Ray Burst Afterglows. *Astrophys. J.* **2003**, *590*, 379. [CrossRef]
21. Ghirlanda, G.; Ghisellini, G.; Lazzati, D. The Collimation-corrected Gamma-Ray Burst Energies Correlate with the Peak Energy of Their $\nu F\nu$ Spectrum. *Astrophys. J.* **2004**, *616*, 331. [CrossRef]
22. Zhang, B.; Mészáros, P. Gamma-Ray Burst Beaming: A Universal Configuration with a Standard Energy Reservoir? *Astrophys. J.* **2002**, *571*, 876. [CrossRef]
23. Lipunov, V.M.; Postnov, K.A.; Prokhorov, M.E. Gamma-Ray Bursts as Standard-Energy Explosions. *Astron. Rep.* **2001**, *45*, 236–240. [CrossRef]
24. Rossi, E.; Lazzati, D.; Rees, M.J. Afterglow light curves, viewing angle and the jet structure of γ-ray bursts. *MNRAS* **2002**, *332*, 945–950. [CrossRef]
25. Abbott, B.P.; Abbott, R.; Abbott, T.D.; Acernese, F.; Ackley, K.; Adams, C.; Adams, T.; Addesso, P.; Adhikari, R.X.; Adya, V.B.; et al. GW170817: Observation of Gravitational Waves from a Binary Neutron Star Inspiral. *Phys. Rev. Lett.* **2017**, *119*, 161101, [CrossRef]
26. Abbott, B.P.; Abbott, R.; Abbott, T.D.; Acernese, F.; Ackley, K.; Adams, C.; Adams, T.; Addesso, P.; Adhikari, R.X.; Adya, V.B.; et al. Gravitational Waves and Gamma-Rays from a Binary Neutron Star Merger: GW170817 and GRB 170817A. *Astrophys. J.* **2017**, *848*, L13, [CrossRef]
27. Abbott, B.P.; Abbott, R.; Abbott, T.D.; Acernese, F.; Ackley, K.; Adams, C.; Adams, T.; Addesso, P.; Adhikari, R.X.; Adya, V.B.; et al. Multi-messenger Observations of a Binary Neutron Star Merger. *Astrophys. J.* **2017**, *848*, L12. [CrossRef]
28. SAO/NASA. Astrophysics Data System (ADS). Available online: https://ui.adsabs.harvard.edu/ (accessed on 10 June 2022).
29. Lazzati, D.; Begelman, M.C. Universal GRB Jets from Jet-Cocoon Interaction in Massive Stars. *Astrophys. J.* **2005**, *629*, 903. [CrossRef]
30. Morsony, B.J.; Lazzati, D.; Begelman, M.C. Temporal and Angular Properties of Gamma-Ray Burst Jets Emerging from Massive Stars. *Astrophys. J.* **2007**, *665*, 569–598. [CrossRef]
31. Granot, J.; Kumar, P. Constraining the Structure of Gamma-Ray Burst Jets through the Afterglow Light Curves. *Astrophys. J.* **2003**, *591*, 1086. [CrossRef]
32. Kumar, P.; Granot, J. The Evolution of a Structured Relativistic Jet and Gamma-Ray Burst Afterglow Light Curves. *Astrophys. J.* **2003**, *591*, 1075–1085. [CrossRef]
33. Ghisellini, G.; Tavecchio, F.; Chiaberge, M. Structured jets in TeV BL Lac objects and radiogalaxies - Implications for the observed properties. *Astron. Astrophys. Suppl. Ser.* **2005**, *432*, 401–410. [CrossRef]
34. Racusin, J.L.; Karpov, S.V.; Sokolowski, M.; Granot, J.; Wu, X.F.; Pal'shin, V.; Covino, S.; van der Horst, A.J.; Oates, S.R.; Schady, P.; et al. Broadband observations of the naked-eye gamma-ray burst GRB 080319B. *Nature* **2008**, *455*, 183–188. [CrossRef] [PubMed]
35. van der Horst, A.J.; Rol, E.; Wijers, R.A.M.; Strom, R.; Kaper, L.; Kouveliotou, C. The Radio Afterglow of GRB 030329 at Centimeter Wavelengths: Evidence for a Structured Jet or Nonrelativistic Expansion. *Astrophys. J.* **2005**, *634*, 1166. [CrossRef]
36. Granot, J. Afterglow Light Curves from Impulsive Relativistic Jets with an Unconventional Structure. *Astrophys. J.* **2005**, *631*, 1022. [CrossRef]
37. Salmonson, J.D. Perspective on Afterglows: Numerically Computed Views, Light Curves, and the Analysis of Homogeneous and Structured Jets with Lateral Expansion. *Astrophys. J.* **2003**, *592*, 1002. [CrossRef]
38. Perna, R.; Sari, R.; Frail, D. Jets in Gamma-Ray Bursts: Tests and Predictions for the Structured Jet Model. *Astrophys. J.* **2003**, *594*, 379–384. [CrossRef]
39. Dai, X.; Zhang, B. A Global Test of a Quasi-universal Gamma-Ray Burst Jet Model through Monte Carlo Simulations. *Astrophys. J.* **2005**, *621*, 875–883. [CrossRef]
40. Lloyd-Ronning, N.M.; Dai, X.; Zhang, B. On the Structure of Quasi-universal Jets for Gamma-Ray Bursts. *Astrophys. J.* **2004**, *601*, 371–379. [CrossRef]
41. Nakar, E.; Granot, J.; Guetta, D. Testing the Predictions of the Universal Structured Gamma-Ray Burst Jet Model. *Astrophys. J.* **2004**, *606*, L37–L40.

42. Ghisellini, G.; Lazzati, D. Polarization light curves and position angle variation of beamed gamma-ray bursts. *MNRAS* **1999**, *309*, L7–L11. [CrossRef]
43. Rossi, E.M.; Lazzati, D.; Salmonson, J.D.; Ghisellini, G. The polarization of afterglow emission reveals γ-ray bursts jet structure. *MNRAS* **2004**, *354*, 86–100. [CrossRef]
44. Lazzati, D.; Covino, S.; Gorosabel, J.; Rossi, E.; Ghisellini, G.; Rol, E.; Castro Cerón, J.M.; Castro-Tirado, A.J.; Della Valle, M.; di Serego Alighieri, S.; et al. On the jet structure and magnetic field configuration of GRB 020813. *Astron. Astrophys.* **2004**, *422*, 121–128.
45. Gill, R.; Kole, M.; Granot, J. GRB Polarization: A Unique Probe of GRB Physics. *Galaxies* **2021**, *9*, 82. [CrossRef]
46. Totani, T.; Panaitescu, A. Orphan Afterglows of Collimated Gamma-Ray Bursts: Rate Predictions and Prospects for Detection. *ApJ* **2002**, *576*, 120–134. [CrossRef]
47. Granot, J.; Panaitescu, A.; Kumar, P.; Woosley, S.E. Off-Axis Afterglow Emission from Jetted Gamma-Ray Bursts. *Astrophys. J.* **2002**, *570*, L61–L64. [CrossRef]
48. Nakar, E.; Piran, T.; Granot, J. The Detectability of Orphan Afterglows. *Astrophys. J.* **2002**, *579*, 699–705. [CrossRef]
49. Rau, A.; Greiner, J.; Schwarz, R. Constraining the GRB collimation with a survey for orphan afterglows. *Astron. Astrophys.* **2006**, *449*, 79–88. [CrossRef]
50. Perna, R.; Loeb, A. Constraining the Beaming of Gamma-Ray Bursts with Radio Surveys. *Astrophys. J.* **1998**, *509*, L85–L88. [CrossRef]
51. Rossi, E.M.; Perna, R.; Daigne, F. 'Orphan' afterglows in the Universal structured jet model for γ-ray bursts. *MNRAS* **2008**, *390*, 675–682.
52. Ghirlanda, G.; Burlon, D.; Ghisellini, G.; Salvaterra, R.; Bernardini, M.G.; Campana, S.; Covino, S.; D'Avanzo, P.; D'Elia, V.; Melandri, A.; et al. GRB Orphan Afterglows in Present and Future Radio Transient Surveys. *PASA* **2014**, *31*, e022. [CrossRef]
53. Ghirlanda, G.; Salvaterra, R.; Campana, S.; Vergani, S.D.; Japelj, J.; Bernardini, M.G.; Burlon, D.; D'Avanzo, P.; Melandri, A.; Gomboc, A.; et al. Unveiling the population of orphan γ-ray bursts. *Astron. Astrophys.* **2015**, *578*, A71. [CrossRef]
54. Granot, J. The Structure and Dynamics of GRB Jets. *arXiv* **2007**, arXiv:astro-ph/0610379
55. Goldstein, A.; Veres, P.; Burns, E.; Briggs, M.S.; Hamburg, R.; Kocevski, D.; Wilson-Hodge, C.A.; Preece, R.D.; Poolakkil, S.; Roberts, O.J.; et al. An Ordinary Short Gamma-Ray Burst with Extraordinary Implications: Fermi-GBM Detection of GRB 170817A. *Astrophys. J.* **2017**, *848*, L14. [CrossRef]
56. Savchenko, V.; Ferrigno, C.; Kuulkers, E.; Bazzano, A.; Bozzo, E.; Brandt, S.; Chenevez, J.; Courvoisier, T.J.L.; Diehl, R.; Domingo, A.; et al. INTEGRAL Detection of the First Prompt Gamma-Ray Signal Coincident with the Gravitational-wave Event GW170817. *Astrophys. J.* **2017**, *848*, L15. [CrossRef]
57. Lamb, G.P.; Kobayashi, S. Electromagnetic counterparts to structured jets from gravitational wave detected mergers. *Mon. Not. R. Astron. Soc.* **2017**, *472*, 4953–4964.
58. D'Avanzo, P.; Campana, S.; Salafia, O.S.; Ghirlanda, G.; Ghisellini, G.; Melandri, A.; Bernardini, M.G.; Branchesi, M.; Chassande-Mottin, E.; Covino, S.; et al. The evolution of the X-ray afterglow emission of GW 170817/ GRB 170817A in XMM-Newton observations. *Astron. Astrophys. Suppl. Ser.* **2018**, *613*, L1. [CrossRef]
59. Alexander, K.D.; Margutti, R.; Blanchard, P.K.; Fong, W.; Berger, E.; Hajela, A.; Eftekhari, T.; Chornock, R.; Cowperthwaite, P.S.; Giannios, D.; et al. A Decline in the X-ray through Radio Emission from GW170817 Continues to Support an Off-axis Structured Jet. *Astrophys. J. Lett.* **2018**, *863*, L18. [CrossRef]
60. Margutti, R.; Alexander, K.D.; Xie, X.; Sironi, L.; Metzger, B.D.; Kathirgamaraju, A.; Fong, W.; Blanchard, P.K.; Berger, E.; MacFadyen, A.; et al. The Binary Neutron Star Event LIGO/Virgo GW170817 160 Days after Merger: Synchrotron Emission across the Electromagnetic Spectrum. *Astrophys. J.* **2018**, *856*, L18.
61. Troja, E.; Piro, L.; Ryan, G.; van Eerten, H.; Ricci, R.; Wieringa, M.H.; Lotti, S.; Sakamoto, T.; Cenko, S.B. The outflow structure of GW170817 from late-time broad-band observations. *Mon. Not. R. Astron. Soc.* **2018**, *478*, L18–L23. [CrossRef]
62. Lazzati, D.; Perna, R.; Morsony, B.J.; Lopez-Camara, D.; Cantiello, M.; Ciolfi, R.; Giacomazzo, B.; Workman, J.C. Late Time Afterglow Observations Reveal a Collimated Relativistic Jet in the Ejecta of the Binary Neutron Star Merger GW170817. *Phys. Rev. Lett.* **2018**, *120*, 241103. [PubMed]
63. Resmi, L.; Schulze, S.; Ishwara-Chandra, C.H.; Misra, K.; Buchner, J.; De Pasquale, M.; Sánchez-Ramírez, R.; Klose, S.; Kim, S.; Tanvir, N.R.; et al. Low-frequency View of GW170817/GRB 170817A with the Giant Metrewave Radio Telescope. *Astrophys. J.* **2018**, *867*, 57.
64. Wu, Y.; MacFadyen, A. Constraining the Outflow Structure of the Binary Neutron Star Merger Event GW170817/GRB170817A with a Markov Chain Monte Carlo Analysis. *Astrophys. J.* **2018**, *869*, 55. [CrossRef]
65. Ryan, G.; van Eerten, H.; Piro, L.; Troja, E. Gamma-Ray Burst Afterglows in the Multimessenger Era: Numerical Models and Closure Relations. *Astrophys. J.* **2020**, *896*, 166. [CrossRef]
66. Mooley, K.P.; Deller, A.T.; Gottlieb, O.; Nakar, E.; Hallinan, G.; Bourke, S.; Frail, D.A.; Horesh, A.; Corsi, A.; Hotokezaka, K. Superluminal motion of a relativistic jet in the neutron-star merger GW170817. *Nature* **2018**, *561*, 355–359. [PubMed]
67. Ghirlanda, G.; Salafia, O.S.; Paragi, Z.; Giroletti, M.; Yang, J.; Marcote, B.; Blanchard, J.; Agudo, I.; An, T.; Bernardini, M.G.; et al. Compact radio emission indicates a structured jet was produced by a binary neutron star merger. *Science* **2019**, *363*, 968–971. [CrossRef]
68. Margutti, R.; Chornock, R. First Multimessenger Observations of a Neutron Star Merger. *ARA&A* **2021**, *59*, 155–202. [CrossRef]

69. Nakar, E. The electromagnetic counterparts of compact binary mergers. *Phys. Rep.* **2020**, *886*, 1–84. [CrossRef]
70. Blandford, R.D.; Znajek, R.L. Electromagnetic extraction of energy from Kerr black holes. *Mon. Not. R. Astron. Soc.* **1977**, *179*, 433–456. [CrossRef]
71. Usov, V.V. Millisecond pulsars with extremely strong magnetic fields as a cosmological source of γ-ray bursts. *Nature* **1992**, *357*, 472–474. [CrossRef]
72. Blandford, R.D.; Rees, M.J. A "twin-exhaust" model for double radio sources. *MNRAS* **1974**, *169*, 395–415. [CrossRef]
73. Marti, J.M.; Mueller, E.; Ibanez, J.M. Hydrodynamical simulations of relativistic jets. *A&A* **1994**, *281*, L9–L12.
74. Matzner, C.D. Supernova hosts for gamma-ray burst jets: Dynamical constraints. *MNRAS* **2003**, *345*, 575–589. [CrossRef]
75. Bromberg, O.; Nakar, E.; Piran, T.; Sari, R. The Propagation of Relativistic Jets in External Media. *Astrophys. J.* **2011**, *740*, 100. [CrossRef]
76. Falle, S.A.E.G. Self-similar jets. *MNRAS* **1991**, *250*, 581–596. [CrossRef]
77. Shapiro, P.R. Relativistic blast waves that accelerate. *Astrophys. J.* **1980**, *236*, 958–969. [CrossRef]
78. Tan, J.C.; Matzner, C.D.; McKee, C.F. Trans-Relativistic Blast Waves in Supernovae as Gamma-Ray Burst Progenitors. *Astrophys. J.* **2001**, *551*, 946–972. [CrossRef]
79. Gottlieb, O.; Nakar, E.; Piran, T. The cocoon emission - an electromagnetic counterpart to gravitational waves from neutron star mergers. *MNRAS* **2018**, *473*, 576–584. [CrossRef]
80. Meszaros, P.; Laguna, P.; Rees, M.J. Gasdynamics of Relativistically Expanding Gamma-Ray Burst Sources: Kinematics, Energetics, Magnetic Fields, and Efficiency. *Astrophys. J.* **1993**, *415*, 181. [CrossRef]
81. Baring, M.G.; Harding, A.K. The Escape of High-Energy Photons from Gamma-Ray Bursts. *Astrophys. J.* **1997**, *491*, 663–686. [CrossRef]
82. Rees, M.J.; Mészáros, P. Dissipative Photosphere Models of Gamma-Ray Bursts and X-ray Flashes. *Astrophys. J.* **2005**, *628*, 847–852. [CrossRef]
83. Kobayashi, S.; Piran, T.; Sari, R. Hydrodynamics of a Relativistic Fireball: The Complete Evolution. *Astrophys. J.* **1999**, *513*, 669–678. [CrossRef]
84. Rees, M.J.; Meszaros, P. Unsteady Outflow Models for Cosmological Gamma-Ray Bursts. *Astrophys. J.* **1994**, *430*, L93. [CrossRef]
85. Zhang, B.; Yan, H. The Internal-collision-induced Magnetic Reconnection and Turbulence (ICMART) Model of Gamma-ray Bursts. *Astrophys. J.* **2011**, *726*, 90. [CrossRef]
86. Drenkhahn, G.; Spruit, H.C. Efficient acceleration and radiation in Poynting flux powered GRB outflows. *Astron. Astrophys.* **2002**, *391*, 1141–1153. [CrossRef]
87. McKinney, J.C.; Uzdensky, D.A. A reconnection switch to trigger gamma-ray burst jet dissipation. *MNRAS* **2012**, *419*, 573–607. [CrossRef]
88. van Marle, A.J.; Langer, N.; Achterberg, A.; García-Segura, G. Forming a constant density medium close to long gamma-ray bursts. *Astron. Astrophys.* **2006**, *460*, 105–116. [CrossRef]
89. Paczynski, B.; Rhoads, J.E. Radio Transients from Gamma-Ray Bursters. *Astrophys. J.* **1993**, *418*, L5. [CrossRef]
90. Sari, R.; Piran, T. Hydrodynamic Timescales and Temporal Structure of Gamma-Ray Bursts. *Astrophys. J.* **1995**, *455*, L143. [CrossRef]
91. Kobayashi, S.; Sari, R. Optical Flashes and Radio Flares in Gamma-Ray Burst Afterglow: Numerical Study. *Astrophys. J.* **2000**, *542*, 819–828. [CrossRef]
92. Blandford, R.D.; McKee, C.F. Fluid dynamics of relativistic blast waves. *Phys. Fluids* **1976**, *19*, 1130–1138. [CrossRef]
93. Zhang, W.; MacFadyen, A. The Dynamics and Afterglow Radiation of Gamma-Ray Bursts. I. Constant Density Medium. *Astrophys. J.* **2009**, *698*, 1261–1272. [CrossRef]
94. van Eerten, H.J.; Leventis, K.; Meliani, Z.; Wijers, R.A.M.J.; Keppens, R. Gamma-ray burst afterglows from transrelativistic blast wave simulations. *MNRAS* **2010**, *403*, 300–316. [CrossRef]
95. Granot, J.; Piran, T. On the lateral expansion of gamma-ray burst jets. *MNRAS* **2012**, *421*, 570–587. [CrossRef]
96. Blandford, R.; Meier, D.; Readhead, A. Relativistic Jets in Active Galactic Nuclei. *arXiv* **2018**, arXiv:1812.06025.
97. Molina, E.; del Palacio, S.; Bosch-Ramon, V. A model for high-mass microquasar jets under the influence of a strong stellar wind. *Astron. Astrophys.* **2019**, *629*, A129. [CrossRef]
98. Romero, G.; Boettcher, M.; Markoff, S.; Tavecchio, F. Relativistic Jets in Active Galactic Nuclei und Microquasars. *arXiv* **2016**, arXiv:1611.09507.
99. Contopoulos, I.; Kazanas, D.; Papadopoulos, D.B. The Force-Free Magnetosphere of a Rotating Black Hole. *arXiv* **2012**, arXiv:1212.0320.
100. Tchekhovskoy, A.; Narayan, R.; McKinney, J.C. Black Hole Spin and The Radio Loud/Quiet Dichotomy of Active Galactic Nuclei. *Astrophys. J.* **2010**, *711*, 50–63. [CrossRef]
101. Camilloni, F.; Dias, O.J.C.; Grignani, G.; Harmark, T.; Oliveri, R.; Orselli, M.; Placidi, A.; Santos, J.E. Blandford-Znajek monopole expansion revisited: Novel non-analytic contributions to the power emission. *J. Cosmol. Astropart. Phys.* **2022**, *2022*, 032. [CrossRef]
102. Popham, R.; Woosley, S.E.; Fryer, C. Hyperaccreting black holes and gamma-ray bursts. *Astrophys. J.* **1999**, *518*, 356–374. [CrossRef]

103. Ruffert, M.; Janka, H.T. Gamma-ray bursts from accreting black holes in neutron star mergers. *Astron. Astrophys.* **1999**, *344*, 573–606.
104. Chen, W.X.; Beloborodov, A.M. Neutrino-cooled Accretion Disks around Spinning Black Holes. *Astrophys. J.* **2007**, *657*, 383. [CrossRef]
105. Levinson, A.; Globus, N. Ultra-relativistic, neutrino-driven flows IN gamma-ray bursts: A double transonic flow solution in a schwarzschild spacetime. *Astrophys. J.* **2013**, *770*, 159.
106. Salafia, O.S.; Giacomazzo, B. Accretion-to-jet energy conversion efficiency in GW170817. *Astron. Astrophys.* **2021**, *645*, A93. [CrossRef]
107. Leng, M.; Giannios, D. Testing the neutrino annihilation model for launching GRB jets. *MNRAS* **2014**, *445*, L1–L5. [CrossRef]
108. Thompson, T.A.; Chang, P.; Quataert, E. Magnetar Spin-Down, Hyperenergetic Supernovae, and Gamma-Ray Bursts. *Astrophys. J.* **2004**, *611*, 380–393. [CrossRef]
109. Metzger, B.D.; Giannios, D.; Thompson, T.A.; Bucciantini, N.; Quataert, E. The protomagnetar model for gamma-ray bursts. *MNRAS* **2011**, *413*, 2031–2056. [CrossRef]
110. Lü, H.J.; Zhang, B.; Lei, W.H.; Li, Y.; Lasky, P.D. The Millisecond Magnetar Central Engine in Short GRBs. *Astrophys. J.* **2015**, *805*, 89. [CrossRef]
111. Zalamea, I.; Beloborodov, A.M. Neutrino heating near hyper-accreting black holes. *MNRAS* **2011**, *410*, 2302–2308. [CrossRef]
112. Kawanaka, N.; Piran, T.; Krolik, J.H. Jet Luminosity from Neutrino-dominated Accretion Flows in Gamma-Ray Bursts. *Astrophys. J.* **2013**, *766*, 31. [CrossRef]
113. Just, O.; Obergaulinger, M.; Janka, H.T.; Bauswein, A.; Schwarz, N. Neutron-star Merger Ejecta as Obstacles to Neutrino-powered Jets of Gamma-Ray Bursts. *Astrophys. J.* **2016**, *816*, L30. [CrossRef]
114. Salafia, O.S.; Barbieri, C.; Ascenzi, S.; Toffano, M. Gamma-ray burst jet propagation, development of angular structure, and the luminosity function. *Astron. Astrophys.* **2020**, *636*, A105. [CrossRef]
115. Hamidani, H.; Ioka, K. Jet propagation in expanding medium for gamma-ray bursts. *MNRAS* **2021**, *500*, 627–642. [CrossRef]
116. Gottlieb, O.; Nakar, E.; Bromberg, O. The structure of hydrodynamic γ-ray burst jets. *MNRAS* **2021**, *500*, 3511–3526. [CrossRef]
117. Bromberg, O.; Granot, J.; Lyubarsky, Y.; Piran, T. The dynamics of a highly magnetized jet propagating inside a star. *MNRAS* **2014**, *443*, 1532–1548. [CrossRef]
118. Levinson, A.; Begelman, M.C. Collimation and Confinement of Magnetic Jets by External Media. *Astrophys. J.* **2013**, *764*, 148. [CrossRef]
119. Bromberg, O.; Tchekhovskoy, A. Relativistic MHD simulations of core-collapse GRB jets: 3D instabilities and magnetic dissipation. *MNRAS* **2016**, *456*, 1739–1760. [CrossRef]
120. Begelman, M.C.; Cioffi, D.F. Overpressured Cocoons in Extragalactic Radio Sources. *Astrophys. J.* **1989**, *345*, L21. [CrossRef]
121. Murguia-Berthier, A.; Ramirez-Ruiz, E.; Montes, G.; De Colle, F.; Rezzolla, L.; Rosswog, S.; Takami, K.; Perego, A.; Lee, W.H. The Properties of Short Gamma-Ray Burst Jets Triggered by Neutron Star Mergers. *Astrophys. J.* **2017**, *835*, L34. [CrossRef]
122. Duffell, P.C.; Quataert, E.; Kasen, D.; Klion, H. Jet Dynamics in Compact Object Mergers: GW170817 Likely Had a Successful Jet. *Astrophys. J.* **2018**, *866*, 3. [CrossRef]
123. Nativi, L.; Lamb, G.P.; Rosswog, S.; Lundman, C.; Kowal, G. Are interactions with neutron star merger winds shaping the jets? *MNRAS* **2022**, *509*, 903–913. [CrossRef]
124. Gottlieb, O.; Nakar, E. Jet propagation in expanding media. *arXiv* **2021**, arXiv:2106.03860.
125. Salafia, O.S.; Ghisellini, G.; Pescalli, A.; Ghirlanda, G.; Nappo, F. Structure of gamma-ray burst jets: Intrinsic versus apparent properties. *Mon. Not. R. Astron. Soc.* **2015**, *450*, 3549–3558.
126. Gottlieb, O.; Bromberg, O.; Levinson, A.; Nakar, E. Intermittent mildly magnetized jets as the source of GRBs. *MNRAS* **2021**, *504*, 3947–3955. [CrossRef]
127. Urrutia, G.; De Colle, F.; Murguia-Berthier, A.; Ramirez-Ruiz, E. What determines the structure of short gamma-ray burst jets? *MNRAS* **2021**, *503*, 4363–4371. [CrossRef]
128. Gandel'Man, G.M.; Frank-Kamenetskii, D.A. Shock Wave Emergence at a Stellar Surface. *Sov. Phys. Dokl.* **1956**, *1*, 223.
129. Colgate, S.A. Early Gamma Rays from Supernovae. *Astrophys. J.* **1974**, *187*, 333–336. [CrossRef]
130. Falk, S.W. Shock steepening and prompt thermal emission in supernovae. *Astrophys. J.* **1978**, *225*, L133–L136. [CrossRef]
131. Matzner, C.D.; McKee, C.F. The Expulsion of Stellar Envelopes in Core-Collapse Supernovae. *Astrophys. J.* **1999**, *510*, 379–403. [CrossRef]
132. Nakar, E.; Sari, R. Relativistic Shock Breakouts—A Variety of Gamma-Ray Flares: From Low-luminosity Gamma-Ray Bursts to Type Ia Supernovae. *Astrophys. J.* **2012**, *747*, 88. [CrossRef]
133. Matzner, C.D.; Levin, Y.; Ro, S. Oblique Shock Breakout in Supernovae and Gamma-Ray Bursts. I. Dynamics and Observational Implications. *Astrophys. J.* **2013**, *779*, 60. [CrossRef]
134. Yalinewich, A.; Sari, R. Analytic asymptotic solution to spherical relativistic shock breakout. *Phys. Fluids* **2017**, *29*, 016103. [CrossRef]
135. Linial, I.; Sari, R. Oblique shock breakout from a uniform density medium. *Phys. Fluids* **2019**, *31*, 097102. [CrossRef]
136. Irwin, C.M.; Linial, I.; Nakar, E.; Piran, T.; Sari, R. Bolometric light curves of aspherical shock breakout. *MNRAS* **2021**, *508*, 5766–5785. [CrossRef]

137. Gottlieb, O.; Nakar, E.; Piran, T.; Hotokezaka, K. A cocoon shock breakout as the origin of the γ-ray emission in GW170817. *Mon. Not. R. Astron. Soc.* **2018**, *479*, 588–600. [CrossRef]
138. Nakar, E.; Piran, T. The Observable Signatures of GRB Cocoons. *Astrophys. J.* **2017**, *834*, 28. [CrossRef]
139. Komissarov, S.; Porth, O. Numerical simulations of jets. *New Astron. Rev.* **2021**, *92*, 101610. [CrossRef]
140. Zhang, W.; Woosley, S.E.; MacFadyen, A.I. Relativistic Jets in Collapsars. *Astrophys. J.* **2003**, *586*, 356–371. [CrossRef]
141. López-Cámara, D.; Morsony, B.J.; Lazzati, D. Photospheric emission from long-duration gamma-ray bursts powered by variable engines. *MNRAS* **2014**, *442*, 2202–2207. [CrossRef]
142. Gottlieb, O.; Bromberg, O.; Singh, C.B.; Nakar, E. The structure of weakly magnetized γ-ray burst jets. *MNRAS* **2020**, *498*, 3320–3333. [CrossRef]
143. Aloy, M.A.; Janka, H.T.; Müller, E. Relativistic outflows from remnants of compact object mergers and their viability for short gamma-ray bursts. *Astron. Astrophys.* **2005**, *436*, 273–311. [CrossRef]
144. Kathirgamaraju, A.; Barniol Duran, R.; Giannios, D. Off-axis short GRBs from structured jets as counterparts to GW events. *MNRAS* **2018**, *473*, L121–L125. [CrossRef]
145. Kathirgamaraju, A.; Tchekhovskoy, A.; Giannios, D.; Barniol Duran, R. EM counterparts of structured jets from 3D GRMHD simulations. *MNRAS* **2019**, *484*, L98–L103. [CrossRef]
146. Fernández, R.; Tchekhovskoy, A.; Quataert, E.; Foucart, F.; Kasen, D. Long-term GRMHD simulations of neutron star merger accretion discs: implications for electromagnetic counterparts. *MNRAS* **2019**, *482*, 3373–3393. [CrossRef]
147. Christie, I.M.; Lalakos, A.; Tchekhovskoy, A.; Fernández, R.; Foucart, F.; Quataert, E.; Kasen, D. The role of magnetic field geometry in the evolution of neutron star merger accretion discs. *MNRAS* **2019**, *490*, 4811–4825. [CrossRef]
148. Ito, H.; Just, O.; Takei, Y.; Nagataki, S. A Global Numerical Model of the Prompt Emission in Short Gamma-ray Bursts. *Astrophys. J.* **2021**, *918*, 59. [CrossRef]
149. Pavan, A.; Ciolfi, R.; Kalinani, J.V.; Mignone, A. Short gamma-ray burst jet propagation in binary neutron star merger environments. *MNRAS* **2021**, *506*, 3483–3498. [CrossRef]
150. Nathanail, A.; Gill, R.; Porth, O.; Fromm, C.M.; Rezzolla, L. 3D magnetized jet break-out from neutron-star binary merger ejecta: afterglow emission from the jet and the ejecta. *MNRAS* **2021**, *502*, 1843–1855. [CrossRef]
151. Gottlieb, O.; Liska, M.; Tchekhovskoy, A.; Bromberg, O.; Lalakos, A.; Giannios, D.; Mösta, P. Black hole to photosphere: 3D GRMHD simulations of collapsars reveal wobbling and hybrid composition jets. *arXiv* **2022**, arXiv:2204.12501.
152. Gottlieb, O.; Moseley, S.; Ramirez-Aguilar, T.; Murguia-Berthier, A.; Liska, M.; Tchekhovskoy, A. On the jet-ejecta interaction in 3D GRMHD simulations of binary neutron star merger aftermath. *arXiv* **2022**, arXiv:2205.01691.
153. Zhang, W.; Woosley, S.E.; Heger, A. The Propagation and Eruption of Relativistic Jets from the Stellar Progenitors of Gamma-Ray Bursts. *Astrophys. J.* **2004**, *608*, 365–377. [CrossRef]
154. Bošnjak, Ž.; Barniol Duran, R.; Pe'er, A. The GRB Prompt Emission: An Unsolved Puzzle. *Galaxies* **2022**, *10*, 38. [CrossRef]
155. Kouveliotou, C.; Meegan, C.A.; Fishman, G.J.; Bhat, N.P.; Briggs, M.S.; Koshut, T.M.; Paciesas, W.S.; Pendleton, G.N. Identification of Two Classes of Gamma-Ray Bursts. *Astrophys. J.* **1993**, *413*, L101. [CrossRef]
156. Ghirlanda, G.; Ghisellini, G.; Celotti, A. The spectra of short gamma-ray bursts. *Astron. Astrophys. Suppl. Ser.* **2004**, *422*, L55–L58. [CrossRef]
157. Ghirlanda, G.; Ghisellini, G.; Nava, L. Short and long gamma-ray bursts: Same emission mechanism? *Mon. Not. R. Astron. Soc. Lett.* **2011**, *418*, L109–L113. [CrossRef]
158. Sakamoto, T.; Barthelmy, S.D.; Barbier, L.; Cummings, J.R.; Fenimore, E.E.; Gehrels, N.; Hullinger, D.; Krimm, H.A.; Markwardt, C.B.; Palmer, D.M.; et al. The First Swift BAT Gamma-Ray Burst Catalog. *Astrophys. J.* **2008**, *175*, 179–190. [CrossRef]
159. Qin, Y.; Liang, E.W.; Liang, Y.F.; Yi, S.X.; Lin, L.; Zhang, B.B.; Zhang, J.; Lü, H.J.; Lu, R.J.; Lü, L.Z.; et al. A comprehensive analysis of fermi gamma-ray burst data. III. energy-dependent T90 distributions of GBM GRBS and instrumental selection effect on duration classification. *Astrophys. J.* **2012**, *763*, 15. [CrossRef]
160. Guidorzi, C.; Dichiara, S.; Amati, L. Individual power density spectra of Swift gamma-ray bursts. *Astron. Astrophys. Suppl. Ser.* **2016**, *589*, A98. [CrossRef]
161. Beloborodov, A.M.; Stern, B.E.; Svensson, R. Power density spectra of gamma-ray bursts. *Astrophys. J.* **2000**, *535*, 158–166. [CrossRef]
162. Guidorzi, C.; Margutti, R.; Amati, L.; Campana, S.; Orlandini, M.; Romano, P.; Stamatikos, M.; Tagliaferri, G. Average power density spectrum of Swift long gamma-ray bursts in the observer and in the source-rest frames. *Mon. Not. R. Astron. Soc.* **2012**, *422*, 1785–1803. [CrossRef]
163. MacLachlan, G.A.; Shenoy, A.; Sonbas, E.; Dhuga, K.S.; Cobb, B.E.; Ukwatta, T.N.; Morris, D.C.; Eskandarian, A.; Maximon, L.C.; Parke, W.C. Minimum variability time-scales of long and short GRBs. *Mon. Not. R. Astron. Soc.* **2013**, *432*, 857–865. [CrossRef]
164. Zach Golkhou, V.; Butler, N.R. Uncovering the Intrinsic Variability of Gamma-ray Bursts. *arXiv* **2014**, arXiv:1403.4254.
165. Norris, J.P.; Bonnell, J.T.; Kazanas, D.; Scargle, J.D.; Hakkila, J.; Giblin, T.W. Long-lag, wide-pulse gamma-ray bursts. *Astrophys. J.* **2005**, *627*, 324–345. [CrossRef]
166. Bernardini, M.G.; Ghirlanda, G.; Campana, S.; Covino, S.; Salvaterra, R.; Atteia, J.L.; Burlon, D.; Calderone, G.; Ghisellini, G.; Heussaff, V.; et al. Comparing the spectral lag of short and long gamma-ray bursts and its relation with the luminosity. *Mon. Not. R. Astron. Soc.* **2015**, *446*, 1129–1138. [CrossRef]

167. Kocevski, D.; Ryde, F.; Liang, E. Search for relativistic curvature effects in gamma-ray burst pulses. *Astrophys. J.* **2003**, *596*, 389–400. [CrossRef]
168. Bhat, P.N.; Briggs, M.S.; Connaughton, V.; Kouveliotou, C.; van der Horst, A.J.; Paciesas, W.; Meegan, C.A.; Bissaldi, E.; Burgess, M.; Chaplin, V.; et al. Temporal deconvolution study of long and short gamma-ray burst light curves. *Astrophys. J.* **2011**, *744*, 141. [CrossRef]
169. Burlon, D.; Ghirlanda, G.; Ghisellini, G.; Lazzati, D.; Nava, L.; Nardini, M.; Celotti, A. Precursors in Swift Gamma Ray Bursts with Redshift. *Astrophys. J.* **2008**, *685*, L19. [CrossRef]
170. Troja, E.; Rosswog, S.; Gehrels, N. Precursors of short gamma-ray bursts. *Astrophys. J.* **2010**, *723*, 1711. [CrossRef]
171. Pescalli, A.; Ronchi, M.; Ghirlanda, G.; Ghisellini, G. From the earliest pulses to the latest flares in long gamma-ray bursts. *Astron. Astrophys. Suppl. Ser.* **2018**, *615*, A80. [CrossRef]
172. Burlon, D.; Ghirlanda, G.; Ghisellini, G.; Greiner, J.; Celotti, A. Time resolved spectral behavior of bright BATSE precursors. *Astron. Astrophys. Suppl. Ser.* **2009**, *505*, 569–575. [CrossRef]
173. Ghirlanda, G.; Celotti, A.; Ghisellini, G. Extremely hard GRB spectra prune down the forest of emission models. *Astron. Astrophys. Suppl. Ser.* **2003**, *406*, 879–892. [CrossRef]
174. Ghirlanda, G.; Pescalli, A.; Ghisellini, G. Photospheric emission throughout GRB 100507 detected by Fermi. *Mon. Not. R. Astron. Soc.* **2013**, *432*, 3237–3244. [CrossRef]
175. Ryde, F.; Pe'er, A. Quasi-Blackbody Component and radiative efficiency of the prompt emission of gamma-ray bursts. *Astrophys. J.* **2009**, *702*, 1211–1229. [CrossRef]
176. Guiriec, S.; Kouveliotou, C.; Daigne, F.; Zhang, B.; Hascoët, R.; Nemmen, R.S.; Thompson, D.J.; Bhat, P.N.; Gehrels, N.; Gonzalez, M.M.; et al. Toward a better understanding of the GRB phenomenon: A new model for GRB prompt emission and its effects on the new Lint– Epeak,Irest,NT Relation. *Astrophys. J.* **2015**, *807*, 148. [CrossRef]
177. Ghisellini, G.; Celotti, A.; Lazzati, D. Constraints on the emission mechanisms of gamma-ray bursts. *Mon. Not. R. Astron. Soc.* **2000**, *313*, L1–L5. [CrossRef]
178. Nakar, E.; Ando, S.; Sari, R. Klein-Nishina Effects on Optically Thin Synchrotron and Synchrotron Self-Compton Spectrum. *Astrophys. J.* **2009**, *703*, 675–691. [CrossRef]
179. Derishev, E.V.; Kocharovsky, V.V.; Kocharovsky, V.V. Physical parameters and emission mechanism in gamma-ray bursts. *Astron. Astrophys.* **2001**, *372*, 1071–1077. [CrossRef]
180. Daigne, F.; Bošnjak, Ž.; Dubus, G. Reconciling observed gamma-ray burst prompt spectra with synchrotron radiation? *Astron. Astrophys.* **2011**, *526*, A110, [CrossRef]
181. Toffano, M.; Ghirlanda, G.; Nava, L.; Ghisellini, G.; Ravasio, M.E.; Oganesyan, G. The slope of the low-energy spectrum of prompt gamma-ray burst emission. *Astron. Astrophys. Suppl. Ser.* **2021**, *652*, A123. [CrossRef]
182. Oganesyan, G.; Nava, L.; Ghirlanda, G.; Celotti, A. Characterization of gamma-ray burst prompt emission spectra down to soft X-rays. *Astron. Astrophys. Suppl. Ser.* **2018**, *616*, A138. [CrossRef]
183. Oganesyan, G.; Nava, L.; Ghirlanda, G.; Melandri, A.; Celotti, A. Prompt optical emission as a signature of synchrotron radiation in gamma-ray bursts. *Astron. Astrophys. Suppl. Ser.* **2019**, *628*, A59. [CrossRef]
184. Ravasio, M.E.; Oganesyan, G.; Ghirlanda, G.; Nava, L.; Ghisellini, G.; Pescalli, A.; Celotti, A. Consistency with synchrotron emission in the bright GRB 160625B observed by Fermi. *Astron. Astrophys. Suppl. Ser.* **2018**, *613*, A16. [CrossRef]
185. Ravasio, M.E.; Ghirlanda, G.; Nava, L.; Ghisellini, G. Evidence of two spectral breaks in the prompt emission of gamma-ray bursts. *Astron. Astrophys. Suppl. Ser.* **2019**, *625*, A60. [CrossRef]
186. Ronchi, M.; Fumagalli, F.; Ravasio, M.E.; Oganesyan, G.; Toffano, M.; Salafia, O.S.; Nava, L.; Ascenzi, S.; Ghirlanda, G.; Ghisellini, G. Rise and fall of the high-energy afterglow emission of GRB 180720B. *Astron. Astrophys. Suppl. Ser.* **2020**, *636*, A55. [CrossRef]
187. Michael Burgess, J.; Bégué, D.; Greiner, J.; Giannios, D.; Bacelj, A.; Berlato, F. Gamma-ray bursts as cool synchrotron sources. *Nat. Astron.* **2019**, *4*, 174–179. [CrossRef]
188. Ghisellini, G.; Ghirlanda, G.; Oganesyan, G.; Ascenzi, S.; Nava, L.; Celotti, A.; Salafia, O.S.; Ravasio, M.E.; Ronchi, M. Proton–synchrotron as the radiation mechanism of the prompt emission of gamma-ray bursts? *Astron. Astrophys. Suppl. Ser.* **2020**, *636*, A82. [CrossRef]
189. Uhm, Z.L.; Zhang, B. Fast-cooling synchrotron radiation in a decaying magnetic field and γ-ray burst emission mechanism. *Nat. Phys.* **2014**, *10*, 351–356. [CrossRef]
190. Florou, I.; Petropoulou, M.; Mastichiadis, A. A marginally fast-cooling proton-synchrotron model for prompt GRBs. *arXiv* **2021**, arXiv:2102.02501.
191. Amati, L.; Frontera, F.; Tavani, M.; in 't Zand, J.J.M.; Antonelli, A.; Costa, E.; Feroci, M.; Guidorzi, C.; Heise, J.; Masetti, N.; et al. Intrinsic spectra and energetics of BeppoSAX Gamma–Ray Bursts with known redshifts. *Astron. Astrophys. Suppl. Ser.* **2002**, *390*, 81–89. [CrossRef]
192. Yonetoku, D.; Murakami, T.; Nakamura, T.; Yamazaki, R.; Inoue, A.K.; Ioka, K. Gamma-Ray Burst Formation Rate Inferred from the Spectral Peak Energy-Peak Luminosity Relation. *Astrophys. J.* **2004**, *609*, 935. [CrossRef]
193. D'Avanzo, P.; Salvaterra, R.; Bernardini, M.G.; Nava, L.; Campana, S.; Covino, S.; D'Elia, V.; Ghirlanda, G.; Ghisellini, G.; Melandri, A.; et al. A complete sample of bright Swift short gamma-ray bursts. *Mon. Not. R. Astron. Soc.* **2014**, *442*, 2342–2356. [CrossRef]
194. Ghirlanda, G.; Bernardini, M.G.; Calderone, G.; D'Avanzo, P. Are short Gamma Ray Bursts similar to long ones? *J. High Energy Astrophys.* **2015**, *7*, 81–89. [CrossRef]

195. Calderone, G.; Ghirlanda, G.; Ghisellini, G.; Bernardini, M.G.; Campana, S.; Covino, S.; Melandri, A.; Salvaterra, R.; Sbarufatti, B.; Tagliaferri, G. There is a short gamma-ray burst prompt phase at the beginning of each long one. *Mon. Not. R. Astron. Soc.* **2015**, *448*, 403–416. [CrossRef]
196. Salafia, O.S.; Pescalli, A.; Nappo, F.; Ghisellini, G.; Ghirlanda, G.; Salvaterra, R.; Tagliaferri, G. Gamma-ray burst jets: Uniform or structured? *arXiv* **2015**, arXiv:1503.01131.
197. Ioka, K.; Nakamura, T. Spectral puzzle of the off-axis gamma-ray burst in GW170817. *MNRAS* **2019**, *487*, 4884–4889. [CrossRef]
198. Beniamini, P.; Nakar, E. Observational constraints on the structure of gamma-ray burst jets. *MNRAS* **2019**, *482*, 5430–5440. [CrossRef]
199. Salafia, O.S.; Ghirlanda, G.; Ascenzi, S.; Ghisellini, G. On-axis view of GRB 170817A. *A&A* **2019**, *628*, A18. [CrossRef]
200. Zhang, B.; Mészáros, P. An Analysis of Gamma-Ray Burst Spectral Break Models. *Astrophys. J.* **2002**, *581*, 1236–1247. [CrossRef]
201. Salafia, O.S.; Ghisellini, G.; Pescalli, A.; Ghirlanda, G.; Nappo, F. Light curves and spectra from off-axis gamma-ray bursts. *MNRAS* **2016**, *461*, 3607–3619. [CrossRef]
202. Barbieri, C.; Salafia, O.S.; Perego, A.; Colpi, M.; Ghirlanda, G. Light-curve models of black hole - neutron star mergers: Steps towards a multi-messenger parameter estimation. *Astron. Astrophys.* **2019**, *625*, A152. [CrossRef]
203. Patricelli, B.; Bernardini, M.G.; Mapelli, M.; D'Avanzo, P.; Santoliquido, F.; Cella, G.; Razzano, M.; Cuoco, E. Prospects for multimessenger detection of binary neutron star mergers in the fourth LIGO-Virgo-KAGRA observing run. *MNRAS* **2022**, *513*, 4159–4168. [CrossRef]
204. Matsumoto, T.; Nakar, E.; Piran, T. Constraints on the emitting region of the gamma-rays observed in GW170817. *MNRAS* **2019**, *483*, 1247–1255. [CrossRef]
205. Matsumoto, T.; Nakar, E.; Piran, T. Generalized compactness limit from an arbitrary viewing angle. *MNRAS* **2019**, *486*, 1563–1573. [CrossRef]
206. Pescalli, A.; Ghirlanda, G.; Salafia, O.S.; Ghisellini, G.; Nappo, F.; Salvaterra, R. Luminosity function and jet structure of Gamma-Ray Burst. *Mon. Not. R. Astron. Soc.* **2015**, *447*, 1911–1921. [CrossRef]
207. Ascenzi, S.; Oganesyan, G.; Salafia, O.S.; Branchesi, M.; Ghirlanda, G.; Dall'Osso, S. High-latitude emission from the structured jet of γ-ray bursts observed off-axis. *A&A* **2020**, *641*, A61. [CrossRef]
208. Cano, Z.; Wang, S.Q.; Dai, Z.G.; Wu, X.F. The Observer's Guide to the Gamma-Ray Burst Supernova Connection. *Adv. Astron.* **2017**, *2017*, 8929054. [CrossRef]
209. Lü, H.J.; Lan, L.; Zhang, B.; Liang, E.W.; Kann, D.A.; Du, S.S.; Shen, J. Gamma-Ray Burst/Supernova Associations: Energy Partition and the Case of a Magnetar Central Engine. *Astrophys. J.* **2018**, *862*, 130. [CrossRef]
210. Wanderman, D.; Piran, T. The luminosity function and the rate of Swift's gamma-ray bursts. *Mon. Not. R. Astron. Soc.* **2010**, *406*, 1944–1958. [CrossRef]
211. Wanderman, D.; Piran, T. The rate, luminosity function and time delay of non-Collapsar short GRBs. *Mon. Not. R. Astron. Soc.* **2015**, *448*, 3026–3037. [CrossRef]
212. Ghirlanda, G.; Salafia, O.S.; Pescalli, A.; Ghisellini, G.; Salvaterra, R.; Chassande-Mottin, E.; Colpi, M.; Nappo, F.; D'Avanzo, P.; Melandri, A.; et al. Short gamma-ray bursts at the dawn of the gravitational wave era. *Astron. Astrophys. Suppl. Ser.* **2016**, *594*, A84. [CrossRef]
213. Lien, A.; Sakamoto, T.; Barthelmy, S.D.; Baumgartner, W.H.; Cannizzo, J.K.; Chen, K.; Collins, N.R.; Cummings, J.R.; Gehrels, N.; Krimm, H.A.; et al. The Third Swift Burst Alert Telescope Gamma-Ray Burst Catalog. *Astrophys. J.* **2016**, *829*, 7. [CrossRef]
214. Nava, L.; Salvaterra, R.; Ghirlanda, G.; Ghisellini, G.; Campana, S.; Covino, S.; Cusumano, G.; D'Avanzo, P.; D'Elia, V.; Fugazza, D.; et al. A complete sample of bright Swift long gamma-ray bursts: Testing the spectral-energy correlations. *Mon. Not. R. Astron. Soc.* **2012**, *421*, 1256–1264. [CrossRef]
215. Eichler, D.; Levinson, A. An Interpretation of the $h\nu_{peak}$-E_{iso} Correlation for Gamma-Ray Bursts. *Astrophys. J.* **2004**, *614*, L13–L16. [CrossRef]
216. Ghirlanda, G.; Nava, L.; Ghisellini, G.; Celotti, A.; Burlon, D.; Covino, S.; Melandri, A. Gamma-ray bursts in the comoving frame. *MNRAS* **2012**, *420*, 483–494. [CrossRef]
217. The LIGO Scientific Collaboration.; the Virgo Collaboration.; the KAGRA Collaboration.; Abbott, R.; Abbott, T.D.; Acernese, F.; Ackley, K.; Adams, C.; Adhikari, N.; Adhikari, R.X.; et al. The population of merging compact binaries inferred using gravitational waves through GWTC-3. *arXiv* **2021**, arXiv:2111.03634.
218. Tagliaferri, G.; Goad, M.; Chincarini, G.; Moretti, A.; Campana, S.; Burrows, D.N.; Perri, M.; Barthelmy, S.D.; Gehrels, N.; Krimm, H.; et al. An unexpectedly rapid decline in the X-ray afterglow emission of long gamma-ray bursts. *Nature* **2005**, *436*, 985–988. [CrossRef]
219. O'Brien, P.T.; Willingale, R.; Osborne, J.; Goad, M.R.; Page, K.L.; Vaughan, S.; Rol, E.; Beardmore, A.; Godet, O.; Hurkett, C.P.; et al. The Early X-ray Emission from GRBs. *Astrophys. J.* **2006**, *647*, 1213–1237.
220. Burrows, D.N.; Romano, P.; Falcone, A.; Kobayashi, S.; Zhang, B.; Moretti, A.; O'brien, P.T.; Goad, M.R.; Campana, S.; Page, K.L.; et al. Bright x-ray flares in gamma-ray burst afterglows. *Science* **2005**, *309*, 1833–1835. [CrossRef] [PubMed]
221. Nousek, J.A.; Kouveliotou, C.; Grupe, D.; Page, K.L.; Granot, J.; Ramirez-Ruiz, E.; Patel, S.K.; Burrows, D.N.; Mangano, V.; Barthelmy, S.; et al. Evidence for a Canonical Gamma-Ray Burst Afterglow Light Curve in the Swift XRT Data. *Astrophys. J.* **2006**, *642*, 389. [CrossRef]

222. Chincarini, G.; Mao, J.; Margutti, R.; Bernardini, M.G.; Guidorzi, C.; Pasotti, F.; Giannios, D.; Valle, M.D.; Moretti, A.; Romano, P.; et al. Unveiling the origin of X-ray flares in gamma-ray bursts. *Mon. Not. R. Astron. Soc.* **2010**, *406*, 2113–2148.
223. Oganesyan, G.; Ascenzi, S.; Branchesi, M.; Salafia, O.S.; Dall'Osso, S.; Ghirlanda, G. Structured Jets and X-ray Plateaus in Gamma-Ray Burst Phenomena. *Astrophys. J.* **2020**, *893*, 88.
224. Kumar, P.; Panaitescu, A. Afterglow Emission from Naked Gamma-Ray Bursts. *Astrophys. J.* **2000**, *541*, L51–L54. [CrossRef]
225. Duque, R.; Beniamini, P.; Daigne, F.; Mochkovitch, R. Flares in gamma-ray burst X-ray afterglows as prompt emission from slightly misaligned structured jets. *arXiv* **2021**, arXiv:2112.02917.
226. Mészáros, P.; Rees, M.J. Optical and Long-Wavelength Afterglow from Gamma-Ray Bursts. *Astrophys. J.* **1997**, *476*, 232–237. [CrossRef]
227. Sari, R.; Piran, T.; Narayan, R. Spectra and Light Curves of Gamma-Ray Burst Afterglows. *Astrophys. J.* **1998**, *497*, L17–L20. [CrossRef]
228. Panaitescu, A.; Kumar, P. Analytic Light Curves of Gamma-Ray Burst Afterglows: Homogeneous versus Wind External Media. *Astrophys. J.* **2000**, *543*, 66–76. [CrossRef]
229. Lamb, G.P.; Kobayashi, S. Reverse shocks in the relativistic outflows of gravitational wave-detected neutron star binary mergers. *MNRAS* **2019**, *489*, 1820–1827. [CrossRef]
230. Fermi, E. On the Origin of the Cosmic Radiation. *Phys. Rev.* **1949**, *75*, 1169–1174. [CrossRef]
231. Bell, A.R. The acceleration of cosmic rays in shock fronts - I. *MNRAS* **1978**, *182*, 147–156. [CrossRef]
232. Blandford, R.; Eichler, D. Particle acceleration at astrophysical shocks: A theory of cosmic ray origin. *Phys. Rep.* **1987**, *154*, 1–75. [CrossRef]
233. Spitkovsky, A. Particle Acceleration in Relativistic Collisionless Shocks: Fermi Process at Last? *Astrophys. J.* **2008**, *682*, L5. [CrossRef]
234. Kobayashi, S. Light Curves of Gamma-Ray Burst Optical Flashes. *Astrophys. J.* **2000**, *545*, 807–812. [CrossRef]
235. Resmi, L.; Zhang, B. Gamma-ray Burst Reverse Shock Emission in Early Radio Afterglows. *Astrophys. J.* **2016**, *825*, 48. [CrossRef]
236. Salafia, O.S.; Ravasio, M.E.; Yang, J.; An, T.; Orienti, M.; Ghirlanda, G.; Nava, L.; Giroletti, M.; Mohan, P.; Spinelli, R.; et al. Multiwavelength View of the Close-by GRB 190829A Sheds Light on Gamma-Ray Burst Physics. *Astrophys. J.* **2022**, *931*, L19. [CrossRef]
237. Rees, M.J.; Mészáros, P. Refreshed Shocks and Afterglow Longevity in Gamma-Ray Bursts. *Astrophys. J.* **1998**, *496*, L1–L4. [CrossRef]
238. Beniamini, P.; Duque, R.; Daigne, F.; Mochkovitch, R. X-ray plateaus in gamma-ray bursts' light curves from jets viewed slightly off-axis. *MNRAS* **2020**, *492*, 2847–2857. [CrossRef]
239. Nakar, E.; Piran, T. Implications of the radio and X-ray emission that followed GW170817. *MNRAS* **2018**, *478*, 407–415. [CrossRef]
240. Nava, L.; Sironi, L.; Ghisellini, G.; Celotti, A.; Ghirlanda, G. Afterglow emission in gamma-ray bursts - I. Pair-enriched ambient medium and radiative blast waves. *MNRAS* **2013**, *433*, 2107–2121. [CrossRef]
241. Lu, W.; Beniamini, P.; McDowell, A. Deceleration of relativistic jets with lateral expansion. *arXiv* **2020**, arXiv:2005.10313.
242. Cannizzo, J.K.; Gehrels, N.; Vishniac, E.T. A Numerical Gamma-Ray Burst Simulation Using Three-Dimensional Relativistic Hydrodynamics: The Transition from Spherical to Jetlike Expansion. *Astrophys. J.* **2004**, *601*, 380–390. [CrossRef]
243. De Colle, F.; Granot, J.; López-Cámara, D.; Ramirez-Ruiz, E. Gamma-Ray Burst Dynamics and Afterglow Radiation from Adaptive Mesh Refinement, Special Relativistic Hydrodynamic Simulations. *Astrophys. J.* **2012**, *746*, 122. [CrossRef]
244. Taylor, G. The Dynamics of the Combustion Products behind Plane and Spherical Detonation Fronts in Explosives. *Proc. R. Soc. Lond. Ser. A* **1950**, *200*, 235–247. [CrossRef]
245. Sedov, L.I. *Similarity and Dimensional Methods in Mechanics*; Academic Press: New York, NY, USA, 1959.
246. Ostriker, J.P.; McKee, C.F. Astrophysical blastwaves. *Rev. Mod. Phys.* **1988**, *60*, 1–68. [CrossRef]
247. Wei, D.M.; Jin, Z.P. GRB afterglow light curves from uniform and non-uniform jets. *A&A* **2003**, *400*, 415–419. [CrossRef]
248. Gill, R.; Granot, J. Afterglow imaging and polarization of misaligned structured GRB jets and cocoons: Breaking the degeneracy in GRB 170817A. *MNRAS* **2018**, *478*, 4128–4141. [CrossRef]
249. Beniamini, P.; Granot, J.; Gill, R. Afterglow light curves from misaligned structured jets. *Mon. Not. R. Astron. Soc.* **2020**, *493*, 3521–3534.
250. Lamb, G.P.; Alexander Kann, D.; Fernández, J.J.; Mandel, I.; Levan, A.J.; Tanvir, N.R. GRB jet structure and the jet break. *arXiv* **2021**, arXiv:2104.11099 .
251. Beniamini, P.; Gill, R.; Granot, J. Robust Features of Off-Axis Gamma-Ray Burst Afterglow Lightcurves. *arXiv* **2022**, arXiv:2204.06008.
252. Achterberg, A.; Gallant, Y.A.; Kirk, J.G.; Guthmann, A.W. Particle acceleration by ultrarelativistic shocks: Theory and simulations. *MNRAS* **2001**, *328*, 393–408. [CrossRef]
253. Cunningham, V.; Cenko, S.B.; Ryan, G.; Vogel, S.N.; Corsi, A.; Cucchiara, A.; Fruchter, A.S.; Horesh, A.; Kangas, T.; Kocevski, D.; et al. GRB 160625B: Evidence for a Gaussian-shaped Jet. *Astrophys. J.* **2020**, *904*, 166. [CrossRef]
254. Taub, A.H. Relativistic Rankine-Hugoniot Equations. *Phys. Rev.* **1948**, *74*, 328–334. [CrossRef]
255. Medvedev, M.V.; Loeb, A. Generation of Magnetic Fields in the Relativistic Shock of Gamma-Ray Burst Sources. *Astrophys. J.* **1999**, *526*, 697–706. [CrossRef]

256. Dermer, C.D.; Chiang, J.; Mitman, K.E. Beaming, Baryon Loading, and the Synchrotron Self-Compton Component in Gamma-Ray Bursts. *Astrophys. J.* **2000**, *537*, 785–795. [CrossRef]
257. Sari, R.; Esin, A.A. On the Synchrotron Self-Compton Emission from Relativistic Shocks and Its Implications for Gamma-Ray Burst Afterglows. *Astrophys. J.* **2001**, *548*, 787–799. [CrossRef]
258. Ghisellini, G.; George, I.M.; Fabian, A.C.; Done, C. Anisotropic inverse Compton emission. *MNRAS* **1991**, *248*, 14–19. [CrossRef]
259. Fan, Y.; Piran, T. Sub-GeV flashes in γ-ray burst afterglows as probes of underlying bright far-ultraviolet flares. *MNRAS* **2006**, *370*, L24–L28. [CrossRef]
260. Mei, A.; Banerjee, B.; Oganesyan, G.; Sharan Salafia, O.; Giarratana, S.; Branchesi, M.; D'Avanzo, P.; Campana, S.; Ghirlanda, G.; Ronchini, S.; et al. GeV emission from a compact binary merger. *arXiv* **2022**, arXiv:2205.08566.
261. Sari, R. The Observed Size and Shape of Gamma-Ray Burst Afterglow. *Astrophys. J.* **1998**, *494*, L49–L52, [CrossRef]
262. Granot, J.; Piran, T.; Sari, R. Images and Spectra from the Interior of a Relativistic Fireball. *Astrophys. J.* **1999**, *513*, 679–689. [CrossRef]
263. Granot, J.; Sari, R. The Shape of Spectral Breaks in Gamma-Ray Burst Afterglows. *Astrophys. J.* **2002**, *568*, 820–829. [CrossRef]
264. Panaitescu, A.; Mészáros, P. Rings in Fireball Afterglows. *Astrophys. J.* **1998**, *493*, L31–L34. [CrossRef]
265. Rybicki, G.B.; Lightman, A.P. *Radiative Processes in Astrophysics*; Wiley: New York, NY, USA, 1979.
266. Granot, J.; Piran, T.; Sari, R. The Synchrotron Spectrum of Fast Cooling Electrons Revisited. *Astrophys. J.* **2000**, *534*, L163–L166. [CrossRef] [PubMed]
267. Granot, J.; Gill, R.; Guetta, D.; De Colle, F. Off-axis emission of short GRB jets from double neutron star mergers and GRB 170817A. *MNRAS* **2018**, *481*, 1597–1608. [CrossRef]
268. Granot, J.; De Colle, F.; Ramirez-Ruiz, E. Off-axis afterglow light curves and images from 2D hydrodynamic simulations of double-sided GRB jets in a stratified external medium. *MNRAS* **2018**, *481*, 2711–2720. [CrossRef]
269. Hotokezaka, K.; Nakar, E.; Gottlieb, O.; Nissanke, S.; Masuda, K.; Hallinan, G.; Mooley, K.P.; Deller, A.T. A Hubble constant measurement from superluminal motion of the jet in GW170817. *Nat. Astron.* **2019**, *3*, 940–944. [CrossRef]
270. Fernández, J.J.; Kobayashi, S.; Lamb, G.P. Lateral spreading effects on VLBI radio images of neutron star merger jets. *MNRAS* **2022**, *509*, 395–405. [CrossRef]
271. Lin, H.; Totani, T.; Kiuchi, K. Non-thermal afterglow of the binary neutron star merger GW170817: A more natural modelling of electron energy distribution leads to a qualitatively different new solution. *MNRAS* **2019**, *485*, 2155–2166. [CrossRef]
272. Lamb, G.P.; Lyman, J.D.; Levan, A.J.; Tanvir, N.R.; Kangas, T.; Fruchter, A.S.; Gompertz, B.; Hjorth, J.; Mandel, I.; Oates, S.R.; et al. The Optical Afterglow of GW170817 at One Year Post-merger. *Astrophys. J.* **2019**, *870*, L15. [CrossRef]
273. Takahashi, K.; Ioka, K. Inverse reconstruction of jet structure from off-axis gamma-ray burst afterglows. *MNRAS* **2020**, *497*, 1217–1235. [CrossRef]
274. Takahashi, K.; Ioka, K. Diverse jet structures consistent with the off-axis afterglow of GRB 170817A. *MNRAS* **2021**, *501*, 5746–5756. [CrossRef]
275. Hallinan, G.; Corsi, A.; Mooley, K.P.; Hotokezaka, K.; Nakar, E.; Kasliwal, M.M.; Kaplan, D.L.; Frail, D.A.; Myers, S.T.; Murphy, T.; et al. A radio counterpart to a neutron star merger. *Science* **2017**, *358*, 1579–1583. [CrossRef] [PubMed]
276. Alexander, K.D.; Berger, E.; Fong, W.; Williams, P.K.G.; Guidorzi, C.; Margutti, R.; Metzger, B.D.; Annis, J.; Blanchard, P.K.; Brout, D.; et al. The Electromagnetic Counterpart of the Binary Neutron Star Merger LIGO/Virgo GW170817. VI. Radio Constraints on a Relativistic Jet and Predictions for Late-time Emission from the Kilonova Ejecta. *Astrophys. J.* **2017**, *848*, L21. [CrossRef]
277. Dobie, D.; Kaplan, D.L.; Murphy, T.; Lenc, E.; Mooley, K.P.; Lynch, C.; Corsi, A.; Frail, D.; Kasliwal, M.; Hallinan, G. A Turnover in the Radio Light Curve of GW170817. *Astrophys. J.* **2018**, *858*, L15. [CrossRef]
278. Corsi, A.; Hallinan, G.W.; Lazzati, D.; Mooley, K.P.; Murphy, E.J.; Frail, D.A.; Carbone, D.; Kaplan, D.L.; Murphy, T.; Kulkarni, S.R.; et al. An Upper Limit on the Linear Polarization Fraction of the GW170817 Radio Continuum. *Astrophys. J.* **2018**, *861*, L10. [CrossRef]
279. Mooley, K.P.; Nakar, E.; Hotokezaka, K.; Hallinan, G.; Corsi, A.; Frail, D.A.; Horesh, A.; Murphy, T.; Lenc, E.; Kaplan, D.L.; et al. A mildly relativistic wide-angle outflow in the neutron-star merger event GW170817. *Nature* **2018**, *554*, 207–210. [CrossRef] [PubMed]
280. Balasubramanian, A.; Corsi, A.; Mooley, K.P.; Brightman, M.; Hallinan, G.; Hotokezaka, K.; Kaplan, D.L.; Lazzati, D.; Murphy, E.J. Continued Radio Observations of GW170817 3.5 yr Post-merger. *Astrophys. J.* **2021**, *914*, L20. [CrossRef]
281. Margutti, R.; Berger, E.; Fong, W.; Guidorzi, C.; Alexander, K.D.; Metzger, B.D.; Blanchard, P.K.; Cowperthwaite, P.S.; Chornock, R.; Eftekhari, T.; et al. The Electromagnetic Counterpart of the Binary Neutron Star Merger LIGO/Virgo GW170817. V. Rising X-ray Emission from an Off-axis Jet. *Astrophys. J.* **2017**, *848*, L20. [CrossRef]
282. Haggard, D.; Nynka, M.; Ruan, J.J.; Kalogera, V.; Cenko, S.B.; Evans, P.; Kennea, J.A. A Deep Chandra X-ray Study of Neutron Star Coalescence GW170817. *Astrophys. J.* **2017**, *848*, L25. [CrossRef]
283. Troja, E.; Piro, L.; van Eerten, H.; Wollaeger, R.T.; Im, M.; Fox, O.D.; Butler, N.R.; Cenko, S.B.; Sakamoto, T.; Fryer, C.L.; et al. The X-ray counterpart to the gravitational-wave event GW170817. *Nature* **2017**, *551*, 71–74. [CrossRef]
284. Nynka, M.; Ruan, J.J.; Haggard, D.; Evans, P.A. Fading of the X-ray Afterglow of Neutron Star Merger GW170817/GRB 170817A at 260 Days. *Astrophys. J.* **2018**, *862*, L19. [CrossRef]

285. Piro, L.; Troja, E.; Zhang, B.; Ryan, G.; van Eerten, H.; Ricci, R.; Wieringa, M.H.; Tiengo, A.; Butler, N.R.; Cenko, S.B.; et al. A long-lived neutron star merger remnant in GW170817: Constraints and clues from X-ray observations. *MNRAS* **2019**, *483*, 1912–1921. [CrossRef]
286. Hajela, A.; Margutti, R.; Alexander, K.D.; Kathirgamaraju, A.; Baldeschi, A.; Guidorzi, C.; Giannios, D.; Fong, W.; Wu, Y.; MacFadyen, A.; et al. Two Years of Nonthermal Emission from the Binary Neutron Star Merger GW170817: Rapid Fading of the Jet Afterglow and First Constraints on the Kilonova Fastest Ejecta. *Astrophys. J.* **2019**, *886*, L17. [CrossRef]
287. Lyman, J.D.; Lamb, G.P.; Levan, A.J.; Mandel, I.; Tanvir, N.R.; Kobayashi, S.; Gompertz, B.; Hjorth, J.; Fruchter, A.S.; Kangas, T.; et al. The optical afterglow of the short gamma-ray burst associated with GW170817. *Nat. Astron.* **2018**, *2*, 751–754. [CrossRef]
288. Fong, W.; Blanchard, P.K.; Alexander, K.D.; Strader, J.; Margutti, R.; Hajela, A.; Villar, V.A.; Wu, Y.; Ye, C.S.; Berger, E.; et al. The Optical Afterglow of GW170817: An Off-axis Structured Jet and Deep Constraints on a Globular Cluster Origin. *Astrophys. J.* **2019**, *883*, L1. [CrossRef]
289. Makhathini, S.; Mooley, K.P.; Brightman, M.; Hotokezaka, K.; Nayana, A.J.; Intema, H.T.; Dobie, D.; Lenc, E.; Perley, D.A.; Fremling, C.; et al. The Panchromatic Afterglow of GW170817: The Full Uniform Data Set, Modeling, Comparison with Previous Results, and Implications. *Astrophys. J.* **2021**, *922*, 154. [CrossRef]
290. Kasliwal, M.M.; Nakar, E.; Singer, L.P.; Kaplan, D.L.; Cook, D.O.; Van Sistine, A.; Lau, R.M.; Fremling, C.; Gottlieb, O.; Jencson, J.E.; et al. Illuminating gravitational waves: A concordant picture of photons from a neutron star merger. *Science* **2017**, *358*, 1559–1565. [CrossRef] [PubMed]
291. Salafia, O.S.; Ghisellini, G.; Ghirlanda, G.; Colpi, M. Interpreting GRB170817A as a giant flare from a jet-less double neutron star merger. *Astron. Astophys.* **2018**, *619*, A18. [CrossRef]
292. Nathanail, A.; Porth, O.; Rezzolla, L. Magnetically Inspired Explosive Outflows from Neutron-star Mergers. *Astrophys. J.* **2019**, *870*, L20. [CrossRef]
293. Nakar, E.; Piran, T. Afterglow Constraints on the Viewing Angle of Binary Neutron Star Mergers and Determination of the Hubble Constant. *Astrophys. J.* **2021**, *909*, 114. [CrossRef]
294. Pian, E.; D'Avanzo, P.; Benetti, S.; Branchesi, M.; Brocato, E.; Campana, S.; Cappellaro, E.; Covino, S.; D'Elia, V.; Fynbo, J.P.U.; et al. Spectroscopic identification of r-process nucleosynthesis in a double neutron-star merger. *Nature* **2017**, *551*, 67–70. [CrossRef]
295. LIGO Scientific Collaboration.; Aasi, J.; Abbott, B.P.; Abbott, R.; Abbott, T.; Abernathy, M.R.; Ackley, K.; Adams, C.; Adams, T.; Addesso, P.; et al. Advanced LIGO. *Class. Quantum Gravity* **2015**, *32*, 074001. [CrossRef]
296. Acernese, F.; Agathos, M.; Agatsuma, K.; Aisa, D.; Allemandou, N.; Allocca, A.; Amarni, J.; Astone, P.; Balestri, G.; Ballardin, G.; et al. Advanced Virgo: A second-generation interferometric gravitational wave detector. *Class. Quantum Gravity* **2015**, *32*, 024001. [CrossRef]
297. Somiya, K. Detector configuration of KAGRA-the Japanese cryogenic gravitational-wave detector. *Class. Quantum Gravity* **2012**, *29*, 124007. [CrossRef]
298. Abbott, B.P.; Abbott, R.; Abbott, T.D.; Abraham, S.; Acernese, F.; Ackley, K.; Adams, C.; Adya, V.B.; Affeldt, C.; Agathos, M.; et al. Prospects for observing and localizing gravitational-wave transients with Advanced LIGO, Advanced Virgo and KAGRA. *Living Rev. Relativ.* **2020**, *23*, 3. [CrossRef]
299. Abbott, R.; Abbott, T.D.; Acernese, F.; Ackley, K.; Adams, C.; Adhikari, N.; Adhikari, R.X.; Adya, V.B.; Affeldt, C.; Agarwal, D.; et al. Search for Gravitational Waves Associated with Gamma-Ray Bursts Detected by Fermi and Swift during the LIGO-Virgo Run O3b. *Astrophys. J.* **2022**, *928*, 186. [CrossRef]
300. Petrov, P.; Singer, L.P.; Coughlin, M.W.; Kumar, V.; Almualla, M.; Anand, S.; Bulla, M.; Dietrich, T.; Foucart, F.; Guessoum, N. Data-driven Expectations for Electromagnetic Counterpart Searches Based on LIGO/Virgo Public Alerts. *Astrophys. J.* **2022**, *924*, 54, [CrossRef]
301. Colombo, A.; Salafia, O.S.; Gabrielli, F.; Ghirlanda, G.; Giacomazzo, B.; Perego, A.; Colpi, M. Multi-messenger observations of binary neutron star mergers in the O4 run. *arXiv* **2022**, arXiv:2204.07592.
302. Howell, E.J.; Ackley, K.; Rowlinson, A.; Coward, D. Joint gravitational wave - gamma-ray burst detection rates in the aftermath of GW170817. *MNRAS* **2019**, *485*, 1435–1447. [CrossRef]
303. Mochkovitch, R.; Daigne, F.; Duque, R.; Zitouni, H. Prospects for kilonova signals in the gravitational-wave era. *Astron. Astophys.* **2021**, *651*, A83. [CrossRef]
304. Duque, R.; Daigne, F.; Mochkovitch, R. Radio afterglows of binary neutron star mergers: A population study for current and future gravitational wave observing runs. *Astron. Astophys.* **2019**, *631*, A39, [CrossRef]
305. Saleem, M. Prospects of joint detections of neutron star mergers and short GRBs with Gaussian structured jets. *MNRAS* **2020**, *493*, 1633–1639. [CrossRef]
306. Zhu, J.P.; Wu, S.; Yang, Y.P.; Zhang, B.; Song, H.R.; Gao, H.; Cao, Z.; Yu, Y.W. Kilonova and Optical Afterglow from Binary Neutron Star Mergers. II. Optimal Search Strategy for Serendipitous Observations and Target-of-opportunity Observations of Gravitational-wave Triggers. *arXiv* **2021**, arXiv:2110.10469.
307. Yu, J.; Song, H.; Ai, S.; Gao, H.; Wang, F.; Wang, Y.; Lu, Y.; Fang, W.; Zhao, W. Multimessenger Detection Rates and Distributions of Binary Neutron Star Mergers and Their Cosmological Implications. *Astrophys. J.* **2021**, *916*, 54. [CrossRef]
308. Salafia, O.S.; Colombo, A.; Gabrielli, F.; Mandel, I. Constraints on the merging binary neutron star mass distribution and equation of state based on the incidence of jets in the population. *arXiv* **2022**, arXiv:2202.01656.

galaxies

MDPI

Review

The GRB Prompt Emission: An Unsolved Puzzle

Željka Bošnjak [1,*,†], Rodolfo Barniol Duran [2,*,†] and Asaf Pe'er [3,*,†]

1 Faculty of Electrical Engineering and Computing, University of Zagreb, Unska ul. 3, 10000 Zagreb, Croatia
2 Department of Physics and Astronomy, California State University, Sacramento, 6000 J Street, Sacramento, CA 95819, USA
3 Physics Department, Bar Ilan University, Ramat-Gan 52900, Israel
* Correspondence: Zeljka.Bosnjak@fer.hr (Ž.B.); barniolduran@csus.edu (R.B.D.); asaf.peer@biu.ac.il (A.P.)
† These authors contributed equally to this work.

Abstract: The recent multi-messenger and multi-wavelength observations of gamma-ray bursts (GRBs) have encouraged renewed interest in these energetic events. In spite of the substantial amount of data accumulated during the past few decades, the nature of the prompt emission remains an unsolved puzzle. We present an overview of the leading models for their prompt emission phase, focusing on the perspective opened by future missions.

Keywords: gamma-ray burst; prompt emission; relativistic jets

Citation: Bošnjak, Ž.; Barniol Duran, R.; Pe'er, A. The GRB Prompt Emission: An Unsolved Puzzle. *Galaxies* **2022**, *10*, 38. https://doi.org/10.3390/galaxies10020038

Academic Editors: Elena Moretti and Francesco Longo

Received: 13 December 2021
Accepted: 4 February 2022
Published: 22 February 2022

1. Introduction

In the past few years, new observations have led to several breakthroughs in the field of high energy astrophysics. The first detection of the binary neutron star merger event GW170817 by the Laser Interferometer Gravitational-Wave Observatory (LIGO) and the Virgo Consortium coinciding with a short-duration gamma-ray burst (GRB) [1–3] was a watershed moment in astronomy. For the first time, both gravitational waves and electromagnetic waves were detected from the same astrophysical source. Furthermore, this detection firmly placed the merger of neutron star binaries as progenitors of (at least, some) short GRBs. This event was accompanied by a "kilonova", also robustly establishing neutron star mergers as critical contributors of the production of heavy elements in the Universe [4,5]. These exciting observations have reinvigorated the interest of the astronomical community in understanding the underlying physics of gamma-ray bursts, their associated jets, and progenitors.

A second major breakthrough was the detection of the very high energy (>100 GeV) emission from GRBs by the Major Atmospheric Gamma Imaging Cherenkov (MAGIC) telescopes and High Energy Stereoscopic System (H.E.S.S.) [6,7]. These discoveries provided crucial data for relativistic jets models in which gamma-ray bursts are produced, as well as the nature of high energy radiation processes. On the other hand, neutrinos from GRBs are expected following the interactions of energetic protons that may be accelerated in the GRB environment, however no neutrinos from GRBs have been firmly detected yet [8]. As a result of this lack of detection, one critical piece of information regarding the possible GRB radiation mechanism is still missing. With the advent of new multi-messenger observations, it is becoming increasingly important to revise theoretical models to understand the physics in the vicinity of black holes and neutron stars, the nature of relativistic jets, and the origin of GRBs as the most energetic events in the Universe.

These recent observations add and extend the knowledge gained in the past several decades about the nature of GRBs. Observationally, we know that the vast majority of GRBs have the following common features: (i) Most GRBs consist of highly variable pulses of gamma-ray photons typically lasting dozen of seconds, having a non-thermal spectrum peaking at $\sim$a few 100 keV. (ii) The occurrence rate is approximately once per day from random directions in the sky [9–11]. (iii) The prompt emission is followed by the afterglow emission detected at lower energies (X-ray, optical, and radio) lasting for days, weeks, months, and (in radio band) years after the main event. (iv) For a number of GRBs, long lasting gamma-ray photons with energy $>$100 MeV have been observed during the afterglow phase.

The extreme nature of these events—short variability time scales $\sim$10 ms, extreme energy of up to (isotropic equivalent) 10^{55} ergs [12], emission over a broad energy scale, from optical to TeV, and the connection of the origin of these explosions with black-hole formation, have posed a challenge for the theoretical modeling of these events. In this review we will focus on the prompt emission of gamma-ray bursts, and provide a short summary of some of the most recent results, and of the proposed models for this emission episode. For more extensive reviews see, e.g., [13–17].

2. Prompt GRB Emission: Key Observational Properties

2.1. Spectral Properties: Sub-MeV Emission

Currently, the wealth of observations on prompt gamma-ray burst emission in the keV/GeV energy range comes from the *Fermi*, *Swift*, *INTEGRAL*, and Konus–*Wind* satellites. The spectrum in sub-MeV energy range is commonly fitted by the so-called Band function [11], which is an empirical function consisting of low- and high-energy power laws, smoothly connected around the peak energy at which most of the energy is emitted. The observed photon spectra indices, α and β of the low- and high-energy power-laws, respectively, may serve to distinguish different radiative mechanisms and properties of the electron distribution (that emit synchrotron radiation, if it is the dominant radiative process, see below). The most recent *Fermi* GBM (Gamma-ray Burst Monitor [18]; covering $\sim$8 keV to 40 MeV) gamma-ray burst spectral catalogue [19] provided α values for time-integrated ("fluence") spectra. When selecting only the models with spectral curvature, the low-energy index values are distributed around $\alpha \sim -1.1$, which is in agreement with previous findings [20,21]. Somewhat steeper low energy spectra $\alpha \sim -0.7$ have been reported for a *Fermi* GBM time-resolved spectral analysis of brightest bursts [22] (excluding the values obtained for simple power-law fits).

Recent works (e.g., [23,24]) provided fits to the gamma-ray burst prompt emission spectra below the spectral peak with not a single, but rather two power laws, connecting at a characteristic low energy spectral break. The break energy below which the spectrum hardened was found to be at (80–280 keV) for a sample of *Fermi* bright long GRBs [24], while it was at lower energies (3–22 keV) for a sample of GRBs contemporaneously observed by *Swift* BAT+XRT [23] (in the latter sample also *Fermi* GBM data were included when available). The importance of these fits lies in the obtained slopes, -0.6 and -1.5 below and above the break, respectively, that are consistent with the prediction of the synchrotron emission theory. A low-energy spectrum having two breaks thus may be a general property of GRB prompt emission though possibly not easily observable with present instruments. On the other hand, studies of the proposed measure of the spectral sharpness, namely the width of the spectral peak [25], showed that a large fraction of the observed GRB prompt spectra is not consistent with the theoretically expected synchrotron model under various assumptions (e.g., delta-function distribution of electrons, and Maxwellian or power-law electron distribution). This result therefore suggests emission mechanisms other than the optically thin synchrotron radiation [26].

A viable alternative is that of a thermal emission, predicted as the first signal arriving from the relativistically expanding fireball, e.g., [27,28]. The thermal spectral component was fitted in the early time-resolved spectra [29,30], or the entire time-integrated spectrum was fitted with a blackbody spectrum [31]. Several authors proposed the fit of a blackbody superimposed on the power-law component in order to fully describe the low energy portion of the spectrum [32–35]. The thermal component exhibited temporal evolution, with a characteristic rise and subsequent decay of the thermal flux. Recent works stress the importance of considering the temporal evolution of the photospheric emission: At earlier times, ~50% of the analyzed pulses were preferably fit with the photospheric emission [36,37].

High Energy Emission

The first observations of GeV emission from GRBs were obtained by EGRET (the Energetic Gamma-Ray Experiment Telescope, [38]) on board the *Compton Gamma Ray Observatory* in flight 1991–2000 [39,40]. The duration of high energy emission was often longer than the emission at keV, and showed a distinct temporal evolution [41,42].

The *Fermi* LAT instrument (the Large Area Telescope; [43]) is sensitive to γ-rays in the energy range ~30 MeV to ≥300 GeV. Since its launch in 2008, it asserted several new observational characteristics at energies >100 MeV [44]: (i) Many of the bright GRBs could not be fitted with commonly used models consisting of the low- and high-energy power law, and an additional power law component was required to fit the high-energy portion of the spectrum abd (ii) the emission above 100 MeV tends to be delayed with respect to emission at lower (sub-MeV) energies. When high energy emission was detected, it started during the prompt phase in >60% of the cases. Given the *Fermi* LAT field of view, this fraction may be even higher; (iii) the high energy emission lasts systematically longer than the sub-MeV prompt emission, and the high energy flux often follows a power law decay $\sim t^{-1}$. Recently, the two Imaging Atmospheric Cherenkov Telescopes, MAGIC and H.E.S.S. telescopes, reported the observations of the very high energy emission [6,7]. The γ-rays from GRB 190114C were observed in the energy range 0.2–1 TeV starting 57 s after the burst onset. The prompt emission duration of this event was ~116 s by *Fermi* GBM and ~362 s by *Swift* BAT. The observed very high energy emission was associated with the inverse Compton component in the afterglow phase, however the contribution from the prompt emission at early times could not be excluded [45].

2.2. Light Curve Properties

Prompt GRB light curves show erratic behavior, and so far no common model has been accepted that would fully describe the observed behavior. The duration of emission is associated with the timescale on which the inner engine producing a GRB operates, while the temporal variability reflects its variations in time [46] (though other sources of the observed variability have been proposed, e.g., local relativistic turbulence [47], see below). Broadly, we distinguish two classes of events, short and long GRBs with the dividing line at $T_{90} \sim 2$ s [48], where T_{90} refers to the time in which 5% to 95% of the counts in the 50–300 keV band is accumulated.

It has been recognized that the dividing line of $T_{90} \sim 2$ s depends on the specific gamma-ray detector used, thus additional information must be used to determine if a GRB is "long" or "short" (e.g., [49,50]). This has a theoretical implication: there is strong evidence that "long" GRBs are associated with the collapse of a massive star (the so-called "collapsar" model [51,52]). This evidence is based on the association of long GRBs with core-collapse supernova and thus massive star progenitors [53,54]. Short GRBs, on the other hand, are believed to be associated with the merger of two compact objects [55]. This idea has been proved by the association of the gravitational wave event GW170817 with a GRB (although this GRB may be atypical [1–3]).

In a small number of short GRBs, there is evidence of an extended emission lasting tens of seconds after the short initial spike [56–58], whose origin is still debated. Extended emission from short GRBs was also observed by the *Fermi* LAT at energies >100 MeV, e.g., in GRB 090510 or GRB 170127C [44].

The observed intrinsic variability during the prompt GRB emission can be rather short, down to ~tens of millisecond timescale or lower [59,60]. It poses a major constraint on prompt emission models, as the short timescale on which the observed signal can vary in the simplest models is given by $\delta T \sim R/(c\Gamma^2)$ [61] (R is a typical radius of the emitting region and the Γ is the jet bulk Lorentz factor). For GRBs with LAT detection, short timescale variability during the prompt phase can be found in a handful of bursts, e.g., GRB 131108A [44] and GRB 170214A [62].

2.3. Polarization

The leading models of the non-thermal emission, namely synchrotron emission and Compton scattering, both produce highly polarized emission [63]. However, in order to observe such a polarized signal, one has to break the spherical symmetry, which seems easier during the later time afterglow phase, due to lateral expansion of the slowing-down jet. Indeed, the first claimed detection of polarization signal was during the afterglow phase [64,65]. For a recent comprehensive study of polarization during the prompt phase for different scenarios see, e.g., [66].

High degree of linear polarization was claimed for several bursts, detected by different instruments: RHESSI, BATSE, and Integral [67–72]. Significant linear polarization was detected by the GAP instrument on *board IKAROS satellite* [73,74] for several GRBs: 100826A ($\Pi = 27 \pm 11\%$), 110301A ($\Pi = 70 \pm 22\%$), and GRB100826A ($\Pi = 84^{+16}_{-28}\%$), in all cases with more than 2.9 σ confidence.

In recent years, there have been dedicated missions to study GRB polarization, such as the Indian-led *ASTROSAT*, which reported several highly polarized signals detected by the CZTI instrument [75–77]. A second dedicated instrument is the POLAR detector [78]. The key result is that, while in many GRBs the time-integrated polarized signal is very low, there are rapid changes in the polarized signal, indicating the need for a time-resolved analysis, in which the signal is much more pronounced.

3. Theories of GRB Prompt Emission

Several current models can successfully interpret some of the spectral and temporal features of GRB emission. The main unknowns in the models for prompt emission are the nature of the energy reservoir and the subsequent energy dissipation, details of the particle acceleration mechanism, and the dominant radiative process. Within current leading models, the observed non-thermal spectrum is interpreted as either [i] a synchrotron and synchrotron self-Compton radiation from a population of relativistic electrons accelerated during the energy dissipation in the outflow (e.g., [79,80]), or [ii] as a Comptonized quasi-thermal emission from the photosphere (e.g., [81–84]). We give an overview and outline the main problems for the several leading models.

3.1. Hot Fireball Model

The hot fireball model assumes the expansion of a fireball composed mostly of photons, electron-positron pairs, and neutrinos [85–87], where magnetic field is energetically sub-dominant. As the fireball expands adiabatically from a very small radius, the energy of photons and pairs is transferred to protons, which are accelerated to large Lorentz factors [88]. At large distances from the central engine, the kinetic energy of the jet is transformed back to thermal energy, and gamma-rays are produced [89].

Using conservation of energy and entropy, it can be shown (e.g., [15]) that the acceleration of the jet is linear with distance from the base of the fireball R_0, namely $\Gamma(r) \propto r$. The acceleration proceeds until the outflow reaches the saturation radius $R_s = R_0\Gamma_s$, where Γ_s is the terminal Lorentz factor. This is true as long as the photons are coupled to electrons in the outflow, therefore, the photospheric radius $R_{\rm ph}$ plays an important role. (i) If $R_s < R_{\rm ph}$, then the jet reaches Γ_s at R_s. (ii) On the other hand, if $R_{\rm ph} < R_s$, then the acceleration mostly stops at $R_{\rm ph}$ [90,91].

The luminosity of the photospheric component in the hot fireball model depends on cases (i) and (ii) mentioned above. For case (i), since the photospheric radius is larger than the saturation radius, the photon temperature decreases due to adiabatic cooling beyond the saturation radius and the thermal luminosity is expected to be lower than in case (ii). The observed photospheric emission is expected to be at a few MeV [85–88]. The emerging spectrum would not be as simple as $f_\nu \propto \nu^2$ below the peak, where ν is the observed frequency and f_ν is the flux density, however integration from different radii would flatten it slightly to approximately $f_\nu \propto \nu^{1.4-1.5}$ [83,92–94] and an angular dependence of the jet Lorentz factor can flatten it to $\propto \nu^0$ [95], see below. The presence of a strong thermal component in the gamma-ray spectrum would point to the hot fireball model. There is some evidence for a photospheric component in a number of GRB spectra, e.g., [96]. On the other hand, the lack of this clear component has been used to support the magnetically dominated jet model [97], which is described below.

3.2. Particle Acceleration

Following the dissipation of kinetic or magnetic energy, particles are accelerated to high energies. These particles, then, emit the high-energy, non-thermal radiation observed. Modeling this radiation (e.g., as synchrotron emission) provides an indirect evidence for particle acceleration to non-thermal distribution. This was first done in the context of the GRB afterglow [98,99].

The theory of test particle acceleration (i.e., assuming a fixed background) has been well established for many decades [100–105]. In the past 10–15 years, advances in parallel computation, in particular, particle-in-cell (PIC) simulations, have enabled the modeling and studying of this process from first principles [106–109], under various conditions (e.g., magnetization, etc.) [110–113]. There have been several attempts to extend the theory beyond the test particle to include the feedback on the surrounding plasma [114,115]. Alternative theories, such as stochastic turbulence acceleration have also been considered [116].

In recent years, there has been a considerable interest in the theory of magnetized outflows. When the magnetic field is energetically dominant, it may convert its energy to kinetic energy by reconnection of the magnetic field lines, namely, a topological change in the magnetic field structure. Using PIC simulations, many authors have demonstrated that efficient acceleration of particles take place in such reconnection layers [113,117]. Furthermore, the accelerated particles obtain a power law distribution, similar to the expectation from a Fermi-like acceleration [111,113,118–125].

3.3. Internal Shock Model

In the context of GRBs, the first ideas for interpreting the observed highly irregular temporal pattern of radiation came soon after establishing the extragalactic origin of GRBs [10]. It was suggested that energy and matter injection by the compact central object ($<10^7$ cm) does not occur at a steady rate. The resulting outflow would in that case consist of a sequence of "shells" with fluctuating Lorentz factors ([126,127]). In the interaction of a faster shell and a slower one emitted earlier, a shock would develop, which would accelerate electrons to relativistic velocities.

Emission from internal shocks in a relativistic wind with varying Lorentz factors has been studied extensively, e.g., [128–132]. The initial kinetic energy is dissipated in collisions of a series of successive shells emitted from the central engine, having a non-uniform distribution of Lorentz factors $\Gamma(t)$. In the model described by [129], shells interact only by direct collisions, and one shock wave is discretized by the series of shocks (for a comparison with a detailed hydrodynamical calculation, see e.g., [133,134]). The dynamic phase is described by the following parameters:The total duration of the energy ejection by the central engine, the distribution $\Gamma(t)$, and the injected kinetic power during the ejection phase.

For each collision, one can calculate the radius, collision time, Lorentz factor of the shocked material, and the energy dissipated in the collision. The advantage of this model is that the variability time of the energy injection roughly translates into the observed variability time in the GRB lightcurve [135]. The fraction of the thermal energy dissipated in collisions is deposited in electrons in the two colliding shells, while the remaining energy goes into proton acceleration and magnetic field amplification. The efficiency of the energy dissipation process is typically low, $\lesssim 15\%$ [129,136,137], which is the main drawback of the internal shock model.

The microphysics related to a shocked medium is usually parametrized by assuming that a fraction ϵ_e of the dissipated energy is given to the ambient electrons [126,129,131,135]. The electrons are assumed to be accelerated to a power-law, $n(\gamma_e) \propto \gamma_e^{-p}$, above their initial thermal distribution (typical Lorentz factor denoted by γ_m). The slope of the electron distribution p depends on the details of the acceleration process [129]. Under the assumption that the leading radiative process is synchrotron emission from these power-law distributed electrons, the observed high energy photon spectral index β of the "Band"-fitting function, provides the indication for the steepness of the particle distribution, $p \approx 2.5$ [138]. For typical parameters, the synchrotron emission produced by the accelerated electrons in a magnetic field B would occur at observed energy $\mathrm{E_{syn}} = 50(\Gamma_*/300)(B/1000\ G)(\gamma_e/100)^2$ eV [129]. To obtain higher electron Lorentz factors (10^3–10^4) in order to reach an observed peak energy at a ~few × 100 keV, several authors have suggested that only a fraction $\zeta \sim 10^{-3}$ of electrons are accelerated [129,131,139].

Note that there is a large uncertainty in the value of the magnetic field. During the prompt phase, there can be two sources of magnetic field: (i) a strong magnetic field may be associated with the central engine (e.g., [140]). Its strength will decay with distance, however it may still be considerable if the source is highly magnetized, and the dissipation does not occur at too large a distance. (ii) In addition, the magnetic field may be generated at the shock front, obtaining an uncertain fraction (referred to as ϵ_B) of the dissipated energy at the shock.

The accelerated relativistic electrons cool mainly by the synchrotron process, and the associated inverse Compton radiation. The high energy portion of the spectrum is attenuated by photon-photon annihilation, and by the EBL (extragalactic background light) absorption. The low energy portion of the spectrum has a steep cutoff due to self-absorption. The temporal profiles of the prompt emission can be obtained when the contributions from all collisions are taken into account. One example of such study is shown in Figure 1. Here the calculation was performed neglecting the interaction between photons emitted in a shocked region and electrons/photons present in another region (see, e.g., [141]); in addition, the possible contribution of the shock accelerated protons was not considered.

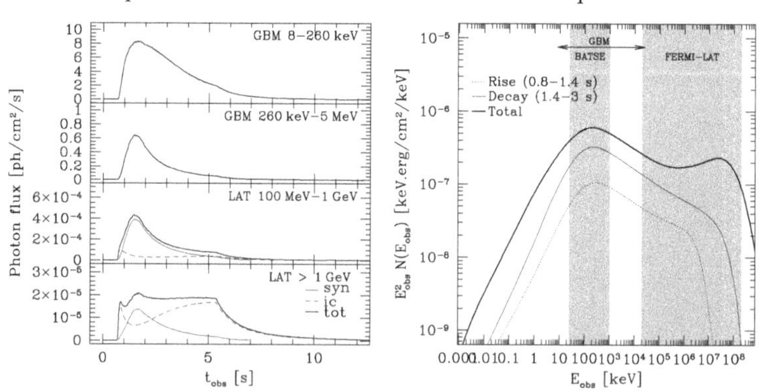

Figure 1. A single pulse burst: the main emission peak is due to the synchrotron radiation. The microphysics parameters used in the simulations are $\epsilon_B = 5\times 10^{-3}$, $\epsilon_e = 1/3$, $\zeta = 2 \times 10^{-3}$, $p = 2.5$, and $\mathrm{dE/dt} = 5 \times 10^{53}$ erg s^{-1}. As the assumed magnetic field is low, the non-negligible signatures of inverse Compton scatterings are favored in the *Fermi* LAT (the Large Area Telescope) energy range LAT energy range. The process included in calculation are the following: adiabatic cooling, synchrotron emission and synchrotron-self absorption, inverse Compton scatterings, and $\gamma\gamma$-annihilation. The effects of secondary pairs were not taken into account. *Left*: observed light curves in *Fermi*-GBM (Gamma-ray Burst Monitor and the LAT range. The synchrotron (thin solid line) and inverse Compton (thin dashed line) components are shown. *Right*: observed time-integrated spectrum during the rise, early decay, and whole duration of the pulse. From Credit: [131], reproduced with permission ©ESO [131].

Some authors have pointed out difficulties within the internal shock model when applied to the 'naked-eye' burst GRB080319B for which variable prompt optical emission is present [142,143]. The main issue seems to be that the observations point to a very large radius of emission: at these large distances, the gamma-ray flux would be much smaller than observed. These difficulties have served as motivation for alternative models [142].

3.4. The Role of Neutrons in the GRB Jet

It is possible that the GRB jet is also composed of a population of neutrons [144–147]. These neutrons may change the GRB jet dynamics and have an effect on the resulting prompt emission phase [92]. As mentioned before, in the case of the hot fireball model without neutrons, the fireball accelerates as long as the photons are coupled to the electrons. Due to the smaller proton-neutron cross section, when neutrons are present in GRB jet they decouple the protons at a smaller distance from the central engine than the Thomson photosphere. If the decoupling radius is also smaller than the radius R_s where protons attain their maximum speed, then neutrons attain a Lorentz factor $\Gamma_n < \Gamma_s$. This two-fluid state or "compound" state of the jet, similarly to the internal shock model, extracts the kinetic energy of internal motions of the jet. More specifically, it extracts the energy of the streaming of plasma through the neutron component throughout a volume instead of being solely confined to the shock front as in internal shocks [92].

Since this jet is prone to collisions between neutrons and protons, it creates multiple $e^{\pm}$ pairs, which can have an effect on the emerging gamma-ray spectrum, by cooling via synchroton and inverse Compton. These cooled pairs form a thermalized pair population which is Coulomb-heated by collisions with protons. This mechanism is able to produce a peak near 1 MeV and a "Band" spectrum with $f_\nu \propto \nu^{1.4}$ and $f_\nu \propto \nu^{-1.5}$ below and above the peak, respectively [92].

Magnetic fields change the spectrum below the peak by significantly cooling the pairs produced in the neutron-proton collisions via the synchroton process [148]. This does not significantly alter the peak of the spectrum, but does flatten the spectrum below the peak, see Figure 2. It also steepens the spectrum above the peak since inverse Compton emission by pairs becomes less important above the peak in lieu of stronger synchrotron emission below it, which may be in tension with spectral observations [14].

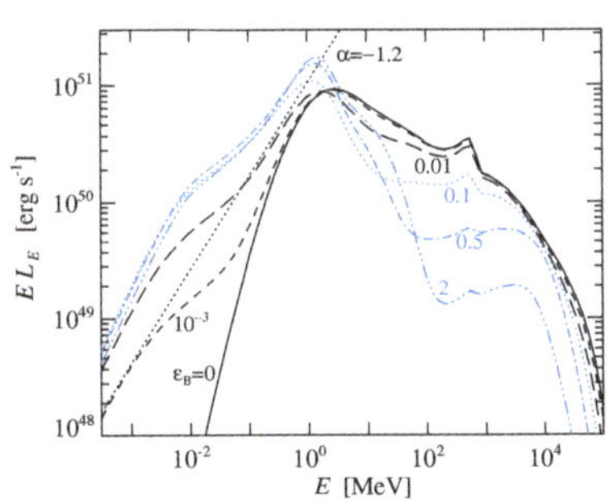

Figure 2. Spectrum of the magnetized, collisionally heated jet. The solid, short-dashed, long-dashed, dotted, dot-dashed, and triple dot-dashed curves correspond to magnetizations of 0, 10^{-3}, 0.01, 0.1, 0.5, and 2, respectively. From ©AAS. Reproduced with permission [148].

3.5. Magnetically Dominated Jet

Some authors have considered models in which the GRB jet is magnetically dominated, that is, where magnetic fields dominate the jet luminosity at the base of the jet [149–152]. In these models, the jet is accelerated as it converts its magnetic energy to bulk kinetic energy. At larger distances from the central engine, the kinetic energy of the jet is transformed to thermal energy, commonly by magnetic reconnection instead of shocks (shocks in this scenario seems to be too inefficient, e.g., [117,153]), and gamma-rays are produced. For general radiation properties in magnetically-dominated jets, see [154].

Jet acceleration in a magnetic model can occur due to the dissipation of the magnetic field in a striped configuration (such as that of a pulsar wind). This "striped jet" model invokes a magnetized jet with small-scale field reversals or "stripes" [120,155–160], where magnetic reconnection is able to start from small distances and continue as the jet accelerates and collimates. In the case of a black hole central engine, the alternating magnetic fields can be produced by the magneto-rotational instability in the innermost regions of the accretion flow [160]. For the case of a magnetar central engine, the alternating fields can be produced by an oblique dipole rotator [155,156]. Recently, the possibility of a distribution of stripe sizes in a magnetized jet has been considered [160].

In the striped jet, the jet accelerates (as the magnetization drops) up to a saturation radius $R_{\rm sat}$, where the magnetization reaches $\sim$1. Jet acceleration proceeds, not linearly with the radius as in the hot fireball model, but as $\Gamma(r) \propto r^{1/3}$, e.g., [155,156]. Magnetic reconnection energizes particles and their emission spectrum will depend on the location of the Thomson photosphere compared to $R_{\rm sat}$. In this model, the observed gamma-ray prompt spectrum can be dominated by a Comptonized thermal spectrum [158,161,162]. Depending on the particle energy injection, the photospheric emission can be subdominant and a non-thermal spectrum can develop. The details of the non-thermal component in this model depend then on the particle energy injection, and several possibilities can be considered [158,163,164].

Jet acceleration can also occur by adiabatic expansion of the outflow [165,166]. In this case, the jet accelerates also as $\Gamma(r) \propto r^{1/3}$ [166]. While there is no magnetic reconnection in this picture, energy dissipation can be driven by internal shocks within the outflow [167,168].

In all models that attempt to explain the prompt gamma-ray emission, reproducing the variability of the observed gamma-ray light curves is crucial. In the case of magnetic reconnection models, including the striped jet model mentioned above, a promising way to explain the light curve variability is to consider small reconnection regions that move relativistically in the co-moving frame of the jet with Lorentz factor $\sim$ few—10 as considered in the "minijets" or "jet in jet" model, relativistic turbulence model, and ICMART (Internal Collision-induced MAgnetic Reconnection and Turbulence) model [47,142,152,169–172]. It is likely that the directions of motion of these small reconnection regions, instead of being isotropically distributed in the comoving frame of the jet, are primarily perpendicular to the direction of the flow [171,172], and this would explain several of the observed prompt GRB temporal and spectral properties [171]. In this particular scenario, the prompt emission would be delayed with respect to the isotropic case, which would allow for the peak of the GRB afterglow to occur during the prompt emission phase in contrast to the simple isotropic model [172], see Figure 3. It would also explain the observed very steep X-ray emission, which is even steeper than the decay expected in the isotropic case [172].

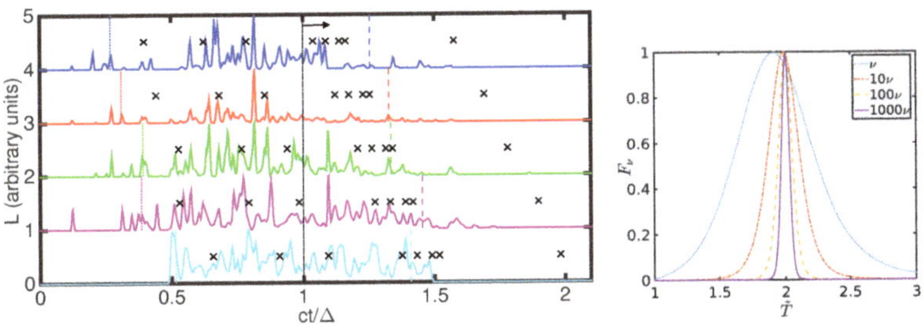

Figure 3. Left: typical prompt emission light curves in the "minijets" model. Degree of anisotropy of minijets' directions increases from top to bottom. Different light curves (normalized) have been shifted vertically for displaying purposes. Anisotropy shifts the overall light curves to later times. The vertical dotted (dashed) lines for each light curve correspond to the T_{90} duration. We include the observed peaks of the GeV light curves (black crosses) for the sample in [173], scaled for each of the simulated light curves. As the level of anisotropy increases, the peaks of the GeV light curves also shift to later times, making most of them consistent with $t \geq \Delta/c$, where Δ is the shell thickness. In the simplest GRB afterglow model, the deceleration time will occur at times $t \geq \Delta/c$ (see small black arrow) and if the peak of the GeV light curves correspond to the deceleration time, then anisotropic minijets' directions alleviate the problem of having them at times much less then Δ/c. From [172]. Right: light curves of a single pulse at different frequencies. The pulse clearly becomes narrower with lower frequencies in some magnetic reconnection models as observed in the prompt emission phase. From [171]. Reprinted (and modified) permission of Oxford University Press on behalf of the Royal Astronomical Society.

3.6. *Radiative Processes*

3.6.1. Synchrotron Emission and Inverse Compton Scatterings

Among the non-thermal radiative processes, synchrotron emission from relativistic electrons has been considered an important mechanism in the context of the prompt emission of GRBs [79,135,174–176]. Several authors have taken a general approach and determined the source properties (e.g., jet bulk Lorentz factor, electrons' Lorentz factor, distance from the central engine to location where gamma-ray are produced) by assuming that the peak of the gamma-ray emission spectrum is produced by the synchrotron process (e.g., [31,130,132,177]; for the case of a magnetic jet, see, e.g., [178]). The major challenge for the synchrotron model is posed by the observed hard low energy spectrum that is in apparent contradiction with the predictions of the simple synchrotron model (e.g., [99,138,179]). The flux $f_\nu \propto \nu^{-1/2}$ below the peak of the spectrum is expected when electron's radiative time scales are much shorter than the dynamical times ('fast-cooling regime') [99]. We define γ_c as the Lorentz factor of electrons whose synchrotron loss timescale is equal to the adiabatic cooling timescale $t_{\rm ex}$, $\gamma_c = 6\pi m_e c/(\sigma_T B^2 t_{\rm ex})$, where m_e is the electron's mass, c is the speed of light, and σ_T is the Thomson cross-section. The synchrotron fast-cooling regime is then characterized by $\gamma_c < \gamma_m$. This regime is favorable for prompt gamma-ray emission as it has a high radiative efficiency. There have been several studies reconciling the observed spectrum with the synchrotron emission, and proposing solutions for harder spectral slope: The pitch-angle distribution [138], the small scale structure of the magnetic field [180,181], or processes that involve the appearance of a quasi-thermal component in addition to non-thermal synchrotron [27,182].

It is also possible to have the synchrotron mechanism responsible for the GRB prompt phase, however modified by including an additional source of cooling due to inverse Compton scatterings [183–186]. The soft low-energy spectral slope of the photon spectrum $\alpha = -1.5$, resulting from the assumption of fast cooling synchrotron spectrum, could be hardened if a sub-dominant radiative process (like inverse Compton scatterings) transferred around 20–40% of the energy from the synchrotron component to higher energies [185]. There are two parameters that control the importance of inverse Compton scatterings:

$w_m = \gamma_m \epsilon_m$, where $\epsilon_m = h\nu_m/m_e c^2$ and $\nu_m = \nu(\gamma_m)$, determine whether the scatterings occur in the Thomson regime ($w_m \ll 1$) or if Klein–Nishina effects need to be taken into account; another parameter is Y_{Th}, which determines the intensity of the inverse Compton component peaking at high energies. When Klein–Nishina corrections are important ($w_m \gtrsim 1$), the cross section and the energy boost are reduced so that the ratio of the total energy in the inverse Compton component over the total energy in the synchrotron component becomes $E_{ic}/E_{syn} \ll Y_{Th}$ [131]. It has been shown that the physical conditions in the emitting region allow for a synchrotron component peak at $\sim$ a few 100 keV, and a moderately efficient inverse Compton scatterings in the Klein–Nishina regime. In particular, in the internal shock scenario [185], a large fraction of the dissipated energy $\epsilon_e \sim 0.1$–$1/3$ should be injected in a small fraction of electrons $\zeta \lesssim 0.01$ and the fraction of the energy injected in the magnetic field should remain low, $\epsilon_B \lesssim 10^{-3}$. Additionally, the 'marginally fast cooling regime' was proposed by [185], considering that electrons are in the fast cooling regime but not deeply in this regime (i.e., $\gamma_c \lesssim \gamma_m$ rather than $\gamma_c \ll \gamma_m$). When the cooling frequency becomes close to the frequency ν_m, the observed photon index can become very close to the value $-2/3$ below the cooling frequency, even in the fast cooling regime. This solution requires collisions at small radii and/or low magnetic fields. However, in this context, and focusing on conditions where synchrotron cooling is balanced by a continuous source of heating, one naturally finds solutions consistent with those of the minijets model in the magnetically dominated jet described above [187], where dissipation occurs far from the central engine.

High energy gamma-rays in the prompt GRB phase could be also produced by inverse Compton scattering of synchrotron photons: "synchrotron-self-Compton" SSC emission. However, in its simplest form, this mechanism either produces a more energetic component in very high energy gamma-rays or would require a more energetic component as a low-energy synchrotron seed, which is inconsistent with observations [143,188]. On the other hand, the SSC it is defined in the beginning of the paragraph origin of GRB prompt emission seems to work well in the context of the relativistic turbulence model [142].

3.6.2. Comptonized Thermal Radiation

Photospheric (thermal) emission is inherent to the "fireball" model as, following the initial explosion, the plasma is optically thick, and photons cannot escape. Rather, they are coupled to the expanding gas, converting their internal energy to kinetic energy of the expanding gas. Only when the gas sufficiently expands does the optical depth decrease such that the photons escape. It is therefore of no surprise that the very first cosmological GRB models considered photospheric emission as a leading radiative process [81,86,87,189]. However, the fact that the prompt spectra appears non-thermal has led to focus on other broad-band models, in particular synchrotron.

Renewed interest in this model resumed in the early 2000s, with the realization that the synchrotron model appears too broad to explain the steep low energy spectral slope (the 'synchrotron line of death') [190]. Several authors considered a possible contribution from photospheric photons to the observed spectra [27,82,90,91,191,192]. It was realized that the observed spectrum of photons originating from the photosphere did not necessarily resemble a "Planck" function, due to two complementary effects. The first is possibly sub-photospheric energy dissipation, e.g., by lateral shock waves at the boundary between the relativistic jet and collapsing star, or reconnection of magnetic field lines, which heats the electrons in the plasma [158,161,193,194]. The dissipation heats the electrons, which then serve as seeds for inverse-Compton scattering. When such events occur below the photosphere, the original 'Planck' spectrum can be heavily modified, and the result depends on the details of the energy exchange between the particles and photon fields; this is demonstrated in Figure 4.

A second, independent effect is the aberration of light, which is essentially the relativistic version of the well-known limb darkening effect from solar observations. Due to the probabilistic nature of the scattering process, the photosphere is in fact 'vague', namely the last scattering location of photons can occur in various spatial locations (as opposed to a single surface) [83,92,93,95,195,196]. This location is angle-dependent: at high angles, it occurs, on the average, at larger radii than at angles $< 1/\Gamma$ (the jet Lorentz factor). In a spherical explosion, this aberration leads to a modification (mainly) of the Rayleigh–Jeans part of the spectrum. However, the jets are not spherical, but have some lateral shape (angle-dependent Lorentz factor). In this case this effect becomes very pronounced and affects both the low as well as the high energy spectral slopes making both of them shallower than the naively expected Rayleigh–Jeans shape [95,197,198], although steeper than the expected from synchrotron radiation, making the spectral slopes consistent with the data [36,164,199,200].

An interesting version of the photospheric model is the 'back scattering' dominated model [201,202]. In this model, the jet drills a funnel through the stellar envelope, and accelerates a 'cork' made of stellar material ahead of it. The photons originate from $e^{\pm}$ pair annihilation close to the central engine, across the virtually empty jet before being back-scattered from the cork material ahead of them (if the cork does not disintegrate too rapidly). Although in the cork frame they are scattered backward, they will be detected by an observed located off axis, due to the relativistic angle change between the cork and observer's frame. It was recently demonstrated [203,204] that the resulting spectra in this setup is in excellent agreement with the observed, both at low and high energy. Furthermore, this model naturally explained the observed peak energy—total energy relation (known as "Amati" correlation; [205]) without the need to invoke any additional assumption.

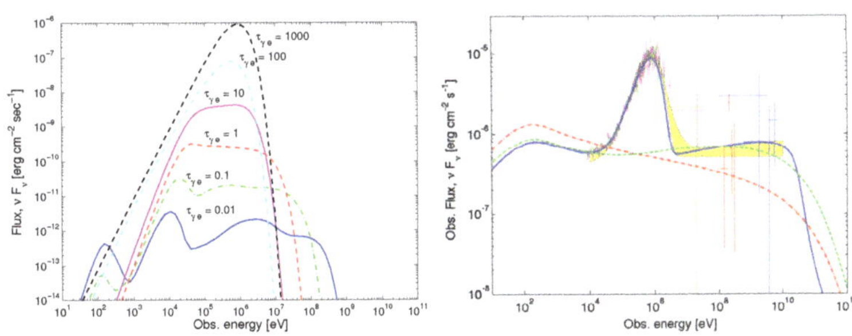

Figure 4. Left: time averaged broadband spectra expected following kinetic energy dissipation at various optical depths. For low optical depth, the two low energy bumps are due to synchrotron emission and the original thermal component, and the high energy bumps are due to inverse Compton. At high optical depth, $\tau \geq 100$, a Wien peak is formed at 10 keV and is blue-shifted to the MeV range by the bulk Lorentz factor of ∼100 expected in GRBs. In the intermediate regime, $0.1 \lesssim \tau \lesssim 100$, a flat energy spectrum above the thermal peak is obtained by multiple Compton scatterings. Figure taken from [194]. ©AAS. Reproduced with permission. Right: spectral decomposition of GRB 090902B (taken 9.6–13.0 s after the GBM trigger) enables clear identification of the physical origin of the emission. The dash-dotted (red) curve shows the spectrum that would have been obtained if synchrotron radiation was the only source of emission. The dashed (green) curve shows the resulting spectrum from synchrotron and synchrotron self-Compton SSC, and the solid (blue) curve shows the spectrum with the full radiative ingredients (synchrotron, SSC, the MeV thermal peak, and Comptonization of the thermal photons). From [206]. Reprinted (and modified) permission of Oxford University Press on behalf of the Royal Astronomical Society.

3.6.3. Hadronic Processes

Hadronic processes refer to radiation from protons that are present in the GRB outflow [207–214]. These protons could in principle radiate via the proton synchrotron process and produce the observed gamma-ray prompt emission. However, even if these protons do not produce the observed gamma-ray spectrum, if present, they could potentially interact with photons and decay to pions through the delta resonance: photopion processes (see, e.g., [207]). Pions decay into neutrinos, making GRBs possible sources of neutrinos; into leptons, which could in turn undergo synchrotron emission, and neutral pions decay directly to two high energy photons. Proton-photon interactions can also generate electron-positron pairs directly via the Bethe–Heitler process. Several authors have used the observed 100 MeV LAT prompt emission to constrain hadronic models [209–211,213]. To explain the 100 MeV LAT photons during the prompt phase, (i) the photopion and Bethe–Heitler processes require energy in protons larger than the observed gamma-ray energy by a factor of a thousand or more and (ii) the proton synchrotron mechanism requires protons to have a minimum Lorentz factor of $\sim 10^6$, which is much larger than expected if the protons are accelerated in shocks [213]. This makes hadronic processes less energetically viable than leptonic models.

4. Discussion: A Look into the Future

With the advent of observations at very high (GeV/TeV) energies by *Imaging Atmospheric Cherenkov Telescopes* such as MAGIC [6] and H.E.S.S. [7], and the perspective being opened by the future multi-messenger environment for gamma-ray bursts, the premise of radiation models will inevitably be revisited. At very high energies, the future Cherenkov Telescope Array (CTA) will provide improved sensitivity to up to an order of magnitude with respect to current IACTs [215,216]. Although the recently provided detection rate is modest (during the prompt phase, it is expected to be $\lesssim 1$ per year [217,218]), if CTA provides GRB observations with high photon statistics, it will help constrain emission models (e.g., the properties in the emission site of high-energy photons). The observed variability could help differentiate between emission mechanisms [217]. As the *Fermi* LAT spectra often displays a hard power-law spectrum extending to GeV energies, the observations of the high energy part could provide the information on the total radiated energy, and the bulk Lorentz factor can be constrained if the high-energy spectral cutoff due to pair production is identified [217]. At the low energy end, the *SVOM* (Space-based multi-band astronomical Variable Objects Monitor) mission aims to survey the high-energy sky and follow-up transients at optical and X-ray wavelengths [219]. Its main goals are observations of the high-redshift GRBs ($z > 5$), and faint/soft nearby events. It will also likely be the alert facility for CTA, opening e.g., the possibility of detecting low luminosity events which are not triggered by the current missions [220]. Other future multi-messenger facilities for GRB-related science include, e.g., the third-generation gravitational-wave observatory Einstein Telescope [221], the development of the extension of the IceCube Neutrino Observatory IceCube-Gen2 [222], and ATHENA [223] satellite for the X-ray domain. Upper limit of neutrino flux from GRBs [224] as well as the observations at longer wavelengths (e.g., using the upcoming Vera Rubin Observatory or the Square Kilometer Array-SKA) could provide information on jet composition—baryonic or magnetic jet. The observational advances need however to be followed by theoretical effort, i.e., numerical simulations of the processes involved in the production of prompt emission, such as energy dissipation and particle acceleration, in order to fully understand the extreme conditions in which gamma-ray bursts are produced.

Author Contributions: All authors Ž.B., R.B.D. and A.P. have contributed to writing—review and editing. All authors have read and agreed to the published version of the manuscript.

Funding: R.B.D. acknowledges support from the National Science Foundation under grants 1816694 and 2107932. A.P. was partially supported by the EU via the ERC consolidating grant 773062 (O.M.J.). Ž.B. acknowledges support from the Croatian Science Foundation (HrZZ) Project IP-2016-06-9782.

Institutional Review Board Statement: Not applicable.

Informed Consent Statement: Not applicable.

Data Availability Statement: Not applicable

Acknowledgments: We thank P. Beniamini, D. Giannios and P. Kumar for useful discussions related to this manuscript.

Conflicts of Interest: The authors declare no conflict of interest.

References

1. Abbott, B.P.; Abbott, R.; Abbott, T.D.; Acernese, F.; Ackley, K.; Adams, C.; Adams, T.; Addesso, P.; Adhikari, R.X.; Adya, V.B.; et al. GW170817: Observation of Gravitational Waves from a Binary Neutron Star Inspiral. *Phys. Rev. Lett.* **2017**, *119*, 161101. [CrossRef]
2. Abbott, B.P.; Abbott, R.; Abbott, T.D.; Acernese, F.; Ackley, K.; Adams, C.; Adams, T.; Addesso, P.; Adhikari, R.X.; Adya, V.B.; et al. Multi-messenger Observations of a Binary Neutron Star Merger. *Astrophys. J. Lett.* **2017**, *848*, L12. [CrossRef]
3. Kasliwal, M.M.; Nakar, E.; Singer, L.P.; Kaplan, D.L.; Cook, D.O.; Van Sistine, A.; Lau, R.M.; Fremling, C.; Gottlieb, O.; Jencson, J.E.; et al. Illuminating gravitational waves: A concordant picture of photons from a neutron star merger. *Science* **2017**, *358*, 1559–1565. [CrossRef]
4. Kasen, D.; Metzger, B.; Barnes, J.; Quataert, E.; Ramirez-Ruiz, E. Origin of the heavy elements in binary neutron-star mergers from a gravitational-wave event. *Nature* **2017**, *551*, 80–84. [CrossRef]
5. Smartt, S.J.; Chen, T.W.; Jerkstrand, A.; Coughlin, M.; Kankare, E.; Sim, S.A.; Fraser, M.; Inserra, C.; Maguire, K.; Chambers, K.C.; et al. A kilonova as the electromagnetic counterpart to a gravitational-wave source. *Nature* **2017**, *551*, 75–79. [CrossRef]
6. MAGIC Collaboration; Acciari, V.A.; Ansoldi, S.; Antonelli, L.A.; Arbet Engels, A.; Baack, D.; Babić, A.; Banerjee, B.; Barres de Almeida, U.; Barrio, J.A.; et al. Teraelectronvolt emission from the γ-ray burst GRB 190114C. *Nature* **2019**, *575*, 455–458. [CrossRef]
7. Abdalla, H.; Adam, R.; Aharonian, F.; Ait Benkhali, F.; Angüner, E.O.; Arakawa, M.; Arcaro, C.; Armand, C.; Ashkar, H.; Backes, M.; et al. A very-high-energy component deep in the γ-ray burst afterglow. *Nature* **2019**, *575*, 464–467. [CrossRef]
8. Deoskar, K.; Coppin, P.; Friedman, E. Searches for Neutrinos from Precursors and Afterglows of Gamma-ray Bursts using the IceCube Neutrino Observatory. *arXiv* **2021**, arXiv:2107.08870.
9. Klebesadel, R.W.; Strong, I.B.; Olson, R.A. Observations of Gamma-Ray Bursts of Cosmic Origin. *Astrophys. J.* **1973**, *182*, L85. [CrossRef]
10. Meegan, C.A.; Fishman, G.J.; Wilson, R.B.; Paciesas, W.S.; Pendleton, G.N.; Horack, J.M.; Brock, M.N.; Kouveliotou, C. Spatial distribution of γ-ray bursts observed by BATSE. *Nature* **1992**, *355*, 143–145. [CrossRef]
11. Band, D.; Matteson, J.; Ford, L.; Schaefer, B.; Palmer, D.; Teegarden, B.; Cline, T.; Briggs, M.; Paciesas, W.; Pendleton, G.; et al. BATSE Observations of Gamma-Ray Burst Spectra. I. Spectral Diversity. *Astrophys. J.* **1993**, *413*, 281. [CrossRef]
12. Abdo, A.A.; Ackermann, M.; Arimoto, M.; Asano, K.; Atwood, W.B.; Axelsson, M.; Baldini, L.; Ballet, J.; Band, D.L.; Barbiellini, G.; et al. Fermi Observations of High-Energy Gamma-Ray Emission from GRB 080916C. *Science* **2009**, *323*, 1688–1693. [CrossRef]
13. Zhang, B. Gamma-Ray Burst Prompt Emission. *Int. J. Mod. Phys. D* **2014**, *23*, 30002. [CrossRef]
14. Kumar, P.; Zhang, B. The physics of gamma-ray bursts and relativistic jets. *Phys. Rep.* **2015**, *561*, 1–109. doi:10.1016/j.physrep.2014.09.008. [CrossRef]
15. Pe'er, A. Physics of Gamma-Ray Bursts Prompt Emission. *Adv. Astron.* **2015**, *2015*, 907321. [CrossRef]
16. Dai, Z.; Daigne, F.; Mészáros, P. The Theory of Gamma-Ray Bursts. *Space Sci. Rev.* **2017**, *212*, 409–427. [CrossRef]
17. Beloborodov, A.M.; Mészáros, P. Photospheric Emission of Gamma-Ray Bursts. *Space Sci. Rev.* **2017**, *207*, 87–110. [CrossRef]
18. Meegan, C.; Lichti, G.; Bhat, P.N.; Bissaldi, E.; Briggs, M.S.; Connaughton, V.; Diehl, R.; Fishman, G.; Greiner, J.; Hoover, A.S.; et al. The Fermi Gamma-ray Burst Monitor. *Astrophys. J.* **2009**, *702*, 791–804. [CrossRef]
19. Poolakkil, S.; Preece, R.; Fletcher, C.; Goldstein, A.; Bhat, P.N.; Bissaldi, E.; Briggs, M.S.; Burns, E.; Cleveland, W.H.; Giles, M.M.; et al. The Fermi GBM Gamma-Ray Burst Spectral Catalog: 10 Years of Data. *arXiv* **2021**, arXiv:2103.13528.
20. Goldstein, A.; Burgess, J.M.; Preece, R.D.; Briggs, M.S.; Guiriec, S.; van der Horst, A.e.J.; Connaughton, V.; Wilson-Hodge, C.A.; Paciesas, W.S.; Meegan, C.A.; et al. The Fermi GBM Gamma-Ray Burst Spectral Catalog: The First Two Years. *Astrophys. J. Suppl. Ser.* **2012**, *199*, 19. [CrossRef]
21. Gruber, D.; Goldstein, A.; Weller von Ahlefeld, V.; Narayana Bhat, P.; Bissaldi, E.; Briggs, M.S.; Byrne, D.; Cleveland, W.H.; Connaughton, V.; Diehl, R.; et al. The Fermi GBM Gamma-Ray Burst Spectral Catalog: Four Years of Data. *Astrophys. J. Suppl. Ser.* **2014**, *211*, 12. [CrossRef]
22. Yu, H.F.; Preece, R.D.; Greiner, J.; Narayana Bhat, P.; Bissaldi, E.; Briggs, M.S.; Cleveland, W.H.; Connaughton, V.; Goldstein, A.; von Kienlin, A.; et al. The Fermi GBM gamma-ray burst time-resolved spectral catalog: Brightest bursts in the first four years. *Astron. Astrophys.* **2016**, *588*, A135. [CrossRef]
23. Oganesyan, G.; Nava, L.; Ghirlanda, G.; Celotti, A. Characterization of gamma-ray burst prompt emission spectra down to soft X-rays. *Astron. Astrophys.* **2018**, *616*, A138. [CrossRef]
24. Toffano, M.; Ghirlanda, G.; Nava, L.; Ghisellini, G.; Ravasio, M.E.; Oganesyan, G. The slope of the low-energy spectrum of prompt gamma-ray burst emission. *Astron. Astrophys.* **2021**, *652*, A123. [CrossRef]
25. Yu, H.F.; van Eerten, H.J.; Greiner, J.; Sari, R.; Narayana Bhat, P.; von Kienlin, A.; Paciesas, W.S.; Preece, R.D. The sharpness of gamma-ray burst prompt emission spectra. *Astron. Astrophys.* **2015**, *583*, A129. [CrossRef]
26. Acuner, Z.; Ryde, F.; Pe'er, A.; Mortlock, D.; Ahlgren, B. The Fraction of Gamma-Ray Bursts with an Observed Photospheric Emission Episode. *Astrophys. J.* **2020**, *893*, 128. [CrossRef]
27. Daigne, F.; Mochkovitch, R. The expected thermal precursors of gamma-ray bursts in the internal shock model. *Mon. Not. R. Astron. Soc.* **2002**, *336*, 1271–1280. [CrossRef]

28. Hascoët, R.; Daigne, F.; Mochkovitch, R. Prompt thermal emission in gamma-ray bursts. *Astron. Astrophys.* **2013**, *551*, A124. [CrossRef]
29. Ghirlanda, G.; Celotti, A.; Ghisellini, G. Extremely hard GRB spectra prune down the forest of emission models. *Astron. Astrophys.* **2003**, *406*, 879–892. [CrossRef]
30. Acuner, Z.; Ryde, F.; Yu, H.F. Non-dissipative photospheres in GRBs: Spectral appearance in the Fermi/GBM catalogue. *Mon. Not. R. Astron. Soc.* **2019**, *487*, 5508–5519. [CrossRef]
31. Bosnjak, Z.; Celotti, A.; Ghirlanda, G. GRB 990413: Insight into the thermal phase evolution. *Mon. Not. R. Astron. Soc.* **2006**, *370*, L33–L37. [CrossRef]
32. Ryde, F.; Pe'er, A. Quasi-blackbody Component and Radiative Efficiency of the Prompt Emission of Gamma-ray Bursts. *Astrophys. J.* **2009**, *702*, 1211–1229. [CrossRef]
33. Ryde, F.; Pe'er, A.; Nymark, T.; Axelsson, M.; Moretti, E.; Lundman, C.; Battelino, M.; Bissaldi, E.; Chiang, J.; Jackson, M.S.; et al. Observational evidence of dissipative photospheres in gamma-ray bursts. *Mon. Not. R. Astron. Soc.* **2011**, *415*, 3693–3705. [CrossRef]
34. Guiriec, S.; Mochkovitch, R.; Piran, T.; Daigne, F.; Kouveliotou, C.; Racusin, J.; Gehrels, N.; McEnery, J. GRB 131014A: A Laboratory for Studying the Thermal-like and Non-thermal Emissions in Gamma-Ray Bursts, and the New $L^{nTh}{}_{i}$-$E^{nTh,rest}{}_{peak,i}$ Relation. *Astrophys. J.* **2015**, *814*, 10. [CrossRef]
35. Guiriec, S.; Gehrels, N.; McEnery, J.; Kouveliotou, C.; Hartmann, D.H. Photospheric Emission in the Joint GBM and Konus Prompt Spectra of GRB 120323A. *Astrophys. J.* **2017**, *846*, 138. [CrossRef]
36. Li, L. Thermal Components in Gamma-Ray Bursts. I. How Do They Affect Nonthermal Spectral Parameters? *Astrophys. J. Suppl. Ser.* **2019**, *245*, 7. [CrossRef]
37. Li, L.; Ryde, F.; Pe'er, A.; Yu, H.F.; Acuner, Z. Bayesian Time-resolved Spectroscopy of Multipulse GRBs: Variations of Emission Properties among Pulses. *Astrophys. J. Suppl. Ser.* **2021**, *254*, 35. [CrossRef]
38. Kanbach, G.; Bertsch, D.L.; Fichtel, C.E.; Hartman, R.C.; Hunter, S.D.; Kniffen, D.A.; Hughlock, B.W.; Favale, A.; Hofstadter, R.; Hughes, E.B. The project EGRET (energetic gamma-ray experiment telescope) on NASA's Gamma-Ray Observatory GRO. *Space Sci. Rev.* **1989**, *49*, 69–84. [CrossRef]
39. Schneid, E.J.; Bertsch, D.L.; Fichtel, C.E.; Hartman, R.C.; Hunter, S.D.; Kanbach, G.; Kniffen, D.A.; Kwok, P.W.; Lin, Y.C.; Mattox, J.R.; et al. EGRET detection of high energy gamma rays from the gamma-ray burst of 3 May 1991. *Astron. Astrophys.* **1992**, *255*, L13.
40. Sommer, M.; Bertsch, D.L.; Dingus, B.L.; Fichtel, C.E.; Fishman, G.J.; Harding, A.K.; Hartman, R.C.; Hunter, S.D.; Hurley, K.; Kanbach, G.; et al. High-Energy Gamma Rays from the Intense 1993 January 31 Gamma-Ray Burst. *Astrophys. J.* **1994**, *422*, L63. [CrossRef]
41. Hurley, K.; Dingus, B.L.; Mukherjee, R.; Sreekumar, P.; Kouveliotou, C.; Meegan, C.; Fishman, G.J.; Band, D.; Ford, L.; Bertsch, D.; et al. Detection of a γ-ray burst of very long duration and very high energy. *Nature* **1994**, *372*, 652–654. [CrossRef]
42. González, M.M.; Dingus, B.L.; Kaneko, Y.; Preece, R.D.; Dermer, C.D.; Briggs, M.S. A γ-ray burst with a high-energy spectral component inconsistent with the synchrotron shock model. *Nature* **2003**, *424*, 749–751. [CrossRef] [PubMed]
43. Atwood, W.B.; Abdo, A.A.; Ackermann, M.; Althouse, W.; Anderson, B.; Axelsson, M.; Baldini, L.; Ballet, J.; Band, D.L.; Barbiellini, G.; et al. The Large Area Telescope on the Fermi Gamma-Ray Space Telescope Mission. *Astrophys. J.* **2009**, *697*, 1071–1102. [CrossRef]
44. Ajello, M.; Arimoto, M.; Axelsson, M.; Baldini, L.; Barbiellini, G.; Bastieri, D.; Bellazzini, R.; Bhat, P.N.; Bissaldi, E.; Blandford, R.D.; et al. A Decade of Gamma-Ray Bursts Observed by Fermi-LAT: The Second GRB Catalog. *Astrophys. J.* **2019**, *878*, 52. [CrossRef]
45. MAGIC Collaboration; Acciari, V.A.; Ansoldi, S.; Antonelli, L.A.; Engels, A.A.; Baack, D.; Babić, A.; Banerjee, B.; Barres de Almeida, U.; Barrio, J.A.; et al. Observation of inverse Compton emission from a long γ-ray burst. *Nature* **2019**, *575*, 459–463. [CrossRef]
46. Sari, R.; Piran, T. Variability in Gamma-Ray Bursts: A Clue. *Astrophys. J.* **1997**, *485*, 270–273. [CrossRef]
47. Narayan, R.; Kumar, P. A turbulent model of gamma-ray burst variability. *Mon. Not. R. Astron. Soc.* **2009**, *394*, L117–L120. [CrossRef]
48. Kouveliotou, C.; Meegan, C.A.; Fishman, G.J.; Bhat, N.P.; Briggs, M.S.; Koshut, T.M.; Paciesas, W.S.; Pendleton, G.N. Identification of two classes of gamma-ray bursts. *Astrophys. J.* **1993**, *413*, L101–L104. [CrossRef]
49. Bromberg, O.; Nakar, E.; Piran, T.; Sari, R. Short versus Long and Collapsars versus Non-collapsars: A Quantitative Classification of Gamma-Ray Bursts. *Astrophys. J.* **2013**, *764*, 179. [CrossRef]
50. Ahumada, T.; Singer, L.P.; Anand, S.; Coughlin, M.W.; Kasliwal, M.M.; Ryan, G.; Andreoni, I.; Cenko, S.B.; Fremling, C.; Kumar, H.; et al. Discovery and confirmation of the shortest gamma-ray burst from a collapsar. *Nat. Astron.* **2021**, *5*, 917–927. [CrossRef]
51. Woosley, S.E. Gamma-ray bursts from stellar mass accretion disks around black holes. *Astrophys. J.* **1993**, *405*, 273–277. [CrossRef]
52. MacFadyen, A.I.; Woosley, S.E. Collapsars: Gamma-Ray Bursts and Explosions in "Failed Supernovae". *Astrophys. J.* **1999**, *524*, 262–289. [CrossRef]
53. Hjorth, J.; Sollerman, J.; Møller, P.; Fynbo, J.P.U.; Woosley, S.E.; Kouveliotou, C.; Tanvir, N.R.; Greiner, J.; Andersen, M.I.; Castro-Tirado, A.J.; et al. A very energetic supernova associated with the γ-ray burst of 29 March 2003. *Nature* **2003**, *423*, 847–850. [CrossRef]

54. Campana, S.; Mangano, V.; Blustin, A.J.; Brown, P.; Burrows, D.N.; Chincarini, G.; Cummings, J.R.; Cusumano, G.; Della Valle, M.; Malesani, D.; et al. The association of GRB 060218 with a supernova and the evolution of the shock wave. *Nature* **2006**, *442*, 1008–1010. [CrossRef] [PubMed]
55. Eichler, D.; Livio, M.; Piran, T.; Schramm, D.N. Nucleosynthesis, neutrino bursts and gamma-rays from coalescing neutron stars. *Nature* **1989**, *340*, 126–128. [CrossRef]
56. Norris, J.P.; Bonnell, J.T. Short Gamma-Ray Bursts with Extended Emission. *Astrophys. J.* **2006**, *643*, 266–275. [CrossRef]
57. Genet, F.; Butler, N.R.; Granot, J. The long rapid decay phase of the extended emission from the short GRB 080503. *Mon. Not. R. Astron. Soc.* **2010**, *405*, 695–700. [CrossRef]
58. Perley, D.A.; Metzger, B.D.; Granot, J.; Butler, N.R.; Sakamoto, T.; Ramirez-Ruiz, E.; Levan, A.J.; Bloom, J.S.; Miller, A.A.; Bunker, A.; et al. GRB 080503: Implications of a Naked Short Gamma-Ray Burst Dominated by Extended Emission. *Astrophys. J.* **2009**, *696*, 1871–1885. [CrossRef]
59. Bhat, P.N.; Fishman, G.J.; Meegan, C.A.; Wilson, R.B.; Brock, M.N.; Paciesas, W.S. Evidence for sub-millisecond structure in a γ-ray burst. *Nature* **1992**, *359*, 217–218. [CrossRef]
60. Walker, K.C.; Schaefer, B.E.; Fenimore, E.E. Gamma-Ray Bursts Have Millisecond Variability. *Astrophys. J.* **2000**, *537*, 264–269. [CrossRef]
61. Piran, T. Gamma-ray bursts and the fireball model. *Phys. Rep.* **1999**, *314*, 575–667. [CrossRef]
62. Tang, Q.W.; Wang, X.Y.; Liu, R.Y. Evidence of an Internal Dissipation Origin for the High-energy Prompt Emission of GRB 170214A. *Astrophys. J.* **2017**, *844*, 56. [CrossRef]
63. Rybicki, G.B.; Lightman, A.P. *Radiative Processes in Astrophysics*; Wiley-Interscience: New York, NY, USA, 1979; p. 393.
64. Covino, S.; Lazzati, D.; Ghisellini, G.; Saracco, P.; Campana, S.; Chincarini, G.; di Serego, S.; Cimatti, A.; Vanzi, L.; Pasquini, L.; et al. GRB 990510: Linearly polarized radiation from a fireball. *Astron. Astrophys.* **1999**, *348*, L1–L4.
65. Wijers, R.A.M.J.; Vreeswijk, P.M.; Galama, T.J.; Rol, E.; van Paradijs, J.; Kouveliotou, C.; Giblin, T.; Masetti, N.; Palazzi, E.; Pian, E.; et al. Detection of Polarization in the Afterglow of GRB 990510 with the ESO Very Large Telescope. *Astrophys. J.* **1999**, *523*, L33–L36. [CrossRef]
66. Gill, R.; Granot, J.; Kumar, P. Linear polarization in gamma-ray burst prompt emission. *MNRAS* **2020**, *491*, 3343–3373. [CrossRef]
67. Coburn, W.; Boggs, S.E. Polarization of the prompt γ-ray emission from the γ-ray burst of 6 December 2002. *Nature* **2003**, *423*, 415–417. [CrossRef]
68. Willis, D.R.; Barlow, E.J.; Bird, A.J.; Clark, D.J.; Dean, A.J.; McConnell, M.L.; Moran, L.; Shaw, S.E.; Sguera, V. Evidence of polarisation in the prompt gamma-ray emission from GRB 930131 and GRB 960924. *Astron. Astrophys.* **2005**, *439*, 245–253. [CrossRef]
69. Kalemci, E.; Boggs, S.E.; Kouveliotou, C.; Finger, M.; Baring, M.G. Search for Polarization from the Prompt Gamma-Ray Emission of GRB 041219a with SPI on INTEGRAL. *Astrophys. J. Suppl. Ser.* **2007**, *169*, 75–82. [CrossRef]
70. McGlynn, S.; Clark, D.J.; Dean, A.J.; Hanlon, L.; McBreen, S.; Willis, D.R.; McBreen, B.; Bird, A.J.; Foley, S. Polarisation studies of the prompt gamma-ray emission from GRB 041219a using the spectrometer aboard INTEGRAL. *Astron. Astrophys.* **2007**, *466*, 895–904. [CrossRef]
71. Götz, D.; Laurent, P.; Lebrun, F.; Daigne, F.; Bošnjak, Ž. Variable Polarization Measured in the Prompt Emission of GRB 041219A Using IBIS on Board INTEGRAL. *Astrophys. J.* **2009**, *695*, L208–L212. [CrossRef]
72. Götz, D.; Covino, S.; Fernández-Soto, A.; Laurent, P.; Bošnjak, Ž. The polarized gamma-ray burst GRB 061122. *Mon. Not. R. Astron. Soc.* **2013**, *431*, 3550–3556. [CrossRef]
73. Yonetoku, D.; Murakami, T.; Gunji, S.; Mihara, T.; Toma, K.; Sakashita, T.; Morihara, Y.; Takahashi, T.; Toukairin, N.; Fujimoto, H.; et al. Detection of Gamma-Ray Polarization in Prompt Emission of GRB 100826A. *Astrophys. J.* **2011**, *743*, L30. [CrossRef]
74. Yonetoku, D.; Murakami, T.; Gunji, S.; Mihara, T.; Toma, K.; Morihara, Y.; Takahashi, T.; Wakashima, Y.; Yonemochi, H.; Sakashita, T.; et al. Magnetic Structures in Gamma-Ray Burst Jets Probed by Gamma-Ray Polarization. *Astrophys. J.* **2012**, *758*, L1. [CrossRef]
75. Rao, A.R.; Chand, V.; Hingar, M.K.; Iyyani, S.; Khanna, R.; Kutty, A.P.K.; Malkar, J.P.; Paul, D.; Bhalerao, V.B.; Bhattacharya, D.; et al. AstroSat CZT Imager Observations of GRB 151006A: Timing, Spectroscopy, and Polarization Study. *Astrophys. J.* **2016**, *833*, 86. [CrossRef]
76. Chand, V.; Chattopadhyay, T.; Oganesyan, G.; Rao, A.R.; Vadawale, S.V.; Bhattacharya, D.; Bhalerao, V.B.; Misra, K. AstroSat-CZTI Detection of Variable Prompt Emission Polarization in GRB 171010A. *Astrophys. J.* **2019**, *874*, 70. [CrossRef]
77. Sharma, V.; Iyyani, S.; Bhattacharya, D.; Chattopadhyay, T.; Rao, A.R.; Aarthy, E.; Vadawale, S.V.; Mithun, N.P.S.; Bhalerao, V.B.; Ryde, F.; et al. Time-varying Polarized Gamma-Rays from GRB 160821A: Evidence for Ordered Magnetic Fields. *Astrophys. J.* **2019**, *882*, L10. [CrossRef]
78. Kole, M.; De Angelis, N.; Berlato, F.; Burgess, J.M.; Gauvin, N.; Greiner, J.; Hajdas, W.; Li, H.C.; Li, Z.H.; Produit, N.; et al. The POLAR Gamma-Ray Burst Polarization Catalog. *arXiv* **2020**, arXiv:2009.04871.
79. Mészáros, P.; Rees, M.J.; Papathanassiou, H. Spectral properties of blast-wave models of gamma-ray burst sources. *Astrophys. J.* **1994**, *432*, 181–193. [CrossRef]
80. Daigne, F.; Mochkovitch, R. The physics of pulses in gamma-ray bursts: Emission processes, temporal profiles and time-lags. *Mon. Not. R. Astron. Soc.* **2003**, *342*, 587–592. [CrossRef]
81. Thompson, C. A Model of Gamma-Ray Bursts. *Mon. Not. R. Astron. Soc.* **1994**, *270*, 480–498. [CrossRef]

82. Rees, M.J.; Mészáros, P. Dissipative Photosphere Models of Gamma-Ray Bursts and X-Ray Flashes. *Astrophys. J.* **2005**, *628*, 847–852. [CrossRef]
83. Pe'er, A. Temporal Evolution of Thermal Emission from Relativistically Expanding Plasma. *Astrophys. J.* **2008**, *682*, 463–473. [CrossRef]
84. Toma, K.; Wu, X.F.; Mészáros, P. Photosphere-internal shock model of gamma-ray bursts: Case studies of Fermi/LAT bursts. *Mon. Not. R. Astron. Soc.* **2011**, *415*, 1663–1680. [CrossRef]
85. Cavallo, G.; Rees, M.J. A qualitative study of cosmic fireballs and γ-ray bursts. *Mon. Not. R. Astron. Soc.* **1978**, *183*, 359–365. [CrossRef]
86. Paczynski, B. Gamma-ray bursters at cosmological distances. *Astrophys. J.* **1986**, *308*, L43–L46. [CrossRef]
87. Goodman, J. Are gamma-ray bursts optically thick? *Astrophys. J.* **1986**, *308*, L47–L50. [CrossRef]
88. Shemi, A.; Piran, T. The Appearance of Cosmic Fireballs. *Astrophys. J.* **1990**, *365*, L55. [CrossRef]
89. Meszaros, P.; Rees, M.J. Relativistic fireballs and their impact on external matter—Models for cosmological gamma-ray bursts. *Astrophys. J.* **1993**, *405*, 278–284. [CrossRef]
90. Mészáros, P.; Rees, M.J. Steep Slopes and Preferred Breaks in Gamma-Ray Burst Spectra: The Role of Photospheres and Comptonization. *Astrophys. J.* **2000**, *530*, 292–298. [CrossRef]
91. Mészáros, P.; Ramirez-Ruiz, E.; Rees, M.J.; Zhang, B. X-Ray-rich Gamma-Ray Bursts, Photospheres, and Variability. *Astrophys. J.* **2002**, *578*, 812–817. [CrossRef]
92. Beloborodov, A.M. Collisional mechanism for gamma-ray burst emission. *Mon. Not. R. Astron. Soc.* **2010**, *407*, 1033–1047. [CrossRef]
93. Pe'er, A.; Ryde, F. A Theory of Multicolor Blackbody Emission from Relativistically Expanding Plasmas. *Astrophys. J.* **2011**, *732*, 49. [CrossRef]
94. Deng, W.; Zhang, B. Low Energy Spectral Index and E_p Evolution of Quasi-thermal Photosphere Emission of Gamma-Ray Bursts. *Astrophys. J.* **2014**, *785*, 112. [CrossRef]
95. Lundman, C.; Pe'er, A.; Ryde, F. A theory of photospheric emission from relativistic, collimated outflows. *Mon. Not. R. Astron. Soc.* **2013**, *428*, 2430–2442. [CrossRef]
96. Abdo, A.A.; Ackermann, M.; Ajello, M.; Asano, K.; Atwood, W.B.; Axelsson, M.; Baldini, L.; Ballet, J.; Barbiellini, G.; Baring, M.G.; et al. Fermi Observations of GRB 090902B: A Distinct Spectral Component in the Prompt and Delayed Emission. *Astrophys. J.* **2009**, *706*, L138–L144. [CrossRef]
97. Zhang, B.; Pe'er, A. Evidence of an Initially Magnetically Dominated Outflow in GRB 080916C. *Astrophys. J.* **2009**, *700*, L65–L68. [CrossRef]
98. Wijers, R.A.M.J.; Rees, M.J.; Meszaros, P. Shocked by GRB 970228: The afterglow of a cosmological fireball. *Mon. Not. R. Astron. Soc.* **1997**, *288*, L51–L56. [CrossRef]
99. Sari, R.; Piran, T.; Narayan, R. Spectra and Light Curves of Gamma-Ray Burst Afterglows. *Astrophys. J.* **1998**, *497*, L17–L20. [CrossRef]
100. Fermi, E. On the Origin of the Cosmic Radiation. *Phys. Rev.* **1949**, *75*, 1169–1174. [CrossRef]
101. Fermi, E. Galactic Magnetic Fields and the Origin of Cosmic Radiation. *Astrophys. J.* **1954**, *119*, 1. [CrossRef]
102. Bell, A.R. The acceleration of cosmic rays in shock fronts. I. *Mon. Not. R. Astron. Soc.* **1978**, *182*, 147–156. [CrossRef]
103. Blandford, R.D.; Ostriker, J.P. Particle acceleration by astrophysical shocks. *Astrophys. J.* **1978**, *221*, L29–L32. [CrossRef]
104. Blandford, R.; Eichler, D. Particle acceleration at astrophysical shocks: A theory of cosmic ray origin. *Phys. Rep.* **1987**, *154*, 1–75. [CrossRef]
105. Jones, F.C.; Ellison, D.C. The plasma physics of shock acceleration. *Space Sci. Rev.* **1991**, *58*, 259–346. [CrossRef]
106. Silva, L.O.; Fonseca, R.A.; Tonge, J.W.; Dawson, J.M.; Mori, W.B.; Medvedev, M.V. Interpenetrating Plasma Shells: Near-equipartition Magnetic Field Generation and Nonthermal Particle Acceleration. *Astrophys. J.* **2003**, *596*, L121–L124. [CrossRef]
107. Nishikawa, K.I.; Hardee, P.; Richardson, G.; Preece, R.; Sol, H.; Fishman, G.J. Particle Acceleration in Relativistic Jets Due to Weibel Instability. *Astrophys. J.* **2003**, *595*, 555–563. [CrossRef]
108. Spitkovsky, A. Particle Acceleration in Relativistic Collisionless Shocks: Fermi Process at Last? *Astrophys. J.* **2008**, *682*, L5. [CrossRef]
109. Haugbølle, T. Three-dimensional Modeling of Relativistic Collisionless Ion-electron Shocks. *Astrophys. J.* **2011**, *739*, L42. [CrossRef]
110. Sironi, L.; Spitkovsky, A. Particle Acceleration in Relativistic Magnetized Collisionless Pair Shocks: Dependence of Shock Acceleration on Magnetic Obliquity. *Astrophys. J.* **2009**, *698*, 1523–1549. [CrossRef]
111. Sironi, L.; Spitkovsky, A. Particle Acceleration in Relativistic Magnetized Collisionless Electron-Ion Shocks. *Astrophys. J.* **2011**, *726*, 75. [CrossRef]
112. Summerlin, E.J.; Baring, M.G. Diffusive Acceleration of Particles at Oblique, Relativistic, Magnetohydrodynamic Shocks. *Astrophys. J.* **2012**, *745*, 63. [CrossRef]
113. Sironi, L.; Spitkovsky, A. Relativistic Reconnection: An Efficient Source of Non-thermal Particles. *Astrophys. J.* **2014**, *783*, L21. [CrossRef]
114. Lemoine, M.; Vanthieghem, A.; Pelletier, G.; Gremillet, L. Physics of relativistic collisionless shocks. II. Dynamics of the background plasma. *Phys. Rev. E* **2019**, *100*, 033209. [CrossRef] [PubMed]

115. Nagar, Y.; Keshet, U. Diffusive acceleration in relativistic shocks: Particle feedback. *Mon. Not. R. Astron. Soc.* **2021**, *501*, 329–336. [CrossRef]
116. Kundu, S.; Vaidya, B.; Mignone, A. Numerical Modeling and Physical Interplay of Stochastic Turbulent Acceleration for Nonthermal Emission Processes. *Astrophys. J.* **2021**, *921*, 74. [CrossRef]
117. Sironi, L.; Petropoulou, M.; Giannios, D. Relativistic jets shine through shocks or magnetic reconnection? *Mon. Not. R. Astron. Soc.* **2015**, *450*, 183–191. [CrossRef]
118. Zenitani, S.; Hoshino, M. The Generation of Nonthermal Particles in the Relativistic Magnetic Reconnection of Pair Plasmas. *Astrophys. J.* **2001**, *562*, L63–L66. [CrossRef]
119. Zenitani, S.; Hoshino, M. Particle Acceleration and Magnetic Dissipation in Relativistic Current Sheet of Pair Plasmas. *Astrophys. J.* **2007**, *670*, 702–726. [CrossRef]
120. McKinney, J.C.; Uzdensky, D.A. A reconnection switch to trigger gamma-ray burst jet dissipation. *Mon. Not. R. Astron. Soc.* **2012**, *419*, 573–607. [CrossRef]
121. Guo, F.; Li, H.; Daughton, W.; Liu, Y.H. Formation of Hard Power Laws in the Energetic Particle Spectra Resulting from Relativistic Magnetic Reconnection. *Phys. Rev. Lett.* **2014**, *113*, 155005. [CrossRef]
122. Guo, F.; Liu, Y.H.; Daughton, W.; Li, H. Particle Acceleration and Plasma Dynamics during Magnetic Reconnection in the Magnetically Dominated Regime. *Astrophys. J.* **2015**, *806*, 167. [CrossRef]
123. Sironi, L.; Giannios, D.; Petropoulou, M. Plasmoids in relativistic reconnection, from birth to adulthood: First they grow, then they go. *Mon. Not. R. Astron. Soc.* **2016**, *462*, 48–74. [CrossRef]
124. Werner, G.R.; Uzdensky, D.A. Nonthermal Particle Acceleration in 3D Relativistic Magnetic Reconnection in Pair Plasma. *Astrophys. J.* **2017**, *843*, L27. [CrossRef]
125. Kagan, D.; Nakar, E.; Piran, T. Physics of the saturation of particle acceleration in relativistic magnetic reconnection. *Mon. Not. R. Astron. Soc.* **2018**, *476*, 3902–3912. [CrossRef]
126. Rees, M.J.; Meszaros, P. Unsteady outflow models for cosmological gamma-ray bursts. *Astrophys. J.* **1994**, *430*, L93–L96. [CrossRef]
127. Paczynski, B.; Xu, G. Neutrino bursts from gamma-ray bursts. *Astrophys. J.* **1994**, *427*, 708–713. [CrossRef]
128. Kobayashi, S.; Piran, T.; Sari, R. Can Internal Shocks Produce the Variability in Gamma-Ray Bursts? *Astrophys. J.* **1997**, *490*, 92. [CrossRef]
129. Daigne, F.; Mochkovitch, R. Gamma-ray bursts from internal shocks in a relativistic wind: Temporal and spectral properties. *Mon. Not. R. Astron. Soc.* **1998**, *296*, 275–286. [CrossRef]
130. Kumar, P.; McMahon, E. A general scheme for modelling γ-ray burst prompt emission. *Mon. Not. R. Astron. Soc.* **2008**, *384*, 33–63. [CrossRef]
131. Bošnjak, Ž.; Daigne, F.; Dubus, G. Prompt high-energy emission from γ-ray bursts in the internal shock model. *Astron. Astrophys.* **2009**, *498*, 677–703. [CrossRef]
132. Bošnjak, Ž.; Daigne, F. Spectral evolution in gamma-ray bursts: Predictions of the internal shock model and comparison to observations. *Astron. Astrophys.* **2014**, *568*, A45. [CrossRef]
133. Daigne, F.; Mochkovitch, R. Gamma-ray bursts from internal shocks in a relativistic wind: A hydrodynamical study. *Astron. Astrophys.* **2000**, *358*, 1157–1166.
134. Mimica, P.; Aloy, M.A.; Müller, E. Internal shocks in relativistic outflows: Collisions of magnetized shells. *Astron. Astrophys.* **2007**, *466*, 93–106. [CrossRef]
135. Sari, R.; Piran, T. Cosmological gamma-ray bursts: Internal versus external shocks. *Mon. Not. R. Astron. Soc.* **1997**, *287*, 110–116. [CrossRef]
136. Kumar, P. Gamma-Ray Burst Energetics. *Astrophys. J.* **1999**, *523*, L113–L116. [CrossRef]
137. Spada, M.; Panaitescu, A.; Mészáros, P. Analysis of Temporal Features of Gamma-Ray Bursts in the Internal Shock Model. *Astrophys. J.* **2000**, *537*, 824–832. [CrossRef]
138. Lloyd, N.M.; Petrosian, V. Synchrotron Radiation as the Source of Gamma-Ray Burst Spectra. *Astrophys. J.* **2000**, *543*, 722–732. [CrossRef]
139. Bykov, A.M.; Meszaros, P. Electron Acceleration and Efficiency in Nonthermal Gamma-Ray Sources. *Astrophys. J.* **1996**, *461*, L37. [CrossRef]
140. Spruit, H.C.; Daigne, F.; Drenkhahn, G. Large scale magnetic fields and their dissipation in GRB fireballs. *Astron. Astrophys.* **2001**, *369*, 694–705. [CrossRef]
141. Gruzinov, A.; Mészáros, P. Photon Acceleration in Variable Ultrarelativistic Outflows and High-Energy Spectra of Gamma-Ray Bursts. *Astrophys. J.* **2000**, *539*, L21–L24. [CrossRef]
142. Kumar, P.; Narayan, R. GRB 080319B: Evidence for relativistic turbulence, not internal shocks. *Mon. Not. R. Astron. Soc.* **2009**, *395*, 472–489. [CrossRef]
143. Zou, Y.C.; Piran, T.; Sari, R. Clues from the Prompt Emission of GRB 080319B. *Astrophys. J.* **2009**, *692*, L92–L95. [CrossRef]
144. Derishev, E.V.; Kocharovsky, V.V.; Kocharovsky, V.V. The Neutron Component in Fireballs of Gamma-Ray Bursts: Dynamics and Observable Imprints. *Astrophys. J.* **1999**, *521*, 640–649. [CrossRef]
145. Mészáros, P.; Rees, M.J. Multi-GEV Neutrinos from Internal Dissipation in Gamma-Ray Burst Fireballs. *Astrophys. J.* **2000**, *541*, L5–L8. [CrossRef]
146. Beloborodov, A.M. Nuclear Composition of Gamma-Ray Burst Fireballs. *Astrophys. J.* **2003**, *588*, 931–944. [CrossRef]

147. Beloborodov, A.M. Hyper-accreting black holes. In *Cool Discs, Hot Flows: The Varying Faces of Accreting Compact Objects*; Axelsson, M., Ed.; American Institute of Physics Conference Series; 2008; Volume 1054, pp. 51–70. Available online: https://ui.adsabs.harvard.edu/abs/2008AIPC.1054...51B (accessed on 1 December 2021).
148. Vurm, I.; Beloborodov, A.M.; Poutanen, J. Gamma-Ray Bursts from Magnetized Collisionally Heated Jets. *Astrophys. J.* **2011**, *738*, 77. [CrossRef]
149. Usov, V.V. Millisecond pulsars with extremely strong magnetic fields as a cosmological source of γ-ray bursts. *Nature* **1992**, *357*, 472–474. [CrossRef]
150. Mészáros, P.; Rees, M.J. Poynting Jets from Black Holes and Cosmological Gamma-Ray Bursts. *Astrophys. J.* **1997**, *482*, L29–L32. [CrossRef]
151. Spruit, H.C. Gamma-ray bursts from X-ray binaries. *Astron. Astrophys.* **1999**, *341*, L1–L4.
152. Lyutikov, M.; Blandford, R. Gamma Ray Bursts as Electromagnetic Outflows. *arXiv* **2003**, arXiv:astro–ph/0312347.
153. Narayan, R.; Kumar, P.; Tchekhovskoy, A. Constraints on cold magnetized shocks in gamma-ray bursts. *Mon. Not. R. Astron. Soc.* **2011**, *416*, 2193–2201. [CrossRef]
154. Kumar, P.; Crumley, P. Radiation from a relativistic Poynting jet: Some general considerations. *Mon. Not. R. Astron. Soc.* **2015**, *453*, 1820–1828. [CrossRef]
155. Drenkhahn, G. Acceleration of GRB outflows by Poynting flux dissipation. *Astron. Astrophys.* **2002**, *387*, 714–724. [CrossRef]
156. Drenkhahn, G.; Spruit, H.C. Efficient acceleration and radiation in Poynting flux powered GRB outflows. *Astron. Astrophys.* **2002**, *391*, 1141–1153. [CrossRef]
157. Giannios, D.; Spruit, H.C. Spectra of Poynting-flux powered GRB outflows. *Astron. Astrophys.* **2005**, *430*, 1–7. [CrossRef]
158. Giannios, D. Prompt GRB emission from gradual energy dissipation. *Astron. Astrophys.* **2008**, *480*, 305–312. [CrossRef]
159. Bégué, D.; Pe'er, A.; Lyubarsky, Y. Radiative striped wind model for gamma-ray bursts. *Mon. Not. R. Astron. Soc.* **2017**, *467*, 2594–2611. [CrossRef]
160. Giannios, D.; Uzdensky, D.A. GRB and blazar jets shining through their stripes. *Mon. Not. R. Astron. Soc.* **2019**, *484*, 1378–1389. [CrossRef]
161. Giannios, D. Prompt emission spectra from the photosphere of a GRB. *Astron. Astrophys.* **2006**, *457*, 763–770. [CrossRef]
162. Giannios, D. The peak energy of dissipative gamma-ray burst photospheres. *Mon. Not. R. Astron. Soc.* **2012**, *422*, 3092–3098. [CrossRef]
163. Beniamini, P.; Giannios, D. Prompt gamma-ray burst emission from gradual magnetic dissipation. *MNRAS* **2017**, *468*, 3202–3211. [CrossRef]
164. Gill, R.; Granot, J.; Beniamini, P. GRB spectrum from gradual dissipation in a magnetized outflow. *Mon. Not. R. Astron. Soc.* **2020**, *499*, 1356–1372. [CrossRef]
165. Contopoulos, J. A Simple Type of Magnetically Driven Jets: An Astrophysical Plasma Gun. *Astrophys. J.* **1995**, *450*, 616. [CrossRef]
166. Granot, J.; Komissarov, S.S.; Spitkovsky, A. Impulsive acceleration of strongly magnetized relativistic flows. *Mon. Not. R. Astron. Soc.* **2011**, *411*, 1323–1353. [CrossRef]
167. Granot, J. The effects of sub-shells in highly magnetized relativistic flows. *Mon. Not. R. Astron. Soc.* **2012**, *421*, 2467–2477. [CrossRef]
168. Komissarov, S.S. Shock dissipation in magnetically dominated impulsive flows. *Mon. Not. R. Astron. Soc.* **2012**, *422*, 326–346. [CrossRef]
169. Lazar, A.; Nakar, E.; Piran, T. Gamma-Ray Burst Light Curves in the Relativistic Turbulence and Relativistic Subjet Models. *Astrophys. J.* **2009**, *695*, L10–L14. [CrossRef]
170. Zhang, B.; Yan, H. The Internal-collision-induced Magnetic Reconnection and Turbulence (ICMART) Model of Gamma-ray Bursts. *Astrophys. J.* **2011**, *726*, 90. [CrossRef]
171. Beniamini, P.; Granot, J. Properties of GRB light curves from magnetic reconnection. *Mon. Not. R. Astron. Soc.* **2016**, *459*, 3635–3658. [CrossRef]
172. Barniol Duran, R.; Leng, M.; Giannios, D. An anisotropic minijets model for the GRB prompt emission. *Mon. Not. R. Astron. Soc.* **2016**, *455*, L6–L10. [CrossRef]
173. Ackermann, M.; Ajello, M.; Asano, K.; Axelsson, M.; Baldini, L.; Ballet, J.; Barbiellini, G.; Bastieri, D.; Bechtol, K.; Bellazzini, R.; et al. The First Fermi-LAT Gamma-Ray Burst Catalog. *Astrophys. J. Suppl. Ser.* **2013**, *209*, 11. [CrossRef]
174. Rees, M.J.; Mészáros, P. Relativistic fireballs: Energy conversion and time-scales. *Mon. Not. R. Astron. Soc.* **1992**, *258*, 41P–43P. [CrossRef]
175. Meszaros, P.; Laguna, P.; Rees, M.J. Gasdynamics of relativistically expanding gamma-ray burst sources—Kinematics, energetics, magnetic fields, and efficiency. *Astrophys. J.* **1993**, *415*, 181–190. [CrossRef]
176. Papathanassiou, H.; Meszaros, P. Spectra of Unsteady Wind Models of Gamma-Ray Bursts. *Astrophys. J.* **1996**, *471*, L91. [CrossRef]
177. Beniamini, P.; Piran, T. Constraints on the Synchrotron Emission Mechanism in Gamma-Ray Bursts. *Astrophys. J.* **2013**, *769*, 69. [CrossRef]
178. Beniamini, P.; Piran, T. The emission mechanism in magnetically dominated gamma-ray burst outflows. *Mon. Not. R. Astron. Soc.* **2014**, *445*, 3892–3907. [CrossRef]
179. Katz, J.I. Low-frequency spectra of gamma-ray bursts. *Astrophys. J.* **1994**, *432*, L107–L109. [CrossRef]

180. Pe'er, A.; Zhang, B. Synchrotron Emission in Small-Scale Magnetic Fields as a Possible Explanation for Prompt Emission Spectra of Gamma-Ray Bursts. *Astrophys. J.* **2006**, *653*, 454–461. [CrossRef]
181. Zhao, X.; Li, Z.; Liu, X.; Zhang, B.b.; Bai, J.; Mészáros, P. Gamma-Ray Burst Spectrum with Decaying Magnetic Field. *Astrophys. J.* **2014**, *780*, 12. [CrossRef]
182. Bhattacharya, M.; Kumar, P. On explaining prompt emission from GRB central engines with photospheric emission model. *arXiv* **2021**, arXiv:2110.14792.
183. Derishev, E.V. Synchrotron emission in the fast cooling regime: Which spectra can be explained? *Astrophys. Space Sci.* **2007**, *309*, 157–161. [CrossRef]
184. Nakar, E.; Ando, S.; Sari, R. Klein-Nishina Effects on Optically Thin Synchrotron and Synchrotron Self-Compton Spectrum. *Astrophys. J.* **2009**, *703*, 675–691. [CrossRef]
185. Daigne, F.; Bošnjak, Ž.; Dubus, G. Reconciling observed gamma-ray burst prompt spectra with synchrotron radiation? *Astron. Astrophys.* **2011**, *526*, A110. [CrossRef]
186. Barniol Duran, R.; Bošnjak, Ž.; Kumar, P. Inverse-Compton cooling in Klein-Nishina regime and gamma-ray burst prompt spectrum. *Mon. Not. R. Astron. Soc.* **2012**, *424*, 3192–3200. [CrossRef]
187. Beniamini, P.; Barniol Duran, R.; Giannios, D. Marginally fast cooling synchrotron models for prompt GRBs. *Mon. Not. R. Astron. Soc.* **2018**, *476*, 1785–1795. [CrossRef]
188. Piran, T.; Sari, R.; Zou, Y.C. Observational limits on inverse Compton processes in gamma-ray bursts. *Mon. Not. R. Astron. Soc.* **2009**, *393*, 1107–1113. [CrossRef]
189. Paczynski, B. Super-Eddington winds from neutron stars. *Astrophys. J.* **1990**, *363*, 218–226. [CrossRef]
190. Preece, R.D.; Briggs, M.S.; Mallozzi, R.S.; Pendleton, G.N.; Paciesas, W.S.; Band, D.L. The Synchrotron Shock Model Confronts a "Line of Death" in the BATSE Gamma-Ray Burst Data. *Astrophys. J.* **1998**, *506*, L23–L26. [CrossRef]
191. Eichler, D.; Levinson, A. A Compact Fireball Model of Gamma-Ray Bursts. *Astrophys. J.* **2000**, *529*, 146–150. [CrossRef]
192. Pe'er, A.; Waxman, E. Prompt Gamma-Ray Burst Spectra: Detailed Calculations and the Effect of Pair Production. *Astrophys. J.* **2004**, *613*, 448–459. [CrossRef]
193. Pe'er, A.; Mészáros, P.; Rees, M.J. Peak Energy Clustering and Efficiency in Compact Objects. *Astrophys. J.* **2005**, *635*, 476–480. [CrossRef]
194. Pe'er, A.; Mészáros, P.; Rees, M.J. The Observable Effects of a Photospheric Component on GRB and XRF Prompt Emission Spectrum. *Astrophys. J.* **2006**, *642*, 995–1003. [CrossRef]
195. Beloborodov, A.M. Radiative Transfer in Ultrarelativistic Outflows. *Astrophys. J.* **2011**, *737*, 68. [CrossRef]
196. Vereshchagin, G.V. Physics of Nondissipative Ultrarelativistic Photospheres. *Int. J. Mod. Phys. D* **2014**, *23*, 30003. [CrossRef]
197. Ito, H.; Nagataki, S.; Ono, M.; Lee, S.H.; Mao, J.; Yamada, S.; Pe'er, A.; Mizuta, A.; Harikae, S. Photospheric Emission from Stratified Jets. *Astrophys. J.* **2013**, *777*, 62. [CrossRef]
198. Lundman, C.; Pe'er, A.; Ryde, F. Polarization properties of photospheric emission from relativistic, collimated outflows. *Mon. Not. R. Astron. Soc.* **2014**, *440*, 3292–3308. [CrossRef]
199. Meng, Y.Z.; Liu, L.D.; Wei, J.J.; Wu, X.F.; Zhang, B.B. The Time-resolved Spectra of Photospheric Emission from a Structured Jet for Gamma-Ray Bursts. *Astrophys. J.* **2019**, *882*, 26. [CrossRef]
200. Parsotan, T.; Lazzati, D. Monte Carlo Simulations of Photospheric Emission in Gamma Ray Bursts. *arXiv* **2021**, arXiv:2110.13038.
201. Eichler, D.; Manis, H. Spectral Lags Explained as Scattering from Accelerated Scatterers. *Astrophys. J.* **2008**, *689*, L85. [CrossRef]
202. Eichler, D. Cloaked Gamma-Ray Bursts. *Astrophys. J.* **2014**, *787*, L32. [CrossRef]
203. Vyas, M.K.; Pe'er, A.; Eichler, D. A Backscattering-dominated Prompt Emission Model for the Prompt Phase of Gamma-Ray Bursts. *Astrophys. J.* **2021**, *908*, 9. [CrossRef]
204. Vyas, M.K.; Pe'er, A.; Eichler, D. Predicting Spectral Parameters in the Backscattering-dominated Model for the Prompt Phase of GRBs. *Astrophys. J.* **2021**, *918*, L12. [CrossRef]
205. Amati, L.; Frontera, F.; Tavani, M.; in't Zand, J.J.M.; Antonelli, A.; Costa, E.; Feroci, M.; Guidorzi, C.; Heise, J.; Masetti, N.; et al. Intrinsic spectra and energetics of BeppoSAX Gamma-Ray Bursts with known redshifts. *Astron. Astrophys.* **2002**, *390*, 81–89. [CrossRef]
206. Pe'er, A.; Zhang, B.B.; Ryde, F.; McGlynn, S.; Zhang, B.; Preece, R.D.; Kouveliotou, C. The connection between thermal and non-thermal emission in gamma-ray bursts: General considerations and GRB 090902B as a case study. *Mon. Not. R. Astron. Soc.* **2012**, *420*, 468–482. [CrossRef]
207. Gupta, N.; Zhang, B. Prompt emission of high-energy photons from gamma ray bursts. *Mon. Not. R. Astron. Soc.* **2007**, *380*, 78–92. [CrossRef]
208. Fan, Y.Z.; Piran, T. High-energy γ-ray emission from gamma-ray bursts—Before GLAST. *Front. Phys. China* **2008**, *3*, 306–330. [CrossRef]
209. Asano, K.; Inoue, S.; Mészáros, P. Prompt High-Energy Emission from Proton-Dominated Gamma-Ray Bursts. *Astrophys. J.* **2009**, *699*, 953–957. [CrossRef]
210. Wang, X.Y.; Li, Z.; Dai, Z.G.; Mészáros, P. GRB 080916C: On the Radiation Origin of the Prompt Emission from keV/MeV TO GeV. *Astrophys. J.* **2009**, *698*, L98–L102. [CrossRef]

211. Razzaque, S.; Dermer, C.D.; Finke, J.D. Synchrotron Radiation from Ultra-High Energy Protons and the Fermi Observations of GRB 080916C. *Open Astron. J.* **2010**, *3*, 150–155. [CrossRef]
212. Asano, K.; Mészáros, P. Delayed Onset of High-energy Emissions in Leptonic and Hadronic Models of Gamma-Ray Bursts. *Astrophys. J.* **2012**, *757*, 115. [CrossRef]
213. Crumley, P.; Kumar, P. Hadronic models for Large Area Telescope prompt emission observed in Fermi gamma-ray bursts. *Mon. Not. R. Astron. Soc.* **2013**, *429*, 3238–3251. [CrossRef]
214. Ghisellini, G.; Ghirlanda, G.; Oganesyan, G.; Ascenzi, S.; Nava, L.; Celotti, A.; Salafia, O.S.; Ravasio, M.E.; Ronchi, M. Proton-synchrotron as the radiation mechanism of the prompt emission of gamma-ray bursts? *Astron. Astrophys.* **2020**, *636*, A82. [CrossRef]
215. Schüssler, F. The Transient program of the Cherenkov Telescope Array. *arXiv* **2019**, arXiv:1907.07567.
216. Fioretti, V.; Ribeiro, D.; Humensky, T.B.; Bulgarelli, A.; Maier, G.; Moralejo, A.; Nigro, C. The Cherenkov Telescope Array sensitivity to the transient sky. *arXiv* **2019**, arXiv:1907.08018.
217. Inoue, S.; Granot, J.; O'Brien, P.T.; Asano, K.; Bouvier, A.; Carosi, A.; Connaughton, V.; Garczarczyk, M.; Gilmore, R.; Hinton, J.; et al. Gamma-ray burst science in the era of the Cherenkov Telescope Array. *Astropart. Phys.* **2013**, *43*, 252–275. [CrossRef]
218. Kakuwa, J.; Murase, K.; Toma, K.; Inoue, S.; Yamazaki, R.; Ioka, K. Prospects for detecting gamma-ray bursts at very high energies with the Cherenkov Telescope Array. *Mon. Not. R. Astron. Soc.* **2012**, *425*, 514–526. [CrossRef]
219. Wei, J.; Cordier, B.; Antier, S.; Antilogus, P.; Atteia, J.L.; Bajat, A.; Basa, S.; Beckmann, V.; Bernardini, M.G.; Boissier, S.; et al. The Deep and Transient Universe in the SVOM Era: New Challenges and Opportunities—Scientific prospects of the SVOM mission. *arXiv* **2016**, arXiv:1610.06892.
220. Rudolph, A.; Bošnjak, Ž.; Palladino, A.; Sadeh, I.; Winter, W. Multi-wavelength radiation models for low-luminosity GRBs, and the implications for UHECRs. *arXiv* **2021**, arXiv:2107.04612.
221. Punturo, M.; Abernathy, M.; Acernese, F.; Allen, B.; Andersson, N.; Arun, K.; Barone, F.; Barr, B.; Barsuglia, M.; Beker, M.; et al. The Einstein Telescope: A third-generation gravitational wave observatory. *Class. Quantum Gravity* **2010**, *27*, 194002. [CrossRef]
222. Clark, B.A.; IceCube-Gen2 collaboration. The IceCube-Gen2 Neutrino Observatory. *J. Instrum.* **2021**, *16*, C10007. [CrossRef]
223. Piro, L.; Ahlers, M.; Coleiro, A.; Colpi, M.; de Oña Wilhelmi, E.; Guainazzi, M.; Jonker, P.G.; Mc Namara, P.; Nichols, D.A.; O'Brien, P.; et al. Multi-messenger-Athena Synergy White Paper. *arXiv* **2021**, arXiv:2110.15677.
224. Pitik, T.; Tamborra, I.; Petropoulou, M. Neutrino signal dependence on gamma-ray burst emission mechanism. *J. Cosmol. Astropart. Phys.* **2021**, *2021*, 034. [CrossRef]

galaxies

MDPI

Review

GRB Polarization: A Unique Probe of GRB Physics

Ramandeep Gill [1,2,3,*], **Merlin Kole** [4,*] **and Jonathan Granot** [3,2,1,*]

1 Department of Physics, The George Washington University, Washington, DC 20052, USA
2 Astrophysics Research Center of the Open University (ARCO), The Open University of Israel, P.O. Box 808, Ra'anana 43537, Israel
3 Department of Natural Sciences, The Open University of Israel, P.O. Box 808, Ra'anana 43537, Israel
4 Department of Nuclear and Particle Physics, University of Geneva, 24 Quai Ernest-Ansermet, 1205 Geneva, Switzerland
* Correspondence: rsgill.rg@gmail.com (R.G.); merlin.kole@unige.ch (M.K.); granot@openu.ac.il (J.G.)

Abstract: Over half a century from the discovery of gamma-ray bursts (GRBs), the dominant radiation mechanism responsible for their bright and highly variable prompt emission remains poorly understood. Spectral information alone has proven insufficient for understanding the composition and main energy dissipation mechanism in GRB jets. High-sensitivity polarimetric observations from upcoming instruments in this decade may help answer such key questions in GRB physics. This article reviews the current status of prompt GRB polarization measurements and provides comprehensive predictions from theoretical models. A concise overview of the fundamental questions in prompt GRB physics is provided. Important developments in gamma-ray polarimetry including a critical overview of different past instruments are presented. Theoretical predictions for different radiation mechanisms and jet structures are confronted with time-integrated and time-resolved measurements. The current status and capabilities of upcoming instruments regarding the prompt emission are presented. The very complimentary information that can be obtained from polarimetry of X-ray flares as well as reverse-shock and early to late forward-shock (afterglow) emissions are highlighted. Finally, promising directions for overcoming the inherent difficulties in obtaining statistically significant prompt-GRB polarization measurements are discussed, along with prospects for improvements in the theoretical modeling, which may lead to significant advances in the field.

Keywords: gamma-ray bursts; polarization; radiation mechanisms; jet structure; instruments & methods

Citation: Gill, R.; Kole, M.; Granot, J. GRB Polarization: A Unique Probe of GRB Physics. *Galaxies* **2021**, *9*, 82. https://doi.org/10.3390/galaxies9040082

Academic Editors: Elena Moretti and Francesco Longo

Received: 28 August 2021
Accepted: 19 October 2021
Published: 27 October 2021

1. Introduction

Gamma-ray bursts (GRBs) are one of the most energetic, and electromagnetically the brightest, transient phenomena in the Universe. They are the ideal test beds for understanding nature at its extreme that involves an explosive release of energy over a short timescale, producing a burst of γ-rays with an isotropic-equivalent luminosity of $L_{\gamma,\rm iso} \sim 10^{51} - 10^{54}\,{\rm erg\,s^{-1}}$. It is now well established that most GRBs are cosmological sources and that they are powered by ultrarelativistic (with bulk Lorentz factors $\Gamma \gtrsim 100$) bipolar beamed outflows driven by a central engine–a compact object. The identity of the central engine, which could be either a black hole (BH) or a millisecond magnetar, is not entirely clear as the highly variable emission is produced far away from it at a radial distance of $R \sim 10^{12} - 10^{16}$cm. The most luminous phase of the burst, referred to as the "prompt" phase is short lived with a bimodal duration distribution, where the short GRBs have typical durations of $t_{\rm GRB} \sim 10^{-1}$ s and the long GRBs typically last for $t_{\rm GRB} \sim 30$ s while the dividing line sits at $t \sim 2$ s [1]. These two classes of GRBs are also distinct spectrally, with the short GRBs being spectrally harder as compared to the long GRBs that produce softer γ-rays. Other clues, e.g., the association of long-soft GRBs with star-forming regions [2] and type-Ib/c supernovae [3–5] and that of the short-hard GRBs with early type galaxies [6,7] lead to the identification of two distinct progenitors. The long-soft GRBs are associated with the core-collapse of massive ($\gtrsim (20-30)M_\odot$) Wolf–Rayet stars [8], whereas the short-hard GRBs were theorized to originate in compact object mergers, namely, that of two neutron stars (NSs) or a NS-BH pair [9,10]. The unequivocal proof of the latter association had to wait until the gravitational wave (GW) detectors, LIGO and Virgo, became operational, which led to the coincident detection of GWs from the merger of two NSs and a short-hard GRB by *Fermi*-GBM and the INTEGRAL-ACS from GW 170817/GRB 170817A [11,12].

Although the global picture is fairly clear, the details of the energy dissipation process, the exact radiation mechanism, and the transfer of radiation in the highly dynamical flow remain poorly understood. All of these different processes combine to produce a non-thermal spectrum that is often well described by the Band function [13], an empirical fit to the spectrum featuring a smoothly broken power law. In νF_ν space, which indicates the observed energy flux around the frequency ν with F_ν being the spectral flux density, this break manifests as a peak at the mean photon energy $\langle E_{\rm br}\rangle \simeq 250$ keV, which also represents the energy at which most of the energy of the burst is released, and the asymptotic power-law photon indices below and above the break energy have mean values of $\langle\alpha_{\rm Band}\rangle \simeq -1$ and $\langle\beta_{\rm Band}\rangle \simeq -2.3$, respectively [14,15]. After decades of spectral modeling of the prompt emission, the basic questions of GRB physics remain unanswered, and it is becoming challenging to advance our understanding with spectral modeling alone.

An exciting opportunity was presented by the claimed detection of high levels of linear polarization, with $\Pi = 80\% \pm 20\%$, in GRB 021206 [16]. Although this result had a detection significance of 5.7σ, further scrutiny by other works [17,18] cast irrevocable doubts and ultimately refuted the final result. Nevertheless, this one result initiated a vigorous theoretical effort to understand the polarization of prompt GRB emission with the expectation that highly sensitive measurements will be able to resolve many of the outstanding questions of GRB physics. Over the past several years, the number of prompt GRB polarization measurements (in some cases time-resolved) have grown; however, the main results remain inconclusive due to inherent difficulties in obtaining highly statistically significant measurements. Therefore, it is hoped that the next generation of γ-ray polarimeters that will be launched in this decade will provide further important clues.

The main objectives of this review were to provide a concise yet comprehensive overview of the current status of theoretical developments as well as observations in the field of prompt GRB polarization and also to highlight the need for developing more sensitive instruments and better analysis tools, which are hoped to yield statistically significant measurements in the coming decade. Many points presented here have also been covered in earlier reviews on the topic e.g., [19–24]. This review begins with a summary of the fundamental questions in GRB physics (Section 2) that can be addressed with measurements of linear polarization along with insights gained from prompt GRB spectral modeling. These include the outflow composition and dynamics (Section 2.1), energy dissipation mechanisms (Section 2.2), radiation mechanisms (Section 2.3), and the angular structure of the outflow (Section 2.4). An overview of γ-ray polarimetry is presented in Section 3, which includes the fundamental principles of γ-ray polarization measurement (Section 3.1) and a summary of the different detectors that have been used for GRB polarimetry (Section 3.3). The theory of GRB polarization is presented next in Section 4, which covers several topics, such as polarization from uniform (Section 4.1) and structured (Section 4.2) jets with different radiation mechanisms, temporal evolution of polarization (Section 4.3), polarization arising from multiple overlapping pulses (Section 4.4), the most likely polarization for a given radiation mechanism (Section 4.5), and the energy dependence of polarization (Section 4.6). The current status of prompt GRB polarization measurements is presented next, which includes time-integrated (Section 5.1), time-resolved (Section 5.2), and energy-resolved (Section 5.3) measurements. The importance of polarization measurements from the other phases of the burst, namely, X-ray flares, reverse-shock emission (optical flash and radio flare), and forward-shock emission, which also probe the properties of the GRB outflow, is emphasized in Section 6. Finally, Section 7 touches upon the outlook for this decade, which will see the launch of more sensitive instruments (Section 7.1). The predicted performance of some is compared in Section 7.2. This review concludes by offering some suggestions for improvements in the polarization data analysis (Section 7.3) and its theoretical modeling (Section 7.4).

2. Key Questions That Can Be Addressed with GRB Polarization

Measurements of the prompt GRB polarization may help shed light on many critical aspects of the relativistic outflow whose knowledge has evaded us so far. Below, we summarize key open questions in GRB physics, which can be probed with spectro-polarimetric observations. More detailed reviews and discussions of these topics can be found in other review articles (e.g., [25–29]). Theoretical modeling of prompt GRB polarization and comprehensive results are provided in Section 4.

2.1. *What Are the Outflow Composition and Dynamics?*

The main dissipation and radiation mechanisms that produce the GRB prompt emission are dictated by the composition of the outflow. The two most widely discussed scenarios invoke an outflow that is either kinetic-energy-dominated (KED) [30] or Poynting-flux-dominated (PFD) [31,32]. In the former, most of the energy is initially thermal (fireball) and is eventually transferred to the kinetic energy of the cold baryons, while in the latter the main energy reservoir is the (likely ordered) magnetic field that drives the expansion and acceleration of the flow. If the radiation mechanism is indeed synchrotron (see Section 2.3), then the level of polarization in both types of flows depends on the structure of the magnetic field that is either generated in situ, e.g., in internal shocks in a KED flow, or survives at large distances from the central engine, which could happen in both types of flows. Our theoretical understanding of the B-field structure in the emission region in a given type of flow is still limited and rather speculative. Any measurement of polarization will put strong constraints on the B-field structure. Therefore, in combination with polarization measurements, spectral and temporal (pulse profiles) modeling will allow us to constrain the composition.

The distinction between a KED and PFD flow can be conveniently parameterized using the magnetization parameter,

$$\sigma \equiv \frac{w'_B}{w'_m} = \frac{B'^2}{4\pi\left(\rho' c^2 + \frac{\hat{\gamma}}{\hat{\gamma}-1}P'\right)} \xrightarrow[cold]{} \frac{B'^2}{4\pi\rho' c^2}, \tag{1}$$

which is defined as the ratio of the comoving (all quantities measured in the comoving/fluid frame are primed) magnetic field enthalpy density, $w'_B = B'^2/4\pi$, to that of matter, $w'_m = \rho' c^2 + \frac{\hat{\gamma}}{\hat{\gamma}-1}P'$ or $w'_m = \rho' c^2$ when it is cold ($P' \ll \rho' c^2$). Here B' is the magnetic field strength, and we assumed here for simplicity that the baryons dominate the total rest mass with density $\rho' = m_p n'$, where n' is the particle number density, m_p is the proton mass, and c is the speed of light. The baryons were assumed to be cold with an adiabatic index $\hat{\gamma} = 5/3$ ($\hat{\gamma} = 4/3$ for a relativistic fluid) and negligible pressure P' when compared with the particle inertia. A KED flow will have $\sigma < 1$; magnetic fields, if present, are weak and randomly oriented with short coherence length scales and are unimportant in governing the dynamics of the outflow. On the other hand, a PFD flow will have $\sigma > 1$, and the magnetic field is much more ordered where it is responsible for accelerating the flow.

A prime example of a KED flow is the standard "fireball" scenario [33,34], in which total energy E is released close to the central engine, launching a radiation-dominated and optically thick outflow, with Thomson optical depth $\tau_T \gg 1$. The temperature at the base of the flow is typically $k_B T \gtrsim$ MeV, which leads to copious production of $e^{\pm}$-pairs via $\gamma\gamma$-annihilation that further enhances the optical depth. The enormous radiation pressure causes the flow to expand adiabatically, thereby converting the radiation field energy to the kinetic energy of baryons, which are inefficient radiators due to their large Thomson cross-sections. The bulk Lorentz factor (LF) of the fireball grows linearly with the radius, $\Gamma(R_0 < R < R_s) \approx R/R_0$, where $R = R_0$ is the launching radius, while its comoving temperature declines as $T'(R) \propto R^{-1}$. The amount of baryon loading, i.e., the amount of baryons with total mass M_b entrained in the flow of given energy E, determines the terminal LF, $\Gamma_\infty = E/M_b c^2$, which is attained at the saturation radius $R = R_s \sim \Gamma_\infty R_0$ at which point the growth in the bulk Γ saturates and the flow simply coasts at $\Gamma = \Gamma_\infty$. The kinetic energy of the baryons is tapped at a large distance ($R > R_s$) from the central engine via internal shocks (see below).

In a PFD flow, large-scale magnetic fields propagate outwards from the central engine with an angular coherence scale $\theta_B > 1/\Gamma$, where $1/\Gamma$ represents the characteristic angular scale over which the flow is causally connected and, as discussed later, also the angular scale into which the emitted radiation is beamed towards the observer from a relativistic flow. While the fireball scenario is well agreed upon and has enjoyed many successes since it is fairly robust, no such *standard model* exists for a magnetized outflow to explain GRB properties. In several works (e.g., [35–39]), ideal-MHD models for a steady-state, axisymmetric, and non-dissipative outflow have been developed in which the flow expands adiabatically due to magnetic stresses. The flow is launched highly magnetized near the light cylinder radius, R_L, with $\sigma(R_L) = \sigma_0 \gg 1$ and bulk LF $\Gamma(R_L) = \Gamma_0 \sim 1$. As the flow expands, its magnetization declines with radius, and in the case of a radial wind (i.e., unconfined, with a negligible external pressure) the flow is limited to a terminal LF of $\Gamma_\infty \sim \sigma_0^{1/3}$ where the corresponding magnetization of the flow is $\sigma \sim \sigma_0^{2/3}$ [40]. For weak external confinement (an external pressure profile $p_{\rm ext} \propto z^{-\kappa}$ with $\kappa > 2$, where $z \approx R = (z^2 + r_{\rm cyl}^2)^{1/2}$ is the distance from the central source along the jet's symmetry axis and $r_{\rm cyl}$ is the cylindrical radius), the acceleration saturates at a terminal LF of $\Gamma_\infty \sim \sigma_0^{1/3}\theta_j^{-2/3}$ and magnetization $\sigma_\infty \sim (\sigma_0\theta_j)^{2/3} \sim (\Gamma_\infty\theta_j)^2 \gg 1$ where θ_j is the jet's asymptotic half-opening angle [41]. For strong external confinement ($p_{\rm ext} \propto z^{-\kappa}$ with $\kappa < 2$), the jet maintains lateral causal contact and equilibrium, leading $\Gamma \sim r_{\rm cyl}/R_L \sim (z/R_L)^{\kappa/4}$, which saturates at $\Gamma_\infty \sim \sigma_0$, $\sigma_\infty \sim 1$, and $\Gamma_\infty\theta_j \sim 1$. Since prompt GRB observations demand the dissipation region to be expanding ultrarelativistically with $\Gamma_\infty \gtrsim 100$, to avoid the *compactness problem* [25,42], and afterglow observations suggest that typically $\theta_j \gtrsim 0.05 - 0.1$, which implies $\Gamma_\infty\theta_j \gtrsim 10$ in GRBs. This suggests that the weakly confined regime is most relevant for GRBs; however, it implies $\sigma_\infty \sim (\Gamma_\infty\theta_j)^2 \gg 1$, which suppresses internal shocks. It has been pointed out [43,44] that the sharp drop in the surrounding (lateral) pressure as the jet exits the progenitor star in long GRBs can lead to $\Gamma_\infty\theta_j \gg 1$ along with a more modest asymptotic magnetization $\sigma_\infty \gtrsim 1$, but even then internal shocks remain inefficient.

When the steady-state assumption is relaxed, alternative models that consider an impulsive and highly variable flow yield a much larger terminal LF with $\Gamma_\infty \sim \sigma_0$ and may achieve $\sigma_\infty < 1$ or even $\sigma_\infty \ll 1$ under certain conditions [45,46]. In this scenario, a thin shell of initial width ℓ_0 is accelerated due to magnetic pressure gradients that causes its bulk LF to grow as $\Gamma \sim (\sigma_0 R/R_0)^{1/3}$, where $R_0 \approx \ell_0$, while its magnetization drops as $\sigma \sim \sigma_0^{2/3}(R/R_0)^{-1/3}$. The bulk LF of the shell saturates at $R_s \sim \sigma_0^2 R_0$ at which point its $\Gamma \sim \Gamma_\infty \sim \sigma_0$ and $\sigma \sim 1$. For $R > R_s$, the magnetization continues to drop further as $\sigma \sim (R/R_s)^{-1}$ as the shell starts to spread radially. For a large number of shells initially separated by $\ell_{\rm gap}$, the radial expansion is limited as neighboring shells collide and one expects an asymptotic mean magnetization of $\sigma_\infty \sim \ell_0/\ell_{\rm gap}$. This scenario offers the dual possibility of magnetic energy dissipation via MHD instabilities when $\sigma > 1$ at $R < R_s$ as well as kinetic energy dissipation via internal shocks when $\sigma < 1$ at $R > R_s$.

Similar outflow dynamics were obtained in a popular model that makes the magnetohydrodynamic (MHD) approximation and features a striped-wind magnetic field structure [47–51], in which magnetic field lines reverse polarity over a characteristic length scale $\lambda \sim \pi R_L = \pi c/\Omega = cP/2 = 1.5 \times 10^7 P_{-3}\,{\rm cm}$. Here, $\Omega = 2\pi/P$ is the central engine's angular frequency with P being its rotational period. Close to the central engine the flow may be accelerated by magneto-centrifugal, and to some extent, thermal acceleration. At distances larger than the Alfvén radius, where $R_A \gtrsim R_L$, these effects are negligible, and when collimation-induced acceleration is ineffective then the properties of the flow can be described using radial dynamics. If a reasonable fraction (the usual assumption is approximately half) of the dissipated energy in the flow goes towards its acceleration, conservation of the total specific energy, while ignoring any radiative losses, yields the relation $\Gamma(R)[1+\sigma(R)] = \Gamma_0[1+\sigma_0]$ for a cold flow, which simplifies to $\Gamma(R)\sigma(R) \approx \Gamma_0\sigma_0$ for $\sigma(R) \gg 1$. At the Alfvén radius, the four velocity of the flow is $u_A = \Gamma_A\beta_A = \sigma_A^{1/2} \approx \Gamma_A \approx \Gamma_0\sigma_0/\sigma_A \approx \sigma_0/\sigma_A$, which implies that $\sigma_A \approx \sigma_0^{2/3}$ and $\Gamma_A \approx \sigma_0^{1/3}$. The terminal LF is achieved at the saturation radius R_s when $\sigma(R_s) \sim 1$, at which point $\Gamma_\infty \approx \Gamma_0\sigma_0 \approx \sigma_0 = \sigma_A^{3/2}$. In this scenario, the saturation radius is given by $R_s = \Gamma_\infty^2\lambda/6\epsilon = 1.7\times10^{13}\Gamma_{\infty,3}^2(\lambda/\epsilon)_8$ cm, where $\epsilon = v_{\rm in}/v_A \sim 0.1$ is a measure of the reconnection rate where it quantifies the plasma inflow velocity $v_{\rm in}$ into the reconnection layer in terms of the Alfvén speed. For magnetized flows, $v_A = c\sqrt{\sigma/(1+\sigma)}$, which approaches the speed of light for $\sigma \gg 1$. Beyond the Alfvén radius, the bulk LF grows as a power law in radius, with $\Gamma(R) = \Gamma_\infty(R/R_s)^{1/3}$, while the magnetization declines as $\sigma(R) = (R/R_s)^{-1/3}$.

In the regime of high magnetization ($\sigma \gg 1$), an alternative model that does not make the MHD approximation was considered by Lyutikov and Blandford [32] and Lyutikov [52].

2.2. *How and Where Is the Energy Dissipated?*

The composition of the outflow has a strong impact on the dominant energy dissipation channel. To produce the prompt GRB emission, the baryonic electrons as well as any $e^\pm$-pairs, which are the primary radiators, cannot be cold, and they need to be accelerated or heated to raise their internal energy. The observed photon energy spectrum is not only shaped by the underlying radiation mechanism but also the radial location in the flow where energy is dissipated. If most of the energy is dissipated much below the photospheric radius, at $R \ll R_{\rm ph}$, where the Thomson optical depth of the flow is $\tau_T \gg 1$ and where the radiation field and particles are tightly coupled via Compton scattering (baryons are coupled with the leptons via Coulombic interactions) and assume a thermal distribution, the final outcome is a quasi-thermal spectrum [31,33,34]. The observed spectrum in this case is not a perfect blackbody, due to the observer seeing different parts of the jet with different Doppler boosts, but close to one with a low-energy (below the spectral peak energy) photon index $\alpha_{\rm ph} = d\ln N_\gamma/d\ln E \approx 0.4$, which is softer from $\alpha_{\rm ph} = 1$ expected for a Rayleigh–Jeans thermal spectrum [53,54]. If instead most of the energy is dissipated in the optically thin ($\tau_T < 1$) parts of the flow, then a non-thermal spectrum emerges. When the flow is continuously heated across the photosphere, the final spectrum is a combination of two components: quasi-thermal and non-thermal.

If the flow is uniform (i.e., quasi-spherical with negligible angular dependence within angles of $\lesssim 1/\Gamma$ around the line of sight), then any thermal component will show negligible polarization as there is no preferred direction for the polarization vector to align with. Even if different parts of the flow may be significantly polarized at the photosphere [55], the net polarization averages out to zero after integrating over the GRB image on the sky. However, angular structure in the flow properties can lead to modest ($\Pi \lesssim 20\%$) polarization [24,56–58]. The polarization of the non-thermal spectral component ultimately depends on the radiation mechanism, discussed in Section 2.3.

In a KED flow, after an initial phase of rapid acceleration of the fireball when the bulk LF saturates, the particles are cold in the comoving frame with negligible pressure ($P' \ll \rho' c^2$). The energy of the flow is dominated by the kinetic energy of the baryons, which is very *ordered*. To produce any radiation, particle motion needs to be *randomized*. A simple and robust method to achieve that is via shocks. The canonical model of internal shocks [30,59–61] posits that the central engine accretes intermittently and ejects shells of matter that are initially separated by a typical length scale $\sim ct_v/(1+z)$ and have fluctuations in their bulk LFs of order $\Delta\Gamma \sim \Gamma$, with Γ being the mean bulk LF. Here, t_v is the observed variability of the prompt emission lightcurve, and z is the redshift of the source. Typically, $R_0 \sim 10^7$ cm and $\Gamma_\infty \sim 10^2 - 10^3$ so that the acceleration saturates at $R_s \sim \Gamma_\infty R_0 \sim 10^9 - 10^{10}$ cm. For $R > R_s$, faster-moving shells catch up from behind with slower ones and collide to dissipate their kinetic energy at internal shocks occurring at the dissipation radius of $R_{\rm dis} = 2\Gamma_\infty^2 c t_v/(1+z) = 6 \times 10^{13}(1+z)^{-1}\Gamma_{\infty,2}^2 t_{v,-1}$ cm.

When the shells collide, a double-shock structure forms with a forward shock going into the slower shell and accelerating it while a reverse shock goes into the faster shell and decelerates it. These shocks heat a fraction ξ_e of the electrons into a power-law energy distribution, with $dN_e/d\gamma_e \propto \gamma_e^{-p}$ for $\gamma_e > \gamma_m$, where these electrons hold a fraction ϵ_e of the total internal energy density behind the shock. The LF of the minimal energy electrons, $\gamma_m = [(p-2)/(p-1)](\epsilon_e/\xi_e)(m_p/m_e)(\Gamma_{\rm ud} - 1)$ (for $p > 2$), depends on the relative bulk LF, $\Gamma_{\rm ud}$, of the upstream to downstream matter across the relevant shock. A fraction ϵ_B of the internal energy density behind the shock is held by the shock-generated magnetic field of strength $B' \sim 10^2 - 10^3$ G. More generally, one can express the comoving magnetic field in terms of the radius and outflow Lorentz factor and magnetization at that radius, as well as the observed isotropic equivalent γ-ray luminosity, $L_{\gamma,\rm iso}$, and the γ-ray emission efficiency, ϵ_γ (i.e., fraction of the total outflow energy channeled into gamma-rays), $B' = 1.8 \times 10^5 \Gamma_2^{-1} R_{14}^{-1} (\frac{\sigma}{1+\sigma})^{1/2} L_{\gamma,\rm iso,52}^{1/2} \epsilon_{\gamma,-1}^{-1/2}$ G. The exact structure of the magnetic field is still an open question, but it has been argued that streaming instabilities [62–66], e.g., the relativistic two-stream and/or Weibel (filamentation) instability, are responsible for generating a small-scale field with coherence scale on the order of the electron and/or proton skin depth, $c/\omega_p' = c(4\pi n' e^2/m)^{-1/2}$ where ω_p' is the plasma frequency, which depends on the particle number density n'; mass m is the particle mass; and e is the elementary charge. Since the coherence length of the shock-generated field is much smaller than the angular size of the beaming cone ($\theta_B \ll 1/\Gamma$), the net polarization is limited to $\Pi \lesssim 30\%$.

Alternatively, interactions of the shock with density inhomogeneities in the upstream can cause macroscopic turbulence in the downstream (e.g., excited by the Richtmyer–Meshkov instability), which can in turn amplify a shock-compressed large-scale upstream magnetic field to near-equipartition with the downstream turbulent motions [67–72]. The dynamo-amplified magnetic field is expected to form multiple mutually incoherent patches (with angular sizes up to a fraction of the visible region) in which the field is largely ordered. The expected polarization, after averaging over such patches in the observed region, is typically expected to be small, with $\Pi \lesssim 2\%$ [69].

As mentioned earlier, in a PFD flow, the main dissipation channel is magnetic reconnection and /or MHD instabilities. Both of these are non-ideal effects that cannot be treated in an ideal MHD formalism. Magnetic field energy is dissipated when opposite polarity field lines reconnect, which leads to acceleration of electrons that then cool by either emitting synchrotron radiation outside of the reconnection sites or inverse-Compton scattering of either synchrotron photons or a pre-existing radiation field advected with the flow. Exactly where dissipation commences depends on the initial magnetic field geometry in the flow as the field lines expand outward from the central engine to larger distances [73]. If the flow is axisymmetric and is not permeated by polarity-switching field lines, magnetic energy can still be dissipated due to current-driven instabilities, e.g., the kink instability [74–77]. Such an instability may also occur at the interface between the jet and the confining medium, e.g., the stellar interior of a Wolf–Rayet star in long-soft GRBs [78] and the dynamically ejected wind during a binary neutron star merger in short-hard GRBs. Magnetic field lines that reverse polarity on some characteristic length scale λ can be embedded into the outflow in a variety of ways [79]. These can indeed be injected at the base of the flow where field polarity reversals are obtained in the accretion disk due to the magnetorotational instability, as demonstrated in several shearing-box numerical MHD simulations [80] as well as in global simulations of black hole accretion [81]. Depending on how particles are heated/accelerated when magnetic energy is dissipated, as the flow becomes optically-thin, as discussed in the next section, the polarization will be energy dependent and can be $\Pi \lesssim 60\%$ if synchrotron emission dominates and the B-field angular coherence length near the line of sight is $\theta_B \gtrsim 1/\Gamma$.

In the striped-wind model [49,50,82], magnetic dissipation commences beyond the Alfvén radius and becomes the dominant contributor towards flow acceleration. Below the Alfvén radius the flow is accelerated due to magneto-centrifugual effects as well as collimation provided by the confining medium [39,83]. If the confining medium has a sharp outer boundary (e.g., the edge of the massive star progenitor for a long GRB), then as the jet breaks out of the confining medium, the flow becomes conical and expands ballistically. The sudden loss of pressure also leads to some further acceleration via the mechanism of *rarefaction acceleration* [44] that operates in PFD relativistic jets. While these ideal MHD processes may continue to operate at length scales relevant for prompt GRB emission, magnetic reconnection in a striped wind provides a source for gradual acceleration out to the saturation radius R_s. Beyond this radius magnetic reconnection subsides, and therefore acceleration ceases and the flow starts to coast. When the prompt emission is produced in an accelerating flow, the degree of polarization is not affected. Instead, the duration of the pulses becomes shorter in comparison to that obtained in a coasting flow—see, e.g., [84].

Other variants of the PFD model, as presented above, include the internal-collision-induced magnetic reconnection and turbulent (ICMART) model [85], in which high-σ shells are intermittently ejected by the central engine that dissipate their energy at $R \sim 10^{15} - 10^{16}$ cm, where collision-induced magnetic reconnection and turbulence radiates away the magnetic energy and reduces the initially high magnetization of the ejecta to order unity. The expected polarization from an ICMART event has been presented in Deng et al. [86] using 3D numerical MHD simulations where they also find a 90° change in polarization angle.

2.3. *What Radiation Mechanism Produces the Band-like GRB Spectrum?*

Few radiation mechanisms have been proposed to explain the Band-like spectrum of prompt emission, the most popular being synchrotron and inverse-Compton. Below, we present a concise summary of the different proposed mechanisms and show the expected polarization in Figure 1.

2.3.1. Optically-Thin Synchrotron Emission

Relativistic electrons gyrating around magnetic field lines cool by emitting synchrotron photons. When the energy distribution of these electrons is described by a power law, e.g., that obtained at collisionless internal shocks due to Fermi acceleration, the emerging synchrotron spectrum is described by multiple power-law segments that join at characteristic break energies [87,88]. These correspond to the synchrotron frequency, $E_m = \Gamma(1+z)^{-1}h\nu'_m = \Gamma(1+z)^{-1}\gamma_m^2(\hbar eB'/m_ec)$, of minimal-energy electrons with LF γ_m and the cooling frequency, $E_c = \Gamma(1+z)^{-1}h\nu'_c = 36\pi^2(1+z)^{-1}(\hbar e m_e c^3/\sigma_T^2)(\Gamma^3\beta^2/B'^3R^2)$, of electrons that are cooling at the dynamical time, $t'_{\rm cool} = t'_{\rm dyn} = R/\Gamma\beta c$. Here, B' is the comoving magnetic field, and σ_T is the Thomson cross-section. The high radiative efficiency of prompt emission demands that the electrons be in the fast-cooling regime for which $E_c < E_m$ and the νF_ν spectrum peaks at $E_{\rm pk} = E_m$. In this case, the spectrum below the peak energy has a photon index $\alpha_{\rm ph} = -2/3$ for $E < E_c$ and $\alpha_{\rm ph} = -3/2$ for $E_c < E < E_m$. Above the peak energy, the photon index is $\alpha_{\rm ph} = -(p+2)/2$ where p is the power-law index of the electron distribution.

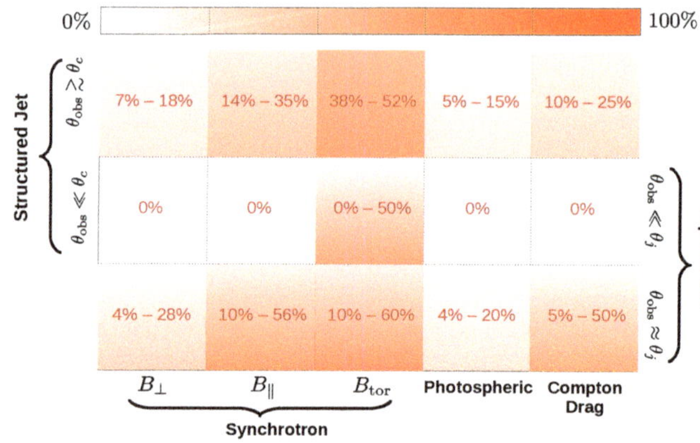

Figure 1. Approximate degree of polarization for different radiation mechanisms and jet structures [24]. If the emission is synchrotron then polarization for different B-field configurations is given (assuming $\Pi_{max} = 70\%$). For each jet structure, a distinction is made between two cases: (i) when the observer's viewing angle (θ_{obs}) is much smaller than the half-opening angle (θ_j) of a top-hat jet or, in the case of a structured jet, if it is much smaller than the core angle (θ_c) and (ii) when θ_{obs} is close to θ_j, i.e., the edge of the jet. For a structured jet, θ_{obs} can exceed θ_c by an order unity factor before the fluence starts to drop significantly. When $\theta_{obs} \approx \theta_j$, the minimum value of polarization can be zero in all cases, except for B_{tor}, for a pulse with a given $\xi_j = (\Gamma\theta_j)^2$, where $\xi_j^{1/2}$ is the ratio of the angular sizes of the jet and of the beaming cone. Different pulses may have slightly different ξ_j (typically with a similar θ_j but different Γ), which on average would yield a finite polarization. The quoted lower range reflects this mean value (see [24] for more details). For the B_{tor} case, $\Pi = 0\%$ when $\theta_{obs} = 0$ due to symmetry and $|\Pi| > 0\%$ otherwise, while $\Pi \approx 50\%$ at $1/\Gamma < \theta_{obs} < \theta_j$.

While synchrotron emission is still regarded as the default emission mechanism, the basic "vanilla" model has been argued to be not as robust as previously thought. First, its predictions have been challenged by a small fraction of GRBs that showed harder low-energy ($E < E_{pk}$) spectral slopes with $\alpha_{Band} > -2/3$ [89–92], often identified as the synchrotron *line of death*. Some possible alternatives that have been suggested to resolve this discrepancy include anisotropic electron pitch angle distribution and synchrotron self-absorption [93], jitter radiation [94], and photospheric emission [95]. The line-of-death violation is generally derived by fitting the empirical Band-function to the observed spectrum. When synthetic synchrotron spectra (after having convolved with the energy response of a detector) are fit with the Band-function, an even softer $\langle\alpha_{Band}\rangle = -0.8$ is found due to the detector's limited energy range (e.g., *Fermi*/GBM [96]), which does not quite probe the asymptotic value of α_{ph}. Since a significant fraction of GRBs show low-energy spectral indices that are harder than this value, it might indicate that another spectral component is possibly contributing at low energies and offsetting the spectral slope. Second, the spectral peak energy in the cosmological rest-frame of the source is given by $E_m(1+z)$, which depends on a combination of Γ, γ_m, and B' to yield the measured peak energy in the range $200\,\mathrm{keV} \lesssim [E_{pk,z} = E_{pk}(1+z)] \lesssim 1\,\mathrm{MeV}$ [97] with a possible peak around $E_{pk,z} \sim m_e c^2$ [98]. Given that all of these quantities can vary substantially between different bursts, the synchrotron model does not offer any characteristic energy scale at which most of the energy is radiated in the event that the $E_{pk,z}$ distribution indeed narrowly peaks around $\sim m_e c^2$ [99]. Third, the synchrotron model predicts wider spectral peaks than that obtained by fitting the Band-function to observations [100]. This issue has now been demonstrated for a large sample of GRBs where the spectral widths obtained with the simplest synchrotron model yielding the narrowest spectral peak, e.g., a slow-cooling Maxwellian distribution of electrons, is inconsistent with most of the GRBs [101,102].

Moreover, it is rather easy to get a wider spectral peak by having, e.g., fast-cooling particles, variable magnetic fields, etc., but it is much harder to obtain narrower peaks.

Several works that find the synchrotron model to be inconsistent with observations invariably use empirical models, e.g., the Band-function, a smoothly-broken power law, to determine low-energy spectral slopes and peak widths. This may become a problem in instances where such models are unable to capture the intrinsic complexity of the underlying data. Therefore, an arguably better approach is to directly fit physical models to the raw data to derive spectral parameters and remove any bias [103–106]. Such an approach has led to alleviating some of the issues encountered by the optically thin synchrotron model, where it was shown that direct spectral fits (in count space rather than energy space) with synchrotron emission from cooling power-law electrons can explain the low-energy spectral slopes as well as the spectral width of the peak [107].

Magnetic Field Structure

If the coherence length of the magnetic field is larger than the gyroradius of particles, the structure of the magnetic field in the dissipation region does not affect the spectrum or the pulse profile. However, it significantly affects the level of polarization. Therefore, spectro-polarimetric observations that strongly indicate synchrotron emission as the underlying radiation mechanism can be used to also determine the magnetic field structure. At least four physically motivated axisymmetric magnetic field structures, and the emergent synchrotron polarization, have been discussed in the literature [108–111]:

1. $\boldsymbol{B}_{\mathbf{ord}}$: An ordered magnetic field with angular coherence length $1/\Gamma \lesssim \theta_B \ll \theta_j$, where $1/\Gamma$ is the angular size of the beaming cone. It is envisioned that the jet surface is filled with several small radiating patches of angular size much smaller than the jet aperture and that these are pervaded by mutually incoherent ordered magnetic fields. In this way, such a field configuration as a whole remains axisymmetric in a statistical sense (despite having a local preferred direction for a given line of sight, namely, the ordered field direction at that line of sight) and also different from a globally ordered B-field. This type of field structure was motivated by the high-polarization claim of $\Pi = 80\% \pm 20\%$ [16] in GRB 021206 and by the notion that the local synchrotron polarization can be very high with $\Pi_{\rm max} \sim 75\%$. Magnetic fields with sufficiently large coherence lengths that are not globally ordered can be advected with the flow from the central engine where their length scale is altered en route to the emission site due to hydromagnetic effects.
2. $\boldsymbol{B}_{\perp}$: A random magnetic field (i.e., with $\Gamma\theta_B \ll 1$) confined to the plane transverse to the local velocity vector of the fluid element in the flow. In many cases, the flow is assumed to be expanding radially, which is a good approximation when the prompt emission is generated since no significant lateral motion is expected at that time. This field structure is motivated by the theoretical predictions of small-scale magnetic fields generated by streaming instabilities at collisionless shocks [62–66].
3. $\boldsymbol{B}_{\parallel}$: An ordered field aligned along the local velocity vector of the outflow. This field structure presents the opposite extreme of $B_\perp$, and in reality the shock-generated field may likely be (at least its emissivity-weighted mean value over the emitting region downstream of the shock) more isotropic than anisotropic whereby it would be a distribution in the $B_\perp - B_\parallel$ parameter space (see, e.g., [109,112] in the context of afterglow collisionless shocks).
4. $\boldsymbol{B}_{\mathbf{tor}}$: A globally ordered toroidal field symmetric around the jet symmetry axis. Such a field configuration naturally arises in a high magnetization flow in which the dynamically dominant field is anchored either to the rotating central engine or in the accretion disk. The azimuthal motion of the magnetic footpoints tightly winds up the field around the axis of rotation, which is also the direction along which the relativistic jet is launched. Due to magnetic flux conservation, the poloidal component declines ($B_p \propto R^{-2}$) more rapidly as compared to the toroidal component ($B_\phi \propto R^{-1}$)

as the flow expands. Therefore, at large distances from the central engine the toroidal field component dominates.

2.3.2. Inverse-Compton Emission

If the energy density of the (isotropic) radiation field ($U'_\gamma = 3L_\gamma/16\pi R^2\Gamma^2 c$, where L_γ is the isotropic-equivalent luminosity) advected with the flow is much larger than that of the magnetic field ($U'_B = B'^2/8\pi$), relativistic particles with LF γ_e cool predominantly by inverse-Compton upscattering softer seed photons, with energy E'_s, to higher energies with a mean value (for an isotropic seed photon field in the comoving frame), $E' = (4/3)\gamma_e^2 E'_s$. When the Thomson optical depth of the flow is $\tau_T > 1$, these seed photons undergo multiple Compton scatterings, where the process is usually referred to as *Comptonization*, until they are able to stream freely when $\tau_T < 1$. Comptonization has been argued as a promising alternative to optically thin synchrotron emission where it is able to explain a broader range of low-energy spectral slopes, provide a characteristic energy scale for the peak of the emission, and yield narrower spectral peaks [99,113] It is the main radiation mechanism in a general class of models known as *photospheric emission* models in which the outflow is heated across the photosphere due to some internal dissipation.

At the base of the flow, where $\tau_T \gg 1$, the radiation field is thermalized and assumes a Planck spectrum. If the outflow remains non-dissipative the Planck spectrum is simply advected with the flow, cooled due to adiabatic expansion, and then released at the photosphere [33,34]. However, only a few GRBs show a clearly thermally dominated narrow spectral peak [114], whereas most have a broadened non-thermal spectrum with a low-energy photon index ($\alpha_{\rm ph} < 1$) softer than that obtained for the Planck spectrum ($\alpha_{\rm ph} = 1$). In many cases, a sub-dominant thermal component in addition to the usual Band function has been identified [115,116]. These observations imply that photospheric emission plays an important role [117], but the pure thermal spectrum must be modified by dissipation across the photosphere [54,95,118–120]. Several theoretical works tried to understand the thermalization efficiency of different radiative process, e.g., Bremmstrahlung, cyclo-synchrotron, and double Compton, below the photosphere to explain the location of the spectral peak and the origin of the low-energy spectral slope e.g., [121–123].

While sub-photospheric dissipation and Comptonization is able to yield the typical low-energy slope, further dissipation near and above the photosphere is needed to generate the high-energy spectrum above the thermal spectral peak. This can be achieved by inverse-Compton scattering of the thermal peak photons by mildly relativistic electrons [24,100,124–127]. If the flow is uniform, the net polarization of the Comptonized spectrum is negligible due to random orientations of the polarization vector at each point of the flow, which, upon averaging over the visible part, adds up to zero polarization. Alternatively, if the flow has an angular structure, particularly in the bulk-Γ profile, then net polarization as large as $\Pi \lesssim 20\%$ can be obtained [24,57].

2.3.3. Dissipative Jet: Hybrid Spectrum

If the jet is dissipative across the photosphere, a hybrid spectrum can emerge where the spectral peak is dominated by a quasi-thermal component but- the low and high-energy wings are dominated by non-thermal emission either from synchrotron or Comptonization. The final outcome depends on the nature of the dissipation and how that leads to particle acceleration/heating. Gill et al. [128], who carried out numerical simulations, and Beniamini and Giannios [129], who performed semi-analytic calculations, considered a steady PFD striped wind outflow, which is heated due to magnetic dissipation commencing at radii when the flow is optically thick to Thomson scattering with initial $\tau_{T0} = 100$. At higher τ_T, and equivalently lower radii, the flow maintains thermal equilibrium while it is being accelerated due to gradual magnetic dissipation. Localized reconnecting layers accelerate the baryonic electrons, as well as any produced $e^\pm$-pairs, into a relativistic power-law distribution. In this instance, since the flow is strongly magnetized with $\sigma > 1$, the relativistic particles are predominantly cooled by synchrotron emission. The development of the spectrum as the flow expands is shown in the left panel of Figure 2 as a function of the total τ_T. The final observed spectrum is indeed Band-like, but it is different from the optically thin synchrotron spectrum even though by the end of the radially extended dissipation the total spectrum (energetically) is synchrotron dominated.

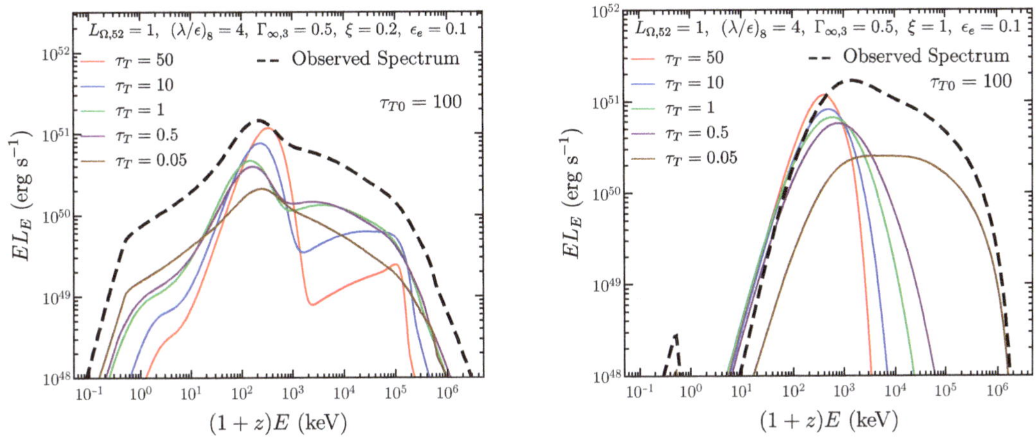

Figure 2. Spectral evolution in a dissipative steady PFD striped wind flow, shown as a function of the Thomson optical depth as the jet is heated accross the photosphere. The spectra are shown for two different particle heating scenarios: (**Left**)—relativistic power-law particles produced by magnetic reconnection, and (**Right**)—mildly relativistic particles forming an almost mono-energetic distribution due to distributed heating and Compton cooling. The flow was evolved from initial $\tau_{T0}=100$ until the total optical depth of baryonic electrons plus produced $e^{\pm}$-pairs was much less than unity. The observed spectrum is effectively a sum over the optically thin spectra. See [128] for more details.

Alternatively, particle heating can occur in a distributed manner [31,124,126,130,131] throughout the whole causal region due to MHD instabilities. In this case, particles remain only mildly relativistic. Their mean energy is governed by a balance between (gradual and continuous) heating and Compton cooling, which leads to a mono-energetic distribution. The spectral evolution as the flow expands is shown in the right panel of Figure 2. In this case the high-energy spectrum is again Band-like, but unlike the previous case it is completely formed through Comptonization [124,126,131]. The mildly relativistic particles do produce some synchrotron emission but only at energies $(1+z)E \lesssim 1$ keV.

Both particle energization mechanims can give rise to a Band-like spectra; however, they can produce completely different energy-dependent polarization. In both, if the jet is uniform and can be approximated as part of a spherical flow (i.e., away from the jet edge in a top-hat jet), then no polarization is expected near the spectral peak, as it is dominated by the quasi-thermal component. In such a scenario, away from the peak, where the spectrum is dominated by non-thermal emission, it is possible to measure high polarization ($\Pi \lesssim 50\%$) if the emission is synchrotron and the flow has a large scale ordered magnetic field, e.g., a $B_{\rm tor}$ field. Other field configurations, namely, $B_\perp$ and $B_\parallel$, will yield vanishingly small net polarization. Alternatively, if the non-thermal component is produced by Comptonization, then the expected polarization is again almost zero. On the other hand, if the LOS passes near the sharp edge of a top-hat jet or the edge of the almost uniform core in a structured jet, then the entire spectrum with non-thermal emission from Comptonization can produce $\Pi \lesssim 20\%$. Similarly, the non-thermal wings coming from synchrotron emission can now yield significant polarization with $4\% \lesssim \Pi \lesssim 28\%$ for $B_\perp$ and $10\% \lesssim \Pi \lesssim 56\%$ for $B_\parallel$, while $B_{\rm tor}$ again yields higher levels with $10\% \lesssim \Pi \lesssim 60\%$.

2.3.4. Other Proposed Mechanisms

(a) ***Compton Drag***

This model envisions the propagation of the relativistic outflow in a dense bath of seed photons with energy $E_{\rm seed}$ that provide the drag for the expanding outflow whereby cold electrons in the outflow Compton upscatter soft photons [132,133]. The seed photons can be provided either by radiation from the associated supernova remnant that exploded a time $\Delta t \simeq$ few hours before the outflow is launched, or if Δt is negligibly small then by the walls of the funnel that has been cleared in the massive star progenitor's envelope by the jet-driven bow shock post core-collapse. These requirements limit the applicability of this model to only long-soft GRBs and do not explain how such an emission would arise in short-hard GRBs. This scenario presents an entirely non-dissipative flow, which is insensitive to the magnetization but yields a high ($\lesssim 50\%$) radiative efficiency. To produce the variability, the flow is required to be unsteady. The required $\tau_T \gtrsim 1$ in this model may make it difficult to produce prompt high-energy emission due to opacity to pair production.

When the prompt GRB emission originates inside the funnel, it is assumed that the funnel is pervaded by a blackbody radiation field emitted by the funnel walls. The spectral peak of the observed prompt emission is then simply the inverse-Compton scattered peak at energy $E_{IC} \sim \Gamma^2 E_{\rm seed}$, where Γ is the bulk LF of the outflow. Inhomogeneity in the funnel temperature and bulk-Γ of subsequent shells, which could also collide to produce internal shocks, gives rise to a Band-like broadened spectrum. The local polarization, i.e., from a given point on the outflow surface, can be as high as 100%; however, the net observable polarization, e.g., in a top-hat jet, is reduced to $\Pi \lesssim 50\%$ for a jet with $(\Gamma\theta_j)^2 > 10$ [24]. If the jet is narrower than this with $(\Gamma\theta_j)^2 < 10$, then the net polarization can be much higher with $\Pi \lesssim 95\%$ [134]. However, such high polarization requires highly idealized assumptions that are hard to meet in reality.

(b) ***Jitter Radiation***

If the magnetic field coherence length is much smaller than the gyroradius of particles, then synchrotron radiation is not the correct description of the radiative mechanism by which relativistic electrons cool, as it assumes homogeneous fields. In this case, the particles experience small pitch-angle scattering where their motion is deflected by magnetic field inhomogeneities by angles that are smaller than the beaming cone of the emitted radiation ($1/\gamma_e$). This scenario has been proposed as a viable alternative to synchrotron radiation [94], where it has been shown to yield harder spectral slopes that cannot be obtained in optically-thin synchrotron emission. In addition, it can produce sharper spectral peaks as compared to synchrotron radiation, which agrees better with observations. The small-scale magnetic fields needed in this scenario may potentially be produced in relativistic collisionless shocks via the Weibel instability (although this may not be easy to achieve in practice; see e.g., [135]). The polarization when this small-scale field is confined to a slab normal to the local fluid velocity is calculated in [136,137], where it is shown that the maximum degree of polarization is obtained only at large viewing angles when the slab is viewed almost edge-on. For small viewing angles that apply to distant GRBs, the polarization is indeed very weak.

(c) ***Synchrotron Self-Compton***

Synchrotron self-Compton (SSC) emission has been considered in some works as a mechanism that could yield low-energy spectral slopes with photon indices as hard as $\alpha_{\rm ph} = 0$, a change of $\Delta\alpha_{\rm ph} = 2/3$ over the synchrotron line of death [138]. This is facilitated by the fact that for typical values of the model parameters in the internal shock scenario, optically thin synchrotron emission peaks at much lower energies (at a few eV when the SSC peak is at $\sim$100 keV) and is mostly self-absorbed. One of the major drawbacks of this radiation mechanism is that it requires the synchrotron emission in the optical, which is the seed for the harder inverse-Compton emission, to be much (by a factor of $\gtrsim 10^3$) brighter than observed (or upper limits [139]). Otherwise, it requires the Compton-y parameter to exceed unity by the same factor, which is hard to accommodate while not strongly violating the total energy budget of the burst [140,141]. The energy-dependent local polarization for SSC in an ultrarelativistic spherical flow for two different ordered B-field configurations, one parallel and the other transverse to the local velocity vector, is calculated in [142], where they found that the local polarization can be as high as $\Pi \lesssim 25\%$ under simplifying assumptions.

2.4. *What's the Angular Structure of the Outflow?*

The angular structure of the relativistic outflow in GRBs affects a number of important observables, such as prompt GRB pulse structure [84], polarization [24], afterglow lightcurve [143], and the detectability of distant GRBs [144,145]. Outflows in GRBs are collimated into narrowly beamed bipolar jets that have an angular scale $\Gamma\theta_j \sim 10$, where θ_j in the simplest model of a uniform conical jet, also referred to as a **top-hat** jet, represents a sharp edge. The notion of narrowly collimated jets in GRBs was first proposed by Rhoads [146], and it was later verified by observations of achromatic **jet-breaks** in the afterglow lightcurve that yielded $\theta_j \simeq 0.05 - 0.4$ (e.g., [147–150]). Since $\Gamma \gtrsim 10^2$ during the prompt emission phase and assuming that θ_j remains approximately the same, this yields $\Gamma\theta_j \sim 5 - 40$. This geometric beaming futher implies that the true radiated energy of these bursts is much smaller [149] with $E_\gamma = f_b E_{\gamma,\rm iso} \sim 10^{48} - 10^{52}$ erg, where $f_b = (1 - \cos\theta_j) \simeq \theta_j^2/2$ is the geometric beaming fraction with the last equality valid for $\theta_j \ll 1$, $E_{\gamma,\rm iso} = 4\pi d_L^2 S_\gamma (1+z)^{-1} \sim 10^{48} - 10^{55}$ erg is the isotropic-equivalent radiated energy, S_γ [erg cm^{-2}] is the burst fluence, and d_L is the luminosity distance. Since f_b is much smaller than 4π, the solid angle into which radiation from a spherical source is emitted, only observers whose line-of-sight (LOS) intersects the surface of the jet or passes very close to the jet edge can detect the GRB, which implies that the true rate of GRBs is enhanced by $\langle f_b^{-1} \rangle \sim 500$ [149] over the observed rate.

A top-hat jet is clearly an idealization even though it is able to explain several features of the afterglow lightcurve. Numerical simulations of jets breaking out of the progenitor star for the long-soft GRBs [151–155] and that from the dynamical ejecta for the short-hard GRBs [155–160] find that these jets naturally develop angular structures by virtue of their interaction with the confining medium. If the true energy reservoir lies in a narrow range and the scatter in $E_{\gamma,\rm iso}$ is instead caused by different viewing angles, then either the jet half-opening angle of a top-hat jet must be different in different GRBs or the jets are not uniform and must have an underlying angular profile for both the energy per unit solid angle, $\epsilon(\theta) = E_{\rm iso}(\theta)/4\pi$, and the (initial) bulk LF, $\Gamma = \Gamma(\theta)$. Such jets are commonly referred to as **structured jets** [161–164] and can be parameterized quite generally as a power law with $\epsilon(\theta) \propto \Theta^{-a}$ and $\Gamma(\theta) - 1 \propto \Theta^{-b}$ where $\Theta = \sqrt{1 + (\theta/\theta_c)^2}$ with θ_c being the core angle. A constant true jet energy among a sample of GRBs implies that $a = 2$, a model referred to as a **universal structured jet** (USJ) [161,165–167], where it corresponds to equal energy per decade in θ and therefore reproduces jet breaks similar to those for a top-hat jet with $\theta_j(\text{top-hat}) \sim \theta_{\rm obs}(\text{USJ})$. This angular profile was used as an alternative model to the top-hat jet to explain the $E_{\gamma,\rm iso} \propto S_\gamma \propto t_b^{-1}$ correlation [149] for the afterglow emission where t_b is the jet-break time [161,167]. Other useful parameterizations of a structured jet include a **Gaussian jet** with $\epsilon(\theta) \propto \Gamma(\theta) - 1 \propto \exp(-\theta^2/2\theta_c^2)$, which is a slightly smoother (around the edges) and more realistic version of the top-hat jet.

The large distances of GRBs have precluded direct confirmation and constraints of the outflow's angular structure. The main difficulty being the rather severe drop in fluence when they are observed outside of the almost uniform core. This changed recently with the afterglow observations of GRB 170817A [12], the first-ever short-hard GRB detected coincidentally with GWs (GW 170817; [11]) from the merger of two neutron stars. Helped by its nearby distance of $D \simeq 40$ Mpc and an impressive broadband (from radio to X-rays) observational campaign (e.g., [168–170]), the afterglow observations led to the first direct and significant constraint on the angular structure of the relativistic jet (e.g., [170–177]). The afterglow from this source showed a peculiar shallow rise ($F_\nu \propto t^{0.8}$) to the lightcurve peak at $t_{\rm pk} \simeq 150$ days, after which point it declined steeply ($F_\nu \propto t^{-2.2}$). Several useful lessons were learned. First, it was shown that a top-hat jet can only explain the afterglow lightcurve near and after the lightcurve peak [177] and not the shallow rise for which a structured jet is needed. Second, both power-law- and Gaussian-structured jets can explain the afterglow of GW 170817, where for a power-law jet the angular structure profile requires $a \sim 4.5$ and $b \gtrsim 1.2$ to explain all the observations [178].

While power-law- and Gaussian-structured jets remain most popular, a few other angular profiles have received some attention. Among them is the two-component jet model [150,179–183] that features a narrow uniform core with initial bulk LF $\Gamma_0 \gtrsim 100$ surrounded by a wider uniform jet with $\Gamma_0 \sim 10-30$. Nothing really guarantees or demands the outflow to be axisymmetric and uniform, in which case an outflow with small variations on small ($\ll 1/\Gamma$) angular scales can be envisioned in the form of a "patchy shell" [184] or an outflow consisting for "mini-jets" [185], with the caveat that significant variations on such causally connected angular scales are rather easily washed out and hard to maintain. In case such variations do indeed persist, it could have important consequences for the time-resolved polarization and PA. For example, patches or mini-jets can have different polarization and/or PA due to mutually incoherent B-field configurations, which can lead to smaller net polarization and PA evolution.

3. Gamma-Ray Polarimetry

Despite the wealth of information that can be obtained from prompt GRB polarization, only a few measurements with modest statistical significance exist. Moreover, many of the results presented in the past were refuted by follow-up studies. A detailed overview of many of these measurements and their respective issues is provided in [186]. The two most recent measurements, by POLAR [187] and Astrosat CZT [188], furthermore appear to be incompatible with one another, indicating probable issues in at least one of these results as well. The lack of detailed measurements, and the many issues with them, result from both the difficulty in measuring γ-ray polarization as well as challenging data analysis at these energies. Below, we discus first the measurement principle, which causes many of the encountered issues. This is followed by a description of the different instruments that have been able to perform measurements to date.

3.1. Measurement Principles

The polarization of X-ray or γ-ray photons can be measured by studying the properties of the particles created during their interaction within the detector. For all the three possible interaction mechanisms, namely, the photo-electric effect, Compton scattering, and pair production, a dependency exists of the orientation of the outgoing products on the polarization vector of the incoming photon. This is illustrated in Figure 3 for the three processes. For the photo-electric effect, it is the azimuthal direction of the outgoing electron that shows a dependency on the polarization vector of the incoming photon; for Compton scattering, it is the azimuthal scattering angle of the photon; and for pair-production, it is the plane defined by the electron-positron pair.

The differential cross section for photo-absorption (via the photo-electric effect) has a dependency on ϕ, which is defined as the azimuthal angle between the polarization vector of the incoming photon $\vec{p}$, as shown in Figure 3, and the projection of the velocity vector of the final state electron $\vec{\beta} = \vec{v}/c$ (where $\hat{\beta} = \hat{k}_2$) on to the plane normal to the momentum vector $\vec{k}_1$ of the photon,

$$\frac{d\sigma}{d\Omega} \propto \cos^2\phi \quad , \quad \phi = \cos^{-1}\left(\frac{\vec{\beta}\cdot\vec{p}}{\beta p \sin\theta}\right) = \cos^{-1}\left(\frac{\vec{k}_2\cdot\vec{p}}{k_2 p \sin\theta}\right), \tag{2}$$

where $d\Omega = \sin\theta d\theta d\phi$ is the unit solid angle, and the polar angle θ is given by $\cos\theta = \hat{\beta}\cdot\hat{k}_1 = \hat{k}_2\cdot\hat{k}_1$. Similarly, for the differential cross section of Compton scattering the dependence on ϕ, here the angle between the polarization vector of the incoming photon

$\vec{p}$ and the projection of the momentum vector of the outgoing photon $\vec{k}_2$ on to the plane normal to the momentum vector $\vec{k}_1$ of the incoming photon, where $\phi = \cos^{-1}\left(\frac{\vec{k}_2 \cdot \vec{p}}{k_2 p \sin\theta}\right)$ as in Equation (2), is

$$\frac{d\sigma}{d\Omega} = \frac{r_0^2}{2}\frac{E'^2}{E^2}\left(\frac{E'}{E} + \frac{E}{E'} - 2\sin^2\theta\cos^2\phi\right). \tag{3}$$

Here, $r_0 = e^2/m_e c^2$ is the classical electron radius with e being the elementary charge, E is the initial photon energy, E' the final photon energy, and $\theta = \cos^{-1}(\hat{k}_2 \cdot \hat{k}_1)$ is the polar scattering angle.

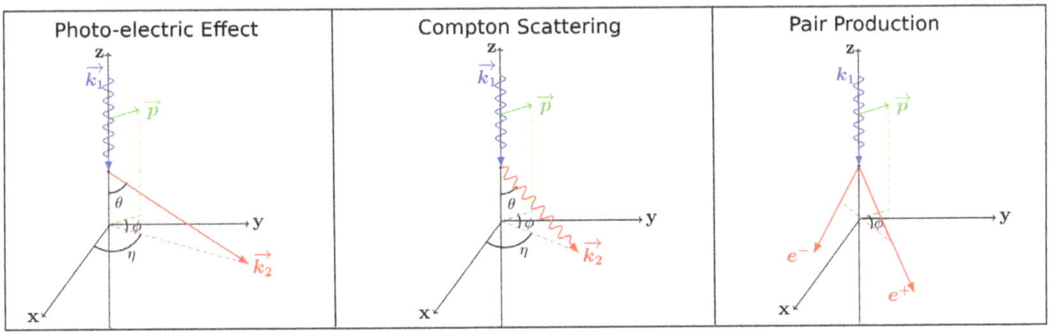

Figure 3. Illustration of the angular dependency of the interaction product on the polarization vector of the incoming photon for the three interaction mechanisms: photo-electric effect (**left**), Compton scattering (**middle**), and pair production (**right**). The incoming photon is shown in blue, its polarization vector in green, and the secondary product(s) in red. The θ angle (as used in Equations (2) and (3)) is defined as the angle between the incoming photon direction and its secondary product. The ϕ angle (again as used in Equations (2) and (3)) is defined as the angle between the projections of the polarization vector and the momentum vector of the secondary product(s) onto the x-y plane. The η angle is the azimuthal angle between the x-axis and the projection of the momentum vector of the secondary particle onto the x–y plane. The θ and η angles can be directly measured in a detector, while ϕ is measured indirectly.

Finally, for pair production the differential cross section is $d\sigma/d\Omega \propto 1 + A(\cos 2\phi)$, where A is the polarization asymmetry of the conversion process (which has dependencies on the photon energy and properties of the target), and ϕ is the angle between the polarization vector of the incoming photon $\vec{p}$ and the plane defined by the momentum vectors of the electron–positron pair, $\vec{k}_{\pm}$.

The general concept for polarimetry in the three energy regimes where these cross sections dominate is therefore similar: one needs to detect the interaction itself and subsequently track the secondary particle, be it an electron, photon, or electron–positron pair. This requirement indicates the first difficulty in polarimetry: simply absorbing the incoming photon flux, as is the case in, for example, standard spectrometry, is not sufficient. The requirement to track the secondary product significantly reduces the efficiency of the detector.

After measuring the properties of the secondary particles, a histogram of ϕ can be made, which shows a sinusoidal variation with a period of 180° referred to as a modulation curve. The amplitude of this is proportional to the polarization degree (PD) and the phase related to the polarization angle (PA). As can be derived from, for example, the Compton scattering cross section, the amplitude for a 100% polarized beam will depend on the energy of the incoming photons as well as on the polar scattering angle. Whereas the energy depends on the source, the polar scattering angle is indirectly influenced by the instrument design. For example, using a detector with a thin large surface perpendicular to the incoming flux, it is more likely to detect photons scattering with a polar angle of 90°, which have a larger sensitivity to polarization than those scattering forward or backward. The relative amplitude, meaning the ratio of the amplitude of the sinusoidal over its mean, is directly proportional to the PD. The relative amplitude one detects for a 100% polarized beam is known as the M_{100}, and it depends on the source spectrum, source location in the sky, the instrument design, and the analysis. Although for specific circumstances the M_{100} can be measured on the ground using, for example, mono-energetic beams with a specific incoming angle w.r.t the detector, its dependency on the energy, incoming angle, and instrument conditions, such as its temperature, implies that the M_{100} required in the analysis of real sources can only be achieved using simulations. The large dependency on simulations provides a source for potential systematic errors in the analysis, which can easily dominate the statistical error in the measurements.

Additionally, it should be noted that in practice retrieving the polarization is significantly more complicated as both instrumental and geometrical effects (such as the incoming angle of the photons w.r.t. the detector and the presence and orientation of materials around the detector) are added to the polarization-induced signal in the modulation curve. In order to retrieve the polarization signal one can, for example, divide it by a simulated modulation curve for an unpolarized signal as illustrated in the first column of Figure 4. This method is often used, for example in [188]. A second option is to model these effects together with the signal and fit the uncorrected curve with this simulated response, as was for example done in [189]. In either case, it requires a highly detailed understanding of the instrument.

In polarization analysis, any imperfections in modelling the instrument will likely result in an overestimation of the polarization. As illustrated in Figure 4, for a modulation curve resulting from an unpolarized flux, removing any instrumental effects from the modulation curve should result in a perfectly flat distribution. This is illustrated in the middle column of this figure. Any error in the model of the instrumental or geometrical effects will, however, result in a non-flat distribution, which, when fitted with a harmonic function, will result in some level of polarization to be detected. It is therefore in practice impossible to measure a PD of 0% as it would require both an infinite amount of statistics, and more importantly, a perfect modelling of all the instrumental effects. On the other extreme, for a PD of 100%, imperfections in the modelling can result in a lower amplitude, but can still also increase it further resulting in measuring a nonphysical PD. Overall, due to the nature of the measurement, both statistical and systematic errors tend to inflate the PD rather than decrease it. Since it is not possible to test the modelling of the instruments when in orbit, as there are no polarization calibration sources, this issue exists for all measurements and can only be minimized by extensive testing of the instrument both on the ground and in-orbit.

A final figure of merit often used in polarimetry is the minimal detectable polarization (MDP) [190]. For GRBs the MDP is best expressed as

$$\mathrm{MDP} = \frac{2\sqrt{-\ln(1 - C.L.)}}{M_{100}C_s}\sqrt{C_s + C_b}\,. \tag{4}$$

Here, $C.L.$ is the confidence level, C_s is the number of signal events, and C_b the number of background events. The MDP expresses the minimum level of polarization of the source that can be distinguished from being unpolarized for a given confidence level. It can therefore be seen as a sensitivity of a given polarimeter for a given observation. Whereas this is highly useful for polarimeters observing point sources, for GRB polarimeters, there is an issue related to the M_{100}. For wide-field-of-view instruments, such as polarimeters designed for GRB observations, the value of M_{100} can start to depend on the PA of the source. For example, in POLAR, the M_{100} was found to depend on the PA for GRBs with a large off-axis incoming angle [187]. This is a result of only being able to resolve two dimensions of the scattering interactions in the detector, making it insensitive (so $M_{100} = 0$) to certain values of PA when the γ-ray photons enter the detector perpendicular to the readout plane [187]. As in such cases the MDP becomes dependent on PA, it loses its use as a figure of merit. However, as the MDP remains highly used in the community and remains the best measure of sensitvity for polarimeters, we used it here in this work as well, although with a small adaptation. In order to remove the PA dependence, we used the mean MDP where the M_{100} is averaged over all possible values of the a priori unknown PA.

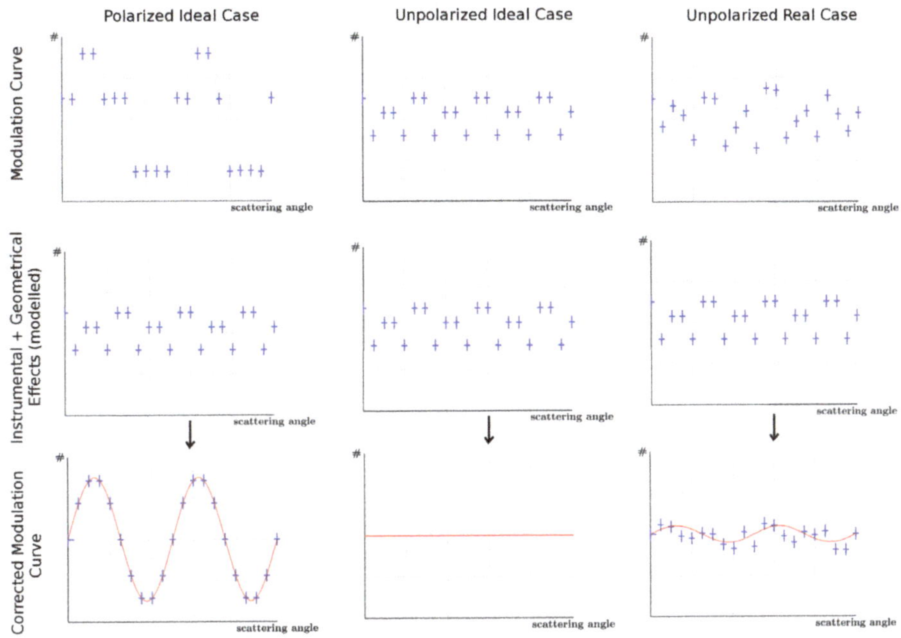

Figure 4. Illustration of recovering the polarization signal from a raw modulation curve. The left column illustrates the ideal case with a high PD value, with a raw measured modulation curve (**top**), abd the perfectly simulated instrumental and geometrical effects (**middle**), which pollute the raw modulation curve. The (**bottom**) panel shows the modulation curve after correction from the instrumental and geometrical effects, which results in a perfect harmonic function. The middle column illustrates the same but for an unpolarized signal resulting in a flat distribution. The right column shows the same for an unpolarized signal; however, random small errors were added to the instrumental and geometrical effects, thereby simulating a non-perfect understanding of the instrument. The result is a non-flat distribution, which, when fitted, shows a low level of polarization.

3.2. Detection Principles

To date, the only GRB polarization measurements performed have made use of Compton scattering in the detector. The majority of these measurements were performed by making use of a segmented detector concept, for example a detector consisting of many relatively small scintillators, e.g., for GAP [191] and POLAR [192], or a segmented semiconductor, such as INTEGRAL-SPI [193] and AstroSAT CZT [194]. In either design, the Compton scattering interaction can be detected in one segment of the detector while an additional interaction of the photon in a second segment can be used to reconstruct the azimuthal Compton scattering angle. This concept is illustrated in Figure 5.

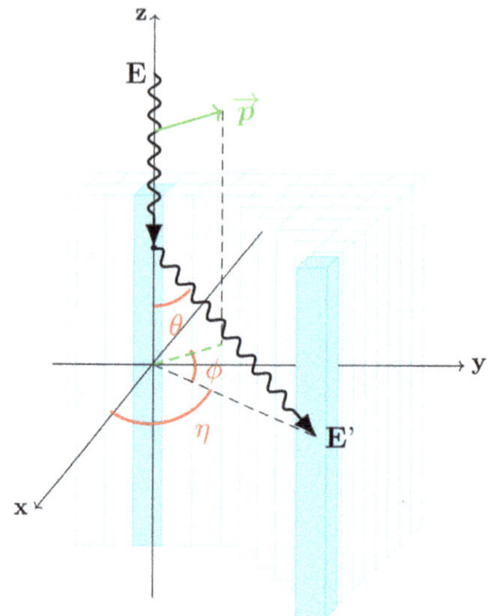

Figure 5. Illustration of the measurement principle of a polarimeter using Compton scattering. The incoming γ-ray Compton scatters in one of the detector segments followed by a photo-absorption (or second Compton scattering) interaction in a different segment. Using the relative position of the two detector segments, the Compton scattering angle can be calculated from which, in turn, the polarization angle can be deduced.

At energies below approximately 10 keV the cross section for photo-absorption dominates. Although no successful GRB polarization measurements have been performed using the photo-electric effect, the detection method has been successfully used recently to perform polarization measurements of the Crab Nebula in the 3–4.5 keV energy range using the PolarLight cubesat [195]. Several large-scale polarimeter ideas have been developed in the past, such as the Low-energy Polarimeter, which was part of the proposed POET mission, which was dedicated to GRBs [196]. Currently, several missions that use the same concept are currently under development [197,198]. In these X-ray polarimeters, the photo-absorption takes place in a thin gas detector. As the produced electron travels through the gas it releases secondary electrons as it ionizes the gas. These secondary electrons can be detected using finely segmented pixel detectors in order to track the path of the electron released in the photo-absorption interaction, allowing to reconstruct its emission angle.

Polarimetry in the pair-production regime is arguably the most challenging as the photon flux is low, and the detection method requires highly precise trackers capable of separating the tracks of the electron and the positron. In spectrometry, the low photon flux is often compensated by using large detectors with a high stopping power. For example, by combining tungsten layers with silicon detectors. Here, the silicon serves to measure the tracks while the tungsten is used to enforce pair production in the detector. However, the use of high Z (atomic number) materials, like tungsten, significantly increases multiple scattering of the electron and positron. Multiple scatterings quickly change the momentum of both products, thereby making it challenging to reconstruct their original emission direction. To overcome this issue, detectors that use silicon both for conversion and detection, have been proposed in the past such as PANGU [199]. Although technically possible, the large number of silicon detectors required to achieve a high sensitivity, with minimal structural material and a potential magnet, which helps to separate the electron–positron pair, make such detectors both costly and challenging to develop for space. A second option is to use gas-based detectors, such as in the HARPO design [200]. This detector, which was successfully tested on the ground [201], allows for precise tracking but has a low stopping power for the incoming γ-rays and therefore a low detection efficiency. This can be compensated with a large volume. However, as the gas volume obviously needs to be pressurized, producing and launching such an instrument for use in space is highly challenging. Despite these challenges, several projects that follow this design are still ongoing, such as the potential future space mission AdEPT [202], which aims to use a time projection chamber to measure polarization in the 5–200 MeV energy range as well as the balloon borne SMILE missions [203].

Although no dedicated pair–production polarimeters are currently in orbit, it should be noted that both the Fermi-LAT [204] and the AMS-02 [205] instruments could, in theory, be used to perform polarization measurements in this energy range. For Fermi-LAT, which is a dedicated γ-ray spectrometer, consisting of silicon strip detectors combined with tungsten conversion layers, the aforementioned multiple scattering induced distortion is again a challenge [206]. The polarization capabilities of Fermi-LAT, which has detected many GRBs to this day, has been studied [206], but no results from actual data have been published to date. For AMS-02, which does not suffer from the use of tungsten layers and additionally contains a magnet which separates the pairs, measurements could be easier. However, as the instrument is designed as a charged particle detector, it remains non-optimized for this purpose, and so far no results have been published by this collaboration.

3.3. GRB Polarimeters

In this section, we aim to provide a summary of the different instruments that have performed GRB polarization to date. For a detailed overview of each individual measurement (up to 2016), the reader is referred to [186].

As mentioned earlier, all polarization measurements of the prompt GRB emission have been performed by making use of Compton scattering. While in the majority of cases the Compton scattering takes place in the detector, there is one exception. The attempts at performing polarization measurements with data from the BATSE detector made use of Compton scattering from the Earth's atmosphere [207,208]. The BATSE detector consisted of several scintillator-based detectors and by itself had no capability to directly perform polarimetry [209]. Instead, it used several detectors pointing towards the Earth, each at different relative angles, to measure the relative intensity of photons scattering off different parts of the Earth's atmosphere. As the probability for photons to scatter off the atmosphere towards different detectors depends on their polarization properties, such a measurement is possible for any detector with an Earth-facing sensitive surface. It does however require a highly detailed modelling of the Earth's atmosphere, software capable of simulating the scattering effects properly, and detailed understanding of the detector response as well as the location and spectra of the GRB. The large number of sources for systematic errors resulted in inconclusive measurements of GRB 930131 [208]. Despite the initial lack of success, improvements have been made since then regarding Compton-scattering models in software such as Geant4 [210]. Furthermore, instruments such as *Fermi*-GBM have measured 1000 s of GRBs over the last decade, and similar studies using this data could prove to be successful in the future.

Systematic errors are a major issue not only for the creative polarization measurement solution used in BATSE but in all GRB polarization results published thus far from different instruments. It is especially important for measurements performed using detectors not originally designed to perform polarimetry such as RHESSI [211] and the SPI and IBIS detectors on board INTEGRAL. Both RHESSI and SPI make use of a segmented detector consisting of germanium detectors and thereby allow to study Compton scattering events by looking for coincident events between different detectors. The IBIS instrument [212] uses two separate sub-detectors instead, namely, the ISGRI detector consisting of 16384 CdTe detectors and the Pixellated Imaging CsI Telescope (PICsIT), an array of 4096 CsI scintillator detectors. Since, similar to RHESSI and SPI, IBIS was not originally designed to perform polarization measurements, the trigger logic in the instrument was not setup to keep coincidence events in the PICsIT or ISGRI alone. Rather, only coincidence events between the PICsIT and the ISGRI are kept, which although lowering the statistics for polarization measurements, still allows for such measurements [213].

Since all three instruments were not designed as polarimeters, one immediate downside of using them as such is the lack of sensitivity. A clear example of this is the non-optimized trigger logic of IBIS. In the case of RHESSI, different analyses of the same GRB [16] resulted in vastly different results, in part due to the difficulty in selecting valid coincident events between different germanium detector channels, again a result of a non-optimized online event selection. The relatively imprecise time measurement of each event prompted a large coincidence window to be set in one of the analyses, which resulted in chance coincidence events induced by different photons or background particles instead of the Compton scattering event [17,18]. If the instrument had been designed and tested on the ground as a polarimeter, the coincidence trigger logic and time measurement would likely have been optimized and event selection methods tested during the calibration phase.

The lack of on-ground calibration for polarization additionally makes verification of the detector response models difficult and prone to errors—for example, dead material around the detector can affect the polarization of the incoming flux when it interacts with it. While such issues are important in spectrometers as well, it can be argued that it is more important in a polarimeter. Imperfect modelling of certain detector channels for a spectral measurement can cause issues. However, if on average the channels are modelled correctly, having a few badly modelled channels will not greatly affect the final flux or spectral result, as over- and under-performing channels can cancel each other out. In a polarimeter it is the difference in the number of events between the detector channels that provides the final measurement and not, as in a simple spectrometer, the average of all the channels. For a polarimeter, however, one single over-performing detector channel would see a larger number of scattering events than expected, causing certain scattering angles to be favoured and thereby faking a polarization signal.

Similarly, dead material in front of the detector channels can easily obscure certain channels more than others causing a similar effect. Understanding all these issues during on-ground calibration is therefore crucial to reduce systematic errors. As a result of such difficulties, the polarization results published by the SPI collaboration clearly mention the possibility of significant systematic errors not taken into account in the analysis, which can affect the results [214,215].

In order to overcome such issues, more recent instruments, such as GAP [191] and POLAR [192], employ small coincidence windows and trigger logics optimized for polarization measurements. Most importantly, such detectors were calibrated prior to launch with polarized photons in different configurations, such as different photon energies and incoming angles [191,216,217]. GAP was the first dedicated GRB polarimeter. It made use of plastic scintillators used to detect Compton scattering photons together with 12 CsI scintillators used to detect the photon after scattering. The instrument flew for several years on the IKAROS solar sail mission during which it detected a few GRBs for which polarization measurements were possible. The POLAR detector also uses plastic scintillators, 1600 in total, to detect the Compton scattering interaction but uses the same scintillators to detect the secondary interaction. As a result, the instrument is less efficient for detecting the secondary interaction and has a poorer energy resolution. However, it allows for a larger scalable effective area as well as a larger field of view, which in the case of GAP is restricted by the CsI detectors that shield the plastic scintillators from a far off-axis source. The POLAR detector took data for six months on board the Tiangong-2 space laboratory, which resulted in the publication of 14 GRB polarization measurements [187].

Two other detectors, which although not fully optimized for polarimetry, were calibrated on the ground for such measurements. They are COSI [218] and the CZTI on Astrosat [194]. The balloon-borne COSI detector uses two layers of germanium double-sided strip detectors allowing for precise measurements of the interaction locations in the instrument. During its long-duration balloon flight in 2017, COSI saw one bright GRB for which a polarization measurement could be performed [219]. The CZT Imager on board the AstroSAT satellite uses, as the name suggests, a CZT semiconducter detector. As this detector is segmented, it allows to look for Compton scattering events. The detector was calibrated with polarized beams prior to launch to study the instrument response to on-axis sources [194]. AstroSAT CZTI has detected a large number of GRBs since its launch and has published polarization measurements of 13 of these to date while it continues to be operational.

4. Theoretical Models of Prompt GRB Polarization

The focus of this section is to present polarization predictions for the popular prompt GRB emission mechanisms, as highlighted in Section 2. Since GRBs are cosmological sources (of modest physical size in astrophysical standards), they remain spatially unresolved. Consequently, the measured polarization is the effective average value over the entire image of the burst on the plane of the sky. Therefore, the obtained polarization is affected by several effects, such as the intrinsic level of polarization at every point on the observed part of the outflow; the geometry of the outflow; i.e., the angular profile of the emissivity and bulk Γ; and the observer's LOS. Even though GRBs are intrinsically very luminous, their large distances drastically reduce the observed flux, making them photon starved. This forces observers to integrate either over the entire pulse or large temporal segments of a given emission episode to increase the photon count. This causes additional averaging—time averaging over the instantaneous polarization from the whole source, which in many cases significantly evolves even within a single spike in the prompt GRB lightcurve.

Before presenting the model predictions for time-resolved polarization in Section 4.3, pulse-integrated polarization is discussed first. In the latter, any radial dependence of the flow properties is ignored for simplicity (but without affecting the accuracy of the calculation). As a result, pulse-integrated polarization ultimately amounts to integrating over a single pulse emitted at a fixed radius, where it is not important what that radius is as it does not enter any of the calculations.

Polarization is most conveniently expressed using the Stokes parameters $(I,\ Q,\ U,\ V)$, where I is the total intensity, Q and U are the polarized intensities that measure linear polarization, and V measures the level of circular polarization. In GRB prompt emission, the circular polarization is typically expected to be negligible compared to the linear polarization ($V^2 \ll Q^2 + U^2$; this is usually expected to hold also for the reverse shock and afterglow emission) and therefore we concentrated here on the linear polarization. The local linear polarization (all local quantities are shown with a "bar") from a given fluid element on the emitting surface of the flow is given by e.g., [220]

$$\bar{\Pi}' = \frac{\sqrt{\bar{Q}'^2 + \bar{U}'^2}}{\bar{I}'} = \frac{\sqrt{\bar{Q}^2 + \bar{U}^2}}{\bar{I}} = \bar{\Pi}\,, \tag{5}$$

where

$$\frac{\bar{U}}{\bar{I}} = \bar{\Pi}\sin 2\bar{\theta}_p\,, \qquad \frac{\bar{Q}}{\bar{I}} = \bar{\Pi}\cos 2\bar{\theta}_p\,, \qquad \bar{\theta}_p = \frac{1}{2}\arctan\left(\frac{\bar{U}}{\bar{Q}}\right), \tag{6}$$

and $\bar{\theta}_p$ is the local polarization position angle (PA). When moving from the comoving frame of the jet to the observer frame, both the Stokes parameters and the direction of the polarization unit vector ($\hat{\bar{\Pi}}' = (\hat{n}' \times \hat{B}')/|\hat{n}' \times \hat{B}'|$, where $\hat{n}'$ and $\hat{B}'$ are the unit vectors in the comoving frame pointing along the observer's LOS and direction of the local B-field, respectively) undergo a Lorentz transformation (e.g., Equation (13) of Gill et al. [24]). The degree of polarization (magnitude of the polarization vector), however, remains invariant (since $\bar{Q}'/\bar{Q} = \bar{U}'/\bar{U} = \bar{I}'/\bar{I}$). The local polarization is different from the *global* one, $\Pi = \sqrt{Q^2 + U^2}/I$ (all global parameters are denoted without a bar), which is derived from the global Stokes parameters. It is the global polarization that is ultimately measured for a spatially unresolved source. For an incoherent radiation field, meaning the emission from the different fluid elements is not in phase, which is also true for most astrophysical sources, the Stokes parameters are additive. Therefore, each global Stokes parameter is obtained by integration of the corresponding local Stokes parameter over the image of the GRB jet on the plane of the sky, such that

$$\left\{ \begin{array}{c} U/I \\ Q/I \end{array} \right\} \to \left\{ \begin{array}{c} U_\nu/I_\nu \\ Q_\nu/I_\nu \end{array} \right\} = \frac{\int d\Omega \left\{ \begin{array}{c} \bar{U}_\nu = \bar{I}_\nu \bar{\Pi} \sin(2\bar{\theta}_p) \\ \bar{Q}_\nu = \bar{I}_\nu \bar{\Pi} \cos(2\bar{\theta}_p) \end{array} \right\}}{\int d\Omega \bar{I}_\nu} = \frac{\int dF_\nu \left\{ \begin{array}{c} \bar{\Pi} \sin(2\bar{\theta}_p) \\ \bar{\Pi} \cos(2\bar{\theta}_p) \end{array} \right\}}{\int dF_\nu}, \tag{7}$$

where $dF_\nu \cong I_\nu d\Omega = I_\nu dS_\perp/d_A^2$ is the flux contributed by a given fluid element, of observed solid angle $d\Omega$ and area $dS_\perp$ on the plane of the sky, and d_A is the angular distance to the distant source. We worked with the Stokes parameters per unit frequency for convenience, such as the specific intensity $\bar{I}_\nu = d\bar{I}/d\nu$. For simplicity, we ignored the radial structure of the outflow and assumed that the emission originates from an infinitely "thin shell". This approximation is valid if the time-scale over which particles cool and contribute to the observed radiation is much smaller than the dynamical time. It implies that the emission region is a thin cooling layer of width (in the lab-frame) $\Delta \ll R/\Gamma^2$. In this approximation, the differential flux density from each fluid element radiating in the direction $\hat{n}$, i.e., the direction of the observer, when the radiating shell is at radius R (radial dependence included here for the general expression) can be expressed as [221]

$$dF_\nu(t_{\rm obs}, \hat{n}, R) = \frac{(1+z)}{16\pi^2 d_L^2} \delta_D^3 L'_{\nu'}(R) d\tilde{\Omega}, \tag{8}$$

where z and d_L are the redshift and luminosity distance of the source, respectively; $L'_{\nu'}$ is the comoving spectral luminosity of the fluid element; and $d\tilde{\Omega} = d\tilde{\mu}d\tilde{\varphi}$ is its solid angle; $\tilde{\mu} = \cos\tilde{\theta}$ with the polar angle $\tilde{\theta}$ measured from the LOS; and $\tilde{\varphi}$ is the azimuthal angle around the LOS. The Doppler factor of the fluid element moving with velocity $\vec{\beta} = \vec{v}/c$ is given by $\delta_D(R) = [\Gamma(1 - \vec{\beta} \cdot \hat{n})]^{-1} = [\Gamma(1 - \beta\tilde{\mu})]^{-1}$ (where the second expression holds for a radial outflow where $\hat{\beta} = \hat{r}$). In order to calculate the Stokes parameters using the differential flux density, the angular structure of the outflow needs to be specified, as was done next.

4.1. Polarization from Uniform Jets

In uniform axisymmetric jets, the comoving spectral luminosity, $L'_{\nu'}$ and the bulk-Γ do not vary with polar angle θ measured from the jet axis, e.g., in a top-hat jet,

$$\frac{L'_{\nu'}(\theta)}{L'_0} = \frac{\Gamma(\theta)}{\Gamma_0} = \begin{cases} 1, & \theta \le \theta_j \\ 0, & \theta > \theta_j \,. \end{cases} \tag{9}$$

It is further assumed that Γ, θ_j, $\theta_{\rm obs}$, and the spectrum (assumed here to be a power law) remain constant with the radius during emission of the prompt GRB (while $L'_{\nu'}$ can vary with the radius). Since the emission arises in an ultrarelativistic jet ($\Gamma \gg 1$), it is strongly beamed along the direction of motion primarily into a cone of angular size $1/\Gamma$. Consequently, most of the observed radiation arrives from angles $\tilde{\theta} \lesssim 1/\Gamma$ around the LOS. If the LOS intersects the jet surface and is more than a beaming cone away from the edge of the jet, i.e., if $\theta_{\rm obs} \lesssim \theta_j - \Gamma^{-1}$ or equivalently if $q \equiv \theta_{\rm obs}/\theta_j \lesssim 1 - \xi_j^{-1/2}$ where $\xi_j \equiv (\Gamma\theta_j)^2$, then the observer remains unaware of the jet's edge (however, see Section 4.3), and the emission can be approximated as if arising from a spherical flow. In this instance, after averaging over the GRB image on the plane of the sky, a finite net polarization will only be obtained if the direction of polarization is not axisymmetric around the LOS. Hence, it becomes necessary to break this symmetry in order to obtain any net polarization. This naturally happens if the LOS lies near the edge of the jet. Therefore, in such cases a special alignment between the flow direction and the observer is needed. This and other effects that break the symmetry and yield finite net polarization are highlighted below.

4.1.1. Synchrotron Emission from Different Magnetic Field Structures

Synchrotron emission is generally partially linearly polarized. The local polarization emerging from a given point on the outflow depends on the geometry of the local B-field and distribution of the emitting electrons, both in energy, $\gamma_e m_e c^2$, and pitch angle, $\chi' = \arccos(\hat{B}' \cdot \hat{\beta}'_e)$, where $\hat{\beta}'_e$ is a unit vector pointing in the direction of the electron velocity. In the case of power-law electrons, with distribution $n_e(\gamma_e) \propto \gamma_e^{-p}$ for $\gamma_e > \gamma_{\rm min}$, and with isotropic velocity distribution so that all pitch-angles are sampled during the emission, the maximum local polarization for a locally ordered B-field depends on the spectrum [108,220]

$$\Pi_{\rm max} = \frac{\alpha+1}{\alpha+5/3} = \frac{p_{\rm eff}+1}{p_{\rm eff}+7/3}\,. \tag{10}$$

Here $\alpha(\nu) = -d\log F_\nu/d\log\nu$ is the *local* spectral index, and $p_{\rm eff} = 2\alpha+1$ is the effective power-law index of the electron distribution. Since the local value of α (and therefore also of $p_{\rm eff}$) smoothly varies with ν, the maximum polarization, $\Pi_{\rm max}$, also varies smoothly with ν across the spectral breaks of the synchrotron spectrum. The *asymptotic* spectral index is different for different power-law segments (PLSs) of the well-studied [87,88] broken power-law synchrotron spectrum. We have $\alpha = 1/2$, $p_{\rm eff} = 2$, and $\Pi_{\rm max} = 9/13 \approx 0.692$ for $\nu_c < \nu < \nu_m$ (fast cooling); $\alpha = (p-1)/2$, $p_{\rm eff} = p$ and $\Pi_{\rm max} = (p+1)/(p+7/3)$ for $\nu_m < \nu < \nu_c$ (slow cooling); $\alpha = p/2$, $p_{\rm eff} = p+1$ and $\Pi_{\rm max} = (p+2)/(p+10/3)$ for $\nu > \max(\nu_c,\nu_m)$ (either slow or fast cooling). For $\nu < \min(\nu_m,\nu_c)$, there is no $p_{\rm eff}$ since emission in this PLS arises from all cooling electrons that are emitting below their typical (optically thin) synchrotron frequency. In this case, $\alpha = -1/3$ and $\Pi_{\rm max} = 1/2$, the lowest local polarization obtained from synchrotron emission. On the other hand, shock-acceleration theory suggests that $2 \lesssim p \lesssim 3$, which means that the maximum local polarization in synchrotron is limited to $\Pi_{\rm max} \lesssim 75\%$.

When the magnetic fields are tangled or switch direction on angular scales $\ll 1/\Gamma$, e.g., in the $B_\perp$ case, the local polarization must be averaged over different B-field orientations. This has been calculated for an infinitely thin ultrarelativistic shell, while assuming $\alpha = 1$, for a tangled B-field [109,222,223]

$$\frac{\bar{\Pi}_{\rm rnd}}{\Pi_{\rm max}} = \frac{(b-1)\sin^2\tilde{\theta}'}{2+(b-1)\sin^2\tilde{\theta}'} = \begin{cases} \frac{-\sin^2\tilde{\theta}'}{1+\cos^2\tilde{\theta}'} & (b=0, B\to B_\perp) \\ 1 & (b=\infty, B\to B_\parallel)\,, \end{cases} \tag{11}$$

where $\tilde{\theta}'$ is the polar angle measured from the LOS in the comoving frame (this holds for a radial flow and more generally $\tilde{\theta}' \to \arccos(\hat{n}' \cdot \hat{n}'_{\rm sh})$). The level of anisotropy of the B-field is quantified by the parameter $b = 2\langle B_\parallel^2\rangle/\langle B_\perp^2\rangle$, which represents the ratio of the average energy densities in the two field orientations. The factor of two simply reflects the two independent directions of the $B_\perp$ component, such that $b=1$ for a field that is isotropic in three dimensions.

The polarization map over the GRB image on the plane of the sky is shown in Figure 6 for different B-field structures (for $\Gamma \gg 1$). Only the area contained within the beaming cone, shown by the red circle, contributes dominantly to the emission. Outside of it, the intensity is strongly suppressed by relativistic beaming, which scales as a power of the Doppler factor. This effect is reflected by the decrease with the angle $\tilde{\theta}$ from the LOS (shown by the red "+" symbol) in the size of the black arrows, which correspond to the magnitude of the polarized intensity. When the jet possesses axial symmetry (and for synchrotron emission the same requirement holds also for the global magnetic field structure), then the image and polarization map are symmetric to reflection along the line connecting the jet symmetry axis to the LOS. Therefore, it is natural to choose a reference direction for measuring the local PA $\tilde{\theta}_p$ either along this line or transverse to it (in the figure, $\tilde{\theta}_p$ as well as

θ_p are measured from the latter, i.e., the horizontal direction). For such a choice, $U = 0$ i.e., the local Stokes parameter $\bar{U} \propto \sin(2\bar{\theta}_p)$ vanishes when integrated over the GRB jet image, and therefore the global polarized intensity is entirely given by Stokes Q, i.e., the integration of $\bar{Q} \propto \cos(2\bar{\theta}_p)$ over the image, where the sign of $\bar{Q}$ for each fluid element depends on the local PA $\bar{\theta}_p$. The different B-field configurations produce completely different polarization maps, with distinct patterns of regions contributing predominantly either to polarization along the line connecting the jet symmetry axis to the LOS (orange–yellow, with local polarization $\bar{\Pi} < 0$) or transverse to it (blue–white, with local polarization $\bar{\Pi} > 0$), as shown by the color map. When averaged over the entire GRB image, these are the only two directions of polarization that can be obtained in an axisymmetric flow in which the magnetic field also possesses axial symmetry about the jet axis, such that it would represent a change of 90° in the PA when the direction of polarization switches from one to the other.

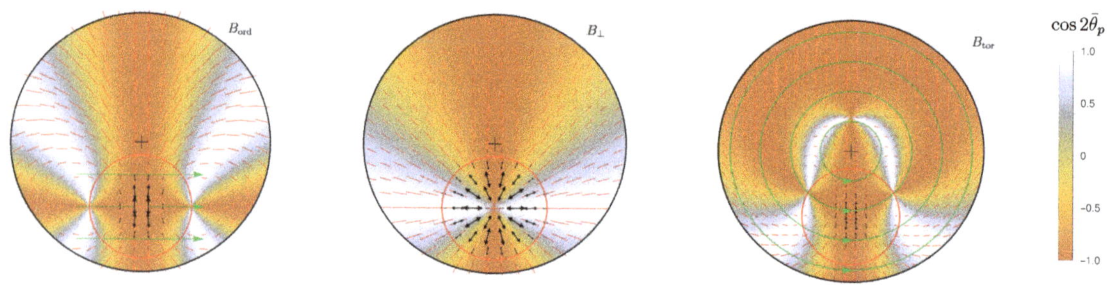

Figure 6. Polarization map for different B-field configurations shown on the surface of a top-hat jet (for $\Gamma \gg 1$). The jet symmetry axis marked with a black "+" symbol and the observer's LOS is marked with a red "+" symbol. The region where the LOS is within the beaming cone of the local emission (i.e., from which the radiation is beamed towards us) is within the red circle, outside of which the the polarized intensity (as shown by the size of the black arrows) declines sharply. The red line segments show the direction and polarized intensity, now without the de-beaming suppression. Green lines show the orientation of the magnetic field lines (in the cases $B_{\rm ord}$ and $B_{\rm tor}$ where it is locally ordered). The color map shows $\bar{Q} \propto \cos(2\bar{\theta}_p)$, with $\bar{\theta}_p$ being the local polarization angle measured counter-clockwise from the horizontal axis, which corresponds to the level at which each point is polarized either along the line connecting the LOS with the jet symmetry axis (orange-yellow dominated) or transverse to it (blue-white dominated).

An example of a B-field configuration that does not possess such axial symmetry is $B_{\rm ord}$. When $B_{\rm ord}$ is not oriented along the line connecting the jet symmetry axis to the LOS or perpendicular to it, then this breaks the symmetry of the image polarization map, thereby enabling other directions of net global polarization to occur, and the corresponding PA can vary continuously with a finite Π [109].

The level of net polarization after averaging over the GRB image depends on the level of symmetry of the polarization map around the LOS. In the case of $B_\perp$, and likewise for $B_\parallel$, the polarization map is symmetric around the LOS and therefore averaging over the GRB image would yield zero net polarization ($\Pi = 0$) due to complete cancellation for a spherical flow (or well within a top-hat jet, $\Gamma(\theta_j - \theta_{\rm obs}) \gg 1$). This symmetry is naturally broken in $B_{\rm ord}$ and $B_{\rm tor}$ where the local B-field is ordered and provides a particular direction (transverse to the local B-field direction and to the propagation direction of the photon) along which the polarization vector aligns. Another way to break the symmetry is by having the LOS close to the edge of the jet, with $\theta_j - \Gamma^{-1} \lesssim \theta_{\rm obs} \lesssim \theta_j + \Gamma^{-1} \Leftrightarrow \Gamma|\theta_{\rm obs} - \theta_j| \lesssim 1$ ($1 - \xi_j^{-1/2} \lesssim q \lesssim 1 + \xi_j^{-1/2} \Leftrightarrow \xi_j^{1/2}|q - 1| \lesssim 1$), so that some part of the beaming cone lies outside of the jet surface. The missing emission, which would otherwise contribute towards cancellation, leads to only partial cancellation and yields a net finite polarization, $|\Pi| > 0$. The sign of net polarization is decided by whichever region, either blue–white or orange–yellow, makes the dominant contribution to the polarized flux. In Figure 6, $\Pi < 0$ for both $B_{\rm ord}$ and $B_{\rm tor}$, whereas $\Pi \approx 0$ for $B_\perp$.

Pulse-integrated polarization as a function of q is shown in Figure 7 for different B-field configurations and different ξ_j, where the latter describes how wide or narrow the jet aperture is compared to the beaming cone. The polarization curves look very different for the three different field configurations, but there are some features that are worth pointing out. The polarization vanishes when the observer is looking down the jet axis, i.e., when $\theta_{\rm obs} = 0$ ($q = 0$), in all cases due to complete cancellation (such a cancellation would not occur for $B_{\rm ord}$, which is not shown in Figure 7). For $q > 0$, polarization grows rapidly for $B_{\rm tor}$ (for which it saturates at $\xi_j^{-1/2} \lesssim q \lesssim 1 - \xi_j^{-1/2} \Leftrightarrow \Gamma^{-1} \lesssim \theta_{\rm obs} \lesssim \theta_j - \Gamma^{-1}$) but slowly for both $B_\perp$ and $B_\parallel$. It reaches a local maxima when the LOS is close to the jet edge, i.e., as before, when $|q-1| \lesssim \xi_j^{-1/2} \Leftrightarrow \Gamma|\theta_{\rm obs} - \theta_j| \lesssim 1$, and declines sharply for $B_\perp$ and $B_{\rm tor}$ when the LOS exceeds one beaming cone outside of the jet, i.e., $\theta_{\rm obs} \gtrsim \theta_j + \Gamma^{-1}$ ($q > 1 + \xi_j^{-1/2}$). The $B_\parallel$ case yields a different behavior where Π becomes maximal when the jet is viewed from outside its edge. In all cases, when $q > 1 + \xi_j^{-1/2} \Leftrightarrow \Gamma(\theta_{\rm obs} - \theta_j) > 1$ the fluence drops off very sharply for a top-hat jet. So, even though a large Π is expected for $B_\parallel$, it will be challenging to detect. Finally, a change in the PA by 90° occurs when $\theta_{\rm obs} \approx \theta_j$ ($q \approx 1$) for $B_\perp$ and $B_\parallel$, at which point $\Pi = 0$.

It is clear from Figure 7 that only the $B_{\rm tor}$ case, an ordered field scenario, yields high levels of polarization when the LOS passes within the aperture of the jet. Since all distant GRBs must be viewed with $q < 1$, otherwise they will be too dim to detect, a measurement of $50\% \lesssim \Pi \lesssim 65\%$ will strongly indicate the presence of an ordered field component. On the other hand, if the B-field configuration is more like $B_\perp$ or $B_\parallel$, then most GRBs will show negligible polarization.

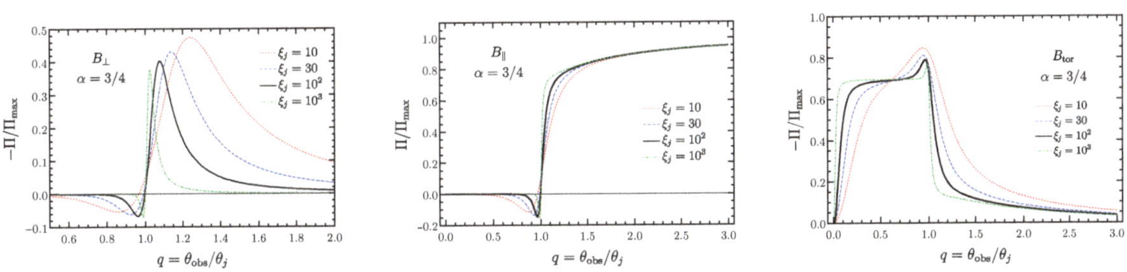

Figure 7. Pulse-integrated polarization of synchrotron emission for different B-field configurations shown for different LOSs (q) and size of the beaming cone w.r.t to the jet aperture ($\xi_j = (\Gamma\theta_j)^2$). The spectral index was fixed to $\alpha = 3/4$, where a larger α produced a larger Π. Figure adapted from [24] but originally produced in [108].

4.1.2. Photospheric Emission from a Uniform Jet

A photospheric spectral component can arise and even dominate the spectral peak in scenarios where energy is dissipated below the photosphere. At the photosphere, radiation decouples from matter and is able to stream freely towards the observer. However, in a matter-dominated flow in which the baryon rest mass energy density, $\rho' c^2$, is much larger than that of the radiation field, U'_γ, the radiation field becomes highly anisotropic at the photosphere [55]. At the last scattering surface, this produces significant *local* polarization at each point of the observed part of the flow. Nevertheless, upon averaging over the GRB image the net polarization is expected to be negligible in an axisymmetric uniform flow since there is no preferred direction for the polarization vector. To obtain finite net polarization, an inhomogeneous outflow with gradients in bulk-Γ (and to a lesser extent in comoving emissivity $L'_{\nu'}$) across the beaming cone are needed. This scenario is discussed in Section 4.2.

Alternatively, if the flow is radiation-dominated, i.e., $U'_\gamma \gg \rho' c^2$, as shown by Beloborodov [55], the comoving angular distribution of the radiation field is preserved in the ultrarelativistic limit as the flow goes from being optically thick to thin. This occurs due to the fact that radiation always tries to push the plasma to an equilibrium Lorentz frame in which the radiative force on the plasma vanishes. As a result, the radiation field accelerates the plasma to a bulk LF $\Gamma(R) \propto R$, which is a special Lorentz frame in which the (comoving) direction of freely streaming photons w.r.t the local radial direction remains unchanged in between successive scatterings. This means that an isotropic radiation field remains isotropic. Since scattering an isotropic radiation field only produces another isotropic field, the flow behaves (to leading order) as if no scatterings took place. Since the radiation field was necessarily isotropic when the flow was optically thick at smaller radii, leading to zero local polarization, it must yield the same (to leading order) when it becomes optically thin, as shown below.

The radiation field is able to accelerate the flow to $\Gamma(R) \propto R$ only if $U'_\gamma/\rho' c^2 \gg 1$ and while the matter maintains a small lag, $\Delta\Gamma = \Gamma_{\rm rad} - \Gamma_{\rm matter}$, which is initially $\ll \Gamma_{\rm matter} \equiv \Gamma \approx \Gamma_{\rm rad}$ (where these LFs are of the respective local center of momentum frames) corresponding to a relative velocity $\beta_{\gamma{\rm m}} \sim \Delta\Gamma/\Gamma \sim \rho' c^2/U'_\gamma \ll 1$, but it gradually increases until it eventually becomes comparable to the two near the saturation radius R_s where $U'_\gamma \tau_T \sim \rho' c^2$ (for $\tau_T < 1$), the point beyond which matter stops accelerating and starts coasting, while the scaling $\Gamma_{\rm rad} \propto R$ remain valid as the radiation-free streams in increasingly more radial directions. In a steady radiation-dominated spherical flow the comoving radiation energy density scales as $U'_\gamma \propto V'^{-4/3} \propto [R^2\Gamma(R)]^{-4/3}$, and the rest mass energy density of the particles scale as $\rho' \propto n'_e \propto V'^{-1} \propto [R^2\Gamma(R)]^{-1}$, where V' is the comoving volume. This yields $U'_\gamma/\rho' c^2 \propto [R^2\Gamma(R)]^{-1/3}$, which for $\Gamma(R) \propto R$ gives $U'_\gamma/\rho' c^2 \propto R^{-1}$ and $\tau_T = n'_e \sigma_T R/\Gamma(R) \propto [R\Gamma^2(R)]^{-1} \propto R^{-3}$. This further yields $U'_\gamma \tau_T/\rho' c^2 = (U'_{\gamma,{\rm ph}}/\rho'_{\rm ph} c^2)(R/R_{\rm ph})^{-4}$ so that $R_s \sim R_{\rm ph}(U'_{\gamma,{\rm ph}}/\rho'_{\rm ph} c^2)^{1/4} \sim R_{\rm ph}\beta_{\gamma{\rm m,ph}}^{-1/4}$ and $\beta_{\gamma{\rm m}} \sim \min[1, (R/R_s)^4]$. Near R_s the comoving radiation anisotropy becomes significant ($\beta_{\gamma{\rm m}} \sim 1$) and therefore so does the polarization of the radiation scattered at $R \sim R_s$, but this is only a fraction $\sim \tau_T(R_s) \sim (U'_{\gamma,{\rm ph}}/\rho'_{\rm ph} c^2)^{-3/4} \sim \beta_{\gamma{\rm m,ph}}^{3/4} \ll 1$ of the photons, and therefore the overall local (i.e., from a particular fluid element) polarization is of the same order, i.e., very small.

4.1.3. Compton Drag

Inverse-Compton scattering of anisotropic radiation yields high levels of polarization for the scattered radiation field with $\Pi \leq 100\%$. This is very different from Comptonization since the polarization vector of the scattered photon can now be aligned with a particular direction, which is transverse to the plane containing the wave vectors, $\vec{k}''_1$ and $\vec{k}''_2$, of the incoming and scattered photons, respectively, in the rest frame of the electron (hence the double primes). If the scattering angle is $\theta''_{\rm sc} = \arccos(\vec{k}''_1 \cdot \vec{k}''_2)$, then Thomson scattering of radiation imparts local polarization

$$\bar{\Pi} = \frac{1-\cos^2\theta''_{\rm sc}}{1+\cos^2\theta''_{\rm sc}} \xrightarrow[\text{electrons}]{\text{cold}} \frac{1-\cos^2\theta'_{\rm sc}}{1+\cos^2\theta'_{\rm sc}} \xrightarrow[\text{flow}]{\text{radial}} \frac{1-\cos^2\tilde{\theta}'}{1+\cos^2\tilde{\theta}'} \tag{12}$$

to the outgoing photon. Indeed, if $\theta''_{\rm sc} = \pi/2$, then $\bar{\Pi} = 100\%$. Here it was assumed that the electrons are cold and therefore their rest frame is the fluid frame ($\theta''_{\rm sc} = \theta'_{\rm sc}$) that is moving with velocity $\vec{v}$, and if it is moving everywhere in the radial direction ($\hat{v} = \hat{r}$), then $\theta'_{\rm sc} = \tilde{\theta}'$. In general, the local polarization depends on the angle θ'_0 between the wave vector of the incoming photon and the velocity vector of the electron. If the electrons have a finite internal energy density, which means that they have a velocity distribution, then the local polarization is obtained by performing a weighted integral over all—see θ'_0 [224] for details.

The expected polarization when assuming cold electrons in the comoving frame of an ultrarelativistic top-hat jet is shown in Figure 8. The polarization curves are very similar to that obtained for synchrotron emission for the $B_\perp$ field configuration, but for Compton drag the normalization is (nearly exactly) higher by $\Pi_{\rm max}^{-1}(\alpha)$ as given by Equation (10). Similar results were first obtained by Lazzati et al. [134] for narrower jets with $\xi_j \leq 25$

where they showed that when $\xi = 0.04$ very high polarization with $\Pi \lesssim 95\%$ can be obtained with Compton drag.

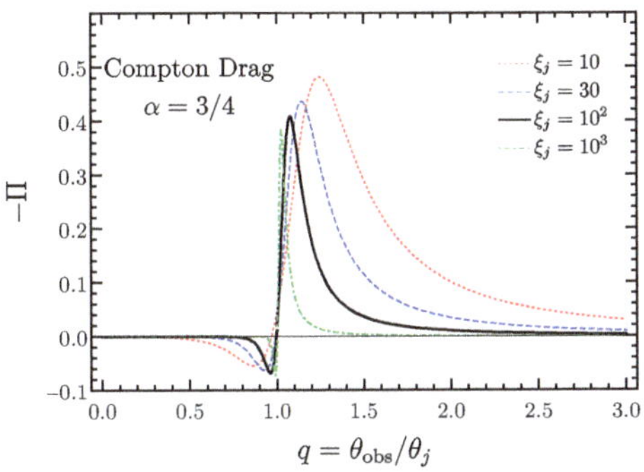

Figure 8. Pulse-integrated polarization of prompt GRB radiation generated by the Compton drag mechanism. The electrons were assumed to be cold in the comoving frame. Figure adapted from [24], but also see [134] for results for a narrower top-hat jet.

4.2. Polarization from Structured Jets

The angular structure of the relativistic jet in GRBs becomes particularly important for relatively nearby events, e.g., GRB170817A ($D \simeq 40\,$Mpc), which can be detected with the current cadre of instruments when the observer is relatively far off-axis and the emission is dim. For distant GRBs, as mentioned earlier, it is challenging to detect emission from significantly off-axis jets. Still, there will be some events in which the LOS is just outside the quasi-uniform core that may not be sharp, as found otherwise in a top-hat jet, but instead be smoother. Then, it becomes important to model the angular structure and compare polarization measurements with accurate theoretical models.

The first level of correction for an idealized top-hat jet model is the consideration of smooth wings of comoving spectral luminosity while the bulk-Γ remains uniform [225]. Like the top-hat jet, $L'_{\nu'} = L'_{\nu',0}$ for $\xi \le \xi_j$ ($\theta \le \theta_j$), but outside of this uniform core the spectral luminosity can have either exponential or power-law wings:

$$\frac{L'_{\nu'}}{L'_{\nu',0}} = \begin{cases} \exp[(\sqrt{\xi_j} - \sqrt{\xi})/\Delta], & \xi > \xi_j \quad \text{(exponential wings)} \\ \left(\frac{\xi}{\xi_j}\right)^{-\delta/2}, & \xi > \xi_j \quad \text{(power-law wings)}\,. \end{cases} \tag{13}$$

Here again it is assumed that Γ, θ_j, $\theta_{\rm obs}$, and the spectrum do not have any radial dependence.

In a more realistic structured jet the core is no longer uniform. Instead, the spectral luminosity as well as the bulk-Γ depend on polar angle θ. In general, the properties of the flow can also depend on the azimuthal angle ϕ, but here the discussion makes the simplifying and physically reasonable assumption of axisymmetric jets. Two different types of structured jets are considered here:

$$\frac{L'_{\nu'}(\theta)}{L'_{\nu',0}} = \frac{\Gamma(\theta)-1}{\Gamma_c - 1} = \exp\left(-\frac{\theta^2}{2\theta_c^2}\right) \qquad \text{(Gaussian Jet)} \tag{14}$$

$$\frac{L'_{\nu'}(\theta)}{L'_{\nu',0}} = \Theta^{-a}, \quad \frac{\Gamma(\theta)-1}{\Gamma_c - 1} = \Theta^{-b}, \quad \Theta = \sqrt{1+\left(\frac{\theta}{\theta_c}\right)^2} \qquad \text{(Power-Law Jet)} \tag{15}$$

Here, $L'_{\nu',0}$ and Γ_c are the core spectral luminosity and bulk-Γ at $\theta = 0$.

4.2.1. Synchrotron Emission from Structured Jets

The polarization curves for a smooth top-hat jet are presented in Figure 9 for different B-field configurations as well as for different levels of smoothness of the edges. The behavior is similar for $\theta_{\rm obs} < \theta_j$, but significant differences between the top-hat jet case appear for $\theta_{\rm obs} > \theta_j$. Now that the spectral luminosity does not fall off so sharply for off-axis observers in the latter case, there is always some emission beamed along the LOS. For B-field configurations that show a larger degree of symmetry of the direction of polarization vectors around the LOS (e.g., $B_\perp$ and $B_\parallel$), the net polarization starts to decline as the edges of the jet are made smoother. This occurs due to the increase in symmetry that was broken sharply in the top-hat jet. A completely opposite behavior is seen in ordered B-field configurations, where the polarization increases with increasing smoothness. This arises since for a very sharp edge the observed flux is dominated by the core and once most of it has a similar weight (i.e., beaming and Doppler factor) then a significant amount of canceling occurs, while for a very smooth or gradual edge the flux is dominated by the region near the line of sight where the B-field is ordered, resulting in very little averaging out of the polarization.

The right column of Figure 9 shows the polarization curves for structured jets. When compared with polarization curves from top-hat jets or even smooth top-hat jets, these are broadly similar. Note that the $\delta = 2$ smooth top-hat jet (left panel of Figure 9) is broadly similar in structure to ($a = 2$, $b = 0$) structured jet (right panel), where both show similar polarization behavior, and therefore a $\delta = 2$ smooth top-hat jet can also be considered a structured jet. In all cases, the curves are now stretched towards larger viewing angles. This means that appreciable polarization can now be measured when the LOS falls outside of the brighter core. In addition to that, the drop in fluence for viewing angles outside of the core is not so severe, as was found for the top-hat jet. Therefore, depending on the exact angular profile, off-axis observers with $q = \theta_{\rm obs}/\theta_c \lesssim$ few to several can still detect the GRB and measure high levels of polarization. This is demonstrated in Figure 9 using a dotted line where the solid to dotted line transition occurs when the off-axis ($\theta_{\rm obs} > 0$) to on-axis ($\theta_{\rm obs} = 0$) fluence ratio has dropped to 1%. Nevertheless, there are additional constraints on the detectability of such off-axis bursts. For example, when the bulk-Γ is non-uniform and declines with θ, the viewing angle out to which the prompt emission can be observed may be limited by compactness e.g., [24,145,226]. This is shown using a thick dot in the figure beyond which the Thomson optical depth of the $e^\pm$-pairs (τ_T) produced due to $\gamma\gamma$-annihilation becomes greater than 10. As a result, the polarization is rather limited to $\Pi \lesssim 20\%$ for $B_\perp$ and $B_\parallel$, but it can be much higher for the ordered field in $B_{\rm tor}$.

4.2.2. Photospheric Emission from Structured Jets

Photospheric emission yields negligible polarization in a uniform jet unless the viewing angle is less than one beaming cone away from the edge of the jet, i.e., $|q-1| \lesssim \xi_j^{-1/2} \Leftrightarrow \Gamma|\theta_{\rm obs} - \theta_j| \lesssim 1$. One way to obtain finite net polarization is by having a structured jet (see Figure 10). This was initially demonstrated in Monte Carlo (MC) simulations of photospheric emission emerging from axisymmetric relativistic outflows [56,57] that featured sheared layers outside of the uniform core with gradients in bulk-Γ as a function of the polar angle θ. It was shown that narrow jets with $\Gamma\theta_c \approx 1$ and steep gradients in bulk-Γ with $\Gamma(\theta) \propto \theta^{-p}$ for $\theta > \theta_c$ (some works use the symbol θ_j instead of θ_c to refer to the half-opening angle of the uniform core) and $p \sim 4$ can yield polarization $\Pi \lesssim 40\%$ for $q = \theta_{\rm obs}/\theta_c \gtrsim 1$. A more realistic scenario would have $\Gamma\theta_c \approx 10$ in which case $\Pi \lesssim 10\%$ is expected. A similar conclusion is reached by carrying out a radial integration of the radiation transfer equations for the Stokes parameters in a steady flow having angular structure in the comoving emissivity and bulk-Γ [24]. The results of this work are shown in the bottom-left panel of Figure 10, and even here it was realized that steep gradients in the bulk-Γ profile are required to achieve significant polarization with $\Pi \lesssim 15\%$.

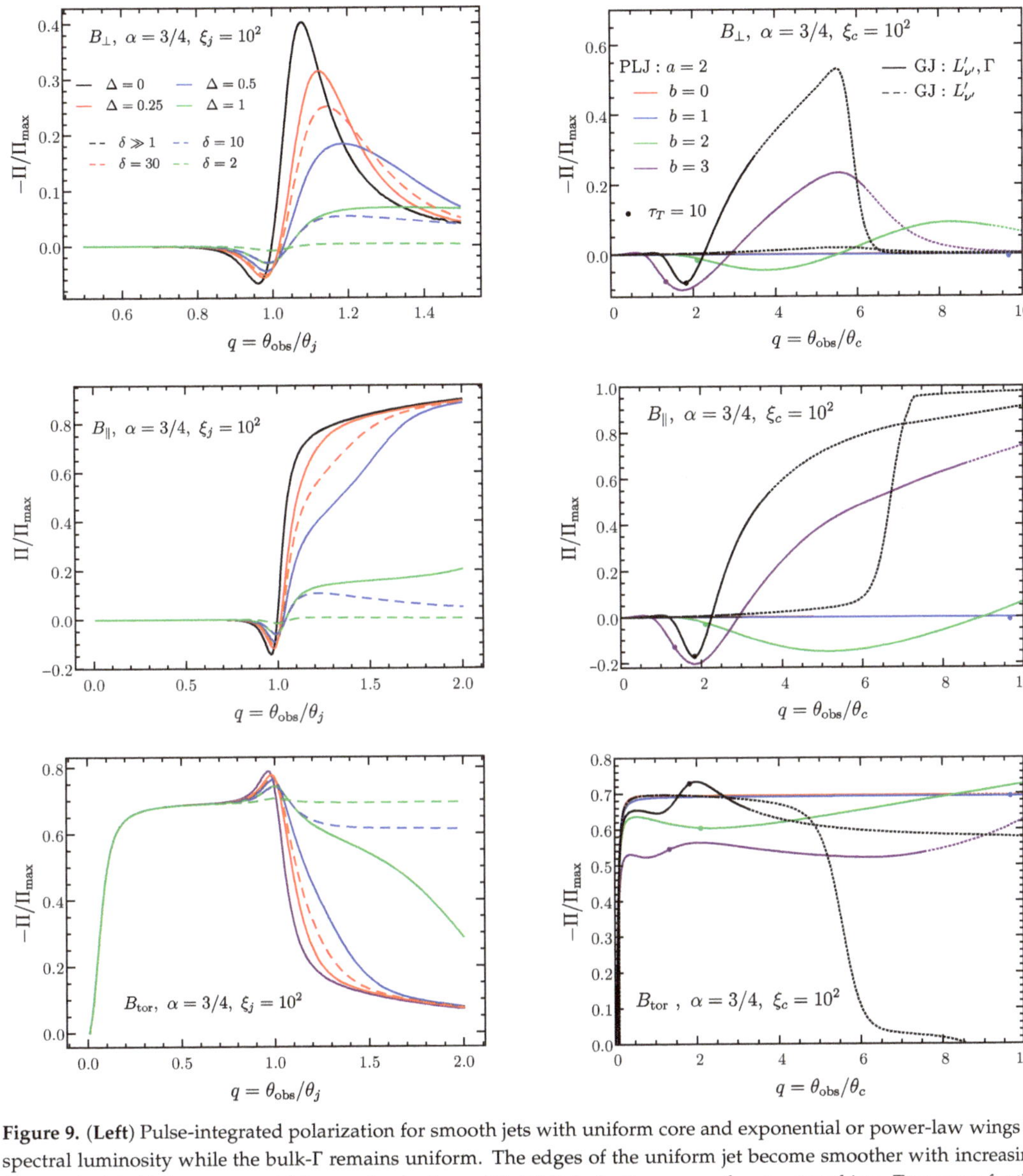

Figure 9. (**Left**) Pulse-integrated polarization for smooth jets with uniform core and exponential or power-law wings in spectral luminosity while the bulk-Γ remains uniform. The edges of the uniform jet become smoother with increasing (decreasing) Δ (δ) for exponential (power law) wings. (**Right**) Polarization curves for structured jets. Two cases for the Gaussian jet (GJ) are shown, where in one both $L'_{\nu'}$ and Γ vary with θ and in the other Γ is kept uniform. For the power-law jet (PLJ), the power-law index for $L'_{\nu'}$ is fixed ($a=2$), but that for the bulk-Γ (b) is varied. The curve for $b=0$ is mostly overlapped by that of $b=1$. The dotted lines show the polarization curves for viewing angles at which the fluence has declined to values smaller than 1% of that expected at $\theta_{obs}=0$. The thick dots mark critical viewing angles beyond which the emission region becomes too compact to $\gamma\gamma$-annihilation, causing the emission to be optically thick to Thomson scattering of the produced $e^{\pm}$-pairs. Figure adapted from [24] and some results for the smoothed top-hat jets were first presented in [225].

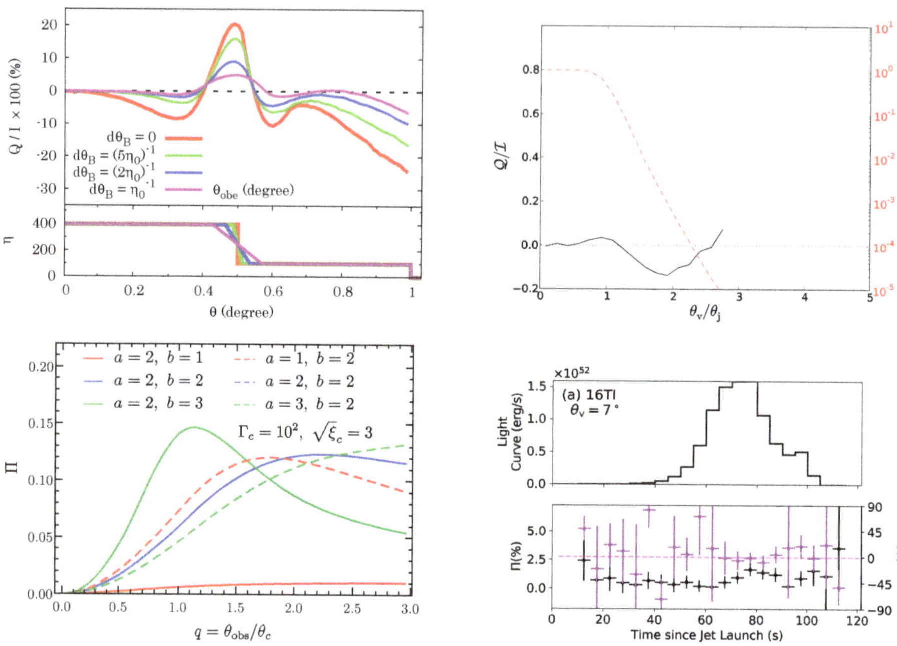

Figure 10. Polarization from non-dissipative photospheric emission model in a structured jet. (**Top-left**) Polarization from the Monte Carlo (MC) simulation of Ito et al. [56] shown for different viewing angles $\theta_{\rm obs}$ and different gradients in bulk-Γ (here η). (**Top-right**) MC simulation results from Lundman et al. [57] featuring a uniform core with half-opening angle θ_j and power-law shear ($\Gamma(\theta) \propto \theta^{-4}$) layer in bulk-Γ. The off-axis spectral luminosity normalized by the on-axis value (viewing angle $\theta_v = 0$) is shown with dashed red line. (**Bottom-left**) Polarization of photospheric emission from a structured jet obtained from semi-analytic radiation transfer calculation of Gill et al. [177] that features angular structure in both the comoving emissivity ($L'_{\nu'}(\theta) \propto \Theta^{-a}$, see Equation (10)) and bulk-Γ ($\Gamma(\theta) \propto \Theta^{-b}$) with $\sqrt{\xi_c} = \Gamma_c\theta_c = 3$ where θ_c is the core angle. The solid lines fix $a = 2$ and dotted lines set $b = 2$ to disentangle the effect of the two profiles. (**Bottom-right**) Polarization derived from a MC simulation with outflow properties obtained from a 2D special relativistic hydrodynamic simulation of a jet launched inside a Wolf-Rayet star (from Parsotan et al. [58]). The top-panel shows the lightcurve and the bottom panel shows the temporal evolution of Π and position angle χ.

A more realistic scenario was explored in Parsotan et al. [58] who carried out two-dimensional (2D) special relativistic hydrodynamic simulations of a jet launched inside a Wolf Rayet star. The flow dynamics and angular structure thus obtained from the simulation were then used with a MC code to obtain the polarization of photospheric emission at the last scattering surface. The results are shown in the bottom-right panel of Figure 10 that shows the lightcurve and temporal evolution of the polarization and PA, with the conclusion that $\Pi \lesssim 2.5\%$ and PA remained steady within the uncertainties. In other cases, where the outflow showed more structure, a slightly larger time-resolved polarization of $\Pi \lesssim 5\%$ and time-variable PA was obtained.

4.3. Temporal Evolution of Polarization

The earlier sections only discuss the pulse-integrated polarization, which is relevant for most GRBs that are not bright enough to be able to yield any time-resolved polarimetric results. However, with the upcoming more sensitive gamma-ray polarimeters in the next decade time-resolved polarimetry of prompt GRB emission will become possible. Therefore, in anticipation of such a development, it is prudent to also construct accurate theoretical model predictions to compare with time-resolved polarization measurements.

When discussing time-resolved polarization it becomes important to include the radial dependence of the flow properties, which were ignored for the pulse-integrated discussion. We first describe a simple and very general pulse model of an accelerating, coasting, or decelerating flow (see, e.g., [227,228]), which is then used to calculate the time-resolved polarization. Consider a thin ultra-relativistic shell that starts to emit prompt GRB photons at radius $R = R_0$. The emission continues over a radial extent ΔR and terminates at $R_f = R_0 + \Delta R$. During this time, the comoving spectrum, with $\nu' L'_{\nu'}$ spectral peak frequency $\nu'_{\rm pk}$, and spectral luminosity evolve as a power law with radius,

$$L'_{\nu'}(R,\theta) = L'_0\left(\frac{R}{R_0}\right)^a S\left(\frac{\nu'}{\nu'_{\rm pk}}\right)f(\theta) \qquad \text{with} \qquad \nu'_{\rm pk} = \nu'_0\left(\frac{R}{R_0}\right)^d, \tag{16}$$

where $L'_0 = L'_{\nu'}(R_0)$ and $\nu'_0 = \nu'_{\rm pk}(R_0)$ are the normalizations. The factor $f(\theta)$ describes the angular profile of $L'_{\nu'}$ where it is normalized to unity at the jet-symmetry axis with $f(0) = 1$, for a uniform spherical flow $f(\theta) = 1$ and for a top-hat jet $f(\theta) = \mathcal{H}(\theta_j - \theta)$ with $\mathcal{H}$ being the Heaviside function and θ_j the jet half-opening angle. The comoving spectrum is described by the function $S(x)$, which is considered here to be the Band function, where $x = \nu'/\nu'_{\rm pk}$. The dynamics of the thin shell are given by the radial profile of the bulk-Γ, such that $\Gamma^2(R) = \Gamma_0^2(R/R_0)^{-m}$ where $\Gamma_0 = \Gamma(R_0)$. The shell is coasting when $m = 0$ and accelerating (decelerating) for $m < 0$ ($m > 0$). Once the power law indices a and d for $L'_{\nu'}$ are provided, one has complete information of the temporal evolution of the pulse. These indices depend on the details of the underlying prompt GRB model, e.g., on the composition and dissipation mechanism. If the prompt GRB spectrum is assumed to be of synchrotron origin, then it can be shown [84] that for a KED flow, where energy is dissipated at internal shocks ($m = 0$), $a = 1$ and $d = -1$. Alternatively, if the flow is PFD with a striped wind B-field structure and energy is dissipated due to magnetic reconnection, which also accelerates the flow with $m = -2/3$, then it is found that $a = 4/3$ and $d = -2$.

The pulse profile and temporal evolution of polarization for a KED flow coasting at $\Gamma_0 \gg 1$ is shown in Figure 11 for an ordered B-field ($B_{\rm ord}$). The different curves are shown for observed frequency $\nu = x_0\nu_0$, which is a fraction x_0 of the peak frequency $\nu_0 = 2\Gamma_0\nu'_0$ of the first photons emitted along the LOS at radius R_0. The apparent arrival time of these first photons is given by $t_{0,z} \equiv t_0/(1+z) = R_0/2(1+m)\Gamma_0^2 c$, which is the characteristic radial delay time between the shell to arrive at radius R_0 and the hypothetical photon that was emitted by the engine at the same time as the shell. For $m = 0$, this is also the angular time over which radiation from within the beaming cone around the LOS arrives at the observer. Depending on x_0, the pulse profile changes and shows a peak at different times with the latest peak occurring at $\tilde{t}_f \equiv t_f/t_0 = \hat{R}_f^{1+m} = (R_f/R_0)^{1+m} = (1 + \Delta R/R_0)^{1+m}$, the arrival time of last photons emitted along the LOS from radius R_f. At $\tilde{t} > \tilde{t}_f$, the flux density declines rapidly, and the pulse becomes dominated by high-latitude emission that originates from outside of the beaming cone, i.e., from angles larger than $1/\Gamma_0$ from the LOS.

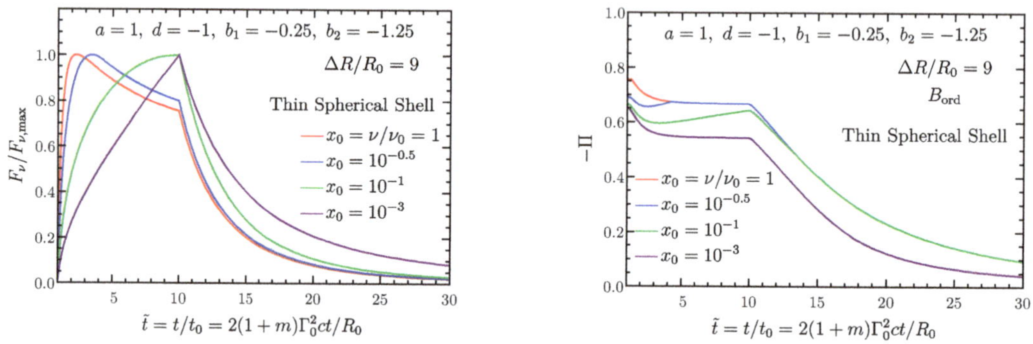

Figure 11. Pulse profile (**left**) and temporal evolution of polarization (**right**) for a coasting ($m = 0$) ultrarelativistic ($\Gamma_0 \gg 1$) thin spherical shell with an ordered field ($B_{\rm ord}$). Here, energy is dissipated in internal shocks in a KED flow and the emission is synchrotron, which is modeled using a Band function with asymptotic spectral indices b_1 and b_2. The shell starts to radiate at $R = R_0$ and terminates at radius $R_f = R_0(1 + \Delta R/R_0)$. The comoving spectral luminosity and spectral peak evolve as a power law in radius with indices a and d, respectively (see Equation (16)). The different curves show the trend at the observed frequency $\nu = x_0\nu_0$ where ν_0 is the νF_ν-peak frequency of the first photons emitted along the observer's LOS from radius R_0, which then arrive at the apparent time $t = t_0$. The emission is assumed to have a Band function spectrum with asymptotic power-law spectral indices b_1 and b_2 below and above the spectral peak energy, respectively. Figure adapted from [84].

The polarization curves show maximal polarization initially, corresponding to $\Pi_{\max}(\alpha)$ depending on the local value of the spectral index α for the Band function as set by x_0. For $\tilde{t} < \tilde{t}_f$, the polarization first declines and then saturates, which reflects the averaging of local polarization over the beaming cone as seen on the plane of the sky, which tends to yield a net polarization lower than $\Pi_{\max}$. For $\tilde{t} > \tilde{t}_f$, like the pulse profile, the polarization also declines rapidly when high-latitude emission becomes dominant. The polarization curves at different x_0 merge at $\tilde{t} = \tilde{t}_{\rm cross}(x_0)$, the crossing time of the break frequency across the observed frequency as the entire spectrum drifts towards softer energies over time. The merging of the polarization curves occurs due to the fact that after time $\tilde{t}_{\rm cross}$ all photons at the observed frequency ν are harder than the Band-function break frequency beyond which the Band function features a strict power law with a given spectral index. Therefore, the level of polarization for all photons sampling the power law is also the same as dictated by $\Pi_{\max}(\alpha)$.

The polarization is not always maximal at the start of the pulse if the magnetic field is not ordered. This is demonstrated in Figure 12 that shows the pulse profile and temporal evolution of synchrotron polarization for different B-field configurations in a top-hat jet. As argued earlier, in B-field configurations, e.g., $B_\perp$ and $B_\parallel$, that produce axisymmetric polarization maps around the LOS the net polarization vanishes. This symmetry is only broken when the observer becomes aware of the jet edge, e.g., in a top-hat jet. It is at that instant the magnitude of polarization begins to grow above zero. The polarization curves for the three B-field configurations also show a change in the PA by $\Delta\theta_p = 90^\circ$ when the curves cross zero. Interestingly, this happens more than once for $B_{\rm tor}$. The reason for this can be understood from the polarization maps shown in Figure 6 where the 90° change in the PA occurs when the net polarization begins to be dominated by emission polarized along the line connecting the jet symmetry axis and the observer's LOS over that polarized in the transverse direction or vice versa. At late times, the observed emission vanishes after the arrival time of the last photons from the edge of the jet furthest from the LOS. Since the flux declines very rapidly at $\tilde{t} > \tilde{t}_f$, the changes in the PA are challenging to detect in practice.

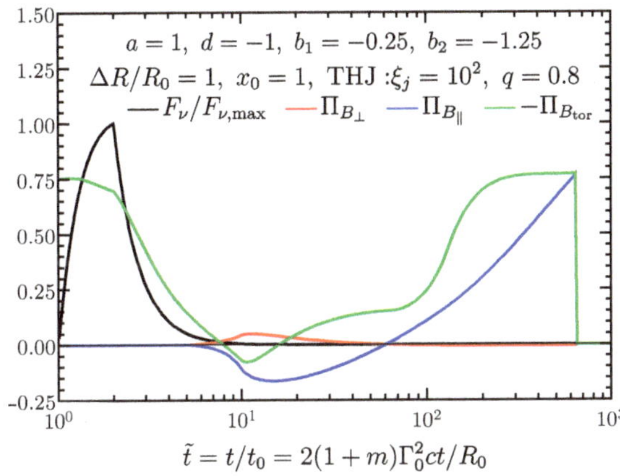

Figure 12. Pulse profile (black) and temporal evolution of synchrotron polarization in a top-hat jet (THJ, with $\xi_j = (\Gamma\theta_j)^2$ and $q = \theta_{\rm obs}/\theta_j$) for different B-field configurations. See caption of Figure 11 for explanation of different symbols and parameters. Figure adapted from [84].

4.4. *Polarization from Multiple Overlapping Pulses*

Since GRBs are generally photon-starved, the only hope of obtaining a statistically significant polarization measurement often relies on integrating over broad segments of the prompt GRB lightcurve. Due to the highly variable nature of the prompt GRB emission, a given emission episode consists of multiple overlapping pulses. The properties of the emission region, e.g., bulk-Γ, B-field configuration, can change between different pulses and improper accounting of these changes in calculating the time-integrated polarization can lead to erroneous results.

In the simplest scenario, multiple pulses are produced by distinct patches or mini-jets within the observed region of size R/Γ of the outflow surface. These patches can be permeated by an ordered B-field the orientation of which is also mutually distinct among the different patches. A broadly similar B-field structure can also be obtained in both internal and external shocks due to macroscopic turbulence excited by, e.g., the Richtmyer–Meshkov instability, which arises in the interaction of shocks and upstream density inhomogeneities [68–70,229]. In the case of mini-jets, the bulk-Γ of the different jets can also be different by a factor of order unity, which will affect the size of the individual beaming cones. Since the Stokes parameters are additive for incoherent emission the time-integrated net polarization of N_p incoherent patches (in the visible region of angular size $1/\Gamma$ around the line of sight) is obtained from [63] (where the motivation was afterglow emission from a shock-generated field rather than incoherent patches or mini-jets).

$$\Pi = \frac{Q}{I} = \frac{\sum_{i=1}^{N_p} Q_i}{\sum_{i=1}^{N_p} I_i} \sim \frac{\Pi_{\max}}{\sqrt{N_p}}. \tag{17}$$

The net polarization is significantly reduced for increasingly large numbers of patches due to the fact that the PA are randomly oriented, and when added together some cancellation occurs. This essentially represents a random walk for the polarized intensity Q while the total intensity adds up coherently. When multiple time-integrated segments of an emission episode are compared, the net polarization and PA will vary between them (the latter is possible as this is a non-axisymmetric global configuration). Alternatively, instead of ordered B-field patches, one can have a shock-produced B-field (e.g., $B_\perp$) with a patchy shell or mini-jets that give different weights to different parts of the image and thereby produce a net polarization (see, e.g., [109,230]).

Another scenario that is worth considering is when multiple overlapping pulses are produced by episodic energization of the emission region, e.g., in the collision of multiple shells in the internal shock scenario where the ejection time of subsequent shells is different, such that the ejection time of the ith shell in the engine frame is $t_{\rm ej,i,z} = t_{\rm ej,i}/(1+z)$. The onset time of each pulse is then given by $t_{\rm onset,i,z} = t_{\rm ej,i,z} + t_{0,z}$. The scenario of multiple pulses from a smooth top-hat jet is demonstrated in Figure 13 using simplifying assumptions, where all pulses have the same R_0 and $\Gamma(R_0)$ (so that the radial delay time $t_{0,z}$ for emission arising from different pulses is the same) and radial extent ΔR. In this case, the onset times of pulses is simply dictated by the different ejection times of the shells. The left panel shows the pulse profile, and the right panel shows the polarization calculated for the $B_{\rm tor}$ field. Time-resolved polarization obtained from multiple temporal segments, where the emission episode is divided into one, two, or three equal duration segments, is shown to demonstrate the different levels of polarization obtained when using the multi-pulse or the single-pulse model. Therefore, when the emission consists of multiple overlapping pulses, it is important to compare the measurement with model predictions that account for multiple pulses.

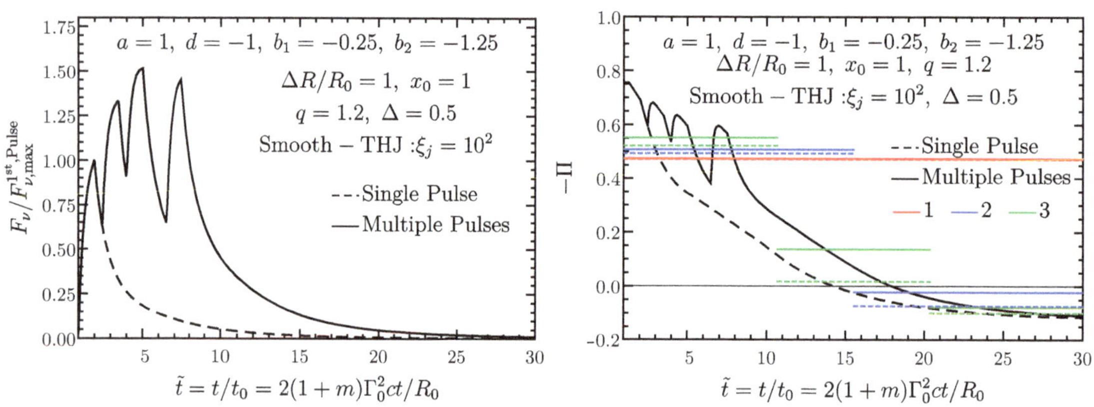

Figure 13. (**Left**) Pulse profile of multiple overlapping pulses in an emission episode, shown here for a KED smooth top-hat jet. A single pulse is also shown for comparison. (**Right**) Temporal evolution of the polarization for a toroidal magnetic field ($B_{\rm tor}$) shown for both the single pulse and multiple pulses. Temporal segments over which polarization is obtained are calculated by dividing the pulse into one (red), two (blue), or three (green) part(s). See the caption of Figure 11 for explanation of different symbols. Figure adapted from [84].

4.5. Most Likely Polarization Measurement

As demonstrated in earlier sections, the prompt GRB polarization depends on (i) the underlying radiation mechanism, (ii) B-field structure (for synchrotron emission), (iii) bulk LF Γ (top-hat jet) or Γ_c (structured jet), (iv) θ_j (top-hat jet) or θ_c (structured jet), (v) viewing angle $\theta_{\rm obs}$, and (vi) angular structure, e.g., power-law indices a and b for a power-law structured jet (see Section 4.2). Due to variations in these parameters the polarization can vary between different pulses within the same GRB as well as between different GRBs. For an ultrarelativistic flow, three basic quantities naturally arise that affect the polarization, namely, (a) the normalized jet/core half-opening angle: $\xi_j^{1/2} = \Gamma\theta_j$ (top-hat jet) or $\xi_c^{1/2} = \Gamma_c\theta_c$ (structure jet), (b) the normalized viewing angle: $q = \theta_{\rm obs}/\theta_j$ (top-hat jet) or $q = \theta_{\rm obs}/\theta_c$ (structured jet), and (c) the normalized viewing-angle dependent fluence: $\tilde{f}_{\rm iso}(q,\xi_j) = E_{\gamma,\rm iso}(q,\xi_j)/E_{\gamma,\rm iso}(0,\xi_j)$ (top-hat jet) or $\tilde{f}_{\rm iso}(q,\xi_c)$ (structure jet), which is the ratio of the off-axis to on-axis isotropic-equivalent radiated energy or equivalently the fluence.

For different pulses emitted by the same GRB, it is natural to expect a considerable change in (iii), while the other parameters are likely to remain more or less fixed. In, e.g., a top-hat jet, this will change the parameter ξ_j, and, for a given distribution of ξ_j between several pulses, the total polarization, after integrating over multiple pulses, will be different from that obtained for a single pulse. When adding up the Stokes parameters of different pulses, an appropriate relative weight using, e.g., $E_{\gamma,\rm iso}$ (or more precisely the relative expected number of photons that will be detected), should be applied.

When comparing emission from different GRBs all of the above-mentioned quantities can in principle vary (or at least there is no strong evidence against this in the observed sample of GRBs). In this case, the fluence ratio is important in determining (i) whether for a given $\theta_{\rm obs} > \theta_j$ (top-hat jet) or $\theta_{\rm obs} > \theta_c$ (structured jet) the pulse will be bright enough to be observed by a given detector and (ii) for a given GRB out to which viewing angle it will be fluent enough for performing polarization measurements. For a top-hat jet, the fluence is strongly suppressed due to Doppler de-beaming when $\Gamma(\theta_{\rm obs} - \theta_j) \gtrsim 1$, whereas, for a structured jet, the suppression in fluence is not as severe and emission from $q \lesssim$ few to several can be detected if it is not suppressed due to compactness, as discussed earlier.

A distribution of polarization for a given radiation mechanism, while accounting for variations in the aforementioned quantities between different pulses from the same source and different GRBs, and its comparison with actual measurements can be used to answer some of the key questions of GRB physics. Such a distribution obtained from a Monte Carlo simulation (see [24] for more details) is shown in Figure 14 for a power-law jet and for different radiation mechanisms as well as different B-field configurations. As expected, the $B_{\rm tor}$ field being ordered yields the highest polarization with $45\% \lesssim \Pi \lesssim 60\%$. Therefore, if GRB jets feature a large-scale toroidal field, then most GRBs that are emitting synchrotron radiation will show $\Pi \sim 50\%$. For the other two B-field configurations, $B_\perp$ and $B_\parallel$, the expected polarization is small with $\Pi \lesssim 10\%$, and one is most likely to find GRBs with negligible polarization. The same conclusion can be drawn for the Compton drag and photospheric radiation mechanisms. The polarization in the photospheric emission model can be $\Pi \lesssim 15\%$ when the flow features a much steeper bulk-Γ angular profile with $\sqrt{\xi_c} = \Gamma_c\theta_c \sim$ few (see Figure 14 of [24] for more details). When comparing with observations, some of which have at least 3σ detection significance, no firm conclusions can be drawn at this point. Measurements made by IKAROS-GAP and AstroSat-CZTI find highly polarized GRBs with $\Pi \gtrsim 50\%$, although with large 1σ error bars. On the other hand, the POLAR data appear to indicate that GRBs are more likely to have significantly smaller polarization with most of their sample consistent with unpolarized sources. The apparent discord between the results of these works not only highlights the challenges involved in obtaining a statistically significant polarization measurement but also calls for the need to build more sensitive detectors.

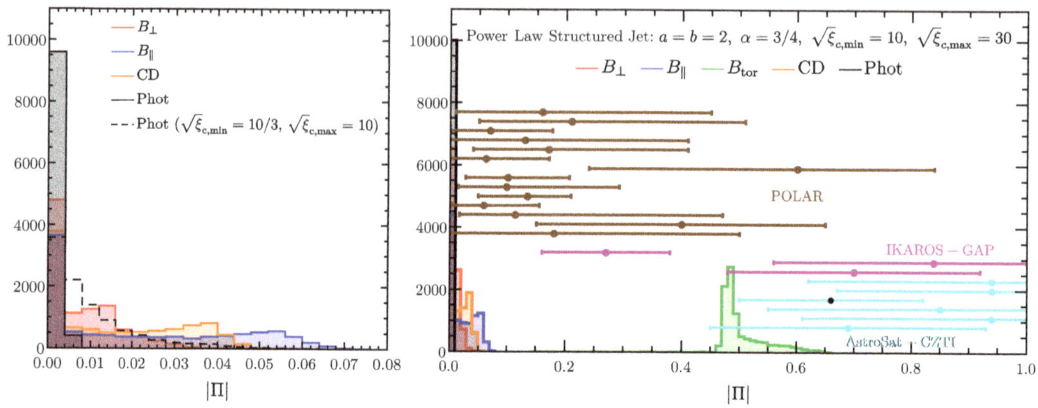

Figure 14. (**Right**): Distribution of polarization from synchrotron emission for different B-field configurations, Compton drag (CD), as well as photospheric (Phot) emission in a power-law-structured jet obtained from a Monte Carlo simulation (with 10^4 samples). Measured polarizations with 1σ error bars from different instruments are shown for comparison. The measurement of $\Pi = 66^{+26}_{-27}\%$ ($\sim 5.3\sigma$) from [231] obtained using AstroSat-CZTI is shown with a black dot with cyan error bars. Figure adapted from [24] where more details can be found. (**Left**): Zoomed-in version of the figure showing the several overlapping distributions for clarity (with a bin size smaller by a factor of 0.4).

4.6. *Energy Dependence of Polarization*

Polarization is energy dependent. This can be easily seen in emission mechanisms where the local polarization depends on the spectral index, e.g., in optically thin synchrotron radiation (see Equation (10)). The energy-dependent spectro-polarimetric evolution in this case is shown in the left panel of Figure 15; temporal evolution of polarization at a given energy and the pulse profile for the same case was shown earlier in Figure 12. The polarization is sensitive to the local spectral index, which, for a Band-like spectrum, changes near the spectral peak and asymptotes far away from it.

Energy-dependent polarization is possible also in emission mechanisms where the local polarization is independent of energy, such as Compton drag in the Thomson regime (where the energy-independent Equation (12) holds). A featureless power-law spectrum will have no energy dependence, but the energy-independent polarization would still depend on the spectral power-law index, $\Pi = \Pi(\alpha)$. This occurs since different α-values give different weights to different parts of the image between which the Doppler factor varies such that the same observed frequency corresponds to different comoving frequencies. For a non-featureless spectrum, the same effect can cause energy dependence in the polarization, e.g., for a Band spectrum, the relative weights of different parts of the image (and therefore also the polarization) will depend on the initial location of the observed frequency relative to the peak frequency along the LOS (i.e., on $x_0 = \nu/\nu_0$).

Alternatively, if multiple spectral components from different radiation mechanisms having different levels of polarization contribute to the observed spectrum, the polarization of the total spectrum will change with energy. This is expected in some photospheric emission models [232] that posit that the spectral peak is dominated by the quasi-thermal photospheric component while the low and/or high energy wings may come from synchrotron emission (see, e.g., Figure 2 and discussion in Section 2.3.3). The right panel of Figure 15 presents such a case, where the polarization grows with decreasing energy owing to the dominance of flux by the synchrotron component. Near the spectral peak, the polarization vanishes. In this way, energy-resolved polarization measurements can be invaluable in understanding the GRB radiation mechanism.

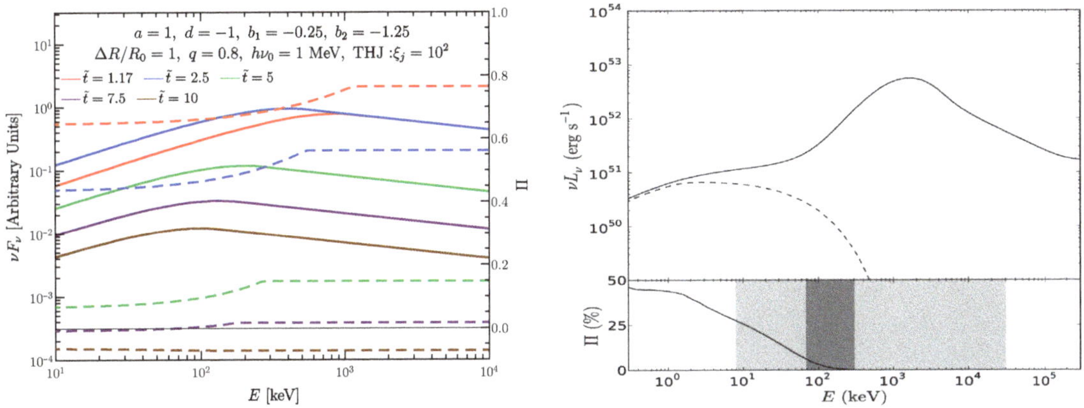

Figure 15. (**Left**): Temporal evolution of the Band-like spectrum (*solid lines*; left y-axis) and the corresponding polarization (*dashed lines*; right y-axis) from synchrotron emission with a $B_{\rm tor}$ field for a KED top-hat jet (THJ) with $\xi_j = (\Gamma\theta_j)^2 = 10^2$ and $q = \theta_{\rm obs}/\theta_j = 0.8$, and $m = 0$. The different colours correspond to different normalized apparent times $\tilde{t} = t/t_0$ where $t_0 = 2(1+m)\Gamma_0^2 ct/R_0$ is the arrival time of the initial photons emitted from radius R_0 along the LOS. The peak frequency of the νF_ν spectrum at this time is given by ν_0. (**Right**): Multi-component GRB spectrum and its energy-resolved polarization. While the photospheric component dominates both spectral peak and at higher energies, the low-energy spectrum is produced by synchrotron emission. As a result, the polarization grows towards lower energies as the fraction of synchrotron photon grows. The light and dark shaded regions correspond to the energy ranges of *Fermi* GBM (NaI + BGO detectors, (8–30 MeV), and GAP (70–300 keV), respectively. Figure from [232].

5. Observations

5.1. Time-Integrated Polarization Measurements

To date, the γ-ray polarization of a total of 31 GRBs has been published. For several GRBs, different analyses have been published, either by different groups using the same data or, in one case, using data from two different instruments. The time- and energy-integrated polarization parameters from these measurements are shown in Table 1, together with the energy range in which they were performed. It is important to note that the energy ranges mentioned here are those stated in the respective publications but that their definitions differ between experiments. The energy ranges stated by SPI for example come from an event selection based on the deposited energy, whereas, for POLAR, which cannot perform measurements of the incoming photon energy directly, the stated range is based on the energy-dependent effective area to polarization.

As can be seen from Table 1, especially for the earliest measurements, at the bottom of the table, the results indicate typically high levels of polarization, although, as explained earlier, this can in some cases be attributed to an error in the analysis. Additionally, publications of GRB polarization measurements have focused on those measurements for which a non-zero PD was found. At least several GRB measurements exist, such as some detected by GAP, for which the PD was found to be compatible with 0%; however, these were not published but only presented at conferences (https://ttt.astro.su.se/groups/head/cost14/talks/Yonetoku.pdf talk accessed on 25 August 2014). This causes an additional bias towards higher PD values found in the list.

Table 1. The list of all GRBs for which a measurement has been published to date. † For GRB 160821A, several analyses were published by members of the AstroSAT collaboration. For this GRB, a time-resolved analysis found high levels of polarization with varying PA as well.

GRB	Instr./Sat.	Pol. (%)	Energy (keV)	Remark
171010A [233]	AstroSAT/CZT	<42	100–300	Significant systematics in mod. curve
170320A [187]	POLAR	18^{+32}_{-18}	50–500	N.A.
170305A [187]	POLAR	40^{+25}_{-25}	50–500	N.A.
170210A [187]	POLAR	$11.4^{+35.7}_{-9.7}$	50–500	N.A.
170207A [187]	POLAR	$5.9^{+9.6}_{-5.9}$	50–500	N.A.
170206A [187]	POLAR	$13.5^{+7.4}_{-8.6}$	50–500	N.A.
170127C [187]	POLAR	$9.9^{+19.3}_{-8.4}$	50–500	N.A.
170114A [187]	POLAR	$10.1^{+10.5}_{-7.4}$	50–500	PA evolution
170101B [187]	POLAR	60^{+24}_{-36}	50–500	N.A.
170101A [187]	POLAR	$6.3^{+10.8}_{-6.3}$	50–500	Hint of PA evolution
161229A [187]	POLAR	17^{+24}_{-13}	50–500	N.A.
161218B [187]	POLAR	13^{+28}_{-13}	50–500	N.A.
161218A [187]	POLAR	$7.0^{+10.7}_{-7.0}$	50–500	N.A.
161217C [187]	POLAR	21^{+30}_{-16}	50–500	N.A.
161203A [187]	POLAR	16^{+29}_{-15}	50–500	N.A.
160910A [188]	AstroSAT/CZTI	94 ± 32	100–300	N.A.
160821A [231]	AstroSAT/CZTI	21^{+24}_{-19}	100–300	Time interval T0 + 115 to T0 + 155 s †
160821A [188]	AstroSAT/CZTI	54 ± 21	100–300	Time interval T0 + 130 to T0 + 149 s
160802A [188]	AstroSAT/CZTI	85 ± 33	100–300	N.A.
160703A [188]	AstroSAT/CZTI	<55	100–300	Best fitted PD > 80% in contour
160623A [188]	AstroSAT/CZTI	<46	100–300	N.A.
160607A [188]	AstroSAT/CZTI	<77	100–300	Best fitted PD > 60% in contour
160530A [219]	COSI	<46	100–1000	N.A.
160509A [188]	AstroSAT/CZTI	<92	100–300	Best fitted PD > 90% in contour
160325A [188]	AstroSAT/CZTI	59 ± 28	100–300	N.A.
160131A [188]	AstroSAT/CZTI	94 ± 33	100–300	N.A.
160106A [188]	AstroSAT/CZTI	69 ± 24	100–300	N.A.
151006A [188]	AstroSAT/CZTI	<84	100–300	Best fitted PD > 80% in contour
140206A [234]	IBIS/INTEGRAL	≥48	200–400	Not calibrated on ground
110721A [235]	GAP/IKAROS	84^{+16}_{-28}	70–300	N.A.
110301A [235]	GAP/IKAROS	70 ± 22	70–300	N.A.
100826A [236]	GAP/IKAROS	27 ± 11	70–300	Pol. Angle evolution
061112 [237]	SPI/INTEGRAL	<60	100–1000	Not calibrated on ground

Table 1. *Cont.*

GRB	Instr./Sat.	Pol. (%)	Energy (keV)	Remark
061112 [238]	IBIS/INTEGRAL	>60	250–800	Not calibrated on ground
041219A [239]	IBIS/INTEGRAL	$\leq$4 and 43 ± 25	200–800	Separated first and second peak
041219A [215]	SPI/INTEGRAL	99 ± 33	100–350	Potential systematic error
041219A [214]	SPI/INTEGRAL	60 ± 35	100–350	Potential systematic error
021206 [16]	RHESSI	80 ± 20	150–2000	Potential systematic errors
021206 [17]	RHESSI	<100	150–2000	Too low signal to background
021206 [18]	RHESSI	41^{+57}_{-44}	150–2000	Potential systematic error
960924 [208]	BATSE/CGRO	$\geq$50	20–1000	Potential systematic errors
930131 [208]	BATSE/CGRO	$\geq$35	20–1000	Potential systematic errors

In recent years, data from GAP, POLAR, and Astrosat CZTI have significantly increased the number of measurements; however, the measured PD shows a large range between the different instruments. POLAR finds results that are mostly compatible with a low or unpolarized flux, whereas Astrosat CZTI reports high levels of polarization in [188], with best fitting PD for 10 out of the 11 GRBs exceeding 50%. Although in numerical form only an upper limit is provided for some of these GRBs by Astrosat CZTI (which are the numbers reported in Table 1) mthe contour plots for these GRBs in Figure 13 of [188] indicate that high levels of PD are favoured for all. In most cases, the best fitting PD is close to the upper limit. The only exception is 160623A where a best fitting PD of approximately 30% is found. It should be noted though that for GRB 160821A, two separate analyses provided different results for the main emission period. The first from [188] indicates a rather high level of polarization, whereas [231] found a time-integrated PD compatible with a lowly or unpolarized flux. The analysis methods used for both analyses were different, while additionally the selected time intervals differed (a period with low fluence was added in [231]). Although the interval selection is not discussed in detail, in [240] it is mentioned that the intervals used in [188] were optimized to maximize the significance of the PD detection, giving a possible explanation. The same analysis as applied in [188] was applied in [241] for GRB 171010A where an upper limit of 42% was reported.

The overall impression given by the Astrosat CZTI results is that GRBs are rather highly polarized. From the POLAR results this is not the case as no significant PD was detected, and all results are compatible with an unpolarized flux within the 99% confidence interval. The POLAR results favors low polarization degrees, with PD values exceeding 50% excluded by five of the brightest GRBs with a 99% confidence level. The results from GAP show both GRBs with a high level of polarization, as well as those with a low level, while COSI, the last of the four detectors, which was well calibrated on ground, additionally excludes high values of PD.

Despite the significant increase in available measurements, no clear conclusion on the PD of GRBs has emerged. It therefore appears that simply continuing to push for more measurements with the current generation of instruments might not be the best way forward. Rather, detailed studies scrutinizing the different results found by different instruments are an easier and more promising way forward. One way to achieve this, which is discussed later on, is the use of more standardized analyses methods as well as by making the polarization data public for an independent analysis by different groups.

5.2. Time-Resolved Measurements

Time-resolved analysis was performed on a range of different GRBs by different collaborations. POLAR only found hints of an evolving PA for two GRBs in their catalog, GRB 170114A and 170101A [187]. Out of the 14 GRBs studied by POLAR, these are the only two with a single fast-rising exponential decay (FRED)-like structure. Such GRBs are of interest as they are typically considered to originate from a single emission region with no contamination from multiple overlapping pulses that complicate the analysis. For any other GRB in the POLAR catalog, all being multi-pulsed, no signs of an evolving PA were found. It is possible that the PA varies between different overlapping pulses, and integrating over different temporal segments of the emission episode results in an approximately fixed PA and also a lower PD due to cancellations. Alternatively, it is equally possible that the PA does not vary between pulses for many GRBs and that the PD is intrinsically low.

The data of POLAR do not allow to determine the nature of the PA evolution for the two GRBs for which hints of it were found. The data are compatible with both random variations as well as a single 90° change [104]. Finer time binning or higher statistics within the time bins are required to fully resolve this.

Both the IBIS, GAP, and AstroSAT CZTI collaborations have reported an evolution of the PA over more complex GRBs consisting either of multiple separated pulses (100826A, 160325A) or overlapping pulses (041219A, 160821A) [231,236,239,242]. For both 100826A and 160821A, the evolution is reported to be compatible with PA changes of 90°. For 160325A, for which a high PD was found in time-integrated analysis presented in [188], the time-resolved analysis found that the first emission episode showed no or low polarization, whereas the second episode showed a PD above 43% with a $1.5\,\sigma$ confidence level. For 041219A, the evolution in PA during the first emission period could explain the low PD observed with time-integrated analysis.

Neither GAP nor AstroSAT has reported any studies of PA evolution for GRBs with FRED-like pulses. Therefore, similar to the time-integrated polarization, currently existing results do not allow to draw any strong conclusions for PA evolution. This is due to the limited number of measurements, lower precision, as well as the disagreement between results found by different groups.

5.3. Energy-Resolved Measurements

To date, no energy-resolved polarization measurements for GRBs have been performed. This is mainly a result of the low statistical significance found for the existing measurements. Dividing the data into energy bins would further reduce the available statistics and therefore not allow for constraining measurements to be performed. A secondary issue with such measurements is the difficulty in the analysis for many of the polarimeters. Unlike in spectrometers, a significant number of the detected photons in a polarimeter are not fully contained in the detector. After a first Compton scattering interaction, a second scattering interaction can follow, after which the photon escapes. As a result, there is a large uncertainty on the incoming photon energy. The analysis is therefore not as simple as dividing the available polarization events based on the energy they deposited in the detector. Instead, one needs to take into account the energy dispersion and use, for example, forward folding methods using an energy-dependent polarization response. Although possible, such methods have not yet been applied to date. It should be noted that for certain instruments, such as COSI, Compton kinematics can be applied to ensure only fully contained photons are selected in the analysis. This significantly reduces the issue of energy dispersion; however, for a proper handling of the data, an energy-dependent polarization response is still required for energy-dependent polarization measurements.

6. Other Polarization Measurements

So far we have concentrated on the polarization of the prompt GRB emission. While this is indeed the main focus of this review, here we briefly outline some of the main features and prospects of polarization measurements from other phases of GRB emission. Such polarization measurements can be very complimentary to prompt GRB polarimetry and provide vital additional constraints on the jet angular structure, our viewing angle, and the magnetic field structure within the GRB outflow or in the shocked external medium. Some of the polarization measurements from these other GRB emission components are performed in the optical, NIR, and sub-mm or radio bands and are therefore technically less challenging and more reliable. We outline the different relevant emission phases in approximate order of increasing time from the GRB onset.

6.1. X-ray Flares

X-ray flares—flaring and re-brightening behavior in the X-ray emission from GRBs—were discovered by the Neil Gehrels *Swift* space observatory [243] and are detected in about a third of *Swift* GRBs [244–249]. They typically display a characteristic shape with a sharp rise in flux followed by a smoother decay, eventually fading back to the pre-flare flux level, and also show a different spectrum (typically harder) compared to the underlying emission. X-ray flares typically occur at times $10^2\,\mathrm{s} \lesssim t \lesssim 10^5\,\mathrm{s}$ after the GRB onset. Their temporal and spectral properties appear to be a smooth continuation of the prompt GRB emission spikes [247,248,250]. While during the prompt GRB emission the typical width or spectrum of the different spikes typically does not show a clear systematic evolution, the X-ray flares gradually become wider (with FWHM Δt satisfying $\Delta t/t \sim 0.1 - 0.3$), less luminous ($\langle L\rangle \propto t^{-2.7\pm0.1}$), and softer with time t. Their overall properties strongly suggest that X-ray flares have a common origin with the prompt GRB emission and likely share similar dissipation and/or emission mechanisms.

Therefore, studying the polarization properties of X-ray flares may provide new insights both for their origin, as well as on the emission and/or dissipation mechanisms that are common with the prompt emission. There are some theoretical predictions for their polarization properties e.g., [251,252], but there is still much room for more detailed and realistic predictions that could be tested against future observations. Their observed similarities to prompt GRB pulses suggests that many of the models for prompt GRB polarization may be generalized to apply also for X-ray flares. The fact that X-ray flares last up to hours or sometimes even days after the GRB onset allows pointed observations by sensitive instruments, while their softer spectrum makes them prime targets for future pointed X-ray polarimeters such as eXTP with a polarimetry focusing array at 2–10 keV energies e.g., [198,253].

6.2. Reverse Shock Emission

As the GRB outflow sweeps up enough external medium, it is decelerated by a reverse shock, while a strong relativistic forward shock propagates into the external medium powering the long-lived afterglow emission. (If the GRB ejecta are still highly magnetized at the deceleration radius $R_{\rm dec}$, $\sigma(R_{\rm dec}) \gtrsim 1$, this may suppress the reverse shock, making it weak or even completely nonexistent.) Most of the outflow's energy is transferred to the shocked external medium when the reverse shock finishes crossing the ejecta shell at the deceleration radius, $R_{\rm dec}$, corresponding to the deceleration (apparent) time, $t_{\rm dec}$, which therefore signals the peak or onset of the afterglow emission e.g., [254–259]. For the "thick shell" case where the reverse shock is at least mildly relativistic, this time is comparable to the prompt GRB duration, $t_{\rm dec} \sim t_{\rm GRB}$, while, for the "thin shell" case (where the reverse shock gradually transitions from Newtonian to mildly relativistic), $t_{\rm dec} > t_{\rm GRB}$. For frequencies that are above the cooling break frequency ν_c of the reverse shock emission at $t_{\rm dec}$, which may include the optical for a sufficiently large $n_{\rm ext}(R_{\rm dec})$ (e.g., as expected for the stellar wind of a massive star progenitor in long GRBs), once the reverse shock finishes crossing the ejecta shell the emission from the LOS sharply drops and the flux decays rapidly ($\sim t^{-3}$), corresponding to high-latitude emission. Otherwise, for frequencies in the range $\max(\nu_a, \nu_m) < \nu < \nu_c$, where ν_a is the break frequency corresponding to synchrotron self-absorption, a slightly less steep flux decay of about t^{-2} is expected, as the emission is dominated by the material along the line of sight where the shocked electrons cool adiabatically. Therefore, the optical emission typically peaks on a timescale of tens of seconds and then sharply drops—the **optical flash** e.g., [183,260–264]. The radio, however, is typically below the self-absorption frequency ν_a at $t_{\rm dec}$ (while $\nu_m < \nu_a$), and its flux keeps rising until ν_a sweeps past the radio band, roughly after a day or so—the **radio flare** (e.g., [148,150,180,265–267]).

In terms of the polarization properties of the reverse shock emission, it is important to keep in mind the following points:

1. The reverse shock emission comes from the shocked ejecta and therefore provides important information about the magnetic field structure within the GRB outflow.
2. In contrast with the prompt GRB emission where the dominant emission mechanism is uncertain, in the reverse shock radio, the optical emission is almost certainly synchrotron radiation (given its large emission radius and broadband SED).
3. Measuring polarization in the optical or radio is generally more reliable than in gamma-ray or X-ray energies, mainly because it is technically less challenging (despite the rapid response robotic telescopes needed for the optical flash).
4. As the ejecta decelerates by sweeping up the external medium, the lower bulk Lorentz factor Γ implies a larger visible region of angle $\sim 1/\Gamma$ around our LOS, in which the structure of the jet and of the magnetic field in the ejecta can be probed.

The **optical flash** emission typically peaks on a timescale of $\sim$10–100 s, and the ejecta Lorentz factor Γ is only somewhat lower than during the prompt GRB emission with $\Gamma \sim 10^2$–$10^{2.5}$. The ejecta are decelerated by the reverse shock, typically reducing Γ down to $\sim \frac{1}{2}\Gamma_\infty$, where Γ_∞ is its value during the coasting phase (it can be lower than this for a highly relativistic reverse shock). However, the prompt GRB emission in photospheric models can arise from $\Gamma < \Gamma_\infty$, at which point the outflow is still accelerating and has not yet reached Γ_∞. The optical flash is therefore expected to probe a comparable (i.e., only somewhat larger) region of angle $\sim 1/\Gamma \sim 10^{-2.5}$–$10^{-2}$ rad around our line of sight. Nonetheless, optical polarization measurements are more reliable than in gamma rays, and the optical flash is almost certainly synchrotron, which enables a cleaner and more robust inference of the ejecta magnetic field structure within this region.

From the observational perspective, since the optical flash usually has significant temporal overlap with the early optical afterglow emission from the shocked external medium, this requires a detailed modeling of both the total flux and the polarized flux as a function of time from these two distinct emission regions in order to properly disentangle between them and derive stronger and more robust constraints on the underlying properties of the GRB ejecta and its magnetic field structure. Most (but not all, e.g., [268]) of the early optical polarimetric observations relevant for the optical flash were done by the RINGO polarimeters on the Liverpool telescope [269–277]. Combining photometric and polarimetric observations [278], they concluded that their data clearly indicates that all epochs in which significant (linear) polarization was measured were dominated by emission from the reverse shock (while the optical afterglow emission from the forward external shock was sub-dominant). Here are a few examples. In GRBs 101112A and 110205A [275], a polarization of $\Pi = 6^{+3}_{-2}\,\%$ and $13^{+13}_{-9}\,\%$, respectively, were measured at the optical peak time of $T_{\rm dec} \sim 299$ s and ~ 1027 s, respectively, which appeared to be dominated by the reverse shock because of the sharp rise to the peak (as $\sim t^{4.2}$ and $\sim t^{4.6}$, respectively). In both GRBs, $T_{\rm dec} \gg T_{\rm GRB}$, indicating a thin shell (with $T_{\rm GRB} \approx T_{90} \sim 9.2$ s and 249 s, respectively). One of the best examples so far is GRB 120308A [277], in which $\Pi = 28\% \pm 4\%$ was detected at $240\,{\rm s} < t < 323\,{\rm s}$, which gradually decreased down to $\Pi = 16^{+4}_{-5}\%$ at $575\,{\rm s} < t < 827\,{\rm s}$, as the emission gradually transitioned from reverse-shock- to forward-shock-dominated (see *left* panel of Figure 16).

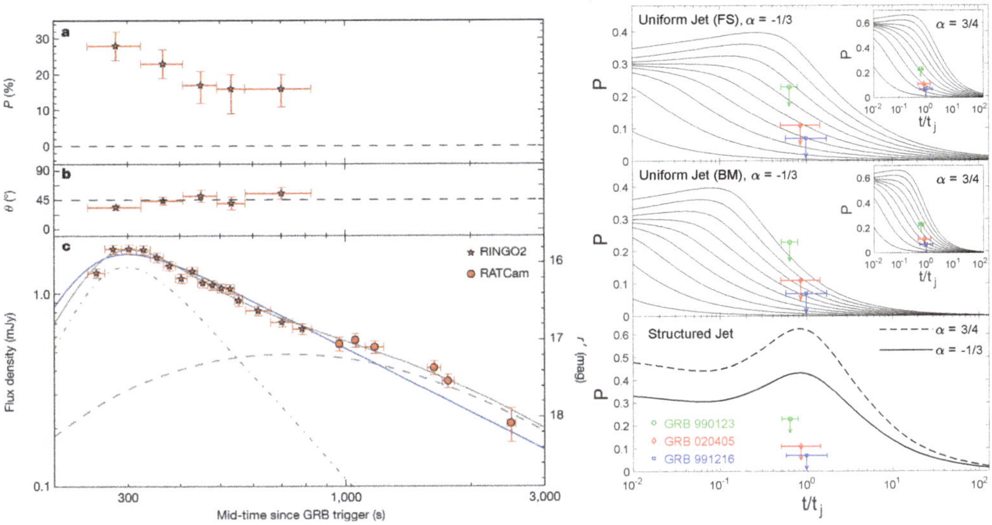

Figure 16. (**Left**): Evolution of optical polarization (degree P (**a**), and position angle θ ((**b**); degrees east of north) and brightness ((**c**) in red (555–690 nm) light using RINGO2 and RATCam, in GRB 120308A (from [277]). (**Right**): $3\,\sigma$ upper limits on the linear polarization of the radio flare emission from three different GRBs overlaid on the theoretical polarization light curves for a toroidal magnetic field in the GRB ejecta (from [111]). The top two panels are for a uniform (top-hat) jet where the different lines, from top to bottom, are for $\theta_{\rm obs}/\theta_j = 0.9, 0.8, ..., 0.1$, while $\alpha = -d\log F_\nu/d\log\nu$ is the spectral index (in the observed radio band) and $\Pi_{\rm max} = (\alpha+1)/(\alpha+5/3)$. In the top panel, the Lorentz factor of the ejecta is assumed to remain equal to that of the freshly shocked fluid just behind the forward shock ("FS"), while in the middle panel it is assumed to follow the Blandford and McKee [279] self-similar solution. The bottom panel is for a "structured" jet, in which the energy per solid angle drops as θ^{-2} outside some small core angle.

The **radio flare** emission, e.g., [148,150,180,265–267], typically peaks on a timescale of a day or so ($\sim 10^5$ s). By this time, the shocked GRB ejecta shell settles in the back of the Blandford and McKee [279] self-similar solution, and its Γ ($\sim$5–10) is smaller by a factor of up to $\sim$1.5–1.8 compared to the material just behind the forward shock that dominates the afterglow emission at the same observed time [111,256]. This corresponds to a visible region of angle $\sim$0.1–0.2 rad around our line of sight, which is significantly larger than during the optical flash. Moreover, it often includes the entire jet (for a simple top-hat jet model) as suggested by the fact that the radio flare peak time is often comparable to the jet break time in the afterglow lightcurve. Granot and Taylor [111] have used VLA data of radio flares from three GRBs (990123, 991216, and 020405) to constrain its polarization, finding only upper limits for both linear and circular polarization. Their best limits are for GRB 991216, for which they found 3σ upper limits on the linear and circular polarization of 7% and 9%, respectively. These limits provide interesting constraints on GRB models and in particular are hard to reconcile with a predominantly ordered toroidal magnetic field in the GRB outflow together with a "structured" jet, where the energy per solid angle drops as the inverse square of the angle from the jet axis (see *right* panel of Figure 16). The polarization of the radio flare may be affected by the location of the observed frequency ν relative to the synchrotron self-absorption break frequency ν_a (polarization is suppressed when $\nu < \nu_a$, during the rising phase of the radio flare) or by Faraday depolarization on the way from the source to us (both are discussed in [111]) and may also be subject to plasma propagation effects within the source (as discussed below, at the end of this section).

Comparing the polarization of the optical flash and radio flare for the same GRB would enable us to study the magnetic field in the GRB ejecta over a wide range of angular scales, probing magnetic structures with a coherence length over this angular range, $10^{-2.5} \lesssim \theta_B \lesssim 10^{-1}$. Measuring the reverse-shock emission polarization at intermediate times and frequencies, such as at sub-mm with ALMA (e.g., [265,267]), would provide a better coverage of this wide range. A particularly interesting example is GRB 190114C, which was also detected at TeV energies [280]. ALMA measured its sub-mm (97.5 GHz) total intensity and linear polarization at 2.2 − 5.2 h after the burst, when the emission was dominated by the reverse shock [267], detecting linear polarization at $\approx 5\sigma$ confidence, decreasing from $\Pi = 0.87\% \pm 0.13\%$ to $\Pi = 0.60\% \pm 0.19\%$, while the position angle evolved from $10^\circ \pm 5^\circ$ to $-44^\circ \pm 12^\circ$. This was the first detection and measurement of the temporal evolution of polarized radio/millimeter emission in a GRB. Using the measured linear polarization, Laskar et al. [267] constrained the coherence scale of tangled magnetic fields in the ejecta to an angular size of $\theta_B \approx 10^{-3}$ rad, while the rotation of the polarization angle rules out the presence of large-scale, ordered axisymmetric magnetic fields and, in particular, a large-scale toroidal field, in the jet.

6.3. *Afterglow Emission*

Linear polarization at the level of a few percent has been detected in the optical or NIR afterglow of about a dozen GRBs [22,281–291]. Higher levels of polarization ($10\% \lesssim \Pi \lesssim 30\%$) have been measured mostly in the very early afterglow, likely being dominated by reverse-shock emission, as discussed above (see, however, [292]). The linear polarization of the afterglow emission was considered as a confirmation that it arises primarily from synchrotron radiation, as was already suggested by its spectral energy distribution.

A variety of **models** have been suggested for GRB afterglow polarization: emission from different patches of uniform but mutually uncorrelated magnetic field, either with microlensing [293] or without it [63], or emission from a random magnetic field within the plane of the afterglow shock together with scintillation in the radio [64] or with a jet viewed not along its symmetry axis [223,294,295], possibly with the addition of an ordered component that pre-exists in the external medium and which is compressed by the afterglow shock and/or a tangled magnetic field that is not purely in the plane of the shock and may even be predominantly in the direction of the shock normal [109,296] or due to clumps in the external medium or a similarly inhomogeneous outflow [109,230].

The most popular models for GRB afterglow polarization feature an axis-symmetric jet viewed not along its symmetry axis along with a tangled shock-produced magnetic field that is symmetric about the local shock normal [109,112,143,222,223,294,295,297–299]. In such models, the only preferred direction on the plane of the sky is that connecting the jet symmetry axis and our LOS, and therefore the net polarization of the unresolved image must lie either along this direction or transverse to it. Indeed, the tell-tale signature of such models for a uniform top-hat jet is a 90° change in the polarization PA θ_p as Π vanishes and reappears rotated by 90°, around the time of the jet break in the afterglow lightcurve [223,294]. On the other hand, for a structured jet viewed from outside of its narrow core, a constant θ_p is expected. Overall, in such models the linear polarization and its temporal evolution depend on: **(i)** the **jet's angular structure**, **(ii)** the local **structure of the shock-generated magnetic field** about the shock normal, and **(iii)** our viewing angle $\theta_{\rm obs}$ from the jet symmetry axis. Therefore, afterglow linear polarization observations can teach us both about the jet's angular structure and about the shock-produced magnetic field structure. However, there is a significant **degeneracy** between the two, which usually requires making large assumptions about one of them in order to significantly constrain the other.

The exceptional case of the short GRB 170817A, which was associated with the first gravitational wave detection of the binary neutron star merger, GW 170817, has allowed us to break this degeneracy. This event was observed from a large off-axis viewing angle, and its low-luminosity prompt gamma-ray emission and subsequent long-lived afterglow emission could be observed thanks to its relatively small distance ($D \approx 40$ Mpc). The combination of an extremely well-monitored afterglow from radio to X-rays e.g., [168–170], and the super-luminal motion of its radio flux centroid ($\langle\beta_{\rm app}\rangle = \langle v_{\rm app}\rangle/c = 4.1 \pm 0.5$

between 75 and 230 days after the burst [300]) has allowed a good determination of our viewing angle and of the jet's angular structure (e.g., [159,160,171–178,301,302]). This has enabled making robust predictions for the linear polarization that depend on the shock-produced magnetic field structure [172]. Shortly thereafter a linear polarization upper limit, $|\Pi| < 12\%$ (99% confidence), was set in the radio (2.8 GHz) at $t = 244$ days [303]. Assuming emission from a two-dimensional surface identified with the afterglow shock front, this has led to a constraint of $0.7 \lesssim b \lesssim 1.5$ on the magnetic field anisotropy parameter, $b \equiv 2\langle B_\parallel^2\rangle/\langle B_\perp^2\rangle$ [172,303], which was introduced by [109], where $B_\parallel$ and $B_\perp$ are the magnetic field compenents parallel and perpendicular to the shock normal direction $\hat{n}_{\rm sh}$, respectively, and $b = 1$ corresponds to an isotropic field in 3D (for which the local and global polarizations vanish). A more detailed analysis [112] accounted for the emission from the whole 3D volume behind the afterglow shock, with the global angular jet structure implied by the GRB 170817A/GW 170817 observation and a local radial hydrodynamic profile set by the Blandford and McKee [279] self-similar solution. The magnetic field was modeled as an isotropic field in 3D that is stretched along $\hat{n}_{\rm sh}$ by a factor $\xi \equiv B_\parallel/B_\perp$, whose initial value $\xi_f = B_{\parallel,f}/B_{\perp,f}$ describes the field that survives downstream on plasma scales $\ll R/\Gamma_{\rm sh}$, and it is evolved downstream according to the [279] solution assuming flux freezing (i.e., no further magnetic dissipation or amplification far downstream of the shock front). In a local coordinate system where $\hat{n}_{\rm sh} = \hat{z}$, in the above definition of b we have $\langle B_\parallel^2\rangle = \langle B_z^2\rangle$ and $\langle B_\perp^2\rangle = \langle B_x^2 + B_y^2\rangle = 2\langle B_x^2\rangle$ due to the B-field's symmetry about $\hat{n}_{\rm sh}$, while here in the definition of ξ, $B_\perp$ represents either B_x or B_y but not $(B_x^2 + B_y^2)^{1/2}$ (while $B_\parallel = B_z$). Gill and Granot [112] found that the shock-produced magnetic field has a finite, but initially sub-dominant, parallel component: $0.57 \lesssim \xi_f \lesssim 0.89$ (see Figure 17).

Circular polarization at the level of $\Pi_{\rm circ} = 0.61\% \pm 0.13\%$ has been reported in the optical afterglow of GRB 121024A [304] at $t = 0.15$ days after the burst, when the linear polarization was $\Pi_{\rm lin} \approx 4\%$, implying a relatively high circular-to-linear polarization ratio of $\Pi_{\rm circ}/\Pi_{\rm lin} \approx 0.15$. Nava et al. [305] performed a detailed analysis of the expected $\Pi_{\rm circ}$ and $\Pi_{\rm lin}$ in GRB afterglows, finding that while ad-hoc configurations may allow large local $\Pi_{\rm circ}$ values, after transformations to the observer frame and integration over the whole visible region are performed, $\Pi_{\rm circ}/\Pi_{\rm lin}$ remains vanishingly small in any realistic optically thin synchrotron afterglow emission model and thus concluding that the origin of the observed $\Pi_{\rm circ}$ in GRB 121024A cannot be intrinsic.

Plasma propagation effects due to the presence of cooler thermal electrons, which are not shock accelerated and represent a fraction $1 - \xi_e$ of the total number, may be important if a significant ordered magnetic field component is present in the emitting region [306–308]. Such effects are most prominent in the early afterglow and around the self-absorption frequency and may therefore potentially affect the reverse shock emission (the "optical flash" or "radio flare"), as well as the forward shock emission in the radio up to a day or so [306–308]. These effects may include Faraday conversion of the linear polarization of the emitted radiation to circular polarization or Faraday depolarization of the emitted linear polarization. For typical GRB afterglow microphysical parameters, the latter effect may strongly suppress the linear polarization in the radio but preserve that in the optical. Therefore, simultaneous observations yielding statistically significant measurements of polarization in both optical and radio can be extremely useful to confirm the population of thermal electrons as well as the existence of an ordered B-field. In some GRBs, this effect may manifest in the sub-mm band where comparison between ALMA and VLA measurements can constrain the value of ξ_e [308]. In fact, Urata et al. [309] argued that the unusually low afterglow polarization ($\Pi = 0.27\% \pm 0.04\%$) of GRB 171205A in the sub-mm band, as compared to the typical late-time optical polarization, may have been the result of Faraday depolarization. Since the true afterglow shock kinetic energy is given by $E' = E/\xi_e$ [310], where E would be the true energy for $\xi_e = 1$, a constraint on ξ_e would lead to better constraints on the burst energetics.

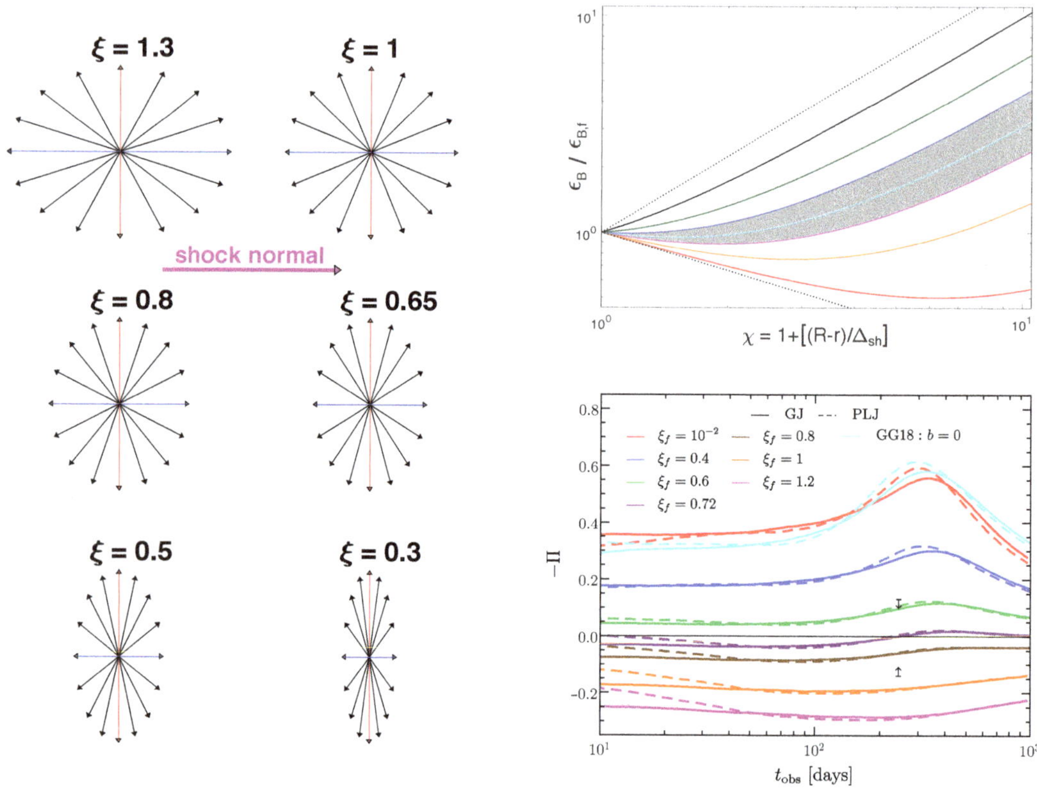

Figure 17. Constraining the magnetic field structure in collisionless relativistic shocks from a radio afterglow linear polarization upper limit in GRB 170817/GW 170817 [112]. (**Left**): Schematic of post-shock magnetic field geometry for different values of the local anisotropy parameter $\xi \equiv B_{\|}/B_{\perp} = \xi_f \chi^{(7-2k)/(8-2k)}$, whose initial value just behind the shock is ξ_f, for an external density profile $\rho_{\rm ext} \propto R^{-k}$, where $\chi = 1 + 2(4-k)\Gamma_{\rm sh}^2(1 - r/R)$ is the Blandford and McKee [279] self-similar variable, r is the radial coordinate, and R and $\Gamma_{\rm sh}$ are the local radius and Lorentz factor of the afterglow shock front, respectively. (**Top Right**): The corresponding evolution of the magnetic field equipartition parameter, ϵ_B, with the distance behind the shock (as parameterized through χ) for $\xi_f = 0, 0.2, 0.4, 0.57, 0.7, 0.89, 1.2, 2, \infty$ (from bottom to top). The two extreme values of $\xi_f = 0, \infty$ are shown as dotted (straight) lines. The light-grey shaded region corresponds to the allowed range found in [112], $0.57 \lesssim \xi_f \lesssim 0.89$. (**Bottom Right**): The linear polarization evolution, $\Pi(t)$, obtained from a volume integration of the flow, shown for different values of ξ_f. The two arrows mark the polarization upper limit, $|\Pi| < 12\%$. Comparison was made between two jet structures–a Gaussian jet (GJ) and a power-law jet (PLJ). The result from [172], which assumed an infinitely thin shell geometry as well as locally isotropic synchrotron spectral emissivity, is also shown (labeled GG18) for the magnetic field anisotropy parameter $b = 0$.

7. Outlook for 2030

The handful of successful γ-ray polarimeters has shown over the previous decade that although challenging, GRB polarization measurements are possible. With this new success, a range of new instruments with not only a higher sensitivity but also a wider energy range are foreseen to be launched over the coming decade. As mentioned earlier, however, simply increasing the number of measurements does not improve our understanding if different instruments provide incompatible results. Below we first discuss the promising advances in detector development for the coming decade. This is followed by a discussion on the need for improvements and standardization of the analysis.

7.1. Future Instruments

7.1.1. POLAR-2 and LEAP

In the Compton energy range of ~10–1000 keV, four instruments are proposed. Both the LEAP [311] and SPHiNX [312] instruments have been proposed for launches in the coming decade, while the POLAR-2 project has already been accepted for launch in 2024 [313]. Additionally, the Daksha mission, a larger-scale full sky monitor follow-up mission based on the Astrosat CZTI is proposed to be launched in the coming decade as well [240]. Out of these four, the POLAR-2 and LEAP projects aim to make the next step in this field by producing instruments with an effective area an order of magnitude larger than the POLAR instrument. The SPHiNX project instead has an effective area similar to that of POLAR and will therefore have to make gains over currently existing measurements by aiming for a longer mission life time. For Daksha, the effective area is planned to be an order of magnitude larger than that of Astrosat CZTI. As the experiment will consist of two satellites, each observing half the sky, this increase in effective area is evenly distributed over the full sky. The design allows for a significant increase in the number of GRBs for which polarization measurements are possible, while also increasing the precision of each such measurement, although not by one full order of magnitude.

The POLAR-2 instrument is similar in design to POLAR with, apart from several minor design improvements, a focus on an improvement in three parts. The first is the size, which is four times larger than POLAR, resulting in a total geometrical area of approximately 2500 cm^2. Secondly, the scintillator readout technology is improved to decrease the low-energy threshold of the instrument from 50 keV to 20 keV, giving a total energy range of 20–800 keV for polarization measurements. Finally, POLAR-2 will be equipped with spectrometers making it independent of other instruments for spectral and location parameters of GRBs, which reduces the systematic error on many GRB measurements. The instrument was approved for launch in early 2024 towards the Chinese Space Station (CSS).

The LEAP instrument is similar to POLAR-2 both in size and in the detection mechanism that uses plastic scintillators. Contrary to POLAR-2, the LEAP instrument will also use high Z scintillators, which increase the absorption cross section. Therefore, the instrument will have a larger sensitivity to polarization and a better spectral response but a reduction in its effective area and field of view. Whereas the total effective area for LEAP that is useful for spectrometry is ~3500 cm^2 at 250 keV, for polarization it is around ~1000 cm^2 [314]. For POLAR-2, the effective area of the polarimeter usable for spectrometry is ~2000 cm^2, and therefore significantly smaller than LEAP. For polarization, however, it is ~1400 cm^2 and therefore larger than LEAP. The reduction in effective area of the polarimeter for spectrometry in POLAR-2 is compensated by separate spectrometers, which will increase this by at least 50%.

The two instruments therefore have different strengths. With a proposed launch in 2025 for LEAP towards the International Space Station (ISS), the combination of both of these instruments in orbit would allow for detailed polarization measurements of the majority of GRBs with fluences (as measured in the 10–1000 keV energy range) above 10^{-6} erg cm^{-2}.

7.1.2. Low-Energy Polarimeters

Apart from adding significant sensitivity in the energy range of ~10–1000 keV, missions are also proposed to perform the first GRB polarization measurements at keV energies.

As previously mentioned, the first polarization measurements at these energies, albeit of point sources, were recently performed by the small scale PolarLite mission [195]. The IXPE mission [197], which uses a similar measurement technology as PolarLite, is planned to be launched in 2021. However, as it is optimized for point sources, it has a narrow FoV. This in combination with a long slewing time makes it unlikely to measure any GRBs. The larger-scale eXTP mission, however, will still be optimized for point sources but is designed to also observe targets of opportunity such as GRBs using a shorter slewing time.

As such, eXTP will be capable of measuring the polarization of the afterglow of GRBs in the 2–10 keV energy range as well as any potential X-ray flares occurring in the afterglow [198]. eXTP is a joint Chinese–European mission and is currently foreseen to be launched in 2028.

The above-mentioned instruments are optimized for point sources and therefore have a small field of view to optimize the signal to noise. In order to measure the polarization of the prompt emission from GRBs, which appears at random positions in the sky, a relatively large field of view is required. A mission under consideration with this capability at keV energies is the Low-energy Polarimetry Detector (LPD) under development at the GuangXi University (Private communication with Prof. Hongbang Liu). The instrument is foreseen to have a sensitivity to polarization in the energy range of 2–30 keV and a maximum effective area of ∼300 cm^2 around 10 keV by using a similar technology as that used by Polarlight with an optimization of the gas for higher energies. The instrument is under consideration to be placed alongside POLAR-2 on the CSS, allowing to perform combined measurements of the prompt emission from 2 keV to 800 keV.

7.1.3. High-Energy Polarimeters

In the MeV energy range, one possible mission to be launched in the coming decade is AMEGO [315]. The AMEGO mission makes use of many layers of silicon placed on top of a calorimeter. This makes it ideal to perform polarization measurements using Compton scattering in the ∼100 keV to 5 MeV energy range. AMEGO will yield polarization measurements for the brightest 1% of GRBs that it will observe.

A second instrument under development is a satellite version of the COSI balloon mission [316]. This instrument will make use of germanium strip detectors capable of measuring the three-dimensional interaction position of incoming photons. The energy range is similar to that of AMEGO (200 keV to 5 MeV). Thanks to its large field of view, it will observe around ∼40 GRBs per year with a fluence exceeding 4×10^{-6} erg cm^{-2} for which it can perform measurements with an MDP of around 50%.

A highly promising instrument concept for polarimetry at MeV energies is the Advanced Particle-astrophysics Telescope [317]. The instrument is designed to maximize the effective area for photons in the MeV to TeV energies without using passive materials for photon conversion. The detector aims to use high Z scintillator crystals for the conversion in combination with scintillating fibres. This allows for a large-scale detector with precise measurements of both electron positron pairs and Compton-scattered photons. The current mission concept would be an order of magnitude more sensitive as a gamma-ray detector than Fermi-LAT and would be capable of performing polarization measurements at MeV energies for GRBs as weak as 170817A, for which an MDP of ∼40% was simulated. The project is in its early stages, and currently a path finder mission is planned for a balloon flight.

Apart from these two instruments the earlier-mentioned HARPO detector [200] will be capable of performing polarization measurements in the MeV energy range using pair production in a gas TPC. Unlike AMEGO and COSI, which are both under consideration for a launch in the coming decade, the HARPO instrument, of which a prototype has been successfully calibrated on ground [201], is currently not under consideration for a launch.

7.2. Performance Predictions

Generally, the coming decade looks promising. In the ∼10–1000 keV energy range a number of new detailed measurements are foreseen, which should be capable of resolving the current differences in PD reported by different groups. This is illustrated in Figure 18, which shows the yearly number of measurements capable of excluding a non-polarized flux as a function of the true polarization degree of GRBs for three different instruments, GAP, POLAR, and POLAR-2. For this figure, the instrument response of POLAR, as used in the POLAR analysis, was used as well as that for POLAR-2 in combination with the Fermi-GBM GRB catalog. For GAP, for which the response is not available, the numbers were produced by scaling the POLAR numbers based on the performance of GAP and

POLAR for respectively detected GRBs, again in combination with the Fermi-GBM GRB catalog. It should be noted that for GAP, for which the detailed response is not known, a fixed M_{100} was used, which, given its design, should be close to the truth.

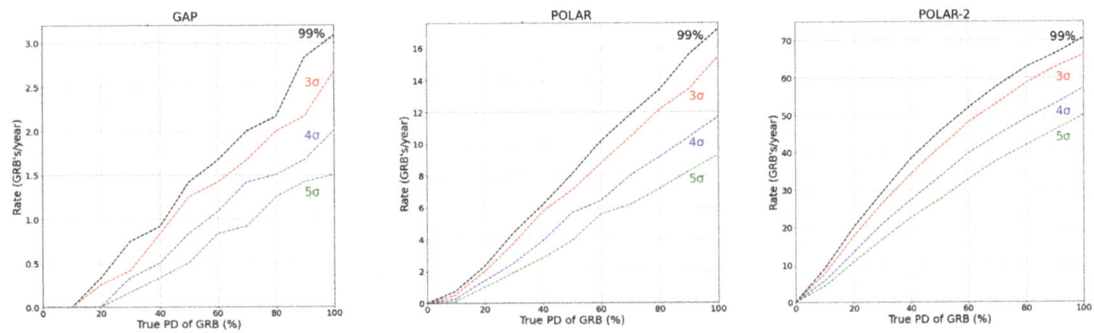

Figure 18. The rate of measurements capable of excluding a non-polarized flux, for different confidence levels, as a function of the true polarization degree (PD) of GRBs for three different instruments, GAP, POLAR, and POLAR-2. Although exact numbers are not available it can be assumed that LEAP will be capable of similar rates as POLAR-2, albeit slightly lower.

It can be seen that with GAP excluding a non-polarized flux was possible for a handful of GRBs per year only in cases where the true PD of the emission is relatively high. For POLAR, the situation improves and, as was the case, with less than a year of data it was able to claim exclusion of polarization levels above ≈50%. It could not, however, effectively probe polarization levels below 30% with a high confidence. With POLAR-2, this region will be probed within a few months, while with 1 year of data it will be capable of determining whether GRB emission is polarized to levels as low as 10%. To illustrate the type of GRBs that can be probed with the different instruments, Figure 19 shows the mean MDP for the three different instruments as a function of the GRB fluence for both short (1 s observed duration) and long (100 s observed duration) GRBs. As an illustration, the fluence of the short and very weak GRB 170817A as well as the long and very bright GRB 190114C are added. It should be noted that the energy ranges used for the different instruments differs, and the energy range of 50–300 keV was used for GAP, 50–500 keV for POLAR and 20–500 keV for POLAR-2. Although no detailed response is available, the performance of LEAP is foreseen to be similar to that of POLAR-2 with a typical effective area ∼30% smaller than that of POLAR-2. A launch of LEAP would therefore further improve the situation, not only regarding the statistics but more importantly regarding the systematics. As for Daksha, not enough details on the instrument are available to make any clear predictions, while the SPHiNX performance would be similar to that of POLAR.

It is evident that the next generation of polarimeters will be capable of almost probing GRBs with fluences as weak as GRB 170817A, a GRB which was hard to even detect with both Fermi-GBM and INTEGRAL-SPI but was important due to its association with a gravitational wave signal [12]. Additionally, for very bright GRBs such as 190114C, which was observed at TeV energies [280,318], highly detailed polarization measurements will become possible, indicating that fine time or energy binning will become an available tool to study such GRBs. It should again be stressed that the mean MDP is simply a figure of merit and that the estimates given here are not exact, as the details will depend not only on the fluence and length of the GRBs but also on its energy spectrum, incoming angle, and position of the polarimeter along its orbit. Additionally, systematic errors, which can be significant, are not taken into account in an MDP calculation. The predictions should therefore be taken only to give an indication of the advancement in the field as well as the possibilities during the coming decade.

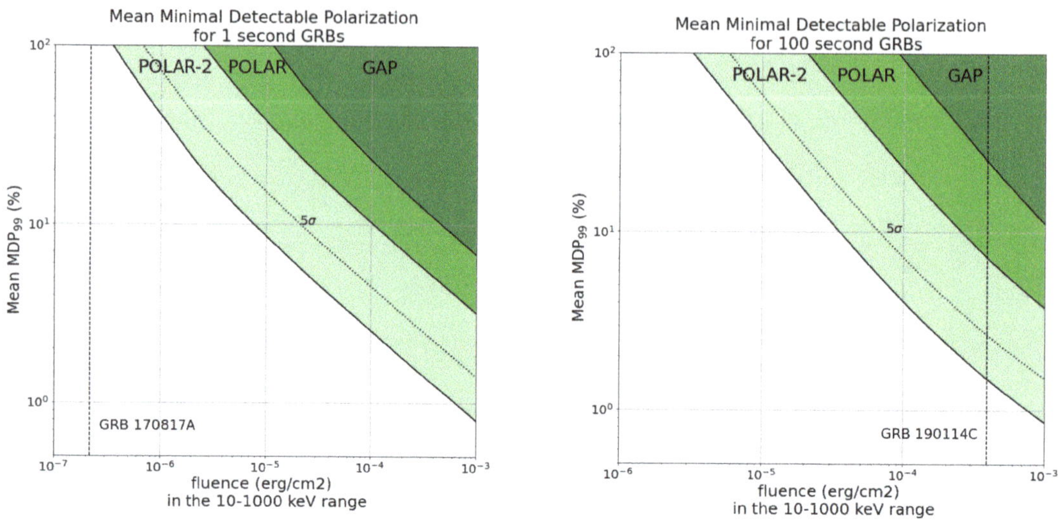

Figure 19. The mean minimal detectable polarization for 99% confidence level (MDP averaged over PA) as a function of the fluence in the 10–1000 keV energy band for GAP, POLAR, and POLAR-2 for both short (1 s observed duration) and long (100 s observed duration) GRBs. Short GRBs with a fluence above 10^{-6} erg/cm^2 occur at a rate of approximately 10 per year on the full sky, whereas, for long GRBs, the rate is about 200 per year. For POLAR-2, the MDP for 5σ confidence is added as well, using a dotted line. The fluences of two well-known GRBs, the weak and short 170817A and the long and bright 190114C, were added as an illustration.

Apart from an improvement in the Compton scattering regime, the first polarization measurements of the prompt emission at MeV energies can be expected towards the end of this decade. There still remains an additional need for energy-dependent polarization measurements. Whereas the eXTP instrument can probe the polarization at keV energies, it is unlikely to detect the prompt emission due to its narrow field of view. An instrument such as the LPD would, especially when placed closed to POLAR-2, allow to provide an energy range of 2–800 keV for many GRBs per year. This would allow to study a potential change in PD in the 10–50 keV energy range, as proposed in some photospheric emission models [232]. In addition, if either COSI or AMEGO will be launched, detailed energy-resolved studies will become possible for bright GRBs in the 2 keV to 5 MeV energy range, thereby fully probing the prompt emission over several orders of magnitude in energy.

7.3. Improvements in Analysis

From the measurement results published to date, it can be seen that increasing the number of measurements alone is likely not enough to provide clear conclusions on the polarization of GRB prompt emission. The clearest example of this is the discrepancy between the results of POLAR and Astrosat CZTI. Out of the 11 GRBs analysed by the Astrosat CZTI collaboration in [188], the six GRBs for which statistically significant measurements (based on the calculation of a Bayes factor required to be above tow) are possible, the polarization levels for all exceeds 50%. For 12 out of the 14 GRB, measurements presented by POLAR the PD were found to be below 25% with the two remaining having a low significance. Although the number of measurements is low, the difference in the results is striking. In order to advance the field, it is prudent to first understand the cause of these differences in these results as well as other earlier published results.

7.3.1. Need for Public Analysis Tools and Data

Polarization analysis is complex, and mistakes can easily lead to high levels of PD being measured. As the field is not yet mature and collaborations are small, every analysis has so far almost exclusively been performed using a tool developed for that specific data. This constant reinvention of the wheel not only allows for mistakes but more importantly results in instrument-specific analysis tools. Such tools are incapable of being applied to other data, and their performance is difficult to verify by a referee or other interested scientists. If, additionally, the code and the data are not public, as is often the case, and publications lack details on the analysis, then it remains nearly impossible to investigate discrepancies with other results.

What is therefore required, arguably more so than more measurements, is a standardized analysis method, which can be adapted to each polarimeter with a public code. Such tools, similar to those widely used in spectrometry such as `Xspec` [319] and `3ML`, would not only allow to understand any potential discrepancies but would also remove the need to reinvent the method by each new collaboration. Furthermore, if additional instrument data and responses exist publicly, it would remove the requirement to have an in-depth understanding of the instrument for being able to perform analysis. This would allow, similar to what happens in spectrometry, for experts in the field of data analysis and statistics to perform the analysis instead of only instrument experts as is now often the case, allowing for more detailed and innovative analyses to be performed.

A first step towards this was produced as part of the 3ML framework [320] for the analysis of the POLAR data. The developed tools aim to provide a framework in which the instrument response and the measurement data are combined to perform the polarization analysis in a transparent way that is usable by anyone. Both for the instrument response and the data format, a standardized format is proposed similar to that used in spectrometry, and the tool can therefore easily be adapted for other polarimeters. The tool has been used first to analyze GRB 170114A [104] in detail using POLAR data, and subsequently to produce the full GRB catalog published by POLAR [187]. The POLAR data used for this analysis is furthermore public https://www.astro.unige.ch/polar/grb-light-curves (accessed on 25 August 2014), allowing further analysis by anyone interested as well as to perform rigorous tests of the validity of the different POLAR results. The public data alone could, for example, already be used by the Astrosat CZTI collaboration using the tools used for the results in [188] to find if their tools provide consistent results are those published in [187]. Although not perfect, such a study would arguably progress the field further than the analysis of additional Astrosat CZTI or POLAR data by the collaborations themselves.

7.3.2. Multi-Instrument Analysis

Thanks to the properties of the 3ML framework, data from different instruments can be combined. So far, this feature was used only to combine the POLAR data with that from *Fermi*-GBM and *Swift*-BAT. This allowed to improve the spectral fits, as the error on the spectrum adds to the systematic error on the polarization measurement, which in turn led to more precise polarization measurements. The 3ML framework additionally allows to fit physical models directly to the data, rather than fitting the data with empirical models and subsequently comparing the results with a parameterized outcome of a theoretical prediction. Although easier, the latter method has, especially in the field of gamma-ray spectrometry, been found to result in over-interpretation of data analysis results and to inconsistent conclusions (see discussion in Section 2.3.1). The fitting of physical models directly to data is especially desirable in the field of polarimetry as it allows to fit these models, potentially unbinned in time and energy, directly both to spectral and polarization data at the same time.

Apart from combining spectral and polarization data in the analysis, in theory, the same can be done using data from two polarimeters in case two different polarimeters observed the same GRB. In fact, several GRBs were observed by both Astrosat and PO-

LAR [187,321]. It would therefore be highly desirable to perform combined analysis of the Astrosat and POLAR data for such GRBs as it would, firstly, allow to study the cause of the likely discrepancy between the results from both instruments. Secondly, it can allow for more detailed measurements of the polarization of these GRBs.

With upcoming instruments sensitive in different energy ranges, such analysis tools will in the future allow to fit physical models to both spectral and polarization data over a broad range in energy by, for example, combining the data of the LPD, LEAP, POLAR-2, and AMEGO or COSI. Whereas with the current level of polarimetry analysis tools, the data has to be studied separately, leaving the full potential of the data unexploited.

As the polarization tool in the 3ML framework discussed here is new and has not been used for the polarization analysis of other instruments, it is to be seen if it will be used by the wider community. However, with the potential of two large-scale polarimeters in LEAP and POLAR-2 launching in the coming years, as well as polarimeters sensitive at keV and at MeV energies, there is a clear need for a collaborative effort between the groups to either further develop this tool or construct a completely new one.

7.4. Improvements in Theoretical Modeling of Prompt GRB Polarization

Pulse-integrated polarization from semi-analytic models of axisymmetric flows with different prompt GRB radiation mechanisms and B-field configurations have been presented in many works [20,24,108–111,134]. The same setup was used to make predictions for the time-dependent polarization for synchrotron emission in some works [84,322,323]. On the other hand, only a few works have attacked the problem using MC simulations [56–58] or radial integration of the transfer equations for the Stokes parameters [55]. Many of these have focused only on photospheric emission.

As the next decade may see the launch of more sensitive instruments to measure GRB polarization with high fidelity, it calls for time- and energy-dependent polarization predictions ($\Pi(E,t)$, $\theta_p(E,t)$) for more realistic outflow models, which would also predict the time-dependent flux density, $F_E(t)$.

One of the weaknesses of current theoretical models is the assumption of an axisymmetric flow, which is usually made for simplicity and convenience. This restricts the change in PA to only $\Delta\theta_p = 90°$, whereas some observations do show, although not so convincingly yet, hints of gradual PA swings. To obtain a change in the PA other than $\Delta\theta_p = 90°$ or to get a gradually changing PA, the condition for axisymmetry must be broken, e.g., the magnetic field configuration/orientation and/or the emissivity can change as a function of (θ, ϕ).

One possibility is that the different pulses that contribute to the emission arise in "mini-jets" within the outflow e.g., [32,85,324–327]. In this case, the different directions of the mini-jets or bright patches w.r.t. the LOS (e.g., [109,230]) would cause the PA to also be different between the pulses even for a field that is locally symmetric w.r.t the local radial direction (e.g., $B_\perp$ or $B_\parallel$) as well as for fields that are axisymmetric w.r.t to the center of each mini-jet (e.g., a local $B_{\rm tor}$ for each mini-jet). Finally, broadly similar results would follow from an ordered field within each mini-jet ($B_{\rm ord}$), which are incoherent between different mini-jets. Time-resolved measurement in such a case would naturally yield a time-varying PA.

Alternatively, as shown by Granot and Königl [109] for GRB afterglow polarization, a combination of an ordered field component (e.g., $B_{\rm ord}$) and a random field, like $B_\perp$, can give rise to a time-varying PA between different pulses (with a different ratio of the two field components) that cab, e.g., arise from internal shocks. The ordered field component here would be that advected from the central enginem and the random field component can be argued to be shock-generated. Notice that the ordered field component should not be axisymmetric in order for the position angle to smoothly vary.

Realistic theoretical predictions can be obtained by coupling radiation transfer modeling with MHD numerical simulations of relativistic jets after they break out of the confining medium. A step towards this direction was taken by Parsotan et al. [58] who used the MHD code `FLASH` to first obtain the jet's angular structure by injecting variable jets into stellar density profiles of Wolf–Rayet stars at core-collapse. They then used an MC code to carry out the radiation transfer of the Stokes parameters and obtain the time-resolved polarization for the photospheric emission (see Figure 10). In another recent work, Ito et al. [328] carried out global neutrino-hydrodynamic simulations of a relativistic jet launched in a binary NS merger scenario. The photospheric emission and polarization from the short GRB was then calculated using a relativistic MC code. While these works focused only on photospheric emission, polarization modeling for other radiation mechanisms performed in the same vein is lacking and can prove to be very fruitful.

MC radiation transfer and MHD numerical simulations of relativistic jets can be computationally expensive. They are nevertheless a useful tool that can be used to calibrate semi-analytic models by delineating the relevant parameter space expected in GRB jets. Ultimately, when high quality observations are made in this decade, fast and computationally inexpensive theoretical models will be required to carry out time-resolved spectro-polarimetric fits in a reasonable amount of time. This further stresses the need for a library of models, akin to `Xspec` [319] that is used routinely for spectral fitting or `boxfit` [329] for GRB afterglow lightcurve modeling, which can be conveniently used by observers. Combining the library of models with the 3ML framework for spectro-polarimetric data analysis will become a very powerful tool for GRB science.

In order to test the different model predictions, e.g., from different radiation mechanisms, on an equal footing, a single underlying theoretical framework should be devised for the jet structure and dynamics, which allows the same freedom in the different model parameters. Such an approach can help to isolate the dominant prompt GRB radiation mechanism when compared with observations.

To conclude, the next decade appears very promising for answering many fundamental questions in GRB physics. With the launch of several dedicated instruments capable of performing high-fidelity γ-ray and X-ray spectro-polarimetry, a larger sample of statistically significant prompt GRB polarization measurements will be obtained. Improvements in polarization data analysis using a single underlying framework that allows simultaneous fitting of both spectrum and polarization from different instruments will yield unbiased and high-quality results. More realistic theoretical models of both time- and energy-dependent polarization based on advanced numerical simulations will allow to better understand the true nature of GRB jets.

Author Contributions: Writing—original draft preparation—review and editing, R.G., M.K. and J.G. All authors have read and agreed to the published version of the manuscript.

Funding: This research was funded in part by the ISF-NSFC joint research program under grant no. 3296/19 (R.G. and J.G.) and the Swiss National Science Foundation (M.K.).

Institutional Review Board Statement: Not Applicable.

Informed Consent Statement: Not Applicable.

Data Availability Statement: No new data were created or analyzed in this study. Data sharing is not applicable to this article.

Acknowledgments: We thank Stefano Covino and Mark McConnell for a thorough read of an earlier version of the manuscript and for their comments and feedback.

Conflicts of Interest: The authors declare no conflict of interest.

References

1. Kouveliotou, C.; Meegan, C.A.; Fishman, G.J.; Bhat, N.P.; Briggs, M.S.; Koshut, T.M.; Paciesas, W.S.; Pendleton, G.N. Identification of Two Classes of Gamma-Ray Bursts. *Astrophys. J.* **1993**, *413*, L101. [CrossRef]
2. Fruchter, A.S.; Levan, A.J.; Strolger, L.; Vreeswijk, P.M.; Thorsett, S.E.; Bersier, D.; Burud, I.; Castro Cerón, J.M.; Castro-Tirado, A.J.; Conselice, C.; et al. Long γ-ray bursts and core-collapse supernovae have different environments. *Nature* **2006**, *441*, 463–468. [CrossRef]

3. Galama, T.J.; Vreeswijk, P.M.; van Paradijs, J.; Kouveliotou, C.; Augusteijn, T.; Böhnhardt, H.; Brewer, J.P.; Doublier, V.; Gonzalez, J.F.; Leibundgut, B.; et al. An unusual supernova in the error box of the γ-ray burst of 25 April 1998. *Nature* **1998**, *395*, 670–672. [CrossRef]
4. Hjorth, J.; Sollerman, J.; Møller, P.; Fynbo, J.P.U.; Woosley, S.E.; Kouveliotou, C.; Tanvir, N.R.; Greiner, J.; Andersen, M.I.; Castro-Tirado, A.J.; et al. A very energetic supernova associated with the γ-ray burst of 29 March 2003. *Nature* **2003**, *423*, 847–850. [CrossRef]
5. Stanek, K.Z.; Matheson, T.; Garnavich, P.M.; Martini, P.; Berlind, P.; Caldwell, N.; Challis, P.; Brown, W.R.; Schild, R.; Krisciunas, K.; et al. Spectroscopic Discovery of the Supernova 2003dh Associated with GRB 030329. *Astrophys. J.* **2003**, *591*, L17–L20. [CrossRef]
6. Gehrels, N.; Sarazin, C.L.; O'Brien, P.T.; Zhang, B.; Barbier, L.; Barthelmy, S.D.; Blustin, A.; Burrows, D.N.; Cannizzo, J.; Cummings, J.R.; et al. A short γ-ray burst apparently associated with an elliptical galaxy at redshift z = 0.225. *Nature* **2005**, *437*, 851–854. [CrossRef]
7. Barthelmy, S.D.; Chincarini, G.; Burrows, D.N.; Gehrels, N.; Covino, S.; Moretti, A.; Romano, P.; O'Brien, P.T.; Sarazin, C.L.; Kouveliotou, C.; et al. An origin for short γ-ray bursts unassociated with current star formation. *Nature* **2005**, *438*, 994–996. [CrossRef]
8. Woosley, S.E. Gamma-Ray Bursts from Stellar Mass Accretion Disks around Black Holes. *Astrophys. J.* **1993**, *405*, 273. [CrossRef]
9. Eichler, D.; Livio, M.; Piran, T.; Schramm, D.N. Nucleosynthesis, neutrino bursts and gamma-rays from coalescing neutron stars. *Nature* **1989**, *340*, 126–128. [CrossRef]
10. Narayan, R.; Paczynski, B.; Piran, T. Gamma-ray bursts as the death throes of massive binary stars. *Astrophys. J.* **1992**, *395*, L83–L86. [CrossRef]
11. Abbott, B.P.; Abbott, R.; Abbott, T.D.; Acernese, F.; Ackley, K.; Adams, C.; Adams, T.; Addesso, P.; Adhikari, R.X.; Adya, V.B.; et al. GW170817: Observation of Gravitational Waves from a Binary Neutron Star Inspiral. *Phys. Rev. Lett.* **2017**, *119*, 161101. [CrossRef]
12. Abbott, B.P.; Abbott, R.; Abbott, T.D.; Acernese, F.; Ackley, K.; Adams, C.; Adams, T.; Addesso, P.; Adhikari, R.X.; Adya, V.B.; et al. Gravitational Waves and Gamma-Rays from a Binary Neutron Star Merger: GW170817 and GRB 170817A. *Astrophys. J.* **2017**, *848*, L13. [CrossRef]
13. Band, D.; Matteson, J.; Ford, L.; Schaefer, B.; Palmer, D.; Teegarden, B.; Cline, T.; Briggs, M.; Paciesas, W.; Pendleton, G.; et al. BATSE Observations of Gamma-Ray Burst Spectra. I. Spectral Diversity. *Astrophys. J.* **1993**, *413*, 281. [CrossRef]
14. Preece, R.D.; Briggs, M.S.; Mallozzi, R.S.; Pendleton, G.N.; Paciesas, W.S.; Band, D.L. The BATSE Gamma-Ray Burst Spectral Catalog. I. High Time Resolution Spectroscopy of Bright Bursts Using High Energy Resolution Data. *Astrophys. J. Suppl. Ser.* **2000**, *126*, 19–36. [CrossRef]
15. Kaneko, Y.; Preece, R.D.; Briggs, M.S.; Paciesas, W.S.; Meegan, C.A.; Band, D.L. The Complete Spectral Catalog of Bright BATSE Gamma-Ray Bursts. *Astrophys. J. Suppl. Ser.* **2006**, *166*, 298–340. [CrossRef]
16. Coburn, W.; Boggs, S.E. Polarization of the prompt γ-ray emission from the γ-ray burst of 6 December 2002. *Nature* **2003**, *423*, 415–417. [CrossRef]
17. Rutledge, R.E.; Fox, D.B. Re-analysis of polarization in the gamma-ray flux of GRB 021206. *Mon. Not. R. Astron. Soc.* **2004**, *350*, 1288–1300. [CrossRef]
18. Wigger, C.; Hajdas, W.; Arzner, K.; Gudel, M.; Zehnder, A. Gamma-ray burst polarization: Limits from rhessi measurements. *Astrophys. J.* **2004**, *613*, 1088–1100. [CrossRef]
19. Lazzati, D. Polarization in the prompt emission of gamma-ray bursts and their afterglows. *New J. Phys.* **2006**, *8*, 131. [CrossRef]
20. Toma, K.; Sakamoto, T.; Zhang, B.; Hill, J.E.; McConnell, M.L.; Bloser, P.F.; Yamazaki, R.; Ioka, K.; Nakamura, T. Statistical Properties of Gamma-Ray Burst Polarization. *Astrophys. J.* **2009**, *698*, 1042–1053. [CrossRef]
21. Toma, K. Polarization of GRB Prompt Emission. *arXiv* **2013**, arXiv:1308.5733.
22. Covino, S.; Gotz, D. Polarization of prompt and afterglow emission of Gamma-Ray Bursts. *Astron. Astrophys. Trans.* **2016**, *29*, 205–244.
23. McConnell, M.L. High energy polarimetry of prompt GRB emission. *New Astron. Rev.* **2017**, *76*, 1–21. [CrossRef]
24. Gill, R.; Granot, J.; Kumar, P. Linear polarization in gamma-ray burst prompt emission. *Mon. Not. R. Astron. Soc.* **2020**, *491*, 3343–3373. [CrossRef]
25. Piran, T. The physics of gamma-ray bursts. *Rev. Mod. Phys.* **2004**, *76*, 1143–1210. [CrossRef]
26. Zhang, B.; Mészáros, P. Gamma-Ray Bursts: Progress, problems & prospects. *Int. J. Mod. Phys. A* **2004**, *19*, 2385–2472. [CrossRef]
27. Mészáros, P. Gamma-ray bursts. *Rep. Prog. Phys.* **2006**, *69*, 2259–2321. [CrossRef]
28. Granot, J.; Ramirez-Ruiz, E. Jets and gamma-ray burst unification schemes. *Gamma-Ray Bursts* **2012**, *51*, 215.
29. Kumar, P.; Zhang, B. The physics of gamma-ray bursts & relativistic jets. *Phys. Rep.* **2015**, *561*, 1–109.
30. Rees, M.J.; Meszaros, P. Unsteady Outflow Models for Cosmological Gamma-Ray Bursts. *Astrophys. J.* **1994**, *430*, L93. [CrossRef]
31. Thompson, C. A Model of Gamma-Ray Bursts. *Mon. Not. R. Astron. Soc.* **1994**, *270*, 480. [CrossRef]
32. Lyutikov, M.; Blandford, R. Gamma Ray Bursts as Electromagnetic Outflows. *arXiv* **2003**, arXiv:astro-ph/0312347.
33. Goodman, J. Are gamma-ray bursts optically thick? *Astrophys. J.* **1986**, *308*, L47. [CrossRef]
34. Paczynski, B. Gamma-ray bursters at cosmological distances. *Astrophys. J.* **1986**, *308*, L43–L46. [CrossRef]

35. Li, Z.Y.; Chiueh, T.; Begelman, M.C. Electromagnetically Driven Relativistic Jets: A Class of Self-similar Solutions. *Astrophys. J.* **1992**, *394*, 459. [CrossRef]
36. Vlahakis, N.; Königl, A. Relativistic Magnetohydrodynamics with Application to Gamma-Ray Burst Outflows. I. Theory and Semianalytic Trans-Alfvénic Solutions. *Astrophys. J.* **2003**, *596*, 1080–1103. [CrossRef]
37. Beskin, V.S.; Nokhrina, E.E. The effective acceleration of plasma outflow in the paraboloidal magnetic field. *Mon. Not. R. Astron. Soc.* **2006**, *367*, 375–386. [CrossRef]
38. Lyubarsky, Y.E. Transformation of the Poynting flux into kinetic energy in relativistic jets. *Mon. Not. R. Astron. Soc.* **2010**, *402*, 353–361. [CrossRef]
39. Komissarov, S.S.; Vlahakis, N.; Königl, A.; Barkov, M.V. Magnetic acceleration of ultrarelativistic jets in gamma-ray burst sources. *Mon. Not. R. Astron. Soc.* **2009**, *394*, 1182–1212. [CrossRef]
40. Goldreich, P.; Julian, W.H. Stellar Winds. *Astrophys. J.* **1970**, *160*, 971. [CrossRef]
41. Lyubarsky, Y. Asymptotic Structure of Poynting-Dominated Jets. *Astrophys. J.* **2009**, *698*, 1570–1589. [CrossRef]
42. Ruderman, M. Theories of gamma-ray bursts. In *Seventh Texas Symposium on Relativistic Astrophysics*; New York Academy of Sciences: New York, NY, USA, 1975, Volume 262, pp. 164–180. [CrossRef]
43. Tchekhovskoy, A.; Narayan, R.; McKinney, J.C. Magnetohydrodynamic simulations of gamma-ray burst jets: Beyond the progenitor star. *New Astron.* **2010**, *15*, 749–754. [CrossRef]
44. Komissarov, S.S.; Vlahakis, N.; Königl, A. Rarefaction acceleration of ultrarelativistic magnetized jets in gamma-ray burst sources. *Mon. Not. R. Astron. Soc.* **2010**, *407*, 17–28. [CrossRef]
45. Granot, J.; Komissarov, S.S.; Spitkovsky, A. Impulsive acceleration of strongly magnetized relativistic flows. *Mon. Not. R. Astron. Soc.* **2011**, *411*, 1323–1353. [CrossRef]
46. Granot, J. The effects of sub-shells in highly magnetized relativistic flows. *Mon. Not. R. Astron. Soc.* **2012**, *421*, 2467–2477. [CrossRef]
47. Lyubarsky, Y.; Kirk, J.G. Reconnection in a Striped Pulsar Wind. *Astrophys. J.* **2001**, *547*, 437–448. [CrossRef]
48. Spruit, H.C.; Daigne, F.; Drenkhahn, G. Large scale magnetic fields and their dissipation in GRB fireballs. *Astron. Astrophys.* **2001**, *369*, 694–705. [CrossRef]
49. Drenkhahn, G. Acceleration of GRB outflows by Poynting flux dissipation. *Astron. Astrophys.* **2002**, *387*, 714–724. [CrossRef]
50. Drenkhahn, G.; Spruit, H.C. Efficient acceleration and radiation in Poynting flux powered GRB outflows. *Astron. Astrophys.* **2002**, *391*, 1141–1153. [CrossRef]
51. Bégué, D.; Pe'er, A.; Lyubarsky, Y. Radiative striped wind model for gamma-ray bursts. *Mon. Not. R. Astron. Soc.* **2017**, *467*, 2594–2611. [CrossRef]
52. Lyutikov, M. The electromagnetic model of gamma-ray bursts. *New J. Phys.* **2006**, *8*, 119. [CrossRef]
53. Pe'er, A. Temporal Evolution of Thermal Emission from Relativistically Expanding Plasma. *Astrophys. J.* **2008**, *682*, 463–473. [CrossRef]
54. Beloborodov, A.M. Collisional mechanism for gamma-ray burst emission. *Mon. Not. R. Astron. Soc.* **2010**, *407*, 1033–1047. [CrossRef]
55. Beloborodov, A.M. Radiative Transfer in Ultrarelativistic Outflows. *Astrophys. J.* **2011**, *737*, 68. [CrossRef]
56. Ito, H.; Nagataki, S.; Matsumoto, J.; Lee, S.H.; Tolstov, A.; Mao, J.; Dainotti, M.; Mizuta, A. Spectral and Polarization Properties of Photospheric Emission from Stratified Jets. *Astrophys. J.* **2014**, *789*, 159. [CrossRef]
57. Lundman, C.; Pe'er, A.; Ryde, F. Polarization properties of photospheric emission from relativistic, collimated outflows. *Mon. Not. R. Astron. Soc.* **2014**, *440*, 3292–3308. [CrossRef]
58. Parsotan, T.; López-Cámara, D.; Lazzati, D. Photospheric Polarization Signatures from Long Gamma-Ray Burst Simulations. *Astrophys. J.* **2020**, *896*, 139. [CrossRef]
59. Papathanassiou, H.; Meszaros, P. Spectra of Unsteady Wind Models of Gamma-Ray Bursts. *Astrophys. J.* **1996**, *471*, L91. [CrossRef]
60. Sari, R.; Piran, T. Variability in Gamma-Ray Bursts: A Clue. *Astrophys. J.* **1997**, *485*, 270–273. [CrossRef]
61. Daigne, F.; Mochkovitch, R. Gamma-ray bursts from internal shocks in a relativistic wind: Temporal and spectral properties. *Mon. Not. R. Astron. Soc.* **1998**, *296*, 275–286. [CrossRef]
62. Weibel, E.S. Spontaneously Growing Transverse Waves in a Plasma Due to an Anisotropic Velocity Distribution. *Phys. Rev. Lett.* **1959**, *2*, 83–84. [CrossRef]
63. Gruzinov, A.; Waxman, E. Gamma-Ray Burst Afterglow: Polarization and Analytic Light Curves. *Astrophys. J.* **1999**, *511*, 852–861. [CrossRef]
64. Medvedev, M.V.; Loeb, A. Generation of Magnetic Fields in the Relativistic Shock of Gamma-Ray Burst Sources. *Astrophys. J.* **1999**, *526*, 697–706. [CrossRef]
65. Bret, A. Weibel, Two-Stream, Filamentation, Oblique, Bell, Buneman...Which One Grows Faster? *Astrophys. J.* **2009**, *699*, 990–1003. [CrossRef]
66. Keshet, U.; Katz, B.; Spitkovsky, A.; Waxman, E. Magnetic Field Evolution in Relativistic Unmagnetized Collisionless Shocks. *Astrophys. J.* **2009**, *693*, L127–L130. [CrossRef]
67. Sironi, L.; Goodman, J. Production of Magnetic Energy by Macroscopic Turbulence in GRB Afterglows. *Astrophys. J.* **2007**, *671*, 1858–1867. [CrossRef]

68. Zhang, W.; MacFadyen, A.; Wang, P. Three-Dimensional Relativistic Magnetohydrodynamic Simulations of the Kelvin-Helmholtz Instability: Magnetic Field Amplification by a Turbulent Dynamo. *Astrophys. J.* **2009**, *692*, L40–L44. [CrossRef]
69. Inoue, T.; Asano, K.; Ioka, K. Three-dimensional Simulations of Magnetohydrodynamic Turbulence Behind Relativistic Shock Waves and Their Implications for Gamma-Ray Bursts. *Astrophys. J.* **2011**, *734*, 77. [CrossRef]
70. Mizuno, Y.; Pohl, M.; Niemiec, J.; Zhang, B.; Nishikawa, K.I.; Hardee, P.E. Magnetic-field Amplification by Turbulence in a Relativistic Shock Propagating Through an Inhomogeneous Medium. *Astrophys. J.* **2011**, *726*, 62. [CrossRef]
71. Mizuno, Y.; Pohl, M.; Niemiec, J.; Zhang, B.; Nishikawa, K.I.; Hardee, P.E. Magnetic field amplification and saturation in turbulence behind a relativistic shock. *Mon. Not. R. Astron. Soc.* **2014**, *439*, 3490–3503. [CrossRef]
72. del Valle, M.V.; Lazarian, A.; Santos-Lima, R. Turbulence-induced magnetic fields in shock precursors. *Mon. Not. R. Astron. Soc.* **2016**, *458*, 1645–1659. [CrossRef]
73. Romanova, M.M.; Lovelace, R.V.E. Magnetic field, reconnection and particle acceleration in extragalactic jets. *Astron. Astrophys.* **1992**, *262*, 26–36.
74. Lyubarskij, Y.E. Energy release in strongly magnetized relativistic winds. *Sov. Astron. Lett.* **1992**, *18*, 356.
75. Eichler, D. Magnetic Confinement of Jets. *Astrophys. J.* **1993**, *419*, 111. [CrossRef]
76. Begelman, M.C. Instability of Toroidal Magnetic Field in Jets and Plerions. *Astrophys. J.* **1998**, *493*, 291–300. [CrossRef]
77. Giannios, D.; Spruit, H.C. The role of kink instability in Poynting-flux dominated jets. *Astron. Astrophys.* **2006**, *450*, 887–898. [CrossRef]
78. Levinson, A.; Begelman, M.C. Collimation and Confinement of Magnetic Jets by External Media. *Astrophys. J.* **2013**, *764*, 148. [CrossRef]
79. McKinney, J.C.; Uzdensky, D.A. A reconnection switch to trigger gamma-ray burst jet dissipation. *Mon. Not. R. Astron. Soc.* **2012**, *419*, 573–607. [CrossRef]
80. Davis, S.W.; Stone, J.M.; Pessah, M.E. Sustained Magnetorotational Turbulence in Local Simulations of Stratified Disks with Zero Net Magnetic Flux. *Astrophys. J.* **2010**, *713*, 52–65. [CrossRef]
81. O'Neill, S.M.; Reynolds, C.S.; Miller, M.C.; Sorathia, K.A. Low-frequency Oscillations in Global Simulations of Black Hole Accretion. *Astrophys. J.* **2011**, *736*, 107. [CrossRef]
82. Giannios, D.; Uzdensky, D.A. GRB and blazar jets shining through their stripes. *Mon. Not. R. Astron. Soc.* **2019**, *484*, 1378–1389. [CrossRef]
83. Tchekhovskoy, A.; McKinney, J.C.; Narayan, R. Simulations of ultrarelativistic magnetodynamic jets from gamma-ray burst engines. *Mon. Not. R. Astron. Soc.* **2008**, *388*, 551–572. [CrossRef]
84. Gill, R.; Granot, J. Temporal Evolution of Prompt GRB Polarization. *arXiv* **2021**, arXiv:2101.06777.
85. Zhang, B.; Yan, H. The Internal-collision-induced Magnetic Reconnection and Turbulence (ICMART) Model of Gamma-ray Bursts. *Astrophys. J.* **2011**, *726*, 90. [CrossRef]
86. Deng, W.; Zhang, H.; Zhang, B.; Li, H. Collision-induced Magnetic Reconnection and a Unified Interpretation of Polarization Properties of GRBs and Blazars. *Astrophys. J.* **2016**, *821*, L12. [CrossRef]
87. Sari, R.; Piran, T.; Narayan, R. Spectra and Light Curves of Gamma-Ray Burst Afterglows. *Astrophys. J.* **1998**, *497*, L17–L20. [CrossRef]
88. Granot, J.; Sari, R. The Shape of Spectral Breaks in Gamma-Ray Burst Afterglows. *Astrophys. J.* **2002**, *568*, 820–829. [CrossRef]
89. Crider, A.; Liang, E.P.; Smith, I.A.; Preece, R.D.; Briggs, M.S.; Pendleton, G.N.; Paciesas, W.S.; Band, D.L.; Matteson, J.L. Evolution of the Low-Energy Photon Spectral in Gamma-Ray Bursts. *Astrophys. J.* **1997**, *479*, L39–L42. [CrossRef]
90. Preece, R.D.; Briggs, M.S.; Mallozzi, R.S.; Pendleton, G.N.; Paciesas, W.S.; Band, D.L. The Synchrotron Shock Model Confronts a "Line of Death" in the BATSE Gamma-Ray Burst Data. *Astrophys. J.* **1998**, *506*, L23–L26. [CrossRef]
91. Preece, R.D.; Briggs, M.S.; Giblin, T.W.; Mallozzi, R.S.; Pendleton, G.N.; Paciesas, W.S.; Band, D.L. On the Consistency of Gamma-Ray Burst Spectral Indices with the Synchrotron Shock Model. *Astrophys. J.* **2002**, *581*, 1248–1255. [CrossRef]
92. Ghirlanda, G.; Celotti, A.; Ghisellini, G. Extremely hard GRB spectra prune down the forest of emission models. *Astron. Astrophys.* **2003**, *406*, 879–892. [CrossRef]
93. Lloyd, N.M.; Petrosian, V. Synchrotron Radiation as the Source of Gamma-Ray Burst Spectra. *Astrophys. J.* **2000**, *543*, 722–732. [CrossRef]
94. Medvedev, M.V. Theory of "Jitter" Radiation from Small-Scale Random Magnetic Fields and Prompt Emission from Gamma-Ray Burst Shocks. *Astrophys. J.* **2000**, *540*, 704–714. [CrossRef]
95. Mészáros, P.; Rees, M.J. Steep Slopes and Preferred Breaks in Gamma-Ray Burst Spectra: The Role of Photospheres and Comptonization. *Astrophys. J.* **2000**, *530*, 292–298. [CrossRef]
96. Burgess, J.M.; Ryde, F.; Yu, H.F. Taking the band function too far: a tale of two α's. *Mon. Not. R. Astron. Soc.* **2015**, *451*, 1511–1521. [CrossRef]
97. Poolakkil, S.; Preece, R.; Fletcher, C.; Goldstein, A.; Bhat, P.N.; Bissaldi, E.; Briggs, M.S.; Burns, E.; Cleveland, W.H.; Giles, M.M.; et al. The Fermi-GBM Gamma-Ray Burst Spectral Catalog: 10 yr of Data. *Astrophys. J.* **2021**, *913*, 60. [CrossRef]
98. Gruber, D.; Goldstein, A.; Weller von Ahlefeld, V.; Narayana Bhat, P.; Bissaldi, E.; Briggs, M.S.; Byrne, D.; Cleveland, W.H.; Connaughton, V.; Diehl, R.; et al. The Fermi GBM Gamma-Ray Burst Spectral Catalog: Four Years of Data. *Astrophys. J. Suppl. Ser.* **2014**, *211*, 12. [CrossRef]
99. Beloborodov, A.M.; Mészáros, P. Photospheric Emission of Gamma-Ray Bursts. *Space Sci. Rev.* **2017**, *207*, 87–110. [CrossRef]

100. Vurm, I.; Beloborodov, A.M. Radiative Transfer Models for Gamma-Ray Bursts. *Astrophys. J.* **2016**, *831*, 175. [CrossRef]
101. Axelsson, M.; Borgonovo, L. The width of gamma-ray burst spectra. *Mon. Not. R. Astron. Soc.* **2015**, *447*, 3150–3154. [CrossRef]
102. Yu, H.F.; van Eerten, H.J.; Greiner, J.; Sari, R.; Narayana Bhat, P.; von Kienlin, A.; Paciesas, W.S.; Preece, R.D. The sharpness of gamma-ray burst prompt emission spectra. *Astron. Astrophys.* **2015**, *583*, A129. [CrossRef]
103. Vianello, G.; Gill, R.; Granot, J.; Omodei, N.; Cohen-Tanugi, J.; Longo, F. The Bright and the Slow—GRBs 100724B and 160509A with High-energy Cutoffs at $\lesssim$100 MeV. *Astrophys. J.* **2018**, *864*, 163. [CrossRef]
104. Burgess, J.M. Is spectral width a reliable measure of GRB emission physics? *Astron. Astrophys.* **2019**, *629*, A69. [CrossRef]
105. Yassine, M.; Piron, F.; Daigne, F.; Mochkovitch, R.; Longo, F.; Omodei, N.; Vianello, G. A new fitting function for GRB MeV spectra based on the internal shock synchrotron model. *Astron. Astrophys.* **2020**, *640*, A91. [CrossRef]
106. Zhang, B.B.; Uhm, Z.L.; Connaughton, V.; Briggs, M.S.; Zhang, B. Synchrotron Origin of the Typical GRB Band Function—A Case Study of GRB 130606B. *Astrophys. Journal* **2016**, *816*, 72. [CrossRef]
107. Burgess, J.M.; Bégué, D.; Greiner, J.; Giannios, D.; Bacelj, A.; Berlato, F. Gamma-ray bursts as cool synchrotron sources. *Nat. Astron.* **2020**, *4*, 174–179. [CrossRef]
108. Granot, J. The Most Probable Cause for the High Gamma-Ray Polarization in GRB 021206. *Astrophys. J.* **2003**, *596*, L17–L21. [CrossRef]
109. Granot, J.; Königl, A. Linear Polarization in Gamma-Ray Bursts: The Case for an Ordered Magnetic Field. *Astrophys. J.* **2003**, *594*, L83–L87. [CrossRef]
110. Lyutikov, M.; Pariev, V.I.; Blandford, R.D. Polarization of Prompt Gamma-Ray Burst Emission: Evidence for Electromagnetically Dominated Outflow. *Astrophys. J.* **2003**, *597*, 998–1009. [CrossRef]
111. Granot, J.; Taylor, G.B. Radio Flares and the Magnetic Field Structure in Gamma-Ray Burst Outflows. *Astrophys. J.* **2005**, *625*, 263–270. [CrossRef]
112. Gill, R.; Granot, J. Constraining the magnetic field structure in collisionless relativistic shocks with a radio afterglow polarization upper limit in GW 170817. *Mon. Not. R. Astron. Soc.* **2020**, *491*, 5815–5825. [CrossRef]
113. Beloborodov, A.M. Regulation of the Spectral Peak in Gamma-Ray Bursts. *Astrophys. J.* **2013**, *764*, 157. [CrossRef]
114. Ryde, F. The Cooling Behavior of Thermal Pulses in Gamma-Ray Bursts. *Astrophys. J.* **2004**, *614*, 827–846. [CrossRef]
115. Guiriec, S.; Connaughton, V.; Briggs, M.S.; Burgess, M.; Ryde, F.; Daigne, F.; Mészáros, P.; Goldstein, A.; McEnery, J.; Omodei, N.; et al. Detection of a Thermal Spectral Component in the Prompt Emission of GRB 100724B. *Astrophys. J.* **2011**, *727*, L33. [CrossRef]
116. Guiriec, S.; Gehrels, N.; McEnery, J.; Kouveliotou, C.; Hartmann, D.H. Photospheric Emission in the Joint GBM and Konus Prompt Spectra of GRB 120323A. *Astrophys. J.* **2017**, *846*, 138. [CrossRef]
117. Ryde, F.; Pe'er, A.; Nymark, T.; Axelsson, M.; Moretti, E.; Lundman, C.; Battelino, M.; Bissaldi, E.; Chiang, J.; Jackson, M.S.; et al. Observational evidence of dissipative photospheres in gamma-ray bursts. *Mon. Not. R. Astron. Soc.* **2011**, *415*, 3693–3705. [CrossRef]
118. Eichler, D.; Levinson, A. A Compact Fireball Model of Gamma-Ray Bursts. *Astrophys. J.* **2000**, *529*, 146–150. [CrossRef]
119. Rees, M.J.; Mészáros, P. Dissipative Photosphere Models of Gamma-Ray Bursts and X-Ray Flashes. *Astrophys. J.* **2005**, *628*, 847–852. [CrossRef]
120. Thompson, C.; Mészáros, P.; Rees, M.J. Thermalization in Relativistic Outflows and the Correlation between Spectral Hardness and Apparent Luminosity in Gamma-Ray Bursts. *Astrophys. J.* **2007**, *666*, 1012–1023. [CrossRef]
121. Vurm, I.; Lyubarsky, Y.; Piran, T. On Thermalization in Gamma-Ray Burst Jets and the Peak Energies of Photospheric Spectra. *Astrophys. J.* **2013**, *764*, 143. [CrossRef]
122. Thompson, C.; Gill, R. Hot Electromagnetic Outflows. III. Displaced Fireball in a Strong Magnetic Field. *Astrophys. J.* **2014**, *791*, 46. [CrossRef]
123. Bégué, D.; Pe'er, A. Poynting-flux-dominated Jets Challenged by their Photospheric Emission. *Astrophys. J.* **2015**, *802*, 134. [CrossRef]
124. Giannios, D. Prompt emission spectra from the photosphere of a GRB. *Astron. Astrophys.* **2006**, *457*, 763–770. [CrossRef]
125. Pe'er, A.; Mészáros, P.; Rees, M.J. The Observable Effects of a Photospheric Component on GRB and XRF Prompt Emission Spectrum. *Astrophys. J.* **2006**, *642*, 995–1003. [CrossRef]
126. Giannios, D. Prompt GRB emission from gradual energy dissipation. *Astron. Astrophys.* **2008**, *480*, 305–312. [CrossRef]
127. Gill, R.; Thompson, C. Non-thermal Gamma-Ray Emission from Delayed Pair Breakdown in a Magnetized and Photon-rich Outflow. *Astrophys. J.* **2014**, *796*, 81. [CrossRef]
128. Gill, R.; Granot, J.; Beniamini, P. GRB spectrum from gradual dissipation in a magnetized outflow. *Mon. Not. R. Astron. Soc.* **2020**, *499*, 1356–1372. [CrossRef]
129. Beniamini, P.; Giannios, D. Prompt gamma-ray burst emission from gradual magnetic dissipation. *Mon. Not. R. Astron. Soc.* **2017**, *468*, 3202–3211. [CrossRef]
130. Ghisellini, G.; Celotti, A. Quasi-thermal Comptonization and Gamma-Ray Bursts. *Astrophys. J.* **1999**, *511*, L93–L96. [CrossRef]
131. Giannios, D.; Spruit, H.C. Spectral and timing properties of a dissipative γ-ray burst photosphere. *Astron. Astrophys.* **2007**, *469*, 1–9. [CrossRef]
132. Lazzati, D.; Ghisellini, G.; Celotti, A.; Rees, M.J. Compton-dragged Gamma-Ray Bursts Associated with Supernovae. *Astrophys. J.* **2000**, *529*, L17–L20. [CrossRef] [PubMed]

133. Ghisellini, G.; Lazzati, D.; Celotti, A.; Rees, M.J. Compton dragged gamma-ray bursts: the spectrum. *Mon. Not. R. Astron. Soc.* **2000**, *316*, L45–L49. [CrossRef]
134. Lazzati, D.; Rossi, E.; Ghisellini, G.; Rees, M.J. Compton drag as a mechanism for very high linear polarization in gamma-ray bursts. *Mon. Not. R. Astron. Soc.* **2004**, *347*, L1–L5. [CrossRef]
135. Sironi, L.; Spitkovsky, A. Synthetic Spectra from Particle-In-Cell Simulations of Relativistic Collisionless Shocks. *Astrophys. J.* **2009**, *707*, L92–L96. [CrossRef]
136. Mao, J.; Wang, J. Application of Jitter Radiation: Gamma-Ray Burst Prompt Polarization. *Astrophys. J.* **2013**, *776*, 17. [CrossRef]
137. Mao, J.; Wang, J. Linear Polarization, Circular Polarization, and Depolarization of Gamma-ray Bursts: A Simple Case of Jitter Radiation. *Astrophys. J.* **2017**, *838*, 78. [CrossRef]
138. Panaitescu, A.; Mészáros, P. Gamma-Ray Bursts from Upscattered Self-absorbed Synchrotron Emission. *Astrophys. J.* **2000**, *544*, L17–L21. [CrossRef]
139. Yost, S.A.; Aharonian, F.; Akerlof, C.W.; Ashley, M.C.B.; Barthelmy, S.; Gehrels, N.; Göğüş, E.; Güver, T.; Horns, D.; Kızıloğlu, Ü.; et al. The Dark Side of ROTSE-III Prompt GRB Observations. *Astrophys. J.* **2007**, *669*, 1107–1114. [CrossRef]
140. Derishev, E.V.; Kocharovsky, V.V.; Kocharovsky, V.V. Physical parameters and emission mechanism in gamma-ray bursts. *Astron. Astrophys.* **2001**, *372*, 1071–1077. [CrossRef]
141. Piran, T.; Sari, R.; Zou, Y.C. Observational limits on inverse Compton processes in gamma-ray bursts. *Mon. Not. R. Astron. Soc.* **2009**, *393*, 1107–1113. [CrossRef]
142. Chang, Z.; Lin, H.N. Gamma-Ray Polarization of the Synchrotron Self-compton Process from a Highly Relativistic Jet. *Astrophys. J.* **2014**, *795*, 36. [CrossRef]
143. Gill, R.; Granot, J. The effect of pair cascades on the high-energy spectral cut-off in gamma-ray bursts. *Mon. Not. R. Astron. Soc.* **2018**, *475*, L1–L5. [CrossRef]
144. Lazzati, D.; Deich, A.; Morsony, B.J.; Workman, J.C. Off-axis emission of short γ-ray bursts and the detectability of electromagnetic counterparts of gravitational-wave-detected binary mergers. *Mon. Not. R. Astron. Soc.* **2017**, *471*, 1652–1661. [CrossRef]
145. Beniamini, P.; Nakar, E. Observational constraints on the structure of gamma-ray burst jets. *Mon. Not. R. Astron. Soc.* **2019**, *482*, 5430–5440. [CrossRef]
146. Rhoads, J.E. How to Tell a Jet from a Balloon: A Proposed Test for Beaming in Gamma-Ray Bursts. *Astrophys. J.* **1997**, *487*, L1–L4. [CrossRef]
147. Harrison, F.A.; Bloom, J.S.; Frail, D.A.; Sari, R.; Kulkarni, S.R.; Djorgovski, S.G.; Axelrod, T.; Mould, J.; Schmidt, B.P.; Wieringa, M.H.; et al. Optical and Radio Observations of the Afterglow from GRB 990510: Evidence for a Jet. *Astrophys. J.* **1999**, *523*, L121–L124. [CrossRef]
148. Kulkarni, S.R.; Frail, D.A.; Sari, R.; Moriarty-Schieven, G.H.; Shepherd, D.S.; Udomprasert, P.; Readhead, A.C.S.; Bloom, J.S.; Feroci, M.; Costa, E. Discovery of a Radio Flare from GRB 990123. *Astrophys. J.* **1999**, *522*, L97–L100. [CrossRef]
149. Frail, D.A.; Kulkarni, S.R.; Sari, R.; Djorgovski, S.G.; Bloom, J.S.; Galama, T.J.; Reichart, D.E.; Berger, E.; Harrison, F.A.; Price, P.A.; et al. Beaming in Gamma-Ray Bursts: Evidence for a Standard Energy Reservoir. *Astrophys. J.* **2001**, *562*, L55–L58. [CrossRef]
150. Berger, E.; Soderberg, A.M.; Frail, D.A.; Kulkarni, S.R. A Radio Flare from GRB 020405: Evidence for a Uniform Medium around a Massive Stellar Progenitor. *Astrophys. J.* **2003**, *587*, L5–L8. [CrossRef]
151. Zhang, W.; Woosley, S.E.; MacFadyen, A.I. Relativistic Jets in Collapsars. *Astrophys. J.* **2003**, *586*, 356–371. [CrossRef]
152. Zhang, W.; Woosley, S.E.; Heger, A. The Propagation and Eruption of Relativistic Jets from the Stellar Progenitors of Gamma-Ray Bursts. *Astrophys. J.* **2004**, *608*, 365–377. [CrossRef]
153. Morsony, B.J.; Lazzati, D.; Begelman, M.C. Temporal and Angular Properties of Gamma-Ray Burst Jets Emerging from Massive Stars. *Astrophys. J.* **2007**, *665*, 569–598. [CrossRef]
154. Mizuta, A.; Ioka, K. Opening Angles of Collapsar Jets. *Astrophys. J.* **2013**, *777*, 162. [CrossRef]
155. Gottlieb, O.; Nakar, E.; Bromberg, O. The structure of hydrodynamic γ-ray burst jets. *Mon. Not. R. Astron. Soc.* **2021**, *500*, 3511–3526. [CrossRef]
156. Aloy, M.A.; Janka, H.T.; Müller, E. Relativistic outflows from remnants of compact object mergers and their viability for short gamma-ray bursts. *Astron. Astrophys.* **2005**, *436*, 273–311. [CrossRef]
157. Lazzati, D.; López-Cámara, D.; Cantiello, M.; Morsony, B.J.; Perna, R.; Workman, J.C. Off-axis Prompt X-Ray Transients from the Cocoon of Short Gamma-Ray Bursts. *Astrophys. J.* **2017**, *848*, L6. [CrossRef]
158. Gottlieb, O.; Nakar, E.; Piran, T. The cocoon emission - an electromagnetic counterpart to gravitational waves from neutron star mergers. *Mon. Not. R. Astron. Soc.* **2018**, *473*, 576–584. [CrossRef]
159. Nathanail, A.; Gill, R.; Porth, O.; Fromm, C.M.; Rezzolla, L. On the opening angle of magnetized jets from neutron-star mergers: the case of GRB170817A. *Mon. Not. R. Astron. Soc.* **2020**, *495*, 3780–3787. [CrossRef]
160. Nathanail, A.; Gill, R.; Porth, O.; Fromm, C.M.; Rezzolla, L. 3D magnetized jet break-out from neutron-star binary merger ejecta: afterglow emission from the jet and the ejecta. *Mon. Not. R. Astron. Soc.* **2021**, *502*, 1843–1855. [CrossRef]
161. Zhang, B.; Mészáros, P. Gamma-Ray Burst Beaming: A Universal Configuration with a Standard Energy Reservoir? *Astrophys. J.* **2002**, *571*, 876–879. [CrossRef]
162. Kumar, P.; Granot, J. The Evolution of a Structured Relativistic Jet and Gamma-Ray Burst Afterglow Light Curves. *Astrophys. J.* **2003**, *591*, 1075–1085. [CrossRef]

163. Granot, J.; Kumar, P. Constraining the Structure of Gamma-Ray Burst Jets through the Afterglow Light Curves. *Astrophys. J.* **2003**, *591*, 1086–1096. [CrossRef]
164. Panaitescu, A.; Kumar, P. The Effect of Angular Structure of Gamma-Ray Burst Outflows on the Afterglow Emission. *Astrophys. J.* **2003**, *592*, 390–400. [CrossRef]
165. Mészáros, P.; Rees, M.J.; Wijers, R.A.M.J. Viewing Angle and Environment Effects in Gamma-Ray Bursts: Sources of Afterglow Diversity. *Astrophys. J.* **1998**, *499*, 301–308. [CrossRef]
166. Lipunov, V.M.; Postnov, K.A.; Prokhorov, M.E. Gamma-Ray Bursts as Standard-Energy Explosions. *Astronomy Reports* **2001**, *45*, 236–240. [CrossRef]
167. Rossi, E.; Lazzati, D.; Rees, M.J. Afterglow light curves, viewing angle and the jet structure of γ-ray bursts. *Mon. Not. R. Astron. Soc.* **2002**, *332*, 945–950. [CrossRef]
168. Abbott, B.P.; Abbott, R.; Abbott, T.D.; Acernese, F.; Ackley, K.; Adams, C.; Adams, T.; Addesso, P.; Adhikari, R.X.; Adya, V.B.; et al. Multi-messenger Observations of a Binary Neutron Star Merger. *Astrophys. J.* **2017**, *848*, L12. [CrossRef]
169. Hallinan, G.; Corsi, A.; Mooley, K.P.; Hotokezaka, K.; Nakar, E.; Kasliwal, M.M.; Kaplan, D.L.; Frail, D.A.; Myers, S.T.; Murphy, T.; et al. A radio counterpart to a neutron star merger. *Science* **2017**, *358*, 1579–1583. [CrossRef]
170. Troja, E.; Piro, L.; van Eerten, H.; Wollaeger, R.T.; Im, M.; Fox, O.D.; Butler, N.R.; Cenko, S.B.; Sakamoto, T.; Fryer, C.L.; et al. The X-ray counterpart to the gravitational-wave event GW170817. *Nature* **2017**, *551*, 71–74. [CrossRef]
171. D'Avanzo, P.; Campana, S.; Salafia, O.S.; Ghirlanda, G.; Ghisellini, G.; Melandri, A.; Bernardini, M.G.; Branchesi, M.; Chassande-Mottin, E.; Covino, S.; et al. The evolution of the X-ray afterglow emission of GW 170817/GRB 170817A in XMM-Newton observations. *Astron. Astrophys.* **2018**, *613*, L1. [CrossRef]
172. Gill, R.; Granot, J. Afterglow imaging and polarization of misaligned structured GRB jets and cocoons: breaking the degeneracy in GRB 170817A. *Mon. Not. R. Astron. Soc.* **2018**, *478*, 4128–4141. [CrossRef]
173. Lamb, G.P.; Kobayashi, S. GRB 170817A as a jet counterpart to gravitational wave triggerGW 170817. *Mon. Not. R. Astron. Soc.* **2018**, *478*, 733–740. [CrossRef]
174. Lazzati, D.; Perna, R.; Morsony, B.J.; Lopez-Camara, D.; Cantiello, M.; Ciolfi, R.; Giacomazzo, B.; Workman, J.C. Late Time Afterglow Observations Reveal a Collimated Relativistic Jet in the Ejecta of the Binary Neutron Star Merger GW170817. *Phys. Rev. Lett.* **2018**, *120*, 241103. [CrossRef] [PubMed]
175. Margutti, R.; Alexander, K.D.; Xie, X.; Sironi, L.; Metzger, B.D.; Kathirgamaraju, A.; Fong, W.; Blanchard, P.K.; Berger, E.; MacFadyen, A.; et al. The Binary Neutron Star Event LIGO/Virgo GW170817 160 Days after Merger: Synchrotron Emission across the Electromagnetic Spectrum. *Astrophys. J.* **2018**, *856*, L18. [CrossRef]
176. Resmi, L.; Schulze, S.; Ishwara-Chandra, C.H.; Misra, K.; Buchner, J.; De Pasquale, M.; Sánchez-Ramírez, R.; Klose, S.; Kim, S.; Tanvir, N.R.; et al. Low-frequency View of GW170817/GRB 170817A with the Giant Metrewave Radio Telescope. *Astrophys. J.* **2018**, *867*, 57. [CrossRef]
177. Gill, R.; Granot, J.; De Colle, F.; Urrutia, G. Numerical Simulations of an Initially Top-Hat Jet and the Afterglow of GW170817 / GRB170817A. *arXiv* **2019**, arXiv:1902.10303.
178. Beniamini, P.; Granot, J.; Gill, R. Afterglow light curves from misaligned structured jets. *Mon. Not. R. Astron. Soc.* **2020**, *493*, 3521–3534. [CrossRef]
179. Panaitescu, A.; Mészáros, P.; Rees, M.J. Multiwavelength Afterglows in Gamma-Ray Bursts: Refreshed Shock and Jet Effects. *Astrophys. J.* **1998**, *503*, 314–324. [CrossRef]
180. Frail, D.A.; Berger, E.; Galama, T.; Kulkarni, S.R.; Moriarty-Schieven, G.H.; Pooley, G.G.; Sari, R.; Shepherd, D.S.; Taylor, G.B.; Walter, F. The Enigmatic Radio Afterglow of GRB 991216. *Astrophys. J.* **2000**, *538*, L129–L132. [CrossRef]
181. Huang, Y.F.; Wu, X.F.; Dai, Z.G.; Ma, H.T.; Lu, T. Rebrightening of XRF 030723: Further Evidence for a Two-Component Jet in a Gamma-Ray Burst. *Astrophys. J.* **2004**, *605*, 300–306. [CrossRef]
182. Peng, F.; Königl, A.; Granot, J. Two-Component Jet Models of Gamma-Ray Burst Sources. *Astrophys. J.* **2005**, *626*, 966–977. [CrossRef]
183. Racusin, J.L.; Karpov, S.V.; Sokolowski, M.; Granot, J.; Wu, X.F.; Pal'Shin, V.; Covino, S.; van der Horst, A.J.; Oates, S.R.; Schady, P.; et al. Broadband observations of the naked-eye γ-ray burst GRB080319B. *Nature* **2008**, *455*, 183–188. [CrossRef]
184. Kumar, P.; Piran, T. Energetics and Luminosity Function of Gamma-Ray Bursts. *Astrophys. J.* **2000**, *535*, 152–157. [CrossRef]
185. Yamazaki, R.; Ioka, K.; Nakamura, T. A Unified Model of Short and Long Gamma-Ray Bursts, X-Ray-rich Gamma-Ray Bursts, and X-Ray Flashes. *Astrophys. J.* **2004**, *607*, L103–L106. [CrossRef]
186. McConnell, M.L.; LEAP Collaboration. LEAP - A LargE Area Burst Polarimeter for the ISS. In *Eighth Huntsville Gamma-Ray Burst Symposium*. LPICo **2016**, *1962*, 4051.
187. Kole, M.; De Angelis, N.; Berlato, F.; Burgess, J.M.; Gauvin, N.; Greiner, J.; Hajdas, W.; Li, H.C.; Li, Z.H.; Produit, N.; et al. The POLAR Gamma-Ray Burst Polarization Catalog. *arXiv* **2020**, arXiv:2009.04871.
188. Chattopadhyay, T.; Vadawale, S.V.; Aarthy, E.; Mithun, N.P.S.; Chand, V.; Ratheesh, A.; Basak, R.; Rao, A.R.; Bhalerao, V.; Mate, S.; et al. Prompt Emission Polarimetry of Gamma-Ray Bursts with the AstroSat CZT Imager. *Astrophys. J.* **2019**, *884*, 123. [CrossRef]
189. Burgess, J.M.; Kole, M.; Berlato, F.; Greiner, J.; Vianello, G.; Produit, N.; Li, Z.H.; Sun, J.C. Time-resolved GRB polarization with POLAR and GBM. Simultaneous spectral and polarization analysis with synchrotron emission. *Astron. Astrophys.* **2019**, *627*, A105. [CrossRef]

190. Weisskopf, M.C.; Elsner, R.F.; O'Dell, S.L. On understanding the figures of merit for detection and measurement of x-ray polarization. *Proc. SPIE* **2016**, *7732*, 98–102. [CrossRef]
191. Yonetoku, D.; Murakami, T.; Gunji, S.; Mihara, T.; Sakashita, T.; Morihara, Y.; Kikuchi, Y.; Takahashi, T.; Fujimoto, H.; Toukairin, N.; et al. Gamma-Ray Burst Polarimeter (GAP) aboard the Small Solar Power Sail Demonstrator IKAROS. *Publ. Astron. Soc. Jpn.* **2011**, *63*, 625–638. [CrossRef]
192. Produit, N.; Bao, T.; Batsch, T.; Bernasconi, T.; Britvich, I.; Cadoux, F.; Cernuda, I.; Chai, J.; Dong, Y.; Gauvin, N.; et al. Design and construction of the POLAR detector. *Nucl. Instruments Methods Phys. Res. Sect. A* **2018**, *877*, 259–268. [CrossRef]
193. Vedrenne, G.; Schönfelder, V.; Albernhe, F.; Borrel, V.; Bouchet, L.; Caraveo, P.; Connell, P.H.; Cordier, B.; Denis, M.; Coszach, R.; et al. The Integral Spectrometer SPI. *Astrophys. Lett. Commun.* **1999**, *39*, 325.
194. Vadawale, S.V.; Chattopadhyay, T.; Rao, A.R. Prospects of hard X-ray polarimetry with Astrosat-CZTI. In Proceedings of the 2013 IEEE Nuclear Science Symposium and Medical Imaging Conference (2013 NSS/MIC), Seoul, Korea, 27 October–2 November 2013; pp. 1–9. [CrossRef]
195. Feng, H.E.A. Re-detection and a possible time variation of soft X-ray polarization from the Crab. *Nat. Astron.* **2020**, *4*, 511–516. [CrossRef]
196. McConnell, M.L. GRB Polarimetry with POET. *AIP Conf. Proc.* **2009**, *1133*, 64. [CrossRef]
197. Weisskopf, M.C.; Ramsey, B.; O'Dell, S.L.; Tennant, A.; Elsner, R.; Soffita, P.; Bellazzini, R.; Costa, E.; Kolodziejczak, J.; Kaspi, V.; et al. The Imaging X-ray Polarimetry Explorer (IXPE). *Results Phys.* **2016**, *6*, 1179–1180. [CrossRef]
198. in't Zand, J.J.M.; Bozzo, E.; Qu, J.; Li, X.D.; Amati, L.; Chen, Y.; Donnarumma, I.; Doroshenko, V.; Drake, S.A.; Hernanz, M.; et al. Observatory science with eXTP. *Sci. China Phys. Mech. Astron.* **2019**, *62*, 29506. [CrossRef]
199. Wu, X.; Walter, R.; Su, M.; Ambrosi, G.; Azzarello, P.; Böttcher, M.; Chang, J.; Chernyakova, M.; Fan, Y.; Farnier, C.; et al. PANGU: a wide field gamma-ray imager and polarimeter. *Proc. SPIE* **2016**, *9905*, 1869–1886. [CrossRef]
200. Bernard, D.; Bruel, P.; Frotin, M.; Geerebaert, Y.; Giebels, B.; Gros, P.; Horan, D.; Louzir, M.; Poilleux, P.; Semeniouk, I.; et al. HARPO: A TPC as a gamma-ray telescope and polarimeter. *Proc. SPIE* **2014**, *9144*, 460–474. [CrossRef]
201. Attié, D.; Amano, S.; Baron, P.; Baudin, D.; Bernard, D.; Bruel, P.; Calvet, D.; Colas, P.; Daté, S.; Delbart, A.; et al. HARPO, prototype of a gamma-ray polarimeter: Results of a polarised photon beam test between 1.7 and 74 MeV. In Proceedings of the 35th International Cosmic Ray Conference—ICRC2017, Busan, Korea, 10–20 July 2017; p. 818. [CrossRef]
202. Hunter, S.D. The advanced energetic pair telescope for gamma-ray polarimetry. *Proc. SPIE* **2018**, *10699*, 652–658.
203. Ueno, K.; Mizumoto, T.; Hattori, K.; Higashi, N.; Iwaki, S.; Kabuki, S.; Kishimoto, Y.; Komura, S.; Kubo, H.; Kurosawa, S.; et al. Development of the balloon-borne sub-MeV gamma-ray Compton camera using an electron-tracking gaseous TPC and a scintillation camera. *J. Instrum.* **2012**, *7*, C01088. [CrossRef]
204. Sadrozinski, H.F.W. GLAST, a Gamma-Ray Large Area Space Telescope. *Nucl. Instruments Methods Phys. Res. A* **2001**, *466*, 292–299. [CrossRef]
205. Ting, S. The Alpha Magnetic Spectrometer on the International Space Station. *Nucl. Phys.-Proc. Suppl.* **2013**, *243-244*, 12–24. [CrossRef]
206. Giomi, M.; Bühler, R.; Sgrò, C.; Longo, F.; Atwood, W.B. Estimate of the Fermi large area telescope sensitivity to gamma-ray polarization. *AIP Conf. Proc.* **2017**, *1792*, 070022. [CrossRef]
207. McConnell, M.; Forrest, D.; Vestrand, W.T.; Finger, M. Using BATSE to measure gamma-ray burst polarization. *AIP Conf. Proc.* **1996**, *384*, 851–855. [CrossRef]
208. Willis, D.R.; Barlow, E.J.; Bird, A.J.; Clark, D.J.; Dean, A.J.; McConnell, M.L.; Moran, L.; Shaw, S.E.; Sguera, V. Evidence of polarisation in the prompt gamma-ray emission from GRB 930131 and GRB 960924. *Astron. Astrophys.* **2005**, *439*, 245. [CrossRef]
209. aciesas, W.S.; Pendleton, G.N.; Lestrade, J.P.; Fishman, G.J.; Meegan, C.A.; Wilson, R.B.; Parnell, T.A.; Austin, R.W.; Berry, F.A., Jr.; Horack, J.M.; et al. Performance Of The Large-Area Detectors For The Burst And Transient Source Experiment (BATSE) On The Gamma Ray Observatory. *Proc. SPIE* **1989**, *1159*, 156–164. [CrossRef]
210. Mizuno, T.; Kamae, T.; Ng, J.; Tajima, H.; Mitchell, J.; Streitmatter, R.; Fernholz, R.; Groth, E.; Fukazawa, Y. Beam test of a prototype detector array for the PoGO astronomical hard X-ray/soft gamma-ray polarimeter. *Nucl. Instruments Methods Phys. Res. Sect. A* **2005**, *540*, 158–168. [CrossRef]
211. The Reuven Ramaty High-Energy Solar Spectroscopic Imager (RHESSI). *Sol Phys* **2002**, *210*, 3–32. [CrossRef]
212. Ubertini, P.; Lebrun, F.; Di Cocco, G.; Bazzano, A.; Bird, A. J.; Broenstad, K.; Goldwurm, A.; La Rosa, G.; Labanti, C.; Laurent, P.; et al. IBIS: The Imager on-board INTEGRAL. *A&A* **2003**, *411*, L131–L139. [CrossRef]
213. Forot, M.; Laurent, P.; Lebrun, F.; Limousin, O. Compton telescope with coded aperture mask: Imaging with the INTEGRAL/IBIS Compton mode. *Astrophys. J.* **2007**, *668*, 1259. [CrossRef]
214. McGlynn, S.; Clark, D.J.; Dean, A.J.; Hanlon, L.; McBreen, S.; Willis, D. R.; McBreen, B.; Bird, A.J.; Foley, S. Polarisation studies of the prompt gamma-ray emission from GRB 041219a using the spectrometer aboard INTEGRAL. *Astron. Astrophys.* **2007**, *466*, 895. [CrossRef]
215. Kalemci, E.; Boggs, S.E.; Kouveliotou, C.; Finger, M.; Barin, M.G. Search for Polarization from the Prompt Gamma-Ray Emission of GRB 041219a with SPI on INTEGRAL. *Astrophys. J. Suppl. Ser.* **2007**, *169*. [CrossRef]
216. Kole, M.; Li, Z.H.; Produit, N.; Tymieniecka, T.; Zhang, J.; Zwolinska, A.; Bao, T.W.; Bernasconi, T.; Cadoux, F.; Feng, M.Z.; et al. Instrument Performance and Simulation Verification of the POLAR Detector. *Nucl. Instrum. Meth. A* **2017**, *872*, 28–40. [CrossRef]

217. Li, Z.H.; Kole, M.; Sun, J.C.; Song, L.M.; Produit, N.; Wu, B.B.; Bao, T.W.; Bernasconi, T.; Cadoux, F.; Dong, Y.W.; et al. In-Orbit Instrument Performance Study and Calibration for POLAR Polarization Measurements. *Nucl. Instrum. Meth. A* **2018**, *900*, 8–24. [CrossRef]
218. Kierans, C.A.; Boggs, S.E.; Chiu, J.-L.; Lowell, A.; Sleator, C.; Tomsick, J.A.; Zoglauer, A.; Amman, M.; Chang, H.-K.; Tseng, C.-H.; et al. The 2016 Super Pressure Balloon flight of the Compton Spectrometer and Imager. *Proc. Int. WorkE* **2016**, *75*, 75. [CrossRef]
219. Lowell, A.W.; Boggs, S.E.; Chiu, J.L.; Kierans, C.A.; Sleator, C.C.; Tomsick, J.A.; Zoglauer, A.C.; Chang, H.K.; Tseng, C.H.; Yang, C.Y.; et al. Polarimetric Analysis of the Long Duration Gamma Ray Burst GRB 160530A With the Balloon Borne Compton Spectrometer and Imager. *Astrophys. J.* **2017**, *848*, 119. [CrossRef]
220. Rybicki, G.B.; Lightman, A.P. *Radiative Processes in Astrophysics*; Wiley-Interscience: New York, NY, USA, 1979.
221. Granot, J. Afterglow Light Curves from Impulsive Relativistic Jets with an Unconventional Structure. *Astrophys. J.* **2005**, *631*, 1022–1031. [CrossRef]
222. Gruzinov, A. Strongly Polarized Optical Afterglows of Gamma-Ray Bursts. *Astrophys. J.* **1999**, *525*, L29–L31. [CrossRef] [PubMed]
223. Sari, R. Linear Polarization and Proper Motion in the Afterglow of Beamed Gamma-Ray Bursts. *Astrophys. J.* **1999**, *524*, L43–L46. [CrossRef]
224. Begelman, M.C.; Sikora, M.; Giommi, P.; Barr, P.; Garilli, B.; Gioia, I.M.; Maccacaro, T.; Maccagni, D.; Schild, R.E. Inverse Compton Scattering of Ambient Radiation by a Cold Relativistic Jet: A Source of Beamed, Polarized X-Ray and Optical Observations of X-Ray–selected BL Lacertae Objects. *Astrophys. J.* **1987**, *322*, 650. [CrossRef]
225. Nakar, E.; Piran, T.; Waxman, E. Implications of the γ-ray polarization of GRB 021206. *J. Cosmol. Astropart* **2003**, *2003*, 5, [CrossRef]
226. Matsumoto, T.; Nakar, E.; Piran, T. Constraints on the emitting region of the gamma-rays observed in GW170817. *Mon. Not. R. Astron. Soc.* **2019**, *483*, 1247–1255. [CrossRef]
227. Genet, F.; Granot, J. Realistic analytic model for the prompt and high-latitude emission in GRBs. *Mon. Not. R. Astron. Soc.* **2009**, *399*, 1328–1346. [CrossRef]
228. Willingale, R.; Genet, F.; Granot, J.; O'Brien, P.T. The spectral-temporal properties of the prompt pulses and rapid decay phase of gamma-ray bursts. *Mon. Not. R. Astron. Soc.* **2010**, *403*, 1296–1316. [CrossRef]
229. Deng, W.; Zhang, B.; Li, H.; Stone, J.M. Magnetized Reverse Shock: Density-fluctuation-induced Field Distortion, Polarization Degree Reduction, and Application to GRBs. *Astrophys. J.* **2017**, *845*, L3. [CrossRef]
230. Nakar, E.; Oren, Y. Polarization and Light-Curve Variability: The "Patchy-Shell" Model. *Astrophys. J.* **2004**, *602*, L97–L100. [CrossRef]
231. Sharma, V.; Iyyani, S.; Bhattacharya, D.; Chattopadhyay, T.; Rao, A.R.; Aarthy, E.; Vadawale, S.V.; Mithun, N.P.S.; Bhalerao, V.B.; Ryde, F.; et al. Time-varying Polarized Gamma-Rays from GRB 160821A: Evidence for Ordered Magnetic Fields. *Astrophys. J.* **2019**, *882*, L10. [CrossRef]
232. Lundman, C.; Vurm, I.; Beloborodov, A.M. Polarization of Gamma-Ray Bursts in the Dissipative Photosphere Model. *Astrophys. J.* **2018**, *856*, 145. [CrossRef]
233. Chand, V.; Chattopadhyay, T.; Oganesyan, G.; Rao, A.R.; Vadawale, S.V.; Bhattacharya, D.; Bhalerao, V.B.; Misra, K. AstroSat-CZTI Detection of Variable Prompt Emission Polarization in GRB 171010A. *Astrophys. J.* **2019**, *874*, 70. [CrossRef]
234. Götz, D.; Laurent, P.; Antier, S.; Covino, S.; D'Avanzo, P.; D'Elia, V.; Melandri, A. GRB 140206A: The most distant polarized gamma-ray burst. *Mon. Not. R. Astron. Soc.* **2014**, *444*, 2776–2782. [CrossRef]
235. Yonetoku, D.; Murakami, T.; Gunji, S.; Mihara, T.; Toma, K.; Morihara, Y.; Takahashi, T.; Wakashima, Y.; Yonemochi, H.; Sakashita, T.; et al. Magnetic Structures in Gamma-Ray Burst Jets Probed by Gamma-Ray Polarization. *Astrophys. J.* **2012**, *758*, L1. [CrossRef]
236. Yonetoku, D.; Murakami, T.; Gunji, S.; Mihara, T.; Toma, K.; Sakashita, T.; Morihara, Y.; Takahashi, T.; Toukairin, N.; Fujimoto, H.; et al. Detection of Gamma-Ray Polarization in Prompt Emission of GRB 100826A. *Astrophys. J.* **2011**, *743*, L30. [CrossRef]
237. McGlynn, S.; Foley, S.; McBreen, B.; Hanlon, L.; McBreen, S.; Clark, D.J.; Dean, A.J.; Martin-Carrillo, A.; O'Connor, R. High energy emission and polarisation limits for the INTEGRAL burst GRB 061122. *Astron. Astrophys.* **2009**, *499*, 465–472. [CrossRef]
238. Gotz, D.; Covino, S.; Fernandez-Soto, A.; Laurent, P.; Bosnjak, Z. The polarized Gamma-Ray Burst GRB 061122. *Mon. Not. Roy. Astron. Soc.* **2013**, *431*, 3550. [CrossRef]
239. Gotz, D.; Laurent, P.; Lebrun, F.; Daigne, F.; Bosnjak, Z. Variable polarization measured in the prompt emission of GRB 041219A using IBIS on board INTEGRAL. *Astrophys. J. Lett.* **2009**, *695*, L208–L212. [CrossRef]
240. Chattopadhyay, T. Hard X-ray Polarimetry—An overview of the method, science drivers and recent findings. *arXiv* **2021**, arXiv:2104.05244.
241. Chand, V.; Chattopadhyay, T.; Iyyani, S.; Basak, R.; Aarthy, E.; Rao, A.R.; Vadawale, S.V.; Bhattacharya, D.; Bhalerao, V.B. Violation of Synchrotron Line of Death by the Highly Polarized GRB 160802A. *Astrophys. J.* **2018**, *862*, 154. [CrossRef]
242. Sharma, V.; Iyyani, S.; Bhattacharya, D.; Chattopadhyay, T.; Vadawale, S.V.; Bhalerao, V.B. Spectropolarimetric analysis of prompt emission of GRB 160325A: Jet with evolving environment of internal shocks. *Mon. Not. R. Astron. Soc.* **2020**, *493*, 5218–5232. [CrossRef]
243. Gehrels, N.; Chincarini, G.; Giommi, P.; Mason, K.O.; Nousek, J.A.; Wells, A.A.; White, N.E.; Barthelmy, S.D.; Burrows, D.N.; Cominsky, L.R.; et al. The Swift Gamma-Ray Burst Mission. *Astrophys. J.* **2004**, *611*, 1005–1020. [CrossRef]
244. Burrows, D.N.; Romano, P.; Falcone, A.; Kobayashi, S.; Zhang, B.; Moretti, A.; O'Brien, P.T.; Goad, M.R.; Campana, S.; Page, K.L.; et al. Bright X-ray Flares in Gamma-Ray Burst Afterglows. *Science* **2005**, *309*, 1833–1835. [CrossRef] [PubMed]

245. Falcone, A.D.; Morris, D.; Racusin, J.; Chincarini, G.; Moretti, A.; Romano, P.; Burrows, D.N.; Pagani, C.; Stroh, M.; Grupe, D.; et al. The First Survey of X-Ray Flares from Gamma-Ray Bursts Observed by Swift: Spectral Properties and Energetics. *Astrophys. J.* **2007**, *671*, 1921–1938. [CrossRef]
246. Chincarini, G.; Mao, J.; Margutti, R.; Bernardini, M.G.; Guidorzi, C.; Pasotti, F.; Giannios, D.; Della Valle, M.; Moretti, A.; Romano, P.; et al. Unveiling the origin of X-ray flares in gamma-ray bursts. *Mon. Not. R. Astron. Soc.* **2010**, *406*, 2113–2148. [CrossRef]
247. Margutti, R.; Bernardini, G.; Barniol Duran, R.; Guidorzi, C.; Shen, R.F.; Chincarini, G. On the average gamma-ray burst X-ray flaring activity. *Mon. Not. R. Astron. Soc.* **2011**, *410*, 1064–1075. [CrossRef]
248. Margutti, R.; Chincarini, G.; Granot, J.; Guidorzi, C.; Berger, E.; Bernardini, M.G.; Gehrels, N.; Soderberg, A.M.; Stamatikos, M.; Zaninoni, E. X-ray flare candidates in short gamma-ray bursts. *Mon. Not. R. Astron. Soc.* **2011**, *417*, 2144–2160. [CrossRef]
249. Yi, S.X.; Xi, S.Q.; Yu, H.; Wang, F.Y.; Mu, H.J.; Lü, L.Z.; Liang, E.W. Comprehensive Study of the X-ray Flares from Gamma-ray Bursts Observed by Swift. *Astrophys. J. Suppl. Ser.* **2016**, *224*, 20. [CrossRef]
250. Krimm, H.A.; Granot, J.; Marshall, F.E.; Perri, M.; Barthelmy, S.D.; Burrows, D.N.; Gehrels, N.; Mészáros, P.; Morris, D. GRB 060714: No Clear Dividing Line between Prompt Emission and X-ray Flares. *Astrophys. J.* **2007**, *665*, 554–568. [CrossRef]
251. Fan, Y.Z.; Zhang, B.; Proga, D. Linearly Polarized X-Ray Flares following Short Gamma-Ray Bursts. *Astrophys. J.* **2005**, *635*, L129–L132. [CrossRef]
252. Geng, J.J.; Huang, Y.F.; Wu, X.F.; Zhang, B.; Zong, H.S. Low-energy Spectra of Gamma-Ray Bursts from Cooling Electrons. *Astrophys. J. Suppl. Ser.* **2018**, *234*, 3. [CrossRef]
253. Zhang, S.N.; Kole, M.; Bao, T.W.; Batsch, T.; Bernasconi, T.; Cadoux, F.; Chai, J.Y.; Dai, Z.G.; Dong, Y.W.; Gauvin, N.; et al. Detailed polarization measurements of the prompt emission of five gamma-ray bursts. *Nat. Astron.* **2019**, *3*, 258–264. [CrossRef]
254. Sari, R.; Piran, T. Hydrodynamic Timescales and Temporal Structure of Gamma-Ray Bursts. *Astrophys. J.* **1995**, *455*, L143. [CrossRef]
255. Sari, R.; Piran, T. Predictions for the Very Early Afterglow and the Optical Flash. *Astrophys. J.* **1999**, *520*, 641–649. [CrossRef]
256. Kobayashi, S.; Sari, R. Optical Flashes and Radio Flares in Gamma-Ray Burst Afterglow: Numerical Study. *Astrophys. J.* **2000**, *542*, 819–828. [CrossRef]
257. Kobayashi, S.; Zhang, B. Early Optical Afterglows from Wind-Type Gamma-Ray Bursts. *Astrophys. J.* **2003**, *597*, 455–458. [CrossRef]
258. Nakar, E.; Piran, T. Early afterglow emission from a reverse shock as a diagnostic tool for gamma-ray burst outflows. *Mon. Not. R. Astron. Soc.* **2004**, *353*, 647–653. [CrossRef]
259. Granot, J. Interaction of a highly magnetized impulsive relativistic flow with an external medium. *Mon. Not. R. Astron. Soc.* **2012**, *421*, 2442–2466. [CrossRef]
260. Akerlof, C.; Balsano, R.; Barthelmy, S.; Bloch, J.; Butterworth, P.; Casperson, D.; Cline, T.; Fletcher, S.; Frontera, F.; Gisler, G.; et al. Observation of contemporaneous optical radiation from a γ-ray burst. *Nature* **1999**, *398*, 400–402. [CrossRef]
261. Sari, R.; Piran, T. GRB 990123: The Optical Flash and the Fireball Model. *Astrophys. J.* **1999**, *517*, L109–L112. [CrossRef]
262. Fox, D.W.; Price, P.A.; Soderberg, A.M.; Berger, E.; Kulkarni, S.R.; Sari, R.; Frail, D.A.; Harrison, F.A.; Yost, S.A.; Matthews, K.; et al. Discovery of Early Optical Emission from GRB 021211. *Astrophys. J.* **2003**, *586*, L5–L8. [CrossRef]
263. Uehara, T.; Toma, K.; Kawabata, K.S.; Chiyonobu, S.; Fukazawa, Y.; Ikejiri, Y.; Inoue, T.; Itoh, R.; Komatsu, T.; Miyamoto, H.; et al. GRB 091208B: First Detection of the Optical Polarization in Early Forward Shock Emission of a Gamma-Ray Burst Afterglow. *Astrophys. J.* **2012**, *752*, L6. [CrossRef]
264. Vestrand, W.T.; Wren, J.A.; Panaitescu, A.; Wozniak, P.R.; Davis, H.; Palmer, D.M.; Vianello, G.; Omodei, N.; Xiong, S.; Briggs, M.S.; et al. The Bright Optical Flash and Afterglow from the Gamma-Ray Burst GRB 130427A. *Science* **2014**, *343*, 38–41. [CrossRef] [PubMed]
265. Laskar, T.; Berger, E.; Zauderer, B.A.; Margutti, R.; Soderberg, A.M.; Chakraborti, S.; Lunnan, R.; Chornock, R.; Chandra, P.; Ray, A. A Reverse Shock in GRB 130427A. *Astrophys. J.* **2013**, *776*, 119. [CrossRef]
266. Perley, D.A.; Cenko, S.B.; Corsi, A.; Tanvir, N.R.; Levan, A.J.; Kann, D.A.; Sonbas, E.; Wiersema, K.; Zheng, W.; Zhao, X.H.; et al. The Afterglow of GRB 130427A from 1 to 10^{16} GHz. *Astrophys. J.* **2014**, *781*, 37. [CrossRef]
267. Laskar, T.; Alexander, K.D.; Gill, R.; Granot, J.; Berger, E.; Mundell, C.G.; Barniol Duran, R.; Bolmer, J.; Duffell, P.; van Eerten, H.; et al. ALMA Detection of a Linearly Polarized Reverse Shock in GRB 190114C. *Astrophys. J.* **2019**, *878*, L26. [CrossRef]
268. Troja, E.; Lipunov, V.M.; Mundell, C.G.; Butler, N.R.; Watson, A.M.; Kobayashi, S.; Cenko, S.B.; Marshall, F.E.; Ricci, R.; Fruchter, A.; et al. Significant and variable linear polarization during the prompt optical flash of GRB 160625B. *Nature* **2017**, *547*, 425–427. [CrossRef] [PubMed]
269. Steele, I.A.; Bates, S.D.; Carter, D.; Clarke, D.; Gomboc, A.; Guidorzi, C.; Melandri, A.; Monfardini, A.; Mottram, C.J.; Mundell, C.G.; et al. RINGO: A novel ring polarimeter for rapid GRB followup. *Proc. SPIE* **2006**, *6269*, 62695M [CrossRef]
270. Mundell, C.G.; Steele, I.A.; Smith, R.J.; Kobayashi, S.; Melandri, A.; Guidorzi, C.; Gomboc, A.; Mottram, C.J.; Clarke, D.; Monfardini, A.; et al. Early Optical Polarization of a Gamma-Ray Burst Afterglow. *Science* **2007**, *315*, 1822. [CrossRef]
271. Steele, I.A.; Mundell, C.G.; Smith, R.J.; Kobayashi, S.; Guidorzi, C. Ten per cent polarized optical emission from GRB090102. *Nature* **2009**, *462*, 767–769. [CrossRef]
272. Steele, I.A.; Bates, S.D.; Guidorzi, C.; Mottram, C.J.; Mundell, C.G.; Smith, R.J. RINGO2: An EMCCD-based polarimeter for GRB followup. *Proc. SPIE* **2010**, *7735*, 773549. [CrossRef]

273. Arnold, D.M.; Steele, I.A.; Bates, S.D.; Mottram, C.J.; Smith, R.J. RINGO3: A multi-colour fast response polarimeter. *Proc. SPIE* **2012**, *8446*, 84462J. [CrossRef]
274. Kopač, D.; Mundell, C.G.; Japelj, J.; Arnold, D.M.; Steele, I.A.; Guidorzi, C.; Dichiara, S.; Kobayashi, S.; Gomboc, A.; Harrison, R.M.; et al. Limits on Optical Polarization during the Prompt Phase of GRB 140430A. *Astrophys. J.* **2015**, *813*, 1. [CrossRef]
275. Steele, I.A.; Kopač, D.; Arnold, D.M.; Smith, R.J.; Kobayashi, S.; Jermak, H.E.; Mundell, C.G.; Gomboc, A.; Guidorzi, C.; Melandri, A.; et al. Polarimetry and Photometry of Gamma-Ray Bursts with RINGO2. *Astrophys. J.* **2017**, *843*, 143. [CrossRef]
276. Jordana-Mitjans, N.; Mundell, C.G.; Kobayashi, S.; Smith, R.J.; Guidorzi, C.; Steele, I.A.; Shrestha, M.; Gomboc, A.; Marongiu, M.; Martone, R.; et al. Lowly Polarized Light from a Highly Magnetized Jet of GRB 190114C. *Astrophys. J.* **2020**, *892*, 97. [CrossRef]
277. Mundell, C.G.; Kopač, D.; Arnold, D.M.; Steele, I.A.; Gomboc, A.; Kobayashi, S.; Harrison, R.M.; Smith, R.J.; Guidorzi, C.; Virgili, F.J.; et al. Highly polarized light from stable ordered magnetic fields in GRB 120308A. *Nature* **2013**, *504*, 119–121. [CrossRef]
278. Kobayashi, S. Polarized Emission from Gamma-Ray Burst Jets. *Galaxies* **2017**, *5*, 80. [CrossRef]
279. Blandford, R.D.; McKee, C.F. Fluid dynamics of relativistic blast waves. *Phys. Fluids* **1976**, *19*, 1130–1138. [CrossRef]
280. MAGIC Collaboration; Acciari, V.A.; Ansoldi, S.; Antonelli, L.A.; Engels, A.A.; Baack, D.; Babić, A.; Banerjee, B.; Barres de Almeida, U.; Barrio, J.A.; et al. Observation of inverse Compton emission from a long γ-ray burst *Nature* **2019**, *575*, 459–463. [CrossRef]
281. Covino, S.; Lazzati, D.; Ghisellini, G.; Saracco, P.; Campana, S.; Chincarini, G.; di Serego, S.; Cimatti, A.; Vanzi, L.; Pasquini, L. GRB 990510: Linearly polarized radiation from a fireball. *arXiv* **1999**, arXiv:astro-ph/9906319.
282. Wijers, R.A.M.J.; Vreeswijk, P.M.; Galama, T.J.; Rol, E.; van Paradijs, J.; Kouveliotou, C.; Giblin, T.; Masetti, N.; Palazzi, E.; Pian, E. Detection of Polarization in the Afterglow of GRB 990510 with the ESO Very Large Telescope. *Astrophys. J.* **1999**, *523*, L33–L36. [CrossRef]
283. Rol, E.; Wijers, R.A.M.J.; Vreeswijk, P.M.; Kaper, L.; Galama, T.J.; van Paradijs, J.; Kouveliotou, C.; Masetti, N.; Pian, E.; Palazzi, E.; et al. GRB 990712: First Indication of Polarization Variability in a Gamma-Ray Burst Afterglow. *Astrophys. J.* **2000**, *544*, 707–711. [CrossRef]
284. Björnsson, G.; Hjorth, J.; Pedersen, K.; Fynbo, J.U. The Afterglow of GRB 010222: A Case of Continuous Energy Injection. *Astrophys. J.* **2002**, *579*, L59–L62. [CrossRef]
285. Masetti, N.; Palazzi, E.; Pian, E.; Simoncelli, A.; Hunt, L.K.; Maiorano, E.; Levan, A.; Christensen, L.; Rol, E.; Savaglio, S.; et al. Optical and near-infrared observations of the GRB020405 afterglow. *Astron. Astrophys.* **2003**, *404*, 465–481. [CrossRef]
286. Covino, S.; Malesani, D.; Ghisellini, G.; Lazzati, D.; di Serego Alighieri, S.; Stefanon, M.; Cimatti, A.; Della Valle, M.; Fiore, F.; Goldoni, P.; et al. Polarization evolution of the GRB 020405 afterglow. *Astron. Astrophys.* **2003**, *400*, L9–L12. [CrossRef]
287. Barth, A.J.; Sari, R.; Cohen, M.H.; Goodrich, R.W.; Price, P.A.; Fox, D.W.; Bloom, J.S.; Soderberg, A.M.; Kulkarni, S.R. Optical Spectropolarimetry of the GRB 020813 Afterglow. *Astrophys. J.* **2003**, *584*, L47–L51. [CrossRef]
288. Rol, E.; Wijers, R.A.M.J.; Fynbo, J.P.U.; Hjorth, J.; Gorosabel, J.; Egholm, M.P.; Castro Cerón, J.M.; Castro-Tirado, A.J.; Kaper, L.; Masetti, N.; et al. Variable polarization in the optical afterglow of GRB 021004. *Astron. Astrophys.* **2003**, *405*, L23–L27. [CrossRef]
289. Lazzati, D.; Covino, S.; di Serego Alighieri, S.; Ghisellini, G.; Vernet, J.; Le Floc'h, E.; Fugazza, D.; Di Tomaso, S.; Malesani, D.; Masetti, N.; et al. Intrinsic and dust-induced polarization in gamma-ray burst afterglows: The case of GRB 021004. *Astron. Astrophys.* **2003**, *410*, 823–831. [CrossRef]
290. Gorosabel, J.; Rol, E.; Covino, S.; Castro-Tirado, A.J.; Castro Cerón, J.M.; Lazzati, D.; Hjorth, J.; Malesani, D.; Della Valle, M.; di Serego Alighieri, S.; et al. GRB 020813: Polarization in the case of a smooth optical decay. *Astron. Astrophys.* **2004**, *422*, 113–119. [CrossRef]
291. Maiorano, E.; Masetti, N.; Palazzi, E.; Savaglio, S.; Rol, E.; Vreeswijk, P.M.; Pian, E.; Price, P.A.; Peterson, B.A.; Jelínek, M.; et al. Physics of the GRB 030328 afterglow and its environment. *Astron. Astrophys.* **2006**, *455*, 423–431. [CrossRef]
292. Bersier, D.; McLeod, B.; Garnavich, P.M.; Holman, M.J.; Grav, T.; Quinn, J.; Kaluzny, J.; Challis, P.M.; Bower, R.G.; Wilman, D.J.; et al. The Strongly Polarized Afterglow of GRB 020405. *Astrophys. J.* **2003**, *583*, L63–L66. [CrossRef]
293. Loeb, A.; Perna, R. Microlensing of Gamma-Ray Burst Afterglows. *Astrophys. J.* **1998**, *495*, 597–603. [CrossRef]
294. Ghisellini, G.; Lazzati, D. Polarization light curves and position angle variation of beamed gamma-ray bursts. *Mon. Not. R. Astron. Soc.* **1999**, *309*, L7–L11. [CrossRef]
295. Rossi, E.M.; Lazzati, D.; Salmonson, J.D.; Ghisellini, G. The polarization of afterglow emission reveals γ-ray bursts jet structure. *Mon. Not. R. Astron. Soc.* **2004**, *354*, 86–100. [CrossRef]
296. Teboul, O.; Shaviv, N. Impact of the ISM magnetic field on GRB afterglow polarization. *arXiv* **2020**, arXiv:2008.10624.
297. Granot, J.; Panaitescu, A.; Kumar, P.; Woosley, S.E. Off-Axis Afterglow Emission from Jetted Gamma-Ray Bursts. *Astrophys. J.* **2002**, *570*, L61–L64. [CrossRef]
298. Shimoda, J.; Toma, K. Multi-wave band Synchrotron Polarization of Gamma-Ray Burst Afterglows. *Astrophys. J.* **2021**, *913*, 58. [CrossRef]
299. Birenbaum, G.; Bromberg, O. Modelling the linear polarization of GRB afterglows across the electromagnetic spectrum. *Mon. Not. R. Astron. Soc.* **2021**, *506*, 4275–4288. [CrossRef]
300. Mooley, K.P.; Deller, A.T.; Gottlieb, O.; Nakar, E.; Hallinan, G.; Bourke, S.; Frail, D.A.; Horesh, A.; Corsi, A.; Hotokezaka, K. Superluminal motion of a relativistic jet in the neutron-star merger GW170817. *Nature* **2018**, *561*, 355–359. [CrossRef]
301. Granot, J.; Gill, R.; Guetta, D.; De Colle, F. Off-axis emission of short GRB jets from double neutron star mergers and GRB 170817A. *Mon. Not. R. Astron. Soc.* **2018**, *481*, 1597–1608. [CrossRef]

302. Troja, E.; Piro, L.; Ryan, G.; van Eerten, H.; Ricci, R.; Wieringa, M.H.; Lotti, S.; Sakamoto, T.; Cenko, S.B. The outflow structure of GW170817 from late-time broad-band observations. *Mon. Not. R. Astron. Soc.* **2018**, *478*, L18–L23. [CrossRef]
303. Corsi, A.; Hallinan, G.W.; Lazzati, D.; Mooley, K.P.; Murphy, E.J.; Frail, D.A.; Carbone, D.; Kaplan, D.L.; Murphy, T.; Kulkarni, S.R. An Upper Limit on the Linear Polarization Fraction of the GW170817 Radio Continuum. *Astrophys. J.* **2018**, *861*, L10. [CrossRef]
304. Wiersema, K.; Covino, S.; Toma, K.; van der Horst, A.J.; Varela, K.; Min, M.; Greiner, J.; Starling, R.L.C.; Tanvir, N.R.; Wijers, R.A.M.J.; et al. Circular polarization in the optical afterglow of GRB 121024A. *Nature* **2014**, *509*, 201–204. [CrossRef]
305. Nava, L.; Nakar, E.; Piran, T. Linear and circular polarization in ultra-relativistic synchrotron sources—Implications to GRB afterglows. *Mon. Not. R. Astron. Soc.* **2016**, *455*, 1594–1606. [CrossRef]
306. Matsumiya, M.; Ioka, K. Circular Polarization from Gamma-Ray Burst Afterglows. *Astrophys. J.* **2003**, *595*, L25–L28. [CrossRef]
307. Sagiv, A.; Waxman, E.; Loeb, A. Probing the Magnetic Field Structure in Gamma-Ray Bursts through Dispersive Plasma Effects on the Afterglow Polarization. *Astrophys. J.* **2004**, *615*, 366–377. [CrossRef]
308. Toma, K.; Ioka, K.; Nakamura, T. Probing the Efficiency of Electron-Proton Coupling in Relativistic Collisionless Shocks through the Radio Polarimetry of Gamma-Ray Burst Afterglows. *Astrophys. J.* **2008**, *673*, L123. [CrossRef]
309. Urata, Y.; Toma, K.; Huang, K.; Asada, K.; Nagai, H.; Takahashi, S.; Petitpas, G.; Tashiro, M.; Yamaoka, K. First Detection of Radio Linear Polarization in a Gamma-Ray Burst Afterglow. *Astrophys. J.* **2019**, *884*, L58. [CrossRef]
310. Eichler, D.; Waxman, E. The Efficiency of Electron Acceleration in Collisionless Shocks and Gamma-Ray Burst Energetics. *Astrophys. J.* **2005**, *627*, 861–867. [CrossRef]
311. McConnell, M.; Baring, M.; Bloser, P.; Briggs, M.; Dwyer, J.; Foucart, F.; Gaskin, J.; Goldstein, A.; Grove, J.; Gunji, S.; et al. LEAP — A Large Area Gamma-Ray Burst Polarimeter for the ISS. *Bull. Am. Astron. Soc.* **2020**, *235*, 373-08.
312. Pearce, M.; Eliasson, L.; Kumar Iyer, N.; Kiss, M.; Kushwah, R.; Larrson, J.; Lundman, C.; Mikhalev, V.; Ryde, F.; Stana, T.A.; et al. Science prospects for SPHiNX—A small satellite GRB polarimetry mission. *Astropart. Phys.* **2019**, *104*, 54–63. [CrossRef]
313. Kole, M. POLAR-2: The First Large Scale Gamma-ray Polarimeter. In Proceedings of the 36th International Cosmic Ray Conference (ICRC2019), Madison, WI, USA, 24 July–1 August 2019; Volume 36, p. 572. [CrossRef]
314. McConnell, M.L.; Baring, M.; Bloser, P.; Briggs, M.S.; Ertley, C.; Fletcher, G.; Gaskin, J.; Gelmis, K.; Goldstein, A.; Grove, E.; et al. The LargE Area burst Polarimeter (LEAP)—A NASA mission of opportunity for the ISS. *Proc. SPIE* **2021**, *11821*, 204–217.
315. McEnery, J.; Barrio, J.A.; Agudo, I.; Ajello, M.; Alvarez, J.M.; Ansoldi, S.; Anton, S.; Auricchio, N.; Stephen, J.B.; Baldini, L.; et al. All-sky Medium Energy Gamma-ray Observatory: Exploring the Extreme Multimessenger Universe. *arXiv* **2019**, arXiv:1907.07558.
316. Tomsick, J.A.; Boggs, S.E.; Zoglauer, A.; Wulf, E.; Mitchell, L.; Philips, B.; Sleator, C.; Brandt, T.; Shih, A.; Robberts, J.; et al. The Compton Spectrometer and Imager. *arXiv* **2019**, arXiv:1908.04334.
317. Chen, W.; Buckley, J.H. The Advanced Particle-astrophysics Telescope:Simulation of the Instrument Performance for Gamma-RayDetection. In Proceedings of the 37th International Cosmic Ray Conference—PoS (ICRC2021), online conference, 12–23 July 2021; Volume 395, p. 590.
318. Ajello, M.; Arimoto, M.; Axelsson, M.; Baldini, L.; Barbiellini, G.; Bastieri, D.; Bellazzini, R.; Berretta, A.; Bissaldi, E.; Blandford, R.D.; et al. Fermi and Swift Observations of GRB 190114C: Tracing the Evolution of High-energy Emission from Prompt to Afterglow. *Astrophys. J.* **2020**, *890*, 9. [CrossRef]
319. Arnaud, K.A. XSPEC: The First Ten Years. *ASPC* **1996**, *101*, 17.
320. Vianello, G.; Lauer, R.J.; Younk, P.; Tibaldo, L.; Burgess, J.M.; Ayala, H.; Harding, P.; Hui, M.; Omodei, N.; Zhou, H. The Multi-Mission Maximum Likelihood framework (3ML). *arXiv* **2015**, arXiv:1507.08343.
321. Kumar, V.; Bhattacharya, D.; Bhalerao, V.; Rao, A.R.; Vadawale, S. GCN CIRCULAR 20351. 2017.
322. Cheng, K.F.; Zhao, X.H.; Bai, J.M. The synchrotron polarization in decaying magnetic field in gamma-ray bursts. *Mon. Not. R. Astron. Soc.* **2020**, *498*, 3492–3502. [CrossRef]
323. Lan, M.X.; Wu, X.F.; Dai, Z.G. Polarization of GRB Prompt Emission and its Application to POLAR's Data. *arXiv* **2020**, arXiv:2008.10746.
324. Shaviv, N.J.; Dar, A. Gamma-Ray Bursts from Minijets. *Astrophys. J.* **1995**, *447*, 863. [CrossRef]
325. Kumar, P.; Narayan, R. GRB 080319B: Evidence for relativistic turbulence, not internal shocks. *Mon. Not. R. Astron. Soc.* **2009**, *395*, 472–489. [CrossRef]
326. Lazar, A.; Nakar, E.; Piran, T. Gamma-Ray Burst Light Curves in the Relativistic Turbulence and Relativistic Subjet Models. *Astrophys. J.* **2009**, *695*, L10–L14. [CrossRef]
327. Narayan, R.; Kumar, P. A turbulent model of gamma-ray burst variability. *Mon. Not. R. Astron. Soc.* **2009**, *394*, L117–L120. [CrossRef]
328. Ito, H.; Just, O.; Takei, Y.; Nagataki, S. A global numerical model of the prompt emission in short gamma-ray bursts. *arXiv* **2021**, arXiv:2105.09323.
329. van Eerten, H.; van der Horst, A.; MacFadyen, A. Gamma-Ray Burst Afterglow Broadband Fitting Based Directly on Hydrodynamics Simulations. *Astrophys. J.* **2012**, *749*, 44. [CrossRef]

galaxies

Article

The Low-Energy Spectral Index of Gamma-ray Burst Prompt Emission from Internal Shocks

Kai Wang [1,*] and **Zi-Gao Dai** [2,3]

[1] Department of Astronomy, School of Physics, Huazhong University of Science and Technology, Wuhan 430074, China
[2] Department of Astronomy, University of Science and Technology of China, Hefei 230026, China; daizg@ustc.edu.cn
[3] School of Astronomy and Space Science, Nanjing University, Nanjing 210023, China
* Correspondence: kaiwang@hust.edu.cn

Abstract: The prompt emission of most gamma-ray bursts (GRBs) typically exhibits a non-thermal Band component. The synchrotron radiation in the popular internal shock model is generally put forward to explain such a non-thermal component. However, the low-energy photon index $\alpha \sim -1.5$ predicted by the synchrotron radiation is inconsistent with the observed value $\alpha \sim -1$. Here, we investigate the evolution of a magnetic field during propagation of internal shocks within an ultrarelativistic outflow, and revisit the fast cooling of shock-accelerated electrons via synchrotron radiation for this evolutional magnetic field. We find that the magnetic field is first nearly constant and then decays as $B' \propto t^{-1}$, which leads to a reasonable range of the low-energy photon index, $-3/2 < \alpha < -2/3$. In addition, if a rising electron injection rate during a GRB is introduced, we find that α reaches $-2/3$ more easily. We thus fit the prompt emission spectra of GRB 080916c and GRB 080825c.

Keywords: gamma rays bursts; radiation mechanisms; non-thermal

Citation: Wang, K.; Dai, Z.-G. The Low-Energy Spectral Index of Gamma-ray Burst Prompt Emission from Internal Shocks. *Galaxies* **2021**, *9*, 68. https://doi.org/10.3390/galaxies9030068

Academic Editors: Elena Moretti and Francesco Longo

Received: 30 July 2021
Accepted: 13 September 2021
Published: 15 September 2021

1. Introduction

The prompt radiation mechanism of gamma-ray bursts (GRBs) is still being debated, even though the prompt spectra can usually be fitted well by the Band function [1], which suggests a smoothly jointed broken power law with the low-energy photon index $\alpha \sim -1$, the high energy photon index $\beta \sim -2.2$ and the peak energy $E_p \sim 250$ keV [2,3]. Currently, neither the possible one-temperature thermal emission from an ultrarelativistic fireball, nor the single synchrotron radiation from shock-accelerated electrons within this fireball, provide an explanation for such a low-energy photon index (see [4,5] for a review).

Generally, there are two mechanisms that explain the low energy photon index $\alpha \sim -1$ of the GRB prompt emission. The first mechanism is the Comptonized quasi-thermal emission from the photosphere of an ultrarelativistic outflow [6–20]. The second mechanism is synchrotron and/or synchrotron self-Compton (SSC) emission in the optically thin region. For fast-cooling synchrotron radiation in the internal shock model, possible solutions include invoking a small-scale rapidly decaying magnetic field [21], a decaying magnetic field with a power-law index in a relativistically-expanding outflow [22–24], a decaying magnetic field in a post-shock region [25], Klein–Nishina (KN) cooling [26,27], an adjustable synchrotron self-absorption frequency [28,29], or the acceleration process [30] and other evolutional model parameters [31]. Alternatively, slow cooling was introduced to understand the low-energy photon index [32]. In addition to the internal shock model, the other energy dissipation mechanisms, such as the ICMART model [33], were proposed to solve the low-energy spectral index issue. In some models (e.g., [34]), the observed prompt emission of GRBs is understood to be dominated by the SSC emission, while the synchrotron radiation is in much lower energy bands.

The fast cooling synchrotron radiation in the internal shock model is generally considered to be a straightforward and leading mechanism to explain the GRB prompt emission spectra, and the most important issue in this model is to explain the low-energy spectral index. An underlying assumption in the traditional synchrotron internal shock model is to calculate the electron cooling without considering the evolution of the magnetic field. In other words, the magnetic field is treated as a constant and its effect in the continuity equation of electrons is usually ignored (e.g., [35]). In the fast cooling case, the predicted low-energy photon index $\alpha \sim -3/2$ is much softer than observed. In this paper, we try to alleviate this problem. We calculate the magnetic field in the realistic internal shock model during a collision of two relativistic thick shells and obtain an evolutional form of the magnetic field, $B' \propto$ constant before the time δt that is nearly equal to the ejection time interval of the two shells, and $B' \propto t^{-1}$ after the time δt. We consider the cooling of electrons accelerated by internal shocks for this evolutional magnetic field, and find the resulting spectral index $\alpha \sim -3/2$ for $B' \propto$ constant and $\alpha \sim -2/3$ for $B' \propto t^{-1}$, by adopting a cooling method similar to that in Ref. [22]. Actually, these two cases may coexist, and the outflow may undergo the first case and then the second case, so theoretically the actual index α will range from $-3/2$ to $-2/3$. Furthermore, below the peak energy E_p there is a gradual process, so that α is only close to -1. In addition, we consider a rising electron injection rate, leading to a larger α, slightly smaller than $-2/3$.

This paper is organized as follows. We calculate the dynamics of a collision between two thick shells in Section 2. In Section 3, we investigate the electron cooling and its synchrotron radiation with an evolutional magnetic field and a rising electron injection rate. In the final section, discussions and conclusions are given.

2. Dynamics of Two-Shell Collision

In the popular internal shock model, an ultrarelativistic fireball consisting of a series of shells with different Lorentz factors can produce prompt emission through collisions among these shells. For the dynamics of two-shell collision, we adopt the same approach as the one in [36]. In order to present one GRB prompt emission component (with duration $\sim$ few seconds), we here consider two thick shell–shell collision to produce a consistent GRB pulse with a duration of few seconds (i.e., the slow pulse) and the fast pulses with a duration of $\sim$0.01 s in GRBs may be caused by the density fluctuation of the shell. Under this assumption, a prior slow thick shell A with bulk lorentz factor γ_A and kinetic luminosity $L_{k,A}$, and a posterior fast thick shell B with bulk lorentz factor γ_B (where $\gamma_B > \gamma_A \gg 1$) and kinetic luminosity $L_{k,B}$ is adopted. The collision of the two shells begins at radius [36]

$$\begin{aligned} R_{col} &= \beta_B c \frac{\beta_A \Delta t_{int}}{(\beta_B - \beta_A)} \simeq \frac{2\gamma_A^2 c \Delta t_{int}}{1-(\gamma_A/\gamma_B)^2} \\ &\equiv 2\gamma_A^2 c\delta t \simeq 5.4\times 10^{14}\gamma_{A,2.5}^2 \delta t_{,-1}\ \mathrm{cm}, \end{aligned} \tag{1}$$

where Δt_{int} is the time interval between the two thick shells, and $\delta t \equiv \Delta t_{int}/[1-(\gamma_A/\gamma_B)^2]$ is a redefined time interval. For $\gamma_A \ll \gamma_B$, $\delta t \simeq \Delta t_{int}$. The conventional expression $Q_{,m} = Q/10^m$ is used. During the collision, there are four regions separated by internal forward-reverse shocks: (1) the unshocked shell A; (2) the shocked shell A; (3) the shocked shell B; and (4) the unshocked shell B, where regions 2 and 3 are separated by a contact discontinuity.

The particle number density of a shell measured in its comoving frame can be calculated as [37]:

$$n_i' = \frac{L_{k,i}}{4\pi R^2 \gamma_i^2 m_p c^3}, \tag{2}$$

where R is the radius of the shell and subscript i can be taken as A or B. As in the literature [36–40], we derive the dynamics of internal forward-reverse shocks. In order to get a high prompt emission luminosity, it is reasonable to assume $\gamma_A \ll \gamma_B$ and $L_{k,A} = L_{k,B} \equiv L_k$. Assuming that γ_1, γ_2, γ_3, and γ_4 are Lorentz factors of regions 1, 2, 3 and 4 respectively, we have $\gamma_1 = \gamma_A$, $\gamma_4 = \gamma_B$, and $n'_1 \gg n'_4$. If a fast shell with low particle number density catches up with a slow shell with high particle number density and then they collide with each other, a Newtonian forward shock (NFS) and a relativistic reverse shock (RRS) may be generated [36,37]. So we can obtain $\gamma_1 \simeq \gamma_2 = \gamma_3 = \gamma \ll \gamma_4$. Then, according to the jump conditions between the two sides of a shock [41], the comoving internal energy densities of the two shocked regions can be calculated following $e'_2 = (\gamma_{21} - 1)(4\gamma_{21} + 3)n'_1 m_p c^2$ and $e'_3 = (\gamma_{34} - 1)(4\gamma_{34} + 3)n'_4 m_p c^2$, where $\gamma_{21} = \frac{1}{2}(\gamma_1/\gamma_2 + \gamma_2/\gamma_1)$ and $\gamma_{34} = \frac{1}{2}(\gamma_3/\gamma_4 + \gamma_4/\gamma_3)$ are the Lorentz factors of region 2 relative to the unshocked region 1, and region 3 relative to region 4, respectively. It is required that $e'_2 = e'_3$ because of the mechanical equilibrium. We have [36,37]

$$\frac{(\gamma_{21} - 1)(4\gamma_{21} + 3)}{(\gamma_{34} - 1)(4\gamma_{34} + 3)} = \frac{n'_4}{n'_1} = \left(\frac{\gamma_1}{\gamma_4}\right)^2 \equiv f. \tag{3}$$

Two relative Lorentz factors can be calculated as $\gamma_{21} \approx \frac{f\gamma_4^2}{7\gamma_1^2} + 1 = \frac{8}{7}$, and $\gamma_{34} = \frac{\gamma_4}{2\gamma_1} \gg 1$. Assuming that t is the observed shell–shell interaction time since the prompt flare onset, the radius of the system during the collision can be written as

$$R = R_{col} + 2\gamma^2 ct \simeq 2\gamma_1^2 c(t + \delta t). \tag{4}$$

During the propagation of the shocks and before the shock crossing time, the instantaneous electron injection numbers (in dt) in regions 2 and 3 can be calculated as follows [38]:

$$dN_{e,2} = 8\pi R^2 n'_1 (\gamma_{21}\beta_{21}/\gamma\beta)\gamma^2 cdt \tag{5}$$

and

$$dN_{e,3} = 8\pi R^2 n'_4 (\gamma_{34}\beta_{34}/\gamma\beta)\gamma^2 cdt, \tag{6}$$

respectively.

3. Synchrotron Radiation with a Decaying Magnetic Field and a Variable Electron Injection Rate

3.1. Synchrotron Radiation with a Decaying Magnetic Field

As usual, we assume that fractions ϵ_B and ϵ_e of the internal energy density in a GRB shock are converted into the energy densities of the magnetic field and electrons, respectively. Thus, using $B'_i = (8\pi\epsilon_B e'_i)^{1/2}$ for i = 2 or 3, we can calculate the strength of the magnetic field before the shock crossing time t_{crs} by

$$B'_2 = B'_3 = \left[\frac{\epsilon_B L_k}{2\gamma_1^6 c^3 (t + \delta t)^2}\right]^{1/2}, \tag{7}$$

and find that the change of the magnetic field before δt can be ignored (i.e., $B'_i \propto$ constant), but after δt the magnetic field B'_i decreases linearly with time t (i.e., $B'_i \propto t^{-1}$). Actually, the evolution of the magnetic field is caused by the expansion of the shocked regions, which is presented in Figure 1. After the shock crossing time t_{crs} (here, t_{crs} is comparable with the peak time of the slow pulse in GRBs), the spreading of the hot materials into the vacuum cannot be ignored and the merged shell undergoes an adiabatic cooling. During this phase, the volume of the merged shell is assumed to expand as $V'_i \propto R^s$, where s is a free parameter and its value is taken to be from 2 to 3. As a result, the particle number density would decrease as $n'_i \propto {V'}_i^{-1} \propto R^{-s}$, the internal energy density as $e'_i \propto {V'}_i^{-4/3} \propto R^{-4s/3}$, and the magnetic field strength as $B'_i \propto (e'_i)^{-1/2} \propto R^{-2s/3} \propto t^{-2s/3}$. Because no additional shock-accelerated electrons are injected after the shock crossing time t_{crs}, we only study the prompt emission before t_{crs} in the remaining part of this paper. What we want to point out is that the redefined time interval δt is not equal to the shock crossing time (t_{crs}), the latter one is dependent on the thickness of the shells. In this paper, the two shells are must be thick enough so that $t_{crs} \gg \delta t$.

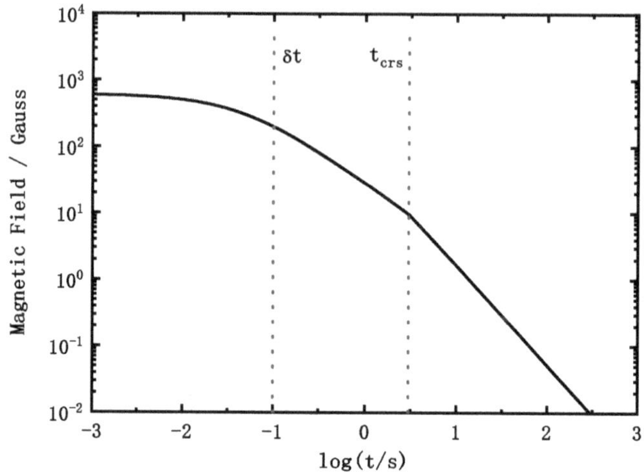

Figure 1. The magnetic field as a function of time. The two blue vertical dotted line represents the redefined interval $\delta t = 0.1$ s and the shock crossing time $t_{crs} = 3$ s, respectively. After the shock crossing time, the merged shell expands adiabatically and $s = 3$ is assumed. The dynamics parameters $L_k = 10^{51}$ erg s^{-1}, $\gamma_1 = 300$, $\gamma_4 = 30{,}000$, $p = 2.5$, $\epsilon_e = 0.3$, $\epsilon_B = 0.3$, and $z = 1$ are taken from numerical calculations.

The electrons accelerated by the shocks are assumed to have a power-law energy distribution, $dN_{e,i}/d\gamma'_{e,i} \propto \gamma_e'^{-p}$ for $\gamma'_{e,i} \geq \gamma'_{e,m,i}$, where $\gamma'_{e,m,i}$ is the minimum Lorentz factor of the accelerated electrons. The following electron cooling discussion is not based on the conventional synchrotron and SSC cooling, which always give us the electron distribution, $\frac{dN_e}{d\gamma'_e} \propto \gamma_e'^{-2}$ for $\gamma'_e < \gamma'_{e,m}$ and $\frac{dN_e}{d\gamma'_e} \propto \gamma_e'^{-p-1}$ for $\gamma'_e > \gamma'_{e,m}$ in the fast cooling case, $\frac{dN_e}{d\gamma'_e} \propto \gamma_e'^{-p}$ for $\gamma'_e < \gamma'_{e,m}$ and $\frac{dN_e}{d\gamma'_e} \propto \gamma_e'^{-p-1}$ for $\gamma'_e > \gamma'_{e,m}$ in the slow cooling case. These electron distributions do not take into account the evolution of the magnetic field. Ref. [22] discussed the electron distribution affected by a decaying magnetic field based on $B' \propto r^{-b}$, where r is the fireball radius and b is the magnetic field decaying index. They considered the electron distribution of a group of plasma in a magnetic field with an arbitrary decaying index b, which is called a "toy box model". Here we consider a more physical process, internal shocks, which generate an evolutional magnetic field and a consistent spectrum with the observed Band spectral shape.

In the comoving frame, the evolution of the Lorentz factor of an electron via synchrotron and SSC cooling and adiabatic cooling can be described by [22]

$$\frac{d}{dt'}\left(\frac{1}{\gamma_e}\right) = \frac{\sigma_T(1+Y_i)}{6\pi m_e c}B_i'^2 - \frac{1}{3}\left(\frac{1}{\gamma_e}\right)\frac{d\ln n'_i}{dt'}, \tag{8}$$

where $Y_i \approx [(4\eta_i\epsilon_e/\epsilon_B + 1)^{1/2} - 1]/2$ is the Compton parameter, which is defined by the ratio of the IC to synchrotron luminosity, with $\eta_i = \min[1, (\gamma'_{e,c,i}/\gamma'_{e,m,i})^{2-p}]$ [42]. $\gamma'_{e,c,i}$ is the cooling Lorentz factor and the comoving time $t' = 2\gamma t$.

The minimum Lorentz factor of the accelerated electrons is $\gamma'_{e,m,i} = \frac{m_p}{m_e}(\frac{p-2}{p-1})\epsilon_e(\gamma_{\rm rel} - 1)$ (where $\gamma_{\rm rel} = \gamma_{21}$ or γ_{34} in region 2 or 3), so it can be written as:

$$\gamma'_{e,m,3} \simeq 1.0 \times 10^4 g_p \epsilon_{e,-1/2}\gamma_{4,4.5}\gamma_{1,2.5}^{-1}, \tag{9}$$

$$\gamma'_{e,m,2} \simeq 30 g_p \epsilon_{e,-1/2}, \tag{10}$$

where $g_p = 3(p-2)/(p-1)$. Moreover, the cooling Lorentz factor $\gamma'_{e,c,i} = 6\pi m_e c/(y_i \sigma_T B_3'^2 \gamma t)$, can be written as

$$\gamma'_{e,c,3} = \gamma'_{e,c,2} \simeq 3.4 \times 10^2 y_{i,0}^{-1} \epsilon_{B,-1/2}^{-1} L_{k,51}^{-1} \gamma_{1,2.5}^{5} \frac{(t+\delta t)_{,0}^2}{t_{,0}}, \tag{11}$$

where $y_i = 1 + Y_i$ is the ratio of the total luminosity to synchrotron luminosity.

From the electron injection rate based on Equations (5) and (6), one can obtain the injected electrons number between t' and $t' + dt'$. Assuming the original electron injection distribution $dN_{e,i}/d\gamma'_{e,i} \propto \gamma_e'^{-p}$ for $\gamma'_{e,i} \geq \gamma'_{e,m,i}$, the injected electrons number between t' and $t' + dt'$ and between γ'_e and $\gamma'_e + d\gamma'_e$ can be derived. So we cut the injected electrons into small pieces in the time space t' and the energy space γ'_e. At the beginning, time $t' = 0$, a number of electrons dN will be injected into the shocked region in a time interval dt' and will be cooled in the initial magnetic field, so one can obtain the change of electron Lorentz factor $\Delta\gamma'_{e,1}$ based on Equation (8) for the electrons between γ'_e and $\gamma'_e + d\gamma'_e$. In the next time interval dt', these electrons with the Lorentz factor between $\gamma'_e + \Delta\gamma'_{e,1}$ and $\gamma'_e + d\gamma'_e + \Delta\gamma'_{e,1}$ will be cooled in the instantaneous magnetic field based on the evolutional magnetic field in Equation (7), and one can obtain another $\Delta\gamma'_{e,2}$ ($\Delta\gamma'_{e,2} \neq \Delta\gamma'_{e,1}$). At the same time, another group electrons are injected and cooled in this instantaneous magnetic field. These processes are continuous before $2\gamma t_{crs}$. The shocked electrons are injected as time and all the electrons are cooled in the instantaneous magnetic field. We sum all electrons at time t' in the energy space, obtain the electron distributions at time t' and present them in Figure 2 ($t_{obs} = t'/2\gamma$). As shown in Figure 2, when $t < \delta t$, the magnetic field does not change significantly (see Figure 1), the electron distribution in the fast cooling case, $\frac{dN_e}{d\gamma'_e} \propto \gamma_e'^{-2}$ for $\gamma'_e < \gamma'_{e,m}$, and $\frac{dN_e}{d\gamma'_e} \propto \gamma_e'^{-p-1}$ for $\gamma'_e > \gamma'_{e,m}$, are expected. However, when $t > \delta t$, the electron distribution below $\gamma'_{e,m}$ would be flattened because of the decaying magnetic field. Due to the magnetic field decay, the electrons injected at later times would cool more slowly than the electrons injected at early times (here, all times are before t_{crs}). In other words, the cooling efficiency would become smaller due to the decaying magnetic field, which induces more electrons accumulating at $\lesssim \gamma'_{e,m}$ than in the invariable magnetic field case. When $t \gg \delta t$ and $t < t_{crs}$, the electron spectral index for $\gamma'_e < \gamma'_{e,m}$ is even flattened to zero.

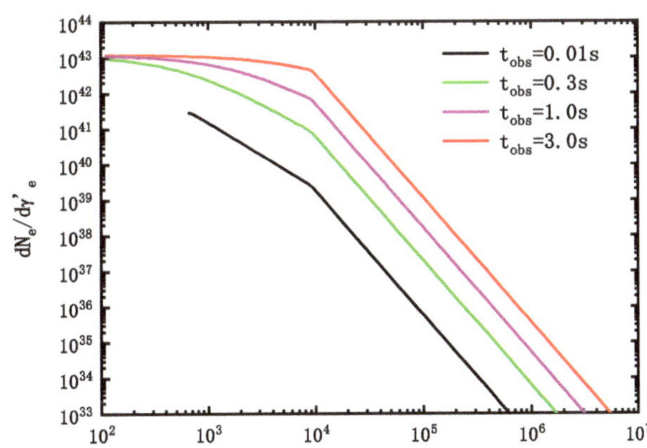

Figure 2. The electron distribution in energy space after cooling time t in the evolutional magnetic field in Figure 1. The same δt, t_{crs}, and dynamics parameters as in Figure 1 are taken in numerical calculations.

In order to find these new results for $t \gg \delta t$ and $t < t_{crs}$, we can evaluate the continuity equation of electrons in energy space, $\frac{\partial}{\partial t'}(dN_{e,\gamma'_e}/d\gamma'_e) + \frac{\partial}{\partial \gamma'_e}[\dot{\gamma}'_e(dN_{e,\gamma'_e}/d\gamma'_e)] = Q(\gamma'_e, t')$, where $dN_{e,\gamma'_e}/d\gamma'_e$ is the instantaneous electron spectrum at time t', and $Q(\gamma'_e, t') = Q_0(t')(\gamma'_e/\gamma'_{e,m})^{-p}$ is the electron injection distribution accelerated by shocks above the minimum injection Lorentz factor $\gamma'_{e,m}$. By ignoring the inconsequential adiabatic cooling term, we can get $\frac{d}{dt'}(\frac{1}{\gamma'_e}) \propto (1+Y_i)B_i'^2 \propto t'^{-2}$, where Y_i is assumed to be a constant before the shock crossing time t_{crs} in the fast cooling case. Then, we can obtain $\gamma'_e \propto t'$, and thus $\dot{\gamma}'_e \propto \gamma_e'^2 t'^{-2} \propto \gamma_e'^0$. For $\gamma'_{e,c} < \gamma'_e < \gamma'_{e,m}$, $Q(\gamma'_e, t') = 0$, to obtain the final and quasi-steady electron spectral shape at the arbitrary time t', by considering a quasi-steady-state system $(\partial/\partial t = 0)$, we can easily find $dN_{e,\gamma'_e}/d\gamma'_e \propto \gamma_e'^0$ below $\gamma'_{e,m}$.

Next, the four characteristic frequencies in regions 2 and 3 that can be calculated from $\nu = \frac{q_e}{2\pi m_e c} B' \gamma_e'^2 \gamma$ are derived as [36]

$$h\nu_{m,2} \simeq 2.1 \times 10^{-4} g_p^2 \epsilon_{e,-1/2}^2 \epsilon_{B,-1/2}^{1/2} L_{k,51}^{1/2} \gamma_{1,2.5}^{-2} (t+\delta t)_{,0}^{-1} \text{ keV}, \quad (12)$$

$$h\nu_{m,3} \simeq 26 g_p^2 \epsilon_{e,-1/2}^2 \epsilon_{B,-1/2}^{1/2} L_{k,51}^{1/2} \gamma_{4,4.5}^2 \gamma_{1,2.5}^{-4} (t+\delta t)_{,0}^{-1} \text{ keV}, \quad (13)$$

and

$$h\nu_{c,2} = h\nu_{c,3} \simeq 3.6 \times 10^{-2} y_{,0}^{-2} \epsilon_{B,-1/2}^{-3/2} L_{k,51}^{-3/2} \gamma_{1,2.5}^{8} \frac{(t+\delta t)_{,0}^3}{t_{,0}^2} \text{ keV}. \quad (14)$$

Here, if $\gamma_1 = 100$ and $\gamma_4 = 10{,}000$, we obtain $h\nu_{m,3} \simeq 186$ keV at time $t = 1$ s, which is approximatively equal to the typical value of E_p of the GRB prompt emission.

We also present the spectrum of region 3 in the top panel of Figure 3 based on the electron distribution shown in Figure 2. However, we do not present the spectrum of region 2 because, from NFS, (1) its photon peak frequency is much smaller than the typical GRB prompt emission E_p, (2) the radiation efficiency can not be high enough as a result of slow cooling, and (3) the flux of region 2 is much lower than that of region 3. The last reason can be evaluated from [35,37]

$$F_{\nu,\max,i} < \frac{N_{e,i}}{4\pi D_L^2} \frac{m_e c^2 \sigma_T}{3 q_e} B_i' \gamma, \quad (15)$$

where D_L is the luminosity distance of the burst and $N_{e,i}$ is the total number of injected electrons until the time t. Since a portion of the electrons have cooled to a much smaller value than γ_{break} (where the break Lorentz factor γ_{break} of an electron distribution, $\gamma_{break} = \gamma_{e,m}$ for fast cooling, and $\gamma_{break} = \gamma_{e,c}$ for slow cooling), the actual number of electrons near γ_{break} is less than $N_{e,i}$ and thus the actual $F_{\nu,\max,i}$ is smaller than the right term of inequality Equation (15). So, we can obtain

$$\nu_{m,3} F_{\nu,\max,3} < 1.5 \times 10^5 g_p^2 \epsilon_{e,-1/2}^2 \epsilon_{B,-1/2} L_{k,51}^2 \gamma_{4,4} \gamma_{1,2}^{-6} \times \frac{t_{,0}}{(t+\delta t)_{,0}^2} D_{L,28}^{-2} \text{ keV cm}^{-2}\text{s}^{-1}, \quad (16)$$

and

$$\nu_{c,2} F_{\nu,\max,2} < 8.7 \times 10^{-1} y_{,0}^{-2} \epsilon_{B-1/2}^{-1} \gamma_{1,2}^{5} \times \frac{(t+\delta t)_{,0}^2}{t_{,0}} D_{L,28}^{-2} \text{ keV cm}^{-2}\text{s}^{-1}, \quad (17)$$

where $\gamma_1 = 100$ and $\gamma_4 = 10{,}000$ are taken.

From Figure 3, we can see that for $t < \delta t$, because of a constant magnetic field, the spectral slope of νF_ν is $1/2$ as described by [35]. However, when $t > \delta t$, the spectral slope will deviate from $1/2$ and become a larger value (even $4/3$). If the electron index

($dN_e/d\gamma_e \propto \gamma_e^{-u}$) is u, the F_ν slope of synchrotron radiation ($F_\nu \propto \nu^{-w}$) would be $w = (u-1)/2$ and the photon spectral index (defined as $dN_\gamma/dE_\gamma = E_\gamma^{-\alpha}$, where E_γ is the photon energy, and N_γ is the photon number flux) would be $\alpha = -(w+1)$. Due to the decaying magnetic field, u tends to be zero, and thus $w = -1/2$ and $\alpha = -1/2$. However, when $\alpha > -2/3$, because of the overlying effect, the low energy photon index of the electrons with $\sim \gamma_{e,m}$ is $-2/3$ and will cover the emission of electrons with smaller Lorentz factors. So, due to the effect of the low-energy radiation tail of electrons with Lorentz factor $\gamma_{e,m}$, α is at most equal to $-2/3$, and we can get $-3/2 < \alpha < -2/3$. This is consistent with the observations [2,3], which suggest $\alpha \sim -1$.

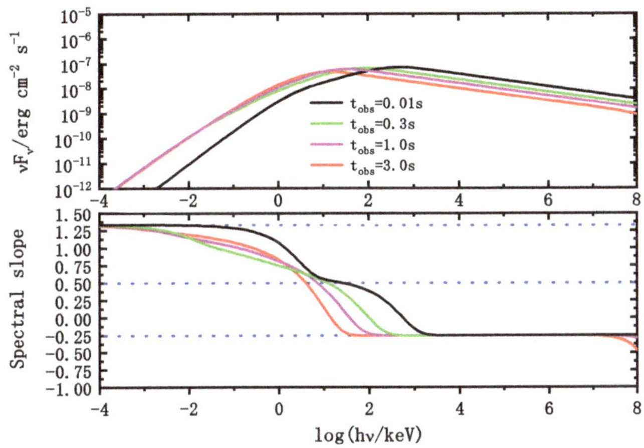

Figure 3. The top panel corresponds to time-resolved spectra in four different t as in Figure 2 and the bottom panel shows corresponding synchrotron spectral slopes. The same δt, t_{crs}, and dynamics parameters as in Figure 1 are taken in numerical calculations.

3.2. The Effect of the Variable Electron Injection Rate

Although theoretically the low-energy photon spectral index α can reach $-2/3$ caused by a decaying magnetic field, since this is a gradual process, α would be softer than $-2/3$ for $E \lesssim E_p$, which can be seen from Figure 3. In fact, $\alpha \sim -1$ can be fitted easily, but fitting α slightly smaller than $-2/3$ is difficult. We here consider a variable electron injection rate, which could induce $\alpha \sim -2/3$. The variable electron injection rate may be suggested by that the actual GRB shell is not homogeneous and presents a density profile, for example, a Gaussian density profile, inducing a rising electron injection rate. Nonetheless, we do not know its growing method clearly. Ref. [22] discussed this effect in their "toy box model", and suggested, because of a rising electron injection rate, α goes from -0.82 to -1.03, which is dependent on the growing power-law index q (where the injection rate $\propto t'^q$ with $q = 1$, 2 or 3). Here we adopt similar expressions of the rising electron injection rate,

$$dN_{e,2} = 8\pi R^2 n_1'(\gamma_{21}\beta_{21}/\gamma\beta)\gamma^2 cdt \times \left(\frac{t}{t_0}\right)^q \tag{18}$$

and

$$dN_{e,3} = 8\pi R^2 n_4'(\gamma_{34}\beta_{34}/\gamma\beta)\gamma^2 cdt \times \left(\frac{t}{t_0}\right)^q, \tag{19}$$

where the factor $(\frac{t}{t_0})$ is to maintain the same electron injection number in the interval t_0 as that for the constant injection rate $q = 0$.

We show the electron distribution for a rising electron injection rate in the top panel of Figure 4. The rising electron injection would increase the electrons injected later, which would cool in a weaker magnetic field and pile up at $\lesssim\gamma'_{e,m}$. This can result in a harder electron distribution and a relative spectrum. The slopes of the spectra are presented in the bottom panel of Figure 4. We can see that the slopes of the spectra tend to reach $-4/3$ more easily than in the constant electron injection case. In addition, a larger q would generate a harder low-energy spectral index.

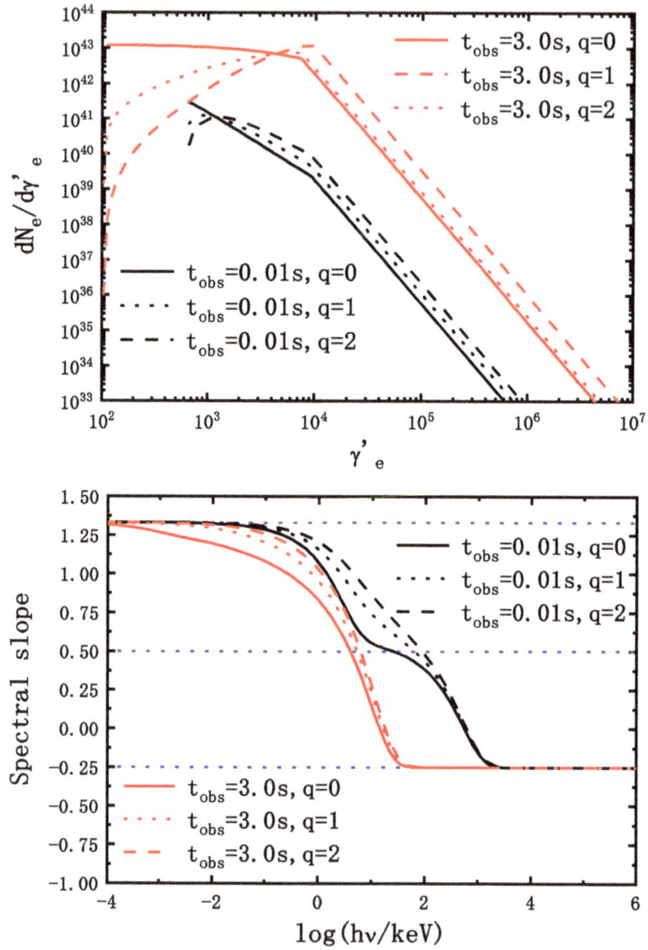

Figure 4. The top panel shows the electron distributions in evolutional magnetic fields and different electron injection rising indices. We adopt the electron injection rising index, $q = 0$ (solid line), $q = 1$ (dotted line) and $q = 2$ (dashed line). The bottom panel shows the corresponding synchrotron spectral slopes for these electron distributions. The same parameters as in Figure 1 are taken for numerical calculations.

4. Application to the Actual GRB Spectra

In order to compare with the actual GRB spectrum, we select the broad band spectrum of GRB 080916c in the interval "b" detected by Gamma-ray Burst Monitor (GBM) and the Large Area Telescope (LAT) aboard the Fermi satellite (see Ref. [43]), from 3.58 s to 7.68 s since the lightcurve during this period is presented as a single and pure pulse. Moreover, its low energy photon index is close to the typical value of the low energy photon index of the GRB, that is, $\alpha \sim -1$, harder than the expectation of synchrotron fast cooling (-1.5). The "b" spectrum of GRB 080916c can be well fitted by the Band function with the low energy photon index of $\alpha = -1.02 \pm 0.02$, the high energy photon spectral slope $\beta = -2.21 \pm 0.03$

and the peak energy $E_p = 1170 \pm 140$ keV [43]. Since the observational data can be well fitted by this Band function with a very small error range, the Band function is precise enough to represent the actual GRB emission. We select some representative points (black points in Figure 5) in this Band function to present the tendency of the actual GRB emission. In addition, more black points around the peak energy in the figure are taken to present the gradual change in behavior there. In Figure 5, by using a time-averaged energy spectrum from $t = 0$ s to $t = 3$ s, the emission of GRB 080916c can be fitted well in our model with the proper parameters, which have been listed in Table 1.

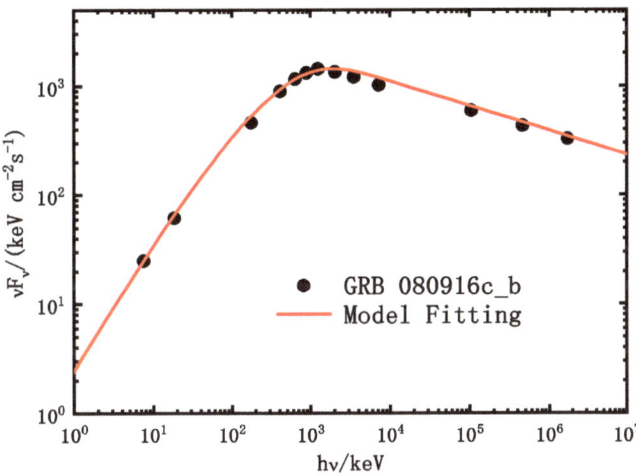

Figure 5. The time-averaged spectrum to fit the interval "b" of GRB 080916c. The black points are selected from the Band function with the low-energy photon index $\alpha = -1.02 \pm 0.02$, the high-energy photon index $\beta = -2.21 \pm 0.03$ and the peak energy $E_p = 1170 \pm 140$ keV provided in Ref. [43], which are precise enough to present the tendency of the actual GRB emission. This spectral duration is from 3.58 s to 7.68 s and we fit it by adopting the time-averaged spectrum from $t = 0$ s to $t = 3$ s ($t_{max} \leq t_{crs}$). The fitting parameters are listed in Table 1.

We select the single pulse spectrum of GRB 080825c in the interval "a" detected by Fermi GBM and LAT (see Ref. [44]), from 0.0 s to 2.7 s, which has a harder photon index, $\alpha \sim -0.76$. The "a" spectrum of GRB 080825c can be well fitted by the Band function with the low energy photon index $\alpha = -0.76 \pm 0.05$, the high energy photon index $\beta = -2.54^{+0.11}_{-0.17}$ and the peak energy $E_p = 291^{+25}_{-22}$ keV [44]. Such a hard photon index could not be approached easily for a constant electron injection rate, that is, $q = 0$, so we consider a rising electron injection rate as suggested in Section 3.2. Some representative points (black points in Figure 6) in this Band function are selected to present the tendency of the actual GRB emission as the same as the treatment for the GRB 080916c. The observational spectrum can be reproduced well in Figure 6 phenomenally by using a time-averaged energy spectrum from $t = 0$ s to $t = 3$ s with an index of rising electron injection rate $q = 2$ and other reasonable parameters (all parameters are listed in Table 1).

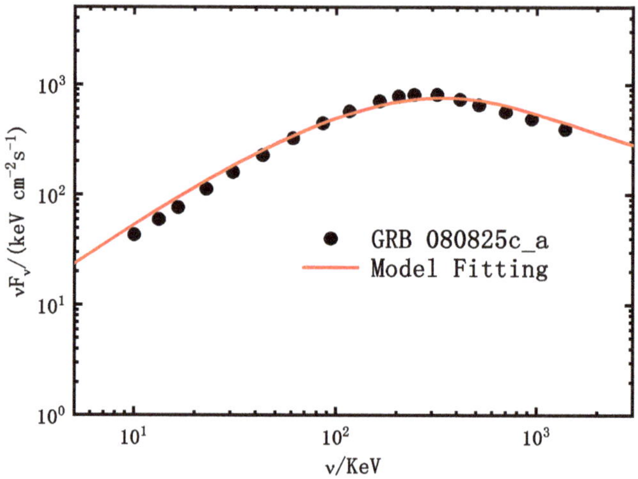

Figure 6. The time-averaged spectrum to fit the interval "a" of GRB 080825c. The black points are selected from the Band function with the low-energy photon index $\alpha = -0.76 \pm 0.05$, the high-energy photon index $\beta = -2.54^{+0.11}_{-0.17}$ and the peak energy $E_p = 291^{+25}_{-22}$ keV provided in Ref. [44], which are precise enough to present the tendency of the actual GRB emission. This spectral duration is from 0.0 s to 2.7 s and we fit it by adopting the time-averaged spectrum from $t = 0$ s to $t = 3$ s ($t_{max} \leq t_{crs}$). The fitting parameters are listed in Table 1.

Table 1. The parameters adopted to fit the spectra of GRB 080916c and GRB 080825c.

Parameters	Symbol	GRB 080916c	GRB 080825c
Redshift	z	4.35	1
Index of electron injection rate	q	0	2
redefined time interval (s)	δt	0.1	0.1
Shock cross time (s)	t_{crs}	3	3
Kenetic luminosity (erg/s)	L_k	3.3×10^{53}	1.2×10^{51}
Bulk Lorentz factor of region 1	γ_1	146	255
Bulk Lorentz factor of region 2	γ_4	3×10^4	3×10^4
Electron injection index	p	2.5	3.2
Electron equipartition factor	ϵ_e	0.3	0.3
Magnetic equipartition factor	ϵ_B	0.3	0.3

The main parameters to effect the final spectrum are listed in Table 1. The dependence of the break energy of the spectrum on the listed parameters could be found in Equation (13) and for the magnitude of peak flux the dependence could be derived roughly in Equation (16). During the model fitting, for simplification, the energy equipartition factors for electrons and the magnetic field, that is,ϵ_e and ϵ_B, and γ_4 are fixed, and then the Lorentz factor γ_1 and kinetic luminosity L_k are adjusted to match the observational peak energy and the peak flux. The electron injection index p is determined by the observational high-energy photon index since the relation between them is $\beta \sim (-p-2)/2$, [1] suggested by the synchrotron radiation. The shock cross time is comparably adopted with the typical duration of the slow pulse of the GRB, namely, $\sim$3 s. Different values of δt could affect the evolutional form of the magnetic field (as shown in Figure 1) and adjust the weight of the cooling in a constant magnetic field and the cooling in a decaying magnetic field. In other words, a smaller δt could make it so that the electron synchroton cooling mainly takes place in a decaying magnetic field and leads the photon index to be harder, while for a larger δt, the electrons are mainly cooling in a constant magnetic field and generating a photon index close to -1.5. As a result, for GRB 080916c with a photon index ~ -1, a relatively small $\delta t = 0.1$ s is adopted. A harder photon index ~ -0.76 for GRB 080825c and a rising electron injection rate with an index $q = 2$ are taken into account as suggested in Section 3.2. Therefore, a certain range of a low-energy photon index from $\sim -3/2$ to $\sim -2/3$ could be approached through the adjustment of δt and the index of the electron injection rate q. However, for a low-energy photon index harder than $-2/3$, this model would become invalid.

5. Discussions and Conclusions

Currently alleviating the tension between the expectation of synchrotron and observations in the GRB prompt regime is a more and more important issue. Two classes of model have been proposed to explain the low-energy photon index of GRB prompt emission, Comptonized quasi-thermal emission from the photosphere within a relativistic outflow and synchrotron and/or SSC emission in the optically thin region. These models can experience difficulties. For Comptonized quasi-thermal emission, the most significant effect to obtain $\alpha \sim -1$ is the equal arrival time effect in this model, which is relevant to the end time of central engine activity, but may not be applicable during the prompt emission phase when a continuous wind is ejected from the central engine [45]. Synchrotron slow cooling in internal shocks may not provide a high radiative efficiency, and a dominant SSC component usually predicts an even more dominant 2nd-order SSC component, which significantly exceeds the total energy budget of GRBs [46,47]. Thus, some evolutional parameters, such as the magnetic field, the fraction of the accelerated electrons, and the energy equipartition factors, were suggested to explain the low-energy index.

In this paper, we have considered a straightforward model, that is, the fast cooling synchrotron radiation in internal shocks. We obtain the magnetic field evolutional form in a practical shell–shell collision, $B' \propto$ constant before δt and $B' \propto t^{-1}$ after this time, and recalculate the electron distribution for this evolutional magnetic field. When $t < \delta t$, the magnetic field is nearly constant, and the fraction of cooling electrons in the invariable magnetic field is high enough so that $\frac{dN_e}{d\gamma'_e} \propto \gamma_e'^{-2}$ for $\gamma'_e < \gamma'_{e,m}$ is expected. However, when $t \gg \delta t$ but $t < t_{crs}$, the fraction of cooling electrons in the evolutional magnetic field is higher than in the invariable magnetic field, so that $\frac{dN_e}{d\gamma'_e}$ will be gradually proportional to $\gamma_e'^{0}$. $2\gamma^2 c\delta t$ and $2\gamma^2 ct$ indicate roughly the collision radius and the propagation distance of a relativistic outflow after the collision takes place but before the shock crossing time t_{crs}, respectively. In other words, if the propagation distance of the outflow is smaller than the collision radius before the shock crossing time, the magnetic field can be treated as a constant and it is not necessary to consider the evolution of the magnetic field when calculating the electron cooling. However, if the propagating distance of the outflow is larger than the collision radius before the shock crossing time, we have to consider the evolution of the magnetic field and can obtain a different electron distribution. Actually, the outflow may undergo the first case and then the second case, so we can obtain a reasonable range of the low-energy photon index α, from $-3/2$ to $-2/3$ theoretically. Since $\frac{dN_e}{d\gamma'_e}$ proportional to $\gamma_e'^{0}$ is a gradual process below E_p, it is usually difficult to get α to be exactly equal to $-2/3$, but this index is only slightly smaller than $-2/3$. Moreover, we also consider a rising electron injection rate, which may exacerbate this situation, inducing α to be closer to $-2/3$.

Ref. [22] considered a decaying magnetic field varying with the distance from the central engine to explore the range of a low energy photon index in the GRB prompt regime. They discussed the radiation spectra of a cloud of plasma in a decaying magnetic field with an arbitrary decaying index b for a simplified model, which is called a "toy box model". Different from their work, we adopt a more physical case for the internal shock by considering the collision of two shells and inducing the decaying form of the magnetic field. As a result, a time-dependent magnetic field is derived (as shown in Figure 1). In fact, a time-dependent magnetic field could be translated to a distance-dependent form due to the propagation of relativistic outflow. For the evolutional magnetic field form obtained from the practical internal shock, we study the influence on the spectral index. In addition to the detailed treatment of shell–shell collision, the kinetic luminosity and the energy equipartition parameters, ϵ_B and ϵ_e are taken into account to obtain the radiation spectra, comparing them with the actual GRB spectra for GRB 080916c and GRB 080825c.

In our model, in order to obtain the high prompt emission luminosity, we assume that $\gamma_4 \gg \gamma_1$. This assumption is reasonable. This is because estimates based on four methods by Ref. [48] show that the mean observed value of the bulk Lorentz factors of GRB outflows is a few hundred, corresponding to $\gamma_2 = \gamma_3 \simeq \gamma_1 \sim 100$ in our model. Furthermore, within the framework of the collapsar model, a prior relativistic jet-like shell (e.g., shell A) first has to propagate through the envelope of a massive star and clean up almost all of the baryons along the propagation direction of this shell, leaving behind a clean passage for a posterior jet-like shell (e.g., shell B). This, therefore, leads to a reasonable possibility that the Lorentz factor of shell B is much greater than that of shell A.

Usually, we have $\gamma_1 \sim 100$, so $\gamma_4 \sim 10^4 \sim \gamma_1^2$ is a universal relationship to obtain the high prompt emission luminosity. Ref. [37] also mentioned that, when $\gamma_4 \sim \gamma_1^2$, the highest luminosity from internal shocks is expected. In fact, this assumption is not a special case. When collisions among a series of shells with different Lorentz factors occur, the highest luminosity from one collision will cover the others. In other words, we always see the brightest. According to Equation (13), if deeming that $\gamma_4\gamma_1^{-2}$ does not vary significantly among bursts, we can easily obtain the so-called "Yonetoku Relation", $E_p \propto L_{iso}^{1/2}$ [49], and the "Amati Relation", $E_p \propto E_{iso}^{1/2}$ [50]. However, this model is also confronted with some issues, for example, the spectrum is somewhat broad near E_p in contrast to the observed data or the Band function [51], which can be seen in Figure 6.

It is important that we should beware of the empirical Band function. The thermal components and more spectral structures are found in the prompt regimes of some GRBs, which deviate from the so-called Band function [52,53]. The thermal emission generated by the photosphere is a natural prediction of the generic fireball scenario. The relative strength of thermal emission and non-thermal emission may depend on the various environments [54,55]. Ref. [53] also claimed that the GRB spectra below the peak energy may present an extra break energy around a few keV, inducing a consistent spectral shape with expectation from the classical synchrotron radiation. More spectral structures of GRBs may make the simple Band function become invalid, and result in an incorrect low-energy spectral index if one forcibly fits them using a Band function. Although our model can present a consistent low-energy spectral index with observations in a certain range, due to the complexities of GRB prompt spectra, more detailed studies are needed.

Author Contributions: All authors have contributed equally to the preparation of this manuscript. All authors have read and agreed to the published version of the manuscript.

Funding: This work was supported by the National Key Research and Development Program of China (grant No. 2017YFA0402600), the National SKA Program of China (grant No. 2020SKA0120300), the National Natural Science Foundation of China under grants No. 11833003, 12003007 and the Fundamental Research Funds for the Central Universities (No. 2020kfyXJJS039).

Institutional Review Board Statement: Not applicable.

Informed Consent Statement: Not applicable.

Data Availability Statement: Not applicable.

Acknowledgments: We thank the anonymous referees for the helpful suggestions and comments.

Conflicts of Interest: The author declares no conflict of interest.

Note

1 The high-energy electron distribution above the break electron energy is $dN_e/d\gamma_e \propto \gamma_e{}^{(-p-1)}$.

References

1. Band, D.; Matteson, J.; Ford, L.; Schaefer, B.; Palmer, D.; Teegarden, B.; Cline, T.; Briggs, M.; Paciesas, W.; Pendleton, G.; et al. BATSE Observations of Gamma-Ray Burst Spectra. I. Spectral Diversity. *Astrophys. J.* **1993**, *413*, 281. [CrossRef]
2. Gruber, D.; Goldstein, A.; von Ahlefeld, V.W.; Bhat, P.N.; Bissaldi, E.; Briggs, M.S.; Byrne, D.; Clevel, W.H.; Connaughton, V.; Diehl, R.; et al. The Fermi GBM Gamma-Ray Burst Spectral Catalog: Four Years of Data. *Astrophys. J. Suppl.* **2014**, *211*, 12. [CrossRef]
3. Preece, R. D.; Briggs, M.S.; Mallozzi, R.S.; Pendleton, G.N.; Paciesas, W.S.; Band, D.L. The BATSE Gamma-Ray Burst Spectral Catalog. I. High Time Resolution Spectroscopy of Bright Bursts Using High Energy Resolution Data. *Astrophys. J.* **2000**, *126*, 19. [CrossRef]
4. Zhang, B. Gamma-ray Burst Prompt Emission. *IJMPD* **2014**, *23*, 1430002. [CrossRef]
5. Kumar, P.; Zhang, B. The Physics of Gamma-Ray Bursts and Relativistic Jets. *Phys. Rep.* **2014**, *561*, 1. [CrossRef]
6. Beloborodov, A.M. Collisional Mechanism for Gamma-Ray Burst Emission. *Mon. Not. R. Astron. Soc.* **2010**, *407*, 1033. [CrossRef]
7. Deng, W.; Zhang, B. Cosmological Implications of Fast Radio Burst/Gamma-Ray Burst Associations. *Astrophys. J.* **2014**, *783*, 35. [CrossRef]
8. Giannios, D. Prompt GRB emission from gradual energy dissipation. *Astron. Astrophys.* **2008**, *480*, 305. [CrossRef]
9. Lazzati, D.; Begelman, M.C.Non-Thermal Emission from the Photospheres of Gamma-Ray Burst Outflows. I. High-Frequency Tails. *Astrophys. J.* **2010**, *725*, 1137. [CrossRef]
10. Lazzati, D.; Morsony, B.J.; Margutti, R.; Begelman, M.C. Photospheric Emission As the Dominant Radiation Mechanism in Long-Duration Gamma-Ray Bursts. *Astrophys. J.* **2013**, *765*, 103 [CrossRef]
11. Lundman, C.; Pe'er, A.; Ryde, F. A Theory of Photospheric Emission from Relativistic, Collimated Outflows. *Mon. Not. R. Astron. Soc.* **2013**, *428*, 2430. [CrossRef]
12. Meszaros, P.; Ramirez-ruiz, E.; Rees, M.J.; Zhang, B. X-Ray-rich Gamma-Ray Bursts, Photospheres, and Variability. *Astrophys. J.* **2002**, *578*, 812. [CrossRef]
13. Meszaros, P.; Rees, M.J. Steep Slopes and Preferred Breaks in Gamma-Ray Burst Spectra: The Role of Photospheres and Comptonization. *Astrophys. J.* **2000**, *530*, 292. [CrossRef]
14. Pe'er, A. Temporal Evolution of Thermal Emission from Relativistically Expanding Plasma. *Astrophys. J.* **2008**, *682*, 463. [CrossRef]
15. Pe'er, A.; Meszaros, P.; Rees, M.J. The Observable Effects of a Photospheric Component on GRB and XRF Prompt Emission Spectrum. *Astrophys. J.* **2006**, *642*, 995. [CrossRef]
16. Pe'er, A.; Ryde, F. A Theory of Multicolor Blackbody Emission from Relativistically Expanding Plasmas. *Astrophys. J.* **2011**, *732*, 49. [CrossRef]
17. Rees, M.J.; Meszaros, P. Dissipative Photosphere Models of Gamma-Ray Bursts and X-Ray Flashes. *Astrophys. J.* **2005**, *628*, 847. [CrossRef]
18. Thompson, C. A Model of Gamma-Ray Bursts. *Mon. Not. R. Astron. Soc.* **1994**, *270*, 480. [CrossRef]
19. Thompson, C.; Meszaros, P.; Rees, M.J. Thermalization in Relativistic Outflows and the Correlation between Spectral Hardness and Apparent Luminosity in Gamma-Ray Bursts. *Astrophys. J.* **2007**, *666*, 1012. [CrossRef]
20. Deng, W.; Zhang, B. Low Energy Spectral Index and Ep Evolution of Quasi-thermal Photosphere Emission of Gamma-Ray Bursts. *Astrophys. J.* **2014**, *785*, 112. [CrossRef]
21. Pe'er, A.; Zhang, B. Synchrotron Emission in Small-Scale Magnetic Fields as a Possible Explanation for Prompt Emission Spectra of Gamma-Ray Bursts. *Astrophys. J.* **2006**, *653*, 454. [CrossRef]
22. Uhm, Z.L.; Zhang, B. Fast-Cooling Synchrotron Radiation in a Decaying Magnetic Field and Γ-Ray Burst Emission Mechanism. *Nat. Phys.* **2014**, *10*, 351. [CrossRef]
23. Zhang, B.B.; Uhm, Z.L.; Connaughton, V.; Briggs, M.S.; Zhang, B. Synchrotron Origin of the Typical GRB Band Function—A Case Study of GRB 130606B. *Astrophys. J.* **2016**, *816*, 72. [CrossRef]
24. Geng, J.J.; Huang, Y.F.; Wu, X.F.; Zhang, B.; Zong, H.S. Low-energy Spectra of Gamma-Ray Bursts from Cooling Electrons. *Astrophys. J.* **2018**, *234*, 3. [CrossRef]
25. Zhao, X.; Li, Z.; Liu, X.; Zhang, B.B.; Bai, J.; Meszaros, P. Gamma-Ray Burst Spectrum with Decaying Magnetic Field. *Astrophys. J.* **2014**, *780*, 12. [CrossRef]
26. Daigne, F.; Bosnjak, Z.; Dubus, G. Reconciling observed gamma-ray burst prompt spectra with synchrotron radiation? *Astron. Astrophys.* **2011**, *526*, 110. [CrossRef]
27. Wang, X.Y.; Li, Z.; Dai, Z.G.; Meszaros, P. GRB 080916C: On the Radiation Origin of the Prompt Emission from keV/MeV TO GeV. *Astrophys. J.* **2009**, *698*, 98. [CrossRef]
28. Lloyd, N.M.; Petrosian, V. Synchrotron Radiation as the Source of Gamma-Ray Burst Spectra. *Astrophys. J.* **2000**, *543*, 722. [CrossRef]
29. Shen, R.F.; Zhang, B. Prompt Optical Emission and Synchrotron Self-Absorption Constraints on Emission Site of GRBs. *Mon. Not. R. Astron. Soc.* **2009**, *398*, 1936. [CrossRef]
30. Xu, S.Y.; Yang, Y.P.; Zhang, B. On the Synchrotron Spectrum of GRB Prompt Emission. *Astrophys. J.* **2018**, *853*, 43. [CrossRef]
31. Bosnjak, Z.; Daigne, F. Spectral Evolution in Gamma-Ray Bursts: Predictions of the Internal Shock Model and Comparison to Observations. *Astron. Astrophys.* **2014**, *568*, A45. [CrossRef]
32. Burgess, J.M. Time-resolved Analysis of Fermi Gamma-Ray Bursts with Fast- and Slow-cooled Synchrotron Photon Models. *Astrophys. J.* **2014**, *784*, 17. [CrossRef]

33. Zhang, B.; Yan, H. The Internal-collision-induced Magnetic Reconnection and Turbulence (ICMART) Model of Gamma-ray Bursts. *Astrophys. J.* **2011**, *726*, 90. [CrossRef]
34. Panaitescu, A.; Meszaros, P.; Gamma-Ray Bursts from Upscattered Self-absorbed Synchrotron Emission. *Astrophys. J.* **2000**, *544*, 17. [CrossRef]
35. Sari, R.; Piran, T.; Narayan, R. Spectra and Light Curves of Gamma-Ray Burst Afterglows. *Astrophys. J.* **1998**, *497*, L17. [CrossRef]
36. Wang, K.; Dai, Z.G. GeV Emission during X-Ray Flares from Late Internal Shocks: Application to GRB 100728A. *Astrophys. J.* **2013**, *772*, 152. [CrossRef]
37. Yu, Y.W.; Dai, Z.G. X-Ray and High Energy Flares from Late Internal Shocks of Gamma-Ray Bursts. *Astrophys. J.* **2009**, *692*, 133. [CrossRef]
38. Dai, Z.G.; Lu, T. Hydrodynamics of Relativistic Blast Waves in a Density-Jump Medium and Their Emission Signature. *Astrophys. J.* **2002**, *565*, L87. [CrossRef]
39. Kumar, P.; Piran, T. Some Observational Consequences of Gamma-Ray Burst Shock Models. *Astrophys. J.* **2000**, *532*, 286. [CrossRef]
40. Sari, R.; Piran, T. Hydrodynamic Timescales and Temporal Structure of Gamma-Ray Bursts. *Astrophys. J.* **1995**, *455*, L143. [CrossRef]
41. Blandford, R.; McKee, C. Fluid Dynamics of Relativistic Blast Waves. *Phys. Fluids* **1976**, *19*, 1130. [CrossRef]
42. Sari, R.; Esin, A.A. On The Synchrotron Self-Compton Emission from Relativistic Shocks and Its Implications for Gamma-Ray Burst Afterglows. *Astrophys. J.* **2001**, *548*, 787. [CrossRef]
43. Abdo, A.A.; Ackermann, M.; Arimoto, M.; Asano, K.; Atwood, W.B.; Axelsson, M.; Baldini, L.; Ballet, J.; Band, D.L.; Barbiellini, G.; et al. Fermi Observations of High-Energy Gamma-Ray Emission from GRB 080916C. *Science* **2009**, *323*, 1688. [PubMed]
44. Abdo, A.A.; Ackermann, M.; Asano, K.; Atwood, W.B.; Axelsson, M.; Baldini, L.; Ballet, J.; B.; D.L.; Barbiellini, G.; Bastieri, D.; et al. Fermi observations of high-energy gamma-ray emission from GRB 080825C. *Astrophys. J.* **2009**, *707*, 580. [CrossRef]
45. Zhang, B. Open Questions in GRB Physics. *Comptes Rendus Phys.* **2011**, *12*, 206. [CrossRef]
46. Derishev, E.V.; Kocharovsky, V.V.; Kocharovsky, V.V. Physical Parameters and Emission Mechanism in Gamma-Ray Bursts. *Astron. Astrophys.* **2011**, *372*, 1071. [CrossRef]
47. Piran, T.; Sari, R.; Zou, Y.C. Observational Limits on Inverse Compton Processes in Gamma-Ray Bursts. *Mon. Not. R. Astron. Soc.* **2009**, *393*, 1107. [CrossRef]
48. Racusin, J.L.; Oates, S.R.; Schady, P.; Burrows, D.N.; De Pasquale, M.; Donato, D.; Gehrels, N.; Koch, S.; McEnery, J.; Piran, T.; et al. Fermi and Swift Gamma-ray Burst Afterglow Population Studies. *Astrophys. J.* **2011**, *738*, 138. [CrossRef]
49. Yonetoku, D.; Murakami, T.; Nakamura, T.; Yamazaki, R.; Inoue, A.K.; Ioka, K. Gamma-Ray Burst Formation Rate Inferred from the Spectral Peak Energy-Peak Luminosity Relation. *Astrophys. J.* **2004**, *609*, 935. [CrossRef]
50. Amati, L.; Frontera, F.; Tavani, M.; Antonelli, A.; Costa, E.; Feroci, M.; Guidorzi, C.; Heise, J.; Masetti, N.; Montanari, E.; et al. Intrinsic Spectra and Energetics of BeppoSAX Gamma–Ray Bursts with Known Redshifts. *Astron. Astrophys.* **2002**, *390*, 81. [CrossRef]
51. Yu, H.F.; Van Eerten, H.J.; Greiner, J.; Sari, R.E.; Bhat, P.N.; Von Kienlin, A.; Paciesas, W.S.; Preece, R.D. The Sharpness of Gamma-Ray Burst Prompt Emission Spectra. *Astron. Astrophys.* **2015**, *583*, 129. [CrossRef]
52. Guiriec, S.; Connaughton, V.; Briggs, M.S.; Burgess, M.; Ryde, F.; Daigne, F.; Meszaros, P.; Goldstein, A.; McEnery, J.; Omodei, N.; et al. Detection of a Thermal Spectral Component in the Prompt Emission of GRB 100724B. *Astrophys. J. Lett.* **2011**, *727*, L33. [CrossRef]
53. Oganesyan, G.; Nava, L.; Ghirlanda, G.; Celotti, A. Detection of Low-energy Breaks in Gamma-Ray Burst Prompt Emission Spectra. *Astrophys. J.* **2017**, *846*, 137. [CrossRef]
54. Daigne, F. Mochkovitch, R.The Expected Thermal Precursors of Gamma-Ray Bursts in the Internal Shock Model. *Mon. Not. R. Astron. Soc.* **2002**, *336*, 1271. [CrossRef]
55. Ryde, F. Is Thermal Emission in Gamma-Ray Bursts Ubiquitous? *Astrophys. J. Lett.* **2005**, *625*, L95. [CrossRef]

galaxies

MDPI

Review

The Detection of GRBs at VHE: A Challenge Lasting for More than Two Decades, What is Next?

Alessio Berti [1,*,†] and Alessandro Carosi [2,*,†]

[1] Max-Planck-Institut für Physik, D-80805 Munich, Germany
[2] INAF-Osservatorio Astronomico di Roma, I-00136 Rome, Italy
* Correspondence: aberti@mpp.mpg.de (A.B.); alessandro.carosi@inaf.it (A.C.)
† These authors contributed equally to this work.

Abstract: Unveiling the mystery of gamma-ray bursts (GRBs) has been the target of many multi-waveband observational and theoretical efforts during the last decades. The results collected by current and past space-based instruments have provided important insights into the mechanisms at the origin of their prompt and afterglow phases. On the other hand, many questions, such as the the origin of the multi-GeV signal observed in a large number of events, remained unanswered. Within this framework, the first firm detections of a very-high-energy (VHE, $E \gtrsim 100\,\mathrm{GeV}$) emission component by MAGIC and H.E.S.S. collaborations represented an important, long-awaited result for the VHE astrophysics community. However, while such discoveries opened a new era in the study of GRBs, they also provided an unexpected complexity due to the differences between the phenomenology of the observed events. This revealed that we still have an incomplete comprehension of GRB physics. In the nearby future, observations by the Cherenkov Telescope Array Observatory (CTAO), with unprecedented sensitivity in the VHE band, will have a key role in the study of these enigmatic objects and their interactions with the surrounding environment. In this review we will cover the recent GRB history, highlighting the efforts of follow-up campaigns by the VHE community that led to the first VHE GRB detection, and outlining what we can expect from future facilities in the next decades.

Keywords: gamma-ray bursts; very high energy; IACTs

Citation: Berti, A.; Carosi, A. The Detection of GRBs at VHE: A Challenge Lasting for More than Two Decades, What is Next? *Galaxies* **2022**, *10*, 67. https://doi.org/10.3390/galaxies10030067

Academic Editors: Elena Moretti, Francesco Longo and Yi-Zhong Fan

Received: 10 March 2022
Accepted: 18 April 2022
Published: 10 May 2022

1. Introduction

In 2019, the announcement of the first detection of VHE gamma-ray emission from GRB 180720B [1], GRB 190114C [2], and GRB 190829A [3] represented a long-awaited result for the astrophysical community and the end of a quest lasting for more than twenty years. The detection of a VHE counterpart of GRBs always posed a major challenge for imaging atmospheric Cherenkov telescopes (IACTs) from both the technical and the scientific point of view see, e.g., [4,5]. On the other hand, catching such a signal has a crucial impact on understanding the poorly-known physics of these objects during the different phases of their emission, motivating the continuous efforts in the VHE observational window. In fact, the observed radiation still has an uncertain origin in many aspects. According to the widely accepted relativistic shock model originally proposed in [6], GRB emission arises from the conversion of the kinetic energy of a relativistic outflow into electromagnetic emission. The details of this conversion remain poorly understood. However, the dissipation might happen in the form of collisionless shocks between the relativistic flow itself (internal shocks, responsible for the prompt phase) or with the circumburst medium (external shocks, responsible for the afterglow emission phase). Alternatively, other dissipation mechanisms have been considered in literature and, noticeably, the possibility of having magnetic reconnection events as the base for particle acceleration; see, e.g., [7,8]. The nature of the possible radiative processes at work is also not firmly established yet. Particles inside the outflow and accelerated towards relativistic regime can emit the observed high-energy photons via many possible non-thermal mechanisms, in particular during the early afterglow phase. In this regard, the prompt-to-early-afterglow phase still remains the least understood in GRB dynamics. Prompt emission spectra have been largely fitted through the so-called Band function [9], an empirical function composed by two smoothly connected power-law functions at a specific break energy. While historically the band function worked quite well in fitting prompt spectra in the 10 keV–1 MeV range for many GRBs, more recent works have showed that extra emission components in the form of an additional power-law and/or a photospheric blackbody component are needed to better fit the observed emission both at lower and at higher energies with respect to the GRB peak energy (see, e.g., [10,11] and references therein). Such components might account for a revision of the theoretical interpretation of the observed radiation as possible synchrotron emission from electrons accelerated within the relativistic outflow. Synchrotron emission has been shown to be in tension with experimental data in many events; see, e.g., [12,13]. Nevertheless, synchrotron is believed to play an essential role in GRB physics and it has been largely considered as the most natural process to explain the GRB sub-MeV emission both during the prompt and afterglow; see, e.g., [14–16]. Furthermore, it has also been suggested that the high-energy photons above ~10 MeV observed by the *Fermi*-LAT (Large Area Telescope) [17], and extending after the end of the prompt emission, might be generated by synchrotron radiation produced in external shocks [18]. However, the observation of an emission component at VHE, as recently detected by current IACTs, challenges the synchrotron-alone emission models and, ultimately, the particle acceleration mechanisms at work in GRBs. In internal/external relativistic shock models, particles can be accelerated up to a maximum Lorentz factor achieved when the comoving acceleration time matches the typical radiative cooling time. The corresponding maximum photon's energy emitted by a synchrotron is around ~50 MeV in the comoving frame corresponding to an observed E_{max}~50 MeV $\times\ \Gamma/(1+z)$ (synchrotron burnoff limit), where Γ is the bulk Lorentz factor of the relativistic outflow and z is the redshift of the source. In the case of GRBs, arguments for the hypothesis of an emitting region moving towards the observer with a bulk Lorentz factor $\Gamma \sim few \times 100$ are known and used to solve inconsistencies between the observed non-thermal emission above the pair production threshold ($\gamma\gamma \rightarrow e^+e^-$) and the time variability observed during the prompt phase; see, e.g., [19]. With the *Fermi* satellite, some firm estimations for Γ have been achieved using the maximum photon's energy detected by *Fermi*-LAT in the GeV band. For some particularly bright GRBs, values exceeding $\approx 10^3$, as in the case of $\Gamma \sim 900$ for GRB 080916C [20] and $\Gamma \sim 1200$ for GRB 090510 [21], have been measured. Although those values dramatically differ from any other relativistic motion observed in other astrophysical sources, they would still appear moderate if considering the signal caught by IACT in the hundreds of GeV or even TeV band. Furthermore, after the end of the prompt phase, Γ decreases with time [22], implying that the maximum energy achievable by synchrotron photons decreases as well. Thus, HE and VHE signals detected deeper in the afterglow phase, such as in the case of GRB 190829A, abundantly exceed the synchrotron burnoff limit, challenging the simple shock acceleration/synchrotron model. The complexity of scenarios provided by the latest IACT results shows a still unsatisfactory level of comprehension of GRB physics and the importance of continuing the observation of GRBs in the VHE band with next-generation IACTs. In the coming decades, the premier facility for VHE astrophysics will be the CTA observatory that will perform observations in the >10 GeV range with unprecedented photon statistics and sensitivity, allowing to investigate the parameter space of a wide range of VHE-transient emitters and their characteristics.

In this paper, we will revise the main experimental results that historically helped in shedding light into the GRB physics in the HE (high energy, $E \gtrsim 100$ MeV) and VHE domains. The paper is organized as follows: in Section 2 we briefly introduce the theoretical emission models used to interpret GRB HE and VHE emission. In Sections 3 and 4 we will summarize the main experimental steps that brought us to the detection of GRBs in the HE and VHE band. Section 5 investigates the open issues that still affect the characterization of GRBs at VHE and that will hopefully be solved by next-generation instruments finally described in Section 6.

2. Models for HE and VHE Emission in GRBs

Although not within the primary scope of this paper, it is important to briefly summarize the main interpretative models able to explain the emission at the highest energies. Many theoretical models were proposed in the last decades to explain the emission from GRBs, with predictions extending to the HE and VHE range. Usually, in these models, the origin of HE and VHE emission can take place in both internal or external shocks. In both cases, either leptonic or hadronic processes might be considered as possible explanation of the observed emission. As already mentioned in the previous section, synchrotron emission is one of the most discussed for the emissions in the keV-MeV band. At higher energy, synchrotron photons might interact through inverse Compton with ultra-relativistic electrons of the outflow. This amplifies the energy of the seed photons by a factor of γ_e^2, where γ_e is the electron's Lorentz factor. Depending on the specific microphysical parameters of the emitting region, this synchrotron self-Compton (SSC) emission can arise and easily produce photons in the HE and VHE ranges. Detailed predictions for such a model are given in [23,24], where the suppression of inverse Compton due to the Klein–Nishina (KN) effect is also widely discussed. This effect can explain the delay observed between the keV and HE emission (see Section 3) if the KN regime is dominant at early times, but then at late times the inverse Compton enters the Thomson regime; see, e.g., [25–27]. Hadronic particles can be also shock-accelerated in the same way as leptons, influencing (potentially) the HE and VHE emission. Hadronic models comprise synchrotron emission from protons or cascade emission (synchrotron) from secondary pairs [28,29]. In the synchrotron scenario, the delay between low- and high-energy emission can be explained as the time required to accelerate protons to high enough energies. However, being a poor emitter compared to leptons, proton energy is mainly lost through p-γ interactions rather than by synchrotron. In this case, the required energy budget to achieve comparable emission level with leptonic processes is normally well above the observed ones ($\gtrsim 10^{55}$ erg), although this requirement can be relaxed with a narrow jet opening angle (<1°). In the case of external shocks, one of the main models considered for HE and VHE emission is SSC at the (external) forward shock. In such a case, a separate component with a second peak at high energies is expected [14,15]. This model was largely used to explain the HE emission in some *Fermi*-LAT bursts (see, e.g., [30–32]), as we will describe further in the next sections. SSC is proposed to produce HE photons also in the reverse shock [33] and it was shown to explain the HE component of some GRBs [34]. Furthermore, in the external shock scenario, hadronic models are also a possible option in order to account for HE and VHE emission. However, as in the internal origin case, hadronic processes suffer the same issue on the required energetics, although a possible non-dominant contribution to the overall HE-VHE emission cannot be completely excluded [35]. We refer to [36] for a more detailed review on theoretical emission models.

3. Gamma-Ray Bursts Observations at High Energies

The first systematic and comprehensive study of GRBs population was carried out by the space-based telescope Compton Gamma-Ray Observatory (CGRO), which operated for about 9 years between April 1991 and June 2000. Thanks to its four onboard instruments and, in particular, the Burst And Transient Source Experiment (BATSE: 25 keV–2 MeV) and the Energetic Gamma-Ray Experiment Telescope (EGRET: 20 MeV–30 GeV), it was possible to have an energy coverage ranging from the soft X-rays to the HE gamma rays. This provided the first meaningful interpretation of the GRB phenomenon. Specifically, thanks to EGRET, it was possible to start studying the properties of the high-energy emission ($\gtrsim$10–20 MeV) of GRBs for the very first time. A notable event was detected by EGRET on 17 February 1994, GRB 940217 [37]. The burst had a duration of 180 s as measured by BATSE. Ten HE photons were detected by EGRET with energy up to ~3 GeV during the prompt emission. Eight other HE photons were detected in the following ∼600 s. After the occultation due to the Earth, EGRET registered another 10 photons more than 4700 s after the burst trigger. The highest-energy photon detected in this observation phase had an energy of 18 GeV (see Figure 1), and for many years it represented the highest energy photon ever detected from a GRB. This delay in high-energy emission was observed in other GRBs detected by EGRET, although not as evident as for GRB 940217 (see [38]). Furthermore, the detection of HE photons pointed out to the possible presence of additional spectral components overlapped to the classical sub-MeV band emission. From a different perspective, the presence of a delayed emission was also considered as an opportunity for TeV detectors, such as IACTs, that needed to be repointed for start follow-up but also for extensive air shower (EAS) arrays.

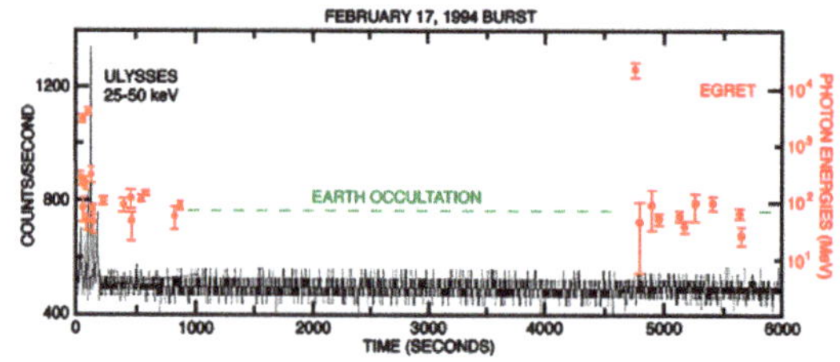

Figure 1. EGRET light curve for the GRB 940217 with the detection of an ∼18 GeV photon that occurred about 90 min after the GRB onset. From https://fermi.gsfc.nasa.gov/science/resources/docs/vwp/. Available online: (accessed on 10 April 2022).

A hint of a distinct emission component in the HE range was found in the case of GRB 941017 [39]. This event showed a $\gtrsim$200 MeV signal rising between 14 and 47 s after the T_0 and lasted for approximately 200 s in addition to the typical GRB emission which peaked at $\lesssim$few hundred keV. The HE component is well fitted by a power law with index close to -1 up to 200 MeV throughout all the burst duration, while the low-energy spectrum was well described by the classical band function. Data were inconsistent with a simple synchrotron model interpretation, and other theoretical emitting scenarios were considered, such as synchrotron self-Compton (SSC) from the reverse shock, created when the GRB ejecta are decelerated by the ambient medium. Additional interpretations were also considered, such as a possible hadronic origin of the HE component, as well as an HE emission taking place in external shocks [33]. Despite these earliest observations that helped significantly in determining some HE properties of GRB, the limited statistics and the large dead time typical of EGRET did not allow to measure precise spectra and study in detail the short timescale variability in the emission, especially during the prompt phase. Many questions were left unanswered after EGRET stopped operations in 2000, mostly related to the jet physics, particle acceleration, and to the nature of the high-energy emission. In this context, there were many expectations for the launch of AGILE (Astro-Rivelatore Gamma a Immagini Leggero) and *Fermi Gamma-ray Space Telescope* (*Fermi* in short). Using silicon trackers, the limitations of the old generation of gamma-ray imagers, such as the small FoV and the large dead time, were partially solved. AGILE, launched in 2007, was the first instrument with this kind of technology, followed by *Fermi* in 2008. These satellites opened a new era in the studies of GRBs in the HE band.

AGILE's onboard instrumentation includes a gamma-ray imaging detector (GRID) sensitive in the 30 MeV–50 GeV band, a hard X-ray monitor (SuperAGILE: 18–60 keV), and a mini-calorimeter (MCAL) non-imaging gamma-ray scintillation detector sensitive in the 350 keV–100 MeV energy range [40]. With the detection of GRB 080514B [41], AGILE confirmed the presence of a delayed and relatively long-lasting high-energy emission, as seen in the EGRET events. The burst was detected by all the instruments onboard AGILE: GRID detected photons from 25 MeV up to 300 MeV, while in the hard X-ray band (SuperAGILE), the 17–50 keV light curve showed a multi-peaked structure with a total duration of 7 s. The high-energy emission did not show any correlation with these peaks, and only three photons above ~30 MeV were detected within 2 s from T_0. All the other high-energy photons were recorded when the X-ray emission had already faded, up to ~30 s after the burst onset.

GRB 100427B [42] is another notable GRB detected by AGILE. Both the MeV and GeV light curves show two bumps where the second peak is broader than the first, with no significant delay with respect to the lower-energy emission in the X-ray band. The second bump resulted harder then the first, and spectral evolution between the bumps and the inter-bump region in MCAL data were detected at the level of 4.0σ. A single power law was shown to be adequate to model the spectrum from 500 keV to 3.5 GeV, given that the spectral index of the MCAL + GRID data and GRID data only were compatible with each other. Even if the redshift was not measured for this GRB, given the highest energy photon of 3.5 GeV, the minimum Lorentz factor during the prompt emission was constrained to be between 50 and 900. For other GRBs observed by AGILE, not detected by GRID, upper limits were derived and found to be consistent with an extrapolation of the band spectrum up to GeV energies; see [43].

Fermi was launched in 2008, approximately one year after AGILE. *Fermi* uses the same detector technology as AGILE and it was designed to be a proficient gamma-ray satellite with improved capabilities with respect to previous-generation gamma-ray detectors. The spacecraft hosts two instruments on board. The gamma-ray burst monitor (GBM) is composed of 14 scintillators (twelve sodium iodide and two bismuth germanate) and covers the energy range from a few keV to ~30 MeV [44]. With a field of view of almost 4π, it is devoted to the detection of GRBs or other burst-like sources and to the quick distribution of GRB localizations. The second instrument is the pair-production telescope LAT [17], operating in the energy range 20 MeV–300 GeV. The adoption of the silicon strips detector technology for the *Fermi*-LAT led to substantial improvements in terms of angular resolution and timing capabilities. Thanks to a big calorimeter, the sensitive energy range of the *Fermi*-LAT extends up to few hundreds of GeV, also providing a good energy resolution. Owing to its efficient design, *Fermi* delivered and is still delivering more detailed results and the highest statistics for studying GRBs in the HE regime. At the same time, it is providing an invaluable overlap with ground-based VHE facilities.

GRB 080825C [45] was the first GRB detected by *Fermi*-LAT, a long burst with $T_{90} = 27$ s. The highest-energy photon was a (572 ± 58) MeV photon detected at $\sim T_0 + 28$ s, just after the low-energy emission measured in *Fermi*-GBM faded almost completely. The spectrum of GRB 080825C in different time bins is well fitted by a band function with a hard-to-soft evolution of the νF_ν spectrum peak energy ($E_{\rm peak}$). In the last time bin, the spectrum is well described by a power law with a harder index -1.95 ± 0.05. This property and the low flux ratio between the first two peaks in the *Fermi*-LAT light curve may suggest a different region of origin for their emission: within the internal and external shock respectively [45].

GRB 080916C [20] is the second *Fermi*-LAT detected burst and one of the brightest in the *Fermi*-LAT GRB sample, with a measured redshift of $z = 4.35 \pm 0.15$ and a total isotropic energy release of 8.8×10^{54} erg. Compared to the signal measured in *Fermi*-GBM, this GRB showed a delayed onset of the LAT pulse and a longer-lived emission in the $\gtrsim$100 MeV band. These features will be confirmed in other GRBs detected at HE. The comparison between the *Fermi*-GBM and the *Fermi*-LAT light curves (Figure 2) showed that the first *Fermi*-GBM peak has no corresponding peak in the *Fermi*-LAT light curve. The first *Fermi*-LAT pulse is instead temporally coincident with the second *Fermi*-GBM peak. A common origin for the two peaks but in spatially different regions is the most likely explanation, with different pairs of colliding shells within the internal shock scenario. The long-lasting emission above 100 MeV was detectable up to $T_0 + 1400$ s, well after the low energy emission faded. The time decay of the high energy flux is well fitted by a power law $t^{-\alpha}$ with $\alpha = -1.2 \pm 0.2$, a value that is typical for other *Fermi*-LAT-detected GRBs. The *Fermi*-GBM flux decays as $t^{-0.6}$ up to $T_0 + 55$ s with a steepening in the index ($\alpha \sim -3.3$) afterward. This might indicate a different nature of the high-energy emission, although no spectral hardening is seen in the *Fermi*-LAT late spectrum, as in the case of GRB 080825C. As in other HE detected burst, GRB 080916C data were used to set a lower limit to the Lorentz factor of the blast-wave, $\Gamma_{\rm min} = 887 \pm 21$. Even if most of the *Fermi*-LAT-detected GRBs belong to the long class, it helped to study the high-energy emission of short GRBs as well. Among them, some interesting cases are GRB 081024B [46] and GRB 090510 [21,47].

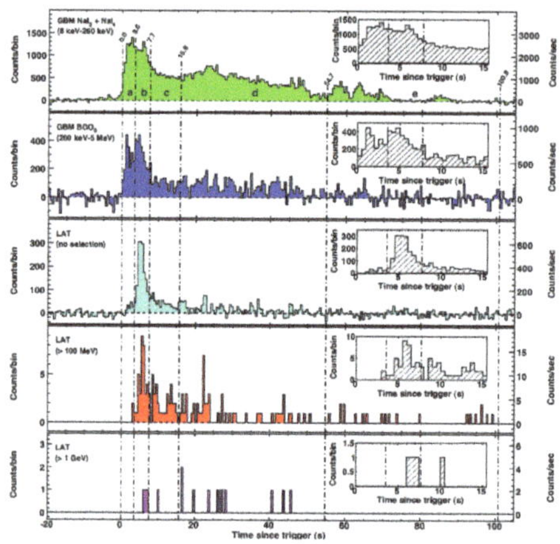

Figure 2. *Fermi*-GBM and *Fermi*-LAT combined light curve for GRB 080916C. Reprinted with permission from Ref. [20].

GRB 081024B is the first short GRB detected by *Fermi*-LAT, with a duration of 0.8 s. In addition for this GRB, the emission above 100 MeV is delayed and is long-lasting ($T_{90} = 2.6$ s above 100 MeV).

GRB 090510 is a short GRB which was detected by both AGILE and *Fermi*. Both instruments confirmed the presence of a $\sim$0.1 s-delayed HE emission component after the onset measured in *Fermi*-GBM. The detection of a 30.5 GeV photon during the prompt phase allowed an evaluation of the bulk Lorentz factor that resulted in a very high lower limit of $\Gamma_{min} \gtrsim 1200$, assuming the estimated redshift $z = 0.903$. However, the most remarkable feature of GRB 090510 is that its time-integrated spectrum for the GBM+LAT prompt emission data cannot be fitted with a simple band function. An additional power law component with index -1.62 ± 0.03, dominant below 20 keV and above 100 MeV, is needed to describe the spectrum (see Figure 3) [21]. In the afterglow phase, a signal has been detected by *Fermi*-LAT up to $\sim T_0 + 150$ s, which prompted several theoretical interpretations for both the prompt and afterglow phases. Some of them consider synchrotron radiation as theoretical interpretation of the low-energy (band) emission while the hard extra-component is generated by the synchrotron photons Compton upscattered by the same electrons accelerated in the shock (synchrotron self-Compton); see, e.g., [48]. This scenario is commonly used to model emission in other VHE sources, such as blazars, and the SSC component results are stronger for a large ratio of non-thermal electron to magnetic-field energy density and low values of Γ. However, in the case of GRB 090510, such an interpretation has difficulties in explaining the delayed onset of the high-energy emission. For example, the SSC model predicts a too-short delay in the assumptions of weak magnetic field [21]. Hadronic scenarios were also proposed but the proton injection isotropic-equivalent energy required is more than two orders higher than the one actually measured for the burst [49]. These observations of short GRBs show that they can be as relativistic as long GRBs and that they seem to have a better efficiency in emitting gamma rays, given that the energy emitted in the high-energy (100 MeV–10 GeV) band is greater than the one at low energy (20 keV–2 MeV). However, the statistic is still limited to few bursts to draw a definitive conclusion.

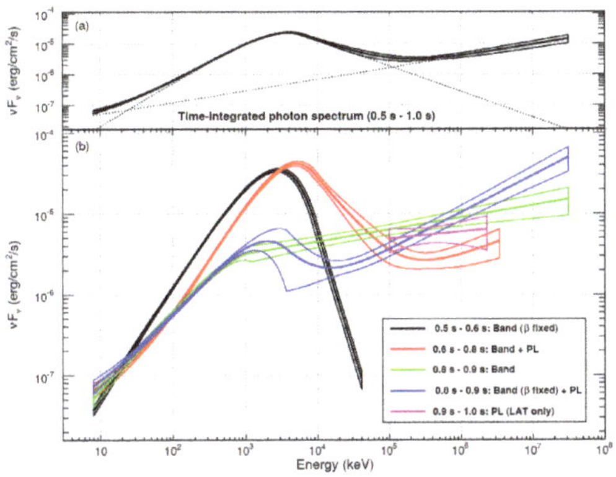

Figure 3. Time−integrated spectrum (upper panel) and model spectra (with $\pm 1\sigma$ error contours) used to fit the emission of GRB 090510 in different time bins. Reprinted with permission from Ref. [21].

As a final example, it is worth to report the *Fermi*-LAT detection of GRB 130427A, one of the most powerful GRBs observed at redshift $z = 0.34$ [50]. The event showed the highest fluence (4.2×10^{-3} erg/cm^2 from 10 keV to 20 MeV), the highest energy photon (95 GeV at $T_0 + 244$ s), and the longest-lasting HE emission extending up to 100 ks after the trigger. It had a total apparent isotropic gamma-ray energy release of $\sim 1.4 \times 10^{54}$ erg. The event showed a delayed emission starting about 10 s after the trigger when the *Fermi*-GBM brightest emission already ended. Therefore, the *Fermi*-LAT emission is temporally distinct from the one in *Fermi*-GBM (see, e.g., Figure 1 in [50]), and this suggests different regions or mechanisms for the two emissions. Having a 95 GeV photon in the early afterglow and a 32 GeV one at $T_0 + 34.4$ ks, it is difficult to accommodate them within the standard synchrotron emission from electrons accelerated in the external shock or in the SSC scenario, at least according to [50]. In [51], the combined X-ray, GeV, and optical data were used to fit the spectrum with a single synchrotron component while authors in [32] proposed an afterglow SSC emission to explain the long-lasting emission. These results show the puzzling interpretative scenarios of GRBs at HE and the lack of a clear physical explanation, both in the prompt and in the afterglow phases.

Summarizing, AGILE and *Fermi* showed that HE emission from GRBs share some common features:

1. The band model is not able to describe the joint low- and high-energy spectra. An additional component (e.g., extra power law) or a cutoff are needed, with no unique solution for all GRBs. Other GRBs may require an additional thermal blackbody component.
2. *Fermi*-LAT-detected GRBs are among the brightest detected by the *Fermi*-GBM. The energy released in high-energy gamma-rays (>100 MeV) in the extended temporal phase is about 10% of the total energy radiated in the prompt phase.
3. The high-energy emission is delayed and longer-lasting with respect to the low-energy one. It might extend in time well after the low-energy emission has faded. The temporal decay is generally consistent with a power law behavior $t^{-\alpha_L}$ with $\alpha_L \sim 1$.

These conclusions are the same as those resulting from the first *Fermi*-LAT GRB catalog, presented in [52], an in-depth systematic study of *Fermi*-LAT-detected GRBs in the first three years of the mission.

4. GRB Observation at VHE: The Story so Far

The field of VHE transient astronomy has been rapidly evolving for the last 30 years, mainly (but not only) due to the development of the imaging atmospheric Cherenkov technique. Towards the end of the last century, the first IACT experiments were built and started operation, proving the robustness and reliability of this detection technique through the first detection of the standard candle VHE emitter, the Crab Nebula [53]. In parallel with the confirmation of the IACT technique, GRBs science was entering for the first time in a phase of systematic population studies thanks to the BATSE and EGRET instruments on board the CGRO (see Section 3) but also to the BeppoSax satellites, launched in 1996 [54]. The latter, thanks to the contemporary presence on the same platform of both wide- and narrow-field instruments, was able to provide, for the first time, arcminute localizations of GRB positions although with $\sim$ hours delay timescale. As reported in the previous section, the discovery of a delayed and persistent HE emission component in some of the EGRET-detected events (see, e.g., [37]) definitively pushed the search for a component also at VHE. Real-time triggers provided through the BATSE Coordinates Distribution Network (BACODINE) and the third Interplanetary Network IPN-[55], although with relative large uncertainties in the localization ($\gtrsim$few degrees), allowed, for the first time, the rapid follow-up by ground-based telescopes including the earliest VHE facilities, such as the first IACTs and EAS arrays. Although not presenting imaging capabilities, EAS arrays were able to cover a wide portion of sky, allowing for offline search of a coincidence signal in the ultra-high-energy (UHE) gamma-ray band ($\gtrsim$100 TeV). Such a search for emission of TeV/PeV gamma rays associated with GRBs has been extensively reported in literature by many different EAS collaborations, such as CYGNUS-I [56], HEGRA-AIROBICC [57], CASA-MIA [58], and EAS-TOP [59]. None of these revealed any convincing evidence for emission in the >100 TeV band. It is important to remark that at the time of these observations, a firm determination of GRB distance was still missing and the detection of $\gtrsim$100 TeV photons represented a concrete possibility and an important insight into the origin (cosmological or local) of these events. A largely discussed, although not conclusive, hint for an emission in the $\sim$TeV band came from the Milagrito experiment [60]. Milagrito was a TeV EAS array based on the water Cherenkov detection technique, a prototype of the larger Milagro detector. The array operated between February 1997 and May 1998 in the 500 GeV–20 TeV energy range, observing 54 BATSE GRBs localized in its field of view. A possible $\sim 3.5\sigma$ evidence of TeV emission was found in the case of GRB 970417A, likely caused by photons of $\gtrsim$650 GeV [61]. This measurement could indicate the first detection of a GRB in VHE regime; however, the weakness of the signal did not allow any spectral analysis of the event. Moreover, no other similar detection was observed by the later Milagro experiment in the same energy range, making the reliability of this observation less constraining.

The first follow-ups by an IACT at lower energies compared to EAS arrays (above $\sim$250 GeV), took place at the beginning of the 1990s, thanks to the Whipple 10 m reflector. These observations represented the first use of the IACT technique in exploring the GRB phenomenon complementing, although not yet overlapping, the band coverage guaranteed by the contemporaneous space-based instrumentation. Whipple reported no significant emission in the VHE band from a sample of nine GRBs observed between May 1994 and December 1995. The obtained upper limits are of the order of that expected for prompt emission if the burst emission extends to TeV energies with a band-like extrapolation without breaks or cutoff [62]. This confirmed the effectiveness of the IACT technique in proving GRB physics while pointing out some of the main difficulties of these follow-ups. Differently from EAS array, being (relatively) narrow field instruments, IACTs need to be repointed to GRB coordinates in order to start the follow-up. This introduced a delay that, for these earliest observations, ranged from 2 to 56 min. Furthermore, due to the large uncertainty in the BATSE localization of the events, the majority of the observations were performed with the source located off-axis (or in some case outside the telescope's field of view), significantly decreasing the sensitivity of the instrument and requiring multiple pointings to scan the burst region (Figure 4). The necessity of having rapid repointing and follow-up observations was the core issue in 2004 of the launch of the *Swift* satellite [63]. *Swift* operates as a multi-band satellite incorporating three different instruments: a large FoV soft-gamma detector for GRBs trigger (BAT, burst alert telescope: 15–150 keV) and two telescopes in the X-ray (XRT, X-ray telescope: 0.3–10 keV) and UV band (UVOT, UV optical telescope) for the low-energy follow-up. These instruments were mounted on an autonomously slewing spacecraft that, using the same driven-logic of BeppoSax, made possible the observation and the precise localization of GRBs within tens of seconds from the event onset. These key features significantly improved the understanding of the early afterglow phase and its connection with the prompt emission [64]. Almost in parallel to the launch of *Swift*, the new generation of IACTs MAGIC (https://magic.mpp.mpg.de/), H.E.S.S. (https://www.mpi-hd.mpg.de/hfm/HESS/), and VERITAS (https://veritas.sao.arizona.edu/) (all websites accessed on 10 April 2022) started operations opening a new phase in GRB study at VHE. Some of these telescopes were explicitly designed to optimize the follow-up observation of GRBs, with the aim to reach the few tens of GeV energy threshold, bridging the observational energy gap between the space-based instrumentation and enlarging the available gamma-ray horizon, one of the critical aspects for high redshift sources such as GRBs. Extensive follow-up campaigns on GRBs were performed by all IACTs collaborations along approximately 15 years of observations and they also progressively bridged the energy coverage gap with AGILE and *Fermi*. However, these extended observations did not report any conclusive evidence of VHE emission from the observed events. We briefly summarize the main outcomes of this first decade of observation.

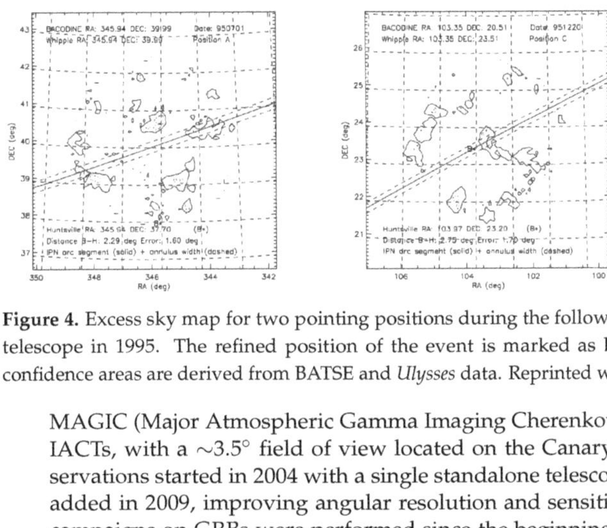

Figure 4. Excess sky map for two pointing positions during the follow-up of two GRBs by Whipple telescope in 1995. The refined position of the event is marked as B* in the plot while the IPN confidence areas are derived from BATSE and *Ulysses* data. Reprinted with permission from Ref. [62].

MAGIC (Major Atmospheric Gamma Imaging Cherenkov) is a system of two 17 m IACTs, with a $\sim$3.5° field of view located on the Canary Island of La Palma. Observations started in 2004 with a single standalone telescope until a second one was added in 2009, improving angular resolution and sensitivity. Extensive follow-up campaigns on GRBs were performed since the beginning of the operations, taking advantage of the instruments low-energy threshold ($\lesssim$50 GeV) combined with a very fast respositioning speed ($\sim$7°/s). Despite the continuous improvement in instrument's reaction to external GRB triggers and in data analysis along the years, no significance evidence of VHE emission was reported during the first $\sim$15 years of observations. However, remarkable results were achieved in terms of performance, such as the first follow-up of GRBs during the prompt emission phase for a bunch of events such as GRB 050713A (Figure 5, left panel), GRB 131030A, GRB 141026A, and GRB 150428B [4,65–67]. Furthermore, within the framework of relativistic shock-wave models, possible emission in the VHE band by synchrotron-self-Compton mechanism in afterglow has been modeled and discussed by the MAGIC collaboration in relation to the obtained upper limits on a few interesting events such as GRB 0804030 [68] and GRB 090102 (Figure 5, right panel), one of the first GRBs with simultaneous data taken with *Fermi*-LAT [5]. Although not particularly constraining, these results showed that IACT performances were mature enough to play an important role in GRB studies.

The High-Energy Stereoscopic System (H.E.S.S.) is an array of IACTs operating in Namibia since 2004. The so-called phase-I included four 12 m diameter telescopes, with an energy threshold of $\sim$100 GeV at zenith, and a 5° field of view. In 2012, a large 28 m diameter telescope was added to the array. This telescope is characterized by a faster repointing and large collection area ($\sim$600 m^2) that guarantee an energy threshold of 50 GeV. Thus, it is a transient-oriented instrument. The introduction of the new telescope marked the beginning of the H.E.S.S. phase-II operations. Despite these improvements, also for H.E.S.S., the first 15 years of observations did not reveal any significant emission for the observed events. Collection of follow -ups and possible interpretation of the obtained upper limits are summarized in different collaboration works, such as in [69–72].

VERITAS (Very Energetic Radiation Imaging Telescope Array System) is an array of four 12 m IACTs located in Arizona operating in the $\gtrsim$100 GeV band. The system is the successor of Whipple and has activated a GRB observing program since the beginning of the operations in 2007. VERITAS did not report any detectable VHE emission from the sample of the observed GRBs; however, in 2013, VERITAS was the only IACT able to follow up GRB 130427A, the first GRB observed at VHE (see Section 3). Unfortunately, VERITAS was only able to perform observations on GRB 130427A approximately 20 h after the event's onset. Although at that time *Fermi*-LAT was still able to detect activity in the HE band, VERITAS did not report a significant emission in the VHE range.The achieved upper limits at $\sim$100 GeV were able to significantly constrain the proposed emission model, pointing out tensions within the Klein–Nishina and Thomson emission regimes [73] (Figure 6).

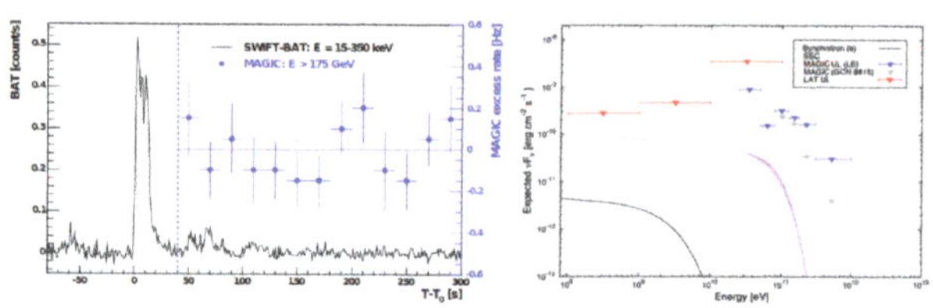

Figure 5. (**Left**) MAGIC excess event rate above the energy threshold of 175 GeV compared with the *Swift*-BAT light curve for GRB 050713A, the first prompt emission followed by an IACT. The vertical line shows the beginning of observations with the MAGIC telescope. Reprinted with permission from Ref. [65]. (**Right**) MAGIC and *Fermi*-LAT overlapping upper limits for GRB 090102. These results are compared to a leptonic synchrotron+SSC afterglow model. From model and data analysis described in [5].

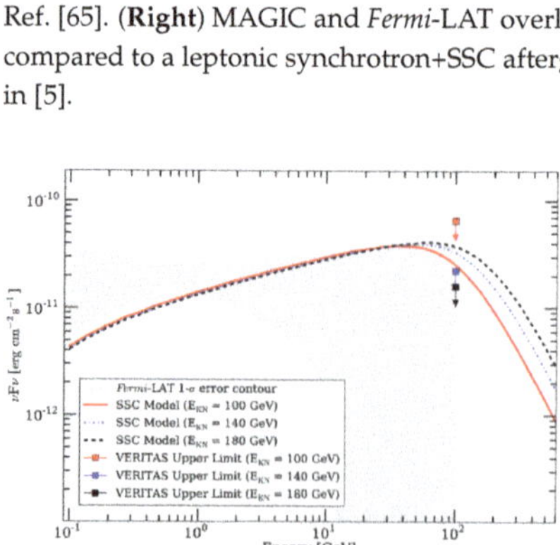

Figure 6. Combined *Fermi*-LAT spectrum including 1σ confidence interval and VERITAS upper limits for the late afterglow of GRB 130427A. Reprinted with permission from Ref. [73].

In parallel to these IACT observations, new EAS facilities, such as ARGO-YBJ, also started taking data in 2004 in the GeV band. No significant VHE emission was reported from any of the events located in the instrument field of view (see, e.g., [74]).

The VHE landscape on GRB study changed dramatically between 2018 and 2019 when the first detections were finally reported by the MAGIC and H.E.S.S. collaborations. These events are described in the following sections and their main parameters are summarized in Table 1.

Table 1. Summary of the main properties of the GRBs detected in the VHE range. GRB 201015A is included due to the strong evidence reported in [75] and is described in Section 4.4. The isotropic energy E_{iso} is calculated in the 50–300 keV range for GRB 180720B, 1–10^4 keV for GRB 190114C, and 10–10^3 keV for GRB 190829A and GRB 201216C. The spectral index α_{obs} is reported for the observed time-integrated spectrum (after absorption due to the EBL) over the whole observation window assuming a power-law model.

Name	T_{90} [s]	Redshift	E_{iso} [erg]	IACT	α_{obs}	E_{max}
180720B	48.9	0.653	6×10^{53}	H.E.S.S.	3.7 ± 1.0	440 GeV
190114C	362	0.4245	3×10^{53}	MAGIC	5.43 ± 0.22	1 TeV
190829A	58.2	0.0785	2×10^{50}	H.E.S.S.	2.59 ± 0.08	3.3 TeV
201216C	48	1.1	5×10^{53}	MAGIC	-	-
201015A	9.8	0.423	10^{50}	MAGIC	-	-

4.1. GRB 190114C

On 14 January 2019, MAGIC detected a very significant (at 50σ level) emission between 300 GeV and 1 TeV from the long GRB 190114C [2]. The event, initially detected by *Swift*-BAT and *Fermi*-GBM, was a bright ($E_{iso} \sim 3 \times 10^{53}$ erg in the 1–10^4 keV energy range), a long ($T_{90} = 362$ s as measured by *Swift*-BAT) and quite nearby ($z = 0.4245$) GRB.

Figure 7 shows the timescale of MAGIC follow-up observation: MAGIC received the alert from *Swift*-BAT 22 s after the GRB onset and started observations about 1 min after the GRB trigger under moderate moon conditions and at a relatively high zenith (58°). The GRB was detected with the MAGIC's real-time analysis with a significance of 20σ in the first 20 min of observations above an approximate threshold of 300 GeV. Later, the signal was confirmed up to 50σ-level in the dedicated offline analyses. The detection was reported as quickly as possible to the astrophysical community to strongly encourage the follow-up of this event at other wavelengths. Due to the timescale of early detection, one of the first questions to be answered was if the emission detected by MAGIC was related to the prompt or to the afterglow phase. While the value of T_{90} may indicate that such emission belongs to the prompt, detailed spectral and temporal studies of the keV–MeV data show that at $\sim T_0 + 25$ s the properties of such low-energy emission are more in agreement with the ones of the afterglow phase. This is additionally confirmed by the similar temporal decay index between the X-ray (from *Swift*-XRT between 0.1 and 1 keV) and the VHE (between 300 GeV and 1 TeV) energy light curves (Figure 8). The intrinsic spectrum of the GRB is compatible with a power law with spectral index of $\alpha_{int} = -2$ between 0.2 and 1 TeV, with no indication of any break or cutoff beyond those energies at the 95% confidence level. Such a flat spectrum shows that the energy output in the VHE range might be considered relevant, turning out to be comparable to the energy release measured at lower energies. Given the high absorption of the VHE flux by the EBL at the redshift of the GRB, the observed spectrum by MAGIC is rather softer and best described by a power law with index $\alpha_{obs} = -5.43 \pm 0.22$. This was tested against different EBL models, resulting in similar spectral indexes compatible within the statistical uncertainties.

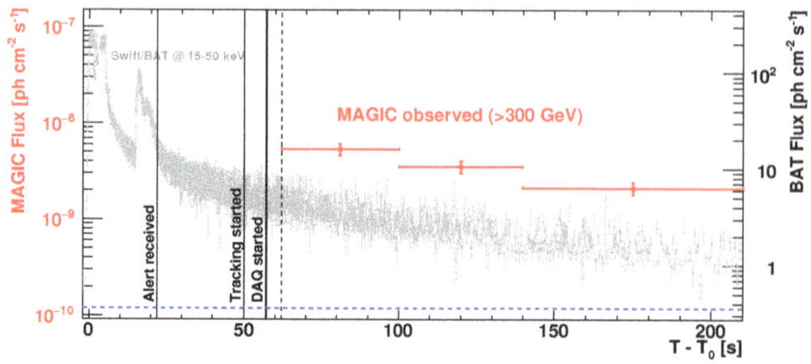

Figure 7. Light curves for MAGIC and *Swift*-BAT. Vertical solid lines show the times of events related to the MAGIC automatic procedure. Extracted from Ref. [2].

The origin of the emission detected by MAGIC is one of the most critical issues. The similarity of the temporal decay in X-ray and VHE energy light curves suggests that the emission processes might be linked and have the same origin. While the simplest hypothesis is that the processes producing the X-ray and VHE photons are the same, namely, synchrotron emission from relativistic electrons accelerated at the external shock in the afterglow of the GRB, this explanation is ruled out if one takes into account that the detected photons largely exceed the synchrotron burn-off limit (see Section 1). Even assuming a Lorentz factor of $\sim$1000, which is not typical for GRBs, the maximum energy of the photons produced by synchrotron is at most around 100 GeV, even considering different density profiles of the interstellar medium density. Therefore, it is reasonable to assume that the VHE emission is due to a different process. In addition, extrapolating the low-energy synchrotron spectrum (from *Fermi*-GBM, *Swift*-XRT, and *Fermi*-LAT data) to VHE range would underestimate the MAGIC flux by approximately one order of magnitude, strengthening the conclusion that the VHE photons are actually produced by a different mechanism. However, the existence of a synchrotron burn-off limit intimately assumes that the radiation came from one single emission region. Having more emission regions might allow synchrotron photons to reach higher energies. In the assumption that the VHE emission is not due to synchrotron, the most simple alternative is the SSC.

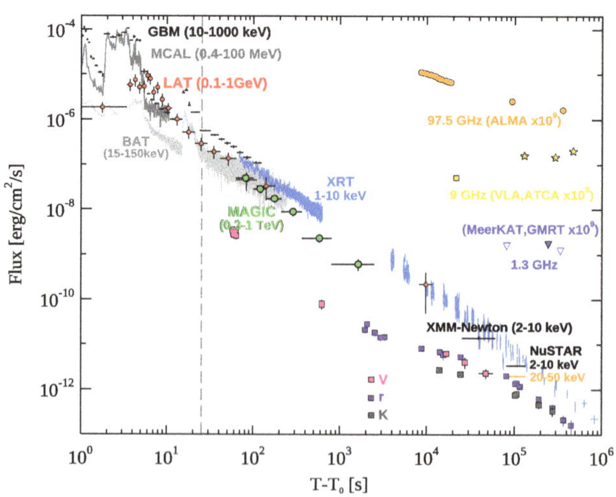

Figure 8. Radio to gamma-ray multi-wavelength light curves of GRB 190114C. The dashed vertical line marks the end of the prompt emission phase. Reprinted with permission from Ref. [76].

The SSC scenario, as commonly observed in other sources such as blazars, foresees a spectral energy distribution (SED) characterized by two distinct emission peaks, one at low energies (X-ray band) due to synchrotron emission, and a second one at higher energies, often in the VHE energy range. The modeling of the GRB 190114C multi-wavelength data with a synchrotron plus SSC emission within the external shock scenario in the afterglow shows exactly these two-peaks features, confirming the presence of an emission component at VHE never observed before (Figure 9). Another remarkable result is the fact that the parameters describing the broadband emission of GRB 190114C have values similar to the ones found in previous studies of GRB afterglows when data only up to the GeV energies were considered. This may hint to the possibility that VHE emission from SSC may be present in all GRBs and that it could be detected by IACTs if favorable conditions apply, i.e., a low enough redshift and good observing conditions. This hypothesis can be confirmed only with the detection of more GRBs in the VHE band.

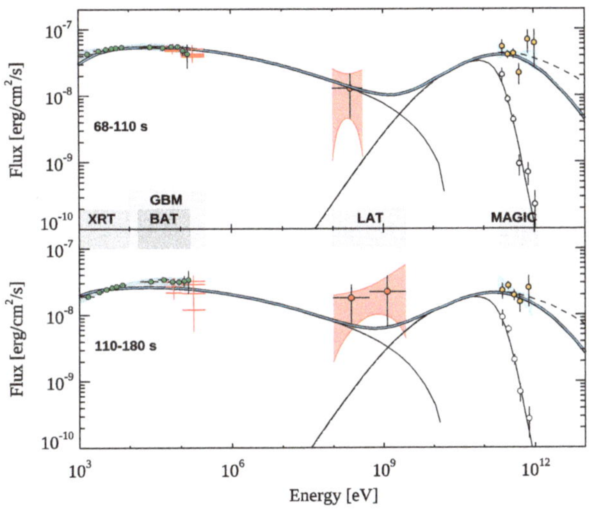

Figure 9. Synchrotron and SSC modeling of the broadband spectra of GRB 190114C in the two time intervals 68–110 s and 110–180 s. Dashed lines represent the SSC emission in the hypothesis of negligible internal $\gamma - \gamma$ opacity. MAGIC points in the VHE band are reported with (empty circles) and without (filled circles) correction for the EBL absorption. From [76].

4.2. GRB 201216C

GRB 201216C was detected by MAGIC [77] after receiving a trigger from *Swift*-BAT [78]. This GRB belongs to the long class as well, and similar to GRB 180720B and GRB 190114C, was very bright, having $E_{\rm iso} \sim 4.7 \times 10^{53}$ erg in the 10 keV to 1 MeV energy range. GRB 201216C was detected also in optical, where large extinction is present, and its redshift has been estimated to be $z = 1.1$ [79]. MAGIC started the observation of the burst about one minute after the *Swift* trigger, for a total of 2.2 h of exposure. The source was detected with a significance of 5.9σ (post trial) using the first 20 min of data, which makes this the farthest source ever detected. The spectrum in the same time interval can be well described by a power law extending from 50 to 200 GeV, while the energy flux light curve decays monotonically with time [80].

4.3. GRB 180720B and GRB 190829A

GRB 190829A was detected by the *Fermi*-GBM on 29 August 2019 at 19:55:53 UTC [81] with a second detection by *Swift* that occurred 51 s later [82]. The measured redshift of $z = 0.0785$ [83] implies a total isotropic energy release of $\approx 10^{50}$ erg during the prompt phase in both the *Fermi*-GBM and the *Swift* energy bands. This relatively low value placed GRB 190829A on the lower edge of the GRB energy distribution. Nonetheless, the very close-by distance made it a relatively bright GRB in the *Swift*-XRT band. H.E.S.S. observations were performed, starting at $T_0 + 4.3$ h, once afterglow phase emission had already taken over. Follow-up was performed with the use of the four smaller telescopes of the H.E.S.S. array and at a starting zenith angle of $\sim 40°$, corresponding to an energy threshold of ~ 170 GeV. The analysis reported a clear detection of a VHE gamma-ray signal during the first night with statistical significance of 21.7σ in 3.6 h of observation. Surprisingly, the GRB was also detected for the two forthcoming nights at $T_0 + 27.2$ h and $T_0 + 51.2$ h, with statistical significance of 5.5σ and 2.4σ, respectively. Figure 10 (left panel) shows the fading VHE signal measured during the three nights of observations. As for GRB 190114C, the X-ray and VHE gamma-ray light curves also show similar decay profiles with a time evolution characterized by a power law of index $\alpha_{VHE} = 1.09 \pm 0.05$ and $\alpha_{XRT} = 1.07 \pm 0.09$ in the H.E.S.S. and *Swift*-XRT bands, respectively. Despite these similarities, the interpretation of the VHE light curve and spectrum of GRB 190829A within the framework of the standard GRB afterglow emission model (such as for GRB 190114C) showed some tensions. The H.E.S.S. data were collected deep in the afterglow phase in a moment in which the bulk Lorentz factor of the outflow was evaluated to be $\Gamma \sim 4.7$ and $\Gamma \sim 2.6$ for the first and second nights of observation, respectively [22,83]. Thus, radiation of few TeV, such as the one measured by H.E.S.S., besides largely exceeding the synchrotron burn-off limit at these specific times, also results in tension with the synchrotron+SSC scenario. With these values of Γ, the electrons producing the VHE emission likely lie in the Klein–Nishina regime. The corresponding reduction in the inverse Compton cross section would introduce a cut-off and a steepening of the flux at VHE. As it appears from H.E.S.S. results (Figure 11), this expected steepening makes it challenging for SSC models to simultaneously reproduce the observed X-ray and VHE spectra [3]. An intriguing possibility is to introduce a leptonic scenario with no limitation placed on the electron maximum energy (and, correspondingly, no synchrotron burn-off limit) that would allow synchrotron emission to produce VHE photons. Although this scenario reproduces the H.E.S.S. data (Figure 11) considerably better, it would require a significant re-evaluation of the relativistic shock-accelerated models. On the other hand, although alternative interpretations have been presented (see, e.g., [84,85]), attempts to model GRB 190829A afterglow using a leptonic synchrotron + SSC emission model have been reported with convincing results and with an obtained set of the shock's microphysical parameters that are similar to those found for GRB 190114C [86].

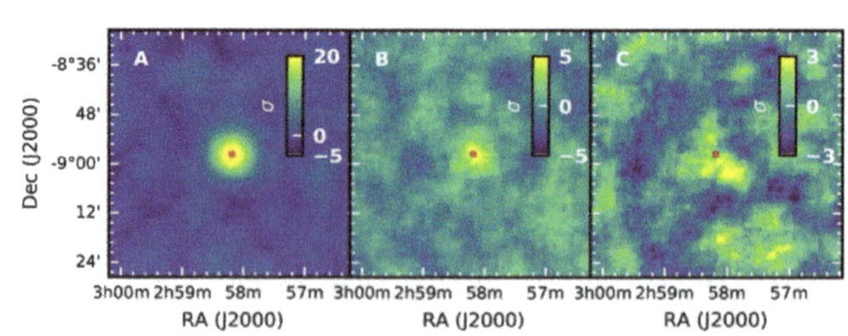

Figure 10. The H.E.S.S. sky map centered at the coordinates of GRB 190829A. The VHE signal was clearly detected for three consecutive nights up to $T_0 + 51$ h. Reprinted with permission from Ref. [3].

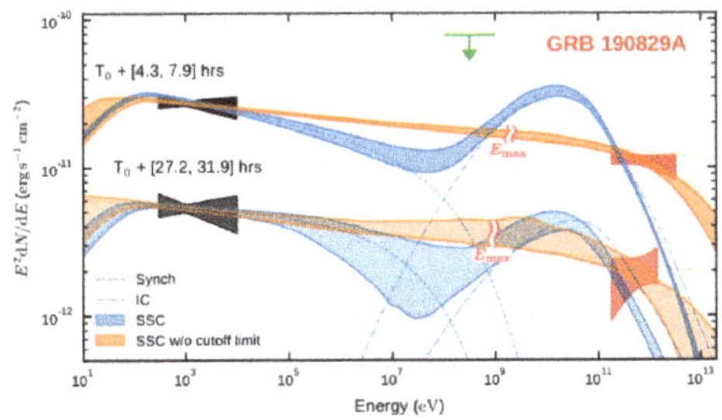

Figure 11. Multi-wavelength models of the first two nights of GRB 190829A H.E.S.S. The black region represents the spectrum and uncertainty of the *Swift*-XRT observations, while the red region is the H.E.S.S. intrinsic spectrum and its uncertainty. The green arrow is the upper limit set by observation during the first night by *Fermi*-LAT. The used model is a standard SSC model with 68% confidence intervals determined from the posterior probability distribution of the model's parameter fitting for the synchrotron component (orange) and the SSC one (blue). No synchrotron cut-off energy (burn-off limit) is considered. Reprinted with permission from Ref. [3].

GRB 180720B was detected by *Fermi*-GBM at 14:21:39.65 UTC [87], and 5 s later by *Swift*-BAT [88]. The event was also detected by the *Fermi*-LAT between T_0 and $T_0 + 700$ s with a maximum photon energy of 5 GeV at $T_0 + 142.4$ s [89]. With a redshift of $z = 0.653$ and an equivalent isotropic energy release of $\sim 6 \times 10^{53}$ erg in the 50–300 keV band, this event is one of the brightest GRBs ever detected by *Fermi*-LAT. The light curves show quite a conventional power-law behavior in both the X and the optical band with a temporal flux decay of index $\alpha_{\rm XRT} = 1.29 \pm 0.01$ and $\alpha_{\rm optical} = 1.24 \pm 0.02$. In the HE band, the flux followed a slightly steeper trend with $\alpha_{\rm LAT} = 0.99 \pm 0.04$, about 1σ from the mean value of the distribution of the decay indices of long GRBs detected by *Fermi*-LAT [1]. Observation by H.E.S.S. started at $T_0 + 10$ h, and the source was detected at $\sim 5\sigma$ level. The detection of a VHE emission at such late times into the afterglow phase implied the presence of very energetic particles accelerated (likely) at the forward shock. Similar to in the case of GRB 190114C, in [1] an SSC-emitting scenario was found to be reasonably in agreement with the observational data, although the marginal level of the significance did not allow more detailed investigation with high statistic and time-resolved spectra and light curve.

4.4. GRB 201015A

GRB 201015A was detected by *Swift*-BAT and followed-up by MAGIC, which reported an excess at the level of $>3\sigma$ in [75,90]. MAGIC observation started from 33 s after T_0, lasting for about 4 h under good weather conditions. The evidence of emission is found above an energy threshold of $\sim$140 GeV. This GRB has some properties similar to GRB 190829A, in particular, the isotropic energy release $E_{\rm iso} \sim 10^{50}$ erg; however, it is located at a much farther distance $z = 0.423$, resulting in a stronger flux attenuation due to the EBL. With the publication of MAGIC data on this GRB, a more detailed comparison with GRB 190829A will be possible, giving more insights into VHE GRBs with rather low luminosity.

The observations performed by MAGIC and H.E.S.S. established the presence of a VHE emission component in both the early and late afterglow phase. For the former, the fast repositioning and reaction of the MAGIC telescopes played a fundamental role. For the latter, the merit is the well-thought strategy based on observational results from other bands. In both cases, these results have proved to be complementary, providing insights into the nature of GRBs and their VHE detectability at different times, constituting an important lesson for the future observation strategies of next-generation VHE facilities.

5. Advances in GRBs Studies and Open Issues at VHE

The detection of a VHE signal from GRB by MAGIC and H.E.S.S. (particularly GRB 190829A) provided a puzzling and unexpected complexity of scenarios, mainly due to the differences between the phenomenology of the observed events. GRB 190114C and GRB 190829A are events that stand on the opposite edges of the GRB energy distribution being on the ~30% sub-sample of more energetic burst for GRB 190114C ($E_{iso} = 3 \times 10^{53}$ erg) and more than three orders of magnitude lower for GRB 190829A ($E_{iso} = 2 \times 10^{50}$ erg). The two events also significantly differ in their temporal profile at VHE, with an extremely bright VHE emission lasting ~15 min for GRB 190114C and a dimmer but much longer-lasting emission for GRB 190829A (up to few days after GRB onset). On the other hand, GRB 180720B, which is more similar to GRB 190114C, was detected several hours after the event's onset. In order to exemplify these differences, we reported in Figure 12 (left panel) the value of the bulk Lorentz factor (Γ_0) at the beginning of the afterglow phase, evaluated for a large sample of GRBs [91] once known their isotropic equivalent energy. The positions of the VHE-detected GRBs are overplotted, from which a different nature of GRB 190829A might be pointed out, although both events lie on the so-called Amati relation (Figure 12 right panel). This is an indication that the observed differences in luminosity and energy are not related to a different geometry of the emission (i.e., GRB 190829A is not an off-axis event). On the other hand, gamma rays of such high energies largely exceed the synchrotron burn-off limit, implying the coexistence of an extra emission component in the VHE band. However, the attempt of broadband modeling of the two events led to a different physical interpretation of the VHE emission. While GRB 190114C has been satisfyingly modeled within a synchrotron + SSC emission scenario, the H.E.S.S.S. collaboration reported an alternative hypothesis for GRB 190829A. In particular, in [3], it was proposed the possibility to interpret the VHE radiation as a synchrotron extending well above the burn-off limit at the time of H.E.S.S. observations. Although intriguing, such an interpretation has been challenged by other works, where, again, a synchrotron+SSC approach seems favorable in modeling the broadband spectrum without requiring peculiar and unconventional choices of the GRB microphysical parameters [86]. Whether an SSC component is at work in all GRBs and which is the maximum energy achievable by different emission mechanisms are still some of the open points that can be addressed with more observations of GRBs at VHE. The tension on the modeling side is also the result of the limited number of VHE GRBs detected up to now, and of the available multi-wavelength (MWL) data collected simultaneously. In particular, *Fermi*-LAT can be of great importance since it covers the energy range where the transition from the synchrotron radiation to the possible SSC component is expected, as exemplified by the case of GRB 190114C. However, such availability of MWL data might not be common, especially if GRBs are detected in the VHE range at late times, when the flux can be below the sensitivity of, e.g., *Fermi*-LAT. Such lack of MWL data can introduce difficulties in the modeling, or lead to degeneracy of the modeling parameters. From this perspective, early VHE follow-up seems to have an advantage, with a higher probability of having more simultaneous MWL data available for later modeling (e.g., the GRB is still bright enough to be detected by instruments such as *Fermi*-LAT). Furthermore, the prompt-to-early-afterglow phase, with the coexistence of forward and reverse shocks in the emitted outflow, could also lead to a large variety of different and interesting emitting scenarios in the VHE band. In this regard, while a deeper understanding of the afterglow phase at VHE is due, one of the next challenges is the detection of VHE emission in the prompt phase. The debate on the physical process at the origin of the prompt emission is still open, with different possibilities missing a clear observational proof. A detection of the prompt emission in the VHE range could resolve such a long-lasting issue, giving a new perspective on this poorly known phase of GRBs. The challenge for IACTs is the short duration of the prompt phase, compared to the delivery times of the alerts from triggering instruments and the time for their reaction. In particular, T_{90} alone is not a good indicator of the duration of the prompt phase and of the nature of a GRB, as already debated within the GRB community (see, e.g., [92,93]). Therefore, long-duration GRBs with T_{90} of the order of hundreds of seconds (e.g., GRB 190114C) can also have prompt phases with a much shorter duration, as shown by the spectral and temporal analysis of the GRB light curves. For this reason, ground-based instruments such as HAWC https://www.hawc-observatory.org/ (that already reported results on GRB observations [94]), LHAASO (http://english.ihep.cas.cn/lhaaso/), and the future SWGO (https://www.swgo.org/SWGOWiki/doku.php) (all websites accessed on 10 April 2022) could be more suited, given their high-duty cycle and sky coverage, with the downside of a higher energy threshold.

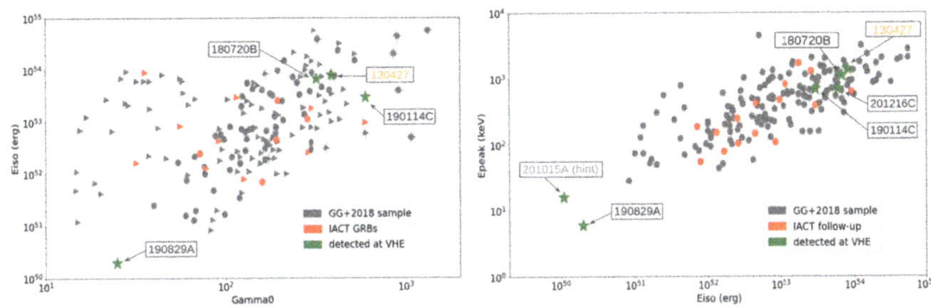

Figure 12. Left panel: Correlation between the bulk Lorentz factor at the beginning of the afterglow phase (Γ_0) and the isotropic equivalent energy E_{iso} for the sample of GRB reported in [91]. **Right panel:** the empirical correlation (Amati relation) between the isotropic equivalent energy E_{iso} and the peak energy of the GRB spectrum for the same sample in [91] and for the events detected in the VHE band. In both panels, GRBs followed-up by IACTs are denoted by green and red symbols, indicating those with and without detection, respectively. Reprinted with permission from Ref. [95].

An additional challenge for IACTs is the detection of short GRBs. While they are located (on average) at smaller redshift with respect to long GRBs, they are also less luminous, making a detection with IACTs difficult. Currently, the strongest evidence for a VHE emission component from short GRBs was reported by MAGIC for GRB 160821B [96]. The telescope's fast response played a major role, despite the adverse observational conditions (reduced atmospheric conditions, relatively high zenith of the observation, and increased night sky background due to the presence of the Moon). A signal at the level of 3σ (post-trial) was found with a flux upper limit of 1.1×10^{-11} cm^{-2} s^{-1} in the first half hour, giving the possibility to perform interesting studies on the expected energy flux at VHE in an MWL context [96]. The SSC model was found to be in tension with the data, nonetheless a firm detection and a higher statistics is needed to rule out this possibility. Such a discovery would be of utmost importance to understand if there are similarities at VHE between long and short GRBs. Moreover, short GRBs are intimately connected with searches of gravitational waves (GWs) from their progenitors. A coincident detection of a short GRB and GWs would allow a comprehensive picture of the system leading to the GRB itself, and of its following evolution.

Finally, the detection of other GRBs at VHE can open up the possibility of interesting studies on more fundamental topics. VHE signal from these events might be extremely valuable for probing external $\gamma - \gamma$ absorption due to the EBL out to larger redshifts than the ones that can be typically reached with other extragalactic sources, such as blazars. Another possibility could involve the searches for Lorentz invariance violation (LIV), where distant GRBs with high-energy photons can considerably improve the sensitivity of the resulting lower limit (see, e.g., [97] for a LIV study using GRB 190114C data). VHE GRB data can also be used for the search of axion-like particles (ALPs), where we expect spectral signatures and a reduction of the optical depth, leading to a lower absorption with respect to the expectation from the EBL (see, e.g., [98]). The GRBs detected so far at VHE confirmed that a low–moderate redshift is still a necessary condition for the detection for current IACTs, especially if the luminosity is towards the low end of the distribution. GRB 201216C, detected at $z = 1.1$, stands as an outsider and can represent an interesting case study for possible EBL or LIV studies. This is indeed a promising result for the next generation of Cherenkov telescopes that, thanks to improved sensitivity and energy threshold, might further extend the gamma-ray horizon of these observations.

6. The Next Decades

The Cherenkov Telescope Array Observatory (CTAO) represents the next-generation ground-based observatory for the study of VHE gamma rays. It will consist of two arrays, one for each hemisphere, made up of IACTs of different size and characteristics. The CTA array will routinely perform follow-up observations of GRB triggers and other transients objects also coming from other cosmic signal, such as neutrino and gravitational waves [99]. The estimation of the detection prospects for such observations are necessarily still preliminary and are dependent on the final array layout and performance. Nonetheless, even starting with simplified assumptions about the GRB emission, the CTA Consortium already reported the possibility of detecting ~hundreds (or more) of photons from moderate to bright GRB, allowing for a significant improvement in the photon statistics and for the possibility to have good-quality time-resolved spectra [100]. The preliminary results reported in such a study show the possibility of detecting up to few GRB per year (considering both arrays) and allowing to move rapidly from the single-case GRB study, such as for current IACT, to a full GRB population study at VHE. In order to confirm these early results and achieve a step forward in the determination of CTA's prospects for GRB follow-ups, the CTA Consortium is currently working on a new study where the potential detection rate is estimated using a theoretical-based approach. Such an approach is based on the *POpulation Synthesis Theory Integrated code for Very high energy Emission* (POSyTIVE) model for GRBs [101]. The aim is to build a GRB population based on few intrinsic properties and assumptions such as E_{peak} and redshift distribution, E_{peak}-E_{iso} correlation (Amati relation) [102], and the bulk Lorentz factor distribution obtained by measured the time of the afterglow onset (providing the bulk Lorentz factor of the event's coasting phase). The population obtained (for both long and short GRBs) is calibrated against a wide dataset of multi-wavelength observations. In order to derive the final expected spectrum, both the prompt and the afterglow emission are simulated according to a standard leptonic synchrotron+SSC emission model [14]. The GRB spectra obtained are then used to simulate the detailed CTA response through the use of dedicated analysis pipelines based on gammapy (https://gammapy.org/) and ctools (http://cta.irap.omp.eu/ctools/) (accessed on 27 April 2022) and making use of the most recent instrument response functions (IRFs). The results of this study are expected by the end of 2022.

In the framework of the CTA, the earliest science operations have recently started thanks to the large-sized telescope prototype (LST-1). LSTs are the largest telescopes designed for CTA, having a 23 m diameter reflector. The first prototype, LST-1 (Figure 13 left panel), is located at the Roque de los Muchachos observatory (28.8° N, 17.8° W, 2200 m a.s.l.), on the Canary Island of La Palma [103], the designed site for the CTA north array. Thanks to the reflective surface of about 400 m^2, the LST-1 will be able to achieve an energy threshold of ≈20 GeV, a value particularly suitable for transients and high-redshift source observations. Furthermore, LSTs are built with a light carbon-fiber structure in order to reduce the total weight of the telescope to about 103 tons and to make possible the fast repositioning (~30 s for 180° azimuth displacement) to catch early emission phases of transient objects. LST-1 was inaugurated in October 2018 and is currently finalizing its commissioning phase. Starting from the first months of 2021, the time allocated for technical observations has been gradually reduced, allowing the first observations of targets of astrophysical interest. Transients follow-up, including GRBs, have the highest priority among LST-1 observed targets. Although a fully automatic procedure that will allow the telescope to react automatically to incoming alerts is still under development, the first observations of GRBs have been performed [95]. Preliminary analysis did not reveal VHE emission associated with any of the observed alerts; however, the continuous effort in improving the telescope's performance and robustness will soon place LST-1 in a key position for VHE observations of those peculiar events, making it a noticeable testbench for the forthcoming full-configured CTA array.

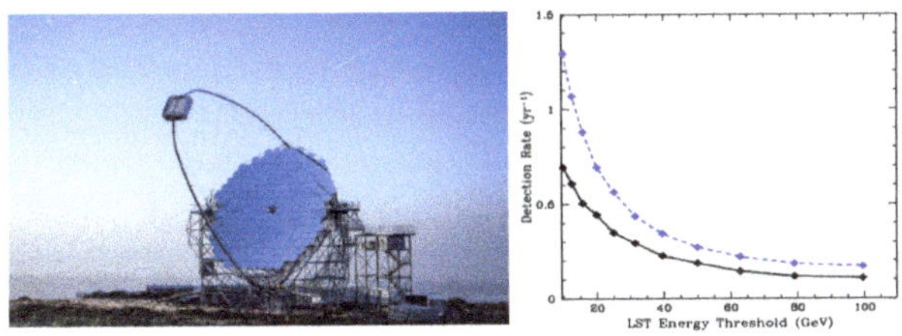

Figure 13. (**Left**) The LST prototype during night operation in La Palma. Picture credit: Tomohiro Inada. (**Right**) GRB detection rate for one of the CTA arrays as a function of the expected LST energy threshold. The curves represent two possible empirical assumptions for GRB spectrum at very high energy; one considering a simple extrapolation of the band emission up to VHE (solid black), the other considering an additional emission component (dashed blue). Reprinted with permission from Ref. [100].

The firm detection of a signal extending up to the multi-TeV band for GRB 190829A has opened new interesting possibilities for observations also with IACTs not specifically designed for transients follow-up, such as the small-sized telescopes (SST) foreseen for the CTA. An interesting example is the case of the ASTRI mini-array, composed of nine imaging atmospheric dual-mirror Cherenkov telescopes at the Teide Observatory site on the Canary Island of La Palma [104]. The telescopes will have a relatively small primary mirror of ~4 m diameter, allowing to detect gamma rays in the 0.5–200 TeV range. Despite the corresponding limited gamma-ray horizon accessible, the authors in [105] proved the feasibility of the ASTRI mini-array to detect bright and nearby GRBs. This would guarantee to cover, with high sensitivity, the extreme edge of the VHE band, complementing the data collected at lower energies with instruments such as the LSTs.

7. Conclusions

VHE observations provide a new channel to study the physics of GRBs in an energy range particularly important for the discrimination of different emitting scenarios and for the constraint of the GRBs' physical parameters in space. The detection of GRBs at VHE represents one of the major breakthroughs for transient astrophysics in the last years. This result was finally achieved thanks to the relentless efforts and the continuous improvements on both the technical and the observational strategy side by current IACTs collaborations. The small sample of detected events shows a large variety of phenomenology that leaves some questions unanswered, creating difficulties in finding a possible common interpretative scenarios. In all detected events, VHE emission has been observed on timescales much longer than the corresponding prompt phase, confirming the results already observed in the GeV band. However, besides the detection of bright and powerful events, that for a long time were assumed to be the best candidates for VHE emission, relatively low-luminosity events also showed long-lasting emission up to the TeV band. This suggests that the detection of these GRBs is likely not unique and the VHE component might be a relatively common feature of many GRBs, although observable only under favorable conditions by IACTs. Whether *all* GRBs have a VHE emission component and whether the parameter space of a possible VHE-emitter GRB is larger than what was previously thought, this will be one of the key issues for the next generation of IACTs, namely, the CTAO. Short timescale transients (including GRBs) have been one of the key motivations when designing the different elements of the CTA, and in particular the LSTs, whose first prototype recently started its operation in the Canary Island of La Palma. Once fully configured, a detection rate of the order of a few bursts per year might be expected, allowing us to build and characterize the GRB population at VHE. Furthermore, the achievable photon statistics would allow CTA to study the spectral and temporal properties of GRBs, shedding light into unresolved issues such as determining the jet formation dynamics and the mechanisms of particle acceleration.

Author Contributions: The authors contributed equally to this work. The authors have read and agreed to the published version of the manuscript.

Funding: This research received no external funding.

Institutional Review Board Statement: Not applicable.

Informed Consent Statement: Not applicable.

Data Availability Statement: Not applicable.

Acknowledgments: The authors acknowledge the anonymous referees for their useful comments.

Conflicts of Interest: The authors declare no conflict of interest.

References

1. Abdalla, H.; Adam, R.; Aharonian, F.; Ait Benkhali, F.; Angüner, E.O.; Arakawa, M.; Arcaro, C.; Armand, C.; Ashkar, H.; Backes, M.; et al. A very-high-energy component deep in the γ-ray burst afterglow. *Nature* **2019**, *575*, 464–467. [CrossRef]
2. MAGIC Collaboration. Teraelectronvolt emission from the γ-ray burst GRB 190114C. *Nature* **2019**, *575*, 455–458. [CrossRef]
3. H.E.S.S. Collaboration; Abdalla, H.; Aharonian, F.; Ait Benkhali, F.; Angüner, E.O.; Arcaro, C.; Armand, C.; Armstrong, T.; Ashkar, H.; Backes, M.; et al. Revealing X-ray and gamma ray temporal and spectral similarities in the GRB 190829A afterglow. *Science* **2021**, *372*, 1081–1085. [CrossRef]
4. Carosi, A.; Antonelli, A.; Becerra Gonzalez, J.; Berti, A.; Covino, S.; Garczarczyk, M.; Gaug, M.; Lombardi, S.; Longo, F.; Moretti, E.; et al. Recent follow-up observations of GRBs in the very high energy band with the MAGIC Telescopes. In Proceedings of the 34th International Cosmic Ray Conference (ICRC2015), The Hague, The Netherlands, 30 July–6 August 2015; International Cosmic Ray Conference; Volume 34, p. 809.
5. Aleksić, J.; Ansoldi, S.; Antonelli, L.A.; Antoranz, P.; Babic, A.; Barres de Almeida, U.; Barrio, J.A.; Becerra González, J.; Bednarek, W.; Berger, K.; et al. MAGIC upper limits on the GRB 090102 afterglow. *Mon. Not. R. Astron. Soc.* **2014**, *437*, 3103–3111. [CrossRef]
6. Paczynski, B. Gamma-ray bursters at cosmological distances. *Astrophys. J. Lett.* **1986**, *308*, L43–L46. [CrossRef]
7. Thompson, C. A model of gamma-ray bursts. *Mon. Not. R. Astron. Soc.* **1994**, *270*, 480–498. [CrossRef]
8. Granot, J. Gamma-Ray Bursts from Magnetic Reconnection: Variability and Robustness of Light Curves. *Astrophys. J. Lett.* **2016**, *816*, L20, [CrossRef]
9. Band, D.; Matteson, J.; Ford, L.; Schaefer, B.; Palmer, D.; Teegarden, B.; Cline, T.; Briggs, M.; Paciesas, W.; Pendleton, G.; et al. BATSE Observations of Gamma-Ray Burst Spectra. I. Spectral Diversity. *Astrophys. J.* **1993**, *413*, 281. [CrossRef]
10. Toffano, M.; Ghirlanda, G.; Nava, L.; Ghisellini, G.; Ravasio, M.E.; Oganesyan, G. The slope of the low-energy spectrum of prompt gamma-ray burst emission. *Astropart. Astrophys.* **2021**, *652*, A123, [CrossRef]
11. Basak, R.; Rao, A.R. Discovery of Smoothly Evolving Blackbodies in the Early Afterglow of GRB 090618: Evidence for a Spine-Sheath Jet? *Astrophys. J.* **2015**, *812*, 156, [CrossRef]
12. Cohen, E.; Katz, J.I.; Piran, T.; Sari, R.; Preece, R.D.; Band, D.L. Possible Evidence for Relativistic Shocks in Gamma-Ray Bursts. *Astrophys. J.* **1997**, *488*, 330–337. [CrossRef]
13. Ghirlanda, G.; Celotti, A.; Ghisellini, G. Extremely hard GRB spectra prune down the forest of emission models. *Astropart. Astrophys.* **2003**, *406*, 879–892. [CrossRef]
14. Sari, R.; Esin, A.A. On the Synchrotron Self-Compton Emission from Relativistic Shocks and Its Implications for Gamma-Ray Burst Afterglows. *Astrophys. J.* **2001**, *548*, 787–799. [CrossRef]
15. Zhang, B.; Mészáros, P. High-Energy Spectral Components in Gamma-Ray Burst Afterglows. *Astrophys. J.* **2001**, *559*, 110–122. [CrossRef]
16. Gupta, N.; Zhang, B. Prompt emission of high-energy photons from gamma ray bursts. *Mon. Not. R. Astron. Soc.* **2007**, *380*, 78–92. [CrossRef]
17. Atwood, W.B.; Abdo, A.A.; Ackermann, M.; Althouse, W.; Anderson, B.; Axelsson, M.; Baldini, L.; Ballet, J.; Band, D.L.; Barbiellini, G.; et al. The Large Area Telescope on the Fermi Gamma-Ray Space Telescope Mission. *Astrophys. J.* **2009**, *697*, 1071–1102. [CrossRef]
18. Ghisellini, G.; Ghirlanda, G.; Nava, L.; Celotti, A. GeV emission from gamma-ray bursts: a radiative fireball? *Mon. Not. R. Astron. Soc.* **2010**, *403*, 926–937. [CrossRef]
19. Piran, T. Gamma-ray bursts and the fireball model. *Phys. Rep.* **1999**, *314*, 575–667. [CrossRef]

20. Abdo, A.A.; Ackermann, M.; Arimoto, M.; Asano, K.; Atwood, W.B.; Axelsson, M.; Baldini, L.; Ballet, J.; Band, D.L.; Barbiellini, G.; et al. Fermi Observations of High-Energy Gamma-Ray Emission from GRB 080916C. *Science* **2009**, *323*, 1688. [CrossRef]
21. Ackermann, M.; Asano, K.; Atwood, W.B.; Axelsson, M.; Baldini, L.; Ballet, J.; Barbiellini, G.; Baring, M.G.; Bastieri, D.; Bechtol, K.; et al. Fermi Observations of GRB 090510: A Short-Hard Gamma-ray Burst with an Additional, Hard Power-law Component from 10 keV TO GeV Energies. *Astrophys. J.* **2010**, *716*, 1178–1190. [CrossRef]
22. Blandford, R.D.; McKee, C.F. Fluid dynamics of relativistic blast waves. *Phys. Fluids* **1976**, *19*, 1130–1138. [CrossRef]
23. Guetta, D.; Granot, J. High-Energy Emission from the Prompt Gamma-Ray Burst. *Astrophys. J.* **2003**, *585*, 885–889. [CrossRef]
24. Guetta, D.; Granot, J. Observational implications of a plerionic environment for gamma-ray bursts. *Mon. Not. R. Astron. Soc.* **2003**, *340*, 115–138. [CrossRef]
25. Daigne, F.; Bošnjak, Ž.; Dubus, G. Reconciling observed gamma-ray burst prompt spectra with synchrotron radiation? *Astron. Astrophys.* **2011**, *526*, A110, [CrossRef]
26. Bošnjak, Ž.; Daigne, F.; Dubus, G. Prompt high-energy emission from gamma-ray bursts in the internal shock model. *Astron. Astrophys.* **2009**, *498*, 677–703. [CrossRef]
27. Daigne, F. *High-Energy Emission from Gamma-Ray Bursts;* International Journal of Modern Physics Conference Series; World Scientific: Singapore, 2012; Volume 8, pp. 196–208. [CrossRef]
28. Razzaque, S. Modeling and Interpretation of High Energy Emission from GRBs Observed with the Fermi Gamma Ray Space Telescope. In Proceedings of the APS April Meeting Abstracts, Denver, Colorado, 2–5 May 2009; p. W8.003.
29. Asano, K.; Inoue, S.; Mészáros, P. Prompt High-Energy Emission from Proton-Dominated Gamma-Ray Bursts. *Astrophys. J.* **2009**, *699*, 953–957. [CrossRef]
30. Beniamini, P.; Nava, L.; Duran, R.B.; Piran, T. Energies of GRB blast waves and prompt efficiencies as implied by modelling of X-ray and GeV afterglows. *Mon. Not. R. Astron. Soc.* **2015**, *454*, 1073–1085. [CrossRef]
31. Wang, X.Y.; He, H.N.; Li, Z.; Wu, X.F.; Dai, Z.G. Klein-Nishina Effects on the High-energy Afterglow Emission of Gamma-ray Bursts. *Astrophys. J.* **2010**, *712*, 1232–1240. [CrossRef]
32. Liu, R.Y.; Wang, X.Y.; Wu, X.F. Interpretation of the Unprecedentedly Long-lived High-energy Emission of GRB 130427A. *Astrophys. J. Lett.* **2013**, *773*, L20, [CrossRef]
33. Wang, X.Y.; Dai, Z.G.; Lu, T. Prompt High-Energy Gamma-Ray Emission from the Synchrotron Self-Compton Process in the Reverse Shocks of Gamma-Ray Bursts. *Astrophys. J. Lett.* **2001**, *546*, L33–L37, [CrossRef]
34. Granot, J.; Guetta, D. Explaining the High-Energy Spectral Component in GRB 941017. *Astrophys. J. Lett.* **2003**, *598*, L11–L14, [CrossRef]
35. Böttcher, M.; Dermer, C.D. High-energy Gamma Rays from Ultra-high-energy Cosmic-Ray Protons in Gamma-Ray Bursts. *Astrophys. J. Lett.* **1998**, *499*, L131–L134, [CrossRef]
36. Nava, L. High-energy emission from gamma-ray bursts. *Int. J. Mod. Phys. D* **2018**, *27*, 1842003, [CrossRef]
37. Hurley, K.; Dingus, B.L.; Mukherjee, R.; Sreekumar, P.; Kouveliotou, C.; Meegan, C.; Fishman, G.J.; Band, D.; Ford, L.; Bertsch, D.; et al. Detection of a γ-ray burst of very long duration and very high energy. *Nature* **1994**, *372*, 652–654. [CrossRef]
38. Merck, M.; Bertsch, D.L.; Dingus, B.L.; Fichtel, C.E.; Hartman, R.C.; Hunter, S.D.; Kanbach, G.; Kniffen, D.A.; Lin, Y.C.; Mayer-Hasselzuander, H.A.; et al., Observations of High-Energy Gamma-ray Bursts with EGRET. In *IAU Colloq. 151: Flares and Flashes;* Springer: Berlin/Heidelberg, Germany 1995; Volume 454, p. 358. [CrossRef]
39. González, M.M.; Dingus, B.L.; Kaneko, Y.; Preece, R.D.; Dermer, C.D.; Briggs, M.S. A γ-ray burst with a high-energy spectral component inconsistent with the synchrotron shock model. *Nature* **2003**, *424*, 749–751. [CrossRef]
40. Tavani, M.; Barbiellini, G.; Argan, A.; Boffelli, F.; Bulgarelli, A.; Caraveo, P.; Cattaneo, P.W.; Chen, A.W.; Cocco, V.; Costa, E.; et al. The AGILE Mission. *Astron. Astrophys.* **2009**, *502*, 995–1013. [CrossRef]
41. Giuliani, A.; Mereghetti, S.; Fornari, F.; Del Monte, E.; Feroci, M.; Marisaldi, M.; Esposito, P.; Perotti, F.; Tavani, M.; Argan, A.; et al. AGILE detection of delayed gamma-ray emission from GRB 080514B. *Astron. Astrophys.* **2008**, *491*, L25–L28, [CrossRef]
42. Del Monte, E.; Barbiellini, G.; Donnarumma, I.; Fuschino, F.; Giuliani, A.; Longo, F.; Marisaldi, M.; Pucella, G.; Tavani, M.; Trifoglio, M.; et al. The AGILE observations of the hard and bright GRB 100724B. *Astron. Astrophys.* **2011**, *535*, A120, [CrossRef]
43. Longo, F.; Moretti, E.; Nava, L.; Desiante, R.; Olivo, M.; Del Monte, E.; Rappoldi, A.; Fuschino, F.; Marisaldi, M.; Giuliani, A.; et al. Upper limits on the high-energy emission from gamma-ray bursts observed by AGILE-GRID. *Astron. Astrophys.* **2012**, *547*, A95. [CrossRef]
44. Meegan, C.; Lichti, G.; Bhat, P.N.; Bissaldi, E.; Briggs, M.S.; Connaughton, V.; Diehl, R.; Fishman, G.; Greiner, J.; Hoover, A.S.; et al. The Fermi Gamma-ray Burst Monitor. *Astrophys. J.* **2009**, *702*, 791–804. [CrossRef]
45. Abdo, A.A.; Ackermann, M.; Asano, K.; Atwood, W.B.; Axelsson, M.; Baldini, L.; Ballet, J.; Band, D.L.; Barbiellini, G.; Bastieri, D.; et al. Fermi Observations of High-energy Gamma-ray Emission from GRB 080825C. *Astrophys. J.* **2009**, *707*, 580–592. [CrossRef]
46. Abdo, A.A.; Ackermann, M.; Ajello, M.; Asano, K.; Atwood, W.B.; Axelsson, M.; Baldini, L.; Ballet, J.; Barbiellini, G.; Bastieri, D.; et al. Fermi Detection of Delayed GeV Emission from the Short Gamma-Ray Burst 081024B. *Astrophys. J.* **2010**, *712*, 558–564. [CrossRef]
47. Giuliani, A.; Fuschino, F.; Vianello, G.; Marisaldi, M.; Mereghetti, S.; Tavani, M.; Cutini, S.; Barbiellini, G.; Longo, F.; Moretti, E.; et al. AGILE Detection of Delayed Gamma-ray Emission From the Short Gamma-Ray Burst GRB 090510. *Astrophys. J.* **2010**, *708*, L84–L88, [CrossRef]

48. Fan, Y.Z.; Piran, T.; Narayan, R.; Wei, D.M. High-energy afterglow emission from gamma-ray bursts. *Mon. Not. R. Astron. Soc.* **2008**, *384*, 1483–1501. [CrossRef]
49. Asano, K.; Guiriec, S.; Mészáros, P. Hadronic Models for the Extra Spectral Component in the Short GRB 090510. *Astrophys. J. Lett.* **2009**, *705*, L191–L194, [CrossRef]
50. Ackermann, M.; Ajello, M.; Asano, K.; Atwood, W.B.; Axelsson, M.; Baldini, L.; Ballet, J.; Barbiellini, G.; Baring, M.G.; Bastieri, D.; et al. Fermi-LAT Observations of the Gamma-Ray Burst GRB 130427A. *Science* **2014**, *343*, 42–47. [CrossRef]
51. Kouveliotou, C.; Granot, J.; Racusin, J.L.; Bellm, E.; Vianello, G.; Oates, S.; Fryer, C.L.; Boggs, S.E.; Christensen, F.E.; Craig, W.W.; et al. NuSTAR Observations of GRB 130427A Establish a Single Component Synchrotron Afterglow Origin for the Late Optical to Multi-GeV Emission. *Astrophys. J. Lett.* **2013**, *779*, L1, [CrossRef]
52. Ackermann, M.; Ajello, M.; Asano, K.; Axelsson, M.; Baldini, L.; Ballet, J.; Barbiellini, G.; Bastieri, D.; Bechtol, K.; Bellazzini, R.; et al. The First Fermi-LAT Gamma-Ray Burst Catalog. *Astrophys. J. Suppl.* **2013**, *209*, 11, [CrossRef]
53. Weekes, T.C.; Cawley, M.F.; Fegan, D.J.; Gibbs, K.G.; Hillas, A.M.; Kowk, P.W.; Lamb, R.C.; Lewis, D.A.; Macomb, D.; Porter, N.A.; et al. Observation of TeV Gamma Rays from the Crab Nebula Using the Atmospheric Cerenkov Imaging Technique. *Astrophys. J.* **1989**, *342*, 379. [CrossRef]
54. Boella, G.; Butler, R.C.; Perola, G.C.; Piro, L.; Scarsi, L.; Bleeker, J.A.M. BeppoSAX, the wide band mission for X-ray astronomy. *Astron. Astrophys. Suppl.* **1997**, *122*, 299–307. [CrossRef]
55. Hurley, K.; Sommer, M.; Boer, M.; Niel, M.; Laros, J.; Fenimore, E.; Klebesadel, R.; Fishman, G.; Kouveliotou, C.; Meegan, C.; et al. ULYSSES precise localizations of gamma-ray bursts. *Astropart. Astrophys.* **1993**, *97*, 39–41.
56. Alexandreas, D.E.; Allen, G.E.; Berley, D.; Biller, S.; Burman, R.L.; Cavalli-Sforza, M.; Chang, C.Y.; Chen, M.L.; Chumney, P.; Coyne, D.; et al. *Search for Ultra High Energy Radiation from Gamma-Ray Bursts;* Gamma-Ray Bursts; Fishman, G.J., Ed.; American Institute of Physics Conference Series; American Institute of Physics: College Park, MD, USA, 1994; Volume 307, pp. 481–485.
57. Padilla, L.; Funk, B.; Krawczynski, H.; Contreras, J.L.; Moralejo, A.; Aharonian, F.; Akhperjanian, A.G.; Barrio, J.A.; Beteta, J.G.; Cortina, J.; et al. Search for gamma-ray bursts above 20 TeV with the HEGRA AIROBICC Cherenkov array. *Astron. Astrophys.* **1998**, *337*, 43–50.
58. Catanese, M.; Chantell, M.; Covault, C.E.; Cronin, J.W.; Fick, B.E.; Fortson, L.F.; Fowler, J.W.; Gibbs, K.G.; Glasmacher, M.A.K.; Green, K.D.; et al. Search for Ultra High Energy (UHE) γ-ray counterparts of BATSE 3B catalog events. In *Gamma-ray Bursts: 3rd Huntsville Symposium*; Kouveliotou, C., Briggs, M.F., Fishman, G.J., Eds.; American Institute of Physics Conference Series; American Institute of Physics: College Park, MD, USA, 1996; Volume 384, pp. 598–602. [CrossRef]
59. Aglietta, M.; Alessandro, B.; Antonioli, P.; Arneodo, F.; Bergamasco, L.; Bertaina, M.; Castagnoli, C.; Castellina, A.; Chiavassa, A.; Cini, G.; et al. Search for high energy GRBs with EASTOP. *Astron. Astrophys.* **1999**, *138*, 595–596. [CrossRef]
60. Atkins, R.; Benbow, W.; Berley, D.; Chen, M.L.; Coyne, D.G.; Delay, R.S.; Dingus, B.L.; Dorfan, D.E.; Ellsworth, R.W.; Espinoza, C.; et al. Milagrito, a TeV air-shower array. *Nucl. Instrum. Methods Phys. Res. A* **2000**, *449*, 478–499. [CrossRef]
61. Atkins, R.; Benbow, W.; Berley, D.; Chen, M.L.; Coyne, D.G.; Dingus, B.L.; Dorfan, D.E.; Ellsworth, R.W.; Evans, D.; Falcone, A.; et al. The High-Energy Gamma-Ray Fluence and Energy Spectrum of GRB 970417a from Observations with Milagrito. *Astrophys. J.* **2003**, *583*, 824–832. [CrossRef]
62. Connaughton, V.; Ackerlof, C.W.; Barthelmy, S.; Biller, S.; Boyle, P.; Buckley, J.; Carter-Lewis, D.A.; Cantanese, M.; Cawley, M.F.; Cline, T.; et al. A Search for TeV Counterparts to BATSE Gamma-Ray Bursts. *Astrophys. J.* **1997**, *479*, 859–867. [CrossRef]
63. Gehrels, N.; Chincarini, G.; Giommi, P.; Mason, K.O.; Nousek, J.A.; Wells, A.A.; White, N.E.; Barthelmy, S.D.; Burrows, D.N.; Cominsky, L.R.; et al. The Swift Gamma-Ray Burst Mission. *Astrophys. J.* **2004**, *611*, 1005–1020. [CrossRef]
64. Gehrels, N.; Razzaque, S. Gamma-ray bursts in the swift-Fermi era. *Front. Phys.* **2013**, *8*, 661–678. [CrossRef]
65. Albert, J.; Aliu, E.; Anderhub, H.; Antoranz, P.; Armada, A.; Baixeras, C.; Barrio, J.A.; Bartko, H.; Bastieri, D.; Becker, J.; et al. MAGIC Upper Limits on the Very High Energy Emission from Gamma-Ray Bursts. *Astrophys. J.* **2007**, *667*, 358–366. [CrossRef]
66. Albert, J.; Aliu, E.; Anderhub, H.; Antoranz, P.; Armada, A.; Asensio, M.; Baixeras, C.; Barrio, J.A.; Bartelt, M.; Bartko, H.; et al. Flux Upper Limit on Gamma-Ray Emission by GRB 050713a from MAGIC Telescope Observations. *Astrophys. J. Lett.* **2006**, *641*, L9–L12, [CrossRef]
67. Berti, A.; Antonelli, L.A.; Bosnjak, Z.; Cortina, J.; Covino, S.; D'Elia, V.; Espiñeira, E.d.S.; Fukami, S.; Inoue, S.; Longo, F.; et al. Searching for GRBs at VHE with MAGIC: the status before CTA. In Proceedings of the 36th International Cosmic Ray Conference (ICRC2019), Madison, WI, USA, 25 July–1 August 2019; International Cosmic Ray Conference; Volume 36, p. 634.
68. Aleksić, J.; Anderhub, H.; Antonelli, L.A.; Antoranz, P.; Backes, M.; Baixeras, C.; Balestra, S.; Barrio, J.A.; Bastieri, D.; Becerra González, J.; et al. MAGIC observation of the GRB 080430 afterglow. *Astron. Astrophys.* **2010**, *517*, A5, [CrossRef]
69. Aharonian, F.; Akhperjanian, A.G.; Barres de Almeida, U.; Bazer-Bachi, A.R.; Behera, B.; Benbow, W.; Bernlöhr, K.; Boisson, C.; Bochow, A.; Borrel, V.; et al. HESS observations of γ-ray bursts in 2003–2007. *Astron. Astrophys.* **2009**, *495*, 505–512. [CrossRef]
70. Aharonian, F.; Akhperjanian, A.G.; Barres DeAlmeida, U.; Bazer-Bachi, A.R.; Behera, B.; Beilicke, M.; Benbow, W.; Bernlöhr, K.; Boisson, C.; Borrel, V.; et al. HESS Observations of the Prompt and Afterglow Phases of GRB 060602B. *Astrophys. J.* **2009**, *690*, 1068–1073. [CrossRef]
71. H.E.S.S. Collaboration; Abramowski, A.; Aharonian, F.; Ait Benkhali, F.; Akhperjanian, A.G.; Angüner, E.; Anton, G.; Balenderan, S.; Balzer, A.; Barnacka, A.; et al. Search for TeV Gamma-ray Emission from GRB 100621A, an extremely bright GRB in X-rays, with H.E.S.S. *Astron. Astrophys.* **2014**, *565*, A16, [CrossRef]

72. Hoischen, C.; Balzer, A.; Bissaldi, E.; Füßling, M.; Garrigoux, T.; Gottschall, D.; Holler, M.; Mitchell, A.; O'Brien, P.; Parsons, R.; et al. GRB Observations with H.E.S.S. II. In Proceedings of the 35th International Cosmic Ray Conference (ICRC2017), Busan, Korea, 12–20 July 2017; International Cosmic Ray Conference; Volume 301, p. 636.
73. Aliu, E.; Aune, T.; Barnacka, A.; Beilicke, M.; Benbow, W.; Berger, K.; Biteau, J.; Buckley, J.H.; Bugaev, V.; Byrum, K.; et al. Constraints on Very High Energy Emission from GRB 130427A. *Astrophys. J. Lett.* **2014**, *795*, L3, [CrossRef]
74. Bartoli, B.; Bernardini, P.; Bi, X.J.; Branchini, P.; Budano, A.; Camarri, P.; Cao, Z.; Cardarelli, R.; Catalanotti, S.; Chen, S.Z.; et al. Search for GeV Gamma-Ray Bursts with the ARGO-YBJ Detector: Summary of Eight Years of Observations. *Astrophys. J.* **2014**, *794*, 82, [CrossRef]
75. Blanch, O.; Gaug, M.; Noda, K.; Berti, A.; Moretti, E.; Miceli, D.; Gliwny, P.; Ubach, S.; Schleicher, B.; Cerruti, M.; et al. MAGIC observations of GRB 201015A: hint of very high energy gamma-ray signal. *GRB Coord. Netw.* **2020**, *28659*, 1.
76. MAGIC Collaboration. Observation of inverse Compton emission from a long γ-ray burst. *Nature* **2019**, *575*, 459–463. [CrossRef]
77. Blanch, O.; Longo, F.; Berti, A.; Fukami, S.; Suda, Y.; Loporchio, S.; Micanovic, S.; Green, J.G.; Pinter, V.; Takahashi, M.; et al. GRB 201216C: MAGIC detection in very high energy gamma rays. *GRB Coord. Netw.* **2020**, *29075*, 1.
78. Beardmore, A.P.; Gropp, J.D.; Kennea, J.A.; Klingler, N.J.; Laha, S.; Lien, A.Y.; Moss, M.J.; Page, K.L.; Palmer, D.M.; Tohuvavohu, A.; et al. GRB 201216C: Swift detection of a burst. *GRB Coord. Netw.* **2020**, *29061*, 1.
79. Vielfaure, J.B.; Izzo, L.; Xu, D.; Vergani, S.D.; Malesani, D.B.; de Ugarte Postigo, A.; D'Elia, V.; Fynbo, J.P.U.; Kann, D.A.; Levan, A.J.; et al. GRB 201216C: VLT X-shooter spectroscopy and potential high redshift of a VHE-emitting GRB. *GRB Coord. Netw.* **2020**, *29077*, 1.
80. Fukami, S.; Berti, A.; Loporchio, S.; Suda, Y.; Nava, L.; Noda, K.; Bošnjak, V.; Asano, K.; Longo, F.; Acciari, V.A.; et al. Very-high-energy gamma-ray emission from GRB 201216C detected by MAGIC. *PoS* **2021**, *ICRC2021*, 788. [CrossRef]
81. Fermi GBM Team. GRB 190829A: Fermi GBM Final Real-time Localization. *GRB Coord. Netw.* **2019**, *25551*, 1.
82. Dichiara, S.; Bernardini, M.G.; Burrows, D.N.; D'Avanzo, P.; Gronwall, C.; Gropp, J.D.; Kennea, J.A.; Klingler, N.J.; Krimm, H.A.; Kuin, N.P.M.; et al. GRB 190829A: Swift detection of a burst consistent with a galaxy at z=0.08. *GRB Coord. Netw.* **2019**, *25552*, 1.
83. Hu, Y.D.; Castro-Tirado, A.J.; Kumar, A.; Gupta, R.; Valeev, A.F.; Pandey, S.B.; Kann, D.A.; Castellón, A.; Agudo, I.; Aryan, A.; et al. 10.4 m GTC observations of the nearby VHE-detected GRB 190829A/SN 2019oyw. *Astron. Astrophys.* **2021**, *646*, A50, [CrossRef]
84. Zhang, B.T.; Murase, K.; Veres, P.; Mészáros, P. External Inverse-Compton Emission from Low-luminosity Gamma-Ray Bursts: Application to GRB 190829A. *Astrophys. J.* **2021**, *920*, 55, [CrossRef]
85. Zhang, L.L.; Ren, J.; Huang, X.L.; Liang, Y.F.; Lin, D.B.; Liang, E.W. Nearby SN-associated GRB 190829A: Environment, Jet Structure, and VHE Gamma-Ray Afterglows. *Astrophys. J.* **2021**, *917*, 95, [CrossRef]
86. Salafia, O.S.; Ravasio, M.E.; Yang, J.; An, T.; Orienti, M.; Ghirlanda, G.; Nava, L.; Giroletti, M.; Mohan, P.; Spinelli, R.; et al. Multi-wavelength view of the close-by GRB~190829A sheds light on gamma-ray burst physics. *arXiv* **2021**, arXiv:2106.07169.
87. Roberts, O.J.; Meegan, C. GRB 180720B: Fermi GBM observation. *GRB Coord. Netw.* **2018**, *22981*, 1.
88. Evans, P.A.; Breeveld, A.A.; Deich, A.; Gropp, J.D.; Kennea, J.A.; Krimm, H.A.; Lien, A.Y.; Marshall, F.E.; Page, K.L.; Siegel, M.H.; et al. GRB 180620B: Swift detection of a burst with an optical counterpart. *GRB Coord. Netw.* **2018**, *22807*, 1.
89. Bissaldi, E.; Racusin, J.L. GRB 180720B : Fermi-LAT detection. *GRB Coord. Netw.* **2018**, *22980*, 1.
90. Suda, Y.; Artero, M.; Asano, K.; Berti, A.; Nava, L.; Noda, K.; Terauchi, K.; Acciari, V.A.; Ansoldi, S.; Antonelli, L.A.; et al. Observation of a relatively low luminosity long duration GRB 201015A by the MAGIC telescopes. *PoS* **2021**, *ICRC2021*, 797. [CrossRef]
91. Ghirlanda, G.; Nappo, F.; Ghisellini, G.; Melandri, A.; Marcarini, G.; Nava, L.; Salafia, O.S.; Campana, S.; Salvaterra, R. Bulk Lorentz factors of gamma-ray bursts. *Astron. Astrophys.* **2018**, *609*, A112, [CrossRef]
92. Zhang, S.; Shao, L.; Zhang, B.B.; Zou, J.H.; Sun, H.Y.; Yao, Y.j.; Li, L.l. A Tight Three-parameter Correlation and Related Classification on Gamma-Ray Bursts. *Astrophys. J.* **2022**, *926*, 170, [CrossRef]
93. Li, Y.; Zhang, B.; Lü, H.J. A COMPARATIVE STUDY OF LONG AND SHORT GRBS. I. OVERLAPPING PROPERTIES. *Astrophys. J. Suppl. Ser.* **2016**, *227*, 7. [CrossRef]
94. Wood, J. Results from the first one and a half years of the HAWC GRB program. *arXiv* **2018**, arXiv:1801.01437.
95. Carosi, A.; Ashkar, H.; Berti, A.; Bordas, P.; de Bony Lavergne, M.; Donini, A.; Dalchenko, M.; Fiasson, A.; Foffano, L.; Fukami, S.; et al. First follow-up of transient events with the CTA Large Size Telescope prototype. In Proceedings of the 37th International Cosmic Ray Conference, Berlin, Germany, 12–23 July 2021.
96. Acciari, V.A.; Ansoldi, S.; Antonelli, L.A.; Arbet Engels, A.; Asano, K.; Baack, D.; Babić, A.; Baquero, A.; Barres de Almeida, U.; Barrio, J.A.; et al. MAGIC Observations of the Nearby Short Gamma-Ray Burst GRB 160821B. *Astrophys. J.* **2021**, *908*, 90, [CrossRef]
97. Acciari, V.A.; Ansoldi, S.; Antonelli, L.A.; Arbet Engels, A.; Baack, D.; Babić, A.; Banerjee, B.; Barres de Almeida, U.; Barrio, J.A.; Becerra González, J.; et al. Bounds on Lorentz Invariance Violation from MAGIC Observation of GRB 190114C. *Phys. Rev. Lett.* **2020**, *125*, 021301, [CrossRef]
98. Abdalla, H.; Abe, H.; Acero, F.; Acharyya, A.; Adam, R.; Agudo, I.; Aguirre-Santaella, A.; Alfaro, R.; Alfaro, J.; Alispach, C.; et al. Sensitivity of the Cherenkov Telescope Array for probing cosmology and fundamental physics with gamma-ray propagation. *J. Cosmol. Astropart. Phys.* **2021**, *2021*, 048, [CrossRef]

99. Carosi, A.; López-Oramas, A.; Longo, F. The Cherenkov Telescope Array transient and multi-messenger program. In Proceedings of the 37th International Cosmic Ray Conference, Berlin, Germany, 12–23 July 2021.
100. Inoue, S.; Granot, J.; O'Brien, P.T.; Asano, K.; Bouvier, A.; Carosi, A.; Connaughton, V.; Garczarczyk, M.; Gilmore, R.; Hinton, J.; et al. Gamma-ray burst science in the era of the Cherenkov Telescope Array. *Astropart. Phys.* **2013**, *43*, 252–275. [CrossRef]
101. Bernardini, M.G.; Bissaldi, E.; Bosnjak, Z.; Carosi, A.; D'Avanzo, P.; Di Girolamo, T.; Inoue, S.; Gasparetto, T.; Ghirlanda, F.; Longo, F.; et al. POSyTIVE-a GRB population study for the Cherenkov Telescope Array. In Proceedings of the 36th International Cosmic Ray Conference (ICRC2019), Madison, WI, USA, 24 July–1 August 2019; International Cosmic Ray Conference; Volume 36, p. 598.
102. Amati, L.; Frontera, F.; Tavani, M.; in't Zand, J.J.M.; Antonelli, A.; Costa, E.; Feroci, M.; Guidorzi, C.; Heise, J.; Masetti, N.; et al. Intrinsic spectra and energetics of BeppoSAX Gamma-Ray Bursts with known redshifts. *Astron. Astrophys.* **2002**, *390*, 81–89. [CrossRef]
103. Cortina, J.; Project, C.L. Status of the Large Size Telescopes of the Cherenkov Telescope Array. In Proceedings of the 36th International Cosmic Ray Conference (ICRC2019), Madison, WI, USA, 24 July–1 August 2019; International Cosmic Ray Conference; Volume 36, p. 653,
104. Scuderi, S. *The ASTRI Mini-Array at the Teide Observatory*; Society of Photo-Optical Instrumentation Engineers (SPIE) Conference Series; Society of Photo-Optical Instrumentation Engineers: Bellingham, WA, USA, 2021; Volume 11822, p. 1182202. [CrossRef]
105. Stamerra, A.; Saturni, F.G.; Green, J.; Nava, L.; Lucarelli, F.; Antonelli, L. TeV Transients with the ASTRI Mini-Array: A case study with GRB 190114C. In Proceedings of the 37th International Cosmic Ray Conference (ICRC2021), Berlin, Germany, 12–23 July 2021; International Cosmic Ray Conference; Volume 395, p. 890.

galaxies

MDPI

Article

Gamma-Ray Bursts Afterglow Physics and the VHE Domain

Davide Miceli [1,2,*] and Lara Nava [3,4,*]

1 Dipartimento di Fisica e Astronomia, Università di Padova, Via Marzolo 8, 35131 Padova, Italy
2 Istituto Nazionale di Fisica Nucleare (INFN), Sez. Padova, Via Marzolo 8, 35131 Padova, Italy
3 INAF—Osservatorio Astronomico di Brera, Via Emilio Bianchi 46, 23807 Merate, Italy
4 Istituto Nazionale di Fisica Nucleare (INFN), Sez. Trieste, Via A. Valerio 2, 34100 Trieste, Italy
* Correspondence: davide.miceli@unipd.it (D.M.); lara.nava@inaf.it (L.N.)

Abstract: Afterglow radiation in gamma-ray bursts (GRB), extending from the radio band to GeV energies, is produced as a result of the interaction between the relativistic jet and the ambient medium. Although in general the origin of the emission is robustly identified as synchrotron radiation from the shock-accelerated electrons, many aspects remain poorly constrained, such as the role of inverse Compton emission, the particle acceleration mechanism, the properties of the environment and of the GRB jet itself. The extension of the afterglow emission into the TeV band has been discussed and theorized for years, but has eluded for a long time the observations. Recently, the Cherenkov telescopes, MAGIC and H.E.S.S., have unequivocally proven that afterglow radiation is also produced above 100 GeV, up to at least a few TeV. The accessibility of the TeV spectral window will largely improve with the upcoming facility CTA (the Cherenkov Telescope Array). In this review article, we first revise the current model for afterglow emission in GRBs, its limitations and open issues. Then, we describe the recent detections of very high energy emission from GRBs and the origin of this radiation. Implications on the understanding of afterglow radiation and constraints on the physics of the involved processes will be deeply investigated, demonstrating how future observations, especially by the CTA Observatory, are expected to give a key contribution in improving our comprehension of such elusive sources.

Keywords: gamma-ray bursts; non-thermal processes; Cherenkov telescopes

Citation: Miceli, D.; Nava, L. Gamma-Ray Bursts Afterglow Physics and the VHE Domain. *Galaxies* **2022**, *10*, 66. https://doi.org/10.3390/galaxies10030066

Academic Editors: Elena Moretti, Francesco Longo and Yi-Zhong Fan

Received: 28 February 2022
Accepted: 29 April 2022
Published: 5 May 2022

1. Introduction

Gamma-ray bursts (GRBs) are observed as transient sources of radiation displaying a distinctive pattern that consists of two different phases. The first phase is dominated by emission in the keV-MeV energy range, lasting from fractions of a second to several minutes, and reaching isotropic equivalent peak luminosities in the range $L{\sim}10^{49}$–10^{53} erg s^{-1}. The bimodal distribution of the prompt emission duration reveals that there are two classes of GRBs, called short and long depending on whether the prompt emission lasts shorter or longer than 2 s [1,2]. The second emission phase, called the afterglow, follows the prompt with a delay of tens of seconds, and is detected on a very wide range of frequencies, from γ-rays to the radio band. The afterglow flux decays smoothly as a power-law in time for weeks or months, and the typical frequency of the radiation moves in time from the X-ray to the radio band. Since 2019, the detection of a few long GRBs between 0.3 and 3 TeV on time-scales from tens of seconds to a few days proved for the first time that GRBs can also be sources of radiation in the TeV band, where they can convey a sizable fraction (20–50%) of the total energy emitted during the afterglow phase [3–5].

All this prompt/afterglow emission is identified with radiation produced as a result of the launch of an ultra-relativistic ($\Gamma{\sim}100$–1000) jet from a newly born compact object. The ejecta first undergoes internal dissipation (through mechanisms such as shocks between different parts of the outflow [6] or magnetic reconnection episodes [7,8]). In a second moment, the ejecta undergoes external dissipation [9], triggered by interactions with the ambient medium (e.g., the interstellar medium or the wind of the progenitor's star [10]). The two different dissipation processes occur at different typical distances from the central engine ($R\sim10^{13-14}$ cm and $R\sim10^{15-20}$ cm) and generate two well distinguished emission phases, identified as the prompt and afterglow emission, respectively.

For long GRBs, it is widely believed that the involved energetics and time-scales and the successful launch of a relativistic jet can find justification in the collapsar model [11,12]. In this model, the core of a massive star collapses into a black hole and the accretion from the surrounding disk powers the launch of two opposite, collimated ($\theta_{jet}\sim$5–10°) outflows. A similar scenario also applies to short GRBs, where the black hole originates from the merger of two neutron stars (as recently proven by the association of a short GRB with a gravitational wave signal [13]) or a neutron star and a black-hole. An alternative model [14–17] considers a millisecond magnetar (i.e., a rapidly rotating neutron star) as the progenitor of long GRBs (or at least a fraction of them). This model has the advantage of more naturally explaining the detections of late time activity (10^2–10^3 s after the prompt onset) in the form of X-ray flares and plateaus, observed in about one third of the population.

Beside the nature of the progenitor's star, another quite pressing open issue in GRB physics concerns the composition of the jet itself, i.e., the nature of the dominant energy stored in the outflow, which can be either magnetic (in the form of Poynting flux [7,14]) or kinetic (i.e., bulk motion of the matter). This uncertainty reflects an uncertainty on the mechanism extracting energy from the jet (i.e., the process converting part of the jet energy into random energy of the particles), which is identified with internal shocks in the latter case, and magnetic reconnection events in the case of a Poynting flux dominated outflows [14]. While internal shocks in a matter-dominated jet have been considered the mainstream model for a long time, tensions between some model predictions and observations have moved the attention in the last decade to a family of models based on magnetic jets [18–20]. In particular, internal shocks are not an efficient mechanism [21,22], and this is in contrast with the evidence that only a relatively small fraction (10–50%) of energy is still in the blast-wave during the afterglow phase, meaning that most of it must have been dissipated and radiated away during the prompt. It must be noted, however, that the estimate of the energy content of the blast during the afterglow is indirect, and contingent upon a proper modeling of the afterglow emission [23]. Investigations that took advantage of GeV emission detected by LAT (the Large Area Telescope onboard the Fermi satellite), reached the conclusion that the blast energy is usually underestimated by studies relying on X-ray emission, and inferred a prompt emission efficiency between 1–10% [24], which is still consistent with internal shocks. The nature and efficiency of the dissipation mechanism in the prompt phase are still matters of intense debate. In any case, the radiation is expected to be produced by the accelerated electrons, which efficiently lose energy via synchrotron cooling [6,25]. Inconsistencies between the expected synchrotron spectrum and the observed spectral shape of the prompt emission [25,26] have also called into question the nature of the radiative process. Recent works have performed major advances towards the comprehension of the radiative mechanism responsible for the prompt emission, supporting the synchrotron interpretation [27–31].

The nature of the afterglow emission is much better understood, at least on its general grounds. The interaction between the jet and the external medium triggers the formation of a forward shock running into the external medium and a reverse shock running into the ejecta [32–34]. These shocks are responsible for the acceleration of particles and for the deceleration of the outflow, eventually down to non-relativistic velocities [35–37]. The observed radiation is the result of synchrotron radiation from electrons accelerated at the forward shock [38]. A contribution from the reverse shock may also be relevant, typically in the radio and optical band [34,39]. Shock formation and particle acceleration in ultra-relativistic shocks are still not completely understood. Very important progresses have been achieved in the last decade on the theoretical side (see [40] for a recent review), especially thanks to numerical particle-in-cell (PIC) simulations [41,42]. Ultra-relativistic shocks in a weakly magnetized medium are found to be efficient particle accelerators, with $\xi_e \sim 1\%$ of the electrons being accelerated into a power-law distribution with spectral index $p \sim 2.5$, carrying about $\epsilon_e \sim 10\%$ of the shock-dissipated energy. A strong magnetic turbulence is self-generated by the accelerated particles counter-streaming in the upstream, ahead of the shock, at a level of magnetization ϵ_B =0.01–0.1 [42]. PIC simulations, however, are currently probing time-scales that are orders of magnitude smaller than the dynamical time-scale of the blast-wave. This implies that results from simulations can only be extrapolated to the relevant time-scales, introducing a certain degree of uncertainty and caution in using the results as inputs for the modeling of GRB afterglows. What is still poorly understood, even though dedicated simulations are starting to give important clues [43], is how the micro-turbulence generated in the shock vicinity evolves (decays) with time. This is particularly important for a proper interpretation of the observations, since it is likely that the particles produce synchrotron and synchrotron-self Compton (SSC) photons in a region of decayed micro-turbulence, and hence feel a magnetic field with $\epsilon_B << 0.01$.

The afterglow emission, its spectral shape from radio to γ-rays, and its temporal evolution from seconds to months, contain a wealth of (convoluted) information on blast dynamics, particle acceleration, magnetic turbulence generation and decay, and external density in the progenitor's surroundings (up to a parsec scale). Nevertheless, since all these ingredients are poorly constrained from theoretical grounds, they enter the afterglow physics as free, unconstrained model parameters. The large degeneracy among different parameters and the small number of observables as compared to model variables are limiting our possibility to extract valuable and robust information from the modeling of the observed afterglow radiation. To go beyond the state-of-the-art, additional efforts are necessary both on the observational and theoretical sides.

An interesting opportunity has recently opened on the observational side, thanks to the discovery that GRBs can be sources of TeV radiation associated with the afterglow phase [3,4]. The characterization of the TeV spectra and light curves offers new observables to further constrain the unknown physics of the afterglow emission. These observations are expected to impact on our current understanding of the environment where GRBs explode (and hence on the nature of their progenitors), of the physics of ultra-relativistic shocks, and of the properties of the jet (e.g., bulk Lorentz factor and energy content). Constraining the jet properties is mandatory for a correct estimate of the prompt mechanism efficiency and then for determining its nature. It is then evident how the opening of this completely new energy window in GRBs is expected to boost the studies in a field that has many connections both with the general understanding of the GRB phenomenon and with topics of general interest, such as star formation and evolution, the last stages of massive stars and their environments and plasma physics under extreme conditions.

Given the impressive amount of new information that VHE observations are going to bring to the field, it is important to revise what is the state-of-the-art, what are the main issues and how we can benefit from the few existing and the upcoming observations in the TeV domain. This review revisits the present understanding of afterglow radiation, the discovery of very-high energy (VHE, > 100 GeV) emission from GRBs and future prospects for the detection of GRBs at VHE with the next generation of Cherenkov telescopes.

For recent and complete reviews on GRB's phenomenology and theoretical interpretation before the TeV era see [44,45]. An overview focused on high-energy emission (0.1–100 GeV) observations and interpretation can be found in [46].

This review is organized as follows. Section 2 presents an overview of the afterglow external shock model, revisiting our common understanding and phenomenological description of (i) the dynamics of the blast-wave, (ii) shock formation, particle acceleration and self-generation of turbulent magnetic field in the shock proximity, and (iii) the main processes shaping the radiative output, on the whole electromagnetic spectrum, from radio to very-high energy γ-rays. In Section 3, we propose a discussion of the main open issues of the afterglow model, outlining which observations are at odds with model predictions, which observed features are missing in the basic scenario and what are the present limitations that prevent us from extracting valuable information from the modeling of multi-wavelength afterglow radiation. In Section 4, we describe the recent discovery that GRBs can be bright TeV emitters. Each GRB with a firm (or with a hint of) detection by MAGIC or H.E.S.S. is discussed in detail. We present multi-wavelength observations and review the proposed interpretations of the detected emission. In Section 5, we compare the general properties of the detected GRBs both among each other and with the general population. We discuss how the TeV emission can help to solve some of the most important issues of the afterglow model. Finally, in Section 6, we discuss the prospects for future studies of TeV emission from GRBs with the next generation of Cherenkov telescopes and their expected impact on GRB physics.

2. The Afterglow Model

Afterglow emission refers to all the broad-band radiation observed from a GRB on longer timescales (minutes to months) as compared to the initial prompt radiation detected in hard X-rays [38,47,48]. Its temporal evolution is usually well described with simple decaying power-laws, in contrast with the short-time (<seconds) variability that characterizes the prompt emission [49–52]. These major differences place the emission region of afterglow radiation at larger radii ($>10^{15}$ cm), pinpointing its origin in the processes triggered by the interaction between the jet and the circumburst medium.

The expansion of the relativistic jet into the external medium is expected to drive two different shocks: the forward shock, running into the external medium, and the reverse shock, running into the jet. The shocked ejecta and the shocked external medium, separated by the contact discontinuity, are both sources of synchrotron radiation from the accelerated electrons [53]. Most of the detected radiation is interpreted as emission from ambient particles energized by the forward shock. Spectra and lightcurves are then shaped by the environment where the GRB explodes, which in turn is strictly connected to the nature of the progenitor. The other player that shapes the properties of afterglow radiation is particle acceleration at relativistic shocks, which is thought to proceed via diffusive shock acceleration, but for which the details of the underlying physics still remain poorly constrained. Moreover, the overall luminosity of the afterglow radiation depends on the energy content of the blast-wave. Such an amount is determined by how efficiently the prompt mechanism has dissipated and released part of the initial explosion energy. Following these considerations, it is evident how the study of afterglow radiation impacts on the general understanding of the GRB phenomenon: the progenitor and its environment, the nature and efficiency of the mechanisms responsible for prompt emission, the properties of the jet, and the micro-physics of relativistic shocks.

In this section, the physics involved in the afterglow scenario is presented, with a particular focus on the forward shock emission and on the radiative output expected at VHE. This section is organized as follows: we revisit the physics of the jet dynamical evolution in its interaction with the ambient medium (Section 2.1), the particle acceleration mechanism (Section 2.2) and the resulting radiative output and its spectral shape (Section 2.3).

2.1. Jet Dynamics

After the reverse shock has crossed the ejecta, the dynamics of the blast-wave enters a self-similar regime ([35], BM76 hereafter). In a thin shell approximation, the reverse shock crossing time corresponds to the time when the blast-wave starts decelerating. The deceleration of the jet, caused by the collision with the external medium, becomes significant at the radius R_{dec}, where the energy transferred to the mass m collected from the external medium ($\sim m(R_{dec})c^2\Gamma_0^2$) is comparable to the initial energy ($E_0 = M_0\Gamma_0 c^2$) carried by the jet. This deceleration radius is typically of the order of $R_{dec}\sim 10^{15}$–10^{16} cm, depending on the density of the external medium, the ejecta mass M_0 and initial bulk Lorentz factor Γ_0. Before reaching this radius, the ejecta expands with constant velocity (coasting phase).

Most analytic estimates of the afterglow evolution with the purpose of modeling data are developed for the deceleration phase, where the self-similar BM76 solution for adiabatic blast-waves is adopted [38,47,54]. Since VHE emission can be detected at quite early times (a few tens of seconds), we are also interested in the description of the coasting phase and in a proper treatment of the transition between coasting and deceleration.

In the following, to derive the evolution of the bulk Lorentz factor, we adopt the approach proposed by [55]. This method allows us to describe the hydrodynamics of a relativistic blast-wave expanding into a medium with an arbitrary density profile $\rho(R)$ and composition (i.e., enriched by pairs), and the transition from the free expansion of the ejecta to the deceleration phase, taking into account the role of radiative and adiabatic losses. The internal structure is neglected (homogeneous shell approximation), and the Lorentz factor Γ considered is the one of the fluids just behind the shock front. In the deceleration phase, the self-similar solutions derived in BM76 are recovered by this method, both for the adiabatic and the fully radiative cases, and for constant and wind-like density profiles of the external medium. The presented approach also allows us to introduce a time-varying radiative efficiency, either resulting from a change with time of ϵ_e or a change in the radiative efficiency of the electrons. Equations reported here are valid after the reverse shock has crossed the ejecta. Corrections to the hydrodynamics before the reverse-shock crossing time can be found in [55].

2.1.1. Equation Describing the Evolution of the Bulk Lorentz Factor

The aim is to derive an equation describing the change $d\Gamma$ of the bulk Lorentz factor of the fluid just behind the shock in response to the collision with a mass $dm(R) = 4\pi R^2\rho(R)dR$ encountered when the shock front moves from a distance R to $R + dR$ and with ρ being the mass density. The change in Γ is determined by dissipation of the bulk kinetic energy, the conversion of internal energy back into bulk motion, and injection of energy into the blast-wave. The latter is sometimes invoked to explain plateau phases in the X-ray early afterglow or to explain flux rebrightenings [56–58]. The following treatment neglects energy injection, which, however, can be easily incorporated in this kind of approach.

To write the equation for energy conservation, from which $d\Gamma/dR$ can be derived, we first need to recall how the energy density transforms under Lorentz transformations. In the following, we denote quantities measured in the frame comoving with the shocked fluid (comoving frame), with a prime, to distinguish them from quantities measured in the frame of the progenitor star (rest frame, without a prime).

The energy density in the comoving frame is $u' = u'_{\rm int} + \rho' c^2$, where $u'_{\rm int}$ is the comoving internal energy and ρ' is the comoving mass density. Applying Lorentz transformations, $u = (u' + p')\Gamma^2 - p'$, where p' is the pressure and is related to the internal energy density by the equation of state, and $p' = (\hat{\gamma} - 1)\, u'_{\rm int} = (\hat{\gamma} - 1)\,(u' - \rho' c^2)$, where $\hat{\gamma}$ is the adiabatic index of the shocked plasma. The energy density is then given by: $u = u'_{\rm int}(\hat{\gamma}\Gamma^2 - \hat{\gamma} + 1) + \rho' c^2\Gamma^2$, which shows how the internal energy and rest mass density transform. The total energy in the progenitor frame will be $E = uV = uV'/\Gamma$, where V is the shell volume in the progenitor frame, and can be expressed as:

$$E = \Gamma M c^2 + \Gamma_{\rm eff} E'_{\rm int}\,, \tag{1}$$

where:

$$\Gamma_{eff} \equiv \frac{\hat{\gamma}\Gamma^2 - \hat{\gamma} + 1}{\Gamma}\,, \tag{2}$$

which properly describes the Lorentz transformation of the internal energy. Here, $M = M_0 + m = \rho' V'$ is the sum of the ejecta mass $M_0 = E_0/\Gamma_0 c^2$ and of the swept-up mass $m(R)$, and $E'_{int} = (u' - \rho' c^2)V'$ is the comoving internal energy. The adiabatic index can be parameterized as $\hat{\gamma} = (4 + \Gamma^{-1})/3$ to obtain the expected limits $\hat{\gamma} \simeq 4/3$ for $\Gamma \gg 1$ and $\hat{\gamma} \simeq 5/3$ for $\Gamma \to 1$. The majority of analytical treatments use Γ instead of Γ_{eff}, which implies an error up to a factor of 4/3 in the ultra-relativistic limit [55].

The blast-wave energy E in Equation (1) can change due to (*i*), the rest mass energy $dm\,c^2$ collected from the medium, (*ii*) radiative losses $dE_{rad} = \Gamma_{eff}\, dE'_{rad}$ and (*iii*) injection of energy. Ignoring possible episodes of energy injections into the blast-wave, the equation of energy conservation in the progenitor frame is:

$$d\left[\Gamma(M_0+m)c^2+\Gamma_{eff}E'_{int}\right]=dm\,c^2+\Gamma_{eff}dE'_{rad}\,. \tag{3}$$

The overall change in the comoving internal energy dE'_{int} results from the sum of three contributions:

$$dE'_{int}=dE'_{sh}+dE'_{ad}+dE'_{rad}\,. \tag{4}$$

The first contribution, $dE'_{sh}=(\Gamma-1)\,dm\,c^2$, is the random kinetic energy produced at the shock as a result of the interaction with an element dm of circum-burst material: as pointed out by BM76, in the post-shock frame, the average kinetic energy per unit mass dE'_{sh}/dm is constant across the shock, and equal to $(\Gamma-1)c^2$. The second term in Equation (4), dE'_{ad}, is the internal energy lost due to adiabatic expansion, that leads to a conversion of random energy back to bulk kinetic energy. The third term, dE'_{rad}, accounts for radiative losses.

From Equation (3), it follows that the variation of the Lorentz factor is:

$$\frac{d\Gamma}{dR}=-\frac{(\Gamma_{eff}+1)(\Gamma-1)\,c^2\frac{dm}{dR}+\Gamma_{eff}\frac{dE'_{ad}}{dR}}{(M_0+m)\,c^2+E'_{int}\frac{d\Gamma_{eff}}{d\Gamma}}\,, \tag{5}$$

from which the evolution $\Gamma(R)$ of the bulk Lorentz factor of the fluid just behind the shock as a function of the shock front radius can be derived.

The term $\Gamma_{\rm eff}\,dE'_{ad}/dR$, accounting for adiabatic losses, allows us to describe the re-acceleration of the fireball: this contribution, usually neglected, becomes important only when the density decreases faster than $\rho\propto R^{-3}$. To evaluate Equation (5), it is necessary to first specify dE'_{ad} and E'_{int}.

2.1.2. Internal Energy and Adiabatic Losses

In specific cases, the adiabatic losses and the internal energy content can be expressed in an analytic form. The following treatment to estimate adiabatic losses and the internal energy content of the blast-wave assumes that, right behind the shock, the freshly shocked electrons instantaneously radiate a fraction $\epsilon_{\rm rad}$ of their internal energy and then they cool only due to adiabatic losses [55]. By assuming that the accelerated electrons promptly radiate at the shock, and then they evolve adiabatically, one is implicitly considering either a fast cooling regime or quasi-adiabatic regime, in which case the radiative losses do not affect the shell dynamics.

Defining $\epsilon_{\rm e}$ as the fraction of energy dE'_{sh} dissipated by the shock and gained by the leptons, the mean random Lorentz factor of post-shock leptons becomes (for a more detailed discussion see Section 2.2):

$$\gamma_{acc,e}-1=(\Gamma-1)\epsilon_e/\mu_e\,. \tag{6}$$

Here, $\mu_e=\rho_e/\rho$ is the ratio between the mass density ρ_e of shocked electrons and positrons (simply "electrons" from now on) and the total mass density of the shocked matter ρ. In the absence of electron-positron pairs $\mu_e=m_e/(m_e+m_p)\simeq m_e/m_p$.

Leptons then radiate a fraction $\epsilon_{\rm rad}$ of their internal energy, i.e., the energy lost to radiation is $dE'_{rad}=-\epsilon_{rad}\epsilon_e\,dE'_{sh}=-\epsilon\,dE'_{sh}$, with $\epsilon\equiv\epsilon_{\rm rad}\epsilon_{\rm e}$ being the overall fraction of the shock-dissipated energy that goes into radiation. After radiating a fraction $\epsilon_{\rm rad}$ of their internal energy, the mean random Lorentz factor of the freshly shocked electrons decreases down to:

$$\gamma_{rad,e}-1=(1-\epsilon_{rad})(\gamma_{acc,e}-1)=(1-\epsilon_{rad})(\Gamma-1)\frac{\epsilon_e}{\mu_e}\,. \tag{7}$$

The assumption of instantaneous radiative losses is verified in the fast cooling regime ($\epsilon_{rad} \sim 1$), which is required (but not sufficient) to have $\epsilon \sim 1$ (i.e., a fully radiative blast-wave). In the opposite case $\epsilon_{rad} \ll 1$, the evolution is nearly adiabatic ($\epsilon \ll 1$), regardless of the value of $\epsilon_{\rm e}$, and the details of the radiative cooling processes are likely to be unimportant for the shell dynamics. The case with intermediate values of $\epsilon_{\rm rad}$ and ϵ is harder to treat analytically, since the electrons shocked at radius R may continue to emit copiously at larger distances as well, affecting the blast-wave dynamics.

A similar treatment can be adopted for protons: if protons gain a fraction $\epsilon_{\rm p}$ of the energy dissipated by the shock (with $\epsilon_p = 1 - \epsilon_e - \epsilon_B$), their mean post-shock Lorentz factor will be:

$$\gamma_{acc,p} - 1 = (\Gamma - 1)\frac{\epsilon_p}{\mu_p}\,, \tag{8}$$

where $\mu_p = \rho_p/\rho$ is the ratio between the mass density of shocked protons ρ_p and the total shocked mass density ρ. In the standard case, when pairs are absent, $\mu_p \simeq 1$. Since the proton radiative losses are negligible, the shocked protons will lose their energy only due to adiabatic cooling.

Adiabatic losses can be computed starting from $dE'_{\rm int} = -p'\,dV'$, where p' is the pressure in the comoving frame. For N particles with Lorentz factor γ, the internal energy density is:

$$u'_{\rm int} = \frac{N(\gamma - 1)m\,c^2}{V'}\,. \tag{9}$$

The radial change of the Lorentz factor, as a result of expansion losses, is:

$$\left(\frac{d(\gamma-1)}{dr}\right)_{ad} = -(\hat{\gamma} - 1)(\gamma - 1)\frac{d\ln V'}{dr}\,. \tag{10}$$

To estimate the adiabatic losses, let us assume that the shell comoving volume scales as $V' \propto R^3/\Gamma$, corresponding to a shell thickness in the progenitor frame $\sim R/\Gamma^2$. This scaling is correct for both relativistic and non-relativistic shocks in the decelerating phase (BM76). For re-accelerating relativistic shocks, Shapiro [59] demonstrated that the thickness of the region containing most of the blast-wave energy is still $\sim R/\Gamma^2$. For the sake of simplicity, changes in the comoving volume due to a time-varying adiabatic index or radiative efficiency are neglected. If the scaling $V' \propto R^3/\Gamma$ is assumed, the equation can be further developed analytically, and reads:

$$\left(\frac{d(\gamma-1)}{dr}\right)_{ad} = -(\hat{\gamma} - 1)(\gamma - 1)\left(\frac{3}{r} - \frac{d\ln\Gamma}{dr}\right)\,. \tag{11}$$

The comoving Lorentz factor at radius R, for a particle injected with $\gamma(r)$ when the shock radius was r, will be

$$\gamma_{\rm ad}(R,r) - 1 = \left(\frac{r}{R}\right)^{3(\hat{\gamma}-1)}\left[\frac{\Gamma(R)}{\Gamma(r)}\right]^{(\hat{\gamma}-1)}(\gamma(r) - 1)\,. \tag{12}$$

where $\gamma(r)$ is given by $\gamma_{rad,e}(r)$ (Equation (7)) for leptons, and by $\gamma_{acc,p}(r)$ (Equation (8)) for protons.

Considering the proton and lepton energy densities separately, the comoving internal energy at radius R will be:

$$E'_{int}(R) = 4\pi c^2\int_0^R dr r^2\left\{\rho_p(r)[\gamma_{ad,p}(R,r) - 1] + \rho_e(r)[\gamma_{ad,e}(R,r) - 1]\right\}\,. \tag{13}$$

With the help of Equation (12), one can explicitly find $E'_{int}(R)$ and insert it in Equation (5).

The other term needed in Equation (5) is dE'_{ad}/dR. First, we have derived $(d\gamma/dR)_{ad}$ for a single particle. Now integrating over the total number of particles, again considering protons and leptons separately, one obtains:

$$\frac{dE'_{ad}(R)}{dR} = -4\pi c^2(\hat{\gamma}-1)\left(\frac{3}{R}-\frac{d\log\Gamma}{dR}\right)\int_0^R dr r^2\Big\{\rho_p(r)(\gamma_{ad,p}-1)+\rho_e(r)(\gamma_{ad,e}-1)\Big\} . \tag{14}$$

In Equations (13) and (14), it is assumed that only the swept-up matter is subject to adiabatic cooling, i.e., that the ejecta particles are cold.

As long as the shocked particles remain relativistic, the equations for the comoving internal energy and for the adiabatic expansion losses assume simpler forms:

$$\begin{aligned} E'_{int}(R) &= 4\pi c^2\int_0^R dr\, r^2\frac{r}{R}\left[\frac{\Gamma(R)}{\Gamma(r)}\right]^{1/3}\Gamma(r)\left\{\frac{\epsilon_p}{\mu_p}\rho_p+(1-\epsilon_{rad})\frac{\epsilon_e}{\mu_e}\rho_e\right\} \\ &= 4\pi c^2\int_0^R dr\, r^2\frac{r}{R}\left[\frac{\Gamma(R)}{\Gamma(r)}\right]^{1/3}\Gamma(r)\,\rho(r)(\epsilon_p+\epsilon_e-\epsilon) \\ \frac{dE'_{ad}(R)}{dR} &= -E'_{int}(R)\left(\frac{1}{R}-\frac{1}{3}\frac{d\log\Gamma}{dR}\right) . \end{aligned} \tag{15}$$

In the absence of significant magnetic field amplification, $\epsilon_p+\epsilon_e\simeq 1$ so that $\epsilon_p+\epsilon_e-\epsilon\simeq 1-\epsilon$, and the radiative processes of the blast-wave are entirely captured by the single efficiency parameter ϵ. In the fast cooling regime $\epsilon_{rad}\sim 1$ and $\epsilon\simeq\epsilon_e$. In this case, the term $\epsilon_p+\epsilon_e-\epsilon$ reduces to $\epsilon_{\rm p}$, meaning that, regardless of the amount of energy gained by the electrons, in the fast cooling regime the adiabatic losses are dominated by the protons, since the electrons lose all their energy to radiation.

Evaluating these expressions for adiabatic blast-waves in a power-law density profile $\rho\propto R^{-s}$, one obtains:

$$E'_{int}(R) = \frac{9-3s}{9-2s}\Gamma mc^2 , \qquad \frac{dE'_{ad}(R)}{dR} = -\frac{(9-s)(3-s)}{2(9-2s)}\frac{\Gamma mc^2}{R} , \tag{16}$$

where $\Gamma\propto R^{-(3-s)/2}$ as in the adiabatic BM76 solution has been used.

In the fully radiative regime $\epsilon=1$, which implies $E'_{int}=0$ and $dE'_{ad}=0$, Equation (5) reduces to:

$$\frac{d\Gamma}{dm} = -\frac{(\Gamma_{eff}+1)(\Gamma-1)}{M_0+m} , \tag{17}$$

which describes the evolution of a momentum-conserving (rather than pressure-driven) snowplow. Replacing $\Gamma_{eff}\to\Gamma$, the solution of this equation coincides with the result by BM76.

Since the model is based on the homogeneous shell approximation, the adiabatic solution does not recover the correct normalization of the BM76 solution. In this treatment, the total energy of a relativistic decelerating adiabatic blast wave in a power-law density profile $\rho(R)\propto R^{-s}$ is

$$E_0\simeq\Gamma_{eff}E'_{int}\simeq\frac{4}{3}\Gamma E'_{int}\simeq\frac{12-4s}{9-2s}\Gamma^2 mc^2 , \tag{18}$$

so that the BM76 normalization can be recovered by multiplying the density of external matter in Equations (5), (13) and (14) by the factor $(9-2s)/(17-4s)$. To smoothly interpolate between the adiabatic regime and the radiative regime, the following correction factor should be adopted:

$$C_{BM76,\epsilon}\equiv\epsilon+\frac{9-2s}{17-4s}(1-\epsilon) . \tag{19}$$

No analytic model properly captures the transition between an adiabatic relativistic blast-wave and the momentum-conserving snowplow, as ϵ increases from zero to unity. The simple interpolation in Equation (19) joins the fully adiabatic BM76 solution with the fully radiative momentum-conserving snowplow.

In summary, Equations (5), (13) and (14), complemented with the correction in Equation (19) (which should by applied to every occurrence of external density and external matter) completely determine the evolution of the shell Lorentz factor Γ as a function of the shock radius R.

2.2. *Relativistic Shock Acceleration*

The spectral shape of the afterglow emission is well described by power-laws over a wide energy range (from radio to GeV-TeV). This is the clear manifestation of the presence of an electron population that has been accelerated in a power-law energy distribution. In GRB afterglows, the main candidate to explain the accelerated non-thermal particles is a Fermi-like mechanism that operates with similar general principles as the non-relativistic diffusive shock acceleration: particles are scattered back and forth across the shock front by magnetic turbulence and gain energy at each shock crossing. The particles themselves are thought to be responsible for triggering the magnetic instability that produces the turbulent field governing their acceleration. The outcome of this acceleration process is determined by the composition of the ambient medium (electron-proton plasma in the case of GRB forward shocks), the fluid Lorentz factor ($\Gamma_{GRB} >> 1$, decreasing to non-relativistic velocity only after several weeks or months) and the magnetization σ (i.e., the ratio between Poynting and kinetic flux in the pre-shocked fluid, $\sigma = B^2/(4\pi\, m_p\, n\, c^2)$), with B being the magnetic field strength. For GRB forward shocks, the magnetization is low, around 10^{-9} in the interstellar medium and in any case below 10^{-5} even for a magnetized circumstellar wind.

In this section, we summarize the present understanding of particle acceleration and magnetic field generation in electron-proton, ultra-relativistic, weakly magnetized shocks. The statements and considerations reported in this section refer specifically to this case (which is the one relevant for forward external shocks in GRBs) and might not be valid for magnetized plasma and/or mildly-relativistic flows and/or electron-positron plasma.

In general, the information that one would extract from theoretical/numerical investigations and compare with observations are: (i) the spectral shape of the emitting electrons (i.e., the minimum and maximum Lorentz factor $\gamma_{min/max}$ and the spectral index p), (ii) the acceleration efficiency (i.e., the fraction of electrons ξ_e and the fraction of energy ϵ_e in the non-thermal population) and (iii) the strength of the self-generated magnetic field, usually quantified in terms of fraction ϵ_B of the shock-dissipated energy conveyed in the magnetic field. In particular, in order to compare with observations, the relevant ϵ_B is the one in the downstream, in the region where radiative cooling takes place and the emission is produced.

After revisiting the state-of-the-art of the theoretical understanding (for recent reviews, see [40,60]), we discuss how particle acceleration and magnetic field amplification are incorporated in GRB afterglow modeling, and then we comment on the constraints on the above-mentioned parameters as inferred from the comparison between the model and the observations.

2.2.1. Inputs from Theoretical Investigations

Analytical approaches and Monte Carlo simulations generally rely on the assumption that electromagnetic waves, providing the scattering centers to regulate and govern the acceleration, are present on both sites of the shock, with a given strength and spectrum, so that the Fermi mechanism can operate. The particle distribution is then evolved under some assumption (such as diffusion in pitch angle) on the scattering process, and considering a test-particle approximation (i.e., the high-energy particles do not modify the properties of the waves).

The main success of these approaches is the verification that under these conditions power-law spectra are indeed produced and the predicted spectral index is in very good agreement with observations of afterglow radiation from GRBs [61]. The spectral index has been calculated for different assumptions on the equation of state, diffusion prescription and for a wide range of shock velocities [61]. A quasi-universal value $p \simeq 2.2 - 2.3$ is found in the ultra-relativistic limit. Figure 1 shows a comparison between analytical and numerical results as a function of the shock velocity for three different types of shocks (see [61] for details). In the ultra-relativistic limit ($\gamma\beta \gg 1$), the estimates of the spectral slopes converge to a universal value $p = s - 2 \sim 2.2$.

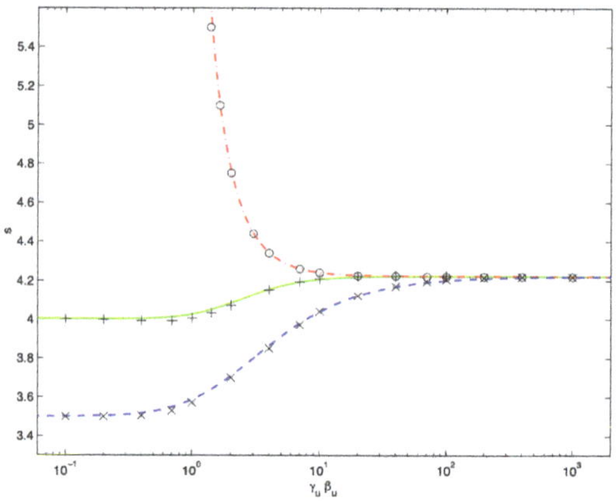

Figure 1. Spectral index ($s_p = p + 2$) of the electrons accelerated in shocks as a function of the shock velocity. Curves refer to the equation derived by [61] under the hypothesis of isotropic, small-angle scattering and is a generalization of the non-relativistic formula. Symbols show the comparison with numerical studies. Different curves refer to different assumptions on the type of shock (see [61] for details): in all cases, the value of the spectral index approaches the same value $s_p \sim 4.2$ (corresponding to $p \sim 2.2$) in the ultra-relativistic limit.

The investigation of relativistic shocks is complemented by particle-in-cell (PIC) simulations, where the non-linear coupling between particles and self-generated magnetic turbulence is captured from first principles.

The limitations of this technique are imposed by the computation time: for accuracy and stability, PIC simulations need to resolve the electron plasma skin depth c/ω_{pe} of the upcoming electrons (where $\omega_{pe} = \sqrt{4\pi e^2 n_e/m_e}$ is the plasma oscillation frequency of the upstream plasma, n_e is the proper density and m_e is the electron mass), which is orders of magnitudes smaller than the scales of astrophysical interest. It is then difficult to follow the evolution on time-scales and length-scales relevant for astrophysics. Low dimensionality (1D or 2D instead of 3D) and small ion-to-electron mass ratios are additional limitations imposed by the computation time. Results of PIC simulations need then to be extrapolated to bridge the gap between the micro-physical scales and the scales of interest. With these caveats in mind, we summarize here the main achievements.

PIC simulations have shown that magnetic turbulence can be efficiently ($\epsilon_B \sim 0.01 - 0.1$) generated by the accelerated particles streaming ahead of the shock (in the so-called precursor region), where they generate strong magnetic waves, which in turn scatter the particles back and fourth across the shock. In particular, in the weakly magnetised shocks discussed in this section, the dominant plasma instability is thought to be the so-called Weibel (or current filamentation) instability [62], generated by the counter-streaming of the accelerated particles against the background plasma in the precursor region [42,63]. PIC simulations have demonstrated that as long as the fluid is ultra-relativistic ($\Gamma > 5$), the main parameter governing the acceleration is the magnetization σ, i.e., the efficiency of the process is insensitive to Γ, as the precursor decelerates the incoming background plasma.

An example of downstream particle spectra derived by PIC simulations is shown in Figure 2 ([42]). The ion and electron spectra are shown for a 2D simulation with $\Gamma = 15$, as an ion-to-electron mass ratio $m_i/m_e = 25$, and $\sigma = 10^{-5}$. The temporal evolution is followed up to $t = 2500\,\omega_{pi}^{-1}$. The formation of a non-thermal tail is clearly visible.

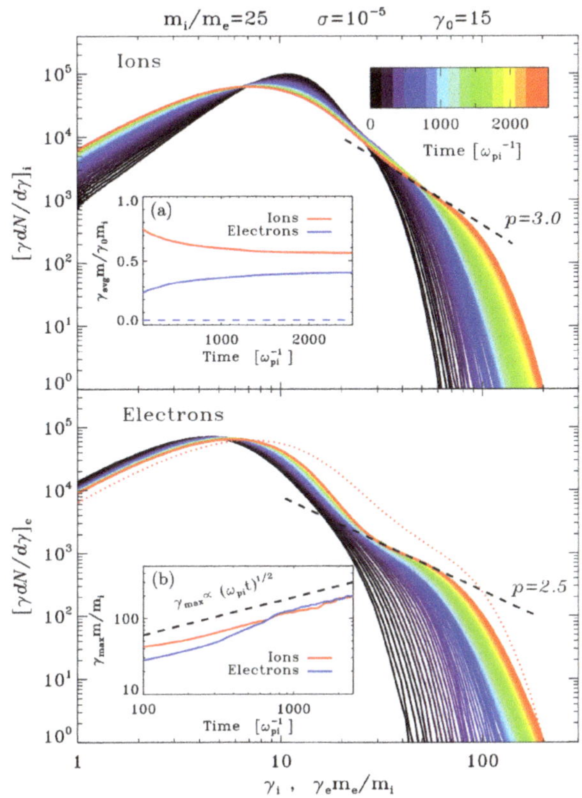

Figure 2. PIC simulations: temporal evolution of the downstream spectrum of ions (**upper** panel) and electrons (**bottom**) for mass ratio $m_i/m_e = 25$, shock Lorentz factor $\Gamma = 15$ and magnetization $\sigma = 10^{-5}$. The evolution is followed until $t = 2500\omega_{pi}^{-1}$. Inset (a): mean post-shock ion (red) and electron (blue) energy (in units of the bulk energy of the upstream flow). The dashed blue line shows the electron energy at injection. Inset (b): temporal evolution of the maximum Lorentz factor of ions (red) and electrons (blue). For comparison, the black dashed line shows the scaling $\gamma_{max} \propto (\omega_{pi}\,t)^{1/2}$. From [42]. ©AAS. Reproduced with permission.

The downstream non-thermal population is found to include around $\xi \simeq 3\%$ of the electrons, carrying $\epsilon_e \simeq 10\%$ of the energy. The spectral index is around $p \sim 2.5$. The acceleration proceeds similarly for electrons and ions, since they enter the shock in equipartition (i.e., their relativistic inertia is comparable) as a result of efficient pre-heating in the self-excited turbulence in the precursor.

The maximum energy γ_{max} increases proportionally to $t^{1/2}$ (see inset in Figure 2), slower than the commonly adopted Bhom rate [64], in which case $\gamma_{max} \propto t$. Extrapolating the γ_{max} behavior to the relevant time-scales and considering that synchrotron cooling will limit the acceleration for high-energy particles, the electron maximal Lorentz factor is found to reach values $\gamma_{max} \sim 10^7$ in the early phase of GRB afterglows, corresponding to synchrotron photon energies around 1 GeV, which is roughly consistent with observations. All these results on the particle spectrum are obtained on time-scales that are too short for the supra-thermal particles to reach a steady-state and their extrapolation to longer time-scales is not trivial.

A still debated open question (because it is computationally demanding) is how the magnetization evolves downstream. PIC simulations have found values of $\epsilon_B \sim 0.1 - 0.01$ in the vicinity of the shock front. How this turbulence evolves on longer time-scales is still a matter of debate. The turbulence is expected to decay rapidly, on time-scales orders of magnitude shorter than the synchrotron cooling time. Magnetization is then predicted to be very different close to the shock and in the region where particle cooling takes place. Electrons would then cool in a region of weak magnetic field [65,66]. These considerations suggest that it might not be correct to define a single magnetization ϵ_B in GRB modeling, infer its value from observations and compare with predictions from PIC simulations referring to the magnetization near the shock front. Magnetization values inferred from observations most likely probe a region downstream, far from the front shock (see Section 3.3 for a discussion).

Theoretical efforts are fundamental to provide physically motivated inputs for the phenomenological parameters included in the afterglow model. The large number of unknown model parameters, coupled with a limited number of constraints provided by observations, implies that constraints from the theory are of paramount importance for a correct interpretation of the emission in GRBs and for grasping the origin of their non thermal emission, from radio to TeV energies. On the other hand, despite the huge progresses in the theoretical understanding of relativistic acceleration, the theory is not quite yet to the point of providing robust inputs for modeling observations. It is then clear how the two approaches must be combined to gain knowledge on the micro-physics of acceleration and magnetic field generation, on the one hand, and on the origin of radiative processes and macro-physics of the emitting region (bulk Lorentz factor and energy content) of the sources, on the other hand.

2.2.2. Description of Shocks in GRB Afterglow Modeling

The theory of relativistic shock acceleration is applied to the GRB afterglow by introducing several unknown parameters in the model. These are the fractions ϵ_e and ϵ_B of dissipated energy gained by the accelerated particles and amplified magnetic field, the spectral index p of the accelerated particle spectrum and the fraction ξ_e of particles, which efficiently enter the Fermi mechanism and populate the non-thermal distribution.

Recalling that the shock-dissipated energy (in the comoving frame) is given by $dE'_{sh} = (\Gamma - 1)dmc^2$ (see Section 2.1), the corresponding energy density is $u'_{\rm sh} = (\Gamma - 1)\rho' c^2$. From shock jump conditions, the density in the comoving frame ρ' is related to the density of the unshocked medium (measured in the rest frame) by the equation:

$$\rho' = 4\Gamma\rho \tag{20}$$

which is valid in both the ultra-relativistic and non-relativistic limits (see e.g., [35]). In the GRB afterglow scenario, it is usually assumed that pairs are unimportant and then the density of protons and electrons is the same: $n_p = n_e = n$. This implies that the mass is dominated by protons: $\rho = nm_p$. In this case, the available energy density that will be distributed to the accelerated particles (electrons and protons) and to the magnetic field can be expressed as:

$$u'_{\rm sh} = 4\Gamma(\Gamma - 1)nm_pc^2 \tag{21}$$

A fraction ϵ_B of this energy will be conveyed to the magnetic field:

$$u'_B = \epsilon_B u'_{\rm sh} = \epsilon_B 4\Gamma(\Gamma - 1)nm_pc^2 \tag{22}$$

from which it follows that the magnetic field strength B' is:

$$B' = \sqrt{32\pi\epsilon_B m_p c^2 n(\Gamma - 1)\Gamma} \tag{23}$$

Similarly, for the accelerated electrons:

$$u'_e = \epsilon_e u'_{sh} = \epsilon_e 4\Gamma(\Gamma - 1)nm_p c^2 = \langle\gamma\rangle m_e c^2 4\Gamma\xi_e n \tag{24}$$

where $\langle\gamma\rangle$ is the average random Lorentz of the accelerated electrons:

$$\langle\gamma\rangle = \frac{\epsilon_e}{\xi_e}\frac{m_p}{m_e}(\Gamma - 1) \tag{25}$$

The accelerated non-thermal electrons are assumed to have a power-law spectrum as a result of shock acceleration. Their energy distribution can be described by a power-law $N(\gamma)d\gamma \propto \gamma^{-p}d\gamma$ for $\gamma_{min} \leq \gamma \leq \gamma_{max}$ where γ_{min} is the *minimum Lorentz factor* of the injected electrons and γ_{max} is the *maximum Lorentz factor* at which electrons can be accelerated. To derive the relation between $\gamma_{\rm min}$, $\gamma_{\rm max}$ and the model parameters, we consider the definition of the average Lorentz factor $\langle\gamma\rangle$:

$$\langle\gamma\rangle = \frac{\int_{\gamma_{min}}^{\gamma_{max}} N(\gamma)\gamma d\gamma}{\int_{\gamma_{min}}^{\gamma_{max}} N(\gamma)d\gamma} \tag{26}$$

and solve the integrals. Equations (25) and (26) leads to (for $p \neq 1$):

$$\begin{aligned} &\left[\frac{\ln\left(\frac{\gamma_{max}}{\gamma_{min}}\right)}{\frac{1}{\gamma_{min}} - \frac{1}{\gamma_{max}}}\right] = \frac{\epsilon_e}{\xi_e}\frac{m_p}{m_e}(\Gamma - 1) \qquad \text{if } p = 2 \\ &\left[\frac{p-1}{p-2}\frac{\gamma_{min}^{-p+2} - \gamma_{max}^{-p+2}}{\gamma_{min}^{-p+1} - \gamma_{max}^{-p+1}}\right] = \frac{\epsilon_e}{\xi_e}\frac{m_p}{m_e}(\Gamma - 1) \qquad \text{if } p \neq 2 \end{aligned} \tag{27}$$

A simplified equation for γ_{min} can be obtained assuming that $\gamma_{max}^{-p+2} \ll \gamma_{min}^{-p+2}$:

$$\gamma_{min} = \frac{\epsilon_e}{\xi_e}\frac{m_p}{m_e}\frac{p-2}{p-1}(\Gamma - 1) \tag{28}$$

Since p is expected to be $2 < p < 3$, this condition is verified for $\gamma_{max} \gg \gamma_{min}$. The minimum Lorentz factor is then not treated as a free parameter of the model, as it is calculated from Equation (28) as a function of the free parameters ϵ_e, ξ_e and p. Concerning the prescription for the value of γ_{max} (for details see Section 3.5), it usually relies on the condition that radiative losses between acceleration episodes are equal to the energy gains, where energy gains proceed at the Bhom rate. As we mentioned in the previous section, PIC simulations, however, have shown that this might not be the case.

A similar treatment can be adopted also for protons simply substituting ϵ_e with ϵ_p, m_e with m_p and assuming a power-law energy distribution with spectral index q. As a result, the minimum Lorentz factor for protons can be derived as:

$$\begin{aligned} &\left[\frac{\ln\left(\frac{\gamma_{max,p}}{\gamma_{min,p}}\right)}{\frac{1}{\gamma_{min,p}} - \frac{1}{\gamma_{max,p}}}\right] = \frac{\epsilon_p}{\xi_p}(\Gamma - 1) \qquad \text{if } q = 2 \\ &\left[\frac{q-1}{q-2}\frac{\gamma_{min,p}^{-q+2} - \gamma_{max p}^{-q+2}}{\gamma_{min,p}^{-q+1} - \gamma_{max,p}^{-q+1}}\right] = \frac{\epsilon_p}{\xi_p}(\Gamma - 1) \qquad \text{if } q \neq 2 \end{aligned} \tag{29}$$

Solving the equations assuming that $\gamma_{max,p} \gg \gamma_{min,p}$ leads to:

$$\gamma_{min,p} = \frac{\epsilon_p}{\xi_p} \frac{q-2}{q-1} (\Gamma - 1) \tag{30}$$

The equations for γ_{min} and Equation (23) for B, coupled with the description of the blast-wave dynamics described in Section 2.1, provides all the necessary equations to derive the radiative output for a jet with energy E and initial bulk Lorentz factor Γ_0 expanding in a medium with density $n(R)$. The derivation of the radiative output is detailed in Section 2.3. To conclude the discussion about particle acceleration, in the next section we anticipate which constraints can be inferred on the physics of particle acceleration from multi-wavelength observations, once the afterglow model is adopted.

2.2.3. Constraints to the Acceleration Mechanism Provided by Observations

Assuming that accelerated particles have a power-law spectrum ($dN^{\rm acc}/d\gamma \propto \gamma^{-p}$) and the cooling is dominated by synchrotron radiation, the spectral slope p can be inferred from observations of the synchrotron spectrum and/or from the temporal decay of the lightcurves if observations are performed at frequencies higher than the typical frequency $\nu_{\rm m}$ of photons emitted by electrons with the Lorentz factor γ_{min} (this is correct in both cases of fast and slow cooling regime). The estimated value of p from the afterglow modeling are spread on a wide range, from $p \sim 2$ to $p \sim 3$, suggesting that the spectrum of injected particles does not seem to have a typical slope, at odds with theoretical predictions. The determination of p, however, suffers from the uncertainties on the spectral index inferred from optical and X-ray observations, where the observed spectra are subject to unknown dust and metal absorption. A derivation of p from the decay rate of the lightcurves is also subject to the correct identification of the spectra regime, and partially also to the assumption on the density profile of the external medium, which is often unconstrained (see Section 3.2).

The typical value of $\epsilon_{\rm e}$ inferred from the afterglow modeling is around 0.1, meaning that 10% of the shock-dissipated energy is gained by the electrons, spanning from 0.01 to large values, such as 0.8. Although this seems a large uncertainty, ϵ_e is perhaps the most well-constrained parameter of the model, and is in good agreement with the values predicted by numerical investigations [42]. For the fraction $\epsilon_{\rm B}$, on the contrary, the inferred values varies in a very wide range, typically from 10^{-5} to 10^{-1} [67–69]. Recent studies that incorporate Fermi-LAT GeV observations [24,70] have demonstrated that the typical values estimated for ϵ_B can be even smaller, in the range $\sim 10^{-7}$–10^{-2}. These values are needed in order to model GeV radiation self-consistently with radiation detected at lower frequencies, with repercussions on the estimates of the other parameters, such as n and E. These small values of the ϵ_B needed to model the radiation have been tentatively interpreted as the sign of turbulence decay in the downstream [65,66]. As a consequence, even though the turbulence is strong ($\epsilon_B \simeq 0.1$) in the vicinity of the shock where the particle is accelerated, it becomes weaker at larger distances, in the region where particles cool (see Section 3.3). Small values of ϵ_B are confirmed by the modeling of recent TeV detections of afterglow radiation from GRBs ([71,72], see Section 4).

Another parameter that one would like to constrain from observations is the fraction of particles ξ_e that are injected into the Fermi process. In the vast majority of the studies, this parameter is not included (i.e., it is implicitly assumed that all the electrons are accelerated, $\xi_e = 1$). This parameter is indeed difficult to constrain, as it is degenerate with all the other parameters [73].

Observations so far have not been able to identify the location of a high-energy cutoff in the synchrotron spectrum that would reveal the maximal energy of the synchrotron photons and then the maximum energy γ_{max} of the accelerated electrons. Observations by Fermi-LAT are in general consistent with a single power-law extending up to at least 1 GeV. Photons with energies in excess of 1 GeV have been detected from several GRBs, the record holder for Fermi-LAT being a 95 GeV photon [74]. These photons cannot be safely associated to synchrotron radiation on the basis of spectral analysis, as their paucity makes it difficult to assess from spectral analysis whether they are consistent with the power-law extrapolation of the synchrotron spectrum or if they are indicative of the rising of a distinct spectral component. In any case, the Fermi-LAT detections are suggesting that synchrotron photons should be produced at least up to a few GeV. This is consistent with the limit commonly invoked for particle acceleration: if the acceleration proceeds at the Bhom rate ($t_{acc} \simeq r_L/c$) with $r_L = E/eB$ being the Larmor radius, and is limited by synchrotron cooling ($t_{syn} \simeq 6\,\pi\, m_e c/\sigma_T B^2 \gamma$) then $\gamma_{max} \sim 10^7 - 10^8$ can be reached. Even though this does not necessarily imply that acceleration must proceed at the Bhom limit, the value of γ_{max} inferred from the detection of GeV photons is quite large and barely consistent with what is found by PIC simulations. Whether or not the observations are in tension with the present derivation of γ_{max} from PIC simulations and theoretical arguments, strongly depends on a clear identification of the origin of photons in the GeV-TeV energy range. Present and future observations with Imaging Atmospheric Cherenkov Telescopes (IACTs) are the main candidates to shed light on this issue.

2.3. Derivation of the Radiative Output

The expected radiative output can be estimated by means of analytical approximations, which provide prescriptions for the location of the synchrotron self-absorption frequency ν_{sa}, the characteristic frequency ν_m emitted by electrons with Lorentz factor γ_{min}, the cooling frequency ν_c emitted by electrons with Lorentz factor γ_c and the overall synchrotron flux [38,47]. In these approaches, the synchrotron spectrum is in general approximated with power-laws connected by sharp breaks, but more sophisticated analytical approximations of numerically derived synchrotron spectra have also been proposed [54]. The associated SSC component in the Thomson regime [48] and corrections to be applied to the synchrotron and SSC spectra to account for the effects of the Klein–Nishina [75] cross section (see Section 2.3.1) are also available in the literature. These prescriptions are usually developed for the deceleration phase, when the Blandford–McKee solution [35] for the blast-wave dynamics applies, i.e., as long as the blast-wave is still relativistic. These models take as input parameters the kinetic energy content of the blast-wave E_k, the external density $n(R) = n_0\, R^{-s}$ (with $s = 0$ or $s = 2$), the fraction of shock-dissipated energy gained by electrons (ϵ_e) and by the amplified magnetic field (ϵ_B), and the spectral slope of the accelerated electrons p. During the deceleration phase, the initial bulk Lorentz factor Γ_0 does not play any role, but its value determines the radius (or time) at which the deceleration begins.

An alternative approach to estimate the expected spectra and their evolution in time consists in numerically solving the differential equation describing the evolution of the particle spectra and estimating the associated emission [76–78]. In this section, we describe a radiative code that simultaneously solves the time evolution of the electron and photon distribution. The code has been adopted, e.g., for the modeling of GRB 190114C presented in [71].

The temporal evolution of the particle distribution is described by the differential equation:

$$\frac{\partial N(\gamma, t')}{\partial t'} = \frac{\partial}{\partial \gamma}\left[\dot{\gamma} N(\gamma, t')\right] + Q(\gamma)\,, \tag{31}$$

where $\dot{\gamma} = \frac{\partial \gamma}{\partial t'}$ is the rate of change of the Lorentz factor γ of an electron caused by adiabatic, synchrotron and SSC losses and energy gains by synchrotron self-absorption. In the SSC mechanism, the synchrotron photons produced by electrons in the emission region act as seed photons that are up-scattered at higher energies by the same population of electrons. Such a scenario will generate a very high energy spectral component, which is the target of searches by IACTs such as MAGIC[1] and H.E.S.S.[2]. In principle, up-scattering of an external population of seed photons can also be considered and included in the cooling term, but here we will ignore this mechanism (external Compton) and focus only on SSC. The source term $Q(\gamma, t') = Q^{acc}(\gamma, t') + Q^{pp}(\gamma, t')$ describes the injection of freshly accelerated particles ($Q^{acc}(\gamma, t') = dN^{acc}/d\gamma\, dt'$) and the injection of pairs $Q^{pp}(\gamma, t')$ produced by photon–photon annihilation.

In the next sections, we explicitly define each one of the terms included in Equation (31) and how to estimate the synchrotron and SSC emission. To solve the equation, an implicit finite difference scheme based on the discretization method proposed by [79] can be adopted.

2.3.1. Synchrotron and SSC Cooling

The synchrotron power emitted by an electron with Lorentz factor γ depends on the pitch angle, i.e., the angle between the electron velocity and the magnetic field line. In the following, we assume that the electrons have an isotropic pitch angle distribution and we use equations that are averaged over the pitch angle (e.g., [80]). The synchrotron cooling rate of an electron with Lorentz factor γ is given by:

$$\dot{\gamma}_{syn} \equiv \left.\frac{d\gamma}{dt'}\right|_{syn} = -\frac{\sigma_T \gamma^2 B'^2}{6\,\pi\, m_e c} \tag{32}$$

The cross section for the inverse Compton mechanism is constant and equal to the Thomson cross section (σ_T) as long as the photon energy in the frame of the electron is smaller than the rest mass electron energy $m_e\, c^2$. For higher photon energies, the cross section decreases as a function of the energy and is described by the Klein-Nishina (KN) cross section. To estimate SSC losses, we adopt the formulation proposed in [81], which is valid for both regimes. Defining the SSC kernel as:

$$K(\gamma, \nu', \tilde{\nu}') = \begin{cases} \frac{\varepsilon}{\tilde{\varepsilon}} - \frac{1}{4\gamma^2} & \frac{\tilde{\varepsilon}}{4\gamma^2} < \varepsilon < \tilde{\varepsilon} \\ 2q\ln q + (1+2q)(1-q) + \frac{1}{2}(1-q)\frac{(4\gamma\tilde{\varepsilon}q)^2}{(1+4\gamma\tilde{\varepsilon}q)} & \tilde{\varepsilon} < \varepsilon < \frac{4\gamma^2\tilde{\varepsilon}}{1+4\gamma\tilde{\varepsilon}}\,, \end{cases} \tag{33}$$

where:

$$\tilde{\varepsilon} = \frac{h\tilde{\nu}'}{m_e c^2} \qquad \varepsilon = \frac{h\nu'}{m_e c^2} \qquad q = \frac{\varepsilon}{4\gamma\tilde{\varepsilon}(\gamma - \varepsilon)}\,. \tag{34}$$

$\tilde{\varepsilon}$ and ε are the energies of the photons (normalized to the rest mass electron energy) before and after the scattering process, respectively. The two terms of Equation (33) account, respectively, for the down-scattering (i.e., $\varepsilon < \tilde{\varepsilon}$) and the up-scattering (i.e., $\varepsilon > \tilde{\varepsilon}$) process. The energy loss term for the SSC can now be calculated with the equation:

$$\dot{\gamma}_{SSC} = \left.\frac{d\gamma}{dt'}\right|_{SSC} = -\frac{3h\sigma_t}{4m_e c\gamma^2}\int d\nu'\nu' \int \frac{d\tilde{\nu}'}{\tilde{\nu}'} n_{\tilde{\nu}'}(t') K(\gamma, \nu', \tilde{\nu}')\,. \tag{35}$$

2.3.2. Adiabatic Cooling

As discussed in Section 2.1, particles lose their energy adiabatically due to the spreading of the emission region. This energy loss term should be inserted in the kinetic equation governing the evolution of the particle distribution. To derive the adiabatic losses, we rewrite Equation (10) as a function of energy losses $d\gamma$ in a comoving time dt':

$$\dot{\gamma}_{ad} = \left.\frac{d\gamma}{dt'}\right|_{ad} = -\frac{\gamma\beta^2}{3}\frac{d\ln V'}{dt'}\,, \tag{36}$$

with β being the random velocity of particles in a unit of c. The comoving volume V' of the emission region can be estimated considering that the contact discontinuity is moving away from the shock at a velocity $c/3$. After a time $t' = \int dR/\Gamma(R)\,c$, the comoving volume is:

$$V' = 4\pi R^2 \frac{ct'}{3}\,, \tag{37}$$

and:

$$\dot{\gamma}_{ad} = \frac{d\gamma}{dt'}\bigg|_{ad} = -\frac{\gamma\beta^2}{3}\left(\frac{2\Gamma c}{R} + \frac{1}{t'}\right). \tag{38}$$

2.3.3. Synchrotron Self-Absorption (SSA)

Electrons can re-absorb low energy photons before they escape from the source region. The absorption coefficient α_ν can be expressed as [80]:

$$\alpha_\nu = -\frac{1}{8\pi\nu'^2 m_e}\int d\gamma P'(\gamma,\nu')\gamma^2\frac{\partial}{\partial\gamma}\left[\frac{N(\gamma)}{\gamma^2}\right] \tag{39}$$

valid for any radiation mechanism at the emission frequency ν', with $P'(\gamma,\nu')$ being the specific power of electrons with Lorentz factor γ at frequency ν' and assuming $h\nu' \ll \gamma m_e c^2$. Thus, the SSA mechanism will mostly affect the low frequency range. This results in a modification of the lower frequency tail of the synchrotron spectrum as:

$$P'_\nu \propto \begin{cases} \nu'^{5/2} & \nu'_i < \nu' < \nu'_{SSA} \\ \nu'^2 & \nu' < \nu'_{SSA} < \nu'_i\,, \end{cases} \tag{40}$$

assuming a power-law distribution of electrons, with $\nu'_i = \min(\nu'_m, \nu'_{cool})$ and ν'_{SSA} the frequency below which the synchrotron flux is self-absorbed and the source becomes optically thick.

2.3.4. Synchrotron and Inverse Compton Emission

Following [82], the synchrotron spectrum emitted by an electron with Lorentz factor γ, averaged over an isotropic pitch angle distribution, is:

$$P'^{syn}_{\nu'}(\nu',\gamma) = \frac{2\sqrt{3}\,e^3 B'}{m_e\,c^2}\,x^2\left[K_{4/3}(x)\,K_{1/3}(x) - 0.6x(K^2_{4/3}(x) - K^2_{1/3}(x))\right], \tag{41}$$

where $x \equiv \nu'\,4\pi\,m_e c/(6\,e\,B'\gamma^2)$ and K_n are the modified Bessel functions of order n. The total power emitted at the frequency ν' is obtained integrating over the electron distribution:

$$P'^{syn}_{\nu'}(\nu') = \int P'^{syn}_{\nu'}(\nu',\gamma)\frac{dN}{d\gamma}d\gamma\,. \tag{42}$$

The SSC radiation emitted by an electron with Lorentz factor γ can be calculated as:

$$P'^{SSC}_{\nu'}(\nu',\gamma) = \frac{3}{4}h\sigma_T c\frac{\nu'}{\gamma^2}\int\frac{d\tilde{\nu}'}{\tilde{\nu}'}n_{\tilde{\nu}'}K(\gamma,\nu',\tilde{\nu}')\,, \tag{43}$$

where $n_{\tilde{\nu}'}$ is the photon density of synchrotron photons and the integration is performed over the entire synchrotron spectrum. Integration over the electron distribution provides the total SSC emitted power at frequency ν'.

2.3.5. Pair Production

Pair production by photon-photon annihilation is particularly important for a correct estimate of the radiation spectrum in the GeV-TeV band. Indeed, some of the emitted VHE photons are lost due to their interaction with photons at lower energies (typically X-ray photons). As a result, the observed flux is attenuated and the resulting spectrum at VHE is modified. Here we follow the treatment presented in [83]. The cross section of the process $\sigma_{\gamma\gamma}$ as a function of β', the centre-of-mass speed of the electron and positron is given by:

$$\sigma_{\gamma\gamma}(\beta') = \frac{3}{16}\sigma_T(1-\beta'^2)\left[(3-\beta'^4)\ln\left(\frac{1+\beta'}{1-\beta'}\right) - 2\beta'(2-\beta'^2)\right] \tag{44}$$

where:

$$\beta'(\omega_t, \omega_s, \mu) = \left[1 - \frac{2}{\omega_t \omega_s (1-\mu)}\right]^{\frac{1}{2}} \tag{45}$$

and $\omega_t = h\nu'_t / m_e c^2$ with ν'_t being the target photon frequency, $\omega_s = h\nu'/m_e c^2$ with ν' being the source photon frequency and $\mu = \cos\phi$, where ϕ is the scattering angle. Then, it is possible to derive the annihilation rate of photons into electron-positron pairs as:

$$R(\omega_t, \omega_s) = c \int_{-1}^{\mu_{max}} \frac{d\mu}{2}(1-\mu)\sigma_{\gamma\gamma}(\omega_t, \omega_s, \mu)\,, \tag{46}$$

where $\mu_{max} = \max(-1, 1 - 2/\omega_s\omega_t)$ coming from the requirement $\beta'^2 > 0$. Considering $x = \omega_t\omega_s$, it is possible to derive asymptotic limits for $R(\omega_t, \omega_s) \equiv R(x)$ in two regimes. For $x \to 1$ (i.e., near the threshold condition) $R(x) \to c\sigma_T/2(x-1)^{3/2}$, while for $x \gg 1$ (i.e., ultra-relativistic limit) $R \to \frac{3}{4}c\sigma_T \ln x/x$. An accurate and simple approximation, which takes into account both regimes, is given by:

$$R(x) \approx 0.652 c\sigma_T \frac{x^2-1}{x^3} \ln(x)\, \mathrm{H}(x-1)\,, \tag{47}$$

where $H(x-1)$ is the Heaviside function [83]. The approximation accurately reproduces the behavior near the peak at $x_{peak} \sim 3.7$ and over the range $1.3 < x < 10^4$, which usually dominates during the calculations. A comparison between Equations (46) and (47) is given in Figure 3, where the goodness of the approximation adopted in the mentioned x range can be observed.

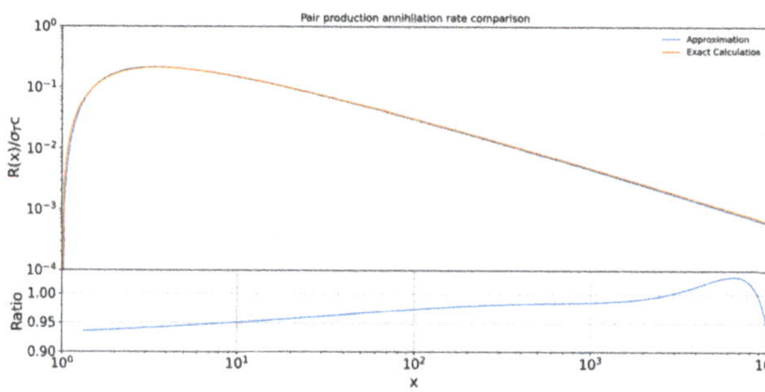

Figure 3. Comparison between the exact annihilation rate (Equation (46)) and the approximated formula (Equation (47)). The ratio between the two curves in the range ($1.3 < x < 10^4$) is shown in the bottom panel. In this range, the ratio is always $\lesssim 7\%$.

The impact of the flux attenuation due to pair production mechanism on the GRB spectra is estimated in terms of the optical depth value $\tau_{\gamma\gamma}$. From its definition:

$$\tau_{\gamma\gamma}(\nu') = \sigma_{\nu'\nu'_t} n'(\nu'_t)\Delta R' \tag{48}$$

where $n'(\nu'_t)$ is the number density of the target photons per unit of volume, $\sigma_{\nu'\nu'_t}$ is the cross section and $\Delta R'$ is the width of the emission region. Introducing the cross section in terms of the annihilation rate $R(x)$ in its approximated formula and integrating over all the possible target photon frequencies:

$$\tau_{\gamma\gamma}(\nu') = \frac{\Delta R'}{c} \int R(\nu', \nu'_t) n'_{\nu'}(\nu'_t) d\nu'_t \tag{49}$$

where ν' and ν'_t are the frequencies of the source and of the target interacting photons. The pair production attenuation factor can be then introduced simply multiplying the flux by a factor $(1 - e^{-\tau_{\gamma\gamma}})/\tau_{\gamma\gamma}$. This attenuation factor will modify the GRB spectrum, giving a non-negligible contribution in the VHE domain, in particular. An example of the modification of a GRB spectrum due to pair production can be observed in Figure 4. Here, the flux emitted in the afterglow external forward shock scenario by synchrotron and SSC radiation and the flux attenuation due to pair production have been calculated with a numerical code. For a set of quite standard afterglow parameters and assuming $\Delta R' = R/\Gamma$, the attenuation of the observed flux due to pair production become relevant above 0.2 TeV, and it reduces the flux by $\sim 30\%$ at 1 TeV and by $\sim 70\%$ at 10 TeV.

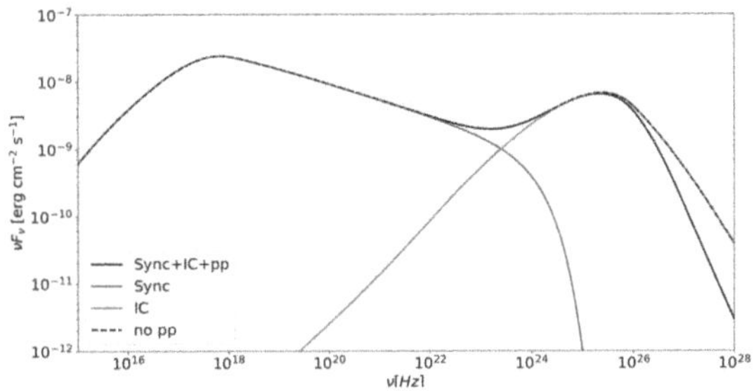

Figure 4. Spectral energy distribution in the GRB afterglow external forward shock scenario estimated with the numerical code presented in this section for $t_{obs} = 170$ s. The effect of the pair production attenuation is clearly evident in the VHE tail. The set of afterglow parameters used are the following: $E_k = 1 \times 10^{53}$ erg, $s = 0$, $A_0 = 1$ cm^{-3}, $\epsilon_e = 0.2$, $\epsilon_B = 0.02$, $\Gamma_0 = 300$, $p = 2.5$ and $z = 0.5$.

Similar considerations can also be conducted for the electron/positron production. Assuming that the electron and positron arises with equal Lorentz factor γ and that $x_{peak} \sim 3.7$, a photon with energy $\omega_s \gg 1$ will mostly interact with a target photon of energy $\omega_t \approx 3/\omega_s$. Then, from the energy conservation condition:

$$2\gamma = \omega_s + \frac{3}{\omega_s} \approx \omega_s = \frac{h\nu'}{m_e c^2} \tag{50}$$

The $e^{\pm}$ production can be observed as an additional source term for the distribution of accelerated particles. As a result, an additional injection term Q_e^{pp} to be inserted in the kinetic equation (Equation (31)) is calculated as:

$$Q_e^{pp}(\gamma, t') = 4\frac{m_e c^2}{h} n_{\nu'}(\frac{2\gamma m_e c^2}{h}, t') \int d\nu_t' n_{\nu'}(\nu_t', t') R(\frac{2\gamma m_e c^2}{h}, \nu_t') \tag{51}$$

2.3.6. Comparison with Analytical Approximations

In order to compare results from the numerical method described in the previous section and analytical prescriptions available in the literature, we give an example in Figure 5. The analytical prescriptions for the synchrotron and the SSC component are calculated following [38,48]. In [38], the synchrotron spectra and light-curves are derived assuming a power-law distribution of electrons in an expanding relativistic shock, cooling only by synchrotron emission. The dynamical evolution is described following BM76 equations for an adiabatic blast-wave expanding in a constant density medium. The resulting emission spectrum (green dashed lines in Figure 5) is described with a series of sharp broken power-laws. The SSC component associated to the synchrotron emission was computed, as a function of the afterglow parameters, in [48]. In this work, calculations are performed assuming that the scatterings occur in Thomson regime. Modifications to the synchrotron spectrum caused by strong SSC electron cooling are also detailed.

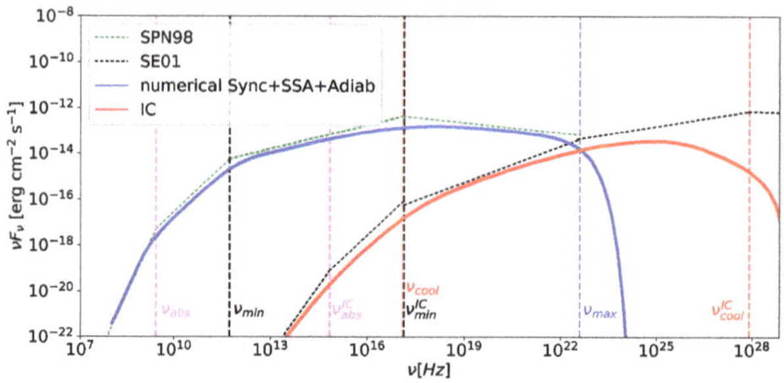

Figure 5. Comparison between spectra estimated with analytical approximations and numerical calculations. For the analytical method, the synchrotron emission (green dashed line) is estimated starting from [38] (SPN98, see legend) while the SSC (black dashed line) is taken from [48] (SE01). Vertical lines mark the break frequencies. The results from the numerical code described in Section 2.3 are shown with solid blue and red lines. This example shows the spectrum calculated at $t = 10^4$ s for $s = 0$, $p = 2.3$, $\epsilon_e = 0.05$, $\epsilon_B = 5 \times 10^{-4}$, $E_k = 10^{52}$ erg, $n_0 = 1$ cm^{-3}, $\Gamma_0 = 400$ and $z = 1$.

From the comparison proposed in Figure 5, it can be clearly observed that analytical and numerical results are in general in good agreement. Both curves follow the same behavior except for the high-energy part of the SSC component. Here, the KN scattering regime, which is not taken into account in the analytical approximation, becomes relevant. As a result, the numerical calculations differ from the analytical ones showing a peak and a cutoff in the SSC spectrum due to the KN effects.

Nevertheless, there are a few minor discrepancies between the two methods. The numerically-derived spectrum is very smooth around the break frequencies, with the result that the theoretically expected slope (e.g., the one predicted by the analytical approximations) is reached only in regions of the spectrum that lie far from the breaks, i.e., is reached only asymptotically. This puts into questions simple methods for discriminating among different regimes and different density profiles based on closure relations, which are relations between the spectral and the temporal decay indices [10,38]. Regarding the flux normalization, there are minor discrepancies between the numerical and analytical results. This is due to the fact that in analytical prescriptions it is assumed that the radiation is entirely emitted at the characteristic synchrotron frequency. On the contrary, in the numerical derivation, the full synchrotron spectrum of a single electron is summed up over the whole electron distribution. Similar considerations apply to the SSC component when comparing with the analytical spectra. Moreover, the discrepancies observed between analytical and numerical SSC spectra are amplified by the differences observed in the target synchrotron spectra.

In general, this comparison shows that the numerical treatment is a powerful tool able to predict the multi-wavelength GRB emission in a more accurate way than the analytical prescriptions. The latter ones, however, are still giving valid approximations of the overall spectral shape. The main limitation of analytical estimates arises when TeV observations are involved. The importance of KN corrections is evident in this band and should be properly treated for a correct interpretation of the TeV spectra, as will be shown in Section 4.

3. Open Questions

As predicted by the basic standard model presented in the previous section, the afterglow emission is the result of particle acceleration and radiative cooling occurring in two different regions: the forward and the reverse shock. The temporal and spectral behavior of the two emission components can be inferred after the jet/blastwave dynamics, acceleration mechanisms and the radiation processes are modeled (Section 2). The general agreement between model predictions and observations convincingly proves that the long-lasting radio-to-GeV radiation is indeed produced in interactions between the ejecta and the external medium. Moreover, the radiative mechanisms involved and the nature of emitting particles are well established, with synchrotron (and possibly SSC) from the accelerated electrons (either at the forward or reverse shock) being the source of the detected radiation.

Despite the general success of the external shock scenario, there are several, long-standing open issues which represent a serious challenge for our present understanding of the afterglow emission and the GRB phenomenon in general. Moreover, even when observations seem to be in qualitative agreement with predictions, the extraction of the model parameters (which would give important feedbacks on our understanding of particle acceleration and GRB environments) is limited by the large degeneracy among parameters and lack of solid inputs from theoretical considerations.

Afterglow emission studies have not experienced relevant progresses in the last years, with observations and techniques that are the same since the launch of the Swift satellite. The recent discovery of TeV radiation from GRBs is opening the possibility to renovate and boost afterglow studies, with major impacts on the general understanding of GRB sources.

In this section, we list and comment on those aspects still lacking a clear explanation, and in particular we selected topics which might largely benefit from observations and detections in the VHE regime.

3.1. X-ray Flares

Observations of the afterglow emission in the X-ray and optical often display behaviors that are not predicted by the standard scenario, and require the inclusion of additional emission components contributing to the detected radiation. In the standard external forward shock scenario, the afterglow light-curves in the X-ray and optical band are expected to decay following a power-law or a broken power-law behavior, where the breaks are interpreted as the cooling or injection frequency crossing the observed band [38,47,54]. The advent of Swift-XRT and the increasing number of optical follow-up observations performed by ground-based robotic telescopes have highlighted the presence, in a good fraction of cases, of unexpected features in the early time afterglow, such as flares and plateaus [49,57].

Flares are episodes of sudden rebrightenings characterized by a very fast rise of the flux, followed by an exponential decay profile. Comprehensive studies of X-ray afterglows demonstrate that an X-ray flare is observed in $\sim$33% of the GRBs [84,85]. The times at which they are observed span a very wide range, from around $\sim$30 s up to $\sim 10^6$ s after the trigger time. The time where the flare peaks is shown in Figure 6 (T_{pk}, x-axis) for a large sample of 468 X-ray flares in long GRBs. Most of the flares occur within 10^3 s, even though there are many cases of flares occurring several hours after the burst. The width of the flare ω is found to evolve linearly with time to larger values following the trend $\omega \sim 0.2T_{pk}$ [84]. The average and peak luminosities L of the flare also display a dependence from T_{pk}, with $L \propto T_{pk}^{-2.7}$ at least for early time ($T_{pk} < 10^3$ s) flares [84,85]. When also including late time flares [86,87], a shallower index is obtained, around ~ -1.2. The energy emitted during flare episodes is quite large and, for early time flares, is around $\sim$10% of the prompt emission or sometimes even comparable [88].

Flares have also been detected in the optical, although the sample of optical flares is far smaller than the X-ray one [89]. A statistical study of optical flares detected by Swift/UVOT demonstrates that most of them correlate with and share similar temporal properties to simultaneous X-ray flares. Nevertheless, there are a few dozen of GRBs for which no X-ray flaring activity is observed simultaneously with optical flares [89].

Flares are believed to have an inner origin and to be associated with a prolonged activity of the GRB central engine [57,90–94]. However, the relatively long timescales on which they are detected represent a challenge for the model. Many questions are still open, such as the location of the emitting region, what is powering the flares, and whether late time flares have a different origin than flares detected at early times.

Speculations about possible signatures of X-ray flares in the GeV-TeV range are present in the literature [95–98]. Assuming that flares have an internal origin and are produced at

$R < R_{dec}$, forward shock electrons will be exposed to the flare radiation, producing an IC emission component by up-scattering the flare photons. Following these estimates, the IC component peaks at $\sim$ 100 GeV and has a flux comparable to the X-ray flux. Alternatively, GeV-TeV radiation associated to flares can be produced by the SSC mechanism, where electrons responsible for X-ray synchrotron flares also upscatter these photons to higher energies. The process is considered less interesting for TeV radiation because the peak of this SSC component is expected to be around 1 GeV [95], due to a relatively low minimum Lorentz factor $\gamma_{min} \sim 60$. Such value is estimated from theoretical considerations where $\gamma_{min} \simeq 60\epsilon_{e,-1}(\Gamma_{IS} - 1)$ for $p = 2.5$, $\epsilon_e = 0.1$ and a relative shock Lorentz factor Γ_{IS} of the order of unity. We notice that the recent estimates of the minimum electron Lorentz factor in the late prompt emission of GRBs [29] may modify these predictions, and place the expected SSC around 100 GeV. The luminosity of this component will strongly depend on the size of the emitting region. As a result, the detection of flares in GeV-TeV band can provide relevant information to identify the properties at the emitting region and the production site of the flaring activity.

To understand what the chances of current and future VHE ground-based instruments are in contributing to the study of flares, we perform some simplified estimates. The MAGIC telescopes observed 138 GRBs in almost ~16.5 years, from 2005 up to June 2021 [99]. More than half of them (74 events) have been observed with delays from shorter than 10^3 s, which means ~4.5 GRBs yr^{-1}, and 37 events observed with delays shorter than 100 s (i.e., 2.2 GRB yr^{-1}). Considering that ~33% of the long GRBs have an X-ray flare and considering the distribution of their peak times (see Figure 6), we estimate that ~1 GRBs/yr is the rate of GRBs with an X-ray flare occurring during MAGIC observations.

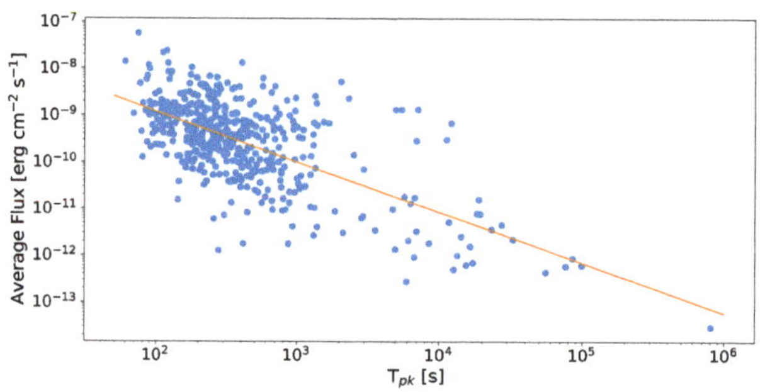

Figure 6. Average fluxes of Swift-XRT flares in the 0.3–10 keV energy range versus peak times T_{pk}. The blue points are the 468 flares observed by Swift from April 2005 to March 2015 (collected from [87]). The orange line is the resulting linear regression, which gives the expression: $\log F_{av} = -6.76 - 1.08 \times \log T_{pk}$.

Let us go a bit further and estimate the detectability of a putative ~10^2 GeV counterpart of X-ray flares. For the flux of the GeV-TeV flare, we consider as reference value the X-ray flux, and discuss what happens if a similar or ten times smaller flux is emitted at ~10^2 GeV.

We collect the X-ray flux of a large sample of flares from the catalog of X-ray flares presented in [87]. The results are shown in Figure 6. The average flux of the flare and the flare peak time correlate, and the orange line represents the best fit. To perform the estimates, we consider two different flare peak times, $T_{pk} = 10^2$ s and $T_{pk} = 10^3$ s. The typical average fluxes at those times are $F = 1 \times 10^{-9}$ erg cm^{-2} s^{-1} and $F = 1 \times 10^{-10}$ erg cm^{-2} s^{-1}, respectively. Assuming that a similar amount of flux is emitted around 100 GeV we can compare these values with the differential sensitivity as a function of the observing time of IACT instruments. Figure 7 [100] shows the sensitivity for several telescopes to the detection of a point-like source at five standard deviations significance as a function of the exposure time and for four selected energies. Considering that the width of the flare is related to the peak time following the relation $\omega \sim 0.2T_{pk}$, we can compare the flare fluxes estimated at $T_{pk} = 10^2$ s and $T_{pk} = 10^3$ s with the differential sensitivity for observing the time of $t_{obs} = 20$ s and $t_{obs} = 200$ s. The flare fluxes lie close to the differential sensitivity of the MAGIC telescopes (for 100 GeV at $t_{obs} = 20$ s is $\sim 1.0 \times 10^{-9}$ erg cm^{-2} s^{-1} and at $t_{obs} = 200$ s is $\sim 5.0 \times 10^{-10}$ erg cm^{-2} s^{-1}). This indicates that MAGIC telescopes can barely detect such a flare. Moreover, Extragalactic Background Light (EBL) attenuation reduces the flux, which is why we are making the estimates at 100 GeV, where the attenuation is still small. We conclude that MAGIC would be able to detect (or place relevant constraints on) only the brightest X-ray flares (as can be observed in Figure 6, the correlation has a large spread, and flares at 10^2 or 10^3 s can easily have fluxes one order of magnitude larger than what is assumed here).

Concerning future instruments, the Cherenkov Telescope Array (CTA[3]) will have a sensitivity which is almost one order of magnitude lower than the MAGIC one and similar slewing capabilities. The same estimates performed for MAGIC can be applied to CTA, with the advantage that CTA will have a northern and southern sites, approximately doubling the possibility to follow GRBs within short time-scales. This is a promising indication that the CTA array will be potentially able to detect possible counterparts at $E \sim 30$ the GeV of X-ray flares, provided that this counterpart has a flux that is no less than ten times smaller than what is detected in X-rays. As a result, it can play a major role in exploring and improving our knowledge of flares and their connection with prompt emission and with the prolonged activity of the central engine.

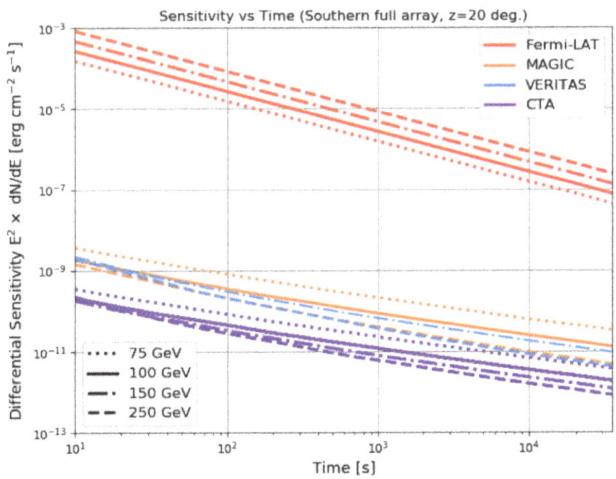

Figure 7. Differential sensitivity as a function of the observation time for several HE and VHE instruments (Fermi-LAT, MAGIC, VERITAS and CTA) at four selected energies (75, 100, 150, 250 GeV). From [100].

3.2. Density Profile of the External Medium

Following the established connection between long GRBs and the core-collapse of massive stars, the jet is expected to produce the afterglow while propagating in the wind of the star in its free-streaming phase. Afterglow radiation of long GRBs should then be produced in the interaction with a medium with a radial density profile $n \propto R^{-2}$. However, several investigations have demonstrated that about half of the long GRBs are better explained if the blast-wave is assumed to run into a medium with constant density. We revise the evidence in support of the constant density medium and discuss the difficulties in reconciling these observations with expectations on the environment surrounding long GRB progenitors.

Long GRBs originate from the core-collapse of massive stars, most likely rapidly rotating, and with a possible evidence of a preference for low-metallicity. The most convincing evidence in support of this paradigm is the association with type Ic supernovae and the proximity of GRBs to young star-forming regions. While the connection of long GRBs (or at least with the bulk of the population) with the core-collapse of massive stars is solid, the role of metallicity and rotation in the launch of the GRB jets, and the identification of the progenitor star, is still uncertain. The progenitor is usually identified with Wolf-Rayet stars, massive stars ($M > 20\,M_\odot$) in the final stages of their evolution, characterized by powerful winds and a high mass loss rate [101]. The wind from the star is expected to interact and deeply modify the environment where the GRB explodes and leaves imprints on its afterglow emission.

More in detail, the interaction between the stellar wind and the ISM four concentric regions with different properties is expected to form. In the inner part (i.e., close to the star), the circumburst medium is permeated by the free-streaming wind, producing a density with radial profile $n \propto R^{-2}$. The density is related to the mass loss rate $\dot{M}$ and to the velocity v_w of the free-streaming stellar wind by:

$$n(R) = \frac{\dot{M}}{4\,\pi\,R^2\,m_p\,v_w}\,. \tag{52}$$

A termination shock separates the unshocked from the shocked wind: the latter forms a hot bubble of thermalized wind material, with a nearly constant density profile, as the formation of pressure and density gradients is prevented by the high sound speed inside the bubble. The hot bubble, in its outer part, is enclosed by a shell of shocked ISM, surrounded by the unshocked ISM. The GRB jet is supposed to trill its way in this stratified medium [10].

To understand where most of the afterglow evolution occurs, we have to estimate the deceleration radius R_d and the non-relativistic radius R_{NR} (i.e., the radius where the blast-wave has decelerated to non-relativistic velocity) and compare them to the termination shock radius. For typical parameters ($\dot{M} = 10^{-5} M_\odot\,\mathrm{yr}^{-1}$ and $v_w = 10^3\,\mathrm{km\,s^{-1}}$), the fit to numerical models of Wolf-Rayet stars [102] give the following relation between the termination shock radius and the density of the unshocked ISM: $R_T = 10\,n_{ISM}^{-1/2}\,\mathrm{pc}$, where n_{ISM} is the density of the unshocked ISM. From the blast-wave dynamics, the deceleration and the non-relativistic radius are $R_d = 6 \times 10^{-5}\,E_{52}\,v_{w,3}/(\dot{M}_{-5}\,\Gamma_{0,2})\,\mathrm{pc}$ and $R_d = 0.6\,E_{52}\,v_{w,3}/\dot{M}_{-5}\,\mathrm{pc}$, respectively. It is evident how the complete evolution of the afterglow radiation occurs well inside the free-streaming region.

In afterglow modeling of long GRBs it is then customary to assume a density profile described by Equation (52), where $\dot{M}$ and v_w are treated as unknown parameters (normalized to the typical values of a Wolf-Rayet star) combined in one single free model parameter $A_\star$: $n(R) = 3 \times 10^{35} A_\star R^{-2}$. Despite this robust prediction, the modeling of afterglow observations shows that in a relevant fraction of cases, observations are better explained by adopting a circumburst medium with a constant density $n = n_0$.

The fraction of this case varies depending on the method and on the selected sample, and is on average about 50% [103–106].

To place the termination shock at least inside the non-relativistic radius, one should invoke a very large density of the ISM, $n \gtrsim 10^5\,\mathrm{cm}^{-3}$, typical of dense cores of molecular clouds: $R_T = 0.03\,(n_{ISM}/10^5\,\mathrm{cm}^{-3})^{-1/2}\,\mathrm{pc}$. Density profiles for different ISM densities are shown in Figure 8, upper panel. Alternatively, one can try to variate the wind parameters. How the termination shock radius changes for different values of $\dot{M}$ and v_w is shown in the bottom panel of Figure 8. A very low mass loss rate $\dot{M} = 10^{-7} M_\odot \mathrm{yr}^{-1}$ (which may find a justification in case of a low-metallicity star) is needed to bring the termination shock radius below 1 pc (for $n_{ISM} = 10\,\mathrm{cm}^{-3}$). With this low mass loss rate, the deceleration and non-relativistic radius increase ($R_d \sim 6 \times 10^{-3}\,\mathrm{pc}$ and $R_{NT} \sim 60\,\mathrm{pc}$), placing the termination shock still after the deceleration radius but well within the non-relativistic radius, allowing for part of the observed emission to develop into a constant density environment. By increasing the blast-wave energy, the deceleration radius can further approach R_T. This suggests that it is more likely for a very energetic GRB to cross the termination shock at early times and then expand in a ISM-like medium, as compared to a faint GRB. An indication of an average larger E_γ in GRBs with a wind-like medium as compared to GRBs with a ISM-like medium has been found in [107], but is in contrast with results from the study performed by [106] on a larger sample.

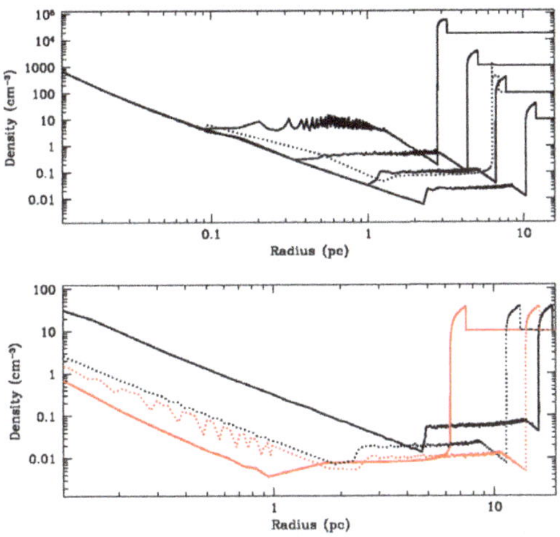

Figure 8. Density profile produced by a star wind as a function of the distance from the central object. The R^{-2} profile characterizes the region where the wind freely streams, which is separated from the shocked wind (with nearly constant density) by the termination shock. **Top** panel: the impact of different values of the ISM density on the termination shock is shown. For all the curves, it is assumed $\dot{M} = 10^{-5} M_\odot\,\mathrm{yr}^{-1}$ and $v_w = 10^3\,\mathrm{km\,s^{-1}}$, while the ISM density varies from 10 to $10^4\,\mathrm{cm}^{-3}$, with the termination shock moving to smaller distances when the density increases. **Bottom** panel: for a fixed ISM density $n_{ISM} = 10\,\mathrm{cm}^{-3}$, the impact of different $\dot{M}$ and v_w is shown. Black-solid: $\dot{M} = 10^{-4} M_\odot\,\mathrm{yr}^{-1}$ and $v_w = 10^3\,\mathrm{km\,s^{-1}}$; red-solid: $\dot{M} = 10^{-6} M_\odot\,\mathrm{yr}^{-1}$ and $v_w = 10^3\,\mathrm{km\,s^{-1}}$; black-dotted: $\dot{M} = 10^{-5} M_\odot\,\mathrm{yr}^{-1}$ and $v_w = 500\,\mathrm{km\,s^{-1}}$; red-dotted: $\dot{M} = 10^{-5} M_\odot\,\mathrm{yr}^{-1}$ and $v_w = 2 \times 10^3\,\mathrm{km\,s^{-1}}$. From [102]. ©AAS. Reproduced with permission.

The parameter space for which part of the afterglow emission can indeed be produced in the ISM-like density profile of the shocked wind is very limited, as it corresponds to the most energetic GRBs, low-metallicity progenitors and high-density ISM, or a combination of these factors [107]. These considerations on the diversity of E_k, $\dot{M}$, v_w and ISM density may not be sufficient to explain the results of the modeling (i.e., the preference for a ISM-like environment). The fraction of GRBs, which might have these peculiar parameters can hardly account for the large fraction of GRBs for which a wind-like profile is excluded by observations. The required conditions are too extreme to be verified in half of the population. However, it is not clear if this percentage has been overestimated by present studies. To quantify the inconsistency, the first step would be to perform a dedicated study of afterglow emission to assess the percentage of long GRB afterglows that are not consistent with a wind-like environment.

Methods based on closure relations may not be valid if the spectrum is modified by Compton scattering in the Klein–Nishina regime (see also [108]). Moreover, these are based on a simple approximation of the synchrotron spectrum into power-law segments, while the wide curvature of the real synchrotron spectra might lead to incorrect estimates of the value of p if the observed frequency is in the vicinity of a synchrotron break frequency. A full modeling is then necessary to really assess the fraction of long GRBs for which an R^{-2} density profile is excluded, and ultimately understand if the paradigm for the environment of GRBs should be drastically modified. Radio observations may be of great help, since the flux temporal behavior does not depend on p and is quite different in the case of constant or wind-like density profile. Similarly, the detection of SSC radiation can help solve this ambiguity.

3.3. Small Values of ϵ_B

For a long time, the typical value of ϵ_B has been considered to variate between 0.01 and 0.1, both on the basis of theoretical considerations on particle acceleration and findings by numerical simulations. Indeed, the present understanding of the micro-physics at weakly magnetized shocks invoke the existence of self-generated micro-turbulence both behind and in front of the shock, at a level corresponding to $\epsilon_B \sim 0.01$–0.1. This layer of intense micro-turbulence is expected on theoretical grounds and recently corroborated by numerical PIC simulations. Inferences of the value of ϵ_B from the early modeling of afterglow radiation were broadly consistent with these numbers, confirming the presence of large self-generated fields in ultra-relativistic weakly magnetized shocks. More recently, several independent methods have provided evidence for significantly lower values.

In particular, several studies on GRBs with GeV temporally extended emission detected by LAT arrive to the same conclusions that in order to explain GeV radiation as part of the synchrotron emission, multi-wavelength observations require $\epsilon_B = 10^{-6}$–10^{-3} [24,109–112]. Similar values have been inferred from studies that are based on radio, optical and/or X-ray emission and do not make use of high energy emission, such as [67–69,113]. A smaller magnetic field in the region where most of the particle cooling occurs might increase the expected relevance of the SSC component, as supported by recent detections of TeV radiation by IACTs.

Such small values of ϵ_B may appear to be problematic [114], because strongly self-generated micro-turbulence must be present to ensure the scattering and acceleration of the particles, which otherwise would be simply advected away.

It was later pointed out that the inferred low values of magnetization might be indicative of a turbulence that is decaying on time-scales comparable with the electron cooling time [65,66]. From a theoretical perspective, indeed, the micro-turbulence is expected to decay beyond some hundreds of skin depths. This picture has been validated by PIC simulations, which however are still far from probing time-scales comparable with the dynamical time-scale of the system. Dedicated simulations the magnetic field does decay behind the shock, on a time-scale much longer than $c\omega_{pi}$. Immediately behind the shock, the magnetic field carries a magnetization $\epsilon_B \sim 0.01$, which decays in time after 10^2–10^3 plasma times. Eventually, the magnetic field will settle to the shock-compressed value $4\Gamma\, B_u$, where B_u is the magnetization of the upstream unperturbed medium. In this scenario, high-energy particles, which produce MeV-GeV photons, feel only the region close to the shock, where the magnetization is large, due to their short cooling time. Particles that cool on longer time-scales (and produce radio, optical and X-ray photons) cool on longer time-scales, and then in a region where the magnetic field has decayed.

The application of cooling in a decaying magnetic turbulence to four GRBs detected by LAT has proved to be very successful and even able to give indications on how fast the turbulence decays, being consistent with a power-law decay $\epsilon_B \propto t^{-\alpha_t}$ with $\alpha_t \sim 0.5$ [66].

To understand and constrain the value of the magnetic field relevant for the particle cooling is of great importance, since an incorrect assumption or prior affects the estimates of all the other afterglow parameters, and in particular the density of the external medium [67,106].

A low value of ϵ_B tends to increase the level of SSC luminosity for a given synchrotron luminosity. The recent detection of bright TeV emission from the afterglow of GRBs is an indication that this might indeed be the case. Existing and future TeV observations will shed a light on this issue, fostering a revision of our prejudice on the value of the magnetic field in the region where particles cool.

3.4. Variation of Micro-Physical Parameters with Time

Thanks to the increasing number of available observations on a wide range of frequencies (from radio to TeV) and times (from seconds to weeks), the basic assumption that micro-physical parameters (such as $\epsilon_e, \epsilon_e, p$ and ξ_e) are constant over the whole afterglow evolution can be tested. We comment on the hints (inferred from afterglow modeling) for temporal evolution of these parameters.

In case of well-sampled multi-wavelength light-curves, the modeling with synchrotron spectra is able not only to identify the location of the spectral breaks at a certain time but also evaluate their evolution in time. As a result, hints that micro-physical parameters ϵ_e and ϵ_B may vary with time have been found in some events with well-detailed multi-wavelength follow-up campaigns.

In [115], broad-band (from near infrared up to X-ray) afterglow data from GRB 091127 were interpreted in the standard external forward shock scenario assuming a constant-density medium. The good quality of the data allows one to identify the breaks in the light-curves and associate them with the synchrotron spectral breaks. As a result, the time evolution of the synchrotron breaks was estimated. In particular, it was calculated that the cooling break frequency ν_{cool} evolves as $\nu_{cool} \propto t^{-1.2}$, which is in contrast with synchrotron predictions for which a less steeper decay $\nu_{cool} \propto t^{-0.5}$ is expected. As a result, some assumptions of the standard model must be relaxed to remove the tension between observations and theoretical predictions. The most likely option able to explain the cooling break observational behavior without affecting the general interpretation of the data is to let the ϵ_B parameter variate with time. Assuming that $\epsilon_B \propto t^{0.49}$ the time evolution of ν_{cool} can be explained successfully.

In [116] for GRB 130427A modeling, in order to explain the observed fast evolution of the injection frequency $\nu_m \propto t^{-1.9}$, a temporal evolution of ϵ_e is claimed. Considering that $\nu_m \propto \epsilon_e^2$, a modest evolution of ϵ_e following the trend $\epsilon_e \propto t^{-0.2}$ is able to satisfactorily describe the observed light-curves.

A time-dependent evolution of the micro-physical parameters has also been proposed in order to explain the features observed in the early afterglow phase, which are not predicted by the external forward shock scenario such as X-ray afterglow plateaus, chromatic breaks, and afterglow rebrightenings [117–120].

Information from TeV observations can be certainly exploited in order to reduce the uncertainty on the values of the micro-physical parameters. The expansion of the broad band afterglow observations to a new spectral window will be a further test and a challenge for the multi-wavelength modeling based on the standard external forward shock scenario. In particular, the time evolution of the different energetic components, also including TeV emission, will give new insights useful to investigating the evolution of the micro-physical parameters. A first proof is provided by the well-sampled multi-wavelength emission observed for GRB 190114C, one of the few GRBs detected so far at TeV energies. The broadband emission can be explained only by invoking the evolution of the micro-physical parameters with time [121], as will be discussed in the next section.

3.5. Maximum Synchrotron Photon Energy

One of the expectations from Fermi-LAT observations of GRB afterglows was the identification of a spectral cutoff in the afterglow synchrotron spectrum marking the maximum energy of synchrotron photons [122,123]. Such a cutoff has not been firmly identified. Its location is directly connected with the shock micro-physics conditions and the maximum energy at which electrons can be accelerated. This maximum energy is typically estimated equating the timescale for synchrotron cooling and the acceleration timescale, where acceleration is assumed to proceed at the Bohm level, considered as the fastest rate. This estimate returns, hence the maximum energy of the accelerated particles. Assuming that the accelerated particles are electrons:

$$t'_L = \frac{r_L}{c} = \frac{\gamma m_e c}{eB'} \quad (53)$$

where r_L is the Larmor radius. For each crossing, the electrons gain energy by a factor ~ 2. On the other hand, the energy losses for synchrotron radiation on this timescale are:

$$\delta E' = t'_L P' = \frac{\gamma m_e c}{eB'} \frac{\sigma_T c \gamma^2 B'^2}{6\pi} = \frac{1}{6\pi} \frac{\sigma_T m_e c^2 B' \gamma^3}{e} \tag{54}$$

The particle stops to gain energy when:

$$\delta E' = \gamma m_e c^2 \tag{55}$$

Therefore, the maximum Lorentz factor for electrons can be derived:

$$\gamma_{max} = \sqrt{\frac{3\pi e}{\sigma_T B'}} \tag{56}$$

The corresponding maximum synchrotron photon energy is:

$$h\nu'_{max} = \frac{eB'\gamma_{max}^2 h}{2\pi m_e c} \tag{57}$$

which for electrons is $\sim$50 MeV in their rest frame.

Similar considerations can also be conducted for protons. Following the same arguments presented below, one obtains:

$$\gamma_{cool,p} = \frac{6\pi m_p^3 c}{\sigma_T m_e^2 B'^2 t'} \tag{58}$$

for the cooling Lorentz factor and:

$$\gamma_{max,p} = \sqrt{\frac{3\pi e m_p^2}{\sigma_T m_e^2 B'}} \tag{59}$$

for the maximum Lorentz factor, which sets a maximum photon energy of $\sim$100 GeV. Synchrotron emission is less efficient for protons so they are less affected by cooling and they can reach higher maximum Lorentz factors than the electrons.

Within this framework, it is expected that observations in the GeV band can be exploited to identify the presence of a cut-off in the HE tail domain. At the current stage, Fermi-LAT observations indicate that the afterglow component of the HE energy emission is usually modeled with a single power-law component with index $\sim$−2 and with no evidence of spectral evolution in time [124] and HE cut-offs.

The absence of the cut-off in the observational data may be explained in several ways. The most likely interpretations are the limited sensitivity of the LAT instrument in the GeV range and the possible contamination due to the rising of the SSC spectral component. As a result, the synchrotron cut-off is hidden behind the VHE spectral component, which can be detected in the GeV-TeV domain. This implies that TeV observations are fundamental in order to disentangle between the two spectral components and infer the cutoff of the synchrotron spectrum.

Another possible interpretation is that the lack of a cutoff in the observational data is genuine. In this case, the synchrotron emission can exceed the limit assumed for the maximum photon energy and extend in the GeV-TeV domain. This interpretation can be tested with VHE observations. The extension of the HE power-law derived by LAT up to the TeV domain should be consistent with VHE data, and no spectral hardening in the GeV-TeV band should be observed. Such a scenario is in contrast with the standard particle acceleration model presented in Section 2. A deep revision of our understanding of acceleration mechanisms is required in order to make TeV emission from synchrotron radiation possible. In particular, a mechanism, able to accelerate electrons up to PeV energies, is needed.

The calculation performed so far assumes the presence of a uniform magnetic field B' throughout the shock-heated plasma. If this assumption is rejected, it is possible to consider a non-uniform magnetic field, stronger close to the shock front and decaying downstream. Following the calculation of [125]. the magnetic field B' can be expressed in terms of the distance from the shock front x as:

$$B'(x) = B'_s\left(\frac{x}{L_p}\right)^{-\eta} + B'_w\,, \tag{60}$$

where B'_s and B'_w are, respectively, the strongest and the weakest magnetic field strengths, η is the power-law decaying index, and L_p is the field decay length scale, which is estimated as [126]:

$$L_p = \left[\frac{m_p\Gamma_s c^2}{4\pi n e^2}\right]^{1/2} = 2.2\times10^7\frac{\Gamma_s}{n}\quad\text{cm} \tag{61}$$

where Γ_s is the shock front Lorentz factor and n is the number density of the accelerated particles in the shocked fluid comoving frame. As a consequence, the Larmor radius r_L increases with the distance from the shock front, since $B'(x)$ becomes weaker and an electron travelling downstream will be likely sent back upstream when $r_L \leq x$. When considering the case $B'_s \gg B'_w$ and $x \gg L_p$ the particles will lose most of their energy in the region of low magnetic field. Therefore, from the condition that losses in the low magnetic field region should be greater than losses in the high magnetic field region, after some algebra the following condition is obtained:

$$\left(\frac{B'_s}{B'_w}\right)^2 \lesssim \frac{r_l}{L_p} \tag{62}$$

valid for $\eta > 1/2$ and $x_0/L_p \gg 1$ where x_0 is the width of the high magnetic field region. Considering that $x_0/L_p \equiv (B'_s/B'_w)^{1/\eta}$, Equation (62) states that the Larmor radius in the high magnetic field region is larger than the actual width of the region and electrons will be barely deflected in such portion of the shocked plasma. As a result, it is possible to calculate the maximum Lorentz factor for electrons that lose most of their energy in the weak magnetic field region following the same conditions presented for the uniform magnetic field case:

$$\gamma_{max} = \sqrt{\frac{3\pi e}{\sigma_T B'_w}} \tag{63}$$

As a result, the maximum synchrotron photon energy is given by:

$$h\nu'_{max} = \frac{e\gamma^2_{max}h}{2\pi m_e c}\left(\frac{B'_s}{B'_w}\right) \tag{64}$$

which is greater than the one calculated in Equation (57) by a factor B'_s/B'_w. Numerical calculations [41] show that this ratio can be larger than $\sim10^2$. As a result, photons of energies $\gtrsim$ 100 GeV can be produced via synchrotron process when assuming a non-uniform magnetic field, which decays downstream of the shock front and with particles losing most of their energies in the weakest field region.

In both interpretations presented here, TeV observations are fundamental in order to investigate with unprecedented details the possible presence or absence of the synchrotron cutoff spectrum. This have also a direct impact on the study of the possible radiation mechanisms responsible for the VHE component in GRBs.

3.6. Prompt Emission Efficiency

The overall efficiency η_γ of the prompt emission mechanism is the result of three processes: the efficiency η_{diss} of the (still unidentified) mechanism responsible for the dissipation of the jet energy, the efficiency ϵ_e of the acceleration mechanism in converting the dissipated energy into random energy of the electrons, and the radiative efficiency ϵ_{rad} of the electrons: $\eta_\gamma = \eta_{diss}\epsilon_e\epsilon_{rad}$. Provided that it is reasonable to assume a fast cooling regime for the prompt emission ($\epsilon_{rad} = 1$), the overall prompt efficiency is limited by the capability of the dissipation mechanism in extracting the kinetic or magnetic energy of the jet and the capability of the particle acceleration process to convey a fraction of this energy into the non-thermal electron population. The value of the efficiency then provides a fundamental clue to placing constraints on the origin of energy dissipation in GRBs, which is still quite uncertain, discriminating between matter and magnetic dominated jets.

From the definition of $\eta_\gamma = E_\gamma/E_0$ (where $E_0 = E_\gamma + E_k$ is the initial explosion energy), we can write the relation $E_k = (1 - \eta_\gamma)/\eta_\gamma E_\gamma$. The parameter η_γ can then be estimated from the comparison of the energy E_γ emitted in the prompt phase and the energy E_k left in the jet after the end of the prompt emission (i.e., at the beginning of the afterglow phase). While the former is directly estimated from observations, the latter can be inferred only indirectly, from the modeling of afterglow radiation.

One of the most adopted methods to infer E_k for large samples of GRBs is to rely on the X-ray luminosity and use it as a proxy for the energy content of the blast-wave [127–131]. This method is solid as long as the X-ray band lies above $max(\nu_m, nu_c)$ and is not affected by inverse Compton cooling. If these two conditions are verified, then the electrons emitting X-ray photons are in a fast cooling regime and their cooling is dominated by synchrotron losses. The luminosity produced is then proportional to the energy content of the accelerated electrons $E_k\epsilon_e$. Assuming a value (typically 0.1) for ϵ_e, then E_k can be estimated. Investigations based on the X-ray emission have inferred very large values of η_γ, between 0.5 and 0.9 [49,129,132,133].

The very same approach can also be applied to 100 MeV-GeV photons detected by the LAT, under the assumption that these are synchrotron photons. A strong correlation between the GeV luminosity and $E_{\gamma,iso}$ has been indeed found, supporting the possibility that GeV photons lie in the high-energy part of the synchrotron spectrum, where the afterglow luminosity is proportional to $E_k\epsilon_e$ and can be used, similarly to the X-ray luminosity, to estimate E_k [134]. A study by [70] revealed that the energetics E_k inferred independently from X-ray and GeV luminosities on a sample of 10 GRBs are inconsistent with each other. The authors show that the inconsistency is solved if $\nu_c > \nu_X$ (where ν_X is the X-ray frequency), or if Compton losses are important in the X-ray band. Full modeling of the GeV, X-ray and optical data support this scenario. In both cases, ϵ_B is required to be much smaller than usually assumed, with values in the range 10^{-7}–10^{-3}. This analysis shows that the GeV band is a much better proxy for E_k, since it is above ν_c and is not affected by inverse Compton cooling, due to Klein–Nishina suppression. Adopting GeV luminosities as a proxy for E_k, the estimated values of E_k increase, also affecting the estimates of η_γ, which are around 5–10% [24].

A correct estimate of η_γ is extremely important, since its value is related to the mechanism dissipating energy in the jet. Since internal shocks can hardly reach values of η_γ larger than 10%, values around 50–90% have been used to argue that internal shocks are not a viable mechanism to explain prompt emission in GRBs, and more efficient mechanisms should be considered (e.g., magnetic reconnection). If the efficiency is, however, smaller than initially estimated, internal shocks may still be a viable solution. Moreover, different estimates of η_γ lead to different estimates on the total initial jet energy $E_0 = E_{\gamma,iso} + E_k$, with repercussions on the energy budget of GRBs and finally on their progenitors and mechanisms for jet launching. Small values of ϵ_B may then relax the problem with very large prompt efficiency, which is definitely unreasonable for internal shocks, but also difficult to attain for magnetic reconnection models (for a discussion, see e.g., [19,20,135]).

A scenario where the magnetic field strength is relatively low in the emitting region implies a stronger SSC emission. Recent TeV detection of GRBs support this scenario, and provide additional observations to constrain the magnetic field. Moreover, as shown by the first detections by IACTs (Section 4), the energy in the TeV component is comparable to the energy in X-rays, providing better estimates for the energy budget in the afterglow phase. Future detections from a larger sample of GRBs can help in assessing the energy budget of the jet more precisely during the afterglow emission, add important information to constrain the efficiency of the prompt emission and favor or exclude some dissipation scenarios.

3.7. Fraction of Accelerated Particles

As described in Section 2, the representation of the relativistic shock acceleration in GRB afterglow relies on some free parameters. These values describe the energy equipartition between particles and magnetic field and the non-thermal accelerated particle distribution.

In particular, the parameter ξ_e is responsible to account for the fraction of particles (here we consider electrons, but the same considerations are also valid for protons) accelerated into a non-thermal distribution. This means that from relativistic shock theory it is expected that a fraction $1-\xi_e$ of electrons is heated into a thermal distribution rather than a non-thermal one (see Figure 9). For simplicity, it is usually assumed in afterglow studies that $\xi_e = 1$, i.e., all the particles are accelerated into a non-thermal distribution. Such an assumption is used in order to avoid the large degeneracy which affects the GRB parameters when including this additional free value. In particular, afterglow modeling predictions obtained assuming $\xi_e = 1$ for parameters E_k, n, ϵ_e and ϵ_B cannot be distinguished from those estimated for any ξ_e in the range $m_e/m_p < \xi_e < 1$ and afterglow parameters $E'_k = E_k/\xi_e$, $n' = n/\xi_e$, $\epsilon'_e = \xi_e\epsilon_e$ and $\epsilon'_B = \xi_e\epsilon_B$ [73]. This can be proven considering the dependencies of the jet dynamics and shock energy equipartition on the model parameters. As shown in BM76 and by previous calculations on the evolution of a relativistic blastwave, in the self-similar regime the bulk Lorentz factor evolve as $\Gamma \propto (E_k/n)^{1/2}$. As a result, the same flow evolution can be obtained with different values of E_k and n as long as their ratio is preserved. The fraction of energy given to the magnetic field is reduced by a factor ξ_e but at the same time the energy density given to the shock is increased by a factor $1/\xi_e$ so that the magnetic field energy density is the same in the scenario when including ξ_e or not. Analogous considerations can be also conducted for the number and the energy density of the electrons so that their values are preserved. As a result, in principle it is not possible to distinguish between the two parameter sets obtained for ξ_e or for any value $m_e/m_p < \xi_e < 1$. This implies that afterglow model parameters, when considering $\xi_e \neq 1$, are estimated with an uncertainty of factor of m_e/m_p, and the afterglow observations do not directly constrain their values (e.g., E_k or ϵ_B) but rather are a fraction of their value multiplied or divided by ξ_e (e.g., E_k/ξ_e or $\epsilon_B\xi_e$).

It is possible to obtain information on the value of ξ_e through PIC simulations or indirect features of the thermal component on the synchrotron afterglow spectra. As mentioned in the previous section, PIC simulations of weakly magnetized shocks have found that the downstream population include around ~3% of the electrons, which are accelerated into a non-thermal distribution. In case the efficiency is low (around ~10% or less) the presence of a large population of thermal electrons may affect the afterglow radiation spectra. The thermal electrons are distributed at lower energies than the non-thermal ones since $\eta\gamma m_e c^2 \ll \gamma m_p c^2$, where $\eta \ll m_p/m_e$ is a factor describing the ignorance on the plasma physics governing electron heating beyond $\gamma m_e c^2$. As a result, the synchrotron radio component emission may be affected through the production of a new emission component from thermal electrons (for $\eta \gg 1$ and moderate $1/\xi_e$) or a large self-absorption optical depth (for $\xi_e \ll 1$), which may be visible in a time scale of few hours. Possible effects of the thermal component are discussed in [136–138].

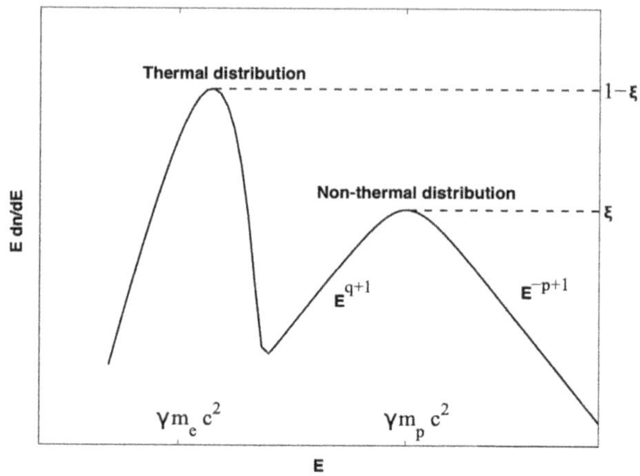

Figure 9. Distribution of accelerated electrons from a relativistic shock. The two bumps correspond, respectively, to the distribution of the $(1-\xi)$ fraction of particles heated into a thermal component with post shock energy $\sim \gamma m_e c^2$ and the fraction ξ of particles accelerated into a non-thermal component with post shock energy $\sim \gamma m_p c^2$. Adapted from [73]. ©AAS. Reproduced with permission.

Information from the TeV component cannot completely solve the degeneracy between afterglow parameters and cannot provide additional clues on the non-thermal electron distribution. However, it can provide unique fundamental data useful to constrain the other afterglow parameters less constrained such as ϵ_B and the density. This will also impact the ξ_e calculation since it can help to reduce the degeneracy between the sets of solutions available in the parameter space. Indeed, the multi-wavelength modeling of GRB 190829A (detected at TeV energies by H.E.S.S.) showed that the only way to explain all the detected radiation, from radio to TeV, is to introduce the parameter ξ_e in the modeling, which is constrained by data to be $\xi_e < 0.13$ [72].

4. Discovery of a TeV Emission Component in GRBs

The robust theoretical framework developed throughout the years to explain the afterglow radiation predicts that, to some extent, GRBs should be TeV emitters (Section 2). Observations in the HE band, and in particular the presence of $\gtrsim$GeV photons with energies up to $\sim$100 GeV, support this possibility. On the other hand, from the observational side, the search for such emission is hampered by several drawbacks. Space-born telescopes, such as Fermi-LAT, sensitive up to few hundred GeV, have an hard time with GRBs due to their low γ-ray photon flux at the highest energies ($\sim 10^2$ GeV), caused by their cosmological distance and strong EBL absorption. These difficulties can be overcome by the much larger effective area of IACTs in the common energy range of sensitivity (50–300 GeV). As a downside, IACTs have a small field of view (a few degrees wide), higher low-energy threshold ($\gtrsim$50–100 GeV), and reduced duty cycle (less than 10%).

In the last decades, IACTs have performed a huge effort to become instruments suitable for GRB observations. In particular, the efforts have been focused in two directions: (i) the development of fast repointing systems to promptly react to GRB alerts and start observations with delays of a few tens of seconds after the trigger time, and (ii) the extension of the energy threshold below 100 GeV, important to reduce the impact of the EBL attenuation on the detection probability of cosmological GRBs.

After a decade of VHE observations resulting in non-detections, the first announcement of GRBs detected by IACTs arrived in 2019, thanks to the MAGIC and H.E.S.S. telescopes [139]. These detections have firmly established that GRBs can be bright sources of TeV radiation. Somewhat unexpectedly, VHE emission was also detected several hours/a few days after the GRB onset, and up to energies of $\gtrsim$3 TeV. The timescales of the detections place the origin of the emission in the afterglow phase. The TeV emission has been studied and interpreted in a multi-wavelength context, in order to evaluate the properties and the nature of the responsible radiation mechanisms. In particular, investigations have focused on SSC, external inverse Compton (EIC), and synchrotron radiation.

In this section, all GRBs for which a detection (significance $> 5\sigma$) or a hint of detection (significance between 3 and 5σ) has been claimed by Cherenkov telescopes, are presented. These are, in total, six events (one short and five long): GRB 160821B (Section 4.1), GRB 180720B (Section 4.2), GRB 190114C (Section 4.3), GRB 190829A (Section 4.4), GRB 201015A (Section 4.5) and GRB 201216C (Section 4.6). For each event, we start with a brief description of the prompt and afterglow multi-wavelength observations. Then, we describe VHE observations and summarize the main results. These detections being a novelty, and some of them laying close to the sensitivity detection threshold of the instrument, we describe in detail the VHE data analysis, the calculation of the significance excess at the GRB position (following the usual prescription used for VHE sources presented in [140]), and the methods adopted for the derivation of the spectral energy distribution (SED) and of the light-curves. For each GRB, we also present the interpretations that have been put forward in the literature. A discussion on the main common properties and differences among this initial population of VHE GRBs and with respect to the whole GRB population is presented in Section 5, where we also address the question of what we have been learning from these few detections.

In this section, all quoted times refer to the time elapsed from the trigger time T_0 of the Swift-BAT or Fermi-GBM instrument, as will be specified. Photon indices are given in the notation $N_\nu \propto \nu^\alpha$, while temporal indices are defined by $F(t) \propto t_T^\beta$.

4.1. GRB 160821B

GRB 160821B is a short GRB at $z = 0.162$ detected by the Swift-BAT [141] on 21 August 2016 at $T_0 = 22:29:13$ UT and by the Fermi-GBM [142]. The analysis of MAGIC observations shows a $\sim 3\sigma$ excess at the GRB position.

4.1.1. General Properties and Multi-Wavelength Observations

The BAT prompt spectrum ($T_{90} = 0.48$ s) is well described by a power-law with index $\alpha = -0.11 \pm 0.88$ and an exponential high-energy cutoff, corresponding to a peak energy with $E_p = 46.3 \pm 6.4$ keV [143]. The GBM prompt spectrum ($T_{90} = 1.088 \pm 0.977$ s) is fitted with a cutoff power-law function as well, with $E_p = 92 \pm 28$ keV. Being located at redshift $z = 0.162$, this is one of the nearest short GRBs detected up to date. Its isotropic-equivalent energy $E_{\gamma,iso} \sim 1.2 \times 10^{49}$ erg is toward the low energy edge of the known distribution for short GRBs [144]).

Afterglow observations are available in the radio, optical, X-ray and (V)HE band. Fermi-LAT observations were performed from the trigger time up to 2315 s and from 5285 s to 8050 s, and did not reveal any significant excess in the 0.3–3 GeV band [145]. Swift-XRT observations [146] started 57 s after the trigger time and revealed a complex behavior of the X-ray afterglow light curve. An initial plateau is followed by a steep decay at around 10^2 s. Then, a power-law decay with index ~ -0.8 is observed after $\sim 10^3$ s [147,148]. Optical observations were performed by several instruments [149–152], revealing the presence of a fading source with a magnitude $r = 22.6 \pm 0.1$ mag 0.95 h after T_0. The identification of the host galaxy allowed the measurement of the spectroscopic redshift $z = 0.162$. The GRB is located in the outskirts of the host spiral galaxy, at $\sim$15 kpc projected distance from its center [148,153]. A fading radio afterglow counterpart was observed at 6 GHz by VLA starting from 3.6 h after the burst trigger [154].

4.1.2. VHE Observations and Results

The MAGIC telescopes started the follow-up of GRB 160821B with a very short delay of 24 s after T_0 and continued observations for about 4 h [145]. The observations were performed with a relatively high Night Sky Background (NSB) (2–8 times brighter than in dark nights) due to the presence of the Moon, and in mid-high zenithal angle conditions (from 34° to 55°). Unfortunately, the first ~1.7 h of the data were strongly affected by clouds. As a result, dedicated and optimized software configurations were used. A more stringent image cleaning with respect to dark conditions was applied to take into account the spurious contribution coming from the high NSB. The analysis required cuts optimized on the Crab Nebula and on Mrk421 observed in similar conditions, and correction factors for the low atmospheric transmission, calculated thanks to the LIDAR facility [155].

The pre-trial analysis showed the presence of a 3.1σ (2.9σ post-trial[4]) significant excess at the GRB position provided by Swift-XRT (see Figure 10). The flux has been estimated for energies above 0.5 TeV assuming a power-law spectrum with photon index $\alpha = -2$. In the first 1.7 h, where data taking was affected by bad atmospheric transmission, only flux upper limits could be derived. This time window has been divided into two time intervals (24–1216 s and 1258–6098 s) and the derived upper limits are, respectively, 1.1×10^{-11} $\text{cm}^{-2}\text{s}^{-1}$ and 5.4×10^{-12} cm^{-2} s^{-1}. In the subsequent 2.2 h (6134–14,130 s) assuming that the signal is real, a flux value could be calculated and is $9.9 \pm 4.8 \times 10^{-13}$ cm^{-2} s^{-1}. For the same time interval, a flux upper limit has also been estimated and gives 3.0×10^{-12} cm^{-2} s^{-1}. All the mentioned fluxes are shown in Figure 11 (red symbols) and refer to the observed values, i.e., without correcting for EBL absorption. Upper limits have been calculated at a 95% confidence level following the prescriptions of [156]. The low ($\sim 3\sigma$) significance estimated did not allow us to obtain an unfolded spectrum. As a result, in the SED (Figure 12) the reconstructed flux in the third bin (6134–14,130 s) over the energy range 0.5–5 TeV is represented as an error box. The statistical error on the photon flux has been taken into account, while, for simplicity, the systematic error for the assumed spectral index was neglected. The flux inferred by MAGIC observations in the 0.5–5 TeV energy range would imply a TeV luminosity at least 5 times larger (when de-absorbed by EBL) than the luminosity emitted in the X-ray band.

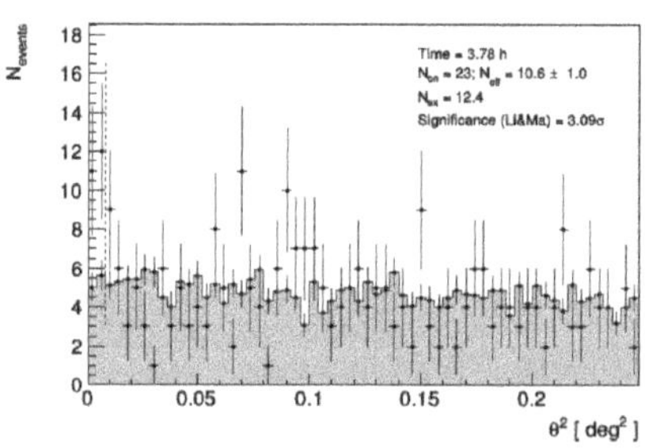

Figure 10. GRB 160821B: angular distance distribution θ^2 between the nominal source position and the reconstructed event arrival directions. The gray histogram represents the background events, while the black points with blue crosses are the γ-like events. The vertical dashed line marks the θ^2 cut value and defines the region in which excess events and signal significance are calculated. From [145]. ©AAS. Reproduced with permission.

4.1.3. Interpretation

A modeling of the multi-wavelength observations, including MAGIC data, has been presented by the MAGIC Collaboration in [145]. The emission is interpreted as the sum of several components, dominating at different times and in different energy bands:

- synchrotron and SSC emission from electrons accelerated at the forward shock; this is in general the dominant emission component;
- synchrotron emission from electrons accelerated by the reverse shock, which is found to dominate the radio band until $t \sim 4.8$ h;
- kilonova emission, powered by freshly synthesized r-process elements released in neutron star mergers; this component is found to dominate the optical/nIR from around 1 to 4 days [148,153];

- an X-ray extended emission component, widely attributed to long-lasting activity of the central engine, here dominating the X-ray band for $t < 10^3$ s.

In performing this multi-component modeling, the synchrotron and SSC forward shock emission have been calculated with a one-zone numerical code (see [71] for details), while the reverse shock and kilonova emission contributions have been taken from [148]. Only X-ray data at $t > 10^3$ s have been included in the modeling, to exclude the extended emission component. The broad-band modeling is shown over-plotted to the light-curves in Figure 11 (solid lines) and to the SED between 1.7 and 4 h in Figure 12.

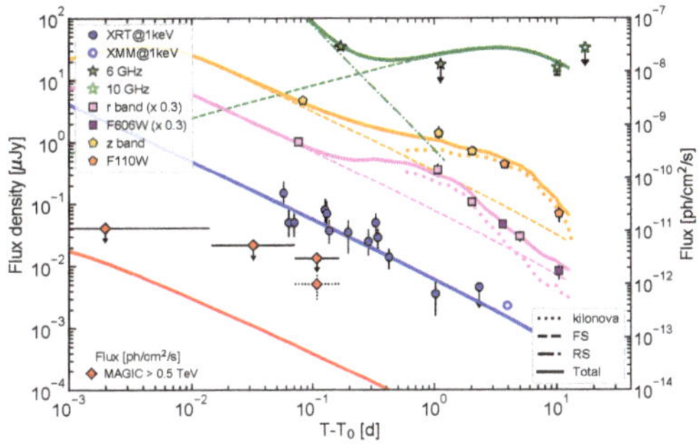

Figure 11. GRB 160821B: multi-wavelength light curves (from radio to TeV) and their modeling according to [145]. ©AAS. Reproduced with permission. The different contributions from the forward shock (FS), reverse shock (RS) and kilonova are shown (see legend).

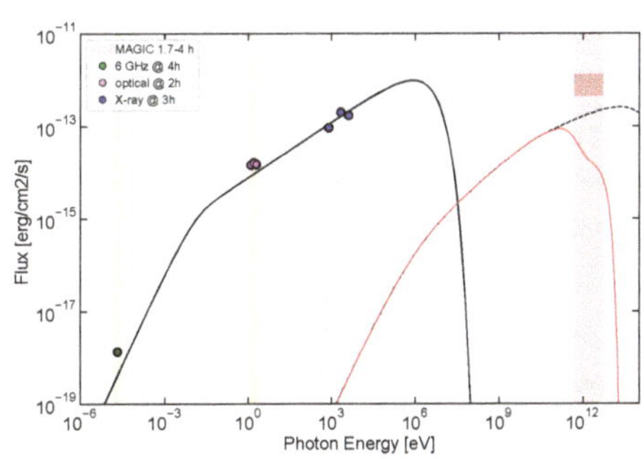

Figure 12. GRB 160821B: modeling of the simultaneous multi-wavelength SED at approximately ~3 h according to [145] (©AAS. Reproduced with permission.), for the same parameters used to model the lightcurves in Figure 11. The shaded areas show the sensitivity energy range of the different instruments. The MAGIC error box on the reconstructed flux is also shown. Synchrotron (solid black line), intrinsic SSC (before EBL absorption, dashed black line) and SSC emission with EBL attenuation (solid red line) estimated from the numerical modeling are shown.

A very good agreement between data and modeling is found in radio (green lines and points), optical (yellow and pink lines and points) and X-rays (blue lines and points). A large degeneracy is present in the parameters, and the data modeling only allows us to identify the ranges for the permitted values of each parameter. These are reported in Table 1 and we note that they are very similar to those estimated in [148] and in a later work by [157]. In the allowed parameter space defined by radio, optical and X-ray observations, different combinations of the parameters predicts different SSC fluxes at 1 TeV are found, reaching at most $F_{SSC}^{(1TeV)} \sim 2 \times 10^{-13}$ erg cm^{-2} s^{-1}. This value, when attenuated by EBL, is at least one order of magnitude fainter than the one inferred from the data analysis of the MAGIC observations. In other words, the parameter space constrained by the observations at lower frequencies is unable to account for such energetic TeV emission, and the SSC forward shock scenario fails to reproduce the observations, provided that the hint of excess found by MAGIC is a real signal from the source.

An alternative scenario that has been explored is the external inverse Compton (EIC) scenario, investigated by [157]. These authors first consider a one-zone SSC model, and reach similar conclusions to those presented by the MAGIC Collaboration [145]: the SSC mechanism predicts a TeV flux around 1–2 orders of magnitude lower than the MAGIC observations (see Figure 13, orange curves). The alternative EIC scenario is then considered by the authors, where the seed photons are provided by the extended X-ray emission and the X-ray plateau. The extended emission and the plateau are fitted using two phenomenological functions. The energy spectrum of the late-prompt emission is described by a broken power-law (see Figure 13, top and bottom panels). For the EIC model, the VHE spectrum is inferred for three different observed times ($t = 1.1, 1.8, 2$ h) and compared to the MAGIC flux averaged between 1.7 and 4 h. As can be observed in Figure 13 (bottom panel), the model flux at 2 h under-predicts the MAGIC flux (the MAGIC observed flux, green shaded area, should be compared with the EBL-absorbed model flux). We conclude that the EIC model is also unable to explain the large TeV flux suggested by MAGIC observations.

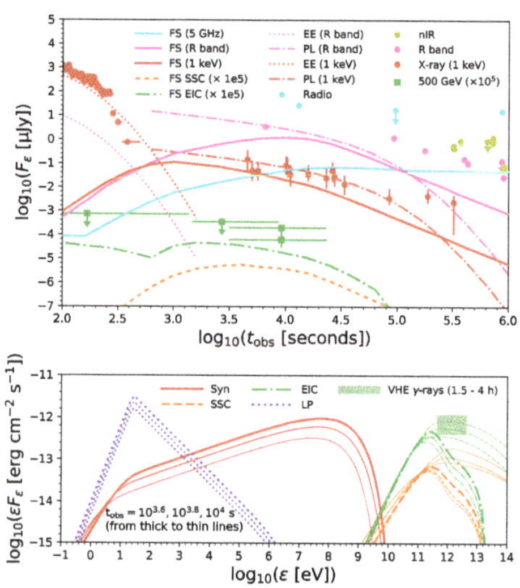

Figure 13. GRB 160821B: modeling of multi-wavelength light curves (top panel) and simultaneous SED (bottom panel) in radio, optical, X-ray and VHE band presented by [157]. ©AAS. Reproduced with permission. Two scenarios are considered: SSC (dashed orange lines) and EIC (dash-dotted green lines). In the top panel, the flux contribution in the different energy bands due to the forward shock (FS) both in the SSC and in the EIC scenario, the extended X-ray emission (EE) and the X-ray plateau emission (PL) are shown. In the bottom panel for the VHE band, both the intrinsic (thin lines) and the absorbed (thick lines) VHE curves are shown for the SSC and EIC scenarios for three different observed times. For the lower energies, the broken power-law of the late prompt emission (dashed violet line) and the expected synchrotron SEDs (solid red line) are displayed. The MAGIC observations (green symbols and green shaded area) are absorbed by EBL.

Table 1. GRB 160821B. List of the best fit parameters inferred from multi-wavelength modeling of the afterglow radiation by different authors.

	E_k erg	$\log(\epsilon_e)$	$\log(\epsilon_B)$	log(n) cm^{-3}	p	ξ_e	θ_j rad
MAGIC Coll.	10^{51}–10^{52}	$[-1; -0.1]$	$[-5.5; -0.8]$	$[-4.85; -0.24]$	2.2–2.35	1	/
Troja + 2019	10^{50}–10^{51}	$[-0.39; -0.05]$	$[-3.1; -1.1]$	$[-4.2; -1.7]$	2.26–2.39	1	0.08–0.50
Zhang + 2021 (SSC)	3×10^{51}	-0.52	-5	-1.3	2.3	0.5	0.15
Zhang + 2021 (EIC)	2×10^{51}	-0.3	-6	-1	2.5	0.1	0.1

4.2. GRB 180720B

GRB 180720B is a long GRB at $z = 0.654$ triggered on 20 July 2018 by the Fermi-GBM [158] at $T_0 = 14:21:39$ UT and 5 s later by the Swift-BAT instrument [159]. The H.E.S.S. telescopes observed and detected GRB 180720B about 11 h after the prompt emission, with a $\sim 5\sigma$ statistical significance.

4.2.1. General Properties and Multi-Wavelength Observations

The extremely bright prompt emission of this event, which is the seventh in brightness among the GRBs detected by the Fermi-GBM until then, lasted for $T_{90} = 48.9 \pm 0.4$ s and released an isotropic-equivalent energy $E_{\gamma,iso} = 6.0 \pm 0.1 \times 10^{53}$ erg in the 50–300 keV range.

Multi-wavelength afterglow observations covered the entire electromagnetic spectrum (see Figure 11). A significant signal was detected by Fermi-LAT from the trigger time up to 700 s, with the highest photon energy of 5 GeV detected 137 s after the burst trigger [160]. The Swift-XRT telescope observed and identified a bright afterglow starting from 90 s. This was still visible almost 30 days after the trigger time. The late-time light curve (from 2×10^3 s to 4×10^6 s) can be modeled with an initial power-law decay with an index $-1.19^{+0.01}_{-0.02}$ followed by a break at $t_{break} = 8 \times 10^4$ s to an index of $-1.55^{+0.04}_{-0.05}$ [5]. Several optical observations [161–170] revealed the presence of a counterpart and allowed us to estimate the redshift value of $z = 0.654$. The optical afterglow was observed to be slowly fading at an almost constant rate from around 10–11 h after the trigger time [171,172] as discussed in [173]. Radio observations (not shown in the figure) were also performed starting from ~ 1.7 days after the burst showing a steep power-law decaying emission [173].

4.2.2. VHE Observations and Results

The H.E.S.S. telescopes followed-up the event for ~ 2 h starting from 10.1 h, revealing the presence of a source with a 5.3σ pre-trial significance (5.0σ post-trial[6]). The observation was performed in standard dark and good weather conditions with a mean zenith angle of 31.5°. Another observation was performed under similar conditions 18 days after the previous one with results consistent with background events. The inferred flux at ~ 11 h and the flux upper limit at 18 d are shown in Figure 14 (red symbols).

The observed spectrum in the 0.1–0.44 TeV energy band has been fitted both with a power-law (panel on the left in Figure 15) and with a power-law with a cutoff $F_{int} = F_{0,int}\left(\frac{E}{E_0}\right)^{-\gamma_{int}}$, to describe an intrinsic power-law spectrum affected by the EBL attenuation $e^{-\tau(E,z)}$ (see Figure 15, panel on the right). With reference to the second fit, the analysis returns a photon index $\gamma_{int} = 1.6 \pm 1.2$ (statistical) ± 0.4 (systematic) and a flux normalization $F_{0,int} = (7.52 \pm 2.03$ (statistical) $^{+4.53}_{-3.84}$) (systematic) $\times 10^{-10}$ TeV^{-1} cm^{-2} s^{-1}, evaluated at an energy $E_{0,int} = 0.154$ TeV.

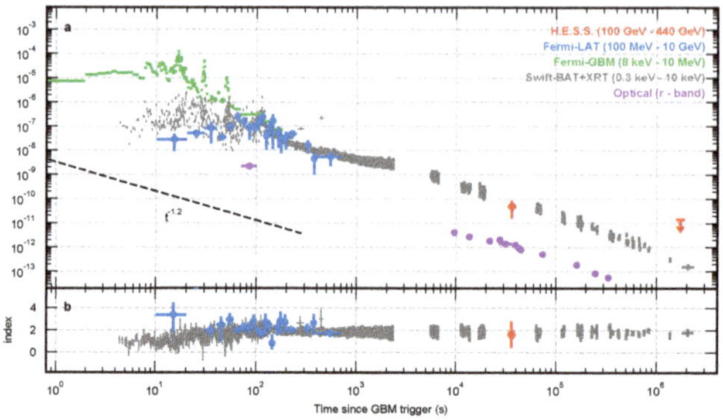

Figure 14. GRB 180720B: broad-band light curves and photon index evolution including optical, X-ray and VHE data. For the late time H.E.S.S. observation (18 days after the trigger) an UL has been derived. A flux temporal decay with index 1.2 (similar to what observed in optical and X-ray) is also shown in black dashed line. From [4].

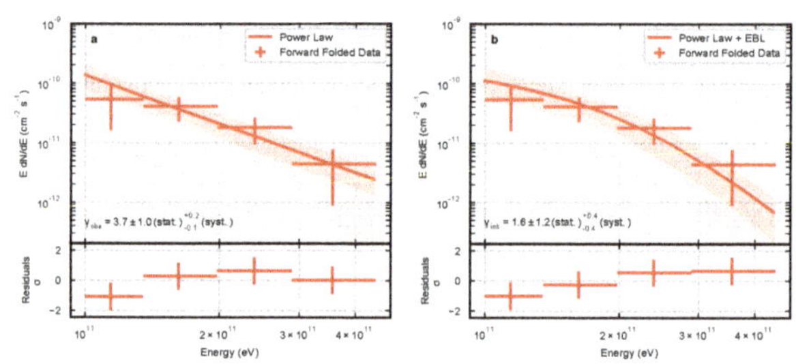

Figure 15. GRB 180720B: spectral fit to the observed emission in the 0.1–0.44 TeV energy range. In the left panel (**a**), the observed spectrum is assumed to be described by a power-law model, while in the panel on the right (**b**), the spectrum is an power-law attenuated by EBL. Statistical and systematic uncertainties at a 1σ confidence level are shown with shaded areas. From [4].

4.2.3. Interpretation

The H.E.S.S. Collaboration explored two possible radiation mechanisms to explain the VHE emission from GRB 180720B [4]: synchrotron emission and SSC radiation. A full modeling is not performed, and the discussion and comparison among the two different scenarios is based on estimates of the typical and maximal electron energy necessary in the two cases and on the comparison between spectral and temporal indices in different energy ranges. A synchrotron spectrum with a flat ($\alpha \sim -2$) slope extending from X-ray to VHE could model the emission with one single broad component and explain the similarity between the H.E.S.S., Fermi-LAT, and Swift-XRT luminosities, and the consistency among their photon index values. The large error on the VHE photon index, however, is not placing strong constraints, leaving open both the possibility of a consistency with the extrapolation of the synchrotron spectrum but also the possibility of a spectral hardening, indicative of a second component. A synchrotron origin of 10^2 GeV photons would require to find a process able to accelerate electrons up to PeV energies, which is in excess of the maximum electron energy achievable in external shocks (for a discussion, see Section 2.3.1). Adopting the standard Bhom limit, >100 GeV emission 10 h after the burst would require a huge bulk Lorentz factor $\Gamma \sim 1000$, which at these late times is really unlikely. As a result, these strong requirements disfavor the synchrotron emission as responsible of the VHE component in GRB 180720B.

The SSC scattering, on the contrary, arises as a natural candidate. A full broad-band modeling of GRB 180720B data in this scenario is presented in [174]. A numerical code reproducing the synchrotron and SSC emission in the afterglow shocks has been used (see [175]). The resulting light-curves and SED in the H.E.S.S. observational time window are shown in Figure 16. The full emission is explained as afterglow forward shock radiation (except for the initial peak in the optical and X-ray curves at $t \sim 10^2$ s, which is attributed to reverse shock emission). In the case of a constant-density ISM environment, the parameters that best reproduce the data are: $E_k = 10^{54}$ erg, $n = 0.1\,\mathrm{cm}^{-3}$, $\epsilon_e = 0.1$, $\epsilon_B = 10^{-4}$, $\Gamma_0 = 300$ and $p = 2.4$. As it can be noticed, the equipartition factor ϵ_B needs to assume quite a low value in order to explain the observations. A stellar wind-like environment is discarded by the authors on the basis of the comparison between the expected flux at ~1–10 GeV, following the prescriptions of [48] ($\gtrsim 2 \times 10^{-6}$ erg cm^{-2} s^{-1} at $t \approx 100$ s) and the one observed by Fermi-LAT (~10^{-8} erg cm^{-2} s^{-1} at $t \approx 100$ s). A low magnetic field equipartition factor is derived from the condition $E_{KN} \gtrsim 0.44$ TeV at $t \approx 10$ h, where E_{KN} is the energy at which the KN scattering becomes relevant. A transition energy between the synchrotron and SSC component of ~1 GeV is derived. Such value falls into the Fermi-LAT energy range and is compatible with a hardening of the spectrum in the VHE band. However, since the Fermi-LAT sensitivity is above the predicted flux of GRB 180720B at 10 h in the GeV band, the data cannot firmly confirm the presence of this transition.

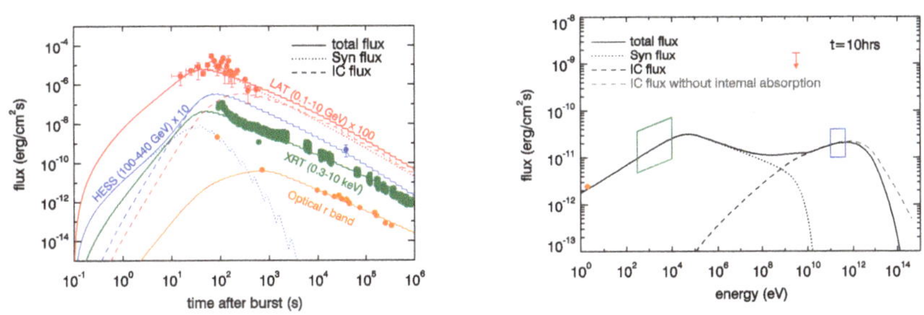

Figure 16. GRB 180720B: modeling of the broad-band light curves (**left** panel) and SED at the time of the H.E.S.S. detection (**right** panel) proposed by [174]. Both the synchrotron and the SSC contribution to the total flux are shown (see legend). In the SED, X-ray and H.E.S.S. data are shown, respectively, with the green and the blue boxes.

4.3. GRB 190114C

GRB 190114C is a long GRB at redshift $z = 0.42$ triggered by the Swift-BAT [176] on 14 January 2019 at $T_0 = 20:57:03$ UT, and by the Fermi-GBM [177]. The event was detected also by several other space γ-ray instruments such as AGILE, INTEGRAL/SPI-ACS, Insight/HXMT, and Konus-Wind [71]. MAGIC detected GRB 190114C starting ~60 s after T_0 with a significance above 50σ.

4.3.1. General Properties and Multi-Wavelength Observations

The duration of the prompt emission is $T_{90} \approx 116$ s as measured by Fermi-GBM and $T_{90} \approx 362$ s by Swift-BAT. However, the prompt light curve showed a multi-peak structure only for about 25 s, suggesting that the remaining activity, which is characterized by a smooth power-law decay and recorded by these instruments, may already be the afterglow emission. Support to such an interpretation is also obtained from a joint spectral and temporal analysis of the Fermi-GBM and Fermi-LAT data [178]. The total radiated prompt energy is $E_{\gamma,iso} = (2.5 \pm 0.1) \times 10^{53}$ erg in the energy range 1–10^4 keV [179].

Extensive follow-up observations from several different instruments from GeV to radio are available. Light-curves are shown in Figure 17. Fermi-LAT observations started since the beginning of the prompt phase. A GeV counterpart was detected from T_0 to 150 s, when the burst left the LAT field of view and remained outside it until 8600 s. When LAT resumed observations, a significant signal was still detected at a flux level $\sim 2 \times 10^{-10}$ erg cm^{-2} s^{-1} (0.1–1 GeV). After ~60 s from the burst trigger, Swift-XRT started follow-up observations, which covered in total $\sim 10^6$ s. The light-curve in the 1–10 keV energy band is consistent with a power-law decay $F \propto t^{\alpha}$ with $\alpha = -1.36 \pm 0.02$ [71]. NuSTAR and XMM-Newton observations are also available around 1–2 days. The NIR, optical and UV data were taken from around $\sim$100 s. The early emission is particularly bright and is interpreted as dominated by the reverse shock component [180]. Afterwards, the decay rate flattens and then steepens again after $\sim 3 \times 10^4$ s (see Figure 17). The Nordic Optical Telescope measured a redshift of $z = 0.4245 \pm 0.0005$ [181], which was then confirmed by Gran Telescopio Canarias [182]. Radio and sub-mm data were taken from $\sim 10^4$ s and exhibit an achromatic behavior, possibly dominated by the reverse shock in the sub-mm range, followed by emission at late times with nearly constant flux.

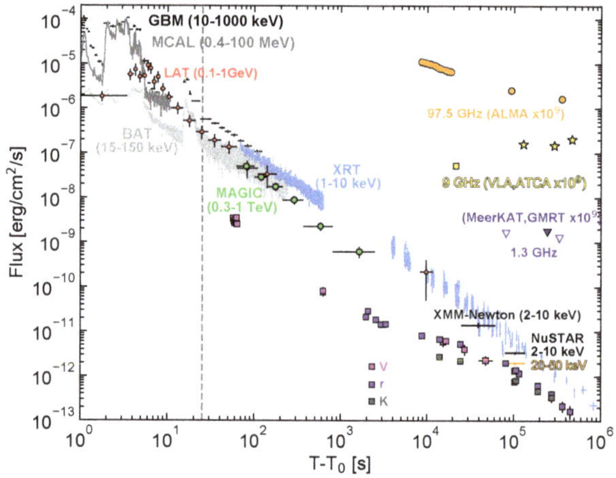

Figure 17. GRB 190114C: light-curves at different frequencies. From [71].

4.3.2. VHE Observations and Results

After receiving (at 22 s after the BAT trigger time) and validating (at 50 s) the GRB alert, the MAGIC telescopes started observing GRB 190114C at 57 s and operated stably from 62 s, starting from a zenith angle of 55.8°. Observations lasted until 15,912 s, when a zenith angle of 81.14° was reached. The observation was performed in good weather conditions but in presence of the moon, resulting in a night sky background approximately six times higher than the standard dark night conditions. The results of the offline analysis demonstrate a clear detection above the 50σ level in the first 20 min of observation [3].

The light-curve (see Figure 18, upper panel) for the intrinsic flux (i.e., corrected for the EBL absorption) in the 0.3–1 TeV range was derived starting from 62 s and up to 2454 s. The TeV light curve is well described by a power-law with temporal decay index $\beta_T = -1.60 \pm 0.07$, steeper than the one exhibited by the X-ray flux. The temporal evolution of the intrinsic spectral photon index α_{int} of the TeV differential photon spectrum is shown in the bottom panel. A constant value of $\alpha_{int} \approx -2$ is consistent with the data, considering the statistical and systematic errors, but there is evidence for a softening of the spectrum with time. The spectral fit in the 0.2–1 TeV energy range for the time-integrated emission (62–2454 s) returns $\alpha_{obs} = -5.34 \pm 0.22$ and $\alpha_{\rm int} = -2.22^{+0.23}_{-0.25}$ for the observed and EBL-corrected spectrum, respectively.

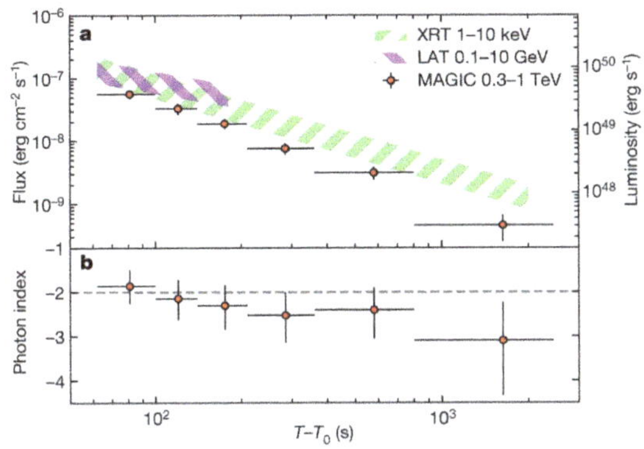

Figure 18. GRB 190114C. Upper panel: MAGIC light-curve (red circles), compared with the XRT (green band) and LAT (red band) emission. Bottom panel: temporal evolution of the intrinsic spectral photon index in the MAGIC data analysis time bins. From [3].

4.3.3. Interpretation

The properties of the VHE light curve and spectrum of GRB 190114C were studied by the MAGIC Collaboration in [3]. The PL behavior, the absence of variability, and the relatively long timescale of the emission support the evidence that the VHE component belongs to the afterglow phase. An estimate of the amount of radiated power in the TeV range can be derived, assuming that the afterglow onset is at ~6 s [178]. In this case, the energy radiated in the TeV band is ~10% of the isotropic-equivalent energy of the prompt emission $E_{\gamma,iso}$ considering the temporal evolution estimated from the MAGIC light-curve.

The energy of the photons observed by the MAGIC telescopes was compared with the maximum energy of synchrotron photons assuming two possible scenarios for the radial profile of the external density, namely constant and wind-like (Figure 19). These estimates of the maximum energy are based on the widely adopted limit on the maximum electron Lorentz factor set by equating the acceleration at Bhom rate with the synchrotron cooling rate (see Equation (57)). Adopting this limit, synchrotron emission can not account for the TeV photons detected by MAGIC, and a different radiation mechanism must be invoked.

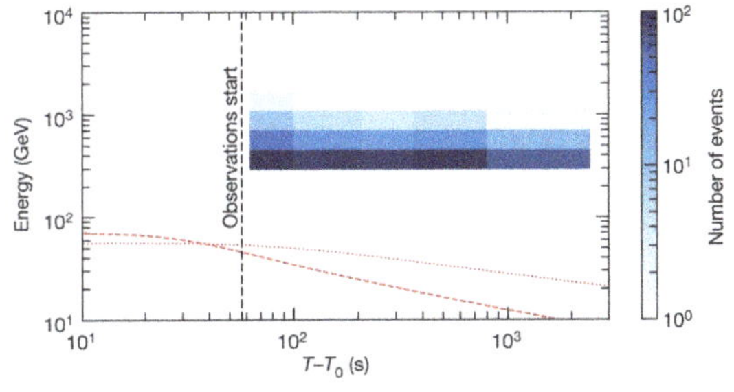

Figure 19. GRB 190114C: comparison between the distribution of the observed γ-ray events binned in time and energy (blue shaded areas) and the limiting curves for synchrotron maximum energy. Two different density radial profiles are considered for the derivation of the theoretical curves: constant (red-dotted curve) and wind-like (red-dashed curve). From [3].

An additional, model independent indication for the presence of a spectral component other than synchrotron is evident after multi-wavelength simultaneous SEDs are built. In [71], the VHE data were rebinned into five time intervals and XRT, BAT, GBM and LAT data were added when available, i.e., in the first two time intervals (Figure 20). The spectrum demonstrates a double-peaked behavior with a first peak in the X-ray band and the second one in the VHE band. The Fermi-LAT data play a particularly important role in revealing the shape of the SED, as they show a dip in the flux, strongly supporting an interpretation of the whole SED as a superposition of two distinct components.

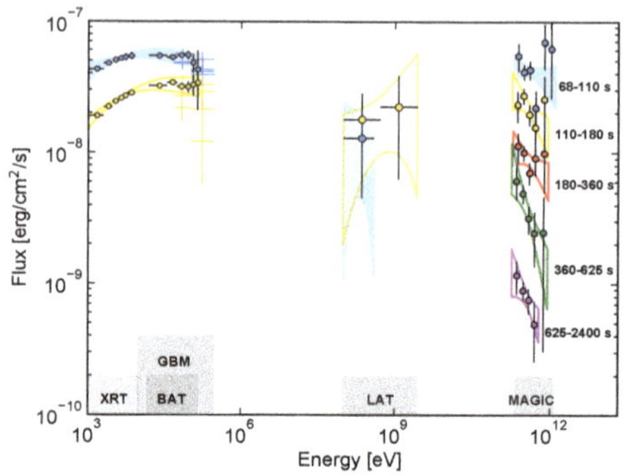

Figure 20. GRB 190114C SEDs from soft X-rays to 1 TeV in five different time intervals. The contour regions for flux uncertainties are also shown. For MAGIC and LAT the 1σ error of their best-fit power-law functions are displayed. For Swift data, the 90% confidence contours for the XRT-BAT joint fit is shown. MAGIC spectra are EBL-corrected assuming the model by [183]. From [71].

Following these considerations, in [71], the SSC mechanism is explored. The broad-band emission is modeled with a numerical code reproducing the synchrotron and SSC radiation in the external forward shock scenario, including the proper KN cross section and the effects of $\gamma - \gamma$ annihilation.

The predicted spectra and lightcurves are compared with the data in Figures 21 and 22. Acceptable modeling of the multi-wavelength afterglow spectra have been found for a constant medium with $E_k \gtrsim 3 \times 10^{53}$ erg, $\varepsilon_e \approx 0.05 - 0.15$, $\varepsilon_B \approx (0.05 - 1) \times 10^{-3}$, $n \approx 0.5 - 5\,\mathrm{cm}^{-3}$ and $p \approx 2.4 - 2.6$. It is found that, the peak of the SSC component being below 200 GeV, the KN suppression and the $\gamma - \gamma$ internal absorption have a non-negligible role in shaping the peak of the VHE spectrum. The modeling reproduces the XRT, LAT and TeV emission very well (solid blue curve in Figure 21 and solid blue, green and red curves in Figure 22), while it overproduces both the optical and radio flux at late times (solid violet, yellow and cyan curves in Figure 22). According to [71], a similar fit is found, also assuming a wind-like profile for the external density. In this case, the parameters are $E_k = 4 \times 10^{53}$ erg, $\varepsilon_e = 0.6$, $\varepsilon_B = 1 \times 10^{-4}$, $A_* = 0.1$ and $p = 2.4$. Very interestingly, the modeling shows that the late LAT observation (around 10^4 s) is completely dominated by SSC emission (red-dashed curve in Figure 22). A different type of modelization is also investigated by [71], under the requirement to model optical data. In this case (dotted curves in Figure 22), the fit is very good for optical, X-ray and LAT observations, but fails in reproducing the MAGIC light-curve.

The values inferred for the GRB afterglow parameters are similar to those used for past GRB afterglow studies at lower frequencies. This is an indication that the SSC component can be a relatively common process for GRB afterglows, since it does not require peculiar values of the parameters to be explained.

Several other successful modelings of GRB 190114C data within the synchrotron and SSC external forward shock scenario have been published in the literature [174,184–186]. A summary of the parameters inferred by different works can be found in Table 2.

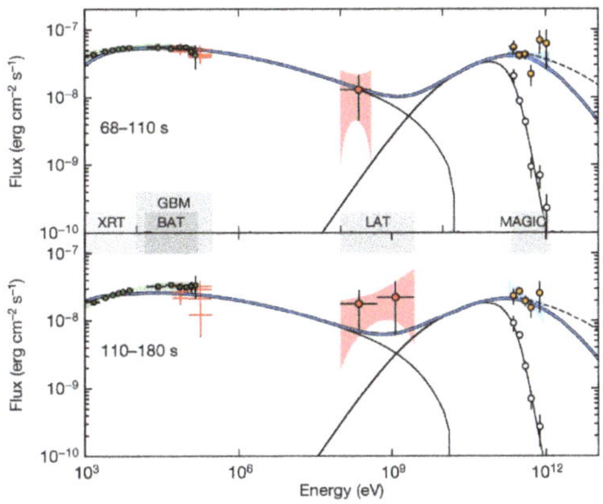

Figure 21. GRB 190114C: modeling of two SEDs in consecutive time intervals. The different curves refer to: the observed spectrum (thin solid line), the EBL deabsorbed spectrum (thick blue line) and the SSC component neglecting the effects of internal γ-γ opacity (dashed line). From [71].

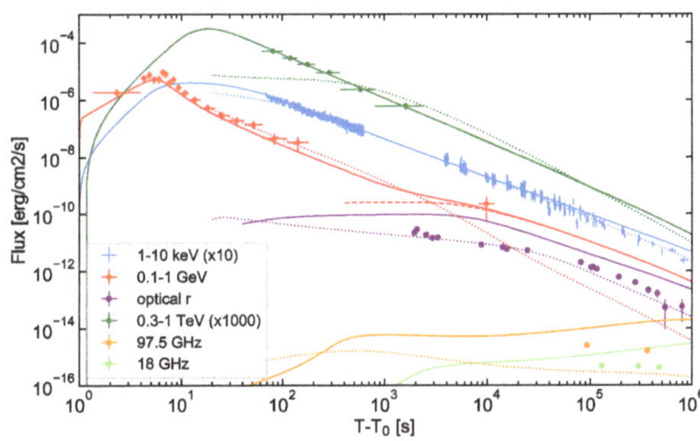

Figure 22. Modeling of broad band GRB 190114C light curves. Description of the different modelings used is given in Section 4.3.3. From [71].

In [174], the X-ray, optical and LAT data before 100 s are attributed to reverse shock emission or prompt contribution. A constant-density environment for the circumburst material is assumed. A time-averaged SED (50–150 s) is estimated, demonstrating that at GeV energies a transition between the synchrotron and SSC component can be identified. From the re-analysis of LAT data, a hard photon index (1.76 ± 0.21) is derived, in agreement with the hardening of the spectrum caused by the rising of the SSC component. Differently from what is observed by [71,184] $\gamma - \gamma$, the absorption does not contribute significantly in shaping the VHE spectrum.

A similar interpretation is given in [185], although the inferred value of ϵ_B is larger ($\epsilon_B \sim 10^{-3}$). A consistent modeling of the multi-wavelength observations as synchrotron and SSC radiation is found in both the ISM and wind-like scenarios. The SED at 80 s (see Figures 2 and 5 in [185]) and the broad-band light curves (Figures 3 and 6 in [185]) are reproduced, despite at 10^3 s, the model predictions in the 0.3–1 TeV band and X-rays are slightly brighter than the observed data.

In [186], analytical approximations are adopted for the description of the synchrotron and SSC components. In addition, the KN cut-off energy and the $\gamma - \gamma$ absorption contribution are calculated and compared with the data. A wind-like environment was used for the circumburst medium. The SEDs in the time intervals 68–110 s and 110–180 s are modeled, and the two values of the KN cut-off energies calculated at these times are

$\sim$3.7 TeV and 2.1 TeV. This implies that the KN effect is relevant only at TeV levels and the VHE data can be modeled assuming that the SSC scattering is in the Thompson regime. The $\gamma - \gamma$ absorption is also considered negligible since the estimated attenuation factor is way lower than the one due to EBL attenuation, and it reaches values around unity only for energies $\gtrsim$ 1 TeV.

In [184], the multi-wavelength data were fitted with a single-zone numerical code with an exact calculation of KN cross-sections as well as the attenuation due to the pair production mechanism. Smoothed analytical approximations for the electron injection function were used. A systematic scan over a 4-dimensional parameter space was performed to search for the best-fit solution at early and later times. The SED calculated at 90 s and 150 s (Figure 3 and Figure 9, respectively in [184]) are found to be well described by a fast cooling regime. The KN effect and the pair production mechanism shape the VHE spectrum significantly. It is estimated that $\simeq$10% of the total emitted power, i.e., $\simeq$25% of initially produced IC power, is absorbed.

Table 2. GRB 190114C: parameters inferred by different authors from the modeling of observations with a synchrotron-SSC scenario.

	E_k erg	ϵ_e	ϵ_B	n cm^{-3}	p	ξ_e
MAGIC Coll.	$\gtrsim 3 \times 10^{53}$	0.05–0.15	$0.05–1 \times 10^{-3}$	0.5–5	2.4–2.6	1
Wang + 2019	6×10^{53}	0.07	4×10^{-5}	0.3	2.5	1
Asano + 2020	10^{54}	0.06	9×10^{-4}	1	2.3	0.3
Asano + 2020	10^{54}	0.08	1.2×10^{-3}	0.1 (wind)	2.35	0.3
Joshi + 2021	4×10^{54}	0.03	0.012	2×10^{-2} (wind)	2.2	1
Derishev + 2021	3×10^{53}	0.1	$2–6 \times 10^{-3}$	2	2.5	1

4.4. GRB 190829A

GRB 190829A is a nearby ($z = 0.078$) long GRB triggered by Swift-BAT [187] and Fermi-GBM [188]. The Fermi-GBM trigger time is $T_0 = 19:55:53.13$ UTC. H.E.S.S. detected this GRB over three consecutive nights, with a significance of 21.7σ during the first night ($\sim$4 h after the GRB trigger).

4.4.1. General Properties and Multi-Wavelength Observations

The prompt emission detected by the two instruments consists of two episodes, with the first one observed in the time interval from the trigger time to 4 s and the second brighter episode from 47 s to 61 s. The two episodes have very different spectral properties: the first one is described by a power-law with index -1.41 ± 0.08 and an exponential high-energy cutoff function with $E_p = 130 \pm 20$ keV, the second one can be described with a Band function with $E_p = 11 \pm 1$ keV, $\alpha = -0.92 \pm 0.62$ and $\beta_T = -2.51 \pm 0.01$ [189]. The (isotropic equivalent) prompt emission energy inferred from the spectral analysis of Fermi-GBM data is $E_{\gamma,iso} \sim 2 \times 10^{50}$ erg.

A multi-wavelength observational campaign of the event was performed covering the entire electromagnetic spectrum. The event was not detected in the HE range by Fermi-LAT. Nevertheless, ULs have been reported in the MeV-GeV band up to 3×10^4 s [190]. The Swift-XRT started observations at 97.3 s and detected a bright X-ray afterglow, which was monitored until $\sim 7.8 \times 10^6$ s [191]. The X-ray light curve in the 0.3–10 keV energy range (observer frame) shows a peculiar behavior with an initial steep decay phase followed by a plateau and a strong flare episode (Figure 23, upper panel, blue points). After the flare, the standard afterglow phase starts with a decay following a power-law with a possible steepening around 10 days. In the UV/optical/NIR band, the event was followed by several instruments. The redshift was estimated to be $z = 0.0785 \pm 0.005$ [192], which makes this event one of the closest GRBs ever detected. Starting from 4.5–5.5 days after the GRB trigger, an associated supernova was reported [193]. Moreover, in the optical data, a flare is observed simultaneously with the one in the X-rays. In the radio band, the detection was reported by several instruments, starting from ~1 day after the trigger [194–197]. The radio flux initially slowly increases and then starts to decay after 20–30 days.

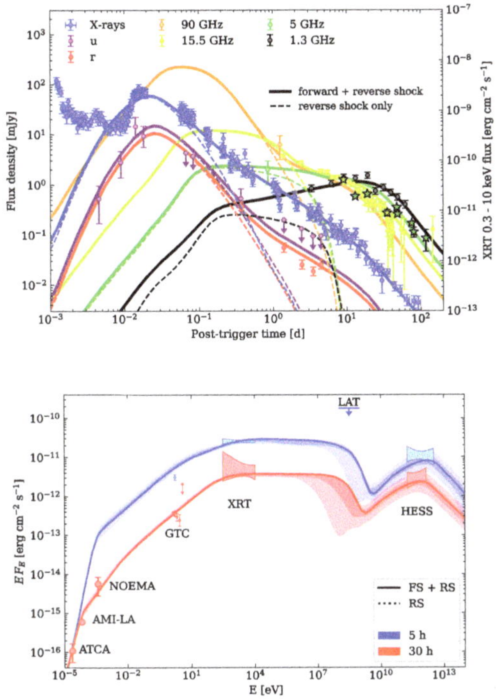

Figure 23. GRB 190829A: modeling of multi-wavelength light curves (**upper** panel) and SED (**bottom** panel) proposed by [72]. The 90% and 50% credible intervals from the fit are shown in lighter shades.

4.4.2. VHE Observations and Results

The afterglow emission of GRB 190829A was followed-up by the H.E.S.S. telescopes starting at 4.3 h and continued for three consecutive nights (up to 55.9 h). The observations were performed using the four medium-size telescopes of H.E.S.S. for a total amount of 13 h divided respectively in 3.6 h (starting at 4.3 h), 4.7 h (starting at 27.2 h) and 4.7 h (starting at 51.2 h). The statistical significance at the GRB position found in the three nights are, respectively, 21.7σ, 5.5σ and 2.4σ.

Spectral analysis was performed for the first two nights fitting the observed photon spectrum with a power-law model. The following values are found: $\alpha_{obs} = -2.59 \pm 0.09$ (stat.) ± 0.23 (syst.) in the 0.18–3.3 TeV (first night) and $\alpha_{obs} = -2.46 \pm 0.22$ (stat.) ± 0.14 (syst.) in the 0.18–1.4 TeV energy range (second night). Fitting a power-law attenuated by EBL, the photon indices inferred for the intrinsic spectrum are: $\alpha_{int} = -2.06 \pm 0.10$ (stat.) ± 0.26 (syst.) in the 0.18–3.3 TeV energy range (first night) and $\alpha_{int} = -1.86 \pm 0.26$ (stat.) ± 0.17 (syst.) in the 0.18–1.4 TeV energy range (second night). The photon indices in each night are consistent and within the systematical uncertainties with those of the simultaneous X-ray emission. Combining all three nights, the photon index is $\alpha_{int} = -2.07 \pm 0.09$ (stat.) ± 0.23 (syst.) in the 0.18–3.3 TeV energy range.

The light-curve in the 0.2–4.0 TeV energy range derived up to 56 h is compared in Figure 24 with the XRT light-curve and the LAT upper limits. The time-evolving flux was satisfactorily modeled with a power-law decay $F(t) \propto t^{\alpha}$ with $\alpha = -1.09 \pm 0.05$. Such a decay index is similar to the X-ray one derived in the same time interval ($\alpha_X = -1.07 \pm 0.09$).

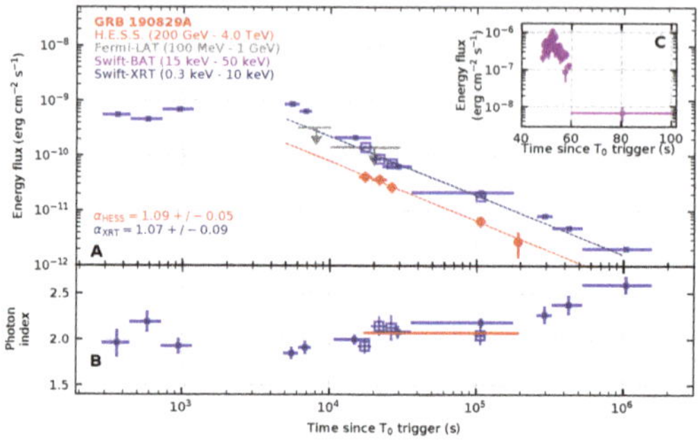

Figure 24. GRB 190829A: multi-wavelength light curves (**A**) (upper panel) and photon index evolution (bottom panel) in the X-ray, HE and VHE band (**B**). The BAT prompt light curve is shown in the inset (**C**). From [5].

4.4.3. Interpretation

The interpretation of the VHE emission from GRB 190829A is debated and different radiation mechanisms including synchrotron, SSC or EIC emission have been proposed so far to explain the origin of the TeV emission.

The H.E.S.S. Collaboration [5] investigated both the synchrotron and the SSC emission in the external forward shock as a responsible radiation mechanism of the observed TeV component. Multi-wavelength data collected simultaneously with H.E.S.S. observations in the first two nights were modeled separately with a time-independent numerical code using the Markov-chain Monte Carlo (MCMC) approach to explore the parameter space. The results of the fitting show that the SSC mechanism fails to explain the VHE emission. The low Lorentz bulk factor predicted by the observations ($\Gamma \lesssim 10$) implies that the SSC emission occurs in KN cross-scattering regime. As a result, a steep spectrum, inconsistent with the observational VHE data, is obtained (see Figure 25, light blue shaded area). Possible improvements between the data and the model foresee a higher Γ, which is in contrast with the observations, or the presence of an additional hard component in the distribution of the accelerated electrons. However, this latter solution implies extreme assumptions on the density of the circumburst medium ($n_0 = 10^{-5}$ cm^{-3} in the case of the strong magnetic field or $n_0 = 10^5$ cm^{-3} for weak magnetic field) and a SED strongly dominated by the SSC component, which is inconsistent with the data. A better fitting of the observational data can be obtained when considering an alternative model where the maximum electron energy set by the radiative losses is ignored. In such scenario, the synchrotron emission is able to extend up to TeV energies and the observational broad-band data are described by a single synchrotron component (see Figure 25, orange shaded area). The SSC contribution is negligible while the $\gamma - \gamma$ absorption shapes the VHE spectrum. The single synchrotron component scenario provides a better fit (>5σ) to the multi-wavelength data. On the other hand, this interpretation requires unknown acceleration processes or non-uniform magnetic field strength in the emission region, as described for GRB 180720B (see Section 4.2).

A complete multi-wavelength modeling of the GRB 190829A data, including contribution of the synchrotron and SSC emission for both the forward and reverse shocks, and considering a constant-density environment, is presented in [72]. The predicted broad-band light curves and the SED at the time of the H.E.S.S. detection are shown in Figure 23. A MCMC approach was adopted in order to estimate the best-fit parameters for the multi-wavelength modeling. The resulting values of the parameters related with the forward shock scenario are shown in Table 3. In contrast with the H.E.S.S. Collaboration results, the VHE emission is well reproduced with the SSC external forward shock scenario. The usual simplified assumption that $\xi_e = 1$ is excluded by the fit, which provides acceptable solutions only for $\xi_e \lesssim 6.5 \times 10^{-2}$. Moreover, an isotropic-equivalent kinetic energy at the afterglow onset $E_k = 2.5^{+1.9}_{-1.3} \times 10^{53}$ erg is estimated. Considering the observed GBM prompt energy, such a value implies that the prompt efficiency is $\eta = 1.2^{+1.0}_{-0.5} \times 10^{-3}$, which is much lower than the typical values derived from the previous GRB studies. The other parameters (n_0, ϵ_e and ϵ_B) are found to be similar to the ones estimated for GRB 190114C.

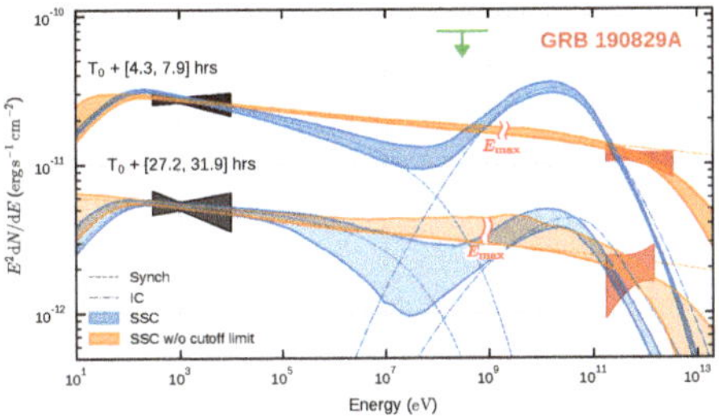

Figure 25. GRB 190829A: modeling of X-ray, LAT and H.E.S.S. data proposed by the H.E.S.S. collaboration for the two time intervals with VHE and X-ray detections. Two scenarios are investigated for the TeV emission: synchrotron and SSC. H.E.S.S. flux contours are displayed considering the statistical uncertainty. The synchrotron and SSC component are shown in dashed and dash-dotted lines, respectively. The shaded areas represent the 68% confidence intervals determined from the posterior probability distribution of the MCMC parameter fitting for the standard SSC model (light blue) and for the model without maximum energy for synchrotron emission (orange). From [5].

A two-component off-axis jet model has also been investigated [198]. Such a model proposes that the GRB jet is observed off-axis ($\theta_{view} = 1.78°$) and it consists of a narrow ($\theta_{jet} = 0.86°$) fast ($\Gamma = 350$) jet and a slow ($\Gamma = 20$) co-axial jet. The former jet component is responsible for the emission of SSC photons in the VHE band. The calculation of the SSC flux at the time of the H.E.S.S. detection is conducted following the prescriptions of [48], considering only the Thompson scattering regime.

An EIC plus SSC scenario has also been proposed for the production of the VHE component [199]. The seed photons belong to the long-lasting X-ray flare observed for GRB 190829A, which can be up-scattered to TeV energies. A numerical calculation of the afterglow dynamics and radiative processes have been used to model the observational data. For $t \sim 10^3$–10^4 s, the EIC component dominates the VHE emission, while for later times ($t \gtrsim 3 \times 10^4$ s) the EIC gradually decays and the SSC component becomes relevant. The initial afterglow kinetic energy used for the modeling ($E_k = 10^{52}$ erg) suggests that GRB 190829A is not a typical low-luminosity GRB but it may have much higher kinetic energy.

Table 3. Parameters for modeling of GRB 190829A.

	E_k erg	ϵ_e	ϵ_B	n cm^{-3}	p	ξ_e	θ_j rad
Hess Coll. (SSC)	2.0×10^{50}	0.91	5.9–7.7 $\times 10^{-2}$	1.	2.06–2.15	1.	/
Hess Coll. (Sync)	2.0×10^{50}	0.03–0.08	≈ 1	1.	2.1	1.	/
Salafia + 2021	1.2–4.4 $\times 10^{53}$	0.01–0.06	1.2–6.0 $\times 10^{-5}$	0.12–0.58	2.01	$<6.5 \times 10^{-2}$	0.25–0.29
Zhang + 2021	9.8×10^{51}	0.39	8.7×10^{-5}	0.09	2.1	0.34	0.1

4.5. GRB 201015A

GRB 201015A is a long GRB at $z = 0.426$ detected by the Swift-BAT [200] on 15 October 2020, at $T_0 = 22:50:13$ UT. The Fermi-GBM instrument did not trigger the event, but the targeted search revealed a transient source consistent with the Swift-BAT location [201]. MAGIC observations show a possible detection with a significance of $\sim 3.5\sigma$.

4.5.1. General Properties and Multi-Wavelength Observations

The (isotropic equivalent) prompt emission energy inferred from spectral analysis of Fermi-GBM data is $E_{\gamma,iso} = (1.1 \pm 0.2) \times 10^{50}$ erg [202]. The prompt duration is $T_{90} =$

9.78 ± 3.47 s (15–350 keV band). The BAT time-average spectrum in the time interval 0–10 s is well fitted by a power-law model with a photon index $\beta_T = -3.03 \pm 0.68$, suggesting a low peak energy $E_p < 10$ keV [203].

Swift-XRT [204] follow-up the event starting only 3214 s after T_0 due to observational constraints. The light curve up to almost 1 day is well described by a power-law with decay index $\alpha = -1.49^{+0.24}_{-0.21}$. Late-time observations performed by the Chandra X-ray Observatory [205] and Swift-XRT [206] from ~8 days up to ~21 days showed a flattening of the X-ray light curve, i.e., a flux level inconsistent (higher) with the extrapolation of the power-law decay rate at early times. Optical observations confirmed the presence of an afterglow counterpart from around 168 s [207]. The optical light-curves showed a clear initial rise with a peak around 200 s followed by a decay [208]. A bright radio counterpart (flux density $\sim 1.3 \times 10^{-4}$ Jy at 6 GHz 1.4 days after the burst) was also detected by several instruments [209–211]. Late-time optical observations identified an associated supernova rising from 5 days after the burst reaching its maximum flux around 12–20 days after T_0 [212,213]. The measurement of the redshift was reported by the GTC ($z = 0.426$) [214] and then confirmed by the NOT ($z = 0.423$) [215] instrument.

4.5.2. VHE Observations and Results

The final results from VHE data analysis of GRB 201015A have not been published yet. Preliminary information reported here have been released in [216,217]. Observations of GRB 201015A were performed by the MAGIC telescopes starting 33 s after the trigger time, under dark conditions, with a zenith angle ranging from 24° up to 48°, and lasted for about 4 h. In the second half of the data, taking the presence of passing clouds affected the observation for ~0.45 h. These data were removed and the remaining ones were analyzed with the standard MAGIC analysis software. Offline analysis showed a possible excess with a 3.5σ significance at the GRB position (see Figure 26) and a significant spot in the sky map. The energy threshold of the analysis is calculated to be 140 GeV from Monte Carlo simulated γ-ray data.

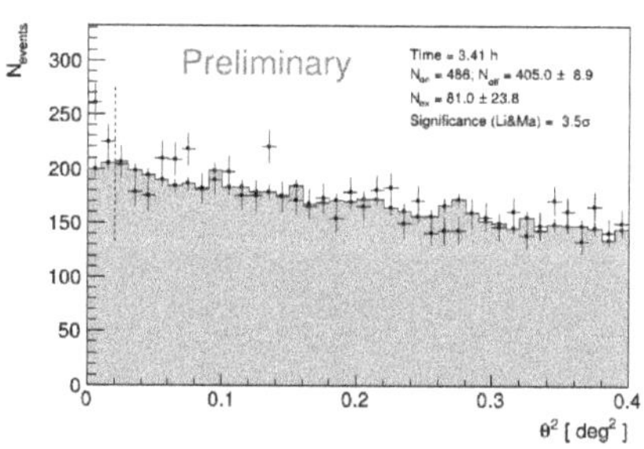

Figure 26. GRB 201015A: distribution of the angular distance θ^2 between the reconstructed event arrival directions and the nominal source position. The gray histogram represents background events while the black point with blue crosses are the γ-like events. The vertical dashed line describes the θ^2 cut value and defines the region in which excess events and signal significance are calculated. From [217].

4.6. GRB 201216C

GRB 201216C is a long GRB at $z = 1.1$ triggered by the Swift-BAT at $T_0 = 23:07:31$ UT on 16 December 2020 [218]. Fermi-GBM also detected the event with a slightly different trigger time (6 second before the Swift-BAT) [219]. MAGIC detected GRB 201216C with a significance of $\sim 6\sigma$.

4.6.1. General Properties and Milti-Wavelength Observations

The duration is estimated as $T_{90} = 48 \pm 16$ s in the 15–350 keV band by Swift-BAT [220] and around 29.9 s in the 50–300 keV band by Fermi-GBM. The light curve shows a multiple peak structure with a main peak around 20 s after the trigger time. The time-averaged GBM spectrum in the first 50 s is best fit by a Band function with $E_p = 326 \pm 7$ keV, $\alpha = -1.06 \pm 0.01$ and $\beta_T = -2.25 \pm 0.03$. The isotropic equivalent energy $E_{\gamma,iso}$ in the 10–1000 keV band is $(4.71 \pm 0.16) \times 10^{53}$ erg, as calculated from the fluence measured by Fermi-GBM.

Fermi-LAT observed the GRB starting from around 3500 s and up to 5500 s. No significant emission was reported [221]. Swift-XRT began the observation at $t = 2966.8$ s due to an observational constrain. A fading source was detected following a broken power-law behavior with decay indices of $1.97^{+0.10}_{-0.09}$ and $1.07^{+0.15}_{-0.10}$ and a break at 9078 s [222]. Optical observations were also performed by several instruments. The r-band light curve made along with VLT data point [223] and inferred data from FRAM-ORM [224] show a power-law decay in flux with the index equal to 1. The Liverpool Telescope observations, performed around 177 s after the trigger time, seems to be around the peak of the optical afterglow [225]. The HAWC observatory followed-up the event but no significant detection was identified in the TeV band [226]. Redshift estimation of $z = 1.1$ was performed by the ESO VLT [227].

4.6.2. VHE Observations and Results

Final results from VHE data analysis of GRB 201216C have not been published yet. Preliminary information reported here have been released in [228,229]. MAGIC observations and data taking of GRB 201216C started with a delay of 56 s after the Swift-BAT trigger time. The observation lasted for 2.2 h and was performed in optimal atmospheric condition and in absence of the moon. The zenith angle ranged from 37° to 68°. The low level of night sky background was also allowed to retain the low energy events and therefore obtain a low energy threshold compared to the other GRBs observed. To keep as many low-energy events as possible, an image cleaning method to extract dimmer Cherenkov showers initiated by gamma rays than the standard method was adopted.

The signal significance was calculated to be 6.0σ pre-trial (5.9σ post-trial[7]) for the first 20 min of observation (see Figure 27). A preliminary time-integrated spectrum for the first 20 min of observation was produced. Due to the strong absorption effect by EBL, a very steep power-law decay was found for the observed spectrum, especially for the events with energies higher than a few hundred of GeV. The intrinsic spectrum, corrected for the EBL absorption, was found to be consistent with a flat single power-law until 200 GeV above, in which no significant spectral points have been derived. A preliminary light curve in the time interval 56 s–2.2 h was also calculated. After 50 min, only upper limits on the emitted flux have been derived, as no significant emission was found after this time. The preliminary results are consistent with a monotonically decaying light curve fitted with a power-law.

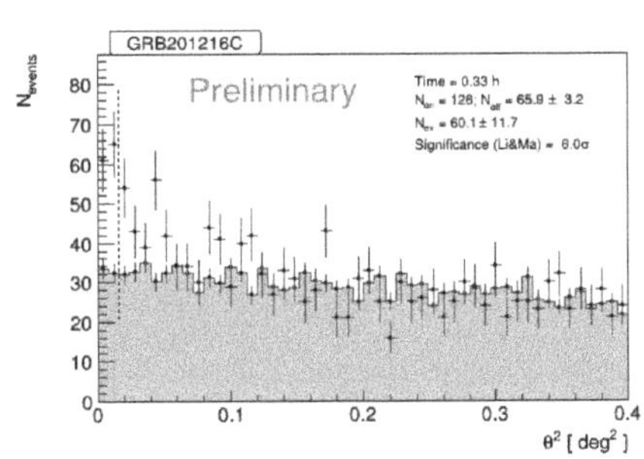

Figure 27. θ^2 angular distance distribution between reconstructed event arrival directions and nominal source position for GRB 201216C. The gray histogram represents background events while the black point with blue crosses are the γ-like events. The vertical dashed line describes the θ^2 cut value and defines the region in which excess events and signal significance are calculated. From [229].

5. The New TeV Spectral Window: Discussion

After decades of searches, MAGIC and H.E.S.S. observations have unequivocally proven that (long) GRBs can be accompanied by a significant amount of TeV emission during the afterglow phase. Table 4 summarizes the main properties of the GRBs detected by IACTs, and presented in detail in the previous section. The list also includes two events (namely GRB 160821B and GRB 201015A), where only a hint of excess (i.e., with a significance at $\sim$3–4 σ) was found. For the other four events, namely GRB 180720B, GRB 190114C, GRB 190829A and GRB 201216C, the detections are robust (>5σ). The table lists several properties, such as duration T_{90} and total emitted energy $E_{\gamma,iso}$ of the prompt emission, redshift, and information on the IACT detection (the starting time T_{delay} of observations elapsed since the trigger time T_0, the energy range where photons have been detected, the name of the telescope and the significance of the excess). GRB 160821B is the only one belonging to the short class; the other five being long GRBs.

In this section, we address the question why these GRBs have been detected, whether they have peculiar properties and whether they show some common behaviors that may be at the basis of the production of TeV radiation. To do that, one should be careful, since these GRBs have been followed-up under very different observational conditions and with very different time delays after the trigger time, and they span quite a large range of redshifts (from 0.078 to 1.1). Keeping in mind these differences, which have a strong impact on the detection capabilities of IACTs, we compare the observed and intrinsic properties of the population of GRBs at VHE, highlighting their similarities and differences, and discuss how they compare to the whole population.

Table 4. List of the GRBs observed by IACTs with a firm detection (significance > 5σ) or a hint of detection (3–4σ) above 100 GeV. The T_{90} and $E_{\gamma,iso}$ refer to the duration and total emitted energy of the prompt emission; the redshift is listed in column 3; T_{delay} is the time delay between the trigger time T_0 and the time when IACT observations started; E_{range} defines the energy range of the detected photons. The name of the telescope which made the observation and the significance of the detection are listed in the last column.

	T_{90} s	$E_{\gamma,iso}$ erg	z	T_{delay} s	E_{range} TeV	IACT (Sign.)
160821B	0.48	1.2×10^{49}	0.162	24	0.5–5	MAGIC (3.1σ)
180720B	48.9	6.0×10^{53}	0.654	3.64×10^4	0.1–0.44	H.E.S.S. (5.3σ)
190114C	362	2.5×10^{53}	0.424	57	0.3–1	MAGIC (>50σ)
190829A	58.2	2.0×10^{50}	0.079	1.55×10^4	0.18–3.3	H.E.S.S. (21.7σ)
201015A	9.78	1.1×10^{50}	0.42	33	0.14	MAGIC (3.5σ)
201216C	48	4.7×10^{53}	1.1	56	0.1	MAGIC (6.0σ)

5.1. Observing Conditions

Low zenith angles, fast repointing, dark nights, low redshift, and highly energetic events have always been considered as optimal, if not necessary, conditions to have some chances for GRB detections with IACTs. On the other hand, these first VHE GRBs have demonstrated that GRBs can have a level of TeV emission large enough to be detected by the current generation of IACTs, even under non-optimal conditions. GRB 190114C was observed with a zenith angle > 55° and in the presence of the moon. Both conditions imply a higher energy threshold (typically $\gtrsim$ 0.2 TeV) and require a dedicated data analysis. Another example is GRB 160821B, that was observed with a NSB 2–8 times higher than the standard dark night conditions. Moreover, significant VHE excess was found not only in case of short delays (less than hundreds of seconds) from the burst trigger but, somewhat surprisingly, also at quite late times, i.e., with delays of several hours or even days, as in the case of GRB 180720B and GRB 190829A, respectively. This showed the importance of pointing a GRB also at relatively late times, in cases fast follow-up observations are not feasible.

Optimal observing conditions and short delays remain, however, crucial to detecting GRBs at a higher redshift, for which the impact of EBL is already large at a few hundreds GeV. This explains how the detection of a GRB has been possible at $z = 1.1$ (GRB 201216C): in this case, optimal observing conditions allowed us to reach a low energy threshold of the sensitivity window (~70 GeV). The excess of signal was indeed found only below 200 GeV (more precisely, between 70 and 200 GeV) where the attenuation by the EBL is still limited.

5.2. Redshift and the Impact of EBL

The redshift of the detected GRBs covers a broad range, from $z = 0.079$ (GRB 190829A) to $z = 1.1$ (GRB 201216C). The impact of the EBL attenuation on the spectrum is severely changing, depending on the redshift value and on the photon energy. For redshift $z \sim 0.4$, the impact becomes relevant for energies $\gtrsim 0.2$ TeV with a flux attenuation of ~50% for 0.2 TeV and almost ~99.5% for 1 TeV [183]. For nearby events ($z \lesssim 0.1$–0.2), the effect of EBL is less severe and becomes relevant only for energies $\gtrsim 0.3$ TeV, reaching an attenuation factor of an order of magnitude only for energies $\gtrsim 2$ TeV. As a result, the GRB observed photon indices and the energy range of detected TeV photons differs significantly between the events. GRBs with redshift $z > 0.4$ such as GRB 190114C or GRB 180720B have very steep photon indices and they are detected in the lower energy range up to 0.44 TeV for GRB 180720B and 1.0 TeV for GRB 190114C. Spectral analysis from GRB 201216C are not yet public but preliminary results indicate that the emission is concentrated in the lower energy band between 0.1–0.2 TeV. Nearby GRBs with redshift $z \lesssim 0.1$–0.2 such as GRB 160821B or GRB 190829A show a less steep photon spectrum (around −2.5) and the TeV detection range extends above 1 TeV. The detection of several GRBs with significant value of redshift ($z > 0.4$) is robust proof that IACTs can overcome the limitations due to the EBL absorption and can expand the VHE detection horizon at the current stage up to $z = 1.1$. On the other hand, it is evident that detection of nearby GRBs is fundamental in order to more robustly explore the spectral shape, unbiased by the EBL effect, which is a non-negligible source of uncertainty for higher redshifts.

5.3. Energetics

In terms of $E_{\gamma,iso}$, the VHE GRB sample spans more than three orders of magnitude from $\sim 10^{49}$ erg up to $\sim 6 \times 10^{53}$ erg. The five long GRBs detected follows the Amati correlation, as shown in Figure 28. GRB 160821B, the only short GRB of the sample, is consistent with the existence of a possible Amati-like correlation for short GRBs, with this event falling in the weak-soft part of the correlation. The detections of GRB 190829A and GRB 201015A show that an event does not need to be extremely energetic in terms of isotropic-equivalent prompt energy in order to produce a TeV emission with (intrinsic) luminosity comparable to the X-ray luminosity. As a result, sources with $E_{\gamma,iso} \sim 10^{50-51}$ erg are not excluded as possible TeV emitters, even though their detection is possible only for relatively low redshift. This reduces the available volume, and hence the detection rate of similar events. In any case, this is relevant also for short GRBs which are less energetic than long ones, with typical isotropic energies falling within the $\sim 10^{49-52}$ erg range [144].

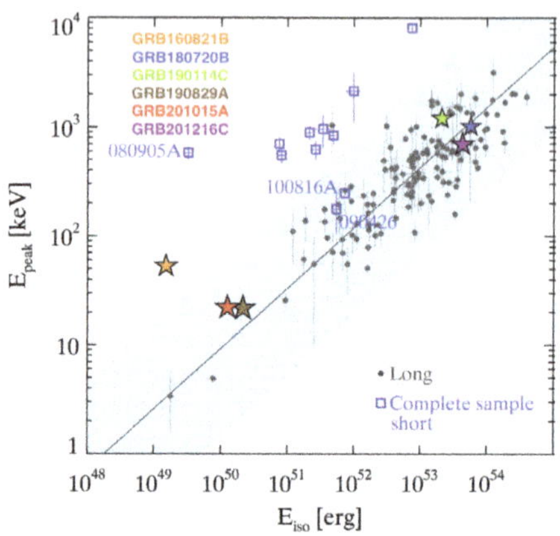

Figure 28. The Amati (E_{peak}–E_{iso}) correlation for a sample of 136 long GRBs (grey dots, from [230]) and a sample of 11 short GRBs (empty blue squares) detected by Swift. The corresponding power-law fit for the sample of long GRBs and the 3σ scatter of the distribution of points around the best fits are shown. The six GRBs detected in VHE are also added in the plot. Adapted from [231].

5.4. X-ray Lightcurves

The comparison between TeV and X-ray light-curves suggests an intimate connection between the emission in these two bands, both in terms of emitted energy and luminosity decay rate. In Figure 29, the XRT afterglow light-curves (luminosity versus rest-frame time) in the 0.3–10 keV energy range are compared with the VHE light-curves (integrated over different energy ranges, depending on the detection window, see Table 4). Different colors refer to the six different GRBs. The VHE luminosity is shown with empty circles.

Considering the X-ray luminosity, the GRB sample can be divided into two groups: GRB 190114C, GRB 180720B and GRB 201216C display large and clustered X-ray luminosity (at $t \sim 10^4$ s their luminosity is around 1–5 $\sim 10^{47}$ erg s^{-1}) and their light curves almost overlap for the entire afterglow phase. The other three GRBs (GRB 190829A, GRB 201015A and GRB 160821B) are much fainter in terms of X-ray luminosity (at least two orders of magnitude at $t \sim 10^4$ s). This is consistent with the fact that they also have a smaller $E_{\gamma,iso}$. The correlation between X-ray afterglow luminosity and prompt $E_{\gamma,iso}$ is found in the bulk of the long GRB population, and these GRBs make no exception.

Observations in the VHE band (empty circles in Figure 29) reveal that the VHE luminosities observed in the afterglow phase are in general smaller but comparable to the simultaneous X-ray luminosity, implying that almost an equal amount of energy is emitted in the two energy bands. Any theory aimed at explaining the origin of the TeV radiation should explain the origin of these similarities. Concerning the decay rate, observations are still not conclusive. The decay rate of the TeV emission is available only for two events. For GRB 190829A, the temporal indices in X-ray and VHE are very similar, while for GRB 190114C, the VHE clearly decays faster than the X-ray emission.

For GRB 190114C at $t \sim 380$ s, the VHE luminosity L_{VHE} is $\sim$1.5–2.5 $\times 10^{48}$ erg s^{-1} and the X-ray one L_X is $\sim$0.6–1.0 $\times 10^{49}$ erg s^{-1}. As a result, the power radiated in the VHE band is about $\sim$25% of the X-ray one. Similarly for GRB 190829A at $t \sim 4.5$ h the VHE luminosity L_{VHE} is $\sim$4.0–8.5 $\times 10^{44}$ erg s^{-1}, which is around $\sim$15–20% of the corresponding X-ray one ($L_X \sim$ 2.0–5.0 $\times 10^{45}$ erg s^{-1}). For GRB 180720B at $t \sim 2 \times 10^4$ s the VHE luminosity L_{VHE} is $\sim$9 $\times 10^{47}$ erg s^{-1} and the X-ray one L_X is $\sim$1.5–2.5 $\times 10^{49}$ erg s^{-1}. In this case, the power radiated in the VHE band is around $\sim$35–60% of the X-ray one.

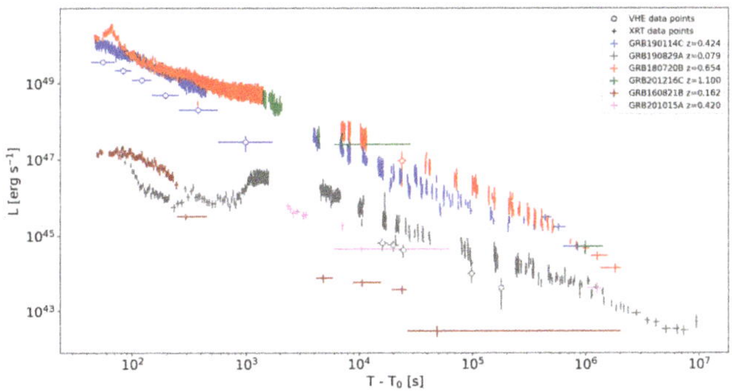

Figure 29. X-ray and VHE luminosity versus rest-frame time for the six GRBs detected in the TeV domain. The X-ray light-curve in the 0.3–10 keV energy range is taken from the Swift Burst Analyzer webpage [8].

5.5. The TeV Contribution to the Multi-Wavelength Modeling

Modeling of multi-wavelength afterglow data provide important insights concerning the GRB afterglow physics. In particular, the VHE data were crucial to investigate (i) the radiation mechanisms responsible for the production of photons between 10–100 GeV already detected by LAT; (ii) the environmental conditions at the GRB site; (iii) the free parameters which describe the shock micro-physics, and in particular, the self-generated magnetic field.

The modelings proposed so far in literature to explain the VHE component have considered two different radiation processes at the origin of the TeV emission: SSC and synchrotron. In the first case, the VHE emission is interpreted as a distinct spectral component from the synchrotron radiation dominating from radio to $\sim$GeV energies, which provides the seed photons that are upscattered at higher energies by the same electron population. In the second scenario, the VHE emission is observed as the extension of the synchrotron spectrum up to TeV energies.

In principle, a simultaneous SED covering the X-ray, HE and VHE range should be sufficient to discriminate among these two different possibilities. A hardening of the spectrum from GeV to TeV energies should be the smoking gun for the presence of a distinct component. In reality, the uncertainties in the spectral slope at VHE (caused by the uncertainty on the EBL and on the narrow energy range of TeV detection) can make the distinction hard to perform. In this case, LAT observations are of paramount importance to reveal the presence of a dip in the SED, which would also prove the need to invoke a different origin for the VHE emission. This seems to be the case for GRB 190114C, for which, at least in one SED, the LAT flux strongly suggests a dip in the GeV flux and hence the presence of the characteristic double bump observed for a synchrotron-SSC emission (Figure 20). For GRB 190829A, LAT provides only an upper limit, which is not constraining for modeling the shape of the SED (Figure 25). In this GRB, an interpretation of the whole SED in terms of synchrotron radiation cannot be excluded, although a modeling as SSC radiation has been proven to be successful [72] (Figure 23). For the other events detected at VHE, either the data do not allow for building a proper SED with simultaneous multi-wavelength observations, or they are not yet public. Despite this, the SSC emission seems to be the most viable mechanism able to explain the TeV data. A firm conclusion on the responsible radiation mechanism has not been reached yet and future detections will be crucial for deeper investigations.

Assuming one of the two scenarios, TeV data coupled with broad band observations at lower energies can be exploited to give additional information on the details of the afterglow external forward shock scenario.

Concerning the shock micro-physics, several modelings have suggested the possibility that the fraction of electrons accelerated in non-thermal distribution, ξ_e, is different from the standard value of one, which is usually assumed. In a few GRBs' modelings, namely for GRB 190114C [185] and GRB 190829A [72,198], the introduction of a $\xi_e < 1$ was essential in order to consistently fit the observational data. In particular in [72], the requirement for a low value of $\xi_e \lesssim 6.5 \times 10^{-2}$ was required to provide an acceptable fit of the data. The other modelings assume a greater value of ξ_e, around $\sim$0.3. Further detections will be exploited in order to verify if such an indication could be present also in other events.

Some considerations can also be drawn for the equipartition parameters ϵ_e and ϵ_B. These values, especially the former one, are usually well unconstrained and can span several orders of magnitudes. The TeV modeling described so far suggest that around $\sim$10% of the energy is given to the electrons, while a lower value (from 10^{-5} to 10^{-3}) is given to the magnetic field. larger values of ϵ_B such as 0.1–0.01, which are considered in an external shock scenario, are excluded. Moreover, some results can also be interpreted as an indication of an evolution in time of these parameters. In Figure 22, the modeling of the broad band light curves of GRB 190114C is shown. Two different modeling are presented: one optimized for the early time X-ray, HE and VHE observations (solid line) and one optimized for the late time lower energy bands (dotted line). This is due to the fact that the model that reproduces the early time data over-predicts the late time optical and radio observations. This result points towards the possibility that some of the fixed parameters of the afterglow theory (e.g., the electron and magnetic field equipartition parameters) may evolve in time. A further clue of the presence of time-dependent shock micro-physics parameters is derived from the low frequency multi wavelength modeling of GRB 190114C presented in [121]. In order to model the optical and the radio data, it is required that the micro-physical parameters evolve with time as $\epsilon_e \propto t^{-0.4}$ and $\epsilon_B \propto t^{0.1}$ in the ISM case and $\epsilon_B \propto t^{0.76}$ for the stellar wind scenario.

An issue that still is not solved by TeV observations is the discrimination between constant and wind-like profile for the ambient density. It is expected that long GRBs occur in wind-like environments. Nevertheless, at the current stage there seems to be no preference between such an environment and a constant ISM one, which is able to well reproduce the observational data. Therefore conclusive answers on the topic cannot be drawn yet.

In conclusion, the current population of GRBs at VHE already demonstrate quite broad properties, spanning more than three orders of magnitude in $E_{\gamma,iso}$ and more than two orders of magnitude in terms of afterglow luminosity and ranging in redshift between 0.079–1.1. The afterglow X-ray and VHE emission have comparable fluxes and decay slopes. The afterglow emitted power in the VHE band seems to constitute from 15% up to 60% of the X-ray one. Data modeling suggest that the responsible VHE radiation mechanism is the SSC emission, although different mechanisms (e.g., synchrotron radiation, EIC) cannot be completely excluded and a conclusive answer cannot be given yet. Multi-wavelength modeling show no preferences concerning the GRB environments between an ISM or wind-like scenario and indicate that shock micro-physics parameters, which seem to be able to reproduce VHE emission, are $\varepsilon_e \sim 0.1$ and $\varepsilon_B \sim 10^{-5}$–$10^{-3}$. Such features can be an indication of the universality of TeV emission in GRBs. It is then expected that a larger sample of GRBs than the current one will be detected in the VHE band, including also short GRBs for which, at the current stage, there are no confirmed detections except for the hint of excesses observed for GRB 160821B.

6. Conclusions and Future Prospects

The recent discoveries performed by the current generation of Cherenkov telescopes in the VHE band have opened a new observational spectral window on GRBs. The presence of a TeV afterglow component has been unequivocally proven and the studies on the currently available samples have demonstrated the potential that such detections have in probing several long-standing open questions in the GRB field. These first studies have focused on the identification of the responsible radiation mechanism, which is the first issue to address, and the comparison of the energetics, luminosity, and temporal behavior of the VHE component with respect to emission at lower frequencies. A modeling of multi-wavelength data covering from radio up to TeV energies was performed, giving interesting insights on the shock micro-physics conditions.

Limitations to the robust use of VHE data for afterglow modeling are imposed by the severe modification of the intrinsic spectrum cased by the energy-dependent flux-attenuation induced by EBL. GRBs with redshift $z > 0.4$, four out of six in the current VHE sample, are strongly affected by EBL absorption starting from hundreds of GeV. This implies large uncertainties on the shape and photon index of the intrinsic VHE spectrum. As a result, firm conclusions on the origin and spectral regime of the TeV component cannot be drawn yet. The low-energy extension of the range of sensitivity of IACTs is then fundamental for reaching a larger rate of detections and a more robust determination of the spectral index of the TeV component.

A full comprehension and exploitation of TeV data is expected to be reached thanks to the next generation of Cherenkov telescopes. The Cherenkov Telescope Array (CTA) will be a huge step forward for the detection of GRBs in the VHE band. The major upgrades with respect to the current generation of Cherenkov telescopes that will impact GRB observations are: (i) a lower energy threshold ($\lesssim$30 GeV), (ii) a larger effective area at multi-GeV energies ($\sim 10^4$ times larger than Fermi-LAT at 30 GeV) and (iii) a rapid slewing capability (180 degrees azimuthal rotation in 20 s). Moreover, its planned mixed-size array of large, medium and small size telescopes (called LST, MST and SST, respectively) situated at two sites in the northern and southern hemispheres will provide a full sky coverage from few tens of GeV up to hundreds of TeV. CTA will have a much better sensitivity and a broader energy range with respect to current ground-based facilities. A comparison is shown in Figure 30. At the present stage, the first prototype of the LSTs has been built and is operative under a commissioning phase[9] in the northern site at the Roque de los Muchachos Observatory in La Palma. Despite these performance improvements, the expected CTA detection rates of GRBs will be influenced anyway by the relatively low duty cycle affecting IACTs and by the synergies with other instruments. Indeed, Cherenkov telescopes' repointing relies on external triggers coming from space satellites. Assuming that currently operating space telescopes will be still operative, GRB alerts will be mostly provided by Swift-BAT and partially by the Fermi-GBM, and in the future by the French-Chinese mission Space-based multi-band astronomical Variable Objects Monitor (SVOM [232]).

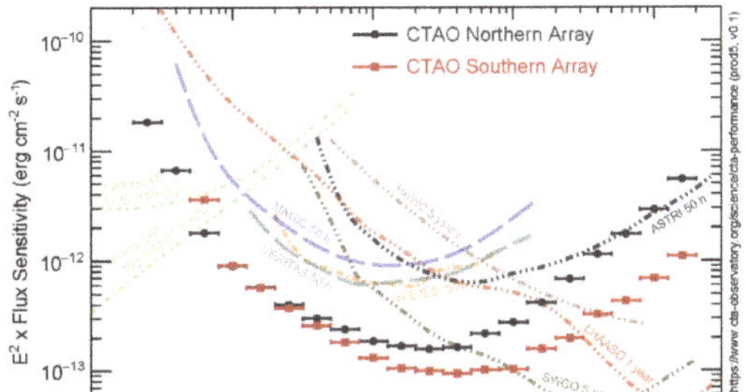

Figure 30. CTAO differential sensitivity[10] (defined as the minimum flux needed to obtain a 5-standard-deviation detection of a point-like source) for 50 h of observations with the Northern and Southern array compared to the sensitivity of several other Cherenkov telescopes and with Fermi-LAT (1 year).

BAT observes around 92 GRBs per year with a typical localization error of a few arcmin [233]. The good localization (later refined by XRT to a few arcsec) is fundamental for Cherenkov telescopes, given their limited field of view (e.g., about 4° for the LSTs and 7° for the MSTs). The GBM provides a much higher number of alerts, around 250 per year but with a larger localization error, from 1–3° up to 10°, which makes follow-up with IACTs very challenging. In the case of such large localization errors, CTA can exploit the so-called divergent mode for observations, which is currently under study [234]. In this pointing strategy, each telescope points to a position in the sky that is slightly offset to extend the field of view. Concerning future instruments, SVOM is expected to provide Swift-like alerts at a rate of $\sim$60–80 GRBs/yr with a localization error $<1^\circ$, including also 10 GRBs/yr with redshift $z < 1$.

Available estimates of the CTA detection rate of GRBs are reported in [235]. These studies were performed before the discovery of TeV emission. They are based on Swift-like alerts (triggered by Swift-BAT or SVOM) and Fermi-GBM alerts. The predicted detection rate is around a few GRBs per year, depending on the energy threshold of the observation and on the observation delay [235]. An updated study that considers current knowledge of TeV emission in the afterglow of GRBs is in progress [236].

Despite for decades GRBs' hunting by Cherenkov telescopes has been primarily focused on reaching low energy thresholds in order to explore the multi-GeV band, these first detections have shown that photons above TeV energies can be produced in GRBs and can be detected. This is mostly valid only for nearby events of redshift below 0.1–0.2, where EBL attenuation is not too severe. The exploration of the GRB emission component above 1 TeV can be of potential interest for SSTs and for the ASTRI Mini-Array. The ASTRI Mini-Array, currently under construction, will be an array of nine imaging atmospheric dual-mirror Cherenkov telescopes at the Teide Observatory site, expected to deliver the first scientific results in 2023. After the detection of GRB 190114C, the capabilities of the ASTRI Mini-Array in detecting and performing spectral studies of an event similar to the MAGIC GRB have been explored [237]. GRB 190114C has been taken as a template to simulate possible GRB emission from a few seconds to hours, and has been extrapolated to 10 TeV on the bases of model predictions.

The results demonstrate that the instrument will be able to detect afterglow TeV emission from an event such as GRB 190114C up to $\sim$ 200 s (see the comparison between the GRB observed flux at 1 TeV and the differential ASTRI Mini-Array sensitivity in Figure 31). By moving the GRB at a smaller redshift (down to $z = 0.078$, the redshift of the TeV GRB 190829A), the time for which the GRB is detectable increases up to $\sim 10^5$ (however, the light-curve in this case should be re-scaled by the lower energetics of nearby events). Nearby GRBs are then potential target of interests for the ASTRI Mini-Array. These are certainly rare events, but their detection will provide a wealth of information, with spectra that can be characterized up to several TeV [237].

In conclusion, after decades of huge efforts, current ground-based VHE facilities have started a new era in the comprehension and study of GRB physics. Their breakthrough detections allow unprecedented studies. As discussed in this review, many open questions in afterglow physics can largely benefit from the inclusion of TeV data. The first detections are providing glimpses of such a huge potential. Luckily, we are at the dawn of the VHE era thanks to the upcoming CTA observatory, which will assure major upgrades in sensitivity, energy range, temporal resolution, and sky coverage. Future observations, if complemented by simultaneous observations in X-rays and at $\sim$ GeV energies, will play a paramount role to improve our knowledge on the physics of GRB during the afterglow phase and hopefully also in the prompt phase. In particular, the afterglow SSC one-zone model will be tested to understand whether it can grasp the main properties of the VHE emission or if a revision of our comprehension on the particle acceleration processes, shock micro-physics and radiation mechanisms is needed.

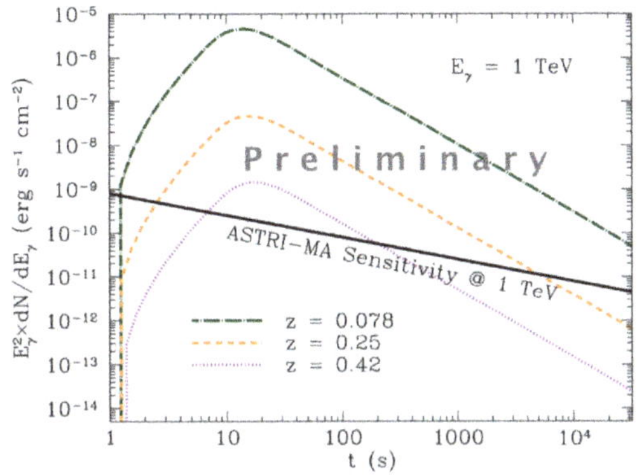

Figure 31. Light-curve of GRB 190114C at 1 TeV (dotted purple curve) compared to the sensitivity of the ASTRI Mini-Array. The yellow dashed and green dot-dashed curves show how GRB 190114C rescaled at redshift $z = 0.25$ and $z = 0.078$, respectively, would appear. From [237].

Author Contributions: All authors, D.M. and L.N., have contributed to writing. All authors have read and agreed to the published version of the manuscript.

Funding: This research received no external funding.

Data Availability Statement: This work made use of data supplied by the UK Swift Science Data Centre at the University of Leicester (https://www.swift.ac.uk/xrt_curves/).

Conflicts of Interest: The authors declare no conflict of interest.

Notes

1. https://magic.mpp.mpg.de/ (accessed on 27 February 2022).
2. https://www.mpi-hd.mpg.de/hfm/HESS/ (accessed on 27 February 2022).
3. https://www.cta-observatory.org (accessed on 27 February 2022).
4. A trial factor of 2 is considered due to the two sets of analysis cuts used for MAGIC data analysis.
5. https://www.swift.ac.uk/xrt_live_cat/00848890/ (accessed on 27 February 2022).
6. post-trial significance is estimated by accounting for the previously well-localized GRBs observed by H.E.S.S. in the same array configuration as of GRB 180720B.
7. See note 4 above.
8. https://www.swift.ac.uk/burst_analyser/ (accessed on 27 February 2022).
9. https://www.lst1.iac.es/ (accessed on 27 February 2022).
10. https://www.cta-observatory.org/science/ctao-performance/ (accessed on 27 February 2022).

References

1. Kouveliotou, C.; Meegan, C.A.; Fishman, G.J.; Bhat, N.P.; Briggs, M.S.; Koshut, T.M.; Paciesas, W.S.; Pendleton, G.N. Identification of Two Classes of Gamma-Ray Bursts. *Astrophys. J.* **1993**, *413*, L101. [CrossRef]
2. Mazets, E.P.; Golenetskii, S.V.; Ilinskii, V.N.; Panov, V.N.; Aptekar, R.L.; Gurian, I.A.; Proskura, M.P.; Sokolov, I.A.; Sokolova, Z.I.; Kharitonova, T.V. Catalog of cosmic gamma-ray bursts from the KONUS experiment data. *Astrophys. Space Sci.* **1981**, *80*, 3–83. [CrossRef]
3. MAGIC Collaboration; Acciari, V.A.; Ansoldi, S.; Antonelli, L.A.; Arbet Engels, A.; Baack, D.; Babić, A.; Banerjee, B.; Barres de Almeida, U.; Barrio, J.A.; et al. Teraelectronvolt emission from the γ-ray burst GRB 190114C. *Nature* **2019**, *575*, 455–458. [CrossRef]
4. Abdalla, H.; Adam, R.; Aharonian, F.; Ait Benkhali, F.; Angüner, E.O.; Arakawa, M.; Arcaro, C.; Armand, C.; Ashkar, H.; Backes, M.; et al. A very-high-energy component deep in the γ-ray burst afterglow. *Nature* **2019**, *575*, 464–467. [CrossRef]
5. HESS Collaboration; Abdalla, H.; Aharonian, F.; Ait Benkhali, F.; Angüner, E.O.; Arcaro, C.; Armand, C.; Armstrong, T.; Ashkar, H.; Backes, M.; et al. Revealing x-ray and gamma ray temporal and spectral similarities in the GRB 190829A afterglow. *Science* **2021**, *372*, 1081–1085. [CrossRef]
6. Rees, M.J.; Meszaros, P. Unsteady Outflow Models for Cosmological Gamma-Ray Bursts. *Astrophys. J.* **1994**, *430*, L93. [CrossRef]
7. Thompson, C. A model of gamma-ray bursts. *Mon. Not. RAS* **1994**, *270*, 480–498. [CrossRef]

8. Spruit, H.C.; Daigne, F.; Drenkhahn, G. Large scale magnetic fields and their dissipation in GRB fireballs *Astron. Astrophys.* **2001**, *369*, 694–705. [CrossRef]
9. Rees, M.J.; Meszaros, P. Relativistic fireballs—Energy conversion and time-scales. *Mon. Not. RAS* **1992**, *258*, 41. [CrossRef]
10. Chevalier, R.A.; Li, Z.Y. Gamma-Ray Burst Environments and Progenitors. *Astrophys. J.* **1999**, *520*, L29–L32. [CrossRef]
11. Woosley, S.E. Gamma-Ray Bursts from Stellar Mass Accretion Disks around Black Holes. *Astrophys. J.* **1993**, *405*, 273. [CrossRef]
12. Paczyński, B. Are Gamma-Ray Bursts in Star-Forming Regions? *Astrophys. J.* **1998**, *494*, L45–L48. [CrossRef]
13. Abbott, B.P.; Abbott, R.; Abbott, T.D.; Acernese, F.; Ackley, K.; Adams, C.; Adams, T.; Addesso, P.E.A. GW170817: Observation of Gravitational Waves from a Binary Neutron Star Inspiral. *Phys. Rev. Lett.* **2017**, *119*, 161101. [CrossRef]
14. Usov, V.V. Millisecond pulsars with extremely strong magnetic fields as a cosmological source of γ-ray bursts. *Nature* **1992**, *357*, 472–474. [CrossRef]
15. Thompson, T.A.; Metzger, B.D.; Bucciantini, N. Proto-Magnetars as GRB Central Engines: Uncertainties, Limitations, & Particulars. In *Deciphering the Ancient Universe with Gamma-ray Bursts*; American Institute of Physics Conference Series; Kawai, N., Nagataki, S., Eds.; American Institute of Physics: College Park, MD, USA, 2010; Volume 1279, pp. 81–88. [CrossRef]
16. Metzger, B.D.; Giannios, D.; Thompson, T.A.; Bucciantini, N.; Quataert, E. The protomagnetar model for gamma-ray bursts. *Mon. Not. RAS* **2011**, *413*, 2031–2056. [CrossRef]
17. Dall'Osso, S.; Stratta, G.; Guetta, D.; Covino, S.; De Cesare, G.; Stella, L. Gamma-ray bursts afterglows with energy injection from a spinning down neutron star. *Astron. Astrophys.* **2011**, *526*, A121. [CrossRef]
18. Giannios, D.; Spruit, H.C. Spectra of Poynting-flux powered GRB outflows. *Astron. Astrophys.* **2005**, *430*, 1–7. [CrossRef]
19. Zhang, B.; Yan, H. The Internal-collision-induced Magnetic Reconnection and Turbulence (ICMART) Model of Gamma-ray Bursts. *Astrophys. J.* **2011**, *726*, 90. [CrossRef]
20. Lazarian, A.; Zhang, B.; Xu, S. Gamma-Ray Bursts Induced by Turbulent Reconnection. *Astrophys. J.* **2019**, *882*, 184. [CrossRef]
21. Lazzati, D.; Ghisellini, G.; Celotti, A. Constraints on the bulk Lorentz factor in the internal shock scenario for gamma-ray bursts. *Mon. Not. RAS* **1999**, *309*, L13–L17. [CrossRef]
22. Sironi, L.; Petropoulou, M.; Giannios, D. Relativistic jets shine through shocks or magnetic reconnection? *Mon. Not. RAS* **2015**, *450*, 183–191. [CrossRef]
23. Fan, Y.; Piran, T. Gamma-ray burst efficiency and possible physical processes shaping the early afterglow. *Mon. Not. RAS* **2006**, *369*, 197–206. [CrossRef]
24. Beniamini, P.; Nava, L.; Piran, T. A revised analysis of gamma-ray bursts' prompt efficiencies. *Mon. Not. RAS* **2016**, *461*, 51–59. [CrossRef]
25. Ghisellini, G.; Celotti, A.; Lazzati, D. Constraints on the emission mechanisms of gamma-ray bursts. *Mon. Not. RAS* **2000**, *313*, L1–L5. [CrossRef]
26. Preece, R.D.; Briggs, M.S.; Mallozzi, R.S.; Pendleton, G.N.; Paciesas, W.S.; Band, D.L. The Synchrotron Shock Model Confronts a "Line of Death" in the BATSE Gamma-Ray Burst Data. *Astrophys. J.* **1998**, *506*, L23–L26. [CrossRef]
27. Oganesyan, G.; Nava, L.; Ghirlanda, G.; Celotti, A. Detection of Low-energy Breaks in Gamma-Ray Burst Prompt Emission Spectra. *Astrophys. J.* **2017**, *846*, 137. [CrossRef]
28. Oganesyan, G.; Nava, L.; Ghirlanda, G.; Celotti, A. Characterization of gamma-ray burst prompt emission spectra down to soft X-rays. *Astron. Astrophys.* **2018**, *616*, A138. [CrossRef]
29. Oganesyan, G.; Nava, L.; Ghirlanda, G.; Melandri, A.; Celotti, A. Prompt optical emission as a signature of synchrotron radiation in gamma-ray bursts. *Astron. Astrophys.* **2019**, *628*, A59. [CrossRef]
30. Ravasio, M.E.; Oganesyan, G.; Ghirlanda, G.; Nava, L.; Ghisellini, G.; Pescalli, A.; Celotti, A. Consistency with synchrotron emission in the bright GRB 160625B observed by Fermi. *Astron. Astrophys.* **2018**, *613*, A16. [CrossRef]
31. Ravasio, M.E.; Ghirlanda, G.; Nava, L.; Ghisellini, G. Evidence of two spectral breaks in the prompt emission of gamma-ray bursts. *Astron. Astrophys.* **2019**, *625*, A60. [CrossRef]
32. Meszaros, P.; Rees, M.J. Relativistic Fireballs and Their Impact on External Matter: Models for Cosmological Gamma-Ray Bursts. *Astrophys. J.* **1993**, *405*, 278. [CrossRef]
33. Mészáros, P.; Rees, M.J. Optical and Long-Wavelength Afterglow from Gamma-Ray Bursts. *Astrophys. J.* **1997**, *476*, 232–237. [CrossRef]
34. Sari, R.; Piran, T. Predictions for the Very Early Afterglow and the Optical Flash. *Astrophys. J.* **1999**, *520*, 641–649. [CrossRef]
35. Blandford, R.D.; McKee, C.F. Fluid dynamics of relativistic blast waves. *Phys. Fluids* **1976**, *19*, 1130–1138. [CrossRef]
36. Sedov, L.I. Propagation of strong shock waves. *J. Appl. Math. Mech.* **1946**, *10*, 241–250.
37. Taylor, G. The Formation of a Blast Wave by a Very Intense Explosion. II. The Atomic Explosion of 1945. *Proc. R. Soc. Lond. Ser. A* **1950**, *201*, 175–186. [CrossRef]
38. Sari, R.; Piran, T.; Narayan, R. Spectra and Light Curves of Gamma-Ray Burst Afterglows. *Astrophys. J.* **1998**, *497*, L17–L20. [CrossRef]
39. Kobayashi, S.; Sari, R. Optical Flashes and Radio Flares in Gamma-Ray Burst Afterglow: Numerical Study. *Astrophys. J.* **2000**, *542*, 819–828. [CrossRef]
40. Vanthieghem, A.; Lemoine, M.; Plotnikov, I.; Grassi, A.; Grech, M.; Gremillet, L.; Pelletier, G. Physics and Phenomenology of Weakly Magnetized, Relativistic Astrophysical Shock Waves. *Galaxies* **2020**, *8*, 33. [CrossRef]

41. Sironi, L.; Spitkovsky, A. Particle Acceleration in Relativistic Magnetized Collisionless Electron-Ion Shocks. *Astrophys. J.* **2011**, *726*, 75. [CrossRef]
42. Sironi, L.; Spitkovsky, A.; Arons, J. The Maximum Energy of Accelerated Particles in Relativistic Collisionless Shocks. *Astrophys. J.* **2013**, *771*, 54. [CrossRef]
43. Keshet, U.; Katz, B.; Spitkovsky, A.; Waxman, E. Magnetic Field Evolution in Relativistic Unmagnetized Collisionless Shocks. *Astrophys. J.* **2009**, *693*, L127–L130. [CrossRef]
44. Kumar, P.; Zhang, B. The physics of gamma-ray bursts & relativistic jets. *Phys. Rep.* **2015**, *561*, 1–109. [CrossRef]
45. Zhang, B. *The Physics of Gamma-Ray Bursts*; Cambridge University Press: Cambridge, UK, 2018; [CrossRef]
46. Nava, L. High-energy emission from gamma-ray bursts. *Int. J. Mod. Phys. D* **2018**, *27*, 1842003. [CrossRef]
47. Panaitescu, A.; Kumar, P. Analytic Light Curves of Gamma-Ray Burst Afterglows: Homogeneous versus Wind External Media. *Astrophys. J.* **2000**, *543*, 66–76. [CrossRef]
48. Sari, R.; Esin, A.A. On the Synchrotron Self-Compton Emission from Relativistic Shocks and Its Implications for Gamma-Ray Burst Afterglows. *Astrophys. J.* **2001**, *548*, 787–799. [CrossRef]
49. Nousek, J.A.; Kouveliotou, C.; Grupe, D.; Page, K.L.; Granot, J.; Ramirez-Ruiz, E.; Patel, S.K.; Burrows, D.N.; Mangano, V.; Barthelmy, S.; et al. Evidence for a Canonical Gamma-Ray Burst Afterglow Light Curve in the Swift XRT Data. *Astrophys. J.* **2006**, *642*, 389–400. [CrossRef]
50. Nardini, M.; Ghisellini, G.; Ghirlanda, G.; Tavecchio, F.; Firmani, C.; Lazzati, D. Clustering of the optical-afterglow luminosities of long gamma-ray bursts. *Astron. Astrophys.* **2006**, *451*, 821–833. [CrossRef]
51. Zaninoni, E.; Bernardini, M.G.; Margutti, R.; Oates, S.; Chincarini, G. Gamma-ray burst optical light-curve zoo: Comparison with X-ray observations. *Astron. Astrophys.* **2013**, *557*, A12. [CrossRef]
52. Bernardini, M.G.; Margutti, R.; Mao, J.; Zaninoni, E.; Chincarini, G. The X-ray light curve of gamma-ray bursts: Clues to the central engine. *Astron. Astrophys.* **2012**, *539*, A3. [CrossRef]
53. Sari, R.; Piran, T. Hydrodynamic Timescales and Temporal Structure of Gamma-Ray Bursts. *Astrophys. J.* **1995**, *455*, L143. [CrossRef]
54. Granot, J.; Sari, R. The Shape of Spectral Breaks in Gamma-Ray Burst Afterglows. *Astrophys. J.* **2002**, *568*, 820–829. [CrossRef]
55. Nava, L.; Sironi, L.; Ghisellini, G.; Celotti, A.; Ghirlanda, G. Afterglow emission in gamma-ray bursts - I. Pair-enriched ambient medium and radiative blast waves. *Mon. Not. RAS* **2013**, *433*, 2107–2121. [CrossRef]
56. Zhang, B.; Mészáros, P. Gamma-Ray Burst Afterglow with Continuous Energy Injection: Signature of a Highly Magnetized Millisecond Pulsar. *Astrophys. J.* **2001**, *552*, L35–L38. [CrossRef]
57. Zhang, B.; Fan, Y.Z.; Dyks, J.; Kobayashi, S.; Mészáros, P.; Burrows, D.N.; Nousek, J.A.; Gehrels, N. Physical Processes Shaping Gamma-Ray Burst X-Ray Afterglow Light Curves: Theoretical Implications from the Swift X-Ray Telescope Observations. *Astrophys. J.* **2006**, *642*, 354–370. [CrossRef]
58. Laskar, T.; Berger, E.; Margutti, R.; Perley, D.; Zauderer, B.A.; Sari, R.; Fong, W.F. Energy Injection in Gamma-Ray Burst Afterglows. *Astrophys. J.* **2015**, *814*, 1. [CrossRef]
59. Shapiro, P.R. Relativistic blast waves that accelerate. *Astrophys. J.* **1980**, *236*, 958–969. [CrossRef]
60. Sironi, L.; Keshet, U.; Lemoine, M. Relativistic Shocks: Particle Acceleration and Magnetization. *Space Sci. Rev.* **2015**, *191*, 519–544. [CrossRef]
61. Keshet, U.; Waxman, E. Energy Spectrum of Particles Accelerated in Relativistic Collisionless Shocks. *Phys. Rev. Lett.* **2005**, *94*, 111102. [CrossRef]
62. Weibel, E.S. Spontaneously Growing Transverse Waves in a Plasma Due to an Anisotropic Velocity Distribution. *Phys. Rev. Lett.* **1959**, *2*, 83–84. [CrossRef]
63. Spitkovsky, A. Particle Acceleration in Relativistic Collisionless Shocks: Fermi Process at Last? *Astrophys. J.* **2008**, *682*, L5. [CrossRef]
64. Böhm, D. *The Characteristics of Electrical Discharges in Magnetic Fields*; Guthrie, A., Wakerling, R.K., Eds.; McGraw-Hill: New York, NY, USA, 1949.
65. Lemoine, M. Synchrotron signature of a relativistic blast wave with decaying microturbulence. *Mon. Not. RAS* **2013**, *428*, 845–866. [CrossRef]
66. Lemoine, M.; Li, Z.; Wang, X.Y. On the magnetization of gamma-ray burst blast waves. *Mon. Not. RAS* **2013**, *435*, 3009–3016. [CrossRef]
67. Santana, R.; Barniol Duran, R.; Kumar, P. Magnetic Fields in Relativistic Collisionless Shocks. *Astrophys. J.* **2014**, *785*, 29. [CrossRef]
68. Wang, X.G.; Zhang, B.; Liang, E.W.; Gao, H.; Li, L.; Deng, C.M.; Qin, S.M.; Tang, Q.W.; Kann, D.A.; Ryde, F.; et al. How Bad or Good Are the External Forward Shock Afterglow Models of Gamma-Ray Bursts? *Astrophys. J. Lett.* **2015**, *219*, 9. [CrossRef]
69. Zhang, B.B.; van Eerten, H.; Burrows, D.N.; Ryan, G.S.; Evans, P.A.; Racusin, J.L.; Troja, E.; MacFadyen, A. An Analysis of Chandra Deep Follow-up Gamma-Ray Bursts: Implications for Off-axis Jets. *Astrophys. J.* **2015**, *806*, 15. [CrossRef]
70. Beniamini, P.; Nava, L.; Duran, R.B.; Piran, T. Energies of GRB blast waves and prompt efficiencies as implied by modelling of X-ray and GeV afterglows. *Mon. Not. RAS* **2015**, *454*, 1073–1085. [CrossRef]
71. MAGIC Collaboration; Veres, P.; Bhat., P.N.; Briggs, M.S.; Cleveland, W.H.; Hamburg, R.; Hui, C.M.; Mailyan, B.; Preece, R.D.; Roberts, O.J.; et al. Observation of inverse Compton emission from a long γ-ray burst. *Nature* **2019**, *575*, 459–463. [CrossRef]

72. Salafia, O.S.; Ravasio, M.E.; Yang, J.; An, T.; Orienti, M.; Ghirlanda, G.; Nava, L.; Giroletti, M.; Mohan, P.; Spinelli, R.; et al. Multi-wavelength view of the close-by GRB~190829A sheds light on gamma-ray burst physics. *arXiv* **2021**, arXiv:2106.07169.
73. Eichler, D.; Waxman, E. The Efficiency of Electron Acceleration in Collisionless Shocks and Gamma-Ray Burst Energetics. *Astrophys. J.* **2005**, *627*, 861–867. [CrossRef]
74. Ackermann, M.; Ajello, M.; Asano, K.; Atwood, W.B.; Axelsson, M.; Baldini, L.; Ballet, J.; Barbiellini, G.; Baring, M.G.; Bastieri, D.; et al. Fermi-LAT Observations of the Gamma-Ray Burst GRB 130427A. *Science* **2014**, *343*, 42–47. [CrossRef] [PubMed]
75. Nakar, E.; Ando, S.; Sari, R. Klein-Nishina Effects on Optically Thin Synchrotron and Synchrotron Self-Compton Spectrum. *Astrophys. J.* **2009**, *703*, 675–691. [CrossRef]
76. Petropoulou, M.; Mastichiadis, A. On the multiwavelength emission from gamma ray burst afterglows. *Astron. Astrophys.* **2009**, *507*, 599–610. [CrossRef]
77. Pennanen, T.; Vurm, I.; Poutanen, J. Simulations of gamma-ray burst afterglows with a relativistic kinetic code. *Astron. Astrophys.* **2014**, *564*, A77. [CrossRef]
78. Bošnjak, Ž.; Daigne, F.; Dubus, G. Prompt high-energy emission from gamma-ray bursts in the internal shock model. *Astron. Astrophys.* **2009**, *498*, 677–703. [CrossRef]
79. Chang, J.S.; Cooper, G. A Practical Difference Scheme for Fokker-Planck Equations. *J. Comput. Phys.* **1970**, *6*, 1–16. [CrossRef]
80. Rybicki, G.B.; Lightman, A.P. *Radiative Processes in Astrophysics*; Wiley: Hoboken, NJ, USA, 1979.
81. Jones, F.C. Calculated Spectrum of Inverse-Compton-Scattered Photons. *Phys. Rev.* **1968**, *167*, 1159–1169. [CrossRef]
82. Ghisellini, G.; Guilbert, P.W.; Svensson, R. The Synchrotron Boiler. *Astrophys. J.* **1988**, *334*, L5. [CrossRef]
83. Coppi, P.S.; Blandford, R.D. Reaction rates and energy distributions for elementary processes in relativistic pair plasmas. *Mon. Not. RAS* **1990**, *245*, 453–507. [CrossRef]
84. Chincarini, G.; Mao, J.; Margutti, R.; Bernardini, M.G.; Guidorzi, C.; Pasotti, F.; Giannios, D.; Della Valle, M.; Moretti, A.; Romano, P.; et al. Unveiling the origin of X-ray flares in gamma-ray bursts. *Mon. Not. RAS* **2010**, *406*, 2113–2148. [CrossRef]
85. Margutti, R.; Bernardini, G.; Barniol Duran, R.; Guidorzi, C.; Shen, R.F.; Chincarini, G. On the average gamma-ray burst X-ray flaring activity. *Mon. Not. RAS* **2011**, *410*, 1064–1075. [CrossRef]
86. Bernardini, M.G.; Margutti, R.; Chincarini, G.; Guidorzi, C.; Mao, J. Gamma-ray burst long lasting X-ray flaring activity. *Astron. Astrophys.* **2011**, *526*, A27. [CrossRef]
87. Yi, S.X.; Xi, S.Q.; Yu, H.; Wang, F.Y.; Mu, H.J.; Lü, L.Z.; Liang, E.W. Comprehensive Study of the X-Ray Flares from Gamma-ray Bursts Observed by Swift. *Astrophys. J. Lett.* **2016**, *224*, 20. [CrossRef]
88. Beniamini, P.; Kumar, P. X-ray flares in GRBs: General considerations and photospheric origin. *Mon. Not. RAS* **2016**, *457*, L108–L112. [CrossRef]
89. Yi, S.X.; Yu, H.; Wang, F.Y.; Dai, Z.G. Statistical Distributions of Optical Flares from Gamma-Ray Bursts. *Astrophys. J.* **2017**, *844*, 79. [CrossRef]
90. Burrows, D.N.; Romano, P.; Falcone, A.; Kobayashi, S.; Zhang, B.; Moretti, A.; O'Brien, P.T.; Goad, M.R.; Campana, S.; Page, K.L.; et al. Bright X-ray Flares in Gamma-Ray Burst Afterglows. *Science* **2005**, *309*, 1833–1835. [CrossRef]
91. Willingale, R.; Genet, F.; Granot, J.; O'Brien, P.T. The spectral-temporal properties of the prompt pulses and rapid decay phase of gamma-ray bursts. *Mon. Not. RAS* **2010**, *403*, 1296–1316. [CrossRef]
92. Fan, Y.Z.; Wei, D.M. Late internal-shock model for bright X-ray flares in gamma-ray burst afterglows and GRB 011121. *Mon. Not. RAS* **2005**, *364*, L42–L46. [CrossRef]
93. Lazzati, D.; Perna, R. X-ray flares and the duration of engine activity in gamma-ray bursts. *Mon. Not. RAS* **2007**, *375*, L46–L50. [CrossRef]
94. Maxham, A.; Zhang, B. Modeling Gamma-Ray Burst X-Ray Flares Within the Internal Shock Model. *Astrophys. J.* **2009**, *707*, 1623–1633. [CrossRef]
95. Wang, X.Y.; Li, Z.; Mészáros, P. GeV-TeV and X-Ray Flares from Gamma-Ray Bursts. *Astrophys. J.* **2006**, *641*, L89–L92. [CrossRef]
96. Ajello, M.; Arimoto, M.; Asano, K.; Axelsson, M.; Baldini, L.; Barbiellini, G.; Bastieri, D.; Bellazzini, R.; Berretta, A.; Bissaldi, E.; et al. Bright Gamma-Ray Flares Observed in GRB 131108A. *Astrophys. J.* **2019**, *886*, L33. [CrossRef]
97. Wang, K.; Dai, Z.G. GeV Emission during X-Ray Flares from Late Internal Shocks: Application to GRB 100728A. *Astrophys. J.* **2013**, *772*, 152. [CrossRef]
98. He, H.N.; Zhang, B.B.; Wang, X.Y.; Li, Z.; Mészáros, P. Origin of the GeV Emission during the X-Ray Flaring Activity in GRB 100728A. *Astrophys. J.* **2012**, *753*, 178. [CrossRef]
99. Berti, A. on behalf of the MAGIC Collaboration. MAGIC View of Gamma-ray Bursts at Very High Energies, Talk at 16th Marcel Grossmann meeting, online, July 2021. Available online: https://indico.icranet.org/event/1/contributions/450/attachments/112/126/berti_MAGIC_GRBs_MG16.pdf(accessedon27February2022).
100. Fioretti, V.; Ribeiro, D.; Humensky, T.B.; Bulgarelli, A.; Maier, G.; Moralejo, A.; Nigro, C. The Cherenkov Telescope Array sensitivity to the transient sky. In Proceedings of the 36th International Cosmic Ray Conference (ICRC2019), Madison, WI, USA, 24 July–1 August 2019; Volume 36, p. 673.
101. Woosley, S.E. Models for Gamma-Ray Burst Progenitors and Central Engines. *arXiv* **2011**, arXiv:1105.4193.
102. Fryer, C.L.; Rockefeller, G.; Young, P.A. The Environments around Long-Duration Gamma-Ray Burst Progenitors. *Astrophys. J.* **2006**, *647*, 1269–1285. [CrossRef]

103. Panaitescu, A.; Kumar, P. Properties of Relativistic Jets in Gamma-Ray Burst Afterglows. *Astrophys. J.* **2002**, *571*, 779–789. [CrossRef]
104. Schulze, S.; Klose, S.; Björnsson, G.; Jakobsson, P.; Kann, D.A.; Rossi, A.; Krühler, T.; Greiner, J.; Ferrero, P. The circumburst density profile around GRB progenitors: A statistical study. *Astron. Astrophys.* **2011**, *526*, A23. [CrossRef]
105. Li, L.; Wu, X.F.; Huang, Y.F.; Wang, X.G.; Tang, Q.W.; Liang, Y.F.; Zhang, B.B.; Wang, Y.; Geng, J.J.; Liang, E.W.; Wei, J.Y.; Zhang, B.; Ryde, F. A Correlated Study of Optical and X-Ray Afterglows of GRBs. *Astrophys. J.* **2015**, *805*, 13. [CrossRef]
106. Gompertz, B.P.; Fruchter, A.S.; Pe'er, A. The Environments of the Most Energetic Gamma-Ray Bursts. *Astrophys. J.* **2018**, *866*, 162. [CrossRef]
107. van Marle, A.J.; Langer, N.; Achterberg, A.; García-Segura, G. Forming a constant density medium close to long gamma-ray bursts. *Astron. Astrophys.* **2006**, *460*, 105–116. [CrossRef]
108. Srinivasaragavan, G.P.; Dainotti, M.G.; Fraija, N.; Hernandez, X.; Nagataki, S.; Lenart, A.; Bowden, L.; Wagner, R. On the Investigation of the Closure Relations for Gamma-Ray Bursts Observed by Swift in the Post-plateau Phase and the GRB Fundamental Plane. *Astrophys. J.* **2020**, *903*, 18. [CrossRef]
109. Kumar, P.; Barniol Duran, R. On the generation of high-energy photons detected by the Fermi Satellite from gamma-ray bursts. *Mon. Not. RAS* **2009**, *400*, L75–L79. [CrossRef]
110. Kumar, P.; Barniol Duran, R. External forward shock origin of high-energy emission for three gamma-ray bursts detected by Fermi. *Mon. Not. RAS* **2010**, *409*, 226–236. [CrossRef]
111. Barniol Duran, R.; Kumar, P. Evidence for mild deviation from power-law distribution of electrons in relativistic shocks: GRB 090902B. *Mon. Not. RAS* **2011**, *417*, 1584–1600. [CrossRef]
112. He, H.N.; Wu, X.F.; Toma, K.; Wang, X.Y.; Mészáros, P. On the High-energy Emission of the Short GRB 090510. *Astrophys. J.* **2011**, *733*, 22. [CrossRef]
113. Barniol Duran, R. Constraining the magnetic field in GRB relativistic collisionless shocks using radio data. *Mon. Not. RAS* **2014**, *442*, 3147–3154. [CrossRef]
114. Piran, T.; Nakar, E. On the External Shock Synchrotron Model for Gamma-ray Bursts' GeV Emission. *Astrophys. J.* **2010**, *718*, L63–L67. [CrossRef]
115. Filgas, R.; Greiner, J.; Schady, P.; Krühler, T.; Updike, A.C.; Klose, S.; Nardini, M.; Kann, D.A.; Rossi, A.; Sudilovsky, V.; et al. GRB 091127: The cooling break race on magnetic fuel. *Astron. Astrophys.* **2011**, *535*, A57. [CrossRef]
116. van der Horst, A.J.; Paragi, Z.; de Bruyn, A.G.; Granot, J.; Kouveliotou, C.; Wiersema, K.; Starling, R.L.C.; Curran, P.A.; Wijers, R.A.M.J.; Rowlinson, A.; et al. A comprehensive radio view of the extremely bright gamma-ray burst 130427A. *Mon. Not. RAS* **2014**, *444*, 3151–3163. [CrossRef]
117. Ioka, K.; Toma, K.; Yamazaki, R.; Nakamura, T. Efficiency crisis of swift gamma-ray bursts with shallow X-ray afterglows: Prior activity or time-dependent microphysics? *Astron. Astrophys.* **2006**, *458*, 7–12. [CrossRef]
118. Kong, S.W.; Wong, A.Y.L.; Huang, Y.F.; Cheng, K.S. Variation of microphysics in wind bubbles: An alternative mechanism for explaining the rebrightenings in Gamma-ray burst afterglows. *Mon. Not. RAS* **2010**, *402*, 409–416. [CrossRef]
119. Huang, L.Y.; Wang, X.G.; Zheng, W.; Liang, E.W.; Lin, D.b.; Zhong, S.Q.; Zhang, H.M.; Huang, X.L.; Filippenko, A.V.; Zhang, B. GRB 120729A: External Shock Origin for Both the Prompt Gamma-Ray Emission and Afterglow. *Astrophys. J.* **2018**, *859*, 163. [CrossRef]
120. Panaitescu, A. Swift gamma-ray burst afterglows and the forward-shock model. *Mon. Not. RAS* **2007**, *379*, 331–342. [CrossRef]
121. Misra, K.; Resmi, L.; Kann, D.A.; Marongiu, M.; Moin, A.; Klose, S.; Bernardi, G.; de Ugarte Postigo, A.; Jaiswal, V.K.; Schulze, S.; et al. Low frequency view of GRB 190114C reveals time varying shock micro-physics. *Mon. Not. RAS* **2021**, *504*, 5685–5701. [CrossRef]
122. Guilbert, P.W.; Fabian, A.C.; Rees, M.J. Spectral and variability constraints on compact sources. *Mon. Not. RAS* **1983**, *205*, 593–603. [CrossRef]
123. de Jager, O.C.; Harding, A.K. The Expected High-Energy to Ultra–High-Energy Gamma-Ray Spectrum of the Crab Nebula. *Astrophys. J.* **1992**, *396*, 161. [CrossRef]
124. Ajello, M.E.A. A Decade of Gamma-Ray Bursts Observed by Fermi-LAT: The Second GRB Catalog. *Astrophys. J.* **2019**, *878*, 52. [CrossRef]
125. Kumar, P.; Hernández, R.A.; Bošnjak, Ž.; Barniol Duran, R. Maximum synchrotron frequency for shock-accelerated particles. *Mon. Not. RAS* **2012**, *427*, L40–L44. [CrossRef]
126. Medvedev, M.V.; Loeb, A. Generation of Magnetic Fields in the Relativistic Shock of Gamma-Ray Burst Sources. *Astrophys. J.* **1999**, *526*, 697–706. [CrossRef]
127. Kumar, P. The Distribution of Burst Energy and Shock Parameters for Gamma-Ray Bursts. *Astrophys. J.* **2000**, *538*, L125–L128. [CrossRef]
128. Berger, E.; Kulkarni, S.R.; Frail, D.A. A Standard Kinetic Energy Reservoir in Gamma-Ray Burst Afterglows. *Astrophys. J.* **2003**, *590*, 379–385. [CrossRef]
129. Lloyd-Ronning, N.M.; Zhang, B. On the Kinetic Energy and Radiative Efficiency of Gamma-Ray Bursts. *Astrophys. J.* **2004**, *613*, 477–483. [CrossRef]
130. Berger, E. The Prompt Gamma-Ray and Afterglow Energies of Short-Duration Gamma-Ray Bursts. *Astrophys. J.* **2007**, *670*, 1254–1259. [CrossRef]

131. D'Avanzo, P.; Salvaterra, R.; Sbarufatti, B.; Nava, L.; Melandri, A.; Bernardini, M.G.; Campana, S.; Covino, S.; Fugazza, D.; Ghirlanda, G.; et al. A complete sample of bright Swift Gamma-ray bursts: X-ray afterglow luminosity and its correlation with the prompt emission. *Mon. Not. RAS* **2012**, *425*, 506–513. [CrossRef]
132. Granot, J.; Königl, A.; Piran, T. Implications of the early X-ray afterglow light curves of Swift gamma-ray bursts. *Mon. Not. RAS* **2006**, *370*, 1946–1960. [CrossRef]
133. Zhang, B.; Liang, E.; Page, K.L.; Grupe, D.; Zhang, B.B.; Barthelmy, S.D.; Burrows, D.N.; Campana, S.; Chincarini, G.; Gehrels, N.; et al. GRB Radiative Efficiencies Derived from the Swift Data: GRBs versus XRFs, Long versus Short. *Astrophys. J.* **2007**, *655*, 989–1001. [CrossRef]
134. Nava, L.; Vianello, G.; Omodei, N.; Ghisellini, G.; Ghirlanda, G.; Celotti, A.; Longo, F.; Desiante, R.; Barniol Duran, R. Clustering of LAT light curves: A clue to the origin of high-energy emission in gamma-ray bursts. *Mon. Not. RAS* **2014**, *443*, 3578–3585. [CrossRef]
135. Beniamini, P.; Granot, J. Properties of GRB light curves from magnetic reconnection. *Mon. Not. RAS* **2016**, *459*, 3635–3658. [CrossRef]
136. Ressler, S.M.; Laskar, T. Thermal Electrons in Gamma-Ray Burst Afterglows. *Astrophys. J.* **2017**, *845*, 150. [CrossRef]
137. Warren, D.C.; Dainotti, M.; Barkov, M.V.; Ahlgren, B.; Ito, H.; Nagataki, S. A Semianalytic Afterglow with Thermal Electrons and Synchrotron Self-Compton Emission. *Astrophys. J.* **2022**, *924*, 40. [CrossRef]
138. Giannios, D.; Spitkovsky, A. Signatures of a Maxwellian component in shock-accelerated electrons in GRBs. *Mon. Not. RAS* **2009**, *400*, 330–336. [CrossRef]
139. Mirzoyan, R.; Noda, K.; Moretti, E.; Berti, A.; Nigro, C.; Hoang, J.; Micanovic, S.; Takahashi, M.; Chai, Y.; Moralejo, A.; et al. MAGIC detects the GRB 190114C in the TeV energy domain. *GRB Coord. Netw.* **2019**, *23701*, 1.
140. Li, T.P.; Ma, Y.Q. Analysis methods for results in gamma-ray astronomy. *Astrophys. J.* **1983**, *272*, 317–324. [CrossRef]
141. Siegel, M.H.; Barthelmy, S.D.; Burrows, D.N.; Lien, A.Y.; Marshall, F.E.; Palmer, D.M.; Sbarufatti, B. GRB 160821B: Swift detection of a short burst. *GRB Coord. Netw.* **2016**, *19833*, 1.
142. Stanbro, M.; Meegan, C. GRB 160821B: Fermi GBM Detection. *GRB Coord. Netw.* **2016**, *19843*, 1.
143. Palmer, D.M.; Barthelmy, S.D.; Cummings, J.R.; Gehrels, N.; Krimm, H.A.; Lien, A.Y.; Markwardt, C.B.; Norris, J.P.; Sakamoto, T.; Siegel, M.H.; et al. GRB 160821B: Swift-BAT refined analysis. *GRB Coord. Netw.* **2016**, *19844*, 1.
144. Berger, E. Short-Duration Gamma-Ray Bursts. *Annu. Rev. Astron Astrophys.* **2014**, *52*, 43–105. [CrossRef]
145. Acciari, V.A.; Ansoldi, S.; Antonelli, L.A.; Arbet Engels, A.; Asano, K.; Baack, D.; Babić, A.; Baquero, A.; Barres de Almeida, U.; Barrio, J.A.; et al. MAGIC Observations of the Nearby Short Gamma-Ray Burst GRB 160821B. *Astrophys. J.* **2021**, *908*, 90. [CrossRef]
146. Evans, P.A.; Goad, M.R.; Osborne, J.P.; Beardmore, A.P. GRB 160821B: Enhanced Swift-XRT position. *GRB Coord. Netw.* **2016**, *19837*, 1.
147. Sbarufatti, B.; Burrows, D.N.; Osborne, J.P.; Page, K.L.; Mingo, B.; Maselli, A.; Melandri, A.; D'Avanzo, P.; Roegiers, T.G.R.; Siegel, M.H. GRB 160821B: Swift-XRT refined Analysis. *GRB Coord. Netw.* **2016**, *19841*, 1.
148. Troja, E.; Castro-Tirado, A.J.; Becerra González, J.; Hu, Y.; Ryan, G.S.; Cenko, S.B.; Ricci, R.; Novara, G.; Sánchez-Rámirez, R.; Acosta-Pulido, J.A.; et al. The afterglow and kilonova of the short GRB 160821B. *Mon. Not. RAS* **2019**, *489*, 2104–2116. [CrossRef]
149. Jeong, S.; Park, I.H.; Hu, Y.; Scarpa, R.; Castro-Tirado, A.J. GRB 160821B: GTC follow-up observation. *GRB Coord. Netw.* **2016**, *19847*, 1.
150. Troja, E.; Tanvir, N.; Cenko, S.B.; Levan, A.; Barnes, J.; Castro-Tirado, A.; Fruchter, A.S.; Gehrels, N.; Greiner, J.; Kawai, N.; et al. GRB 160821B: HST detection of the optical and IR counterpart. *GRB Coord. Netw.* **2016**, *20222*, 1.
151. Xu, D.; Malesani, D.; de Ugarte Postigo, A.; Gafton, E.; Rivero Losada, I. GRB 160821B: NOT optical afterglow candidate. *GRB Coord. Netw.* **2016**, *19834*, 1.
152. Levan, A.J.; Wiersema, K.; Tanvir, N.R.; Malesani, D.; Xu, D.; de Ugarte Postigo, A. GRB 160821B: WHT afterglow confirmation and redshift of candidate host. *GRB Coord. Netw.* **2016**, *19846*, 1.
153. Lamb, G.P.; Tanvir, N.R.; Levan, A.J.; de Ugarte Postigo, A.; Kawaguchi, K.; Corsi, A.; Evans, P.A.; Gompertz, B.; Malesani, D.B.; Page, K.L.; et al. Short GRB 160821B: A Reverse Shock, a Refreshed Shock, and a Well-sampled Kilonova. *Astrophys. J.* **2019**, *883*, 48. [CrossRef]
154. Fong, W.; Alexander, K.D.; Laskar, T. GRB 160821B: VLA 5.0 GHz observations. *GRB Coord. Netw.* **2016**, *19854*, 1.
155. Fruck, C.; Gaug, M.; Zanin, R.; Dorner, D.; Garrido, D.; Mirzoyan, R.; Font, L. A novel LIDAR-based Atmospheric Calibration Method for Improving the Data Analysis of MAGIC. *arXiv* **2014**, arXiv:1403.3591.
156. Rolke, W.A.; López, A.M.; Conrad, J. Limits and confidence intervals in the presence of nuisance parameters. *Nucl. Instrum. Methods Phys. Res. A* **2005**, *551*, 493–503. [CrossRef]
157. Zhang, B.T.; Murase, K.; Yuan, C.; Kimura, S.S.; Mészáros, P. External Inverse-Compton Emission Associated with Extended and Plateau Emission of Short Gamma-Ray Bursts: Application to GRB 160821B. *Astrophys. J.* **2021**, *908*, L36. [CrossRef]
158. Roberts, O.J.; Meegan, C. GRB 180720B: Fermi GBM observation. *GRB Coord. Netw.* **2018**, *22981*, 1.
159. Siegel, M.H.; Burrows, D.N.; Deich, A.; Gropp, J.D.; Kennea, J.A.; Laporte, S.J.; Lien, A.Y.; Moss, M.J.; Page, K.L.; Palmer, D.M.; et al. Correction: Swift trigger #848890 is GRB 180720B (not GRB 180720A). *GRB Coord. Netw.* **2018**, *22975*, 1.
160. Bissaldi, E.; Racusin, J.L. GRB 180720B: Fermi-LAT detection. *GRB Coord. Netw.* **2018**, *22980*, 1.

161. Sasada, M.; Nakaoka, T.; Kawabata, M.; Uchida, N.; Yamazaki, Y.; Kawabata, K.S. GRB 180720B: Kanata 1.5m optical/NIR observation. *GRB Coord. Netw.* **2018**, *22977*, 1.
162. Martone, R.; Guidorzi, C.; Kobayashi, S.; Mundell, C.G.; Gomboc, A.; Steele, I.A.; Cucchiara, A.; Morris, D. GRB 180720B: LCO Haleakala possible bright optical candidate. *GRB Coord. Netw.* **2018**, *22976*, 1.
163. Reva, I.; Pozanenko, A.; Volnova, A.; Mazaeva, E.; Kusakin, A.; Krugov, M. GRB 180720B: TSHAO optical observations. *GRB Coord. Netw.* **2018**, *22979*, 1.
164. Watson, A.M.; Butler, N.; Becerra, R.L.; Lee, W.H.; Roman-Zuniga, C.; Kutyrev, A.; Troja, E. GRB 180720B: COATLI Optical Detection. *GRB Coord. Netw.* **2018**, *23017*, 1.
165. Schmalz, S.; Graziani, F.; Pozanenko, A.; Volnova, A.; Mazaeva, E.; Molotov, I. GRB 180720B: ISON-Castelgrande optical observations. *GRB Coord. Netw.* **2018**, *23020*, 1.
166. Zheng, W.; Filippenko, A.V. GRB 180720B: KAIT Optical Observations. *GRB Coord. Netw.* **2018**, *23033*, 1.
167. Lipunov, V.; Gorbovskoy, E.; Tiurina, N.; Vlasenko, D.; Kornilov, V.; Kuznetsov, A.; Chazov, V.; Gorbunov, I.; Zimnukhov, D.; Kuvshinov, D.; et al. GRB 180720B: MASTER Global Net OT observations. *GRB Coord. Netw.* **2018**, *23023*, 1.
168. Itoh, R.; Murata, K.L.; Tachibana, Y.; Harita, S.; Morita, K.; Shiraishi, K.; Iida, K.; Oeda, M.; Adachi, R.; Niwano, S.; et al. GRB 180720B: MITSuME Akeno optical observations. *GRB Coord. Netw.* **2018**, *22983*, 1.
169. Izzo, L.; Kann, D.A.; de Ugarte Postigo, A.; Thoene, C.C.; Bensch, K.; Blazek, M.; Diaz-Martin, M.C.; Rodriguez-Llano, S. GRB 180720B: OAJ optical observations. *GRB Coord. Netw.* **2018**, *23040*, 1.
170. Covino, S.; Fugazza, D. GRB 180720B: REM photometry. *GRB Coord. Netw.* **2018**, *23021*, 1.
171. Jelinek, M.; Strobl, J.; Hudec, R.; Polasek, C. GRB 180720B: D50 optical observations. *GRB Coord. Netw.* **2018**, *23024*, 1.
172. Kann, D.A.; Izzo, L.; Casanova, V. GRB 180720B: OSN detection, fading. *GRB Coord. Netw.* **2018**, *22985*, 1.
173. Rhodes, L.; van der Horst, A.J.; Fender, R.; Monageng, I.M.; Anderson, G.E.; Antoniadis, J.; Bietenholz, M.F.; Böttcher, M.; Bright, J.S.; Green, D.A.; et al. Radio afterglows of very high-energy gamma-ray bursts 190829A and 180720B. *Mon. Not. RAS* **2020**, *496*, 3326–3335. [CrossRef]
174. Wang, X.Y.; Liu, R.Y.; Zhang, H.M.; Xi, S.Q.; Zhang, B. Synchrotron Self-Compton Emission from External Shocks as the Origin of the Sub-TeV Emission in GRB 180720B and GRB 190114C. *Astrophys. J.* **2019**, *884*, 117. [CrossRef]
175. Liu, R.Y.; Wang, X.Y.; Wu, X.F. Interpretation of the Unprecedentedly Long-lived High-energy Emission of GRB 130427A. *Astrophys. J.* **2013**, *773*, L20. [CrossRef]
176. Gropp, J.D.; Kennea, J.A.; Klingler, N.J.; Krimm, H.A.; Laporte, S.J.; Lien, A.Y.; Moss, M.J.; Palmer, D.M.; Sbarufatti, B.; Siegel, M.H. GRB 190114C: Swift detection of a very bright burst with a bright optical counterpart. *GRB Coord. Netw.* **2019**, *23688*, 1.
177. Hamburg, R.; Veres, P.; Meegan, C.; Burns, E.; Connaughton, V.; Goldstein, A.; Kocevski, D.; Roberts, O.J. GRB 190114C: Fermi GBM detection. *GRB Coord. Netw.* **2019**, *23707*, 1.
178. Ravasio, M.E.; Oganesyan, G.; Salafia, O.S.; Ghirland a, G.; Ghisellini, G.; Branchesi, M.; Campana, S.; Covino, S.; Salvaterra, R. GRB 190114C: From prompt to afterglow? *Astron. Astrophys.* **2019**, *626*, A12. [CrossRef]
179. Ajello, M.e.a. Fermi and Swift Observations of GRB 190114C: Tracing the Evolution of High-energy Emission from Prompt to Afterglow. *Astrophys. J.* **2020**, *890*, 9. [CrossRef]
180. Laskar, T.; Alexander, K.D.; Gill, R.; Granot, J.; Berger, E.; Mundell, C.G.; Barniol Duran, R.; Bolmer, J.; Duffell, P.; van Eerten, H.; et al. ALMA Detection of a Linearly Polarized Reverse Shock in GRB 190114C. *Astrophys. J.* **2019**, *878*, L26. [CrossRef]
181. Selsing, J.; Fynbo, J.P.U.; Heintz, K.E.; Watson, D. GRB 190114C: NOT optical counterpart and redshift. *GRB Coord. Netw.* **2019**, *23695*, 1.
182. Castro-Tirado, A.J.; Hu, Y.; Fernandez-Garcia, E.; Valeev, A.; Sokolov, V.; Guziy, S.; Oates, S.; Jeong, S.; Pandey, S.B.; Carrasco, I.; et al. GRB 190114C: Refined redshift by the 10.4m GTC. *GRB Coord. Netw.* **2019**, *23708*, 1.
183. Domínguez, A.; Primack, J.R.; Rosario, D.J.; Prada, F.; Gilmore, R.C.; Faber, S.M.; Koo, D.C.; Somerville, R.S.; Pérez-Torres, M.A.; Pérez-González, P.; et al. Extragalactic background light inferred from AEGIS galaxy-SED-type fractions. *Mon. Not. RAS* **2011**, *410*, 2556–2578. [CrossRef]
184. Derishev, E.; Piran, T. GRB afterglow parameters in the era of TeV observations: The case of GRB 190114C. *arXiv* **2021**, arXiv:2106.12035.
185. Asano, K.; Murase, K.; Toma, K. Probing Particle Acceleration through Broadband Early Afterglow Emission of MAGIC Gamma-Ray Burst GRB 190114C. *Astrophys. J.* **2020**, *905*, 105. [CrossRef]
186. Joshi, J.C.; Razzaque, S. Modelling synchrotron and synchrotron self-Compton emission of gamma-ray burst afterglows from radio to very-high energies. *Mon. Not. RAS* **2021**, *505*, 1718–1729. [CrossRef]
187. Dichiara, S.; Bernardini, M.G.; Burrows, D.N.; D'Avanzo, P.; Gronwall, C.; Gropp, J.D.; Kennea, J.A.; Klingler, N.J.; Krimm, H.A.; Kuin, N.P.M.; et al. GRB 190829A: Swift detection of a burst consistent with a galaxy at z = 0.08. *GRB Coord. Netw.* **2019**, *25552*, 1.
188. Fermi GBM Team. GRB 190829A: Fermi GBM Final Real-time Localization. *GRB Coord. Netw.* **2019**, *25551*, 1.
189. Chand, V.; Banerjee, A.; Gupta, R.; Dimple; Pal, P.S.; Joshi, J.C.; Zhang, B.B.; Basak, R.; Tam, P.H.T.; Sharma, V.; et al. Peculiar Prompt Emission and Afterglow in the H.E.S.S.-detected GRB 190829A. *Astrophys. J.* **2020**, *898*, 42. [CrossRef]
190. Piron, F.; Longo, F.; Axelsson, M.; Arimoto, M.; Racusin, J.L.; Bissaldi, E.; Fermi-LAT Collaboration. GRB 190829A: Fermi-LAT Upper Limits. *GRB Coord. Netw.* **2019**, *25574*, 1.
191. Osborne, J.P.; Page, K.L.; Capalbi, M.; Perri, M.; D'Elia, V.; Burrows, D.N.; Gropp, J.D.; Tohuvavohu, A.; Beardmore, A.P.; Dichiara, S.; et al. GRB 190829A: Swift-XRT refined Analysis. *GRB Coord. Netw.* **2019**, *25568*, 1.

192. Valeev, A.F.; Castro-Tirado, A.J.; Hu, Y.D.; Fernand ez-Garcia, E.; Sokolov, V.V.; Carrasco, I.; Castellon, A.; Garcia Alvarez, D.; Rivero, M. GRB 190829A: 10.4m GTC spectroscopy. *GRB Coord. Netw.* **2019**, *25565*, 1.
193. Lipunov, V.; Balakin, F.; Gorbovskoy, E.; Kornilov, V.; Tyurina, N.; Balanutsa, P.; Kuznetsov, A.; Vladimirov, V.; Vlasenko, D.; Gorbunov, I.; et al. GRB 190829A: MASTER confirmation of GROND SN. *GRB Coord. Netw.* **2019**, *25652*, 1.
194. Laskar, T.; Bhandari, S.; Schroeder, G.; Fong, W.; Alexander, K.D.; Berger, E.; Chornock, R.; Coppejans, D.; Margutti, R.; Ayache, E.; et al. GRB 190829A: ATCA cm-band detection. *GRB Coord. Netw.* **2019**, *25676*, 1.
195. Monageng, I.; van der Horst, A.J.; Woudt, P.A.; Bottcher, M. GRB 190829A: MeerKAT radio observation. *GRB Coord. Netw.* **2019**, *25635*, 1.
196. de Ugarte Postigo, A.; Bremer, M.; Kann, D.A.; Schulze, S.; Thoene, C.C.; Blazek, M.; Bensch, K.; Agui, J.F.; Izzo, L.; Perley, D.A.; et al. GRB 190829A: NOEMA detection of the mm afterglow. *GRB Coord. Netw.* **2019**, *25589*, 1.
197. Chandra, P. GRB 190829A: Detection of radio afterglow with the uGMRT. *GRB Coord. Netw.* **2019**, *25627*, 1.
198. Sato, Y.; Obayashi, K.; Yamazaki, R.; Murase, K.; Ohira, Y. Off-axis jet scenario for early afterglow emission of low-luminosity gamma-ray burst GRB 190829A. *Mon. Not. RAS* **2021**, *504*, 5647–5655. [CrossRef]
199. Zhang, B.T.; Murase, K.; Veres, P.; Mészáros, P. External Inverse-Compton Emission from Low-luminosity Gamma-Ray Bursts: Application to GRB 190829A. *Astrophys. J.* **2021**, *920*, 55. [CrossRef]
200. D'Elia, V.; Ambrosi, E.; Barthelmy, S.D.; D'Ai, A.; Gropp, J.D.; Klingler, N.J.; Lien, A.Y.; Palmer, D.M.; Sbarufatti, B.; Siegel, M.H.; et al. GRB 201015A: Swift detection of a burst. *GRB Coord. Netw.* **2020**, *28632*, 1.
201. Fletcher, C.; Veres, P.; Fermi-GBM Team. Fermi GBM Sub-Threshold Detection of GRB 201015A. *GRB Coord. Netw.* **2020**, *28663*, 1.
202. Minaev, P.; Pozanenko, A. GRB 201015A: Classification as long GRB. *GRB Coord. Netw.* **2020**, *28668*, 1.
203. Markwardt, C.B.; Barthelmy, S.D.; Cummings, J.R.; D'Elia, V.; Krimm, H.A.; Laha, S.; Lien, A.Y.; Palmer, D.M.; Sakamoto, T.; Stamatikos, M.; et al. GRB 201015A: Swift-BAT refined analysis (a soft short pulse with a tail emission). *GRB Coord. Netw.* **2020**, *28658*, 1.
204. D'Ai, A.; Gropp, J.D.; Kennea, J.A.; Tohuvavohu, A.; Page, K.L.; Beardmore, A.P.; Evans, P.A.; Melandri, A.; D'Elia, V.; Swift-XRT Team. GRB 201015A: Swift-XRT refined Analysis. *GRB Coord. Netw.* **2020**, *28660*, 1.
205. Gompertz, B.; Levan, A.; Tanvir, N.; Fruchter, A.; Cucchiara, A.; Greiner, J.; Hjorth, J.; Kangas, T.; Lamb, G.; Lyman, J.; et al. GRB 201015A: Late X-ray Detections with Chandra. *GRB Coord. Netw.* **2020**, *28822*, 1.
206. D'Elia, V.; Swift Team. GRB 201015A: Swift-XRT late time observations. *GRB Coord. Netw.* **2020**, *28857*, 1.
207. Lipunov, V.; Gorbovskoy, E.; Kornilov, V.; Tyurina, N.; Balanutsa, P.; Kuznetsov, A.; Balakin, F.; Vladimirov, V.; Vlasenko, D.; Gorbunov, I.; et al. Swift GRB 201015A: Global MASTER-Net observations report. *GRB Coord. Netw.* **2020**, *28634*, 1.
208. Grossan, B.; Maksut, Z.; Kim, A.; Krugov, M.; Smoot, G.F.; Linder, E. GRB 201015A NUTTelA-TAO/BSTI Early Measurements (Preliminary). *GRB Coord. Netw.* **2020**, *28674*, 1.
209. Fong, W.; Schroeder, G.; Rastinejad, J.; Hajela, A. GRB 201015A: 6 GHz VLA radio afterglow candidate detection. *GRB Coord. Netw.* **2020**, *28688*, 1.
210. Giarratana, S.; Giroletti, M.; Marcote, B.; Ghirlanda, G.; Ribó, M.; Paredes, J.M. GRB 201015A radio afterglow with e-MERLIN. *GRB Coord. Netw.* **2020**, *28939*, 1.
211. Marcote, B.; Ribó, M.; Paredes, J.M.; Giroletti, M.; Giarratana, S.; Ghirlanda, G. GRB 201015A: 5-GHz radio afterglow detection with the EVN. *GRB Coord. Netw.* **2020**, *29028*, 1.
212. Pozanenko, A.; Belkin, S.; Volnova, A.; Moskvitin, A.; Burhonov, O.; Kim, V.; Krugov, M.; Rumyantsev, V.; Klunko, E.; Inasaridze, R.Y.; et al. GRB 201015A: Optical observations and supernova identification. *GRB Coord. Netw.* **2020**, *29033*, 1.
213. Rossi, A.; Benetti, S.; Palazzi, E.; D'Avanzo, P.; D'Elia, V.; De Pasquale, M.; CIBO Collaboration. GRB 201015A: Evidence of supernova in LBT spectra. *GRB Coord. Netw.* **2021**, *29306*, 1.
214. de Ugarte Postigo, A.; Kann, D.A.; Blazek, M.; Agui Fernandez, J.F.; Thoene, C.; Gomez Velarde, G. GRB 201015A: Redshift from GTC/OSIRIS. *GRB Coord. Netw.* **2020**, *28649*, 1.
215. Izzo, L.; Malesani, D.B.; Zhu, Z.P.; Xu, D.; de Ugarte Postigo, A.; Pursimo, T. GRB 201015A: Redshift confirmation. *GRB Coord. Netw.* **2020**, *28661*, 1.
216. Blanch, O.; Gaug, M.; Noda, K.; Berti, A.; Moretti, E.; Miceli, D.; Gliwny, P.; Ubach, S.; Schleicher, B.; Cerruti, M.; et al. MAGIC observations of GRB 201015A: Hint of very high energy gamma-ray signal. *GRB Coord. Netw.* **2020**, *28659*, 1.
217. Suda, Y.; Artero, M.; Asano, K.; Berti, A.; Nava, L.; Noda, K.; Terauchi, K.; MAGIC Collaboration. Observation of a relatively low luminosity long duration GRB 201015A by the MAGIC telescopes. In Proceedings of the 37th International Cosmic Ray Conference—PoS (ICRC2021), Berlin, Germany, 12 July 2021. [CrossRef]
218. Beardmore, A.P.; Gropp, J.D.; Kennea, J.A.; Klingler, N.J.; Laha, S.; Lien, A.Y.; Moss, M.J.; Page, K.L.; Palmer, D.M.; Tohuvavohu, A.; et al. GRB 201216C: Swift detection of a burst. *GRB Coord. Netw.* **2020**, *29061*, 1.
219. Malacaria, C.; Veres, P.; Meegan, C.; Bissaldi, E.; Fermi GBM Team. GRB 201216C: Fermi GBM detection. *GRB Coord. Netw.* **2020**, *29073*, 1.
220. Ukwatta, T.N.; Barthelmy, S.D.; Beardmore, A.P.; Krimm, H.A.; Laha, S.; Lien, A.Y.; Markwardt, C.B.; Palmer, D.M.; Sakamoto, T.; Stamatikos, M. GRB 201216C: Swift-BAT refined analysis. *GRB Coord. Netw.* **2020**, *29080*, 1.
221. Bissaldi, E.; Omodei, N.; Kocevski, D.; Axelsson, M.; Longo, F.; Moretti, E.; Fermi-LAT Collaboration. GRB 201216C: Fermi-LAT upper limit. *GRB Coord. Netw.* **2020**, *29076*, 1.

222. Evans, P.A.; Osborne, J.P.; Bernardini, M.G.; Ambrosi, E.; Capalbi, M.; Tohuvavohu, A.; Sbarufatti, B.; Burrows, D.N.; Beardmore, A.P.; Swift-XRT Team. GRB 201216C: Swift-XRT refined Analysis. *GRB Coord. Netw.* **2021**, *29278*, 1.
223. Izzo, L.; Malesani, D.B.; Kann, D.A. GRB 201216C: VLT afterglow candidate discovery. *GRB Coord. Netw.* **2020**, *29066*, 1.
224. Jelinek, M.; Strobl, J.; Karpov, S.; Masek, M.; Janecek, P.; Jurysek, J.; Ebr, J.; Cunniffe, R.; Travnicek, P.; Prouza, M. GRB 201216C: FRAM-ORM afterglow confirmation. *GRB Coord. Netw.* **2020**, *29070*, 1.
225. Shrestha, M.; Melandri, A.; Smith, R.; Steele, I.A.; Kobayashi, S.; Mundell, C.; Gomboc, A.; Guidorzi, C. GRB 201216C: Liverpool Telescope First Hour Observations. *GRB Coord. Netw.* **2020**, *29085*, 1.
226. Ayala, H. GRB201216C: No significant detection in HAWC. *GRB Coord. Netw.* **2020**, *29086*, 1.
227. Vielfaure, J.B.; Izzo, L.; Xu, D.; Vergani, S.D.; Malesani, D.B.; de Ugarte Postigo, A.; D'Elia, V.; Fynbo, J.P.U.; Kann, D.A.; Levan, A.J.; et al. GRB 201216C: VLT X-shooter spectroscopy and potential high redshift of a VHE-emitting GRB. *GRB Coord. Netw.* **2020**, *29077*, 1.
228. Blanch, O.; Longo, F.; Berti, A.; Fukami, S.; Suda, Y.; Loporchio, S.; Micanovic, S.; Green, J.G.; Pinter, V.; Takahashi, M.; et al. GRB 201216C: MAGIC detection in very high energy gamma rays. *GRB Coord. Netw.* **2020**, *29075*, 1.
229. Fukami, S.; Berti, A.; Loporchio, S.; Suda, Y.; Nava, L.; Noda, K.; Bosnjak, Z.; Asano, K.; Longo, F.; MAGIC Collaboration. Very-high-energy gamma-ray emission from GRB 201216C detected by MAGIC. In Proceedings of the 37th International Cosmic Ray Conference—PoS (ICRC2021), Berlin, Germany, 12 July 2021; Volume 395, p. 788. [CrossRef]
230. Nava, L.; Salvaterra, R.; Ghirlanda, G.; Ghisellini, G.; Campana, S.; Covino, S.; Cusumano, G.; D'Avanzo, P.; D'Elia, V.; Fugazza, D.; et al. A complete sample of bright Swift long gamma-ray bursts: Testing the spectral-energy correlations. *Mon. Not. RAS* **2012**, *421*, 1256–1264. [CrossRef]
231. D'Avanzo, P.; Salvaterra, R.; Bernardini, M.G.; Nava, L.; Campana, S.; Covino, S.; D'Elia, V.; Ghirlanda, G.; Ghisellini, G.; Melandri, A.; et al. A complete sample of bright Swift short gamma-ray bursts. *Mon. Not. RAS* **2014**, *442*, 2342–2356. [CrossRef]
232. Wei, J.; Cordier, B.; Antier, S.; Antilogus, P.; Atteia, J.L.; Bajat, A.; Basa, S.; Beckmann, V.; Bernardini, M.G.; Boissier, S.; et al. The Deep and Transient Universe in the SVOM Era: New Challenges and Opportunities—Scientific prospects of the SVOM mission. *arXiv* **2016**, arXiv:1610.06892.
233. Lien, A.; Sakamoto, T.; Barthelmy, S.D.; Baumgartner, W.H.; Cannizzo, J.K.; Chen, K.; Collins, N.R.; Cummings, J.R.; Gehrels, N.; Krimm, H.A.; et al. The Third Swift Burst Alert Telescope Gamma-Ray Burst Catalog. *Astrophys. J.* **2016**, *829*, 7. [CrossRef]
234. Donini, A.; Gasparetto, T.; Bregeon, J.; Pierro, F.D.; Longo, F.; Maier, G.; Moralejo, A.; Vuillaume, T. The Cherenkov Telescope Array Performance in Divergent Mode. In Proceedings of the 36th International Cosmic Ray Conference (ICRC2019), Madison, WI, USA, 24 July–1 August 2019; Volume 36, p. 664.
235. Inoue, S.; Granot, J.; O'Brien, P.T.; Asano, K.; Bouvier, A.; Carosi, A.; Connaughton, V.; Garczarczyk, M.; Gilmore, R.; Hinton, J.; et al. Gamma-ray burst science in the era of the Cherenkov Telescope Array. *Astropart. Phys.* **2013**, *43*, 252–275. [CrossRef]
236. Grazia Bernardini, M.; Bissaldi, E.; Bosnjak, Z.; Carosi, A.; D'Avanzo, P.; Di Girolamo, T.; Inoue, S.; Gasparetto, T.; Ghirlanda, G.; Longo, F.; et al. POSyTIVE—A GRB population study for the Cherenkov Telescope Array (ICRC-2019). *arXiv* **2019**, arXiv:1908.01544.
237. Stamerra, A.; Saturni, F.G.; Green, J.G.; Nava, L.; Lucarelli, F.; Antonelli, L.A. TeV Transients with the ASTRI Mini-Array: A case study with GRB 190114C. In Proceedings of the 37th International Cosmic Ray Conference—PoS (ICRC2021), Berlin, Germany, 12 July 2021; Volume 395, p. 890. [CrossRef]

galaxies

MDPI

Article

Distributed Architectures and Constellations for γ-ray Burst Science

Fabrizio Fiore [1,2,*], Norbert Werner [3] and Ehud Behar [4]

[1] INAF-Osservatorio Astronomico di Trieste, Via Tiepolo 11, 34143 Trieste, Italy
[2] Institute for the Fundamental Physics of the Universe, IFPU, Via Beirut, 2, 34151 Trieste, Italy
[3] Department of Theoretical Physics and Astrophysics, Faculty of Science, Masaryk University, Kotlářská 2, 611 37 Brno, Czech Republic; werner@physics.muni.cz
[4] Department of Physics, Technion, Haifa 32000, Israel; behar@physics.technion.ac.il
* Correspondence: fabrizio.fiore@inaf.it

Abstract: The gravitational wave/γ-ray burst GW/GRB170817 event marked the beginning of the era of multi-messenger astrophysics, in which new observations of Gravitational Waves (GW) are combined with traditional electromagnetic observations from the very same astrophysical source. In the next few years, Advanced LIGO/VIRGO and KAGRA in Japan and LIGO-India will reach their nominal/ultimate sensitivity. In the electromagnetic domain, the Vera C. Rubin Observatory and the Cherenkov Telescope Array (CTA) will come online in the next few years, and they will revolutionize the investigation of transient and variable cosmic sources in the optical and TeV bands. The operation of an efficient X-ray/γ-ray all-sky monitor with good localisation capabilities will play a pivotal role in providing the high-energy counterparts of the GW interferometers and Rubin Observatory, bringing multi-messenger astrophysics to maturity. To reach the required precision in localisation and timeliness for an unpredictable physical event in time and space requires a sensor distribution covering the whole sky. We discuss the potential of large-scale, small-platform-distributed architectures and constellations to build a sensitive X-ray/γ-ray all-sky monitor and the programmatic implications of this, including the set-up of an efficient assembly line for both hardware development and data analysis. We also discuss the potential of a constellation of small platforms operating at other wavelengths (UV/IR) that are capable of repointing quickly to follow-up high-energy transients.

Keywords: γ-ray burst; multi-messenger astrophysics; nano-satellites

Citation: Fiore, F.; Werner, N.; Behar, E. Distributed Architectures and Constellations for γ-ray Burst Science. *Galaxies* **2021**, *9*, 120. https://doi.org/10.3390/galaxies9040120

Academic Editors: Elena Moretti and Francesco Longo

Received: 3 November 2021
Accepted: 7 December 2021
Published: 16 December 2021

1. Introduction

The sky is teeming with explosive, energetic transient events, many of which remain hidden from our view. Some of the most exciting transient phenomena are γ-ray bursts (GRBs), discovered in 1967 by the Vela military satellites [1]. About two GRBs are detected per day, and last from a fraction of a second to a few minutes (in exceptional cases a few hours), in the energy range from several keV to a few MeV. They are some of the most extreme explosive events ever observed, momentarily outshining any other phenomena in the sky and are associated with the death of stars and the coalescence of compact objects (e.g., neutron stars) to form a new black hole. Despite great efforts and numerous observations, many open questions about their detailed physics remain. The emergence of multi-messenger astronomy provides a unique opportunity to shed new light on GRB physics [2]. To make progress, we need to perform a sensitive, all-sky monitoring of the high-energy sky, detect and localize the transients simultaneously with other probes, and follow them up rapidly with telescopes which observe other wavelengths.

It may sound like a mockery that the most dramatic events in the cosmos produce among the most luminous objects in the Universe (GRBs) but that all this light is most likely produced quite far from where the action is; far from the newborn event horizon, the accretion disk, and the region where relativistic jets are launched. On the other hand, gravitational waves (GWs), encoding the rapid/relativistic motion of compact objects, allow us to look directly into the innermost regions of these systems, providing precise information on space–time dynamics, and therefore the mass, spin, interior properties and inclination of the systems, as well as accurate distances. Electromagnetic measurements can hardly provide accuracies comparable to GW observations on these quantities, which are key to testing general relativity, the physics of compact objects and the build-up of the most efficient accelerators in the Universe. However, the information carried by GWs can be greatly amplified by identifying the context in which the event occurs. Electromagnetic observations can provide this context, as the GW/GRB170817 event strikingly demonstrated.

Multi-messenger astrophysics can include many more sources and astrophysical context, in addition to compact binary mergers, with their associated GRBs and kilonovae, such as supernovae, binary white dwarfs, coalescence of supermassive black holes, tidal disruption events in the vicinity of supermassive black holes and many others. In this paper, we limit ourselves to a discussion on compact binary coalescences (CBCs). The paper is organized as follows: we first discuss the main scientific goals of the multi-messenger approach to CBCs; we then summarize where we stand today; and, finally, discuss the role of distributed architectures in CBCs multi-messenger research during the present decade and the 2030s.

2. Why Multi-Messenger Astrophysics? What We Want to Learn from Compact Binary Coalescence

Binary systems including two compact objects (black holes, BH, and neutron stars, NS) are unique laboratories for studying physics that is inaccessible in terrestrial laboratories; in particular strong-field general relativity, relativistic acceleration mechanisms, and matter under extreme conditions (density, temperature, magnetic field). It was postulated for many years that the NS-NS and BH-NS coalescence leads to the production of a short GRB, a bright flash of gamma-rays lasting less than a few seconds [3] (see also the reviews) [4,5]. The reason for this can be outlines as follows. The outcome of the coalescence is a BH (or a metastable NS), including most of the mass of the binary system, and an accretion disk including the leftovers (on the order of a few percent of the solar mass). Accretion of this matter onto the newly formed compact object can release 10^{52}–10^{53} ergs of gravitational energy, so even if a small fraction of this large amount can be converted to electromagnetic radiation a GRB can be generated. The duration of the burst is determined by the lifetime of the disk, which is expected to be a fraction of a second at these masses. Accreting compact objects can produce powerful jets, which can transport the energy from the launching site near the inner engine to the photosphere, usually at 10^7–10^8 Schwarzschild radii R_S. The jet launching site is thought to be near the newly formed BH, a potential laboratory for GR in the strong-field and the dynamics of matter under extreme conditions, or even new physics. In other words, are the BH and NS we observe in GW the same as described by GR? GWs are produced during the binary coalescence and settling of their remnants; therefore, they should precede the launching of the jet (the short GRB), while providing precise information on the physical condition of the system just before engine ignition. The physics at play outside the photosphere includes particle acceleration in internal shocks to produce the GRB prompt emission and external shocks to produce the GRB afterglow; both can be probed by electromagnetic observations. Detailed GRMHD simulations express this in a more quantitative framework ([6] and references therein). However, only observations can quantitatively confirm this scenario, connecting the physics near the event horizon of the newly formed BH to the physics of jet formation, collimation, and propagation. For example, the variability of the observed gamma-ray light curve may reflect the energy injection at the base of the jet [7,8]. The following key questions can be addressed by the multi-messenger approach to CBC:

- **What happens during the merger of compact objects?** How frequent is the coincidence with short GRBs; how frequent is the formation of powerful relativistic jets?

Non-detections are, in principle, as important as detections. This question is addressed through simultaneous observations and studies of the GW event and the high-energy emission associated with jet production.
- **What is the nature of the short GRB's central engine?** What powers the most powerful accelerators of the Universe, NS or BH accretion? The study of the gravitational wave form can distinguish the nature of the remnant. The detection of a short GRB would indicate that a powerful accelerator is in place.
- **What is the jet launching mechanism?** The delay time between the GW emission and γ-rays emission (the short GRB) can distinguish between different jet launching scenarios [9].
- **Do jets have a universal structure or does the structure depend on the type of the compact binary?** Perhaps on the mass/spin of the binary components or of the merger remnant? The jet structure can be addressed through both the analysis of the prompt event and its afterglow [10], as well as direct VLBI imaging [11,12]. This requires accurate localizations and follow-up observations from radio to X-rays.
- **What is the role of CBCs in the production of heavy elements in the Universe?** This requires accurate localizations and follow-up spectroscopic observations from UV to IR.

The first three questions require the detection of GW signals from a statistical sample of NS-NS and NS-BH systems and their sensitive high-energy coverage over the full sky. The fourth and fifth questions require the accurate determination of the position of the source of GWs and multiwavelength follow-up observations, from radio waves to gamma-rays, of the electromagnetic counterpart.

3. CBC Multi-Messenger Astrophysics Today

It is often assumed that the multi-messenger revolution has a very precise starting date: 17 August 2017, with the detection of GW170817 by the LIGO/Virgo interferometers and short GRB170817 by the Fermi/GBM and INTEGRAL/SPI. Unfortunately, no other event was detected simultaneously by GWs and gamma-rays (or any electromagnetic radiation) during the third LIGO/VIRGO observing run (O3). Since one swallow does not make a summer, all questions presented in the previous section are still open. Certainly, GW/GRB1708017 has shown how powerful the multi-messenger approach could be. Its full impact, however, measured by the ability to open a long-lasting, brand-new field must still be determined, and it will greatly depend on the capability of collecting statistical samples of GW-electromagnetic events in the near future, that is, during the LIGO/VIRGO O4 and O5 observing runs planned for this decade.

As of now, the LIGO/Virgo interferometers have detected two NS-NS coalescence events (GW170817 and GW190425) and one very likely BH-NS merger (GW190814).

The GW/GRB170817 event does not need to be commented on in detail here; many comprehensive reviews exist [6]. Briefly, Fermi/GBM and INTERGAL/SPI detected a short burst lasting for about 2 s, just 1.7 s after the NS-NS detection by LIGO/Virgo. The LIGO/Virgo error box was about 30 deg^2, a value that did not improve much after including the Inter Planetary Network (IPN) error box. The distance of the source 40^{+8}_{-14}Mpc was determined directly from the GW detection and greatly limited the search for an optical counterpart to only about 50 galaxies. An optical counterpart was discovered in NGC4993 only after about 11 h from the GW/GRB event, which gave rise to an impressive follow-up by nearly one hundred ground-based and space-based telescopes, placing constraints on the optical/IR counterpart of the GW source/GRB and discovering the first confirmed kilonova in history. The GRB detection, which was nearly simultaneous with the GW signal, confirmed that, at least in this case, a merger of two NSs is the origin of short GRBs. The follow-up observations allowed the complex jet structure to be determined, and the jet was observed as off-axis for the first time [10–12]. It was also found that kilonovae were one of the prime sites of heavy element production through the r-processes [6,13].

GW190425 was detected with a high significance by one interferometer only, and thus the uncertainty region included a large fraction of the sky [14]. No GRB was detected at the same time, with an upper limit on the 50–300 keV fluence of $\sim 10^{-6}$ ergs/cm^2 in 1 s [15], corresponding to a few ph/cm^2, by INTEGRAL/SPI. Swift/BAT and Fermi/GBM covered only a fraction of the region, and therefore could not contribute much to the INTEGRAL observation. The distance of the event inferred from the GW signal is 159^{+69}_{-71} Mpc. Therefore, even for a gamma-ray luminosity comparable to that of GRB170817, its flux would need to be 16 times dimmer, <0.1 ph/cm^2; hence, it is relatively difficult to detect with the available detectors. Moreover, the observed gamma-ray flux is a strong function of the inclination at which the jet is observed (a power law with an exponent $-5 \div -6$). Unfortunately, the inclination of this binary system was not well constrained by the interferometers, but assuming similar characteristics of GRB170817, we can infer a lower limit on the inclination between the jet and the line-of-sight of about 10 deg.

GW190814 was detected by all three interferometers, thus providing good constraints on the position of the source (error box of 18.5 deg^2), distance (241^{+41}_{-45} Mpc) and inclination (46 +/− 11 deg), in addition to the masses of the two coalescing objects and of the remnant (this was the GW event with the most unequal mass ratio, 0.11 +\− 0.01 ever detected). No significant electromagnetic counterpart was reported for this event, with INTEGRAL/SPI inferring a three-sigma upper limit for the flux of 3×10^{-7} erg/cm^2/s in the energy range 75–2000 keV [16]. Swift/BAT covered >99% of the error box with >10% partial coding, reporting a five-sigma upper limit of 10^{-7} erg/cm^2 between 15–350 keV in 1 s [17]. The lack of gamma-ray detection is likely due to the high inclination of the system, which corresponds to a huge (factor of 10^6–10^7) reduction in the observed flux. The jet emission at these inclinations is hardly observable by any conceivable all-sky monitor at the moment.

In summary, the lessons learnt from the three NS-NS and BH-NS events collected so far are: (1) the need for an **all-sky monitor** with a sufficient sensitivity to detect off-axis jets with intrinsic luminosities down to 10^{47} ergs, at least to a distance of 100–200 Mpc and an inclination of 10–20 deg (of course smaller inclination values allow for detection over larger distances). In fact, within this decade the LIGO/Virgo interferometers will reach their target sensitivity and the searched volume will become much larger with respect to O1–O3: the horizon for NS-NS merging events detected with a signal-to-noise ratio of 8 will reach ~200 Mpc for LIGO and 100–130 Mpc for Virgo in O4, implying a discovery volume ~100 times larger than in the GW170817 case. The capability to instantaneously cover the whole sky is mandatory, because the number of events will be small; therefore, missing simply one event would mean a considerable loss for scientific research. (2) The capability of determining the position of the transients with uncertainties is smaller than a few degrees. Within the volume defined by this spatial constraint and the distance of the GWE provided by the interferometric measurements, the number of optical transients will be small enough to easily assess the correct transient to associate with the GWE, thus prompting further follow-ups.

4. The Role of Distributed Architectures in Tomorrow's Multi-Messenger Astrophysics

4.1. A Constellation of Nano-Satelliters for High-Energy Transient Detection and Localization

Today, X-ray and gamma-ray monitors dedicated to the search and localization of high-energy transients are mostly monolithic instruments (e.g., NASA Swift/BAT, ESA INTEGRAL/IBIS, INTEGRAL/SPI, ASI AGILE/SuperAgile, AGILE/Microcalorimeter) or multiple detectors hosted by the same large spacecraft (NASA FERMI/GBM). One good example of distributed architecture used for GRB science since the beginning of this enterprise is the Inter-Planetary network (IPN). The accurate timing from X-ray and γ-ray detectors hosted on different spacecrafts for both Low-Earth Orbit (LEO) and High-Earth orbit (HEO) or even inter-planetary routes, is joined together to determine the transient position in the sky. This is achieved by using the delay time of arrival of the transient signal on different detectors, a strategy used since the earliest GRB detections by the VELA satellites at the end of the 1960s.

While the VELA satellites were fully dedicated to the detection of gamma-ray flashes (in this case the main target was the detection of gamma-rays from nuclear tests above the Earth's atmosphere), today, the IPN includes instruments on satellites fully dedicated to high-energy astrophysics, and instruments hosted by spacecrafts with quite different primary purpose. The diverse gamma-ray instrumentation used by the IPN, the poor knowledge of the spacecraft position when outside the Earth's GNSS infrastructure, as well as the difficulty of defining an absolute time with a good precision for all spacecrafts and the delay in communication with remote solar system spacecrafts, significantly limit the ability of the IPN to routinely provide accurate (better than a few degrees) and timely localizations. In other words, systematic errors associated with IPN transient localization are usually much greater than the statistical errors.

All of these difficulties could be overcome by a constellation of satellites in LEO hosting similar, if not identical, X-ray and gamma-ray detectors. The GPS and Galileo infrastructures offers the opportunity to constrain LEO satellite positions to within a few tens of meters, and the absolute time within a few tens of nanoseconds, orders of magnitude better than what is possible to achieve outside of the GNSS infrastructure. Adopting identical detectors ensures similar responses to cosmic events, reducing systematic uncertainties.

Since GRBs, and high-energy transients in general, are relatively bright (the flux from GRB170817 was about 2 ph/cm^2/s in the 50–300 keV band (and this was a faint GRB) seen from a relatively large off-axis jet angle), even a small instrument can be efficiently used for their detection. In fact, the collecting area of the Fermi/GBM modules is about 120 cm^2. Today, this class of instruments can be hosted by compact nano-satellites. Nano-satellites were developed at the end of the 20th century for didactical purposes, but today they are used for the most diverse applications: Earth observation, aircraft and ship tracking, telecommunication, and science. Nano-satellites have several major advantages with respect to traditional satellites. First and foremost, they can be developed on a relatively short timescale (a few years and, in extreme cases, a few months) compared with one to several decades usually required for standard space missions. Second, their cost is a few orders of magnitude smaller than standard large satellites. Both of these advantages imply that modularity can be exploited to its maximum. Modularity can allow us to (a) avoid single (or even multiple) point failures (if one or several units are lost the constellation and the experiment can still be operative); (b) fully test the hardware in orbit with the first launches and then improve it, if needed, with the following launches (iterative development, used in the context of gamma-ray flashes and GRBs early on by the VELA satellites. The second generation included gamma-ray detectors with much better timing capabilities, strongly improving the localization capabilities of the constellation); (c) to build the final mission step by step, gradually increasing its performance while diluting costs and risks.

An all-sky monitor, capable of monitoring the whole sky, or a large fraction of the sky, at all times, requires either a distributed infrastructure on LEO or a dedicated spacecraft far from the Earth. Given the advantages of miniaturized instrumentation hosted by CubeSats, it is relatively natural to propose building a sensitive all-sky monitor based on a constellation of nano-satellites.

In the previous section, in the context of multi-messenger astrophysics, we discussed the scientific relevance of even simple high-energy transient detections at the time of GWEs. Of course, their accurate localization would multiply the scientific return, prompting multi-wavelength (and even multi-messenger) follow-ups.

In the context of multi-wavelength astrophysics, it must be considered that, in a few years, the Vera Rubin Observatory (VRO) and the Cerenkov Array Telescope (CTA) will come online. VRO will have the ability to cover ~1/4 of the sky every night, finding millions of transients per night down to a magnitude r < 24.5 using real-time data analysis. On the one hand, optical counterparts of GRBs and GWEs will likely be serendipitously found in VRO images (covering each about 10 deg^2), providing arcsec positions and immediately prompting multiwavelength follow-up. On the other hand, the detection of X-ray and gamma-ray emission will promptly characterize the VRO transients (magnetars, soft gamma-ray repeaters, tidal distruption events, thermonuclear bursts from accreting NS, novae, AGN jets, etc., in addition to GRBs), thus better focusing the multiwavelength follow-up.

CTA will boost the study of the 20 GeV–300 TeV energy range. Only three GRBs have been detected at TeV energies so far: one by MAGIC [18] and two by HESS [19]. The CTA's fast re-positioning capabilities (20 s) and the improved sensitivity, due to the larger collecting area and lower energy threshold (~20 GeV) compared to MAGIC and HESS, will aid the study of GRB high-energy radiation routine, opening the possibility to accurately derive the jet Lorentz factor, assessing the role of synchrotron and inverse Compton radiation, and constraining the magnetic field strength and configuration. Given the limited field of view (FoV) of CTA at GeV energies (4.3°), an instrument operating during the 20 s and providing the localization of GRBs with errors smaller than the CTA FoV is of paramount importance for triggering CTA follow-up observations.

4.2. A Powerful Combination of Nano- and Micro-/Small Satellites

Four years later, GW170817A remains the only gravitational wave source with a detection of an electromagnetic counterpart. While GW170817 was quickly followed by a short GRB seen at an angle of 19–42 degrees from the jet axis, it is likely that most kilonovae will not emit a GRB observable from the Earth. The prompt γ-ray emission is strongly beamed, and it is estimated that only about 1 in 100 kilonovae will be detectable at high energies [20]. However, GRB170817A is an unusually long and faint short-GRB, and detections of other GRB counterparts for gravitational wave events can be important discoveries. In the previous section, we discussed how a constellation of nano-satellites can be efficiently used to provide a powerful, high-energy, all-sky monitor at a relatively low cost and on short timescales. Here, we discuss how a synergic constellation of micro-satellites could be used for follow-up observations, greatly enhancing the scientific return.

It will require coordination, but to a large extent, several nano- and micro-satellite constellations with different detectors could work together in conjunction, forming one network. The micro-satellites will perform rapid follow-up observations at near-UV/optical and near-IR wavelengths. To follow up kilonovae without GRB counterparts, detected only by GW observatories, these observations shall be triggered directly by their GW emission. This will require the near UV/optical/IR observatories to have large fields of view and fast repointing capabilities, enabling them to locate the electromagnetic counterparts of kilonovae after short mosaicing observations. The early time evolution of the near-UV to near-IR flux ratios will provide the key diagnostics to distinguish between various scenarios of kilonova explosions. No existing or proposed mission provides all-sky monitoring and localization together with rapid multi-wavelength follow-up capabilities.

The rapid follow-up observations at near-UV/optical, near infrared, and X-ray wavelengths are expected to produce real breakthroughs in our understanding of kilonovae. The luminous optical counterpart of GW170817 was initially blue in colour with the emission peaking at near-UV wavelengths. Then, over the course of a few days the emission shifted to the near-IR wavelengths. This fast spectral evolution was unlike that of any previously observed event. However, the optical counterpart was discovered only about 11 h after the gravitational wave signal. A wide-field UV space telescope, able to rapidly slew on-source, could revolutionize our understanding of these exciting events (e.g., Ultrasat, http://space.gov.il/en/node/1129; accessed on 1 December 2021). Theoretical modelling predicts that the first few hours might be dominated by near-UV emission from free neutrons, which do not have time to be captured by the nuclei. Observing this early emission is thus key for the understanding of the nucleosynthesis of kilonovae.

Figure 1 of Fernandez & Mezger 2016 [21] presents the phases of a binary neutron star merger as a function of time, showing the observational signatures, as well as the possible outcomes and the associated physical phenomena. The in-spiral and coalescence of neutron stars, which can be observed through gravitational waves, can result in a hyper-massive neutron star that quickly collapses into a black hole; into a stable, rapidly spinning, highly magnetised neutron star; or directly into a black hole. The merger gives rise to the ejection of 10^{-4}–10^{-2} solar masses of unbound matter, with velocities 0.1–0.3 c from the tidal tails in the equatorial region [20]. The ejected matter that remains bound to the resulting compact object falls back and forms an accretion disc that helps launch the ultra-relativistic jet, which produces the observed short GRB. The equatorial ejecta are expected to be rich in heavier elements, known as lantanides, and produce long-lasting infrared emission. UV and blue emissions are produced early in the kilonova and last for only about a day. This may arise from free neutrons or from lantanide-poor polar ejecta with a higher electron fraction. While the properties of the tidal ejecta are sensitive to the mass ratio of the neutron stars, the properties of the polar and wind ejecta are sensitive to the neutron star radii and to the nature of the merger product. The different ratios between the observed kilonova fluxes obtained by near-UV, optical and near-infrared observations, will allow us to identify and constrain the properties of the different ejecta [13]. In particular, UV observations, performed in the first few hours of the kilonova (unavailable for GW170817), will probe the mass, composition, and thermal content of the fastest ejecta and allow us to constrain its geometry, quantity, and kinematics, as well as the nature of the merger product. Micro-satellites carrying relatively small (with a collecting area around 200 cm^2) near-IR and near-UV space telescopes will probe the emission from kilonovae out to the distance beyond 200 Mpc.

Observations in the near-IR should also allow us to detect afterglows at redshifts $z > 5$, corresponding to the first billion years of the Universe. Long GRBs are mostly observed at cosmological distances, with the most distant GRB detected at the redshift of $z = 9.4$, which corresponds to a look-back time of 13.3 billion years, only 500 million years after the Big Bang. Long GRBs are thus excellent probes for examining the early Universe, when the first massive stars and their host galaxies were being formed, the first heavier elements were produced, and the diffuse interstellar and intergalactic matter was re-ionised. To truly exploit long GRBs as probes of the early Universe, we need to identify more GRBs from the first billion years after the Big Bang. This can only be achieved by rapid near-infrared, follow-up observations from space, capable of imaging the afterglows of GRBs at the edge of the observable Universe. This was the main goal of the Theseus mission [22], which unfortunately was not selected by ESA for realization. CubSats again can provide a contribution here; see, for example, the SkyHopper proposal [23] https://skyhopper.research.unimelb.edu.au; accessed on 1 December 2021.

Next to the near-IR and near-UV telescopes, the constellation would also benefit from a microsatellite carrying a wide-field 10 cm^2–20 cm^2 X-ray telescope observing in the 0.5–8 keV band. Since GRBs are bright in X-rays, a rapidly slewing X-ray telescope can aid the quick arcmin scale identification of the GRB position in the sky. The time and spectral evolution of the early X-ray emission can also provide valuable information about the possible two-step collapse model (through a short-lived massive neutron star) and the jet geometry of the source.

4.3. An Incomplete List of Pathfinder/Precursor Missions

In the past, many projects in the broad area of nano-satellites for high-energy astrophysics were proposed and several were funded, in order to build demonstrators or pathfinders. Among these, we mention the following, without pretending to be complete.

4.3.1. CAMELOT and GRBAlpha

One of the concepts developed to perform all-sky monitoring and timing-based localization of gamma-ray transients is called CAMELOT: Cubesats Applied for MEasuring and LOcalising Transients [24]. It is a constellation of 3U CubeSats equipped with large and thin ($150 \times 75 \times 5$ mm) CsI(Tl) scintillators read out by SiPM detectors, called multi-pixel photon counters (MPPCs), by Hamamatsu. The detectors are placed on two perpendicular walls of the satellites to maximize the effective photon collecting area on a CubeSat of this size.

The detector concept developed for the CAMELOT mission was first demonstrated on a 1U CubeSat, named GRBAlpha [25]. It carries a smaller, $75 \times 75 \times 5$ mm CsI (Tl) scintillator, which provides one-eighth of the expected effective area of the 3U CubeSats envisioned for the CAMELOT mission. GRBAlpha was successfully launched on 22 March 2021 to a sun-synchronous polar orbit with a Soyuz 2.1 rocket from Bajkonur. Following a short commissioning phase, the detector was switched on and regular measurements began. The on-board data acquisition software stack is periodically updated in orbit, continuously improving the capabilities of the science payload. The degradation of the SiPM detectors, which are protected by a 2 mm thick lead shield, is being monitored. About half of the polar orbit is plagued by a high particle background, and thus the duty cycle of the detector cannot be better than 50%, even if it is operated continuously. The satellite communicates on amateur radio frequencies and its ground segment is supported by the radio amateur community. The mission also takes advantage of the SatNOGS network for increased data downlink volume. Figure 1 shows GRBAlpha before the launch.

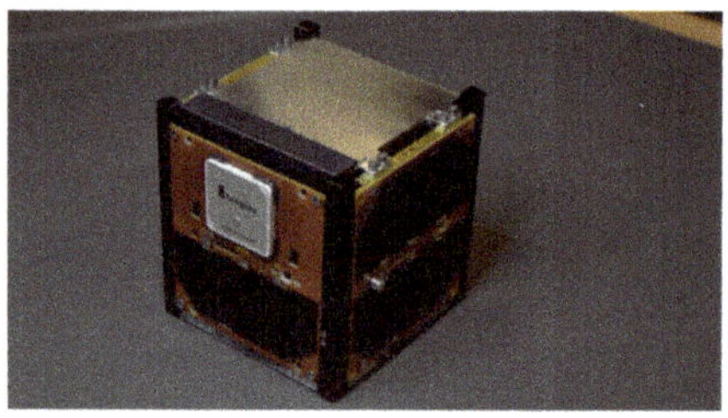

Figure 1. GRBAlpha, a 1U CubeSAT equipped with a gamma-ray detector (visible on the top, see [25] for details).

GRBAlpha became the first nano-satellite to detect multiple confirmed GRBs. At the time of writing, the mission detected five GRBs (four long and one short GRBs [26–30]), two of them over the course of a single night from 18 to 19 October. Figure 2 shows the light curves of GRB211018A observed by GRBAlpha in different energy bands [30]. GRB detectors developed for the CAMELOT mission are also going to be launched on VZLUSAT-2, which is a technological 3U CubeSat built by the Czech Aerospace Research Centre. The satellite will carry two perpendicular detectors the same size as GRBAlpha. VZLUSAT-2 is expected to be launched in January 2022 into a sun-synchronous polar orbit on a Falcon 9 rideshare mission.

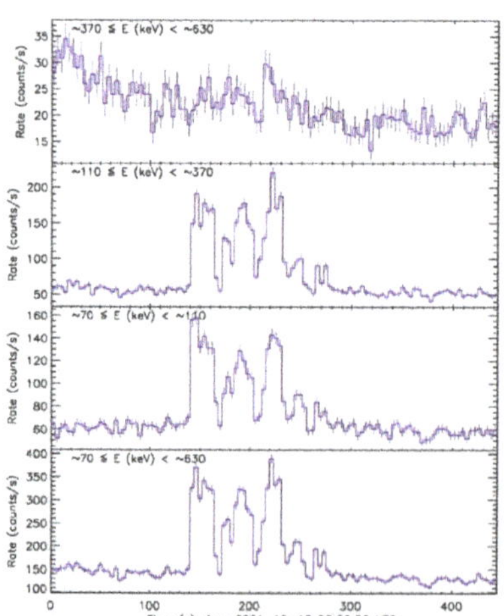

Figure 2. Light curves of GRB 211018A observed by GRBAlpha in different energy bands [30].

4.3.2. HERMES Pathfinder

HERMES-Technologic and Scientific pathfinder (HERMES pathfinder) is an in-orbit demonstration consisting of a constellation of six 3U nano-satellites hosting simple but innovative X-ray detectors for the monitoring of cosmic high-energy transients such as GRBs and the electromagnetic counterparts of GWEs [31]. The main objective of HERMES-TP/SP is to prove that an accurate position of high-energy cosmic transients can be obtained using miniaturized hardware, with a cost at least one order of magnitude smaller than that of conventional scientific space observatories and a development time as short as a few years.

The transient position is obtained by studying the delay time of arrival of the signal to different detectors hosted by nano-satellites on low Earth orbits [32]. To this purpose, particular attention is placed on reaching the best time resolution and time accuracy, with the goal of reaching an overall accuracy of a fraction of a micro-second [33]. The main goals of the project are: (1) join the multi-messenger revolution by providing the first mini-constellation for GRB localization with a total of six units (the first experiment of GRB triangulation with miniaturized instrumentation); (2) develop miniaturized payload technology for breakthrough science; (3) demonstrate COTS applicability to challenging missions, contribute to Space 4.0 goals, push and prepare for high-reliability large constellations.

Figure 3 shows the HERMES pathfinder detector system during integration. The 60 GAGG scintillator crystals can be seen to the right and the 12 10×10 silicon drift detector mosaics used to read out the crystals to the left (see [33] and references therein for a detailed description of the HERMES pathfinder payload).

Figure 3. HERMES pathfinder detector system during integration at Fondazione Bruno Kessler laboratories in Trento, Italy.

The HERMES-TP project is funded by the Italian Ministry for Education, University and Research, and the Italian Space Agency. The HERMES-SP project is funded by the European Union's Horizon 2020 Research and Innovation Programme under Grant Agreement No. 821896.

The consortium started the integration and testing of the first flight unit during the summer of 2021; the proto-flight model and its qualification review is foreseen for Q1 2022. The other five units will be integrated and tested during 2022 and the constellation is set to be launched to a nearly equatorial LEO in 2023.

HERMES pathfinder is intrinsically a modular experiment that can be naturally expanded to provide a global, sensitive, all-sky monitor for high-energy transients. The next step is SpIRIT, a 6U cubesat funded by the Australian Space Agency and managed by the University of Melbourne. SpIRIT will host one HERMES pathfinder payload, and will fly on an SSO at the same time as the HERMES pathfinder, forming a constellation of seven satellites in two different orbits.

4.3.3. GALI

A new concept for identifying the direction of GRBs was suggested recently by Rahin et al. [34]. The concept was named GALI (GAmma-ray burst Localizing Instrument). Its basic idea is to use numerous small scintillators (e.g., $1 \times 1 \times 1$ cm^3 cubes) in a 3D array utilizing their mutual shielding. Consequently, the relative γ-photon count of each scintillator varies strongly with the direction of the burst. In a sense, GALI can be thought of as a coded-mask detector, but where the mask itself has detecting elements. Moreover, the detector (and mask) have no preferred direction, and thus provide full-sky coverage, as opposed to coded-mask instruments. A configuration such as CAMELOT, benefits from the SiPMs, which occupy little volume; hence, they enable the compact packing of the scintillators. As with GRBAlpha/CAMELOT, the SiPMs are radiation-sensitive, and need to be protected. A GALI laboratory prototype was successfully tested, and a flight model to be launched to the International Space Station (ISS) is being built. The laboratory prototype is shown in Figure 4.

Figure 4. Laboratory model of the GALI detector system with 90 scintillators crystals. The individual scintillators in their reflective wrappings can be seen, arranged in a random order. A larger version is currently being built for the ISS.

The GALI concept can be scaled to any size, and thus can fit many platforms. Clearly, larger versions will be more sensitive, and, more importantly, they will provide a better angular resolution. An advantage of the GALI configuration is the reduced sky background on the inner scintllators, which will light up only for GRBs in specific directions. This allows for a high signal-to-background ratio on these scintillators, and the exploitation of soft γ-rays below 50 keV for directionality. The aforementioned flight model will consist of 350 scintillators, occupying a total volume of merely ~1 L. Simulations show that even such a small detector can identify the direction of a burst down to approximately $\pm 2°$ for 1 s GRBs with a 10 keV–1 MeV flux of 10 ph cm^{-2} s^{-1}, $\pm 5°$ for 5 ph cm^{-2} s^{-1}, and $\pm 10°$ for 2.5 ph cm^{-2} s^{-1} [34]. Although GALI can operate onboard a single satellite, it can also be incorporated into a distributed satellite architecture to enhance the sky coverage and directional capabilities of the entire constellation.

4.3.4. Other Projects

BurstCube is a 6U CubeSat developed by NASA, which will detect GRBs using four CsI scintillators, each with an effective area ~90 cm^2 [35]. BurstCube is expected to be launched in 2022.

The Educational Irish Research Satellite 1 (EIRSAT-1), supported by ESA's Fly Your Satellite program, will carry a gamma-ray module (GMOD) to detect gamma-ray bursts [36]. GMOD uses SensL B-series SiPM detectors and a CeBr scintillator. EIRSAT-1 will be launched from the ISS in 2022.

Nanosatellite constellations include the Chinese Gamma-Ray Integrated Detectors (GRID), which will consist of GRB detectors (as secondary payloads) on 10–24 CubeSats [37]. Two GRID units have been launched so far and one GRB has been recently detected by the second unit [38].

Author Contributions: Conceptualization, F.F., N.W. and E.B., writing, F.F., N.W. and E.B. All authors have read and agreed to the published version of the manuscript.

Funding: This research was funded by European Union Horizon 2020 Research and Innovation Framework Programme under grant agreements HERMES-SP n. 821896 and AHEAD2020 n. 871158, by ASI INAF Accordo Attuativo n. 2018-10-HH.1.2020 HERMES—Technologic Pathfinder Attivita' scientifiche, by a Center of Excellence of the Israel Science Foundation, grant No. 2752/19, and by a grant from the Israel Space Agency.

Institutional Review Board Statement: Not applicable.

Informed Consent Statement: Not applicable.

Data Availability Statement: Not applicable.

Acknowledgments: This article benefited from the invaluable input of the HERMES pathfinder, CAMELOT and GALI teams. In particular we would like to thank J. Ripa, A. Sanna, L. Burderi, M. Lavagna, P. Bellutti, Y. Evangelista, M. Trenti for inspiring discussions.

Conflicts of Interest: The authors declare no conflict of interest.

References

1. Klebesadel, R.W.; Strong, I.B.; Olson, R.A. Observations of Gamma-Ray Bursts of Cosmic Origin. *Astrophys. J. Lett.* **1973**, *182*, L85. [CrossRef]
2. Phinney, S.E. Finding and Using Electromagnetic Counterparts of Gravitational Wave Sources. A white paper for the Astro2010 Decadal Review. *arXiv* **2019**, arXiv:0903:0098.
3. Eichler, D.; Livio, M.; Piran, T.; Schramm, D.N. Nucleosynthesis, neutrino bursts and γ-rays from coalescing neutron stars. *Nature* **1989**, *340*, 126–128. [CrossRef]
4. Nakar, E. Short hard Gamma-ray bursts. *Phys. Rep.* **2007**, *442*, 166–236.
5. Berger, E. Short-Duration Gamma-Ray Bursts. *Annu. Rev. Astron. Astrophys.* **2014**, *52*, 43–105.
6. Nakar, E. The electromagnetic counterparts of compact binary mergers. *Phys. Rep.* **2020**, *886*, 1. [CrossRef]
7. Nakar, E.; Piran, T. Gamma-ray bursts light-curves—Another clue on the inner engine. *Astrophys. J. Lett.* **2002**, *572*, L139. [CrossRef]
8. Morsony, B.J.; Lazzati, D.; Begelman, M.C. The origin and propagation of the variability in the outflows of long duration gamma-ray bursts. *Astrophys. J. Lett.* **2010**, *723*, 267. [CrossRef]
9. Lazzati, D.; Ciolfi, R.; Perna, R. Intrinsic Properties of the Engine and Jet that Powered the Short Gamma-ray Burst Associated with GW170817. *Astrophys. J. Lett.* **2020**, *898*, 59.
10. Troja, E.; Piro, L.; Ryan, G.; Van Eerten, H.; Ricci, R.; Wieringa, M.H.; Lotti, S.; Sakamoto, T.; Cenko, S.B. The outflow structure of GW170817 from late-time broad-band observations. *Mon. Not. R. Astron. Soc.* **2018**, *478*, L18. [CrossRef]
11. Mooley, K.P.; Deller, A.T.; Gottlieb, O.; Nakar, E.; Hallinan, G.; Bourke, S.; Frail, D.A.; Horesh, A.; Corsi, A.; Hotokezaka, K. Superluminal motion of a relativistic jet in the neutron-star merger GW170817. *Nature* **2018**, *561*, 355. [CrossRef] [PubMed]
12. Ghirlanda, G.; Salafia, O.S.; Paragi, Z.; Giroletti, M.; Yang, J.; Marcote, B.; Blanchard, J.; Agudo, I.; An, T.; Bernardini, M.G.; et al. Compact radio emission indicates a structured jet was produced by a binary neutron star merger. *Science* **2019**, *363*, 968. [CrossRef] [PubMed]
13. Metzger, B.D. Kilonovae. *Living Rev. Relativ.* **2019**, *23*, 1–59. [CrossRef] [PubMed]
14. LIGO/Virgo Collaborations. GW190425: Observation of a Compact Binary Coalescence with Total Mass$\sim$3.4 $M_\odot$. *Astrophys. J. Lett.* **2020**, *892*, L3. [CrossRef]
15. Chelovekov, I.; Pozanenko, A.; Minaev, P.; Grebenev, S. GCN 2418. 2019. Available online: https://gcn.gsfc.nasa.gov/gcn/ (accessed on 1 December 2021).
16. Molkov, S.; Mereghetti, S.; Savchenko, V.; Ferrigno, C.; Rodi, J.; Coleiro, A. GCN 25323. 2019. Available online: https://gcn.gsfc.nasa.gov/gcn/gcn3/25323.gcn3 (accessed on 1 December 2021).
17. Palmer, M. GCN 25341. 2019. Available online: https://gcn.gsfc.nasa.gov/gcn3/25341.gcn3 (accessed on 1 December 2021).

18. Collaboration, M.A.G.I.C. Teraelectronvolt emission from the γ-ray burst GRB 190114C. *Nature* **2019**, *575*, 455. [CrossRef]
19. Abdalla, H.; Adam, R.; Aharonian, F.; Benkhali, F.A.; Angüner, E.O.; Arakawa, M.; Arcaro, C.; Armand, C.; Ashkar, H.; Backes, M.; et al. A very-high-energy component deep in the γ-ray burst afterglow. *Nature* **2019**, *575*, 464. [CrossRef] [PubMed]
20. Metzger, B.D.; Berger, E. What is the Most Promising Electromagnetic Counterpart of a Neutron Star Binary Merger? *Astrophys. J. Lett.* **2012**, *746*, 48. [CrossRef]
21. Fernandez, R.; Metzger, B.D. Electromagnetic Signatures of Neutron Star Mergers in the Advanced LIGO Era Annual Review of Nuclear and Particle. *Science* **2016**, *66*, 23.
22. Amati, L.; O'Brien, P.; Götz, D.; Bozzo, E.; Tenzer, C.; Frontera, F.; Ghirlanda, G.; Labanti, C.; Osborne, J.P.; Stratta, G.; et al. The THESEUS space mission concept: Science case, design and expected performances. *Adv. Space Res.* **2018**, *62*, 191–244. [CrossRef]
23. Mearns, R.; Trenti, M. Near real-time telecommand solutions for CubeSats: State of the art and applications to the SkyHopper mission. *arXiv* **2018**, arXiv:1808.06746.
24. Werner, N.; Řípa, J.; Pál, A.; Ohno, M.; Tarcai, N.; Torigoe, K.; Tanaka, K.; Uchida, N.; Mészáros, L.; Galgóczi, G.; et al. CAMELOT: Cubesats Applied for MEasuring and LOcalising Transients-Mission Overview. *arXiv* **2018**, arXiv:1806.03681v1.
25. Pal, A.; Ohno, M.; Mészáros, L.; Werner, N.; Řípa, J.; Frajt, M.; Hirade, N.; Hudec, J.; Kapuš, J.; Koleda, M.; et al. GRBAlpha: A 1U CubeSat mission for validating timing-based gamma-ray burst localization. *arXiv* **2020**, arXiv:2012.01298.
26. Ripa, J.; Ohno, M.; Topinka, M.; Pal, A.; Meszaros, L.; Csak, B.; Werner, N.; Topinka, M.; Munz, F.; Hroch, F.; et al. GCN 30624. 2021. Available online: https://gcn.gsfc.nasa.gov/gcn3/30624.gcn3 (accessed on 1 December 2021).
27. Ohno, M.; Ripa, J.; Takahashi, H.; Pal, A.; Meszaros, L.; Csak, B.; Werner, N.; Topinka, M.; Munz, F.; Hroch, F.; et al. GCN 30697. 2021. Available online: https://gcn.gsfc.nasa.gov/gcn3/30697.gcn3 (accessed on 1 December 2021).
28. Ripa, J.; Ohno, M.; Takahashi, H.; Pal, A.; Meszaros, L.; Csak, B.; Werner, N.; Topinka, M.; Munz, F.; Hroch, F.; et al. GCN 30840. 2021. Available online: https://gcn.gsfc.nasa.gov/gcn3/30840.gcn3 (accessed on 1 December 2021).
29. Ripa, J.; Ohno, M.; Takahashi, H.; Pal, A.; Meszaros, L.; Csak, B.; Werner, N.; Topinka, M.; Munz, F.; Hroch, F.; et al. GCN 30946. 2021. Available online: https://gcn.gsfc.nasa.gov/gcn3/30946.gcn3 (accessed on 1 December 2021).
30. Ripa, J.; Ohno, M.; Takahashi, H.; Pal, A.; Meszaros, L.; Csak, B.; Werner, N.; Topinka, M.; Munz, F.; Hroch, F.; et al. GCN 30945. 2021. Available online: https://gcn.gsfc.nasa.gov/gcn3/30945.gcn3 (accessed on 1 December 2021).
31. Fiore, F.; Burderi, L.; Lavagna, M.; Bertacin, R.; Evangelista, Y.; Campana, R.; Fuschino, F.; Lunghi, P.; Monge, A.; Negri, B.; et al. *The HERMES-Technologic and Scientific Pathfinder*; Society of Photo-Optical Instrumentation Engineers (SPIE) Conference Series: Washington, DC, USA, 2020; on line only; p. 11444-168.
32. Sanna, A.; Gambino, A.F.; Burderi, L.; Riggio, A.; Di Salvo, T.; Fiore, F.; Lavagna, M.; Bertacin, R.; Evangelista, Y.; Campana, R.; et al. *Timing Techniques Applied to Distributed Modular High-Energy Astronomy: The H.E.R.M.E.S. Project*; Society of Photo-Optical Instrumentation Engineers (SPIE) Conference Series: Washington, DC, USA, 2020; on line only; p. 11444-251.
33. Evangelista, Y.; Fiore, F.; Fuschino, F.; Campana, R.; Ceraudo, F.; Demenev, E.; Guzman, A.; Labanti, C.; La Rosa, G.; Fiorini, M.; et al. *The Scientific Payload on-Board the HERMES-TP and HERMES-SP CubeSat Missions*; Society of Photo-Optical Instrumentation Engineers (SPIE) Conference Series: Washington, DC, USA, 2020; on line only; p. 11444-168.
34. Rahin, R.; Moleri, L.; Vdovin, A.; Feigenboim, A.; Margolin, S.; Tarem, S.; Behar, E.; Ghelman, M.; Osovizky, A. *GALI a Gamma-ray Burst Localizing Instrument*; Society of Photo-Optical Instrumentation Engineers (SPIE) Conference Series: Washington, DC, USA, 2020; on line only; p. 1444-168.
35. Racusin, J.; Perkins, J.S.; Briggs, M.S.; Nolfo, G.; Krizmanic, J.; Caputo, R.; McEnery, J.E.; Shawhan, P.; Morris, D.; Connaughton, V.; et al. BurstCube: A CubeSat for Gravitational Wave Counterparts. *arXiv* **2017**, arXiv:1708.09292v1.
36. Murphy, D.; Murphy, D.; Flanagan, J.; Thompson, J.; Doyle, M.; Erkal, J.; Gloster, A.; O'Toole, C.; Salmon, L.; Sherwin, D.; et al. EIRSAT-1 The Educational Irish Research Satellite. In Proceedings of the 2nd Symposium on Space Educational Activities, Budapest University of Economics and Technology (BME), Budapest, Hungary, 11–13 April 2018; Volume 8859.
37. Wen, J.; Long, X.; Zheng, X.; An, Y.; Cai, Z.; Cang, J.; Che, Y.; Chen, C.; Chen, L.; Chen, Q.; et al. GRID: A Student Project to Monitor the Transient Gamma-Ray Sky in the Multi-Messenger Astronomy Era. *arXiv* **2019**, arXiv:1907.06842v1. [CrossRef]
38. Wang, Y.; Zheng, X.; Xiao, S.; Yang, J.; Liu, Z.-K.; Yang, J.-H.; Zhang, B.-B.; Zeng, M.; Xiong, S.-L.; Feng, H.; et al. GRB 210121A: A Typical Fireball Burst Detected by Two Small Missions. *Astrophys. J. Lett.* **2021**, *922*, 237. [CrossRef]

galaxies

MDPI

Review

The SVOM Mission

Maria Grazia Bernardini [1,*], Bertrand Cordier [2] and Jianyan Wei [3,†] on behalf of the SVOM Collaboration

[1] INAF—Osservatorio Astronomico di Brera, Via Emilio Bianchi 46, 23807 Merate, Italy
[2] CEA Saclay, DSM/IRFU/Service d'Astrophysique, 91191 Gif-sur-Yvette, France; bertrand.cordier@cea.fr
[3] National Astronomical Observatories, CAS, 20A Datun Road, Beijing 100012, China; wjy@nao.cas.cn
* Correspondence: maria.bernardini@inaf.it
† Membership of the SVOM Consortium is provided here: https://www.svom.eu/en/#filter=.cooperationarticle.portfolio201load.

Abstract: SVOM (Space-based multiband astronomical Variable Objects Monitor) is a sino-french mission that is dedicated to Gamma-Ray Burst (GRB) science, expected to be launched in mid 2023. The mission includes four space-based and three ground-based instruments that, working together, will discover GRBs and provide rapid multi-wavelength follow-up in order to obtain a complete coverage of the GRB emission over seven decades in energy, from the trigger up to the very late phases of the afterglow. Thanks to its characteristics, SVOM will play a crucial role in time-domain and multi-messenger astronomy.

Keywords: SVOM (Space-based multiband astronomical Variable Objects Monitor); gamma-ray bursts; time-domain astronomy

Citation: Bernardini, M.G.; Cordier, B.; Wei, J.; on behalf of the SVOM Collaboration. The SVOM Mission. *Galaxies* **2021**, *9*, 113. https://doi.org/10.3390/galaxies9040113

Academic Editors: Elena Moretti and Francesco Longo

Received: 16 November 2021
Accepted: 1 December 2021
Published: 4 December 2021

1. Overview of the Mission

SVOM (Space-based multiband astronomical Variable Objects Monitor) is a sino-french mission that is dedicated to Gamma-Ray Burst (GRB) science and, more generally, to the discovery and multi-wavelength follow-up of transient sources. It is expected to be launched in mid 2023 for a nominal duration of three years, plus an extended duration of two additional years. The mission includes four space-based and three ground-based instruments (Figure 1) that, working together, will provide multi-wavelength follow-up of the targets in order to obtain a complete coverage of the GRB emission over seven decades in energy, from the trigger up to the very late phases of the afterglow (see Figure 2). In particular, SVOM will be able to trigger and locate GRBs, and distribute alerts and accurate localization in real time. In addition to that, SVOM will adopt an optimal pointing strategy for ground-based follow-up, fostering synergies with other space and ground-based facilities and allowing for a large fraction of GRBs detected and observed with redshift measurements.

SVOM is the result of a collaboration between the Chinese and French national space agencies, CNSA (China National Space Administration) and CNES (Centre national d'études spatiales), with the main contributions from the Institute of Research into the Fundamental Laws of the Universe (Irfu) and the Research Institute of Astrophysics and Planetology (IRAP) for France and the National Astronomical Observatory (NAOC) and the Beijing High Energy Institute (IHEP) for China.

In the following, we describe the instruments and the observational strategy (Section 2) and give an overview of the main science cases that can be addressed with SVOM (Section 3). Finally, we outline the crucial role that SVOM will play thanks to its characteristics and its synergies with the large ground-based facilities that will be operational in the next decade (Section 4).

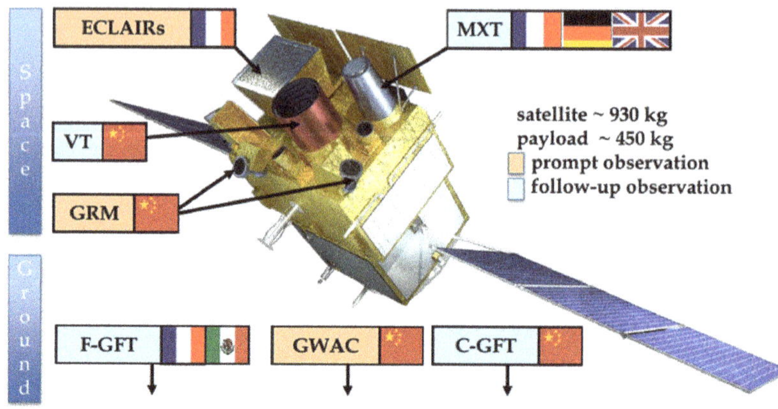

Figure 1. The SVOM (Space-based multiband astronomical Variable Objects Monitor) mission: The satellite with the four instruments on board (ECLAIRs, GRM, MXT, and VT), and the three instruments on ground (GWAC and the two GFTs).

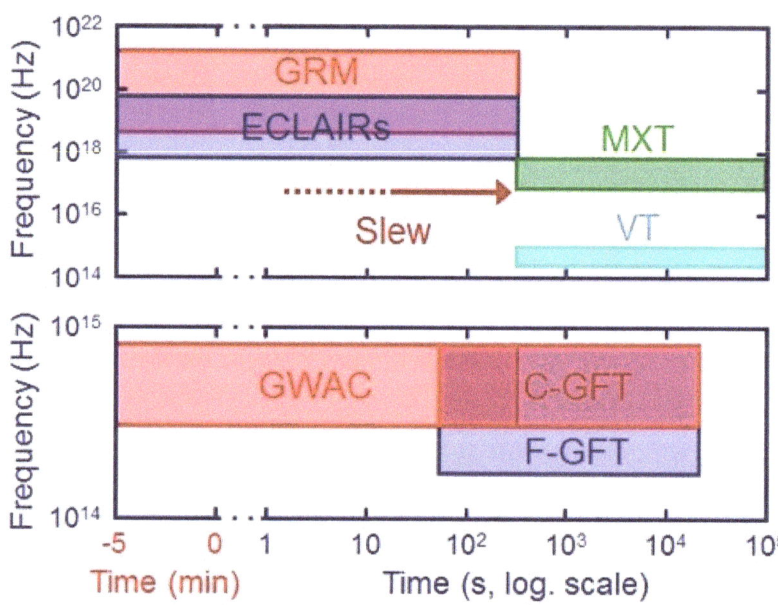

Figure 2. The ranges of energy and the typical timescales covered by the different instruments of the SVOM mission in space (**upper panel**) and on ground (**lower panel**).

2. Instruments and Observational Strategy

The SVOM satellite will be launched by the Chinese Long March 2C rocket from the Xichang launch base, and will be put into an orbit with a 30-degree inclination, an altitude of 625 km and an orbital period of 96 min. It will be equipped with four instruments, two French and two Chinese:

- The telecope ECLAIRs (France), a coded-mask instrument with a field of view of 2 sr, capable of triggering and locating GRBs with a precision of less than 12′ in the 4–120 keV energy band;

- The Gamma-Ray burst Monitor (GRM; China), composed by three units that together cover a larger field of view than ECLAIRs (∼5.6 sr) and extend its energy band up to 5 MeV;
- The Microchannel X-ray Telescope (MXT; France, with Germany and UK participation), that detects GRBs in the 0.2–10. keV energy band and is capable of locating them with a precision <13″;
- The Visible Telescope (VT; China), that observes GRBs in the visible band (400–1000 nm) and can localize them with a precision ∼1″.

These instruments will be supported by three ground-based facilities:

- The Ground–based Wide Angle Camera instrument (GWAC; China) consists of two sets of 16 wide-field cameras. Each set covers a field of view of 2500 deg^2, which is about 30% of the field of view of ECLAIRs. The first set is already operational and installed in the eastern part of China (Xinglong). The second set will be installed in the western part of China (Muztagh);
- Two 1-m class robotic GFTs (Ground Follow-up Telescopes; one French with Mexican contribution, one Chinese), that operate in the visible and Near Infrared (NIR) bands (400–1700 nm and 400–900 nm, respectively).

When a GRB is triggered within the field of view of ECLAIRs, the satellite automatically slews within a few minutes to repoint the two narrow-field instruments on board, MXT and VT, to provide fast detection and accurate localization of the possible X-ray and optical counterparts. The alert messages and the preliminary information collected are transmitted to the ground in about 10 s at Very High Frequency (VHF) in a frequency band between 137 and 138 MHz by an on-board antenna. The messages are then downlinked through a network of radio stations that, to ensure permanent contact with the satellite, will consist of ∼45 VHF antennae homogeneously deployed in the inter-tropical zone around the Earth, between latitudes −30° and +30°. They will relay every alert message from the satellite to the French Science Center (FSC) located at Saclay in France. After a first analysis at the FSC, the messages will be distributed to the whole scientific community through the GCN alert network or other networks available, in particular to the GFTs robotic telescopes that will refine the GRB position and give an initial indication of distance.

The rapid follow-up of SVOM-detected GRBs with ground-based optical telescopes in order to have a redshift measurement for a large fraction of them (∼2/3) is one of the scientific objectives of the mission. In order to avoid targets that are too close to the Sun and are thus not accessible by ground telescopes during their nighttime, an optimal pointing strategy will be adopted such that the optical axis of the instruments on board the satellite point in the direction opposite to the Sun. This choice implies that the Earth will hide the field of view of the instruments once per orbit, up to 50% of the period of revolution.

3. Science Cases

In the following, we give a description of the SVOM main observing programs, the GRB "core" program, the general program, and the Target of Opportunities (ToOs) program. For a complete discussion of all the science cases related to these programs, we refer to Wei, Cordier et al. (2016) [1].

The core program: Thanks to the unique properties of the mission outlined in Section 2, SVOM will provide a unique sample of ∼30–40 GRBs/yr with: (i) Complete coverage of the prompt emission over three decades in energy, thanks to the synergy between ECLAIRs and GRM (see e.g., Figure 3 and [2]), and possibly the detection of the prompt optical counterpart with the GWAC (∼16% of cases); (ii) the X-ray counterpart observed from a few minutes after the trigger up to a few days in X-rays, optical, and NIR; and (iii) a redshift measurement for ∼2/3 of them, allowing for a complete study of their physical properties in the rest frame. The low-energy threshold of ECLAIRs and the extension to the NIR of the GFTs will allow SVOM to tackle many open issues in GRBs science, as high-redshift GRBs ($z > 5$) that allows to probe the primordial universe, or soft GRBs (X-ray rich and X-ray flashes) and ultra-long GRBs ([3,4], see also Figure 3) that will shed light to the nature of the progenitors in the local universe and the physics of explosion. The joint observations of ECLAIRs and GRM will increase the efficiency in detecting short GRBs that, besides their interest per se, are one of the possible counterparts of Gravitational Wave (GW) emitters. Finally, the possible detection (or limits from the non-detection) of the optical counterpart of the GRB prompt emission will open a new window for unexpected discoveries.

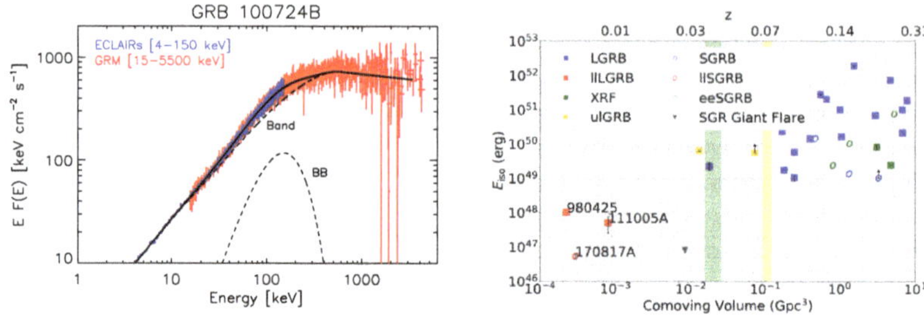

Figure 3. Left panel: The multi-component (band plus blackbody) prompt emission spectrum of the Fermi/GBM detected GRB 100724B, simulated as it would have been seen by ECLAIRs and GRM. The two instruments together will be capable to well characterize the peak energy and all the features of the prompt emission of GRBs. Adapted with permission from ref. [2], Copyright Year: 2017, Copyright Owner: Springer. **Right panel:** ECLAIRs on-axis count Signal-to-Noise ratio (SNR) for short high-energy transients in the local Universe: Long and Short GRBs (LGRB, SGRB), X-ray Flashes (XRF), ultra-long GRBs (ulGRBs), and Soft Gamma-ray Repeater (SGR) giant flares. The orange horizontal band represents the detection threshold of ECLAIRs at SNR = 6.5. The green and yellow bands represent the O4 LIGO distance sensitivity limits for double neutron stars and neutron star-black hole mergers, respectively. The light grey trails represent the evolution of the on-axis count SNR with the redshift. Adapted with permission from ref. [4], Copyright Year: 2020 Copyright Owner: Springer.

The general program: Although SVOM is a mission specifically designed for GRB science, its characteristics will find many applications in the study of many extra-galactic sources as Active Galactic Nuclei (AGN), Ultra-Luminous X-ray sources (ULX), and Tidal Disruption Events (TDE), and galactic sources as accreting systems, pulsars, magnetars, and flaring stars.

Rapid follow-up observations: Thanks to its flexibility, SVOM will be a powerful tool for the multi-wavelength follow-up of transients discovered by external facilities and not triggered by SVOM. This will be possible thanks to a Target of Opportunities (ToOs) program. SVOM will take advantage from wide-field instruments such as ECLAIRs, GRM, and GWAC to search for the counterparts at different wavelengths of events discovered by external facilities, and also of an X-ray telescope (MXT) with a field of view of 1 deg^2 that will allow SVOM to cover larger regions than Swift/XRT. This will be particularly important to identify the Electromagnetic (EM) counterparts of GW or neutrino emitters. Indeed, SVOM will have a specific ToO program dedicated to EM counterpart search in response to a multi-messenger alert. Concerning GW alerts, SVOM already joined the first three observing runs of the Advanced LIGO and Virgo interferometers with part of the ground segment (see e.g., [5]), and will be ready to join the first part of O4 (starting in late 2022) with its ground-based instruments before the launch of the spacecraft.

4. Conclusions

Time-domain, multi-messenger astronomy is a recognized priority in the next decade (see e.g., Pathways to Discovery in Astronomy and Astrophysics for the 2020s the National Academies of Science, Engineering and Medicine's latest decadal survey [1]), and the role of small-size space missions is considered to be crucial to both discover new transients and fully characterize them throughout the EM spectrum. SVOM will play a major role in this context. Thanks to its flexibility, the characteristics of the instruments and the observation strategies, it will: (i) Provide a unique sample of GRBs monitored over seven decades in energy and with redshift measurement, that can be used for population studies; (ii) open new discovery windows for GRB science (soft GRBs in the local universe, optical prompt emission, and high-redshift GRBs); (iii) complement current missions that are aging, as Fermi and Swift; and (iv) be a powerful tool to search for and characterize EM counterparts of external triggers. Concerning this last point, SVOM is specifically designed to foster synergies with the large ground-based facilities, and this aspect will be even more vital when the Cherenkov Telescope Array (CTA), the Vera Rubin Observatory (LSST), or the Square Kilometre Array (SKA) will be operational. Missions like SVOM are also necessary to fully exploit the scientific potential of transient astrophysical phenomena discovered via GWs and neutrinos, since these events also require EM observations across the spectrum for identification and further study.

Author Contributions: Writing—original draft preparation, M.G.B.; writing—review and editing, B.C. and J.W.; project administration, B.C. and J.W. All authors have read and agreed to the published version of the manuscript.

Funding: This research received no external funding.

Institutional Review Board Statement: Not applicable.

Informed Consent Statement: Not applicable.

Data Availability Statement: Not applicable.

Conflicts of Interest: The authors declare no conflict of interest.

Notes

1 https://nap.edu/resource/26141/interactive/ (accessed on 16 November 2021).

References

1. Wei, J.; Cordier, B.; Antier, S.; Antilogus, P.; Atteia, J.-L.; Bajat, A.; Basa, S.; Beckmann, V.; Bernardini, M.G.; Boissier, S.; et al. The Deep and Transient Universe in the SVOM Era: New Challenges and Opportunities—Scientific prospects of the SVOM mission. *arXiv* **2016**, arXiv:1610.06892.
2. Bernardini, M.G.; Xie, F.; Sizun, P.; Piron, F.; Dong, Y.; Atteia, J.-L.; Antier, S.; Daigne, F.; Godet, O.; Cordier, B.; et al. Scientific prospects for spectroscopy of the gamma-ray burst prompt emission with SVOM. *Exp. Astron.* **2017**, *44*, 113–127. [CrossRef]
3. Dagoneau, N.; Schanne, S.; Atteia, Je.; Götz, D.; Cordier, B. Ultra-Long Gamma-Ray Bursts detection with SVOM/ECLAIRs. *Exp. Astron.* **2020**, *50*, 91–123. [CrossRef]
4. Arcier, B.; Atteia, J.L.; Godet, O.; Mate, S.; Guillot, S.; Dagoneau, N.; Rodriguez, J.; Gotz, D.; Schanne, S.; Bernardini, M.G. Detection of short high-energy transients in the local universe with SVOM/ECLAIRs. *Astrophys. Space Sci.* **2020**, *365*, 185. [CrossRef]
5. Turpin, D.; Wu, C.; Han, Xu.; Xin, Li.; Antier, S.; Leroy, N.; Cao, L.; Cai, Ho.; Cordier, B.; Deng, J.; et al. The mini-GWAC optical follow-up of gravitational wave alerts—Results from the O2 campaign and prospects for the upcoming O3 run. *Res. Astron. Astrophys.* **2020**, *20*, 013. [CrossRef]

galaxies

MDPI

Article

Breakthrough Multi-Messenger Astrophysics with the THESEUS Space Mission †

Giulia Stratta [1,2,*], Lorenzo Amati [2], Marica Branchesi [3], Riccardo Ciolfi [4], Nial Tanvir [5], Enrico Bozzo [6], Diego Götz [7], Paul O'Brien [5] and Andrea Santangelo [8]

1 Istituto Nazionale di Astrofisica-Istituto di Astrofisica e Planetologia Spaziali di Roma, Via Fosso del Cavaliere 100, 00133, Rome, Italy
2 Istituto Nazionale di Astrofisica-Osservatorio di Astrofisica e Scienza dello spazio Bologna, Via P. Gobetti 101, 40129, Bologna, Italy; lorenzo.amati@inaf.it
3 Gran Sasso Science Institute, Viale F. Crispi 7, I-67100 L'Aquila, Italy; marica.branchesi@gssi.it
4 INAF, Osservatorio Astronomico di Padova, Vicolo dell'Osservatorio 5, I-35122 Padova, Italy; riccardo.ciolfi@inaf.it
5 School of Physics and Astronomy, University of Leicester, Leicester LE1 7RH, UK; nrt3@leicester.ac.uk (N.T.); pto2@leicester.ac.uk (P.O.)
6 Department of Astronomy, University of Geneva, 1205 Versoix, Switzerland; enrico.bozzo@unige.ch
7 Commissariat à l'Énergie Atomique et aux Énergies Alternatives, Centre National de la Recherche Scientifique, Astrophysics, Instrumentation-Modeling, Université Paris Saclay, F-91190 Gif-sur-Yvette, France; diego.gotz@cea.fr
8 Institut für Astronomie und Astrophysik, Abteilung Hochenergieastrophysik, Kepler Center for Astro and Particle Physics, Eberhard Karls Universitat, Sand 1, D 72076 Tuebingen, Germany; andrea.santangelo@uni-tuebingen.de
* Correspondence: giulia.stratta@inaf.it
† On behalf of the THESEUS Consortium: http://www.isdc.unige.ch/theseus/.

Citation: Stratta, G.; Amati, L.; Branchesi, M.; Ciolfi, R.; Tanvir, N.; Bozzo, E.; Götz, D.; O'Brien, P.; Santangelo, A. Breakthrough Multi-Messenger Astrophysics with the THESEUS Space Mission †. *Galaxies* **2022**, *10*, 60. https://doi.org/10.3390/galaxies10030060

Academic Editor: Elena Moretti, Francesco Longo and Yosuke Mizuno

Received: 28 February 2022
Accepted: 15 April 2022
Published: 21 April 2022

Abstract: The mission concept THESEUS (Transient High Energy Sky and Early Universe Surveyor) aims at exploiting Gamma-Ray Bursts (GRB) to explore the early Universe, as well as becoming a cornerstone of multi-messenger and time-domain astrophysics. To achieve these goals, a key feature is the capability to survey the soft X-ray transient sky and to detect the faint and soft GRB population so far poorly explored. Among the expected transients there will be high-redshift GRBs, nearby low-luminosity, X-ray Flashes and short GRBs. Our understanding of the physics governing the GRB prompt emission will benefit from the 0.3 keV–10 MeV simultaneous observations for an unprecedented large number of hundreds of events per year. In particular the mission will provide the identification, accurate sky localisation and characterization of electromagnetic counterparts to sources of gravitational wave and neutrino sources, which will be routinely detected during the 2030s by the upgraded second generation and third generation Gravitational Wave (GW) interferometers and next generation neutrino detectors.

Keywords: gamma–rays: bursts; cosmology: early universe; multi-messenger astrophysics: gravitational waves; neutrinos; instrumentation: X/gamma–ray astrophysics from space; fundamental physics

1. Introduction

Since their discovery in the late 1960s [1], the attention given by the scientific community to the gamma-ray burst (GRB) phenomenon has continuously increased. Nowadays, GRBs represent one of the most important sources in the Universe for observational cosmology, multi-messenger astronomy as well as extreme physics.

Their brightness in gamma-rays during the prompt phase allows us to detect the so-called "long-duration" GRBs, the GRBs generated by the gravitational collapse of rapidly rotating and massive stars, up to distances where the very first stars and galaxies form. The census of high-redshift GRBs allows us to shed light on the main processes responsible for Universe reionization after Dark Ages.

In the context of multi-messenger astronomy, "short-duration" GRBs are today playing a major role after the confirmation of their association with binary neutron star merger (BNS) systems, in particular for the case of GRB 170817A [2,3]. Electromagnetic counterparts of gravitational wave (GW) sources, are fundamental for accurate sky localization that allows us to improve GW parameter estimation accuracy and to measure the cosmological redshift of the GW sources. The latter has crucial implications for the measurement of cosmological parameters and, being an independent method, can potentially solve current tensions over the value of the Hubble constant [4].

Finally, GRBs provide a unique and extremely powerful benchmark for performing tests of fundamental physics. For instance, the vast photon flux emitted during the prompt emission phase of these phenomena over several orders of magnitude of energy, combined with their extreme cosmological distances, make these phenomena powerful probes for testing the Lorentz Invariance Violation (LIV), which is predicted by different families of Quantum Gravity theories [5].

In all these respects, future GRB missions (such as the proposed THESEUS and Gamow Explorer mission concepts) will provide an ideal synergy with the large electromagnetic facilities of the future like the VRO/LSST, ELT, TMT, SKA, CTA and ATHENA in the electromagnetic domain, and advanced second generation (2G) and third generation (3G) GW detectors and future large neutrino detectors (e.g., KM3NeT) in the non-photonic domain.

2. Multi-Messenger Astrophysics with Gamma-ray Bursts

In 2015 the advanced GW interferometers LIGO and Virgo [6,7] have detected gravitational waves for the first time [8,9]. Nowadays, the GW transient source catalogs published by the LIGO/Virgo collaboration counts several dozens of GW source candidates [10,11], all identified as compact binary coalescences. Among these sources, two are consistent with being BNSs and one has been associated with a short GRB (GRB170817A) [2,3]. In contrast, no neutrino counterpart has been found associated with any bright GRBs, putting stringent constraints on neutrino production in relativistic jets associated with these events [12]. The nearby class of long GRBs with low-luminosity (LL-GRBs) as well as ultra-long duration have been suggested as more promising candidates with respect to the bright GRBs (see e.g., [13] and references therein). The key feature of THESEUS is to independently detect these electromagnetic counterparts at the time of neutrino events and to provide refined sky localizations to allow multi-wavelength prolonged follow-up with other facilities.

2.1. GRB 170817

The first multi-messenger observations of a GRB happened on 17 August 2017 when a short GRB (GRB170817A) was detected with *Fermi* [14] and *INTEGRAL* [15].

The sky position and burst trigger time resulted to be consistent with an independent detection of GW event (GW 170817) achieved with the advanced LIGO and Virgo network [3,16]. The observed gravitational waveform of GW 170817 was consistent with the one expected from a BNS. The association with GRB 170817A has a gaussian-equivalent significance of 5.3σ [3] and marks the first direct evidence of short GRB progenitor. This breakthrough result confirmed a wealth of indirect evidence gathered in almost 20 years of short GRB observations, for instance the lack of any association with core-collapse supernova (contrary to long GRBs) yet plausible evidence for kilonova emission in some cases the mixed-type nature of the host galaxies (early and late) and the GRB sites within the host galaxy (see e.g., [17] and references therein).

Another milestone reached with GRB 170817A was the first direct evidence of a narrow, relativistic expanding jet. Indeed, from both the prompt and afterglow emission properties, it was realized that this burst was observed with a non-null viewing angle with respect to the jet axis. This allowed imaging and monitoring in time of the orthogonal component of the jet with the superb spatial accuracy of the VLBI and the results were consistent with a compact source moving at relativistic velocities [18]. This result in turn confirms that BNSs can produce ultra-relativistic jets.

A major advance was also achieved due to the fact that the gravitational waveform of compact binary coalescences encodes not only information on the masses, but also on the luminosity distance D_L. Indeed, by combining the measured D_L for GW170817 with the redshift measurement of the host galaxy NGC 4993, identified through the electromagnetic counterpart follow-up [2], it was possible to estimate the Universal expansion rate through the Hubble constant (H_0) measurement [4]. Despite the large uncertainties obtained with this single measure, this result showed the feasibility of a new, independent method to measure H_0, with strong implications for the current tension plaguing the outcomes from different probes used to measure this fundamental parameter (see [4] and references therein).

2.2. Next Achievements

In the next years, with the second generation GW interferometer network (i.e., LIGO, Virgo, KAGRA and by 2025 also LIGO-India, see [19]) we expect to detect other cases like GRB 170817A from BNS mergers and possibly short GRBs associated with NS-BH merger systems up to distances $z < 0.1 - 0.2$. The joint detection rate is still very uncertain but possibly limited to a few cases within the 2020s. This low rate is mainly due to the collimated nature of GRB emission that confines simultaneous detections to a tiny fraction of the total GW events associated with BNS or NS-BH, for which GWs are emitted isotropically.

Interesting predictions concern the possible formation of a NS remnant with large magnetic field (magnetar) after a BNS merger. In this case, the dipole radiation, expected mainly in the X-ray band and with low level of collimation, increases the chances of joint detection. The presence of a spinning-down, new-born magnetar is among the possible scenarios invoked to explain the "Extended Emission" detected after some short GRBs, a $\sim$100 s lasting component with softer spectrum than the main short burst, that may represents a potential X-ray counterpart of a BNS merger [20]. Another potential multi-messenger X-ray target expected to be less collimated than short GRBs is represented by the afterglow emission during the so called "plateau" phase, characterized by a nearly constant flux level lasting on a timescale that goes from a few hundreds of seconds up to $\sim$1 day. The origin of the plateaus is still debated: among the possible scenarios is the presence of a magnetar pumping energy into the forward shock. In this case, a long-transient continuous GW emission might be simultaneously detected [21]. An alternative scenario invokes the high-latitude prompt emission or afterglow emission from a structured jet for an observer line of sight slightly offset with respect to the jet axis (e.g., [22,23]). In this case, GW continuous emission is not necessarily expected. We note that X-ray plateaus are observed also for long GRBs: in this case, if a magnetar is the origin of this feature, long-transient continuous GWs may be detected also from this other class of GRBs.

During the 2030s, the third generation GW detectors are expected to be operational, with sensitivity nearly one order of magnitude higher. By that time, the distance up to which a BNS can be detected is $z \gtrsim 1$, thus implying a huge detection rate, of the order of $O(10^5)$ per year [24].

With such large detection rate, the fraction of joint detection as short GRBs will be high and will allow statistical studies on large samples. Among the possible issues that can be tackled with a statistical approach there are: (i) jet launching mechanisms and efficiency, (ii) the universality of the jet structure, (iii) differences/commonalities among BNS and NS-BH systems; (iv) accurate cosmological parameter measurements. The high sensitivity of 3G detectors, in addition, will make the detection of the faint GW emission from cc-SN a realistic proposition, possibly up to Mpc scales, allowing us to gain crucial insights on the still uncertain explosion mechanisms.

3. The THESEUS Mission Concept

The Transient High Energy Sky and Early Universe Surveyor (THESEUS) mission concept (Figure 1), developed in recent years by a large European-led collaboration involving also scientists worldwide, aims to fully exploit the unique and breakthrough potential of GRBs for investigating the Early Universe and substantially advancing multi-messenger astrophysics, while simultaneously vastly increasing the discovery space of most high energy transient phenomena over the entirety of cosmic history and allowing tests of fundamental physics [25]. THESEUS will achieve these ambitious goals through a step change in capabilities for detection and characterisation of GRBs and other transients over a very broad energy band (0.3 keV to 10 MeV) and wide field of view, including on-board near-infrared imaging and spectroscopy, and is designed to be at the forefront of these science fields in the late 2030s.

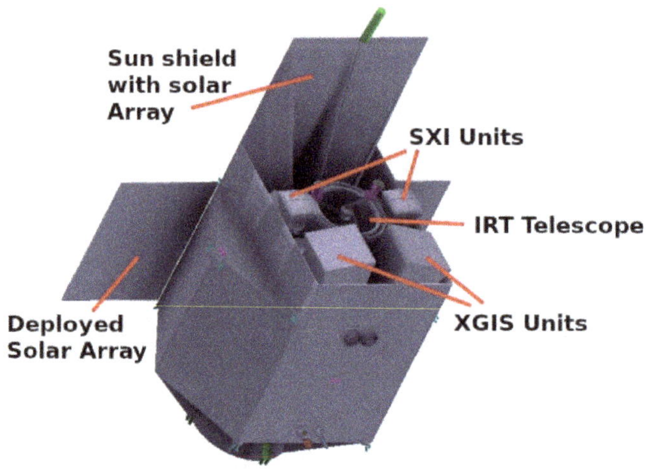

Figure 1. Possible spacecraft design and payload accomodation of THESEUS (Credit: ESA and THESEUS Consortium).

THESEUS will inherently be a mission enabling great synergies with the premier future observatories, providing simultaneous wide sky monitoring, rapid follow-up and real-time alerts (Figure 2). Figure 3 provides a tentative timeline of these observatories with respect to the THESEUS expected operational period. From *ELT* to *ATHENA*, *CTA* to Einstein Telescope, the Vera Rubin Observatory to the Roman Space Telescope, the science returns from combining observations with multiple facilities is a classic case of "the whole being much greater than the sum of the parts" [26]. A broad range of other science programmes will be enabled by THESEUS, including using observations of GRB emission as laboratories of ultra-relativistic matter and, e.g., for testing Lorentz invariance [5], as well as gathering statistics on large populations of other high-energy sources and transients [27]. Thus, THESEUS data will be of interest to a very wide user community, also through its open guest-observer programme.

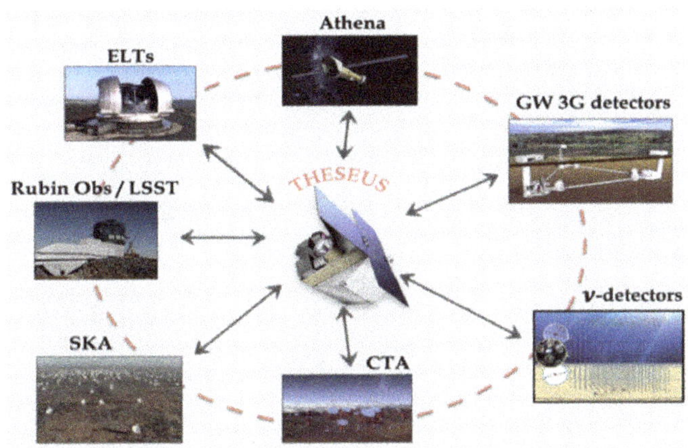

Figure 2. Synergy of THESEUS with next generation very large facilities in the multi-wavelength and multi-messenger domains [26].

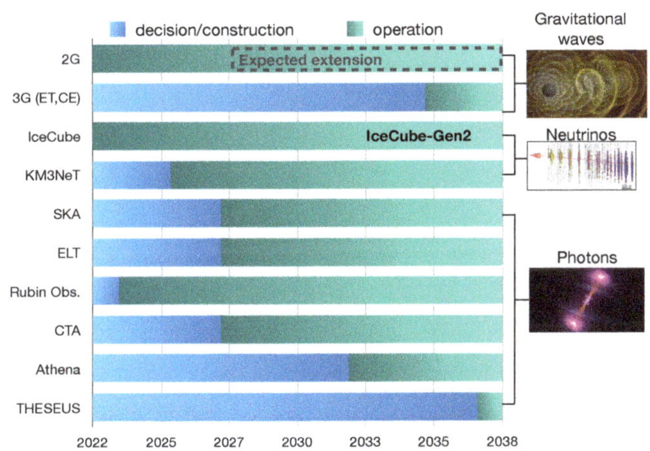

Figure 3. Planned timelines of the main failicities for Multi-Messenger Astronomy in the next decades with which THESEUS will operate in strong synergy (see Figure 2) [26].

3.1. Scientific Goals and Requirements

The scientific goals for the exploration of the early Universe require the detection, identification, and characterization of several tens of long GRBs occurring in the first billion years of the Universe ($z > 6$) within the 4 years of nominal mission lifetime of THESEUS [28]. This would be a giant leap with respect to what has been obtained in the last 20 years (8 GRBs at $z > 6$), using past and current GRB dedicated experiments like *Swift/BAT*, *Fermi/GBM*, *Konus-WIND* combined with intensive follow-up programs from the ground with small robotic and large telescopes (e.g., VLT). This breakthrough performance can be achieved by overcoming the current limitations through an extension of the GRB monitoring passband to the soft X-rays with an increase of at least one order of magnitude in minimum detectable flux with respect to previously-flown wide-field X-ray monitors. As well, a substantial improvement of the efficiency of counterpart detection, spectroscopy and redshift measurement will be enabled through prompt on-board near-infrared (NIR) follow-up observations (Table 1).

At the same time, the goals for multi-messenger astrophysics and time domain astronomy require:(i) a substantial advance in the detection and localization, over a large (>2 sr) FoV of short GRBs as electromagnetic counterparts of GW signals coming from BNS, and possibly NS-BH mergers; (ii) monitoring the high-energy sky with an unprecedented combination of sensitivity, location accuracy and field of view in the soft X-rays; iii) imaging up to the hard X-rays and spectroscopy/timing of the soft gamma-rays [27,29,30].

Table 1. Key science performance requirements of THESEUS[1]. The sensitivity requirements assume a power-law spectrum with a photon index of 1.8 and an absorbing column density of 5×10^{20} cm^{-2}.

SXI sensitivity (3σ)	1.8×10^{-11} erg cm^{-2} s^{-1} (0.3–5 keV, 1500 s) 10^{-10} erg cm^{-2} s^{-1} (0.3–5 keV, 100 s)
XGIS sensitivity (1 s, 3σ)	10^{-8} erg cm^{-2} s^{-1} (2–30 keV) 3×10^{-8} erg cm^{-2} s^{-1} (30–150 keV) 2.7×10^{-7} erg cm^{-2} s^{-1}(150 keV–1 MeV)
IRT sensitivity (imaging, SNR = 5, 150 s)	20.9 (I), 20.7 (Z), 20.4 (Y), 20.7 (J), 20.8 (H)
SXI FoV	0.5 sr – 31×61 deg^2
XGIS FoV ($\geq$20% efficiency)	2 sr (2–150 keV) – 117×77 deg^2 4 sr ($\geq$150 keV)
IRT FoV	$15' \times 15'$
Redshift accuracy ($6 \leq z \leq 10$)	$\leq$ 10%
IRT resolving power	$\geq$400
XGIS background stability	$\leq$10% over 10 min
Field-of-Regard	$\geq$50% of the sky
Trigger broadcasting delay to ground-based networks	$\leq$30 s (65% of the alerts) $\leq$20 min (65% of the alerts)
External alert (e.g., GW or ν events) reaction time	>4–12 h
SXI positional accuracy (0.3–5 keV, 99% c.l.)	$\leq$2 arcmin
XGIS positional accuracy (2–150 keV, 90% c.l.)	$\leq$7 arcmin (50% of triggered short GRBs) $\leq$15 arcmin (90% of triggered short GRBs)
IRT positional accuracy (5σ detections) real time post-processing	 $\leq$5 arcsec $\leq$1 arcsec

3.2. *On-Board Scientific Instruments*

Based on the above mentioned mission scientific requirements and the unique heritage and worldwide leadership of the Consortium in the enabling technologies, the THESEUS payload (Figure 1) will include the following scientific instruments:

- Soft X-ray Imager (SXI, 0.3–5 keV): a set of two "Lobster-eye" telescope units, covering a total FoV of ~0.5 sr with source location accuracy $< 2'$, focusing onto innovative large size X-ray CMOS detectors [31];
- X-Gamma ray Imaging Spectrometer (XGIS, 2 keV–10 MeV): a set of two coded-mask cameras using monolithic SDD+CsI X- and gamma-ray detectors, granting a ~2 sr imaging FoV and a source location accuracy <15 arcmin in 2–150 keV, an energy band from 2 keV up to 10 MeV and few μs timing resolution [32];

- InfraRed Telescope (IRT, 0.7–1.8 μm): a 0.7-m class IR telescope with 15′ × 15′ FoV, with imaging (I, Z, Y, J and H) and moderate spectroscopic (resolving power, R 400, through 2′ × 2′ grism) capabilities [33].

The instruments' Data Handling Units (DHU) will operate in synergy, thus optimizing the capability of detecting, identifying and localizing likely transients in the SXI, XGIS and IRT FoVs, as well as providing the unprecedented capability of on-board autonomous redshift measurements.

3.3. Mission Profile

The baseline launcher / orbit configuration is a launch with a Vega-C to a low inclination (<6°) Low Earth Orbit (LEO, 550–640 km altitude), which has the unique advantages of granting a low and stable background level in the high-energy instruments, allowing the exploitation of the Earth's magnetic field for spacecraft fast slewing and facilitating the prompt transmission of transient triggers and positions to the ground. The mission profile will include: (a) a spacecraft autonomous slewing capability >7°/min; (b) the capability of promptly (within a few tens of seconds at most) transmitting to the ground the trigger time and positions of GRBs (and other transients of interest) through the Trigger Broadcasting Unit (TBU) transmitter (via inter-satellite systems like, e.g., ORBCOMM, Iridium and, in case of US contribution, the NASA/TDRSS) and the THESEUS Burst Alert Ground Segment (TBAGS). The main ground station will be 10 m antenna (X-band receiver) at ASI "Luigi Broglio Space Centre" in Malindi (Kenya). As assessed during ESA/M5 Phase A study through a sophisticated Mission Observation Simulator (MOS), the mission scientific goals could be achieved with a nominal duration of 4 years (about 3.5 years of scientific operations). The Mission Operation Control (MOC) and Science Operations Centre (SOC) will be managed by ESA, while the Science Data Centre (SDC) will be under the responsibility of the Consortium.

The thorough R&D activities carried on by the THESEUS Consortium and ESA during the M5 Phase A study, grant a Technology Readiness Level (TRL) already close to that required for mission adoption for the main payload elements. The technical feasibility of the spacecraft, including payload accommodation and thermal control, the required pointing accuracy and stability, the reliability of reaching TRL at mission adoption, the compatibility of a launch with Vega-C in LEO, as well of the overall mission profile, within the M-class mission boundaries and according to the THESEUS scientific requirements, has also already been successfully assessed by M5 Phase A study.

The baseline mission operation concept includes a Survey mode, during which the monitors are waiting for GRBs and other transients of interest. Following a GRB (or transient of interest) trigger validated by the Data Handling Unit (DHU) system, the spacecraft enters a Burst mode (improved data acquisition and spacecraft slewing), followed by a pre-determined (but flexible) IRT observing sequence (Follow-up and Characterization or Deep Imaging modes). The pointing strategy during the Survey mode will be such as to maximize the combined efficiency of the sky monitoring by SXI and XGIS and that of the follow-up with the IRT. Small deviations (of the order of a few degrees until core science goals are achieved) from the Survey mode pointing strategy will be possible so to point the IRT on sources of interest pre-selected through a Guest Observer (GO) programme. Scientific modes also include an external trigger (or Target of Opportunity) mode, in which the IRT and high-energy monitors will be pointed to the direction of a GRB, transient or, e.g., to the error region of a GW or neutrino signal, provided by an external facility.

4. Multi-Messenger Astrophysics with THESEUS

The large THESEUS/XGIS and SXI field of view and sky localization accuracy will secure independent triggers on the electromagnetic counterparts of several GW and neutrino sources and their localization down to arcmin/arcsec level. The synergies of THESEUS with next generation neutrino and GW observatories will significantly increase the number of multi-messenger detections, enabling unprecedented robust statistical studies of multi-messenger sources [30].

By the end of the 2030s, 3G GW interferometers as the Einstein Telescope [24,34] and the Cosmic Explorer [35], are expected to operate at full sensitivity and likely within a network configuration. In the left panel of Figure 4 we show the THESEUS/XGIS short GRB redshift distribution, where the expected detection rate is of the order of 12/year. These numbers are obtained from simulations of THESEUS pointing strategy, considering all observational constraints, and a random set of short GRB triggers based on the population model of Ghirlanda et al. (2016). This model is built on past short GRBs observed with *Swift* and *Fermi* (i.e., before GRB170817A), considered to be "aligned" (for which the line of sight falls inside the narrow core of the corresponding jet). By taking into account the BNS merger detection efficiency of the 3G interferometers, we also show in each redshift bin the expected fraction that will be jointly detected with ET only and by a network of ET located in Europe plus 2 CEs assumed to be located one in USA and the other in Australia. Expected joint detection rates are quoted in Table 1.

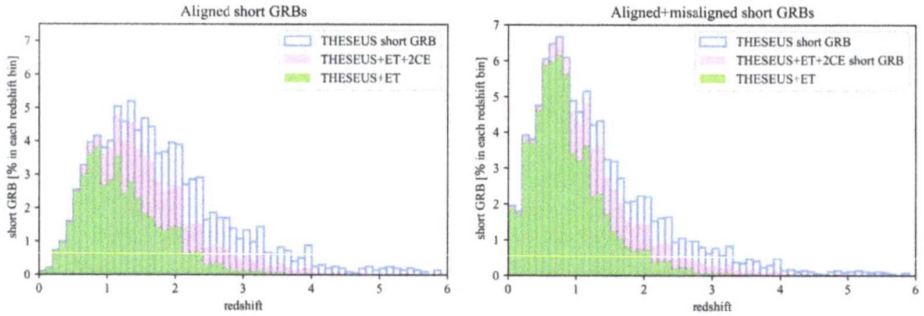

Figure 4. (**Left**): Short GRB redshift distribution detected with THESEUS/XGIS assuming an on-axis configuration (blue) and jointly with ET (green) and a network of 3G composed by 2CE and ET (pink). (**Right**): Same figure but now Short GRB with off-axis configurations are included.

THESEUS/XGIS is suitable to detect soft-faint bursts as those we expect to observe from large viewing angles ("misaligned") [30]: in the right panel of Figure 4 we included "misaligned" short GRBs by assuming a structured jet model from [18,36]. We find that the most nearby events can be detected up to large viewing angles [30]. As a consequence, the number of THESEUS short GRB detections at small redshifts is significantly increased with respect to the "aligned" only case. Since at low redshift the GW interferometers BNS merger detection efficiency is near 100%, this improvement is also reflected in the number of joint detections (see third column of Table 1).

With such joint detection rates, THESEUS will allow us to build statistical samples of multi-messenger sources with which a number of fundamental issues can be investigated [30]. To mention some examples:

- through the detection or non detection of a SGRB simultaneously with a BNS/NS-BH event for which the orbital inclination angle is accurately measured through the GW waveform, the jet formation efficiency in such systems can be estimated as well as a comparison between BNS and NS-BH systems.
- from the temporal gap between the GRB trigger and the GW events, jet launching mechanisms can be deeply investigated.
- from the detection or non detection of continuous GWs after the merger it will be possible to constrain the formation of a NS remnant versus a BH, a crucial issue that encodes information on the NS Equation of State and that will allow us to understand the origin of still unexplained features observed in short GRBs such as the Extended Emission and the plateau that THESEUS/XGIS and SXI will detect (Figure 5).

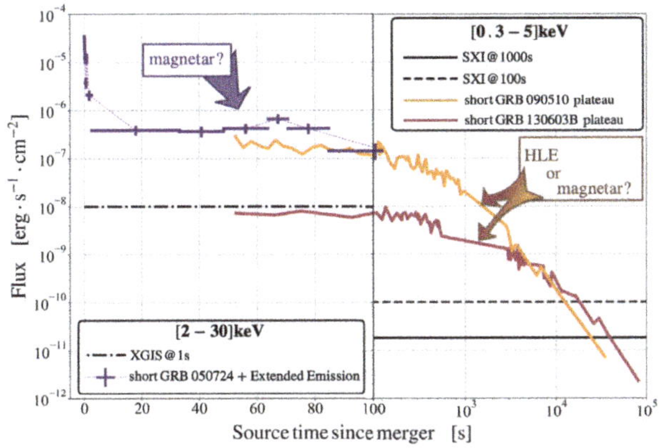

Figure 5. The X-raylong monitoring from prompt emission to the afterglow of relatively bright short GRB with THESEUS/XGIS and SXI. The linear temporal-scale plot on the left shows the short GRB prompt with the "Extended Emission" as it would be detected by THESEUS/XGIS possibly explained by invoking a magnetar remnant. The logarithmic temporal-scale plot on the right shows the X-ray plateau phase with the indication of two possible scenarios invoked for its origin, i.e., High-Latitude Emission of the prompt assuming a structured jet [22] and a spinning-down magnetar pumping energy into the external shock [37] [Credit: S. Vinciguerra].

Last but not least, from a large sample of sources with independent measurement of the cosmological redshift and luminosity distance, the Hubble constant can be measured with the sufficient accuracy to solve the current tensions among different measurements. In Figure 6 we plot the gaussian probability distributions of the current estimates of H_0 and the expected improvement of the GW170817 distribution by assuming a realistic number of 20 joint detected events (Table 2).

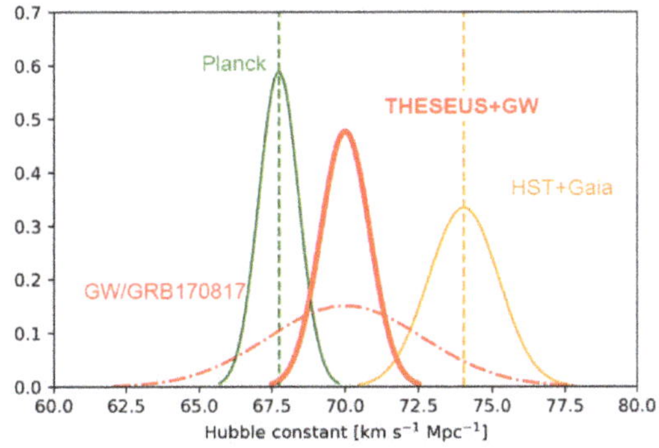

Figure 6. Gaussian probability distribution of the current H_0 measurements, including the one obtained from the BNS GW170817, and the expected one with ~20 joint GW+GRB detections. During the end of the 2030s, 20 joint detection is a conservative estimate for three years of synergies between THESEUS and a network of ET and 2 CE by taking into account the expected fraction of short GRB for which a redshift can be measured.

Table 2. Expected joint detections of short GRBs with THESEUS and the 3G GW interferometers (ET = Einstein Telescope, CE = Cosmic Explorer) expected to be operational in the second half of the 2030s, by assuming 3.45 years of joint observations [30].

GW Detectors in the '30s	Joint Detections (Aligned Short GRBs)	Joint Detections (Aligned + Misaligned Short GRBs)
ET	20	45
ET + CE	25	55
ET + 2CE	30	60

THESEUS will disseminate accurate sky localization (arcmin/arcsecond uncertainties) within seconds/minutes to the astronomical community, thus enabling large ground and space-based telescopes available by the end of 2030s to observe and deeply characterise the nature of large sample of multi-messenger sources [26,38]. Figure 2 illustrates the main large multi-wavelength and multi-messenger facilities expected to operate in synergy with THESEUS and Figure 3 shows the planned timeline.

5. Conclusions and Perspectives

THESEUS mission concept aims to fully exploit the transformative potential of GRBs for investigating the Early Universe and multi-messenger astrophysics, while simultaneously vastly expanding the discovery space of most high energy transient phenomena over the entirety of cosmic history and allowing tests of fundamental physics. THESEUS will achieve these ambitious goals through a step change in capabilities for the detection and characterisation of GRBs and other transients over a very broad energy band (0.3 keV to 10 MeV) and wide field of view, including on-board near-infrared imaging and spectroscopy, and is designed to be at the forefront of these science fields in the late 2030s.

THESEUS will also enable huge synergies with the premier future observatories (e.g., ELT, ATHENA, CTA, the next generation GW and neutrino telescopes), providing simultaneous wide sky monitoring, rapid follow-up and real-time alerts. A broad range of other science programmes will be enabled by THESEUS, including using observations of GRB emission as laboratories of ultra-relativistic matter and, e.g., for testing Lorentz invariance, as well as gathering statistics on large populations of other high energy sources and transients. Thus, THESEUS data will be of interest to a very wide user community, also through its open guest-observer programme.

The European leadership in these scientific and technological area builds on past success and ongoing investment: in addition to the involvement in pioneering satellite missions (BeppoSAX, HETE-2, XMM-Newton, INTEGRAL, Swift, SVOM), Europe has been at the heart of developments in multi-messenger astrophysics (e.g., Virgo, Einstein Telescope, KM3Net) and in electromagnetic follow-up via European-led consortia (e.g., ENGRAVE, VINROUGE, STARGATE). THESEUS is one of the three mission concepts that were selected by ESA in 2018 for a Phase A study as candidates for the M5 mission. The ESA Phase A study as candidate M5 mission, conducted from Fall 2018 to Spring 2021 led to detailed, workable and well qualified solutions for the spacecraft, its payload and operations. It has also demonstrated the technical and programmatic feasibility of accomplishing the core science goals with this mission concept. Based on these great achievements and heritage, the THESEUS project is being further developed for responding to new opportunities for medium-class missions. The later launch schedule with respect to M5 of these new opportunities will further improve the scientific return of THESEUS for multi-messenger astrophysics and time-domain astronomy, allowing for a great synergy with third generation GW detectors (expected to begin operations only in the second half of the 2030s) and improving the synergy with, e.g., the ATHENA and LISA space observatories (expected to be launched in the mid '30s), while maintaining the relevance of the other key science goals.

Author Contributions: Conceptualization, L.A.; writing–original draft preparation, G.S.; formal analysis, G.S., M.B., R.C.; supervision, N.T., E.B., D.G., P.O. and A.S. All authors have read and agreed to the published version of the manuscript.

Funding: This research was funded by ASI and INAF through agreement n 2018-29-HH.0., by the UE H2020 programme through the AHEAD2020 project and by the national space agencies and institutions supporting the THESEUS Consortium.

Institutional Review Board Statement: Not applicable.

Informed Consent Statement: Not applicable.

Data Availability Statement: Not applicable.

Acknowledgments: The authors, on behalf of the entire THESEUS consortium, are grateful to the ESA appointed THESEUS Science Study Team, as well as the entire ESA Study Team, for all their work and support across the mission assessment phase (2018–2021). The authors acknowledge the anonymous referees for their useful comments.

Conflicts of Interest: The authors declare no conflict of interest.

Note

1 from the THESEUS Assessment Study Report https://sci.esa.int/s/8Zb0RB8 (18 February 2021).

References

1. Klebesadel, R.W.; Strong, I.B.; Olson, R.A. Observations of Gamma-Ray Bursts of Cosmic Origin. *Astrophys. J.* **1973**, *182*, L85. [CrossRef]
2. Abbott, B.P.; Bloemen, S.; Canizares, P.; Falcke, H.; Fender, R.P.; Ghosh, S.; Groot, P.; Hinderer, T.; el Hör, J.R.; Jonker, P.G.; et al. Multi-messenger Observations of a Binary Neutron Star Merger*. *Astrophys. J. Lett.* **2017**, *848*, L12. [CrossRef]
3. Abbott, B.P.; Abbott, R.; Abbott, T.D.; Acernese, F.; Ackley, K.; Adams, C.; Adams, T.; Addesso, P.; Adhikari, R.X.; Adya, V.B.; et al. Gravitational Waves and Gamma-Rays from a Binary Neutron Star Merger: GW170817 and GRB 170817A. *Astrophys. J. Lett.* **2017**, *848*, L13. [CrossRef]
4. Abbott, B.P.; Abbott, R.; Abbott, T.D.; Acernese, F.; Ackley, K.; Adams, C.; Adams, T.; Addesso, P.; Adhikari, R.X.; Adya, V.B.; et al. A gravitational-wave standard siren measurement of the Hubble constant. *Nature* **2017**, *551*, 85–88. [CrossRef]
5. Burderi, L.; Sanna, A.; Di Salvo, T.; Riggio, A.; Iaria, R.; Gambino, A.F.; Manca, A.; Anitra, A.; Mazzola, S.M.; Marino, A. Quantum gravity with THESEUS. *Exp. Astron.* **2021**, *52*, 439–452. [CrossRef]
6. Collaboration, T.L.S.; Aasi, J.E.A. Advanced LIGO. *Class. Quantum Gravity* **2015**, *32*, 074001. [CrossRef]
7. Acernese, F.A.; Agathos, M.; Agatsuma, K.; Aisa, D.; Allemandou, N.; Allocca, A.; Amarni, J.; Astone, P.; Balestri, G.; Ballardin, G.; et al. Advanced Virgo: A second-generation interferometric gravitational wave detector. *Class. Quantum Gravity* **2015**, *32*, 024001. [CrossRef]
8. Abbott, B.P.; Abbott, R.; Abbott, T.D.; Abernathy, M.R.; Acernese, F.; Ackley, K.; Adams, C.; Adams, T.; Addesso, P.; Adhikari, R.X.; et al. Observation of Gravitational Waves from a Binary Black Hole Merger. *Phys. Rev. Lett.* **2016**, *116*, 061102. [CrossRef]
9. Abbott, B.P.; Abbott, R.; Abbott, T.D.; Abernathy, M.R.; Acernese, F.; Ackley, K.; Adams, C.; Adams, T.; Addesso, P.; Adhikari, R.X.; et al. Localization and Broadband Follow-up of the Gravitational-Wave Transient GW150914. *Astrophys. J. Lett.* **2016**, *826*, L13. [CrossRef]
10. Abbott, B.P.; Abbott, R.; Abbott, T.D.; Abraham, S.; Acernese, F.; Ackley, K.; Adams, C.; Adhikari, R.X.; Adya, V.B.; Affeldt, C.; et al. GWTC-1: A Gravitational-Wave Transient Catalog of Compact Binary Mergers Observed by LIGO and Virgo during the First and Second Observing Runs. *Phys. Rev. X* **2019**, *9*, 031040. [CrossRef]
11. Abbott, R.; Abbott, T.D.; Abraham, S.; Acernese, F.; Ackley, K.; Adams, A.; Adams, C.; Adhikari, R.X.; Adya, V.B.; Affeldt, C.; et al. GWTC-2: Compact Binary Coalescences Observed by LIGO and Virgo during the First Half of the Third Observing Run. *Phys. Rev. X* **2021**, *11*, 021053. [CrossRef]
12. Aartsen, M.G.; Ackermann, M.; Adams, J.; Aguilar, J.A.; Ahlers, M.; Ahrens, M.; Al Samarai, I.; Altmann, D.; Andeen, K.; Anderson, T.; et al. Extending the Search for Muon Neutrinos Coincident with Gamma-Ray Bursts in IceCube Data. *Astrophys. J.* **2017**, *843*, 112. [CrossRef]
13. Kimura, S.S. Neutrinos from Gamma-ray Bursts. *arXiv* **2022**, arXiv:2202.06480.
14. Goldstein, A.; Veres, P.; Burns, E.; Briggs, M.S.; Hamburg, R.; Kocevski, D.; Wilson-Hodge, C.A.; Preece, R.D.; Poolakkil, S.; Roberts, O.J.; et al. An Ordinary Short Gamma-Ray Burst with Extraordinary Implications: Fermi -GBM Detection of GRB 170817A. *Astrophys. J. Lett.* **2017**, *848*, L14. [CrossRef]
15. Savchenko, V.; Ferrigno, C.; Kuulkers, E.; Bazzano, A.; Bozzo, E.; Brandt, S.; Chenevez, J.; Courvoisier, T.J.L.; Diehl, R.; Domingo, A.; et al. INTEGRAL Detection of the First Prompt Gamma-Ray Signal Coincident with the Gravitational-wave Event GW170817. *Astrophys. J. Lett.* **2017**, *848*, L15. [CrossRef]

16. LIGO Scientific Collaboration and Virgo Collaboration; Abbott, B.P.; Abbott,R.; Abbott, T.D.; Acernese, F.; Ackley, K.; Adams, C.; Adams, T.; Addesso, P.; Adhikari, R. X.; et al. GW170817: Observation of Gravitational Waves from a Binary Neutron Star Inspiral. *Phys. Rev. Lett.* **2017**, *119*, 161101. [CrossRef] [PubMed]
17. Berger, E. Short-Duration Gamma-Ray Bursts. *Annu. Rev. Astron. Astrophys.* **2014**, *52*, 43–105. [CrossRef]
18. Ghirlanda, G.; Salafia, O.S.; Paragi, Z.; Giroletti, M.; Yang, J.; Marcote, B.; Blanchard, J.; Agudo, I.; An, T.; Bernardini, M.G.; et al. Compact Radio Emission Indicates a Structured Jet Was Produced by a Binary Neutron Star Merger. *Science* **2019**, *363*, 968. [CrossRef]
19. Abbott, B.P.; Abbott,R.; Abbott, T.D.; Abraham, S.; Acernese, F.; Ackley, K.; Adams, C.; Adya, V.B.; Affeldt, C.; Agathos, M.; et al. Prospects for observing and localizing gravitational-wave transients with Advanced LIGO, Advanced Virgo and KAGRA. *Living Rev. Relativ.* **2020**, *23*, 3. [CrossRef]
20. Bucciantini, N.; Arons, J.; Amato, E. Modelling spectral evolution of pulsar wind nebulae inside supernova remnants. *Mon. Not. R. Astron. Soc.* **2011**, *410*, 381–398. [CrossRef]
21. Dall'Osso, S.; Stella, L. Millisecond Magnetars. *arXiv* **2021**, arXiv:2103.10878.
22. Oganesyan, G.; Ascenzi, S.; Branchesi, M.; Salafia, O.S.; Dall'Osso, S.; Ghirlanda, G. Structured Jets and X-Ray Plateaus in Gamma-Ray Burst Phenomena. *Astrophys. J.* **2020**, *893*, 88. [CrossRef]
23. Beniamini, P.; Granot, J.; Gill, R. Afterglow Lightcurves from Misaligned Structured Jets. *Mon. Not. R. Astron. Soc.* **2020**, *493*, 3521–3534. [CrossRef]
24. Maggiore, M.; Van Den Broeck, C.; Bartolo, N.; Belgacem, E.; Bertacca, D.; Bizouard, M.A.; Branchesi, M.; Clesse, S.; Foffa, S.; García-Bellido, J.; et al. Science case for the Einstein telescope. *J. Cosmol. Astropart. Phys.* **2020**, *3*, 050. [CrossRef]
25. Amati, L.; O'Brien, P.T.; Götz, D.; Bozzo, E.; Santangelo, A.; Tanvir, N.; Frontera, F.; Mereghetti, S.; Osborne, J.P.; Blain, A.; et al. The THESEUS space mission: Science goals, requirements and mission concept. *Exp. Astron.* **2021**, *52*, 183–218. [CrossRef]
26. Rosati, P.; Basa, S.; Blain, A.W.; Bozzo, E.; Branchesi, M.; Christensen, L.; Ferrara, A.; Gomboc, A.; O'Brien, P.T.; Osborne, J.P.; et al. Synergies of THESEUS with the large facilities of the '30s and GO opportunities. *arXiv* **2021**, arXiv:2104.09535.
27. Mereghetti, S.; Balman, S.; Caballero-Garcia, M.; Del Santo, M.; Doroshenko, V.; Erkut, M.H.; Hanlon, L.; Hoeflich, P.; Markowitz, A.; Osborne, J.P.; et al. Time domain astronomy with the THESEUS satellite. *Exp. Astron.* **2021**, *52*, 309–406. [CrossRef]
28. Tanvir, N.R.; Le Floc'h, E.; Christensen, L.; Caruana, J.; Salvaterra, R.; Ghirlanda, G.; Ciardi, B.; Maio, U.; D'Odorico, V.; Piedipalumbo, E.; et al. Exploration of the high-redshift universe enabled by THESEUS. *Exp. Astron.* **2021**, *52*, 219–244. [CrossRef]
29. Stratta, G.; Ciolfi, R.; Amati, L.; Bozzo, E.; Ghirlanda, G.; Maiorano, E.; Nicastro, L.; Rossi, A.; Vinciguerra, S.; Frontera, F.; et al. THESEUS: A key space mission concept for Multi-Messenger Astrophysics. *Adv. Space Res.* **2018**, *62*, 662–682. [CrossRef]
30. Ciolfi, R.; Stratta, G.; Branchesi, M.; Gendre, B.; Grimm, S.; Harms, J.; Lamb, G.P.; Martin-Carrillo, A.; McCann, A.; Oganesyan, G.; et al. Multi-messenger astrophysics with THESEUS in the 2030s. *Exp. Astron.* **2021**, *52*, 245–275. [CrossRef]
31. O'Brien, P.; Hutchinson, I.; Lerman, H.N.; Feldman, C.H.; McHugh, M.; Lodge, A.; Willingale, R.; Beardmore, A.; Speight, R.; Drumm, P. The soft X-ray imager on THESEUS: The transient high energy survey and early universe surveyor. *arXiv* **2021**, arXiv:2102.08700.
32. Labanti, C.; Amati, L.; Frontera, F.; Mereghetti, S.; Gasent-Blesa, J.L.; Tenzer, C.; Orleanski, P.; Kuvvetli, I.; Campana, R.; Fuschino, F.; et al. The X/Gamma-ray Imaging Spectrometer (XGIS) on-board THESEUS: Design, main characteristics, and concept of operation. *arXiv* **2021**, arXiv:2102.08701.
33. Götz, D.; Basa, S.; Pinsard, F.; Martin, L.; Arhancet, A.; Bozzo, E.; Cara, C.; Escudero Sanz, I.; Frugier, P.A.; Floriot, J.; et al. The Infra-Red Telescope (IRT) on board the THESEUS mission. *arXiv* **2021**, arXiv:2102.08696.
34. Punturo, M.; Abernathy, M.; Acernese, F.; Allen, B.; Andersson, N.; Arun, K.; Barone, F.; Barr, B.; Barsuglia, M.; Beker, M.; et al. The Einstein Telescope: A third-generation gravitational wave observatory. *Class. Quantum Gravity* **2010**, *27*, 194002. [CrossRef]
35. Reitze, D.; Adhikari, R.X.; Ballmer, S.; Barish, B.; Barsotti, L.; Billingsley, G.; Brown, D.A.; Chen, Y.; Coyne, D.; Eisenstein, R.; et al. Cosmic Explorer: The U.S. Contribution to Gravitational-Wave Astronomy beyond LIGO. *Bulletin Am. Astron. Soc.* **2019**, *51*, 35.
36. Salafia, O.S.; Ghirlanda, G.; Ascenzi, S.; Ghisellini, G. On-axis view of GRB 170817A. *Astron. Astrophys.* **2019**, *628*, A18. [CrossRef]
37. Dall'Osso, S.; Stratta, G.; Guetta, D.; Covino, S.; De Cesare, G.; Stella, L. Gamma-ray bursts afterglows with energy injection from a spinning down neutron star. *Astron. Astrophys.* **2011**, *526*, A121. [CrossRef]
38. Piro, L.; Ahlers, M.; Coleiro, A.; Colpi, M.; de Oña Wilhelmi, E.; Guainazzi, M.; Jonker, P.G.; Mc Namara, P.; Nichols, D.A.; O'Brien, P.; et al. Multi-messenger-Athena Synergy White Paper. *arXiv* **2021**, arXiv:2110.15677.

galaxies

MDPI

Review

Probing Gamma-Ray Burst VHE Emission with the Southern Wide-Field-of-View Gamma-Ray Observatory

Giovanni La Mura [1,*,†], Ulisses Barres de Almeida [2], Ruben Conceição [1,3], Alessandro De Angelis [4,5,6], Francesco Longo [7,8,9], Mário Pimenta [1,2], Bernardo Tomé [1,3] and Davide Miceli [10]

1 Laboratório de Instrumentação e Física Experimental de Partículas (LIP), Av. Prof. Gama Pinto 2, 1649-003 Lisboa, Portugal; ruben@lip.pt (R.C.); pimenta@lip.pt (M.P.); bernardo@lip.pt (B.T.)
2 Brazilian Center for Physics Research (CBPF), Rua Dr. Xavier Sigaud 150, Rio de Janeiro 22290-180, Brazil; ulisses@cbpf.br
3 Instituto Superior Técnico (IST), Av. Rovisco Pais 1, 1049-001 Lisboa, Portugal
4 Dipartimento di Fisica e Astronomia—Università di Padova, Via Marzolo 8, 35131 Padova, Italy; alessandro.deangelis@unipd.it
5 Dipartimento di Scienze Matematiche, Informatiche e Fisiche—Università Degli Studi di Udine, Via Palladio 8, 33100 Udine, Italy
6 Istituto Nazionale di Fisica Nucleare sez. Padova (INFN), Via Marzolo 8, 35131 Padova, Italy
7 IFPU—Institute for Fundamental Physics of the Universe, Via Beirut 2, 34014 Trieste, Italy; francesco.longo@ts.infn.it
8 INFN, Sezione di Trieste, Via A. Valerio 2, 34100 Trieste, Italy
9 Dipartimento di Fisica, Università degli Studi di Trieste, Via A. Valerio 2, 34100 Trieste, Italy
10 Laboratoire d'Annecy de Physique des Particules (LAPP), CNRS-IN2P3, 9 Chemin de Bellevue—BP 110, 74941 Annecy, France; davide.miceli@lapp.in2p3.fr
* Correspondence: glamura@lip.pt or swgo_spokespersons@swgo.org
† Current Address: Laboratório de Instrumentação e Física Experimental de Partículas (LIP), Av. Prof. Gama Pinto 2, 1649-003 Lisboa, Portugal

Citation: La Mura, G.; Barres de Almeida, U.; Conceição, R.; de Angelis, A.; Pimenta, M.; Tomé, B.; Miceli, D. Probing Gamma-Ray Burst VHE Emission with the Southern Wide-Field-of-View Gamma-Ray Observatory. *Galaxies* **2021**, *9*, 98. https://doi.org/10.3390/galaxies9040098

Academic Editor: Yi-Zhong Fan

Received: 29 September 2021
Accepted: 5 November 2021
Published: 8 November 2021

Abstract: Recent observations have confirmed that Gamma-Ray Burst (GRB) afterglows produce Very High-Energy radiation (VHE, $E > 100\,\text{GeV}$). This highly anticipated discovery opens new scenarios in the interpretation of GRBs and in their role as probes of Extragalactic Background Light (EBL) and Lorentz Invariance Violation (LIV). However, some fundamental questions about the actual nature of VHE emission in GRBs and its evolution during the burst are still unsolved. These questions will be difficult to address, even with future imaging Cherenkov telescopes, such as the Cherenkov Telescope Array (CTA). Here we investigate the prospects of gamma-ray sky monitoring with Extensive Air Showers arrays (EAS) to address these problems. We discuss the theoretical aspects connected with VHE radiation emission and the implications that its temporal evolution properties have on the interpretation of GRBs. By revisiting the high-energy properties of some *Fermi*-LAT detected GRBs, we estimate the typical fluxes expected in the VHE band and compare them with a range of foreseeable instrument performances, based on the Southern Wide Field-of-view Gamma-ray Observatory concept (SWGO). We focus our analysis on how different instrument capabilities affect the chances to explore the burst onset and early evolution in VHE, providing invaluable complementary information with respect to Cherenkov telescope observations. We show that under the assumption of conditions already observed in historical events, the next-generation ground monitoring detectors can actually contribute to answer several key questions.

Keywords: instrumentation—detectors; gamma rays: general; gamma-ray bursts: general

1. Introduction

The recent observation of Gamma-Ray Bursts (GRB) in the Very High-Energy domain (VHE, $E \geqslant 100\,\text{GeV}$) [1,2] marked an extraordinary milestone in our understanding of these outstandingly powerful transients. From the moment of their first identification as cosmological sources [3–5], it was immediately clear that their luminosity ranged up to values as high as $L \sim 10^{52}\,\mathrm{erg\,s^{-1}}$, making them the brightest sources of electromagnetic radiation known in the Universe and requiring an extremely efficient energy production mechanism. At present it is thought that GRBs arise as the consequence of ultra-relativistic shocks in magnetized plasma jets, which are launched in the fast accretion process that follows the collapse of very massive stars ($M \geqslant 20\,\mathrm{M_\odot}$) or the merger of compact stellar remnants, such as neutron stars (NS) and black holes (BH), to form a *magnetar* or a new BH [6,7]. The conversion of the huge amount of gravitational binding energy and thermal energy into kinetic and radiative power, within fractions of a second, triggers the emission of intense radiation, through mechanisms that naturally lead to the production of high-energy photons.

In general, a GRB is characterized by two emission regimes: an initial pulse of radiation, named *prompt* emission, most frequently observed in the energy range between 100 keV and 1 MeV and characterized by fast and strong variability [8], followed by a smoothly decaying *afterglow* phase, which can be detected from the radio and optical frequencies all the way up to energetic γ rays. The prompt phase lasts only a few seconds and it can be used to distinguish between a class of short GRBs, where the prompt emission takes place for less than 2 s, and one of long GRBs, whose prompt radiation is emitted for longer times [9–11]. These two classes can be fairly well interpreted by different types of source, with the long GRBs being more likely associated with a massive star core-collapse event [12,13], while the short ones can be better reproduced in the compact binary merger scenario [7,14], as confirmed by the Multi-Messenger observation of GRB 170817A in connection with GW170817 [15–17].

Apart from the generally well-understood picture, the nature of GRBs and of their radiation mechanisms still poses many difficult questions. On one side, there are solid theoretical arguments that predict energetic radiation from relativistic magnetized plasmas. On the other hand, the explanation of the spectral and temporal properties is not totally consistent with simple leptonic scenarios. It is very likely that synchrotron emission should be the dominant radiation mechanism for GRBs, as argued, e.g., in [18–20]. However, the variability that is observed down to millisecond timescales in the prompt stage [21] implies compact emission regions, where the magnetic and radiative energy densities are so high that the radiative cooling should consequently be very fast and produce soft spectra [22]. This is inconsistent with the observational evidence [23,24]. Many different possibilities, invoking thermal components, re-acceleration mechanisms or non-leptonic scenarios e.g., [25–27], have been proposed to address this problem.

The observation of VHE photons has a crucial role in the identification of the radiation mechanisms at work. These photons can be produced as a consequence of shocks between relativistic blobs in the jet (internal shocks, probably dominant during the prompt stage), as well as between the jet and the external environment (external shocks, expected to occur in the afterglow). In addition, they require compact sources in relativistic motion, to escape the production site. VHE radiation can be observed at the ground, using either Imaging Atmospheric Cherenkov Telescopes (IACT), such as MAGIC [28] and H.E.S.S. [29], or Extensive Air Shower (EAS) particle detector arrays, such as HAWC [30] and LHAASO [31]. At present, IACTs have been able to firmly detect VHE emission in the afterglow of some powerful GRBs. It is very likely that the next-generation Cherenkov Telescope Array (CTA) [32] will further improve our ability to investigate the VHE signal of GRBs. However, due to their small field of view (FoV) and to the consequent requirement to be alerted and pointed towards the source, these instruments can only track GRBs with a certain delay after their actual onset.

Here we describe the scientific opportunities that can be explored by means of EAS arrays. Thanks to their large FoV, which grants a continuous sky coverage of more than 1 sr, these instruments have a higher chance to probe the early phases of GRB emission, without the necessity of an external trigger. We discuss the issue taking into account the characteristics of an array concept based on Water Cherenkov Detectors (WCD) and investigated by the Southern Wide-field-of-view Gamma-ray Observatory collaboration

(SWGO) [33]. Our work is structured as follows: in §2 we present the theoretical framework of GRB emission; in §3 we describe the known and the expected VHE properties of GRBs; in §4 we discuss the detection opportunities of an instrument such as SWGO; finally, in §5 we summarize our conclusions.

2. Theoretical Framework

Despite several decades of investigation, we do not yet have a complete theory for GRBs. What we know for sure is that the presence of high-energy γ rays with a non-thermal spectrum implies emission from a highly relativistic source. This result stems from the well-known *compactness problem* [34,35]. A relatively simple argument can be used to illustrate the concept. A bright GRB has a time integrated energy flux, or *fluence*, of the order of $F \sim 10^{-7}\,\mathrm{erg\,cm^{-2}}$, which is approximately related to the total emitted energy E by:

$$E = 4\pi D_L^2 F \approx 10^{50}\mathrm{erg}\left(\frac{D_L}{3000\,\mathrm{Mpc}}\right)^2\left(\frac{F}{10^{-7}\,\mathrm{erg\,cm^{-2}}}\right), \tag{1}$$

where D_L is the luminosity distance. The typical variability timescale is $\delta T \approx 10^{-2}\,\mathrm{s}$, implying an emitting region size limit $R \leqslant c\delta T \approx 3000\,\mathrm{km}$. As a result, the source would be characterized by an extremely high radiation density. High-energy photons can produce electron-positron pairs whenever the condition $\sqrt{E_1 E_2} \geqslant 2m_e c^2$ is met. Introducing a probability factor f_p that accounts for the likelihood of the pair production mechanism, we obtain a pair production opacity of:

$$\tau_{\gamma\gamma} = \frac{f_p \sigma_T F D_L^2}{R^2 m_e c^2 (1+z)} \approx \frac{10^{13} f_p}{1+z}\left(\frac{F}{10^{-7}\,\mathrm{erg\,cm^{-2}}}\right)\left(\frac{D_L}{3000\,\mathrm{Mpc}}\right)^2\left(\frac{\delta T}{10^{-2}\,\mathrm{s}}\right)^{-2}, \tag{2}$$

where σ_T denotes the Thomson scattering cross-section. For typical GRB characteristics, the opacity predicted in Equation (2) is very large and should result in a thermal spectrum, in clear contradiction with observational evidence. If we allow the source to be a blob of plasma, approaching in a relativistic motion with bulk Lorentz factor Γ, at small angles with respect to the line of sight, and characterized by a power-law energy spectrum $N(\gamma) \propto \gamma^{-\alpha}$, the energy and the rate of arrival of the observed photons are both a factor Γ higher than the corresponding values in the emission frame. Due to the Doppler effect on frequency, this implies that the photons that we observe at a given frequency are a factor of $\Gamma^{2\alpha}$ denser than what would be seen in the emitting frame. The size of the emitting region is also affected by relativistic contraction, implying that $R_{em} \leqslant \Gamma^2 c\delta T$. Thus, Equation (2) should be corrected to:

$$\tau_{\gamma\gamma} \approx \frac{10^{13} f_p}{(1+z)\Gamma^{(4+2\alpha)}}\left(\frac{F}{10^{-7}\,\mathrm{erg\,cm^{-2}}}\right)\left(\frac{D_L}{3000\,\mathrm{Mpc}}\right)^2\left(\frac{\delta T}{10^{-2}\,\mathrm{s}}\right)^{-2}, \tag{3}$$

that since $1 < \alpha < 3$ and $f_p < 1$, predicts an optically thin regime when the condition $\Gamma \geqslant 10^{13/(4+2\alpha)} \approx 10^2$ is satisfied.

Similar order of magnitude considerations could be drawn to estimate the predicted spectrum. When reproducing GRB spectra starting from theoretical considerations, it is common use to introduce a set of free parameters containing some assumptions due to unknown properties of the detailed structure of the jet, the ongoing acceleration process and the shock micro-physics. In particular, an unknown fraction of the energy dissipated through the shock will go to the particle distribution and to the magnetic field. To account for this effect, two normalization parameters ϵ_e and ϵ_B, expressing, respectively, the fractions of total energy carried by the particles and the magnetic field, are then introduced. When particles are accelerated in shock waves through the Fermi mechanism, we expect a resulting power-law energy distribution in the form of $N(\gamma) \propto \gamma^{-p}$ and magnetic fields that can be as large as $B \approx 10^4\,\mathrm{G}$ [36]. If the radiating species are electrons and positrons, the energy is quickly converted into radiation, through the emission of fast pulses of synchrotron and Compton scattered photons, with a spectral form similar to the one illustrated in Figure 1 [37,38]. As a result, we expect that GRBs should in principle be powerful sources of transient VHE emission, although the most energetic part of the spectrum is prone to pair production opacity on the Extragalactic Background Light (EBL) photons, which implies suppression of the most energetic radiation from sources located at large cosmological distances [39–41].

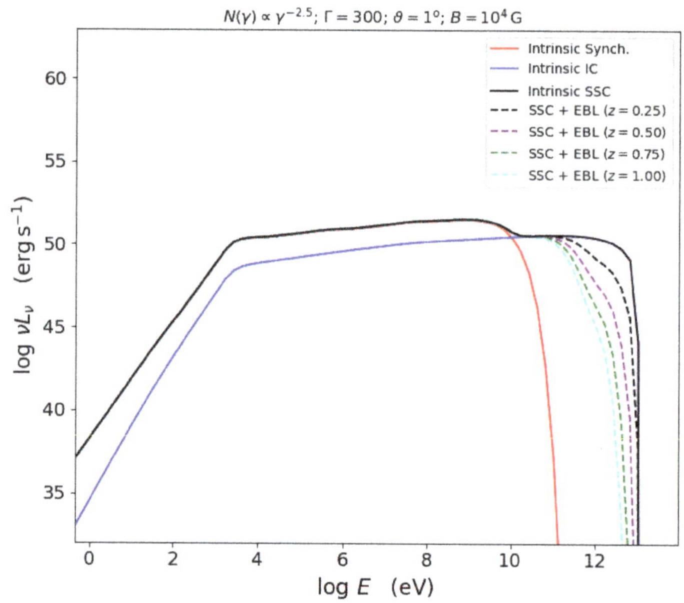

Figure 1. Spectral energy distribution expected for a relativistic blob of plasma, with bulk Lorentz factor $\Gamma = 300$, moving with an inclination of $\vartheta = 1^{\circ}$ from the line of sight and carrying a magnetic field $B = 10^{4}$ G. The radiating particles are assumed to be electron-positron pairs in a power-law distribution $N(\gamma) \propto \gamma^{-2.5}$, with $10 \leqslant \gamma \leqslant 10^{6}$. The red line represents synchrotron emission, while the blue line shows the inverse Compton scattering contribution. The resulting spectrum is represented as a continuous black line. The dashed curves illustrate the effects of $\gamma\gamma$-opacity on the Universe background radiation for redshifts $0.25 \leqslant z \leqslant 1$, in steps of 0.25.

Although the scenario depicted above can in principle justify the energies and the spectra that we see in GRBs, it is nonetheless prone to many important problems. The pulse-like appearance of the prompt stage light curve is consistent with the presence of particle acceleration processes, followed by a rapid cooling. The typical burst duration, however, requires many acceleration events or an effective supply of energetic particles, to match the data. A critical aspect is the onset and the duration of the production of the most energetic photons [42]. If they are emitted as a continuation of the synchrotron spectrum, they should be highly correlated with the low energy radiation. On the contrary, the interaction of several emitting regions, or the presence of non-leptonic contributions, can lead to the prediction of delayed high-energy emission [43]. Although the currently available observations tend to favor a delayed detection of energetic photons, the existence of earlier VHE contributions is not ruled out and it represents a critical factor to discriminate between different possible scenarios. More accurate spectral models would need to take into account the evolution of the system and the probably important effects of non-homogeneity and orientation. We also must consider the possibility that the existence of very energetic photons (up to the TeV scale) in a presumably dense environment can lead to important photo-hadronic interactions and, therefore, to potentially much more elaborated spectral forms. This type of processes is actually expected to occur when the jet plasma collides with the external environment, therefore taking a major role in the afterglow emission. The true nature of GRBs in their initial stages, however, will only be clarified when precise spectral and temporal information on their most energetic emission at early times are obtained.

3. VHE Properties of GRBs

So far, the direct detection of VHE radiation from GRBs has only been possible for a limited number of cases, thanks to ground-based follow-up observations. The monitoring campaign carried out by the *Fermi* Large Area Telescope (*Fermi*-LAT) [44] led to the identification of a high-energy spectral component, coming in the form of power-law emission, which appears to be a common feature of bright GRBs and may even arise very early in the event [45]. Identifying the origin of this component and its relationship with the low energy portion of the spectrum has important implications on the interpretation of GRBs. If it arises as a high-energy extension of the synchrotron spectrum, we would expect a strong degree of correlation between different spectral bands, with direct implications on the energy of the radiating particles. If, on the contrary, it represents an independent contribution, its distribution among GRBs and its spectral characteristics may prove fundamental to understand its origin. The data collected so far, however, do not yet allow the drawing of a conclusive picture.

Due to its limited collecting area, the LAT cannot place strong constraints on the spectral features of short transients at $E \geqslant 100\,\mathrm{GeV}$. In the assumption of a composite synchrotron and inverse Compton spectrum, the upper limits placed above $\sim$30 GeV for the VHE detected GRB 190114C represented an invaluable reference to estimate the transition between the two regimes, though the possibility of alternative interpretations and the limited temporal information still leave room for open questions. In any case, the results of *Fermi*-LAT observations can be used as a starting point to estimate the possible extension of the GRB properties to the VHE domain and therefore evaluate their detection possibility with other instruments.

Assuming, for the sake of simplicity, that the temporal evolution of the spectrum is only limited to a scaling factor, without relevant spectral changes, we can express the high-energy spectrum of a GRB as a function of energy in the form of:

$$\frac{\mathrm{d}N(t)}{\mathrm{d}E} = N_0(t)\left(\frac{E}{E_0}\right)^{-\alpha} \exp[-\tau(E,z)] \quad [\mathrm{photons\,cm^{-2}\,s^{-1}\,GeV^{-1}}], \tag{4}$$

where $N_0(t)$ is the flux of photons per unit energy observed at time t and pivot energy E_0, α is the spectral index, which is often within the range $1.5 \leqslant \alpha \leqslant 3$, with an average value close to 2, and $\tau(E,z)$ is the opacity due to pair production on EBL, given as a function of energy and redshift. The temporal evolution of the flux is typically well represented by a power-law, or a broken power-law, which can be written as:

$$N_0(t) = \begin{cases} N_{peak}\left(\dfrac{t-T_0}{T_{peak}-T_0}\right) & \text{for} \quad T_0 \leqslant t < T_{peak} \\ N_{peak}\left(\dfrac{t}{T_{peak}}\right)^{-\gamma} & \text{for} \quad t \geqslant T_{peak}, \end{cases} \tag{5}$$

where we denoted with T_0 the trigger time, with T_{peak} the time taken to achieve peak emission, with N_{peak} the maximum flux, and γ the temporal evolution index, which is often found to be $1 \leqslant \gamma \leqslant 2$.

Using the second catalog of *Fermi*-LAT detected GRBs (2FLGC) [45], which provides measurements of the observed photon fluxes in the energy range between 100 MeV and 10 GeV, together with information on the spectral index and on the light-curve shape, for a sample of GRBs observed during 10 years of regular monitoring operations, we are able to apply Equation (4), with the inclusion of Equation (5), to estimate the expected high-energy fluxes as a function of time, as illustrated for instance in Figure 2. In principle, we can extend this type of spectra to the VHE domain and, thus, obtain an estimate for the expected fluxes. In practice, this operation is not directly possible, due to the lack of a redshift measurement for most of the LAT detected GRBs, which implies an unknown EBL opacity in Equation (4). Although the effects of EBL are generally negligible for the observed LAT band, they become quickly very important at higher energies, with a typical EBL opacity horizon set by $\tau = 1$ for $z \approx 1$ already at $E = 100\,\mathrm{GeV}$ [41]. For this reason, we combined the spectral and temporal fits, which we obtained from the LAT data, with a set of simulations, aiming at estimating the effects of EBL opacity on the VHE extension of the GRBs that resulted in the observed LAT fluxes.

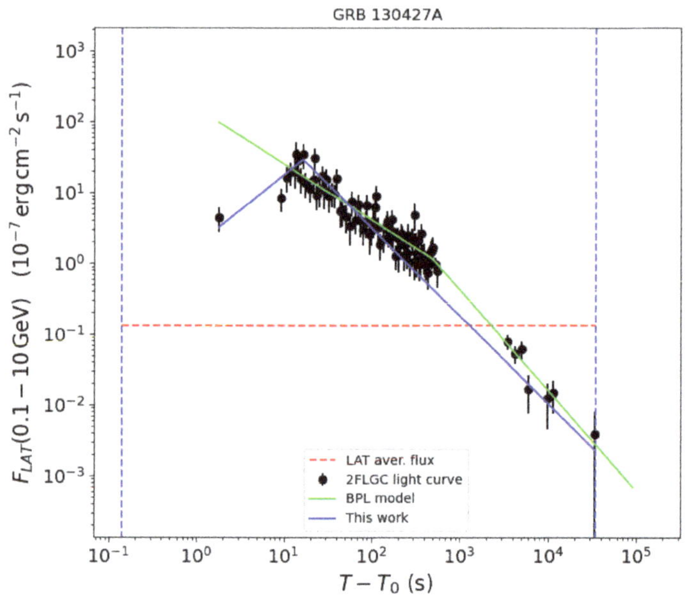

Figure 2. Comparison between the *Fermi*-LAT light curve of GRB 130427A and the model based on Equations (4) and (5). The vertical blue dashed lines mark the temporal window of the LAT signal, the red horizontal line is the average energy flux collected during the emission, the green continuous line is the 2FLGC broken power-law fit to the data, while the blue continuous line is a model using the light curve of Equation (5).

The approach that we adopted in our simulations was to extract the fluences of all the LAT detected GRBs, reported in 2FLGC, and to assign a set of 1000 random redshift values to each GRB without an available redshift measurement. The result of this process is the production of 1000 random GRB redshift distributions, corresponding to an equal number of random luminosity distributions, all of which yield the observed 2FLGC population. As shown in Figure 3, the simulation set provides a distribution of GRB luminosities which is in good agreement with the one followed by the 2FLGC GRBs with a measured redshift. The combination of all the different simulations, therefore, can be used to estimate the likelihood that a GRB with measured spectral and temporal characteristics is associated with a specific redshift range (see [46] for more details on the simulation process).

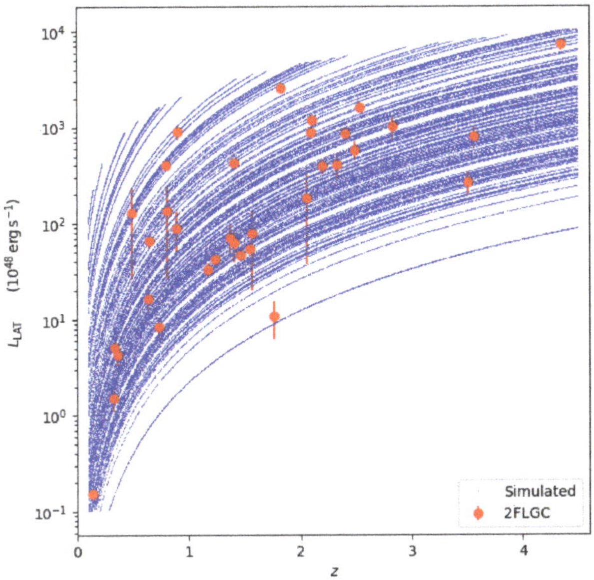

Figure 3. Comparison between the luminosity estimated in the LAT band $0.1\,\text{GeV} \leqslant E \leqslant 10\,\text{GeV}$ for the 2FLGC GRBs with a redshift measurement (large red dots) and the luminosity of the 2FLGC GRBs with 1000 randomly distributed redshifts (small blue points).

4. Monitoring GRBs with SWGO

Both theoretical and observational arguments suggest that VHE emission is an important property of GRBs. This is a well verified feature in the afterglow of some bright events. Evidence for the existence of energetic photons in the prompt phase is much harder to obtain, although there are some *Fermi*-LAT detected bursts that hint in this direction. Solving questions such as the occurrence rate of energetic spectral components in GRB afterglows, the existence of fast VHE pulses associated with the prompt stage, and assess the delay in the onset of VHE components is a problem that requires an extensive monitoring campaign, possibly covering a large sample of GRBs with an instrument characterized by a large collecting area. At present, several experiments are available or are being constructed to provide VHE spectral coverage with different sensitivities and resolution.

The SWGO collaboration is currently investigating the design of a new WCD array-based EAS observatory, to be constructed in the Southern Hemisphere [47]. The target performance domain is summarized in Figure 4, together with the sensitivities achieved by other instruments and with a comparison of the expected limiting fluxes with the ones emitted by GRBs that were detected in the HE and VHE domains. The role of such a new instrument will be to provide constant scanning of a wide FoV in the Southern sky, thus complementing the FoV covered by Northern facilities such as HAWC and LHAASO and providing a triggering and alert system for CTA. Adopting a compact array concept with a collecting area of 80,000 m^2, located in a high-altitude site ($\geqslant$4400 m a.s.l.), this type of instrument has the possibility to probe the flux range that we expect to be characteristic of GRB emission, with a transient localization accuracy $\alpha_{68} \leqslant 1^{\circ}$ below 1 TeV, although the possibility to detect different types of events depends critically on the overall instrument performance and on its ability to probe the lower part of the spectral range, where the effects of EBL opacity are less severe and a larger volume of the Universe is potentially accessible.

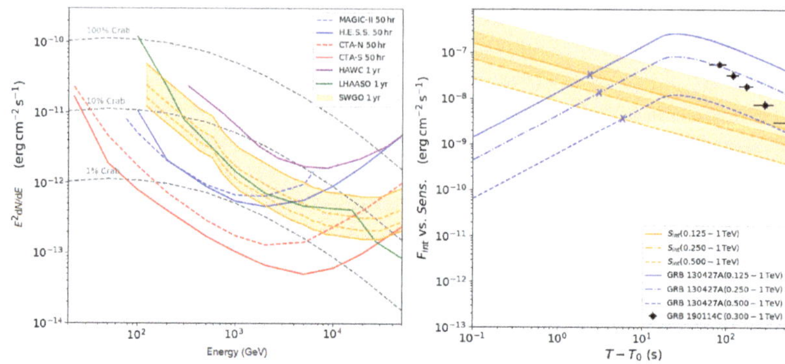

Figure 4. (**Left panel**) Differential sensitivity to the flux of a point-like source located at zenith distance $\vartheta = 20^{\circ}$ for SWGO, HAWC and LHAASO (computed for 1 year of data taking) and for MAGIC, H.E.S.S. and CTA (computed for 50 h of exposure time). Different fractions of the Crab Nebula flux are also shown for comparison. (**Right panel**) Expected detection times for a GRB with the temporal and spectral characteristics of GRB 130427A, located at redshift $z = 0.34$, for different fractions of the optimal SWGO performance, taken integrating the spectrum above thresholds of $E = 125\,\text{GeV}$, 250 GeV and 500 GeV. For each case, the blue crosses mark the time required to accumulate an integrated flux above the corresponding detection threshold. The black points with error bars denote the temporal evolution of the VHE flux detected by MAGIC for $E \geqslant 300\,\text{GeV}$ from GRB 190114C [1].

To test the potential role of SWGO as a monitoring and alert system, we took the sample of GRBs with simulated redshifts, discussed in §3, and we calculated the expected VHE fluxes, integrating Equation (4), with the inclusion of Equation (5), in time and in energy, using the spectral and the temporal characteristics extracted from 2FLGC and applying the $\gamma\gamma$ absorption effects predicted by an EBL opacity model [39]. The results of these calculations are summarized in Figure 5 for different possible performances of the experiment. The plots shown in Figure 5 represent the number of simulated redshift distributions that result in the GRB detections reported on the x-axes out of a total of 1000 simulations. Although we observe that all the explored configurations have some degree of detection chances, we can easily verify that an instrument performing at the optimal sensitivity, down to a low energy threshold of $E_{low} = 125\,\text{GeV}$, has a predicted ability to detect significantly more than 10 GRBs in 10 years in approximately 75% of the simulated scenarios, provided that they occur within a zenith distance of $\vartheta \leqslant 20^{\circ}$. Adopting lower performance solutions, or using a spectral window with a higher limiting threshold, such as 250 GeV or 500 GeV, leads to lower detection chances, due to the loss of many relatively low-flux events and to a more severe effect of EBL opacity, which limits the visible horizon to a smaller volume of the Universe. Incidentally, these calculations approximately correspond to what we would expect at larger zenith distances and, therefore, within increasing fractions of the available FoV, from an optimally performing solution. As a result, even if the FoV covered by a ground-based instrument below 1 TeV is smaller than the one that was available in the case of the *Fermi*-LAT observations, by taking into account the estimated distribution of expected GRB fluxes, we can conclude that the possibility to detect more than 1 event per year is a reasonable expectation. Due to the characteristic shapes of the tested light-curves, which are generally dominated by the peak flux, a fraction of approximately $(78 \pm 16)\%$ of these detections is estimated to occur within the first 10 s of the event, resulting in promisingly good chances to investigate the elusive properties of the prompt emission.

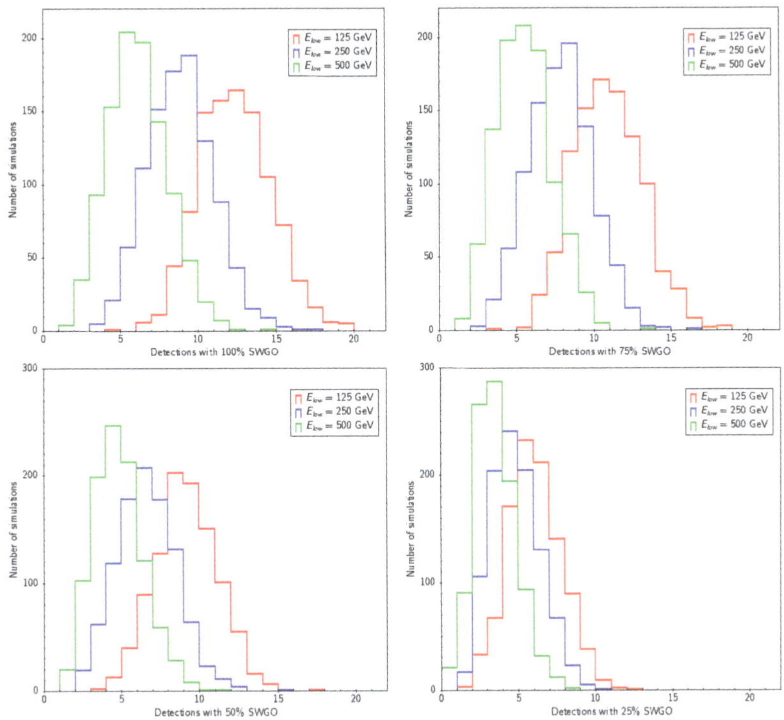

Figure 5. From upper left to lower right: histograms of the number of expected detections of GRBs extracted from 2FLGC and associated with 1000 random redshift values, computed for flux integrations above a low energy threshold of E_{low} = 125 GeV (red histograms), 250 GeV (blue histograms) and 500 GeV (green histograms), with an overall performance reaching up to 100%, 75%, 50% and 25% of the optimal SWGO sensitivity and zenith distance up to $\vartheta = 20^{\circ}$. The optimal SWGO design is expected to be able to detect more than 10 GRBs in an observing period of 10 years in more than 750 simulations out of 1000. Reductions of performance or higher energy thresholds reflect in gradually lower expected detection chances.

5. Conclusions

Investigating the VHE properties of GRB will have fundamental implications in our understanding of these extremely powerful events. The existence of VHE radiation components, particularly if associated with the prompt stage, represents a fundamental piece of information to model the physics of the radiating environment, thanks to the strong implications that VHE spectral and temporal properties have on the radiating species. The degree of correlation, or the occurrence of delays of the energetic components with respect to lower energy emission, depend on the particle energy distribution and on the interactions within the emitting regions, or between these regions and the environment. Testing the distribution of these properties among long and short GRBs will reduce the ambiguity implied by the partial overlap of these classes and further characterize the jets produced in the two cases. In addition, the possibility to verify whether the early VHE emission is dominated by smooth temporal evolution or irregular variability will improve our understanding of the transition between the prompt stage and the afterglow emission, providing invaluable information on the close GRB environment.

In our investigation, we used the HE data from GRBs detected during the first 10 years of *Fermi*-LAT observations, to infer the expected properties of GRBs in the VHE domain. Using the spectral and timing information of the LAT detected events, to estimate the expected fluxes, with the aid of ancillary simulations to derive fiducial redshift distributions for most of the sources with unknown z, we extrapolated the predicted spectra up to the TeV scale, taking into account the effects of EBL opacity. We then calculated the detection prospects of the resulting GRB distributions for different performances of the experiment, based on estimates of the SWGO potential. We found that a new monitoring facility, with the characteristics investigated by the SWGO collaboration, could effectively monitor VHE emission from GRBs, providing localization information at the level of a few square degrees and reaching timescales of less than 10 s for the brightest events. This is a time domain that is very hard to explore with IACT facilities, despite being critical to distinguish whether the high-energy component is a spectral extension, characteristic of powerful bursts, or the result of an additional process, dominated by some external contribution. With the ability to trigger on VHE transient signals, the localization regions of Multi-Messenger source candidates will further improve and the efficiency in the execution of follow-up campaigns will subsequently increase.

Author Contributions: Conceptualization, U.B.d.A. and F.L.; methodology, G.L.M.; software, G.L.M., R.C. and B.T.; validation, F.L.; investigation, M.P. and A.d.A.; writing—original draft preparation, G.L.M.; funding acquisition, M.P. and U.B.d.A.; modeling, D.M. and G.L.M.; All authors have read and agreed to the published version of the manuscript.

Funding: This research was funded by Fundação para a Ciência e Tecnologia, under project PTDC/FIS-PAR/4300/2020 and grant DL57/2016/cP1330/cT0002, by CNPq Productivity Research, Grant no. 311997/2019-8, by Serrapilheira Institute Grant number Serra—1812-26906, and by FAPERJ Young Scientist Fellowship no. E-26/202.818/2019.

Institutional Review Board Statement: Not applicable.

Informed Consent Statement: Not applicable.

Data Availability Statement: There are no new data associated with this article. The data used to estimate the GRB HE properties are provided by the *Fermi*-LAT collaboration and publicly available online at the HEASARC website https://heasarc.gsfc.nasa.gov/W3Browse/fermi/fermilgrb.html (accessed on 20 September 2021).

Acknowledgments: The authors gratefully thank Edna Ruiz-Velasco and Brian Reville for discussion and comments leading to the improvement of the manuscript. The SWGO Collaboration acknowledges the support from the agencies and organizations listed here: https://www.swgo.org/SWGOWiki/doku.php?id=acknowledgements (accessed on 20 September 2021). Part of this work is based on public data provided by the *Fermi*-LAT Collaboration. The *Fermi*-LAT Collaboration acknowledges generous ongoing support from several agencies and institutes that have supported both the development and the operation of the LAT as well as scientific data analysis. These include the National Aeronautics and Space Administration and the Department of Energy in the United States, the Commissariat à l'Energie Atomique and the Centre National de la Recherche Scientifique/Institut National de Physique Nucléaire et de Physique des Particules in France, the Agenzia Spaziale Italiana and the Istituto Nazionale di Fisica Nucleare in Italy, the Ministry of Education, Culture, Sports, Science and Technology (MEXT), High-Energy Accelerator Research Organization (KEK) and Japan Aerospace Exploration Agency (JAXA) in Japan, and the K. A. Wallenberg Foundation, the Swedish Research Council and the Swedish National Space Agency in Sweden. Additional support for science analysis during the operations phase is gratefully acknowledged from the Istituto Nazionale di Astrofisica in Italy and the Centre National d'Études Spatiales in France. This work performed in part under DOE Contract DEAC02-76SF00515.

Conflicts of Interest: The authors declare no conflict of interest.

Abbreviation

The following abbreviations are used in this manuscript:

2FLGC	2nd *Fermi*-LAT Gamma-ray burst Catalogue
BH	Black Hole
CTA	Cherenkov Telescope Array
GRB	Gamma-Ray Burst
GW	Gravitational Wave
EAS	Extensive Air Shower
FoV	Field of View
H.E.S.S.	High-Energy Stereoscopic System
HE	High Energy
HAWC	High-Altitude Water Cherenkov
IACT	Imaging Atmospheric Cherenkov Telescope
LHAASO	Large High-Altitude Air Shower Observatory
MAGIC	Major Atmospheric Gamma-ray Imaging Cherenkov
NS	Neutron Star
SWGO	Southern Wide-field-of-view Gamma-ray Observatory
VHE	Very High Energy
WCD	Water Cherenkov Detector

References

1. MAGIC Collaboration. Teraelectronvolt emission from the γ-ray burst GRB 190114C. *Nature* **2019**, *575*, 455–458. [CrossRef] [PubMed]
2. Abdalla, H.; Adam, R.; Aharonian, F.; Ait Benkhali, F.; Angüner, E.O.; Arakawa, M.; Arcaro, C.; Armand, C.; Ashkar, H.; Backes, M.; et al. A very-high-energy component deep in the γ-ray burst afterglow. *Nature* **2019**, *575*, 464–467. [CrossRef] [PubMed]
3. Fishman, G.J.; Meegan, C.A. Gamma-Ray Bursts. *Annu. Rev. Astron. Astrophys.* **1995**, *33*, 415–458. [CrossRef]
4. Metzger, M.R.; Djorgovski, S.G.; Kulkarni, S.R.; Steidel, C.C.; Adelberger, K.L.; Frail, D.A.; Costa, E.; Frontera, F. Spectral constraints on the redshift of the optical counterpart to the γ-ray burst of 8 May 1997. *Nature* **1997**, *387*, 878–880. [CrossRef]
5. Kulkarni, S.R.; Djorgovski, S.G.; Ramaprakash, A.N.; Goodrich, R.; Bloom, J.S.; Adelberger, K.L.; Kundic, T.; Lubin, L.; Frail, D.A.; Frontera, F.; et al. Identification of a host galaxy at redshift z = 3.42 for the γ-ray burst of 14 December 1997. *Nature* **1998**, *393*, 35–39. [CrossRef]
6. Woosley, S.E. Gamma-Ray Bursts from Stellar Mass Accretion Disks around Black Holes. *Astrophys. J.* **1993**, *405*, 273. [CrossRef]
7. Eichler, D.; Livio, M.; Piran, T.; Schramm, D.N. Nucleosynthesis, neutrino bursts and γ-rays from coalescing neutron stars. *Nature* **1989**, *340*, 126–128. [CrossRef]
8. Nakar, E.; Piran, T. Temporal properties of short gamma-ray bursts. *Mon. Not. R. Astron. Soc.* **2002**, *330*, 920–926. [CrossRef]
9. Mazets, E.P.; Golenetskii, S.V.; Ilinskii, V.N.; Panov, V.N.; Aptekar, R.L.; Gurian, I.A.; Proskura, M.P.; Sokolov, I.A.; Sokolova, Z.I.; Kharitonova, T.V. Catalog of cosmic gamma-ray bursts from the KONUS experiment data. *Astrophys. Space Sci.* **1981**, *80*, 3–83. [CrossRef]
10. Norris, J.P.; Cline, T.L.; Desai, U.D.; Teegarden, B.J. Frequency of fast, narrow γ-ray bursts. *Nature* **1984**, *308*, 434–435. [CrossRef]
11. Kouveliotou, C.; Meegan, C.A.; Fishman, G.J.; Bhat, N.P.; Briggs, M.S.; Koshut, T.M.; Paciesas, W.S.; Pendleton, G.N. Identification of Two Classes of Gamma-Ray Bursts. *Astrophys. J. Lett.* **1993**, *413*, L101. [CrossRef]
12. Galama, T.J.; Vreeswijk, P.M.; van Paradijs, J.; Kouveliotou, C.; Augusteijn, T.; Böhnhardt, H.; Brewer, J.P.; Doublier, V.; Gonzalez, J.-F.; Leibundgut, B.; et al. An unusual supernova in the error box of the γ-ray burst of 25 April 1998. *Nature* **1998**, *395*, 670–672. [CrossRef]
13. Bloom, J.S.; Kulkarni, S.R.; Djorgovski, S.G.; Eichelberger, A.C.; Côté, P.; Blakeslee, J.P.; Odewahn, S.C.; Harrison, F.A.; Frail, D.A.; Filippenko, A.V.; et al. The unusual afterglow of the γ-ray burst of 26 March 1998 as evidence for a supernova connection. *Nature* **1999**, *401*, 453–456. [CrossRef]
14. Li, L.X.; Paczyński, B. Transient Events from Neutron Star Mergers. *Astrophys. J. Lett.* **1998**, *507*, L59–L62. [CrossRef]
15. Abbott, B.P.; Abbott, R.; Abbott, T.D.; Acernese, F.; Adams, C.; Adams, T.; Addesso, P.; Adhikari, R.X.; Adya, V.B.; Affeldt, C.; et al. Multi-messenger Observations of a Binary Neutron Star Merger. *Astrophys. J. Lett.* **2017**, *848*, L12. [CrossRef]
16. Abbott, B.P.; Abbott, R.; Abbott, T.D.; Acernese, F.; Ackley, K.; Adams, C.; Adams, T.; Addesso, P.; Adhikari, R.X.; Adya, V.B.; et al. Gravitational Waves and Gamma-Rays from a Binary Neutron Star Merger: GW170817 and GRB 170817A. *Astrophys. J. Lett.* **2017**, *848*, L13. [CrossRef]
17. Cowperthwaite, P.S.; Berger, E.; Villar, V.A.; Metzger, B.D.; Nicholl, M.; Chornock, R.; Blanchard, P.K.; Fong, W.; Margutti, R.; Soares-Santos, M.; et al. The Electromagnetic Counterpart of the Binary Neutron Star Merger LIGO/Virgo GW170817. II. UV, Optical, and Near-infrared Light Curves and Comparison to Kilonova Models. *Astrophys. J. Lett.* **2017**, *848*, L17. [CrossRef]
18. Katz, J.I. Delayed Hard Photons from Gamma-Ray Bursts. *Astrophys. J. Lett.* **1994**, *432*, L27. [CrossRef]

19. Rees, M.J.; Meszaros, P. Unsteady Outflow Models for Cosmological Gamma-Ray Bursts. *Astrophys. J. Lett.* **1994**, *430*, L93. [CrossRef]
20. Tavani, M. Shock Emission Model for Gamma-Ray Bursts. *Phys. Rev. Lett.* **1996**, *76*, 3478–3481. [CrossRef]
21. Walker, K.C.; Schaefer, B.E.; Fenimore, E.E. Gamma-Ray Bursts Have Millisecond Variability. *Astrophys. J.* **2000**, *537*, 264–269. [CrossRef]
22. Ghisellini, G.; Celotti, A. Quasi-thermal Comptonization and Gamma-Ray Bursts. *Astrophys. J. Lett.* **1999**, *511*, L93–L96. [CrossRef]
23. Band, D.; Matteson, J.; Ford, L.; Schaefer, B.; Palmer, D.; Teegarden, B.; Cline, T.; Briggs, M.; Paciesas, W.; Pendleton, G.; et al. BATSE Observations of Gamma-Ray Burst Spectra. I. Spectral Diversity. *Astrophys. J.* **1993**, *413*, 281. [CrossRef]
24. Preece, R.D.; Briggs, M.S.; Mallozzi, R.S.; Pendleton, G.N.; Paciesas, W.S.; Band, D.L. The Synchrotron Shock Model Confronts a "Line of Death" in the BATSE Gamma-Ray Burst Data. *Astrophys. J. Lett.* **1998**, *506*, L23–L26. [CrossRef]
25. Vietri, M. GeV Photons from Ultrahigh Energy Cosmic Rays Accelerated in Gamma Ray Bursts. *Phys. Rev. Lett.* **1997**, *78*, 4328–4331. [CrossRef]
26. Ghirlanda, G.; Pescalli, A.; Ghisellini, G. Photospheric emission throughout GRB 100507 detected by Fermi. *Mon. Not. R. Astron. Soc.* **2013**, *432*, 3237–3244. [CrossRef]
27. Burgess, J.M.; Preece, R.D.; Ryde, F.; Veres, P.; Mészáros, P.; Connaughton, V.; Briggs, M.; Pe'er, A.; Iyyani, S.; Goldstein, A.; et al. An Observed Correlation between Thermal and Non-thermal Emission in Gamma-Ray Bursts. *Astrophys. J. Lett.* **2014**, *784*, L43. [CrossRef]
28. Aleksić, J.; Ansoldi, S.; Antonelli, L.A.; Antoranz, P.; Babic, A.; Bangale, P.; Barceló. M.; Barrio, J.A.; Becerra González, J.; Bednarek, W. The major upgrade of the MAGIC telescopes, Part I: The hardware improvements and the commissioning of the system. *Astropart. Phys.* **2016**, *72*, 61–75. [CrossRef]
29. Aharonian, F.; Akhperjanian, A.G.; Bazer-Bachi, A.R.; Beilicke, M.; Benbow, W.; Berge, D.; Bernlöhr, K.; Boisson, C.; Bolz, O.; Borrel, V.; et al. Observations of the Crab nebula with HESS. *Astron. Astrophys.* **2006**, *457*, 899–915. [CrossRef]
30. DeYoung, T. The HAWC observatory. *Nucl. Instruments Methods Phys. Res. A* **2012**, *692*, 72–76. [CrossRef]
31. Zhen, C.; Ming-Jun, C.; Song-Zhan, C.; Hong-Bo, H.; Cheng, L.; Ye, L.; Ling-ling, M.; Xin-hua, M.; Xiang-dong, S.; Han-rong, W.; et al. Introduction to Large High Altitude Air Shower Observatory (LHAASO). *Chin. Astron. Astrophys.* **2019**, *43*, 457–478. [CrossRef]
32. CTA Consortium. *Science with the Cherenkov Telescope Array*; World Scientific Publishing Co.: Singapore, 2019. [CrossRef]
33. Barres de Almeida, U.; Giacinti, G.; Longo, F. Benchmarking the Science for the Southern Wide-Field Gamma-ray Observatory (SWGO). In Proceedings of the 37th International Cosmic Ray Conference (ICRC2021), Berlin, Germany, 12–23 July 2021; Volume 395, p. 10.
34. Ruderman, M. Theories of gamma -ray bursts. In *Seventh Texas Symposium on Relativistic Astrophysics*; Bergman, P.G., Fenyves, E.J., Motz, L., Eds.; John Wiley & Sons, Ltd.: Hoboken, NJ, USA, 1975; Volume 262, pp. 164–180. [CrossRef]
35. Schmidt, W.K.H. Distance limit for a class of model γ-ray burst sources. *Nature* **1978**, *271*, 525–527. [CrossRef]
36. Piran, T. Gamma-ray bursts and the fireball model. *Phys. Rep.* **1999**, *314*, 575–667. [CrossRef]
37. Rybicki, G.B.; Lightman, A.P. *Radiative Processes in Astrophysics*; John Wiley & Sons, Ltd.: Hoboken, NJ, USA, 1986. [CrossRef]
38. Xue, R.R.; Tam, P.H.; Wagner, S.J.; Behera, B.; Fan, Y.Z.; Wei, D.M. Very High Energy γ-Ray Afterglow Emission of Nearby Gamma-Ray Bursts. *Astrophys. J.* **2009**, *703*, 60. [CrossRef]
39. Dominguez, A.; Primack, J.R.; Rosario, D.J.; Prada, F.; Gilmore, R.C.; Faber, S.M.; Koo, D.C.; Somerville, R.S.; Pérez-Torres, M.A.; Pérez-González, P.; et al. Extragalactic background light inferred from AEGIS galaxy-SED-type fractions. *Mon. Not. R. Astron. Soc.* **2011**, *410*, 2556–2578. [CrossRef]
40. Franceschini, A.; Foffano, L.; Prandini, E.; Tavecchio, F. Very high-energy constraints on the infrared extragalactic background light. *Astron. Astrophys.* **2019**, *629*, A2. [CrossRef]
41. Saldana-Lopez, A.; Domínguez, A.; Pérez-González, P.G.; Finke, J.; Ajello, M.; Primack, J.R.; Paliya, V.S.; Desai, A. An observational determination of the evolving extragalactic background light from the multiwavelength HST/CANDELS survey in the Fermi and CTA era. *Mon. Not. R. Astron. Soc.* **2021**. [CrossRef]
42. Fan, Y.Z.; Piran, T. High-energy γ-ray emission from gamma-ray bursts—Before GLAST. *Front. Phys. China* **2008**, *3*, 306. [CrossRef]
43. Asano, K.; Mészáros, P. Delayed Onset of High-energy Emissions in Leptonic and Hadronic Models of Gamma-Ray Bursts. *Astrophys. J.* **2012**, *757*, 115. [CrossRef]
44. Atwood, W.B.; Abdo, A.A.; Ackermann, M.; Althouse, W.; Anderson, B.; Axelsson, M.; Baldini, L.; Ballet, J.; Band, D.L.; Barbiellini, G.; et al. The Large Area Telescope on the Fermi Gamma-Ray Space Telescope Mission. *Astrophys. J.* **2009**, *697*, 1071–1102. [CrossRef]
45. Ajello, M.; Arimoto, M.; Axelsson, M.; Baldini, L.; Barbiellini, G.; Bastieri, D.; Bellazzini, R.; Bhat, P.N.; Bissaldi, E.; Blandford, R.D.; et al.; A Decade of Gamma-Ray Bursts Observed by Fermi-LAT: The Second GRB Catalog. *Astrophys. J.* **2019**, *878*, 52. [CrossRef]

46. La Mura, G.; Barres de Almeida, U.; Conceição, R.; De Angelis, A.; Longo, F.; Pimenta, M.; Prandini, E.; Ruiz-Velasco, E.; Tomé, B. Gamma-ray burst detection prospects for next generation ground-based VHE facilities. *Mon. Not. R. Astron. Soc.* **2021**, *508*, 671–679. [CrossRef]
47. Schoorlemmer, H.; Conceição, R.; Smith, A.J. Simulating the performance of the Southern Wide-view Gamma-ray Observatory. In Proceedings of the 37th International Cosmic Ray Conference (ICRC2021). In Proceedings of the 37th International Cosmic Ray Conference (ICRC2021), Berlin, Germany, 12–23 July 2021; Volume 395, p. 9.

galaxies

MDPI

Article

Two Dimensional Clustering of *Swift*/BAT and *Fermi*/GBM Gamma-ray Bursts

Lána Salmon [1,2,*], Lorraine Hanlon [1,2] and Antonio Martin-Carrillo [1,2]

1 School of Physics, University College Dublin, Belfield, D04 V1W8 Dublin, Ireland; lorraine.hanlon@ucd.ie (L.H.); antonio.martin-carrillo@ucd.ie (A.M.-C.)
2 Centre for Space Research, University College Dublin, Belfield, D04 V1W8 Dublin, Ireland
* Correspondence: lana.salmon@ucdconnect.ie

Abstract: Studies of Gamma-ray Burst (GRB) properties, such as duration and spectral hardness, have found evidence for additional classes beyond the short-hard (merger) and long-soft (collapsar) prototypes. Several clustering analyses of the duration-hardness plane identified a third, intermediate duration, class. In this work, Gaussian Mixture Model-based (GMM) clustering is applied to the *Swift*/BAT and *Fermi*/GBM samples of GRBs. The results obtained by the hierarchical combination of Gaussian components (or clusters) based on an entropy criterion are presented. This method counteracts possible overfitting arising from the application of Gaussian models to non-Gaussian underlying data. While the initial GMM clustering of the hardness-duration plane identifies three components (short/intermediate/long) for the *Swift*/BAT and *Fermi*/GBM samples, only two components (short/long) remain once the entropy criterion is applied. The analysis presented here suggests that the intermediate duration class may be the result of overfitting, rather than evidence of a distinct underlying population.

Keywords: gamma-ray burst; clustering; statistical analysis

Citation: Salmon, L.; Hanlon, L.; Martin-Carrillo, A. Two Dimensional Clustering of *Swift*/BAT and *Fermi*/GBM Gamma-ray Bursts. *Galaxies* **2022**, *10*, 77. https://doi.org/10.3390/galaxies10040077

Academic Editors: Elena Moretti and Francesco Longo

Received: 27 April 2022
Accepted: 22 June 2022
Published: 25 June 2022

1. Introduction

The bimodal duration distribution of Gamma-ray Bursts (GRBs) suggests the separation of GRBs at $T_{90} \approx 2$ s into short/hard and long/soft classes [1]. The association of long GRBs with star forming galaxies [2] and Type Ic supernovae (Galama et al. [3], Woosley and Bloom [4]; for a review, see Cano et al. [5]) provides an observational link between long GRBs and the deaths of massive stars, supporting the collapsar scenario [6]. There is substantial evidence to support compact object mergers (neutron star–neutron star or neutron star–black hole) as the progenitors of short GRBs [7,8]. The location offsets of short GRBs from their host galaxies [9,10], their proximity to elliptical galaxies [11], and the association of GRB 170817A, an unusual short GRB, with the neutron star merger event GW 170817 detected by aLIGO [12–14], all support the merger hypothesis for the origin of short GRBs.

Other formation scenarios for short GRBs include the accretion-induced collapse of a white dwarf, double white dwarf mergers, or neutron star–white dwarf mergers [15–17], possibly leading to an unstable magnetar remnant. There are notable exceptions to the short-merger/long-collapsar paradigm, such as the short-collapsar event GRB 200826A [18–20], and GRB 060614, a long GRB without a supernova [21]. It has been suggested that many of the short duration GRBs of high redshift arise from collapsars [22]. Consideration of additional GRB characteristics, such as late X-ray flares in some short GRBs, and the non-detection of a supernova associated with some long GRBs [23], led to the suggestion of a new classification scheme [21], with Type I (massive star origin) and Type II (compact object merger origin) GRBs defined by many multiple observational criteria beyond the traditional duration and hardness [22,24]. Lü et al. [25] suggested a new parameter ϵ, based on the isotropic equivalent energy and peak energy, to classify bursts. Additionally, Donaghy et al. [26] considered 10 observational criteria for HETE-2 bursts, concluding that the best criteria to classify GRBs as 'short population' or 'long population' bursts are host galaxy properties, spectral lag, and the presence of a long-soft bump or gravitational waves.

In view of the diversity in GRB phenomenology, a definitive classification of GRBs based on duration alone is challenging. Several studies have found evidence for an additional 'intermediate' duration class of GRBs, first identified through Gaussian fits to the duration distribution of GRBs in the Third BATSE catalogue [27] and, subsequently, in fits to the GRB duration distributions of BeppoSAX [28], RHESSI [29], and *Swift* [30–35]. This class appears as an additional Gaussian 'component' required for the best-fit solution. However, the observed duration distribution can be recovered by modelling it as two skewed distributions [31,36,37], without requiring a third component.

GRB catalogues provide a set of standard parameters measured for each GRB, including duration (T_{90}), hardness ratio (HR), fluence (S), peak flux (PF), peak energy (E_{peak}), and spectral fit parameters, including the low and high energy spectral indices of the Band function [38], which fits the keV-MeV GRB spectrum, typically denoted in the literature as α and β, respectively. In the case of *Fermi*/GBM, the catalogue contains over 300 parameters for each GRB [39,40]. The availability of such large GRB catalogues allows the application of bivariate and multidimensional analyses to the data.

Table 1 summarises the previous studies, along with the resulting number of components identified for different GRB datasets. Between two and five classes of GRBs are found, depending on the sample, parameters, and methods used. Clustering of the duration-hardness plane of the final BATSE GRB catalogue identified three [41–43] or five [44–46] classes of GRBs separated by their duration, fluence, and hardness. Unsupervised neural network analysis also revealed an intermediate class [47] or two classes [48,49]. However, only two classes were found in the BATSE sample using self-organising maps [50] and fits to the duration-hardness plane with skewed bivariate distributions [51,52].

The clustering of the duration and hardness of *Swift*/BAT GRBs [53,54] and the clustering of light curve shape indicators [55] identified three classes of bursts. Gaussian Mixture Model-based (GMM) clustering applied to the *Fermi*/GBM sample revealed that GRB 170817A fit within the intermediate class in the duration-hardness plane [56], and that five classes could be identified by clustering spectral fit parameters, fluences, and durations [57]. Principal Component Analysis (PCA) also identified three classes in *Fermi*/GBM [58] and BATSE [59] samples.

Table 1. Methods and resulting components identified in clustering, fitting, and dimensionality reduction techniques applied to GRB populations. HR denotes Hardness Ratio, PF denotes Peak Flux, and S represents fluence. Studies which consider intrinsic properties such as redshift-corrected duration and hardness are marked with a *.

Study	Method	Parameters	Components
BATSE			
Horváth [27]	Fit (Gaussian)	T_{90}	3
Mukherjee et al. [41]	Clustering (Hierarchical)	T_{90}, HR, PF, S	3
Hakkila et al. [48]	Supervised pattern recognition	T_{90}, HR, PF, S, E_{peak}, α, β	2
Balastegui et al. [47]	Clustering (Hierarchical), PCA, Neural Network	T_{90}, PF, S	3
Horváth [60]	Fit (log-normal)	T_{90}	3
Rajaniemi and Mähönen [50]	Self-Organising Maps	T_{90}, HR, S	2
Hakkila et al. [49]	Clustering (k-means), Neural Network	T_{90}, HR, S	2
Chattopadhyay et al. [42]	Clustering (k-means, Dirichlet mixture)	T_{90}, HR, PF, S	3
Zitouni et al. [32]	Fit (Gaussian)	T_{90}	2
Zhang et al. [61]	Fit (Gaussian)	T_{90}	2
Bhave et al. [54]	Clustering (Gaussian Mixture-Model)	T_{90}, HR	2
Chattopadhyay and Maitra [44]	Clustering (k-means, Gaussian Mixture-Model)	T_{90}, HR, PF, S	5

Table 1. *Cont.*

Study	Method	Parameters	Components
Swift			
Kulkarni and Desai [35]	Fit (log-normal)	T_{90}	2
Chattopadhyay and Maitra [45]	Clustering (Ellipsoidal Gaussian, t-mixture)	T_{90}, PF, S	5
Tarnopolski [51]	Fit (Skewed bi-variate)	T_{90}, HR	2
Tarnopolski [52]	Fit (Skewed bi-variate)	T_{90}, HR	2
Tóth et al. [46]	Clustering (Gaussian Mixture-Model)	T_{90}, HR, PF, S	5
Modak [43]	Clustering (Fuzzy)	T_{90}, HR, PF, S	3
Horváth et al. [62]	Fit (log-normal)	T_{90}	3
Zhang and Choi [63]	Fit (log-normal)	T_{90}	2
Zhang and Choi [63] *	Fit (log-normal)	$T_{90,rest}$	2
Huja et al. [30]	Fit (Gaussian)	T_{90}	3
Huja et al. [30] *	Fit (Gaussian)	$T_{90,rest}$	1
Horváth et al. [64]	Fit (Gaussian)	T_{90}, HR	3
Veres et al. [53]	Clustering (Hierarchical, k-means)	T_{90}, HR	3
Koen and Bere [31]	Clustering (Gaussian)	T_{90}, HR	3
Tsutsui and Shigeyama [55]	Clustering (Gaussian)	Light curve shape indicators	3
Zitouni et al. [32]	Fit (Gaussian)	T_{90}	3
Zitouni et al. [32] *	Fit (Gaussian)	$T_{90,rest}$	3
Horváth and Tóth [33]	Fit (log-normal)	T_{90}	3
Tarnopolski [34]	Fit (Skew-normal)	T_{90}	3
Tarnopolski [34] *	Fit (Skew-normal)	$T_{90,rest}$	1
Yang et al. [65] *	Clustering (Gaussian Mixture-Model)	$T_{90,rest}$, HR	2
Zhang et al. [61]	Fit (Gaussian)	T_{90}	3
Zhang et al. [61] *	Fit (Gaussian)	$T_{90,rest}$	2
Bhave et al. [54]	Clustering (Gaussian Mixture-Model)	T_{90}, HR	3
Bhave et al. [54] *	Clustering (Gaussian Mixture-Model)	$T_{90,rest}$, HR	3
Kulkarni and Desai [35]	Fit (log-normal)	T_{90}	3
Kulkarni and Desai [35] *	Fit (log-normal)	$T_{90,rest}$	2
Fermi			
Zhang et al. [61]	Fit (Gaussian)	T_{90}	2
Bhave et al. [54]	Clustering (Gaussian Mixture-Model)	T_{90}, HR	2
Kulkarni and Desai [35]	Fit (log-normal)	T_{90}	2
Acuner and Ryde [57]	Clustering (Gaussian Mixture-Model)	T_{90}, S, E_{peak}, α, β	5
Horváth et al. [56]	Clustering (Gaussian Mixture-Model)	T_{90}, HR	3
Zitouni et al. [66]	Fit (Gaussian)	T_{90}	2
Zitouni et al. [66] *	Fit (Gaussian)	$T_{90,rest}$	2
Horváth et al. [58]	Principal Component Analysis	T_{90}, PF, S, E_{peak}, α, β	3
Tarnopolski [51]	Fit (skewed bivariate)	T_{90}, HR	2
BeppoSAX			
Horváth [28]	Fit (log-normal)	T_{90}	3
Kulkarni and Desai [35]	Fit (log-normal)	T_{90}	2
RHESSI			
Řípa et al. [29]	Fit (log-normal)	T_{90}	2
Řípa et al. [29]	Fit (log-normal)	T_{90}, HR	3
Řípa et al. [67]	Clustering (Gaussian Mixture-Model, k-means)	T_{90}, HR	3
INTEGRAL			
Minaev et al. [68]	Fit (log-normal)	T_{90}	2
Konus-Wind			
Svinkin et al. [69]	Fit (log-normal)	T_{50}	2
Svinkin et al. [69]	Clustering (Gaussian Mixture Model)	T_{50}, HR	3
Multiple samples			
Minaev and Pozanenko [70] *	Fit (Skew-normal)	$T_{90,rest}$, E_{iso}, $E_{peak,rest}$	2

Observational bias has been suggested as a possible origin of the putative intermediate class. Bias caused by short temporal trigger windows favours short low-fluence bursts (fluence-duration bias; Hakkila et al. [49]), while the low signal-to-noise ratios of long faint bursts can cause them to be mistaken for short bursts ('tip-of-the-iceberg' effect; Lü et al. [71]). However, neither of these effects have been able to reproduce the third class in simulations. It has been shown that the third class can arise as a consequence of fitting symmetrical models to the GRB duration distribution, which may be skewed rather than symmetrical [31,36,37,51], possibly as a result of the GRB pulse shapes [72].

The significant number of GRBs with measured redshift in the *Swift* and *Fermi* samples has allowed studies of intrinsic properties, which have pointed to the existence of two classes in the *Fermi*/GBM sample [32]. For the *Swift*/BAT sample of bursts, one [30,34], two [35,61,63,65], or three [32,54] classes have been identified. However, cosmological time dilation applied to GRB durations has not been found to transform a rest-frame two-component Gaussian duration distribution to the observed skewed one [73]. While there are now more than 400 *Swift* GRBs with measured redshift, there are only 25 short duration bursts with $T_{90,\mathrm{obs}} < 2\,\mathrm{s}$. The rest-frame studies outlined in Table 1 note that the short duration sample is not statistically significant, and a larger sample is required [54,65].

This paper reports on an updated two-dimensional clustering analysis in the duration-hardness plane of the large *Fermi*/GBM and *Swift*/BAT GRB samples. Advancing previous studies, the analysis presented here makes use of an entropy criterion to identify 'excess' components that may be identified in the standard GMM clustering of data but which arise from the application of Gaussian models to non-Gaussian underlying data [74]. This method has been applied in other astrophysical contexts, for example in the clustering of stars [75]. As the number of short GRBs with redshift has not grown significantly since previous studies, this paper focuses on GMM clustering using observed, rather than intrinsic, properties.

Section 2 outlines the sample construction, while Section 3 provides details of the methods applied to perform clustering. The results and discussion are presented in Sections 4 and 5 respectively, while the conclusions are outlined in Section 6.

2. Datasets and Data Preparation

2.1. *Swift/BAT*

The Third *Swift*/BAT Catalogue [76] contains 1388 bursts detected between 17 December 2004 and 28 August 2020 and provides the durations, spectral fit parameters, fluxes, and fluences calculated in the simple Power-Law (PL) and Cut-off Power-Law (CPL) models. The hardness ratio HR_{32} for each GRB was calculated as the ratio of the fluence in energy range 3 (50–100 keV) to energy range 2 (25–50 keV), given by

$$\mathrm{HR}_{32} = \frac{\int_{50\,\mathrm{keV}}^{100\,\mathrm{keV}} Ef(E)dE}{\int_{25\,\mathrm{keV}}^{50\,\mathrm{keV}} Ef(E)dE}, \tag{1}$$

where $f(E)$ is the photon flux at energy E. For the PL model, this is given by

$$f(E) = K_{50}^{\mathrm{PL}} \left(\frac{E}{50\,\mathrm{keV}} \right)^{\alpha^{\mathrm{PL}}}, \tag{2}$$

where α^{PL} is the PL index, and K_{50}^{PL} is the normalisation factor at 50 keV, with units of photons $\mathrm{cm}^{-2}\,\mathrm{s}^{-1}\,\mathrm{keV}^{-1}$. The CPL model is described as

$$f(E) = K_{50}^{\mathrm{CPL}} \left(\frac{E}{50\,\mathrm{keV}} \right)^{\alpha^{\mathrm{CPL}}} \exp\left(\frac{-E(2+\alpha^{\mathrm{CPL}})}{E_{\mathrm{peak}}} \right), \tag{3}$$

where α^{CPL} is the CPL index, K_{50}^{CPL} is the normalisation factor at 50 keV, with units of photons $\mathrm{cm}^{-2}\,\mathrm{s}^{-1}\,\mathrm{keV}^{-1}$, and E_{peak} is the peak energy in keV of the νF_ν or $E^2 f(E)$ spectrum.

This is the flux density integrated over the energy range, also known as the spectral flux density.

The sample of 1388 bursts was filtered to remove 52 GRBs for which no duration or best-fit model was documented. A further 20 GRBs with duration or hardness errors in excess of 50% of their magnitude were removed, resulting in a final sample of 1316 GRBs for clustering.

2.2. Fermi/GBM

The *Fermi*/GBM catalogue was accessed using the *Fermi*/GBM Data Tools [40] and limited to the period between 10 August 2008 and 17 March 2021, which yielded a sample of 3001 bursts. The hardness ratio was calculated by comparing the counts detected in the 8–50 keV band to the 50–300 keV band. Counts within the T_{90} interval were summed from the 64 ms light curves, generated using Time-Tagged Event (TTE) data in the *Fermi*/GBM Data Tools. Only triggered detectors were used, and the background subtraction was performed using the background intervals defined in the *Fermi*/GBM catalogue. Bursts with no documented duration or incorrect background subtraction were removed, resulting in a sample of 2669 bursts. Prior to clustering, 36 outliers were identified by the R package `HDOutliers` [77] and removed from the sample, leaving a final sample of 2633 bursts for clustering.

3. Clustering Methods

3.1. GMM Clustering

GMM clustering was carried out in R using the MCLUST [78]. GMM clustering assumes that the observed data are generated from a mixture of K components, where the density of each component is described by a multivariate Gaussian distribution. MCLUST fit 14 different models to the data, parameterised by the shape (spherical or ellipsoidal) and volume. In the case of ellipsoidal models, the alignment of the axes and the difference in shape of the fitted ellipsoidals was specified. This is known as Volume-Shape-Orientation (VSO) decomposition. For a given model, the volume, shape, and orientation can be constrained to equal variance, denoted by 'E'. If the variance is free to change, the model is denoted 'V'. Additionally, the orientation of the clusters relative to each other can be constrained to Equal or Varying, or a model can have alignment limited to the coordinate axis, and is labelled 'I'. For example, 'EVI' denotes equal volume components, with variable shapes (i.e., not spherical) and orientation aligned with the axes.

MCLUST makes use of the Bayesian Information Criterion (BIC; Schwarz et al. [79]) to compare mixture models fitted on the data. The best-fit model and number of components are chosen based on the largest BIC value. A difference in BIC value between models of 6–10 is considered significant, while a difference of greater than 2 provides positive evidence for a better fit [80]. This standard GMM fit method is the same as that employed in some previous studies, for example Horváth et al. [56] and Bhave et al. [54].

3.2. Combination of Gaussian Components

In the case where Gaussian components were overlapping or components were suspected to be non-Gaussian, as has been shown for the BATSE and *Fermi*/GBM GRB duration distributions [51,52], the MCLUST function `clustCombi` was used to hierarchically combine components using an entropy criterion [74]. Entropy is a measure of the uncertainty of the observations belonging to a certain cluster or component. Thus, a large decrease in entropy signifies a better fit with smaller uncertainty. For MCLUST, the final number of components was chosen based on the observed 'elbow' in the entropy plot. The number of components at which the elbow occurred pointed to a large decrease in entropy and, therefore, a model with smaller uncertainty.

There are several methods for joining Gaussian Mixture components. In comparison to the entropy criterion, these methods have limitations, for example, requiring spherical components [81] or one-dimensional data [82]. Other suggested methods assume the number of clusters [83] or make use of hard clustering methods, which assigns points to one cluster rather than applying a probabilistic method (e.g., Tantrum et al. [84]). The method employed in this study was a soft-clustering probabilistic method, which is computationally efficient and applicable to multiple dimensions. Hence, it was the chosen method to achieve a robust clustering result for the complex GRB datasets.

4. Results

The results of the initial MCLUST fit and subsequent `clustCombi` method applied to the *Swift*/BAT and *Fermi*/GBM samples are summarised in Table 2.

Table 2. Number of components (K), Bayesian Information Criterion (BIC) values, models, and number of bursts (#) identified in the MCLUST and subsequent `clustCombi` fits to the *Swift*/BAT and *Fermi*/GBM samples.

	Initial MCLUST Fit			`clustCombi` Fit		
	Model	**K**	**BIC**	**K**	**# Short**	**# Long**
Swift/BAT	VEI	3	−720	2	85	1231
Fermi/GBM	VEI	3	−3970	2	295	2338

4.1. *Swift/BAT*

The BIC values for the top three models versus the number of components, resulting from the application of MCLUST to the full *Swift*/BAT sample, are shown in Figure 1a. The 'VEI' model with three components had the largest BIC value. The three components were labelled 'long', 'intermediate', and 'short' according to their durations and are projected onto the hardness-duration plane in Figure 1b. The clear round edge between the intermediate and long components suggests that a Gaussian was being fit to a non-Gaussian component.

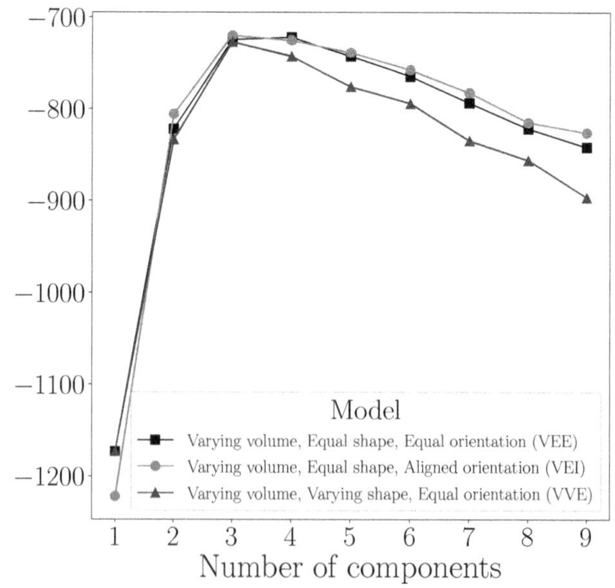

(**a**)

Figure 1. *Cont.*

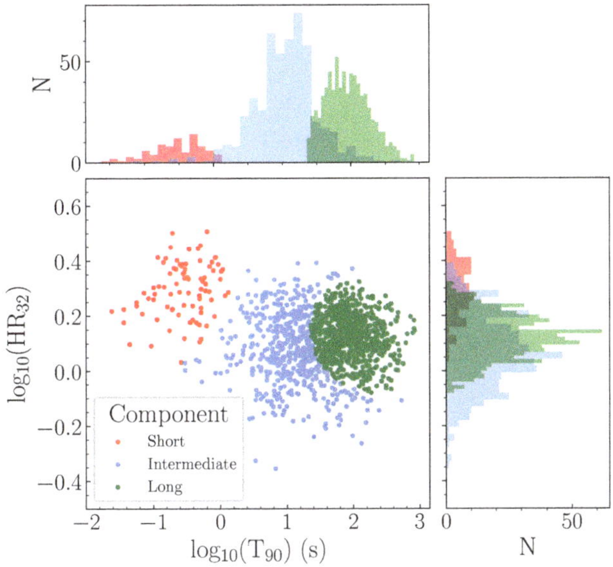

(**b**)

Figure 1. (**a**) BIC values of the top three MCLUST models fit to the *Swift*/BAT sample and (**b**) the resulting duration-hardness plane for the best-fit three-component model (VEI).

After `clustCombi` was applied, the 'intermediate' and 'long' components were combined, producing a large decrease in entropy as shown in Figure 2. The two remaining components or classes were labelled 'long' and 'short', as shown in Figure 3. Table 2 presents the sample size of the classes.

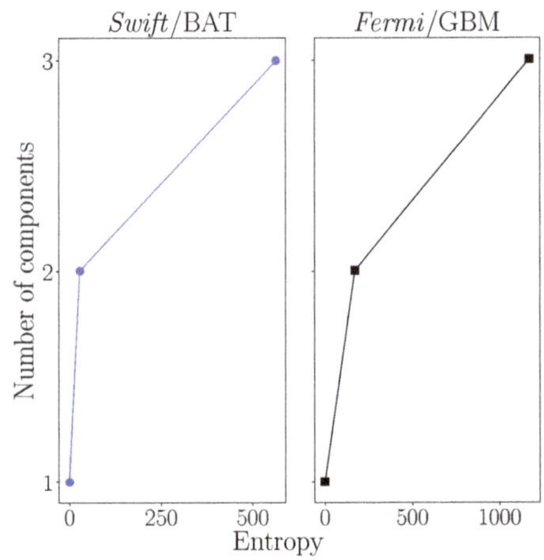

Figure 2. Entropy plots returned by `clustCombi` depicting the entropy of the initial MCLUST fits (three components) and the entropy after combination of the initial MCLUST components for *Swift*/BAT (left) and *Fermi*/GBM (right). An inflection, or elbow, in the entropy plot signifies a model with the optimal number of components.

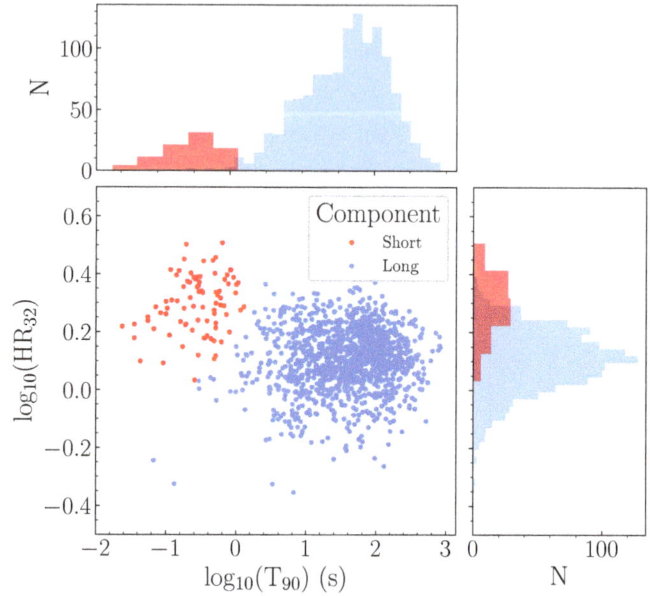

Figure 3. The results of `clustCombi` applied to the components identified by MCLUST in the duration-hardness plane for *Swift*/BAT.

The distributions of the duration (T_{90}) and hardness ratio (HR_{32}) are depicted in the violin plot in Figure 4. The mean, standard deviation, and median values of these parameters for the long and the short classes are presented in Table 3.

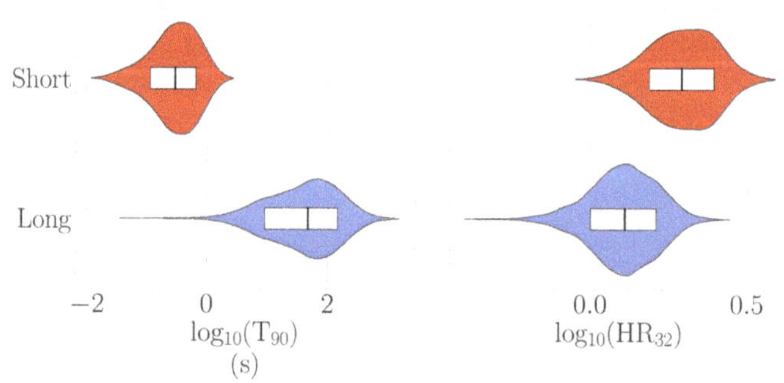

Figure 4. Violin plots showing the distribution of the duration (T_{90}) and hardness ratio (HR_{32}) for the *Swift*/BAT 'short' (red) and 'long' (blue) classes identified by `clustCombi`. The median of each parameter is marked as a black line within the box, which represents the 1 σ interval (i.e., the 16th to 84th percentile).

Table 3. Mean (μ), standard deviation (σ), and median of the properties of the *Swift*/BAT 'long' and 'short' classes identified by `clustCombi`.

	Short			Long		
	μ	σ	**Median**	μ	σ	**Median**
T_{90} (s)	0.39	0.29	0.30	79.8	101.0	47.7
HR_{32}	2.02	0.46	1.96	1.31	0.32	1.28

4.2. Fermi/GBM

For *Fermi*/GBM, the initial MCLUST fit indicated that a three-component fit to the data was preferred. The BIC values of the top three models are shown in Figure 5a. The three best-fit components are depicted in Figure 5b. The three-component 'VEI' model exhibited a BIC value difference of ∼6 compared to the next best model; thus, it was considered a significant result. The classification components were labelled 'short', 'intermediate', and 'long', according to their duration. The boundary between the 'intermediate' and 'long' components exhibited a similar round-edge feature as identified in the results for *Swift*/BAT.

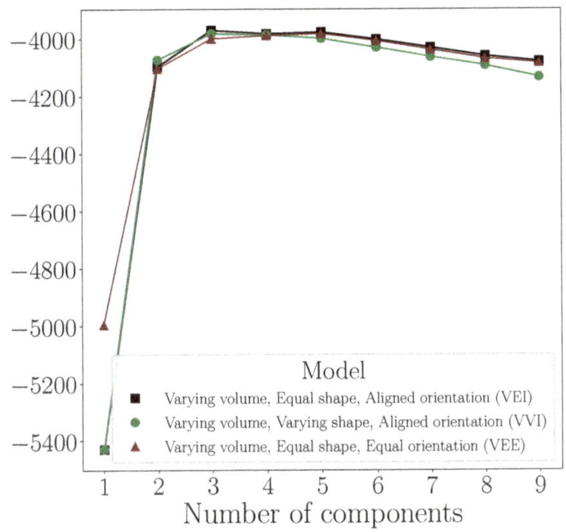

(**a**)

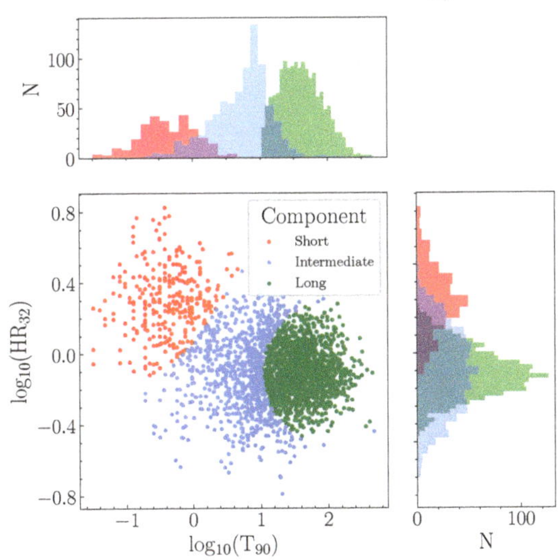

(**b**)

Figure 5. (**a**) BIC values of the top three MCLUST models fit to the *Fermi*/GBM sample and (**b**) the resulting duration-hardness plane for the best-fit three-component model (VEI).

The results obtained from applying `clustCombi` to this sample are shown in Figure 6, indicating that a model consisting of two components or classes, rather than three, provided a better fit to the data, based on a decrease in entropy depicted in the entropy plot in Figure 2. The number of bursts in the 'long' and 'short' classes identified by `clustCombi` is presented in Table 2.

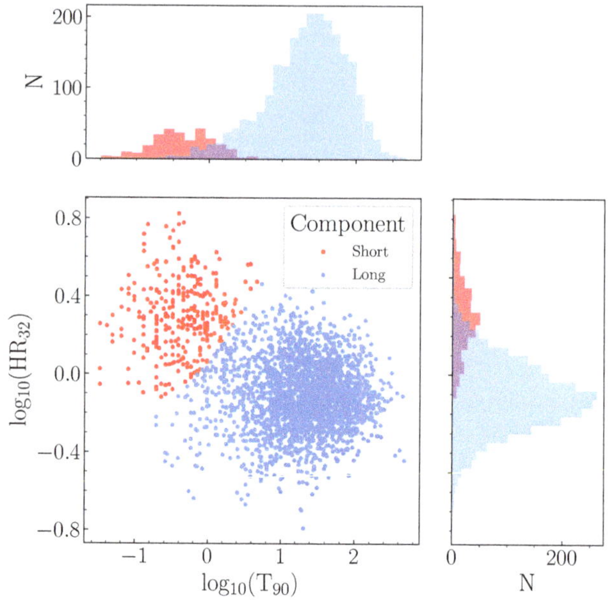

Figure 6. The results of `clustCombi` applied to the components identified by MCLUST in the duration-hardness plane for *Fermi*/GBM.

The violin plot in Figure 7 for the *Fermi*/GBM sample demonstrates the distributions of the duration and hardness ratio, while the summary statistics of these populations are presented in Table 4.

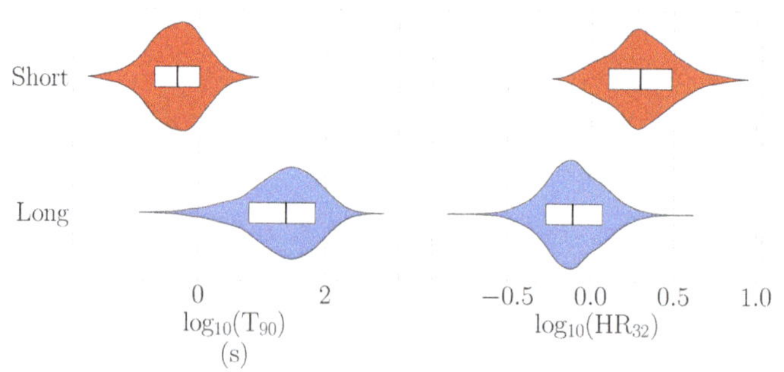

Figure 7. Violin plots showing the distribution of the duration (T_{90}) and hardness ratio (HR_{32}) for the *Fermi*/GBM 'short' (red) and 'long' (blue) classes identified by `clustCombi`. The median of each parameter is marked as a black line within the box, which represents the 1 σ interval (i.e., the 16th to 84th percentile).

Table 4. Mean (μ), standard deviation (σ), and median of the properties of the *Fermi*/GBM 'long' and 'short' classes identified by `clustCombi`.

	Short			Long		
	μ	σ	Median	μ	σ	Median
T_{90} (s)	0.64	0.65	0.45	38.6	23.4	45.4
HR_{32}	1.99	1.53	1.96	0.78	1.49	0.77

5. Discussion

5.1. *Swift/BAT*

For *Swift*/BAT, the three components identified in Figure 1b had a similar size and structure to those identified in the GMM clustering by Bhave et al. [54]. In this analysis, the hardness ratio was computed using the best-fit model for *Swift*, consistent with the method undertaken by Bhave et al. [54], enabling comparison of results. The clear-cut round boundary between the intermediate and long components in Figure 1b was also found by Bhave et al. [54] and is a signature of the application of a Gaussian model to a non-Gaussian underlying distribution.

The result of applying `clustCombi` after the GMM clustering indicated that the intermediate duration component, combined with the long duration component, provided a better fit to the sample of *Swift*/BAT bursts (Figure 3). Thus, the intermediate class was likely identified by the overfitting resulting from GMM clustering applied to the complex distribution of *Swift*/BAT bursts in duration-hardness space.

Figure 4 and Table 3 show that the mean duration of the short class identified by `clustCombi` was $T_{90} \approx 0.3$ s (1 σ standard deviation of 0.29 s), while the long class had a mean $T_{90} \approx 70$ s (1 σ standard deviation of 101 s). This is consistent with the peaks of the *Swift* short ($T_{90} < 2$ s) and long ($T_{90} > 2$ s) duration distributions [85]. The shorter duration class had a larger hardness ratio than the longer duration class, as expected from the traditional short/long paradigm. The separation between the short and long classes occurred at $T_{90} \approx 0.5$–2 s. This is in agreement with the findings of Bromberg et al. [86], whose modelling of the duration distribution of *Swift*/BAT bursts using the Collapsar model suggested a separation at $T_{90} \approx 0.8$ s.

5.2. *Fermi/GBM*

Prior to the removal of the 36 outlier bursts, MCLUST initially suggested a fit with four components in the *Fermi*/GBM sample. The fit included one group of bursts with very high or very low hardness ratios situated in a halo around the three groups in Figure 5b. These outlier bursts were effectively removed using `HDOutliers` (Section 2), following previous studies including Tóth et al. [46], Horváth et al. [56,58] and were likely the result of unsuitable background subtraction. Upon removal of the outliers, MCLUST identified three components, which were similar to the components obtained for *Swift*/BAT (Figure 5b). The intermediate duration component contained more bursts than the class identified by Horváth et al. [56], whose intermediate class only contained bursts with low spectral hardness. This difference can be attributed to their smaller sample size of 1298 bursts.

A signature of a Gaussian component is visible at the sharp boundary between the intermediate and long duration components in Figure 5b, indicating an arbitrary Gaussian component was identified. Consistent with the results obtained for *Swift*/BAT, the intermediate component was disregarded when `clustCombi` was applied, indicating that it was likely an overfitting component identified by the GMM clustering procedure. Thus, a short and long duration class remained.

In this analysis, the hardness ratio was computed using the background-subtracted counts to be consistent with previous *Fermi*/GBM studies and to enable direct comparison with those results. The short and long duration classes in Figure 6 were comparable to the classes found in previous GMM clustering analyses by Bhave et al. [54], Bhat et al. [87], and in skewed bi-variate fits carried out by Tarnopolski [51]. Table 4 and Figure 7 show that the short duration class was spectrally harder than the long duration class, as expected. The mean duration of the classes were 0.64 s (1 σ standard deviation of 0.65 s) and 38.6 s (1 σ standard deviation of 23.4 s) for the short and long classes, respectively. This result is consistent with the mean durations of the short (0.82 s) and long (28.3 s) classes identified in the GMM clustering of the third *Fermi*/GBM catalogue [87].

5.3. Comparison to GRB Subclasses

Groups 1 and 2 of the *Swift*/BAT and *Fermi*/GBM samples resemble the traditional short/hard and long/soft prototypes. The groups can be compared to several subclasses of GRBs, including those with associated supernovae, extended emission episodes, and plateaus. The longer duration Group 2 contained all 49 bursts with an X-ray plateau from the platinum sample identified by Dainotti et al. [88]. Similarly, all bursts in the sample with an optical plateau [89] and those with an associated supernova and a plateau [90] lay in Group 2. The *Swift* sample analysed also contained four ultra-long GRBs from the Gold sample and five possible ultra-long GRBs from the Silver sample of Gendre et al. [91]. All of these bursts resided in Group 2 as expected, given their duration.

Short GRBs with extended emission episodes have challenged the typical duration-based classification scheme of GRBs. The population of *Swift* GRBs with extended emission identified by Gibson et al. [92] contains bursts chosen from the sample in Kaneko et al. [93] and Gompertz et al. [94,95]. The Gibson et al. [92] sample was found to only contain bursts from Group 2 of our *Swift*/BAT results. This is understandable, given that the rebrightening exhibited in their light curves can lead to an increase in the measured T_{90} [70], thus placing them in Group 2. The extended emission episodes are typically softer than the initial spike, dominating the overall detected fluence, thus resulting in a longer duration GRB.

Group 2 resembled the standard long-duration group for both the *Swift* and *Fermi* samples. Thus, bursts with associated supernovae were expected to belong to this group. The sample of supernova-associated GRBs from Cano et al. [5] was updated to include more recent events GRB 161219B/SN 2016jca [96], GRB 171205A/SN 2017htp [97], GRB 180728A/ SN 2018fip [98,99], GRB 190114C/SN 2019jrj and GRB 190829A/AT2019oyw [100], and GRB 200826A [18–20]. There were 25 *Swift* and 15 *Fermi* GRB-SN cases within the sample analysed, all of which resided in Group 2 as expected. The only confirmed GRB with a kilonova, GRB 170817, was also in Group 1 of the *Fermi* sample.

5.4. Selection Effects

Svinkin et al. [69] suggested that T_{50}, the time during which 50% of the counts above background are recorded, is a more robust duration measure than T_{90}, since it may be less affected by detector energy ranges. To eliminate possible selection effects and to verify the two-component solutions, the clustering analysis was repeated using T_{50} as the duration parameter. For *Swift*/BAT, the initial MCLUST fit returned a three-component solution similar to Figure 1b—the short duration group remained the same, while the two long duration groups also exhibited the clear-cut spherical feature identified in the T_{90} analysis. When `clustCombi` was applied, a two group-solution was the best fit. Group 1 and Group 2 were identical to the groups found in the T_{90} analysis. Thus, for *Swift*/BAT, this method did not favour T_{50} over T_{90} as a duration measure, and the results further supported the two-group solution.

For *Fermi*/GBM, the initial MCLUST fit identified an extra long duration group in a four-component solution. The long duration group in Figure 5b was split in two, with the remaining structure matching the results of the T_{90} analysis. `clustCombi` resulted in a two-component fit closely resembling the structure and makeup of Group 1 and Group 2 of the T_{90} fit. However, Group 1 contained ≈ 100 more GRBs than the T_{90} fit. For *Fermi*/GBM, the two-component fit was supported, and while the T_{50} parameter returned slightly different proportions in each group, it did not demonstrate any clear advantage over T_{90} as a duration parameter.

6. Conclusions

GMM clustering with MCLUST identified three Gaussian components of *Swift*/BAT and *Fermi*/GBM bursts in the duration-hardness plane. The third component resembled the intermediate duration group identified in previous studies. However, combining components, based on an entropy criterion, identified a short and long duration class only for both samples.

This study highlights the drawbacks of fitting GRB populations with model-based methods. Similar model-based fitting methods, including the log-normal fit procedures applied to GRB duration distributions, have exhibited components thought to be identified incorrectly due to the inherently skewed distribution of long GRB durations [31,36]. Table 1 highlights the diversity of results from model-based studies.

The lack of consensus regarding a definitive number of GRB classes, both in the analysis presented here and in previous studies of GRB catalogues, is a motivator for a model-independent analysis of GRB light curves. The light curves may also contain more information than the summary data provided by the GRB catalogues. Fourier analysis of the *Swift*/BAT GRB light curves by Jespersen et al. [101] identified two classes of bursts. Following on from the analysis presented in this paper is a wavelet-based feature extraction analysis of GRB light curves from *Swift*/BAT, BATSE and *Fermi*/GBM (Salmon et al. [102], in preparation).

Author Contributions: Conceptualization, L.S., L.H. and A.M.-C.; methodology, L.S., L.H. and A.M.-C.; software, L.S.; validation, L.S., L.H. and A.M.-C.; formal analysis, L.S.; investigation, L.S.; resources, L.S., L.H. and A.M.-C.; data curation, L.S.; writing—original draft preparation, L.S.; writing—review and editing, L.H. and A.M.-C.; visualization, L.S.; supervision, L.H. and A.M.-C. ; project administration, L.H. and A.M.-C.; funding acquisition, L.H. and A.M.-C. All authors have read and agreed to the published version of the manuscript.

Funding: L.S. acknowledges the Irish Research Council Postgraduate Scholarship No GOIPG/2017/1525. L.H. acknowledges support from Science Foundation Ireland (Grant number 19/FFP/6777) and the EU H2020 (Grant agreement 871158).

Institutional Review Board Statement: Not applicable.

Informed Consent Statement: Not applicable.

Data Availability Statement: This paper makes use of the *Swift*/BAT catalogue [76], which can be accessed through the online *Swift*/BAT Gamma-ray Burst Catalogue [1]. The *Fermi*/GBM catalogue is accessed through the *Fermi*/GBM data tools [40].

Acknowledgments: This research made use of the following Python packages: NumPy [103], Matplotlib [104] and pandas [105,106]. The authors are grateful to Brendan Murphy and Michael Fop for their assistance with MCLUST.

Conflicts of Interest: The authors declare no conflict of interest. The funders had no role in the design of the study; in the collection, analyses, or interpretation of data; in the writing of the manuscript, or in the decision to publish the results.

Note

1 https://swift.gsfc.nasa.gov/results/batgrbcat/, (accessed on 16 June 2022).

References

1. Kouveliotou, C.; Meegan, C.A.; Fishman, G.J.; Bhat, N.P.; Briggs, M.S.; Koshut, T.M.; Paciesas, W.S.; Pendleton, G.N. Identification of Two Classes of Gamma-ray Bursts. *Astrophys. J.* **1993**, *413*, L101. [CrossRef]
2. Perley, D.A.; Niino, Y.; Tanvir, N.R.; Vergani, S.D.; Fynbo, J.P.U. Long-Duration Gamma-ray Burst Host Galaxies in Emission and Absorption. *Space Sci. Rev.* **2016**, *202*, 111–142. [CrossRef]
3. Galama, T.J.; Vreeswijk, P.M.; van Paradijs, J.; Kouveliotou, C.; Augusteijn, T.; Böhnhardt, H.; Brewer, J.P.; Doublier, V.; Gonzalez, J.F.; Leibundgut, B.; et al. An unusual supernova in the error box of the γ-ray burst of 25 April 1998. *Nature* **1998**, *395*, 670–672. [CrossRef]

4. Woosley, S.E.; Bloom, J.S. The Supernova Gamma-ray Burst Connection. *Annu. Rev. Astron. Astrophys.* **2006**, *44*, 507–556. [CrossRef]
5. Cano, Z.; Wang, S.Q.; Dai, Z.G.; Wu, X.F. The Observer's Guide to the Gamma-ray Burst Supernova Connection. *Adv. Astron.* **2017**, *2017*, 8929054, [CrossRef]
6. MacFadyen, A.I.; Woosley, S.E. Collapsars: Gamma-ray Bursts and Explosions in "Failed Supernovae". *Astrophys. J.* **1999**, *524*, 262–289. [CrossRef]
7. Eichler, D.; Livio, M.; Piran, T.; Schramm, D.N. Nucleosynthesis, neutrino bursts and γ-rays from coalescing neutron stars. *Nature* **1989**, *340*, 126–128. [CrossRef]
8. Narayan, R.; Paczynski, B.; Piran, T. Gamma-ray Bursts as the Death Throes of Massive Binary Stars. *Astrophys. J.* **1992**, *395*, L83. [CrossRef]
9. Berger, E. A Short Gamma-ray Burst "No-host" Problem? Investigating Large Progenitor Offsets for Short GRBs with Optical Afterglows. *Astrophys. J.* **2010**, *722*, 1946–1961. [CrossRef]
10. Tunnicliffe, R.L.; Levan, A.J.; Tanvir, N.R.; Rowlinson, A.; Perley, D.A.; Bloom, J.S.; Cenko, S.B.; O'Brien, P.T.; Cobb, B.E.; Wiersema, K.; et al. On the nature of the 'hostless' short GRBs. *Mon. Not. R. Astron. Soc.* **2014**, *437*, 1495–1510. [CrossRef]
11. Berger, E. Short-Duration Gamma-ray Bursts. *Annu. Rev. Astron. Astrophys.* **2014**, *52*, 43–105. [CrossRef]
12. Abbott, B.P.; Abbott, R.; Abbott, T.D.; Acernese, F.; Ackley, K.; Adams, C.; Adams, T.; Addesso, P.; Adhikari, R.X.; Adya, V.B.; et al. Multi-messenger Observations of a Binary Neutron Star Merger. *Astrophys. J.* **2017**, *848*, L12. [CrossRef]
13. Goldstein, A.; Veres, P.; Burns, E.; Briggs, M.S.; Hamburg, R.; Kocevski, D.; Wilson-Hodge, C.A.; Preece, R.D.; Poolakkil, S.; Roberts, O.J.; et al. An Ordinary Short Gamma-ray Burst with Extraordinary Implications: Fermi-GBM Detection of GRB 170817A. *Astrophys. J.* **2017**, *848*, L14. [CrossRef]
14. Savchenko, V.; Ferrigno, C.; Kuulkers, E.; Bazzano, A.; Bozzo, E.; Brandt, S.; Chenevez, J.; Courvoisier, T.L.; Diehl, R.; Domingo, A.; et al. INTEGRAL detection of the first prompt gamma-ray signal coincident with the gravitational-wave event GW170817. *Astrophys. J.* **2017**, *848*, L15. [CrossRef]
15. Qin, B.; Wu, X.P.; Chu, M.C.; Fang, L.Z.; Hu, J.Y. The Collapse of Neutron Stars in High-Mass Binaries as the Energy Source for the Gamma-ray Bursts. *Astrophys. J.* **1998**, *494*, L57–L59. [CrossRef]
16. Levan, A.J.; Wynn, G.A.; Chapman, R.; Davies, M.B.; King, A.R.; Priddey, R.S.; Tanvir, N.R. Short gamma-ray bursts in old populations: Magnetars from white dwarf-white dwarf mergers. *Mon. Not. R. Astron. Soc. Lett.* **2006**, *368*, L1–L5. [CrossRef]
17. Metzger, B.D.; Quataert, E.; Thompson, T.A. Short-duration gamma-ray bursts with extended emission from protomagnetar spin-down. *Mon. Not. R. Astron. Soc. Lett.* **2008**, *385*, 1455–1460. [CrossRef]
18. Ahumada, T.; Singer, L.P.; Anand, S.; Coughlin, M.W.; Kasliwal, M.M.; Ryan, G.; Andreoni, I.; Cenko, S.B.; Fremling, C.; Kumar, H.; et al. Discovery and confirmation of the shortest gamma ray burst from a collapsar. *arXiv* **2021**, arXiv:2105.05067.
19. Rossi, A.; Rothberg, B.; Palazzi, E.; Kann, D.A.; D'Avanzo, P.; Klose, S.; Perego, A.; Pian, E.; Savaglio, S.; Stratta, G.; et al. The peculiar short-duration GRB 200826A and its supernova. *arXiv* **2021**, arXiv:2105.03829.
20. Zhang, B.B.; Liu, Z.K.; Peng, Z.K.; Li, Y.; Lü, H.J.; Yang, J.; Yang, Y.S.; Yang, Y.H.; Meng, Y.Z.; Zou, J.H.; et al. A Peculiarly Short-duration Gamma-ray Burst from Massive Star Core Collapse. *arXiv* **2021**, arXiv:2105.05021.
21. Zhang, B.; Zhang, B.B.; Liang, E.W.; Gehrels, N.; Burrows, D.N.; Mészáros, P. Making a Short Gamma-ray Burst from a Long One: Implications for the Nature of GRB 060614. *Astrophys. J.* **2007**, *655*, L25–L28. [CrossRef]
22. Zhang, B.; Zhang, B.B.; Virgili, F.J.; Liang, E.W.; Kann, D.A.; Wu, X.F.; Proga, D.; Lv, H.J.; Toma, K.; Mészáros, P.; et al. Discerning the Physical Origins of Cosmological Gamma-ray Bursts Based on Multiple Observational Criteria: The Cases of z = 6.7 GRB 080913, z = 8.2 GRB 090423, and Some Short/Hard GRBs. *Astrophys. J.* **2009**, *703*, 1696–1724. [CrossRef]
23. Hjorth, J.; Bloom, J.S. The Gamma-ray Burst—Supernova Connection. In *Gamma-ray Bursts*; Cambridge University Press: Cambridge, MA, USA, 2012; Chapter 9, pp. 169–190.
24. Li, Y.; Zhang, B.; Yuan, Q. A Comparative Study of Long and Short GRBs. II. A Multiwavelength Method to Distinguish Type II (Massive Star) and Type I (Compact Star) GRBs. *Astrophys. J.* **2020**, *897*, 154. [CrossRef]
25. Lü, H.J.; Liang, E.W.; Zhang, B.B.; Zhang, B. A New Classification Method for Gamma-ray Bursts. *Astrophys. J.* **2010**, *725*, 1965–1970. [CrossRef]
26. Donaghy, T.Q.; Lamb, D.Q.; Sakamoto, T.; Norris, J.P.; Nakagawa, Y.; Villasenor, J.; Atteia, J.L.; Vanderspek, R.; Graziani, C.; Kawai, N.; et al. HETE-2 Localizations and Observations of Four Short Gamma-ray Bursts: GRBs 010326B, 040802, 051211 and 060121. *arXiv* **2006**, arXiv:astro-ph/0605570.
27. Horváth, I. A Third Class of Gamma-ray Bursts? *Astrophys. J.* **1998**, *508*, 757–759. [CrossRef]
28. Horváth, I. Classification of BeppoSAX's gamma-ray bursts. *Astrophys. Space Sci.* **2009**, *323*, 83–86. [CrossRef]
29. Řípa, J.; Wigger, C.; Huja, D.; Hudec, R. Gamma-ray Burst Classes Found in the Rhessi Data Sample. *Baltic Astron.* **2009**, *18*, 305–310.
30. Huja, D.; Mészáros, A.; Řípa, J. A comparison of the gamma-ray bursts detected by BATSE and Swift. *Astron. Astrophys.* **2009**, *504*, 67–71. [CrossRef]
31. Koen, C.; Bere, A. On multiple classes of gamma-ray bursts, as deduced from autocorrelation functions or bivariate duration/hardness ratio distributions. *Mon. Not. R. Astron. Soc. Lett.* **2012**, *420*, 405–415. [CrossRef]
32. Zitouni, H.; Guessoum, N.; Azzam, W.J.; Mochkovitch, R. Statistical study of observed and intrinsic durations among BATSE and Swift/BAT GRBs. *Astrophys. Space Sci.* **2015**, *357*, 7. [CrossRef]

33. Horváth, I.; Tóth, B.G. The duration distribution of Swift Gamma-ray Bursts. *Astrophys. Space Sci.* **2016**, *361*, 155. [CrossRef]
34. Tarnopolski, M. Analysis of the observed and intrinsic durations of gamma-ray bursts with known redshift. *Astrophys. Space Sci.* **2016**, *361*, 125. [CrossRef]
35. Kulkarni, S.; Desai, S. Classification of gamma-ray burst durations using robust model-comparison techniques. *Astrophys. Space Sci.* **2017**, *362*, 70. [CrossRef]
36. Tarnopolski, M. Distinguishing short and long Fermi gamma-ray bursts. *Mon. Not. R. Astron. Soc. Lett.* **2015**, *454*, 1132–1139. [CrossRef]
37. Tarnopolski, M. Analysis of gamma-ray burst duration distribution using mixtures of skewed distributions. *Mon. Not. R. Astron. Soc.* **2016**, *458*, 2024–2031. [CrossRef]
38. Band, D.L. Gamma-ray Burst Spectral Evolution through Cross-Correlations of Discriminator Light Curves. *Astrophys. J.* **1997**, *486*, 928–937. [CrossRef]
39. Von Kienlin, A.; Meegan, C.A.; Paciesas, W.S.; Bhat, P.N.; Bissaldi, E.; Briggs, M.S.; Burns, E.; Cleveland, W.H.; Gibby, M.H.; Giles, M.M.; et al. The Fourth Fermi-GBM Gamma-ray Burst Catalog: A Decade of Data. *Astrophys. J.* **2020**, *893*, 46. [CrossRef]
40. Goldstein, A.; Cleveland, W.H.; Kocevski, D. Fermi GBM Data Tools: v1.1.1. 2020. Available online: https://fermi.gsfc.nasa.gov/ssc/data/analysis/gbm (accessed on 17 June 2022).
41. Mukherjee, S.; Feigelson, E.D.; Jogesh Babu, G.; Murtagh, F.; Fraley, C.; Raftery, A. Three Types of Gamma-ray Bursts. *Astrophys. J.* **1998**, *508*, 314–327. [CrossRef]
42. Chattopadhyay, T.; Misra, R.; Chattopadhyay, A.K.; Naskar, M. Statistical Evidence for Three Classes of Gamma-ray Bursts. *Astrophys. J.* **2007**, *667*, 1017–1023. [CrossRef]
43. Modak, S. Distinction of groups of gamma-ray bursts in the BATSE catalog through fuzzy clustering. *Astron. Comput.* **2021**, *34*, 100441. [CrossRef]
44. Chattopadhyay, S.; Maitra, R. Gaussian-mixture-model-based cluster analysis finds five kinds of gamma-ray bursts in the BATSE catalogue. *Mon. Not. R. Astron. Soc. Lett.* **2017**, *469*, 3374–3389. [CrossRef]
45. Chattopadhyay, S.; Maitra, R. Multivariate t-mixture-model-based cluster analysis of BATSE catalogue establishes importance of all observed parameters, confirms five distinct ellipsoidal sub-populations of gamma-ray bursts. *Mon. Not. R. Astron. Soc. Lett.* **2018**, *481*, 3196–3209. [CrossRef]
46. Tóth, B.G.; Rácz, I.I.; Horváth, I. Gaussian-mixture-model-based cluster analysis of gamma-ray bursts in the BATSE catalog. *Mon. Not. R. Astron. Soc. Lett.* **2019**, *486*, 4823–4828. [CrossRef]
47. Balastegui, A.; Ruiz-Lapuente, P.; Canal, R. Reclassification of gamma-ray bursts. *Mon. Not. R. Astron. Soc. Lett.* **2001**, *328*, 283–290. [CrossRef]
48. Hakkila, J.; Haglin, D.J.; Pendleton, G.N.; Mallozzi, R.S.; Meegan, C.A.; Roiger, R.J. Gamma-ray Burst Class Properties. *Astrophys. J.* **2000**, *538*, 165–180. [CrossRef]
49. Hakkila, J.; Giblin, T.W.; Roiger, R.J.; Haglin, D.J.; Paciesas, W.S.; Meegan, C.A. How Sample Completeness Affects Gamma-ray Burst Classification. *Astrophys. J.* **2003**, *582*, 320–329. [CrossRef]
50. Rajaniemi, H.; Mähönen, P. Classifying gamma-ray bursts using self-organizing maps. *Astrophys. J.* **2002**, *566*, 202. [CrossRef]
51. Tarnopolski, M. Analysis of the Duration-Hardness Ratio Plane of Gamma-ray Bursts Using Skewed Distributions. *Astrophys. J.* **2019**, *870*, 105. [CrossRef]
52. Tarnopolski, M. Multivariate Analysis of BATSE Gamma-ray Burst Properties Using Skewed Distributions. *Astrophys. J.* **2019**, *887*, 97. [CrossRef]
53. Veres, P.; Bagoly, Z.; Horváth, I.; Mészáros, A.; Balázs, L.G. A Distinct Peak-flux Distribution of the Third Class of Gamma-ray Bursts: A Possible Signature of X-ray Flashes? *Astrophys. J.* **2010**, *725*, 1955–1964. [CrossRef]
54. Bhave, A.; Kulkarni, S.; Desai, S.; Srijith, P.K. Two Dimensional Clustering of Gamma-ray Bursts using durations and hardness. *arXiv* **2017**, arXiv:1708.05668.
55. Tsutsui, R.; Shigeyama, T. On the subclasses in Swift long gamma-ray bursts: A clue to different central engines. *Publ. Astron. Soc. Jpn.* **2014**, *66*, 42. [CrossRef]
56. Horváth, I.; Tóth, B.G.; Hakkila, J.; Tóth, L.V.; Balázs, L.G.; Rácz, I.I.; Pintér, S.; Bagoly, Z. Classifying GRB 170817A/GW170817 in a Fermi duration-hardness plane. *Astrophys. Space Sci.* **2018**, *363*, 53. [CrossRef]
57. Acuner, Z.; Ryde, F. Clustering of gamma-ray burst types in the Fermi GBM catalogue: Indications of photosphere and synchrotron emissions during the prompt phase. *Mon. Not. R. Astron. Soc. Lett.* **2018**, *475*, 1708–1724. [CrossRef]
58. Horváth, I.; Hakkila, J.; Bagoly, Z.; Tóth, L.V.; Rácz, I.I.; Pintér, S.; Tóth, B.G. Multidimensional analysis of Fermi GBM gamma-ray bursts. *Astrophys. Space Sci.* **2019**, *364*, 105. [CrossRef]
59. Modak, S.; Chattopadhyay, A.K.; Chattopadhyay, T. Clustering of Gamma-ray bursts through kernel principal component analysis. *arXiv* **2017**, arXiv:1703.05532.
60. Horváth, I. A further study of the BATSE Gamma-ray Burst duration distribution. *Astron. Astrophys.* **2002**, *392*, 791–793. [CrossRef]
61. Zhang, Z.B.; Yang, E.B.; Choi, C.S.; Chang, H.Y. Classifying gamma-ray bursts with Gaussian Mixture Model. *Mon. Not. R. Astron. Soc. Lett.* **2016**, *462*, 3243–3254. [CrossRef]
62. Horváth, I.; Balázs, L.G.; Bagoly, Z.; Veres, P. Classification of Swift's gamma-ray bursts. *Astron. Astrophys.* **2008**, *489*, L1–L4. [CrossRef]

63. Zhang, Z.B.; Choi, C.S. An analysis of the durations of Swift gamma-ray bursts. *Astron. Astrophys.* **2008**, *484*, 293–297. [CrossRef]
64. Horváth, I.; Bagoly, Z.; Balázs, L.G.; de Ugarte Postigo, A.; Veres, P.; Mészáros, A. Detailed Classification of Swift's Gamma-ray Bursts. *Astrophys. J.* **2010**, *713*, 552–557. [CrossRef]
65. Yang, E.B.; Zhang, Z.B.; Jiang, X.X. Two dimensional classification of the Swift/BAT GRBs. *Astrophys. Space Sci.* **2016**, *361*, 257. [CrossRef]
66. Zitouni, H.; Guessoum, N.; AlQassimi, K.M.; Alaryani, O. Distributions of pseudo-redshifts and durations (observed and intrinsic) of Fermi GRBs. *Astrophys. Space Sci.* **2018**, *363*, 223. [CrossRef]
67. Řípa, J.; Mészáros, A.; Veres, P.; Park, I.H. On the Spectral Lags and Peak Counts of the Gamma-ray Bursts Detected by the RHESSI Satellite. *Astrophys. J.* **2012**, *756*, 44. [CrossRef]
68. Minaev, P.Y.; Pozanenko, A.S.; Loznikov, V.M. Short gamma-ray bursts in the SPI-ACS INTEGRAL experiment. *Astrophys. Bull.* **2010**, *65*, 326–333. [CrossRef]
69. Svinkin, D.S.; Aptekar, R.L.; Golenetskii, S.V.; Frederiks, D.D.; Ulanov, M.V.; Tsvetkova, A.E. Classification of gamma-ray bursts observed with Konus-Wind. *J. Phys. Conf. Ser.* **2019**, *1400*, 022010. [CrossRef]
70. Minaev, P.Y.; Pozanenko, A.S. The $E_{p,I}$-E_{iso} correlation: Type I gamma-ray bursts and the new classification method. *Mon. Not. R. Astron. Soc. Lett.* **2020**, *492*, 1919–1936. [CrossRef]
71. Lü, H.J.; Zhang, B.; Liang, E.W.; Zhang, B.B.; Sakamoto, T. The 'amplitude' parameter of gamma-ray bursts and its implications for GRB classification. *Mon. Not. R. Astron. Soc. Lett.* **2014**, *442*, 1922–1929. [CrossRef]
72. Tarnopolski, M. How does the shape of gamma-ray bursts' pulses affect the duration distribution? *Mon. Not. R. Astron. Soc.* **2021**, *507*, 1450–1457. [CrossRef]
73. Tarnopolski, M. Can the Cosmological Dilation Explain the Skewness in the Gamma-ray Burst Duration Distribution? *Astrophys. J.* **2020**, *897*, 77. [CrossRef]
74. Baudry, J.P.; Raftery, A.E.; Celeux, G.; Lo, K.; Gottardo, R. Combining Mixture Components for Clustering. *J. Comput. Graph. Stat.* **2010**, *19*, 332–353. [CrossRef] [PubMed]
75. Kuhn, M.A.; Hillenbrand, L.A.; Carpenter, J.M.; Avelar Menendez, A.R. The Formation of a Stellar Association in the NGC 7000/IC 5070 Complex: Results from Kinematic Analysis of Stars and Gas. *Astrophys. J.* **2020**, *899*, 128. [CrossRef]
76. Lien, A.; Sakamoto, T.; Barthelmy, S.D.; Baumgartner, W.H.; Cannizzo, J.K.; Chen, K.; Collins, N.R.; Cummings, J.R.; Gehrels, N.; Krimm, H.A.; et al. The Third Swift Burst Alert Telescope Gamma-ray Burst Catalog. *Astrophys. J.* **2016**, *829*, 7. [CrossRef]
77. Fraley, C. *HDoutliers: Leland Wilkinson's Algorithm for Detecting Multidimensional Outliers*, R Package Version 1.0.2. 2020. Available online: https://CRAN.R-project.org/package=HDoutliers (accessed on 17 June 2022).
78. Scrucca, L.; Fop, M.; Murphy, T.B.; Raftery, A.E. mclust 5: Clustering, classification and density estimation using Gaussian finite mixture models. *R J.* **2016**, *8*, 289–317. [CrossRef]
79. Schwarz, G. Estimating the dimension of a model. *Ann. Stat.* **1978**, *6*, 461–464. [CrossRef]
80. Kass, R.E.; Raftery, A.E. Bayes Factors. *J. Am. Stat. Assoc.* **1995**, *90*, 773–795. [CrossRef]
81. Wang, N.; Raftery, A.E. Nearest-neighbor variance estimation (NNVE) robust covariance estimation via nearest-neighbor cleaning. *J. Am. Stat. Assoc.* **2002**, *97*, 994–1019. [CrossRef]
82. Walther, G. Detecting the presence of mixing with multiscale maximum likelihood. *J. Am. Stat. Assoc.* **2002**, *97*, 508–513. [CrossRef]
83. Li, J. Clustering based on a multilayer mixture model. *J. Comput. Graph. Stat.* **2005**, *14*, 547–568. [CrossRef]
84. Tantrum, J.; Murua, A.; Stuetzle, W. Assessment and pruning of hierarchical model based clustering. In Proceedings of the 9th ACM SIGKDD International Conference on Knowledge Discovery and Data Mining, Washington, DC, USA, 24–27 August 2003; pp. 197–205.
85. Sakamoto, T.; Barthelmy, S.D.; Baumgartner, W.H.; Cummings, J.R.; Fenimore, E.E.; Gehrels, N.; Krimm, H.A.; Markwardt, C.B.; Palmer, D.M.; Parsons, A.M.; et al. The Second Swift Burst Alert Telescope Gamma-ray Burst Catalog. *Astrophys. J. Suppl. Ser.* **2011**, *195*, 2. [CrossRef]
86. Bromberg, O.; Nakar, E.; Piran, T.; Sari, R. Short versus Long and Collapsars versus Non-collapsars: A Quantitative Classification of Gamma-ray Bursts. *Astrophys. J.* **2013**, *764*, 179. [CrossRef]
87. Bhat, P.N.; Meegan, C.A.; von Kienlin, A.; Paciesas, W.S.; Briggs, M.S.; Burgess, J.M.; Burns, E.; Chaplin, V.; Cleveland, W.H.; Collazzi, A.C.; et al. The Third Fermi GBM Gamma-ray Burst Catalog: The First Six Years. *Astrophys. J. Suppl. Ser.* **2016**, *223*, 28. [CrossRef]
88. Dainotti, M.G.; Lenart, A.Ł.; Sarracino, G.; Nagataki, S.; Capozziello, S.; Fraija, N. The X-Ray Fundamental Plane of the Platinum Sample, the Kilonovae, and the SNe Ib/c Associated with GRBs. *Astrophys. J.* **2020**, *904*, 97. [CrossRef]
89. Dainotti, M.G.; Livermore, S.; Kann, D.A.; Li, L.; Oates, S.; Yi, S.; Zhang, B.; Gendre, B.; Cenko, B.; Fraija, N. The Optical Luminosity-Time Correlation for More than 100 Gamma-ray Burst Afterglows. *Astrophys. J.* **2020**, *905*, L26. [CrossRef]
90. Dainotti, M.G.; Nagataki, S.; Maeda, K.; Postnikov, S.; Pian, E. A study of gamma ray bursts with afterglow plateau phases associated with supernovae. *Astron. Astrophys.* **2017**, *600*, A98. [CrossRef]
91. Gendre, B.; Joyce, Q.T.; Orange, N.B.; Stratta, G.; Atteia, J.L.; Boër, M. Can we quickly flag ultra-long gamma-ray bursts? *Mon. Not. R. Astron. Soc. Lett.* **2019**, *486*, 2471–2476. [CrossRef]
92. Gibson, S.L.; Wynn, G.A.; Gompertz, B.P.; O'Brien, P.T. Fallback accretion on to a newborn magnetar: Short GRBs with extended emission. *Mon. Not. R. Astron. Soc. Lett.* **2017**, *470*, 4925–4940. [CrossRef]

93. Kaneko, Y.; Bostancı, Z.F.; Göğüş, E.; Lin, L. Short gamma-ray bursts with extended emission observed with Swift/BAT and Fermi/GBM. *Mon. Not. R. Astron. Soc. Lett.* **2015**, *452*, 824–837. [CrossRef]
94. Gompertz, B.P.; O'Brien, P.T.; Wynn, G.A.; Rowlinson, A. Can magnetar spin-down power extended emission in some short GRBs? *Mon. Not. R. Astron. Soc. Lett.* **2013**, *431*, 1745–1751. [CrossRef]
95. Gompertz, B.P.; O'Brien, P.T.; Wynn, G.A. Magnetar powered GRBs: Explaining the extended emission and X-ray plateau of short GRB light curves. *Mon. Not. R. Astron. Soc. Lett.* **2014**, *438*, 240–250. [CrossRef]
96. Ashall, C.; Mazzali, P.A.; Pian, E.; Woosley, S.E.; Palazzi, E.; Prentice, S.J.; Kobayashi, S.; Holmbo, S.; Levan, A.; Perley, D.; et al. GRB 161219B/SN 2016jca: A powerful stellar collapse. *Mon. Not. R. Astron. Soc. Lett.* **2019**, *487*, 5824–5839. [CrossRef]
97. Melandri, A.; Malesani, D.B.; Izzo, L.; Japelj, J.; Vergani, S.D.; Schady, P.; Sagués Carracedo, A.; de Ugarte Postigo, A.; Anderson, J.P.; Barbarino, C.; et al. GRB 171010A/SN 2017htp: A GRB-SN at z = 0.33. *Mon. Not. R. Astron. Soc. Lett.* **2019**, *490*, 5366–5374. [CrossRef]
98. Izzo, L.; Rossi, A.; Malesani, D.; Heintz, K.; Selsing, J.; Schady, P.; Starling, R.; Sollerman, J.; Leloudas, G.; Cano, Z.; et al. GRB 180728A: Discovery of the associated supernova. *GRB Coord. Netw.* **2018**, *23142*, 1.
99. Selsing, J.; Izzo, L.; Rossi, A.; Malesani, D.; Heintz, K.; Schady, P.; Starling, R.; Sollerman, J.; Leloudas, G.; Cano, Z.; et al. GRB 180728A: Classification of the associated SN 2018fip. *GRB Coord. Netw.* **2018**, *23181*, 1.
100. Hu, Y.D.; Castro-Tirado, A.J.; Kumar, A.; Gupta, R.; Valeev, A.F.; Pandey, S.B.; Kann, D.A.; Castellón, A.; Agudo, I.; Aryan, A.; et al. 10.4 m GTC observations of the nearby VHE-detected GRB 190829A/SN 2019oyw. *Astron. Astrophys.* **2021**, *646*, A50. [CrossRef]
101. Jespersen, C.K.; Severin, J.B.; Steinhardt, C.L.; Vinther, J.; Fynbo, J.P.U.; Selsing, J.; Watson, D. An Unambiguous Separation of Gamma-ray Bursts into Two Classes from Prompt Emission Alone. *Astrophys. J.* **2020**, *896*, L20. [CrossRef]
102. Salmon, L.; Martin-Carrillo, A.; Hanlon, L. Two Classes of Gamma-ray Bursts Distinguished within the First Second of Their Prompt Emission. *Galaxies* 2022, *submitted*. [CrossRef]
103. Harris, C.R.; Millman, K.J.; van der Walt, S.J.; Gommers, R.; Virtanen, P.; Cournapeau, D.; Wieser, E.; Taylor, J.; Berg, S.; Smith, N.J.; et al. Array programming with NumPy. *Nature* **2020**, *585*, 357–362. [CrossRef]
104. Hunter, J.D. Matplotlib: A 2D graphics environment. *Comput. Sci. Eng.* **2007**, *9*, 90–95. [CrossRef]
105. Pandas Development Team. *Pandas-Dev/pandas*, Pandas-v1.4.2. 2020. Available online: https://doi.org/10.5281/zenodo.3509134 (accessed on 17 June 2022).
106. Wes McKinney. Data Structures for Statistical Computing in Python. In Proceedings of the 9th Python in Science Conference, Austin, TX, USA, 28 June–3 July 2010; van der Walt, S., Millman, J., Eds.; SciPy: Austin, TX, USA, 2010; pp. 56–61. [CrossRef]

galaxies MDPI

Article

Two Classes of Gamma-ray Bursts Distinguished within the First Second of Their Prompt Emission

Lána Salmon *, Lorraine Hanlon and Antonio Martin-Carrillo

School of Physics and Centre for Space Research, University College Dublin, Belfield, D04 V1W8 Dublin, Ireland; lorraine.hanlon@ucd.ie (L.H.); antonio.martin-carrillo@ucd.ie (A.M.-C.)
* Correspondence: lana.salmon@ucdconnect.ie

Abstract: Studies of Gamma-Ray Burst (GRB) properties, such as duration and spectral hardness, have found evidence for additional classes, beyond the short/hard and long/soft prototypes, using model-dependent methods. In this paper, a model-independent approach was used to analyse the gamma-ray light curves of large samples of GRBs detected by BATSE, *Swift*/BAT and *Fermi*/GBM. All the features were extracted from the GRB time profiles in four energy bands using the Stationary Wavelet Transform and Principal Component Analysis. t-distributed Stochastic Neighbourhood Embedding (t-SNE) visualisation of the features revealed two distinct groups of *Swift*/BAT bursts using the T_{100} interval with 64 ms resolution data. When the same analysis was applied to 4 ms resolution data, two groups were seen to emerge within the first second (T_1) post-trigger. These two groups primarily consisted of short/hard (Group 1) and long/soft (Group 2) bursts, and were 95% consistent with the groups identified using the T_{100} 64 ms resolution data. Kilonova candidates, arising from compact object mergers, were found to belong to Group 1, while those events with associated supernovae fell into Group 2. Differences in cumulative counts between the two groups in the first second, and in the minimum variability timescale, identifiable only with the 4 ms resolution data, may account for this result. Short GRBs have particular significance for multi-messenger science as a distinctive EM signature of a binary merger, which may be discovered by its gravitational wave emissions. Incorporating the T_1 interval into classification algorithms may support the rapid classification of GRBs, allowing for an improved prioritisation of targets for follow-up observations.

Keywords: gamma-ray burst; feature extraction; machine learning

Citation: Salmon, L.; Hanlon, L.; Martin-Carrillo, A. Two Classes of Gamma-ray Bursts Distinguished within the First Second of Their Prompt Emission. *Galaxies* **2022**, *10*, 78. https://doi.org/10.3390/galaxies10040078

Academic Editors: Elena Moretti and Francesco Longo

Received: 27 April 2022
Accepted: 23 June 2022
Published: 26 June 2022

Publisher's Note: MDPI stays neutral with regard to jurisdictional claims in published maps and institutional affiliations.

1. Introduction

Gamma-Ray Bursts (GRBs) are traditionally classified based on their duration and hardness as short/hard or long/soft bursts. These classes are separated at $T_{90} \approx 2$ s, derived from the duration distribution of the Third BATSE catalogue [1]. T_{90} is defined as the duration during which 5–95% of the counts above background are detected. The properties of these classes suggest different progenitors—long GRBs often lie in star-forming galaxies [2] and some long GRBs are associated with Type Ic supernovae [3–6] linking them to the deaths of massive stars [7]. Short GRBs are linked to compact object mergers [8,9], as some short GRBs have been identified near elliptical galaxies [10], and many are offset from their hosts [11,12]. The detection of GRB 170817A [13,14], associated with the neutron star merger GW170817, detected in gravitational waves by LIGO [15], lends further weight to this progenitor theory.

The classification of GRBs based on their duration is affected by the significant overlap between the duration distributions of the long and short groups, and is further complicated by a possible 'intermediate' class of GRBs, first identified through Gaussian fits to the duration distribution of GRBs in the third BATSE catalogue [16]. Clustering of the duration–hardness plane and multi-dimensional analyses of GRB samples from different satellites have also revealed evidence of more than two classes of bursts.

Salmon et al. [17] presents a review of previous studies and reports on an updated clustering analysis of *Swift*/BAT and *Fermi*/GBM bursts which finds that Gaussian models applied to *Swift*/BAT and *Fermi*/GBM GRB samples recover three clusters, including an intermediate-duration one. However, the latter is identified as an excess Gaussian component when an entropy criterion is used and the resulting best-fit solution contains two classes, which are broadly consistent with the typical short- and long-duration groups. A key conclusion of the analysis is that model-based methods may identify spurious components in one-, two- and multi-dimensional analyses of GRB samples and that model-independent analyses of GRBs should be conducted, for example, using GRB light curves.

Short GRBs with extended emissions have been detected, which may form an additional sub-class [18–20] and are possibly associated with a magnetar central engine [21]. These episodes, combined with the late X-ray flares in some short GRBs, and the non-detection of supernovae associated with some long GRBs, led to the suggestion of a new classification scheme by Zhang et al. [22]. Type I (massive star/collapsar origin) and Type II (compact-object merger origin) bursts are defined by multiple observational criteria beyond duration and hardness [23]. Other classification methods, based on afterglow and host galaxy properties [24], minimum variability timescales [25] and prompt emission and energetics, have been defined [26–30]. The instrument, sample size and classification method used can lead to different results [31], and the collapsar/merger fractions for each instrument's sample cannot simply be defined by a $T_{90} = 2$ s threshold [32].

Analysis of GRB light curves in several bands does not rely on summary statistics, such as parameters derived from spectral fits, which could be poorly fit or incorrect. Jespersen et al. [33] extracted features from 64 ms-resolution *Swift*/BAT light curves using Discrete Fourier Transforms and found two groups using t-distributed Stochastic Neighbourhood Embedding (t-SNE). This approach does not assume the underlying distribution of the variables, unlike model-based clustering and distribution fitting.

An alternative to Fourier analysis is wavelet analysis, which has been used to study non-stationary time-series [34]. Wavelet analysis has the advantage of extracting both frequency and temporal information, and for this reason it has been used to compress and de-noise GRB light curves for the study of their time evolution [35–37], to identify peaks [38–41], and to quantify the minimum variability timescale of GRBs [42–46]. Wavelet decomposition has been used to reduce the dimensionality of supernova light curves for classification [47], and has been combined with Principal Component Analysis (PCA) and t-SNE for classification [48,49]. Lochner et al. [48] found that classifiers performed better when supplied with wavelet coefficients of supernova light curves, in contrast to feature extraction using parametric models.

GRB pulses exhibit spectral evolution, including hard-to-soft [50] or intensity-tracking [51] behaviour. Other common features of all GRB pulses include longer-observed durations at lower energies [52] and asymmetric shapes [53,54]. These commonalities suggest that a similar emission mechanism creates GRB pulses, regardless of the progenitor [55,56].

However, pulses in short and long bursts also exhibit some differences. Long GRB pulses are observed to peak earlier at higher energies, but these spectral lags are not typically significant in short GRBs [18,54,57–63]. The minimum variability timescales [44–46] retrieved from wavelet analysis of long and short GRBs are $\sim$200 ms and $\sim$10 ms respectively. Hakkila and Preece [64] found that pulses in short GRBs are shorter and harder than long GRBs, and exhibit more spectral evolution. Coupled with the observation that shorter pulses have a higher peak flux and $\sim$90% of short GRBs consist of a single pulse, compared to 25–40% for long GRBs, the pulse properties are likely to be a distinguishing feature in the first seconds of a burst. In particular, spectral evolution is evident at early times in previous studies of bursts from BATSE [65–70], *Swift* [71] and *Fermi*/GBM [72–74].

Redshift effects have not been observed in GRB light curves, as the standard time dilation of GRB pulses is thought to be masked by a contrasting effect whereby only the shorter, brightest portion of the burst is observed [75]. Therefore, analysis of GRB light curves is unlikely to be strongly affected by cosmological effects [76].

In this work, the light curves of GRBs in four energy bands from three different instruments are analysed, using wavelets as a feature-extraction method. The T_{100} burst intervals, during which 100% of the counts above background is recorded, are studied at 64-ms resolution, and the early phase of GRB emission (first few seconds) at 4-ms resolution. Wavelet coefficients are extracted and reduced and then visualised using PCA and t-SNE. Section 2 outlines the sample construction, while Section 3 provides details of the methods applied to perform feature extraction. Results are presented in Section 4 and consistency checks with other studies and between instrument samples are discussed in Section 5. The classification of notable GRBs is presented in Section 6. Possible signatures in the first second are discussed in Section 7, while conclusions are outlined in Section 8.

2. GRB Light Curves

The analysed GRB samples include bursts detected by the BATSE instrument on the *Compton Gamma-Ray Observatory* [77], the Burst Alert Telescope (BAT) on the *Neil Gehrels Swift Observatory* (hereafter *Swift*/BAT; Gehrels et al. [78]) and the Gamma-ray Burst Monitor (GBM) on the *Fermi Gamma-ray Space Telescope* (hereafter *Fermi*/GBM; Meegan et al. [79]).

2.1. BATSE

The BATSE 64 ms-binned light curves were stored as ascii files on the BATSE Public Data Archive (https://heasarc.gsfc.nasa.gov/FTP/compton/data/batse/trigger/, accessed on 17 February 2021). There were 2704 bursts in the final BATSE catalogue from 21 April 1991 to 17 August 2000, and 1956 light curves at 64 ms resolution were available. Background subtraction was applied via polynomial fits to the 64 ms light curves pre- and post-burst. The BATSE 4 ms-binned light curves were generated using the TTE files for individual bursts in the BATSE Public Data Archive, which stores 1732 TTE files, 1721 of which have successful background subtraction. The count rate was divided by the number of triggered detectors to obtain light curves measured in counts s^{-1} det^{-1}. The resulting light curves were stored in the four standard BATSE bands (20–50 keV, 50–100 keV, 100–300 keV and >300 keV).

2.2. Swift/BAT

The *Swift*/BAT Gamma-Ray Burst Catalogue (https://swift.gsfc.nasa.gov/results/batgrbcat/, accessed on 29 January 2021) hosts ascii files containing the 64 ms- and 4 ms-binned background-subtracted light curves. There were 1388 GRBs detected between 17 December 2004 and 28 August 2020 in this catalogue, which was extended from the Third *Swift* BAT Catalogue [80]. 1273 light curve files were available at 4 ms resolution, containing four background-subtracted light curves, corresponding to four bands (15–25 keV, 25–50 keV, 50–100 keV and 100–350 keV) in units of counts s^{-1} det^{-1}. Twenty-two bursts with no documented duration (T_{90}) were removed from the sample. At 64 ms resolution, light curves were available for the same set of bursts, with three additional GRBs added to the sample.

2.3. Fermi/GBM

Fermi/GBM light curves were generated from TTE data in 64 ms and 4 ms bins using the *Fermi*-GBM Data Tools [81]. A total of 3000 bursts from 10 August 2008 to 17 March 2021 were included, with 2678 successful background subtracted light curves created. Only triggered detectors were used, and the background intervals defined in the *Fermi*/GBM catalogue were used for background subtraction. Count rates were transformed to counts s^{-1} det^{-1} by normalising according to the number of triggered detectors. Unlike *Swift* and BATSE, *Fermi*/GBM does not have defined light curve bands. Thus, they were chosen to capture the energy ranges of the NaI and BGO detectors, and the bands were considered in hardness ratio calculations. Four energy bands were considered: the *Fermi* trigger band (50–300 keV), the lower energy band used in hardness ratio calculations (8–50 keV), the energy range of the NaI detectors (8–1000 keV) and the higher energy range of the BGO detectors (>1000 keV). The effect of the choice of these bands was studied by repeating the analysis of light curves within the four *Swift* bands, which was shown to produce similar results.

3. Feature Extraction

The feature extraction algorithm consists of multiple steps, which are outlined in Figure 1 for the analysis of light curves in the T_{100} interval at 64 ms resolution. Light curves were first pre-processed, before Stationary Wavelet Transform was applied. PCA was used to reduce the dimensionality of the resulting coefficients before visualisation with a t-SNE map. Figure 2 depicts the steps that were followed for the analysis of the first second of prompt emission. This section outlines the details of each step in the feature extraction algorithm.

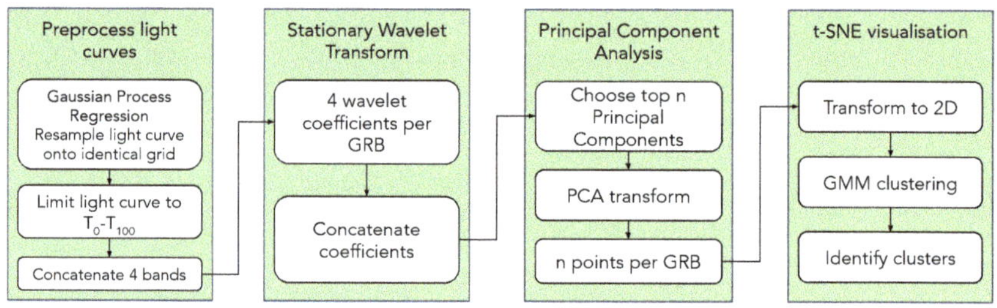

Figure 1. Flowchart of the feature extraction and clustering algorithm for analysis of 64 ms-binned light curves in the interval T_0 to T_{100}.

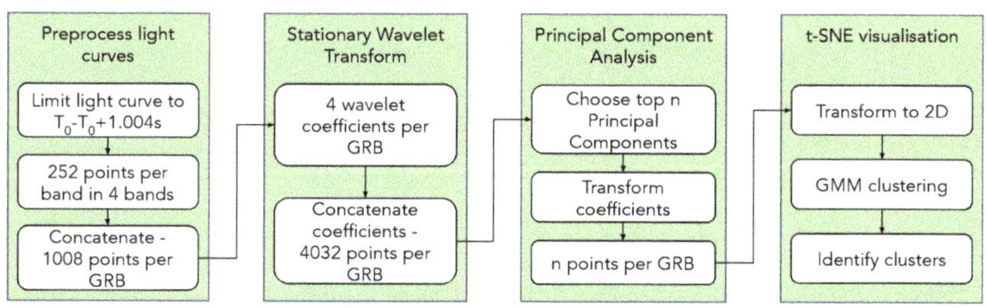

Figure 2. Flowchart of the feature extraction and clustering algorithm for analysis of light curves in the interval T_0 to T_0 + 1.004 s.

3.1. *Light Curve Pre-Processing*

3.1.1. 64 ms Light Curves

To obtain counts recorded at identical times relative to the trigger time for each light curve, BATSE and *Swift*/BAT light curves were modelled and resampled onto an identical grid using Gaussian Process Regression (GPR), a machine learning method that uses the input data to infer the function and explain the observations [82]. Gaussian processes model observations function as joint multivariate normal distributions, which can be fully specified by a mean function and covariance matrix. GPR determines the mean function and the entries of the covariance matrix using a user-specified covariance function (kernel). Hyperparameters of the kernel were optimised to maximise the marginal likelihood of the data under the Gaussian process prior.

The Gaussian Process model was implemented using the GPFlow library in Python [83], which originates from GPy but is built on TensorFlow [84]. A heteroscedastic regression model was used, which incorporates uncertainty in each point into the interpolation process by applying less weight to points with greater uncertainty. The radial-basis function kernel (also known as squared exponential kernel) was used, as it is infinitely differentiable and produces smooth functions. The Adam and natural gradient optimisers were used to converge to the best-fit hyperparameters. The resulting equally spaced, evenly sampled 64 ms light curves were zero-padded beyond T_{100} to ensure noise was discarded. The T_{100} interval was extracted from the GRB catalogues. The four-band light curves were concatenated together and input to the feature extraction algorithm depicted in Figure 1.

3.1.2. 4 ms Light Curves

Wavelet decomposition requires a time series of equal and even lengths. At 4 ms resolution, the light curves were restricted to even-length time intervals starting at T_0, the burst start time documented in the GRB catalogues. Light curves spanning different intervals were created and extended in intervals of 0.1 s until T_0 + 3 s. Many of the BATSE TTE datasets did not extend past 3 s, and BATSE light curves were zero-padded if they did not extend to the specified interval. For each GRB, the light curves in the four energy bands were concatenated together to form one vector to be input to the feature extraction algorithm, depicted in Figure 2 for the case in which light curves between T_0 and T_0 + 1.004 s were studied.

3.2. *Wavelet Decomposition*

Fourier analysis is often used to examine the frequency composition of signals and to extract features from time series (e.g., Jespersen et al. [33]). However, a drawback of Fourier transforms is the loss of temporal information and the stringent sine and cosine basis functions. Wavelets are more suited to the analysis of images, music and transient events, as they overcome the limitations of Fourier analysis by encoding both time and frequency information in the basis function [85]. The Stationary Wavelet Transform (SWT), also known as the Á Trous algorithm [86], is a shift-invariant transform, which convolves a signal with scaled and shifted versions of the basis wavelet function. The shift-invariance feature of the SWT has made it a popular method for pattern recognition [87,88]. The SWT returns two coefficients, known as Approximation and Detail coefficients, of equal length to the input signal. The coefficients are computed using a filter-bank algorithm [34] with low- and high-pass filters, which decomposes the input signal. Multiple levels of decomposition can be performed, whereby the output of the low-pass filter is successively fed to the next decomposition level.

The `pywt.swt` function of the PyWavelets package [89] was applied to the light curve vectors using the symlet family of wavelets, which is a more symmetric version of the Daubechies wavelet family [90], but other wavelet families produce similar results. A two-level decomposition was performed, resulting in four components of equal length to the vector containing the light curves in four bands (Figure 2). These were concatenated into one vector for each GRB prior to dimensionality reduction.

3.3. *Principal Component Analysis*

After performing a two-level wavelet decomposition, the dataset for each GRB increased in length by a factor of four. A dimensionality reduction technique was used to extract only the most significant information encoded in the wavelet coefficients. PCA is a form of decomposition, which extracts uncorrelated Principal Components from correlated data via an orthogonal transformation [91,92]. PCA involves eigenvalue decomposition of the covariance matrix of the input wavelet coefficient data. The eigenvectors are sorted by the magnitude of their eigenvalues. The user must choose how many eigenvectors to keep based on the percentage of variance explained by each eigenvector. The chosen eigenvectors represent the original data in a new PCA reference frame and are known as the Principal Components (PCs). The matrix of PCs is used to project the wavelet coefficients onto the lower-dimensional PCA space.

In this work, PCA was carried out using the `sklearn.decomposition.PCA` function. For *Swift*/BAT, the components whose cumulative variance reached >70% were chosen as the new representation of the dataset, as the number of components required to meet >90% was large. For BATSE and *Fermi*/GBM, the number of retained components ensured that >90% of the variance was captured.

3.4. t-SNE

The chosen PCA components require transformation to a 2D space so that features can be visualised. Stochastic Neighbourhood Embedding (SNE; Hinton and Roweis [93]) provided a 2D visual representation of the components on arbitrary axes by computing the probability that each point is a neighbour of another point. This used a Gaussian probability density and Kullback–Leibler minimisation [94] to ensure that the low-dimensional space adequately represented the high-dimensional space. A user-specified parameter called Perplexity specified the importance of local or global structure. In general, the Perplexity can be considered representative of the number of nearest neighbours of each point.

t-SNE (t-distributed SNE; Maaten and Hinton [95]) used a Student t-distribution with a single degree of freedom, replacing the Gaussian comparison between points. The `sklearn.manifold.TSNE` method was used with a Perplexity, which maximises the separation of clusters in the final representation. In this case, the smaller *Swift*/BAT sample was analysed with a Perplexity of 40, while for the larger samples of BATSE and *Fermi*/GBM, Perplexities of 50 and 70 were used, respectively. The result is a 2D representation of the PCA feature space, in which similar light curves were grouped together.

3.5. GMM Clustering

Finally, Gaussian Mixture Model (GMM)-based clustering was applied to the t-SNE plots to identify clusters using the MCLUST package in R [96,97]. GMM clustering assumes that the observed data are generated from a mixture of K components, where the density of each component is described by a multivariate Gaussian distribution. MCLUST applies 14 different models and chooses the best-fit model and number of clusters based on the Bayesian Information Criterion (BIC; Schwarz et al. [98]). Since the underlying distributions are non-Gaussian, clusters are combined using the `clustCombi` function to converge on the optimum number of clusters, calculated via an entropy criterion [99].

4. Results

The results obtained by analysing GRB light curves, as described in Section 3 for the T_{100} intervals at 64 ms resolution, and for the first three seconds post-trigger at 4 ms resolution, are presented.

4.1. 64 ms Results

The t-SNE plots, coloured by burst duration (T_{90}), are shown in Figure 3 for the T_{100} intervals of bursts from BATSE, *Swift*/BAT and *Fermi*/GBM. t-SNE plots produce a mapping onto an arbitrary space, whereby the scale of the axes have no units or physical meaning. Thus, the t-SNE plots presented in this paper do not label the X and Y axes, and the precise position of points along the axes is not significant. However, the structure within the t-SNE space is significant and is identified. A separate group of shorter-duration bursts is evident in Figure 3b for Swift/BAT, while for BATSE and *Fermi*/GBM, the separation is not as clear.

GMM clustering, applied to the t-SNE map for *Swift*/BAT (Figure 3b), identified four clusters of bursts. However, the distribution is complex and is likely unsuitable for model-based clustering. When coloured by duration, it is clear that two groups of bursts were identified within the T_{100} intervals of *Swift*/BAT light curves: one consisting primarily of short bursts and a larger group of longer duration bursts.

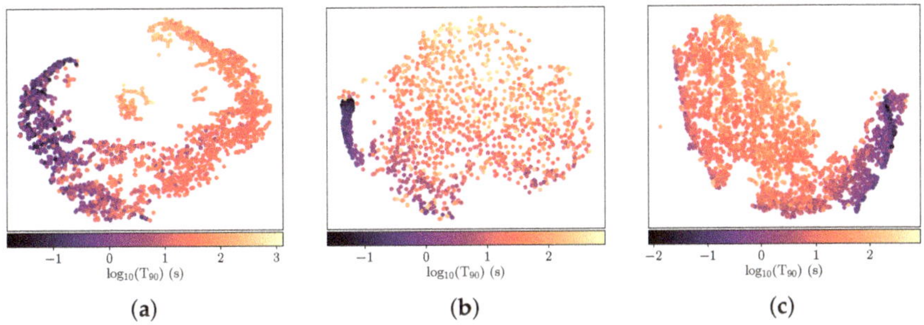

Figure 3. 2D t-SNE representation of the extracted wavelet and PCA features from the 64 ms light curves from T_0 to T_{100}, coloured by burst duration T_{90}, for (**a**) BATSE, (**b**) *Swift*/BAT and (**c**) *Fermi*/GBM.

4.2. 4 ms Results

The t-SNE plots from the analysis of the 4 ms light curves are shown in the animations in Figure 4, coloured by burst duration (T_{90}). The video animations are available to download in the Supplementary Materials. The intervals shown in each iteration of the t-SNE plot increase by 0.1 s, starting from the burst trigger time, T_0. For *Swift*/BAT, a small group of shorter-duration bursts, begins to form and separate from the larger group of longer bursts within T_0 + 0.2 s. This shorter group of bursts grows and detaches from the longer group by T_0 to T_0 + 1.004 s, remaining detached up to the first 3 s post-trigger, which is the maximum interval available at 4 ms resolution. For BATSE and *Fermi*/GBM, the distinction between groups is not as clear, but a similar pattern is observed—a group of shorter bursts begins to form at $\sim T_0$ + 0.2 s and grows, separating itself from the larger, longer-duration group. We conclude that the time at which the two clusters of bursts become clearly separated is T_0 + 1.004 s.

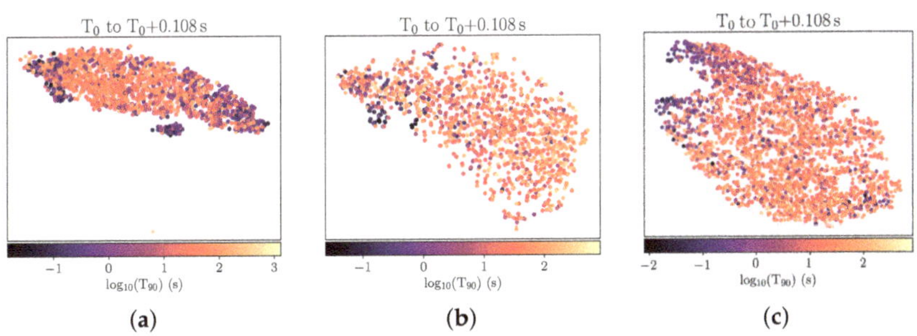

Figure 4. Animation of t-SNE projections for different GRB light-curve time-intervals at 4 ms resolution for (**a**) BATSE (**b**) *Swift*/BAT and (**c**) *Fermi*/GBM, coloured by their T_{90} duration. The title indicated on the top axis of each figure denotes the analysed time interval, since the burst trigger. The video files are available in the Supplementary Materials.

4.3. Properties of GRB Clusters Identified in the First Second Post-Trigger

2D t-SNE representation of the extracted wavelet and PCA features from the first second (T_0 to T_0 + 1.004 s) of GRB light curves, coloured by burst duration T_{90} and hardness ratio HR_{32} for BATSE, *Swift*/BAT and *Fermi*/GBM, are shown in Figure 5. The projections indicate the presence of two groups of bursts, which can be seen clearly in Figure 5b for *Swift*/BAT. For BATSE (Figure 5a) and *Fermi*/GBM (Figure 5c), this separation is less well-defined.

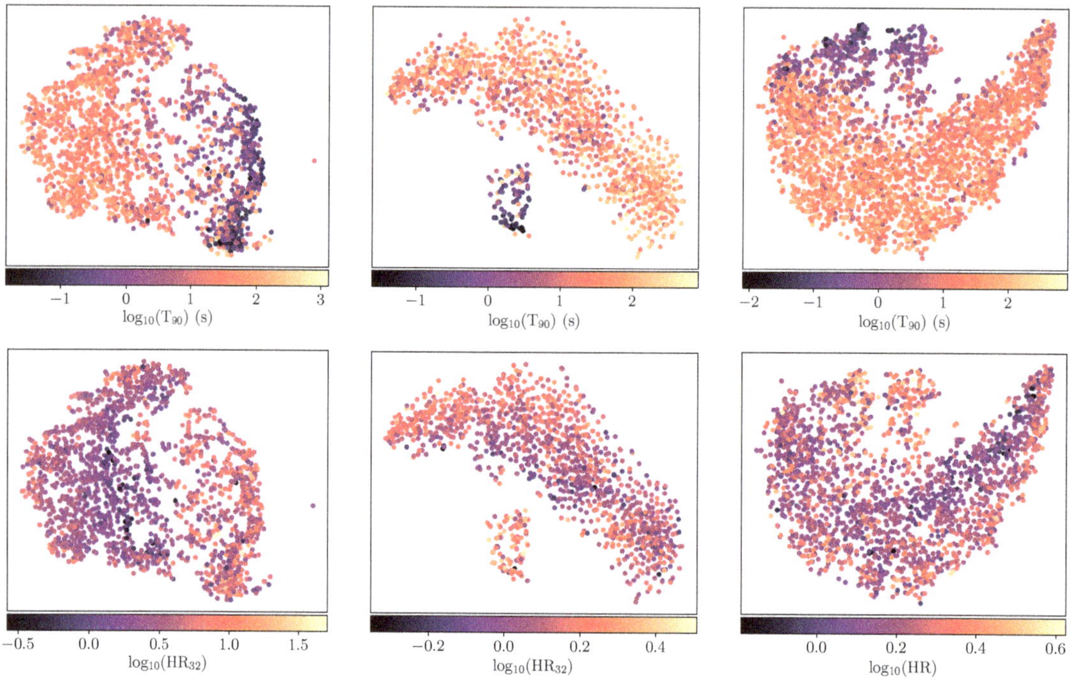

(**a**) (**b**) (**c**)

Figure 5. 2D t-SNE representation of the extracted wavelet and PCA features from the first second (T_0 to T_0 + 1.004 s) of burst light curves, coloured by burst duration T_{90} (top row) and hardness ratio HR_{32} (bottom row) for (**a**) BATSE, (**b**) *Swift*/BAT and (**c**) *Fermi*/GBM. Hardness ratios (HR_{32}) for *Swift*/BAT and BATSE are calculated as the ratio of fluence in Band 3 and Band 2. The hardness ratio of *Fermi*/GBM bursts is defined as the ratio of fluence in the 50–300 keV and 10–50 keV bands, calculated using the best-fit spectral parameters.

GMM clustering applied to the *Swift*/BAT projection identifies two separate groups, shown in Figure 6a. The group consisting of mostly short/hard bursts is labelled Group 1, and the larger, longer-duration group is denoted Group 2. The groups are shown projected onto the duration-hardness plane in Figure 6b.

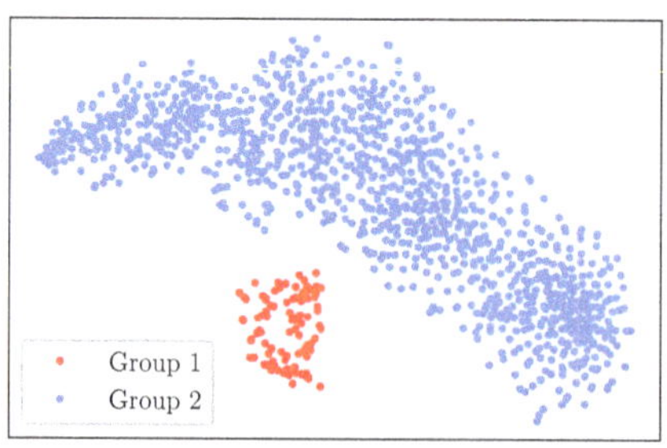

(**a**)

Figure 6. *Cont.*

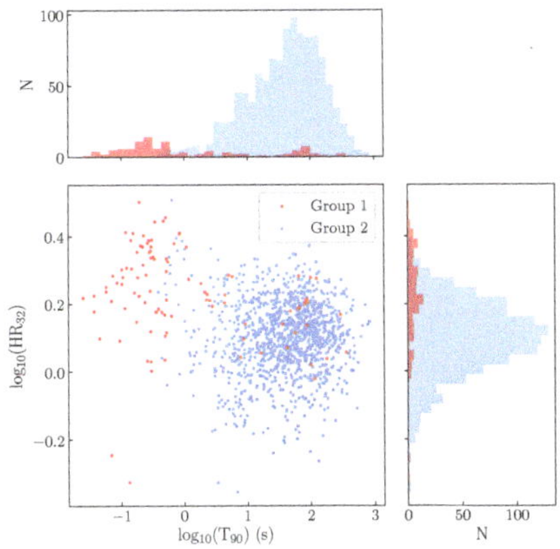

(b)

Figure 6. (**a**) The t-SNE map of *Swift*/BAT bursts derived from the T_0 to T_0 + 1.004 s interval at 4 ms resolution showing 2 clearly separated groups and (**b**) their projection onto the duration-hardness plane. Histograms indicate the distribution of duration and hardness for each group.

Two-dimensional Kolmogorov–Smirnov (KS) tests applied to Group 1 and Group 2 verify that there are statistically significant differences in GRB properties such as the duration (T_{90}), hardness (HR_{32}), peak energy (E_{peak}) and fluence (S) of the two clusters. Table 1 presents the results of the KS test. The probability (p-value) presented in Table 1 indicates the probability that Groups 1 and 2 are drawn from the same distribution. This hypothesis is rejected, as all probabilities are below 1%. Figure 7 demonstrates the distribution of the GRB properties for Group 1 and Group 2.

Table 1. Results of the 2D KS test comparing Group 1 and Group 2 identified within the first second of prompt emission of *Swift*/BAT bursts.

Parameter	p-Value
T_{90}	4.3×10^{-42}
HR_{32}	1.7×10^{-19}
E_{peak}	1.9×10^{-3}
Fluence (15–350 keV)	2.9×10^{-22}

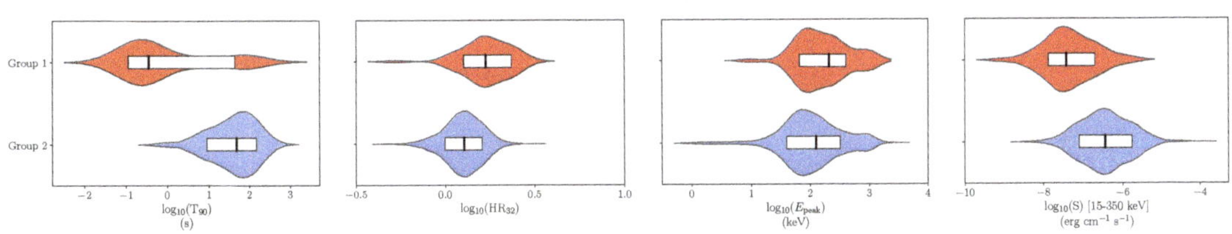

Figure 7. Violin plots showing the distribution of GRB properties for Group 1 (red) and Group 2 (blue) *Swift*/BAT bursts identified in the T_0 to T_0 + 1.004 s light curve interval. The white box plots represent the 1σ interval (i.e., the 16th to 84th percentile), with the median of each parameter marked as a black line.

Table 2 lists the cluster memberships of a subset of the *Swift* GRBs for both the analysis of the T_{100} interval at 64 ms resolution, and the interval from T_0 to T_0 + 1.004 s at 4 ms resolution. The full table is available to download from the Supplementary Materials. When the first 1 s of prompt emission is considered, Group 1 contains 107 bursts, 73 of which are short-duration ($T_{90} < 2$ s). There are 1144 bursts in Group 2, containing 1112 long-duration bursts ($T_{90} > 2$ s). The composition of each group and the properties of GRBs in Groups 1 and 2 are further discussed in Section 5.

Table 2. Group membership of *Swift* GRBs, based on the analysis of the first second of prompt emission at 4 ms resolution (T_1) and the T_{100} interval at 64 ms resolution. The full table is provided in the Supplementary Materials.

GRB	T_1 Group	T_{100} Group	T_{90} (s)	HR_{32}
GRB041220	2	2	5.6	1.3
GRB041223	2	2	109.1	1.8
GRB041224	2	2	177.6	1.2
...	...	...	...	...
GRB200809B	... [1]	2	4.2	1.8
GRB200819A	2	2	26.9	1.0
GRB200829A	... [1]	2	13.1	1.9

[1] No 4 ms light-curve file available.

As with the T_{100} analysis (Section 4.1), the results obtained for the T_0 to $T_0 + 1.004$ s interval at 4 ms resolution for BATSE and *Fermi*/GBM GRBs are not as clear-cut as they are for *Swift*/BAT. In the case of BATSE, GMM clustering with MCLUST identifies six clusters within the t-SNE projection in Figure 5a. However, we can tentatively identify two clusters of bursts for BATSE by eye. These two groups resemble the short/hard and long/soft groups identified for *Swift*/BAT. Similarly to *Swift*/BAT, a KS test applied to the two BATSE groups reveals significant differences in their duration, hardness, peak energy and peak flux. BATSE has a harder energy range than *Swift*/BAT; thus, the BATSE population contains more short/hard bursts. Therefore for BATSE, the short-duration Group 1 contains a larger proportion of bursts compared to *Swift*/BAT.

For *Fermi*/GBM, the projection in Figure 5c indicates two groups, primarily consisting of short/hard and long/soft bursts, but their clustering is not dense enough to allow for a clean separation between them. MCLUST identifies five clusters. This may indicate that the application of a Gaussian model does not adequately represent the underlying complex distributions [17].

5. Consistency Checks

The analysis of the first second of prompt emission (T_1) identifies two groups of bursts within the BATSE, *Swift*/BAT and *Fermi*/GBM samples. We focus this discussion on the more clear-cut results obtained with *Swift*/BAT.

5.1. T_{100} vs. T_1 Analysis

For *Swift*/BAT, two groups are identified using both the T_{100} and T_1 intervals. Table 2 provides the classification results obtained with each interval. The sample sizes of the two groups are shown in Table 3, separated into long- ($T_{90} > 2$ s) and short- ($T_{90} < 2$ s) duration bursts.

Table 3. Sample sizes of short-duration ($T_{90} < 2$ s) and long-duration ($T_{90} > 2$ s) bursts in the *Swift*/BAT sample, and Groups 1 and 2, based on the analysis of the T_1 and the T_{100} intervals.

Sample	Number of Bursts			
	$T_{90} < 2$ s		$T_{90} > 2$ s	
	T_1	T_{100}	T_1	T_{100}
Swift/BAT sample	107	107	1144	1147
Group 1	73	91	32	14
Group 2	34	16	1112	1133

A total of 95% (1185 of the 1251 bursts) of the classifications of *Swift*/BAT bursts determined using the T_1 interval at 4 ms resolution are consistent with those derived using the T_{100} intervals at 64 ms resolution. There are 21 short-duration bursts, which are classified as Group 2 bursts when the T_1 interval is considered, but move to Group 1 when the T_{100} interval is used for the analysis. There are 28 long-duration bursts, which move from Group 1 in the T_1 analysis to Group 2 when the T_{100} interval is considered. These include five bursts in the list of *Swift*/BAT bursts with extended emission episodes compiled by Gibson et al. [100]. The long-duration supernova-accompanied burst GRB 101219B moves to Group 2 in the T_{100} analysis, correctly placing it amongst the other bursts with associated supernovae. The inclusion of the full light-curve data in these cases is important for correct classification. The classification of bursts with associated supernovae is further discussed in Section 6. Some of the movement between groups may reflect the different temporal resolutions used for the T_1 (4 ms) and T_{100} (64 ms) analyses. The minimum variability timescale with short GRBs of order 10 ms would not necessarily be captured by the T_{100} analysis. There may also be cases where there are pre-trigger emissions that are not captured in the current approach, which starts at the trigger time.

5.2. Inter-Comparison of Swift/BAT and Fermi/GBM Results

In all three GRB samples, clear evidence of a separation into two groups by the end of the first second was observed. The clean separation of bursts in the *Swift*/BAT sample indicates that the T_1 interval could potentially be used to classify GRBs independently of their T_{90} duration. The less clear-cut cluster separation found in the BATSE and *Fermi*/GBM samples most likely arose from instrumental differences (e.g., energy ranges, triggering methods and sensitivities). Of the three instruments, *Swift*/BAT has the largest effective area, and detects more spectrally softer, long-duration GRBs [101], and fewer short GRBs, than BATSE or *Fermi*/GBM [80,102].

A total of 293 bursts were analysed, which were detected by both *Fermi*/GBM and *Swift*/BAT. There is excellent (274/293) agreement between the two instruments in the cluster membership of these GRBs using the T_1 interval at 4 ms resolution. Differences in classification can primarily be attributed to the lack of clear separation between groups in the *Fermi*/GBM sample, which makes cluster identification less definitive than it is for *Swift*/BAT. In 6 of the 19 cases where cluster membership is found to disagree between the two detectors, significantly different (>50%) T_{90} durations are recorded by the instruments.

5.3. Time Intervals

The analysis was repeated for different intervals within the bursts to investigate the intervals in which classes may be identified in the *Swift*/BAT sample.

First, the feature extraction analysis with 4 ms resolution light curves was performed for the interval of $T_0 - 1$ s to $T_0 + 1$ s (Figure 8a). The addition of pre-trigger data is shown to produce almost identical results to those obtained by starting at T_0. However, the analysis requires additional Principal Components to explain the variance, indicating that including 1 s of data before the trigger adds more noise than information. Secondly, when the selected interval is between $T_0 + 1.004$ s and $T_0 + 2.008$ s, the separation disappears, as shown in Figure 8b, indicating that the early prompt emission in the first second post-trigger is the key interval for separating the two classes.

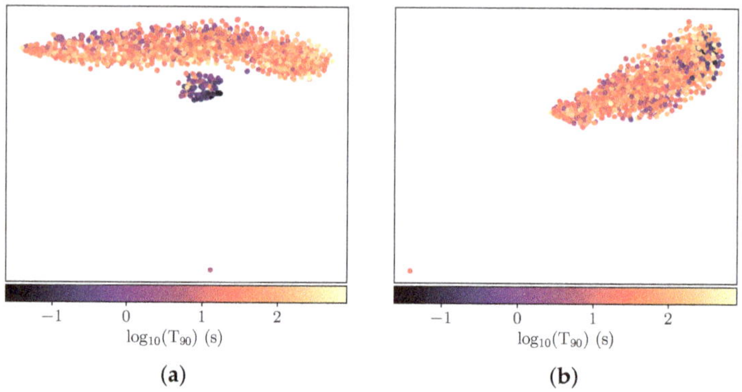

Figure 8. 2D t-SNE representation of the wavelet feature extraction applied to *Swift*/BAT light curves covering time intervals (**a**) $T_0 - 1$ s to $T_0 + 1$ s and (**b**) $T_0 + 1$ s to $T_0 + 2$ s. The plots are coloured by burst duration T_{90}.

5.4. Light Curve Classifications

In Salmon et al. [17], Gaussian Mixture Model-based clustering of the hardness-T_{90} plane identified two classes of bursts in the *Swift*/BAT and *Fermi*/GBM samples. The results suggest that the intermediate duration class may be an artefact of the application of unsuitable models, as was also suggested by Tarnopolski [25], Koen and Bere [103], Tarnopolski [104], Tarnopolski [105]. The results of the model-independent analysis of light curves presented in this paper and by Jespersen et al. [33] lend further support to this conclusion.

Jespersen et al. [33] found two distinct groups of bursts in their t-SNE map obtained from Fourier decomposition of full *Swift*/BAT light curves. The composition of the 'type-S' and 'type-L' groups from Jespersen et al. [33] are compared to the clusters found in this work. For the T_{100} interval, there is an agreement in classification for 96% of bursts, after removing bursts for which no light curve files are available. Although the same burst intervals are considered, Jespersen et al. [33] input flux-normalised light curves into a Fourier-based analysis, potentially leading to small differences in the resulting burst memberships compared to the wavelet-based analysis presented here. The light curves in this study were not flux-normalised, as the resulting t-SNE maps did not separate the groups.

More than 96% of the bursts classified using wavelets and the T_1 interval were found to match the membership assigned by Jespersen et al. [33] based on Fourier decomposition in the T_{100} interval. Eight of the 20 bursts found to be within Group 1 in this study, classified as type-L by Jespersen et al. [33], have extended emission episodes, which are not captured within the first second.

5.5. Collapsar 'Contamination'

The fraction of *Swift*/BAT bursts within Groups 1 and 2 can be compared to the expected distributions that are specific to the *Swift*/BAT detector.

Bromberg et al. [32] quantified the contamination of short GRB samples by collapsar bursts by fitting the duration distribution of *Swift* bursts with a function representing the merger and collapsar duration distributions. The model is based on the plateau in the duration distribution for shorter durations than the jet breakout time, which is predicted by the collapsar model [106]. According to this model $\sim$40% of *Swift*/BAT bursts with durations <2 s are collapsar bursts. *Swift*/BAT is more sensitive to soft GRBs, meaning that low-fluence long GRBs contaminate the short GRB population to a greater extent than they do for BATSE ($\sim$10%) and *Fermi*/GBM ($\sim$15%).

There are 27 bursts for which Bromberg et al. [32] assigns a probability of being a non-Collapsar of >90%. The majority (23) of these are classified as Group 1 in this analysis, indicating that Group 1 primarily consists of bursts arising from mergers. The collapsar (Group 2) contamination of *Swift*/BAT bursts with durations <2 s from our analysis of the first 1 s of prompt emission is 31.8%, or 34/107 bursts (Table 3), consistent within 1σ with the predictions in Bromberg et al. [32]. The collapsar contamination of short-duration bursts for the T_{100} analysis is significantly lower, at $\sim$15%. These results suggest that classification based on the T_1 interval may be useful for identifying collapsar 'imposters' in short GRB samples.

6. Notable GRBs

We discuss cluster membership for some notable GRBs. As a default, the discussion relates to the results obtained with the T_1 interval at 4 ms resolution unless otherwise indicated. We identify cluster membership for GRBs in any of the three analysed samples, which have associated kilonovae or supernovae. We note that cluster membership for the BATSE and *Fermi* samples is not as clear-cut as it is for the *Swift* sample, and the cut in t-SNE space is made by eye.

6.1. GRBs with Associated Supernovae

As discussed in Section 5.5, Group 2 is associated with collapsar bursts. Thus, it is expected that GRBs with associated supernovae will lie within Group 2. The list of 31 *Swift* and 10 *Fermi* Supernova (SN)-GRBs provided in Cano et al. [107] is extended to include additional SN-GRB events GRB 161219B/SN 2016jca [108], GRB 171205A/SN 2017htp [109], GRB 180728A/SN 2018fip [110,111], GRB 190114C/SN 2019jrj and GRB 190829A/AT2019 oyw [112] and the peculiar short GRB 200826A [113–115]. Figure 9 indicates the location of these bursts within the t-SNE plot. The 25 SN-GRBs for which light curve files are available lie in Group 2 of *Swift*/BAT and *Fermi*/GBM, as expected, with the exception of GRB 101219B, which is an outlier to *Swift*/BAT Group 1 but lies within Group 2 of *Fermi*/GBM bursts. The analysis using the T_{100} interval correctly places GRB 101219B in Group 2 for *Swift*/BAT. The shortest collapsar burst detected to date, GRB 200826A, lies within Group 2 of the *Fermi*/GBM sample despite its observed duration of $T_{90} \approx 0.96$ s [115].

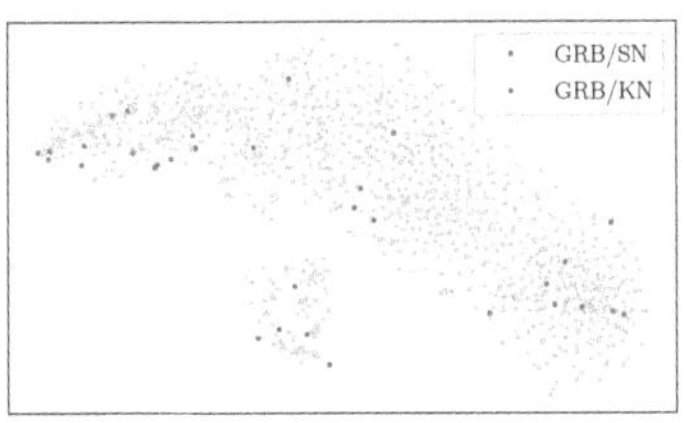

(**a**)

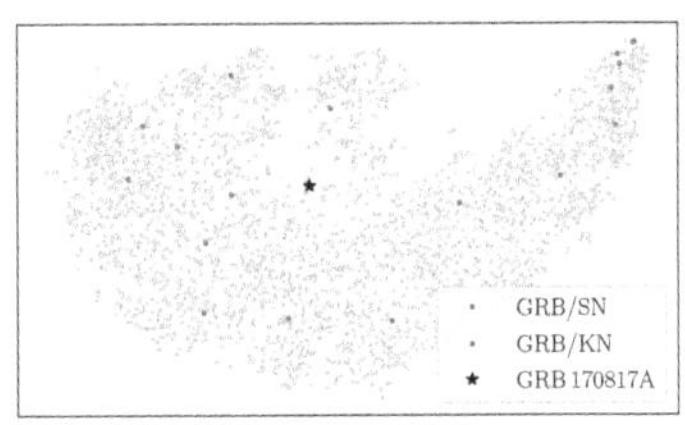

(**b**)

Figure 9. Locations of GRBs with associated Supernovae (GRB/SN) and possible Kilonovae (GRB/KN) within the t-SNE projections of the T_1 interval for (**a**) *Swift*/BAT and (**b**) *Fermi*/GBM. The location of the only confirmed kilonova, associated with GRB 170817A, is indicated with a black star.

The remaining bursts are identified in Group 2, with their classifications unchanged by the interval used in analysis, with the exception of GRB 050824, which migrates to Group 1 when the T_{100} interval is used for analysis.

6.2. GRBs with Possible Kilonovae

The only confirmed kilonova is associated with GRB 170817A, which is in the *Fermi*/GBM sample (Figure 9b) and does not clearly belong to either group. However, GRB 170817A was not a standard 'short' GRB, and would probably have been unremarked on if not for the associated gravitational wave source [13,14]. Near-infrared excesses, similar to kilonova signatures, have been found in the afterglows of a handful of nearby short GRBs. Following detection in GRB 130603B [116,117], reanalysis of GRB 060614 [118,119], GRB 080503 [120] and GRB 050709 [121] revealed similar near-IR components. Since then, GRB 150101B [122], GRB 160821B [123–126] and GRB 200522A [127] have all been suggested as kilonova candidates.

Figure 9 shows that all the kilonova candidate bursts lie within Group 1 of *Swift* and *Fermi* GRBs, except for GRB 050709, for which no light curve file is available, and GRB 060614, which is in Group 2. GRB 060614 is an anomalous GRB with a short pulse followed by a longer period of soft flaring emission. Some properties of this burst are typical of the long GRB population [128], but the lack of supernova detection for this close burst ($z = 0.125$ Price et al. [129], Fugazza et al. [130]) and possible near-infrared excess led to the suggestion that this burst originates from a merger [22], or is within its own subclass [131–135]. Our results agree with the classification by Jespersen et al. [33], who place this burst in the longer-duration, collapsar group. When the T_{100} light-curve interval is considered, the classifications remain unchanged for the kilonova candidates, with the exception of GRB 080503, which moves to Group 2. This is an example of a GRB with a short initial spike and extended emission, which may be the result of a merger rather than a collapsar [120]. The T_1 interval appears to return the more appropriate classification in this case.

7. Discussion

Studies of GRB pulses at early times have revealed that the dominant radiation process is usually photospheric emission [136–141]. These thermal pulses exhibit significant spectral evolution, with bursts usually evolving to be dominated by synchrotron emission [137,139,142]. If this is the case, the first second of all GRBs should be dominated by thermal pulses; therefore, the radiation process is unlikely to be the driver of the observed differences in light curves that appear at early times.

The feature extraction algorithm may identify differences in the spectral evolution and pulse shapes of the two burst groups. The spectral lags of long and short bursts are different, with many short bursts exhibiting zero lag [18,63]. The minimum variability timescales for short and long bursts have also been found to be different [44–46]. For example Golkhou et al. [46] found median minimum variability timescales of 10 ms and 45 ms for short and long bursts, respectively. Hakkila and Preece [64] found that pulses in short GRBs are shorter and harder than those in long GRBs, and exhibit more rapid spectral evolution. Coupled with the observation that shorter pulses have a higher peak flux and $\sim$90% of short GRBs consist of a single pulse, compared to 25–40% for long GRBs, pulse properties are likely to be a distinguishing feature in the first pulses and first seconds of a burst. Short GRBs have shorter pulse durations and their triple peaked substructure shows more intense precursor and decay peaks (on either side of the central peak) than long GRBs Hakkila et al. [56].

The magnitude of the PCA components in the different *Swift*/BAT energy bands indicate that Bands 2 and 3 contain the most variance; therefore, they are the most important for the 4 ms light curves. Figure 10 shows that the results of the feature extraction algorithm only applied to Band 3 data, showing that some segregation of the bursts into two groups is evident using light curves in one energy band. Thus, energy-dependent pulse characteristics are not the sole driver of the classification.

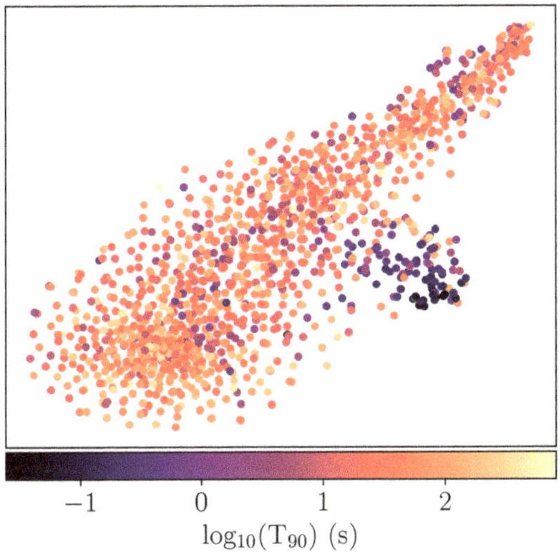

Figure 10. 2D t-SNE representation of the wavelet coefficients and PCA features extracted from the light curves measured in Band 3 for *Swift*/BAT (50–100 keV). The plot is coloured by burst duration T_{90}.

Cumulative Counts

Figure 6 shows that, for the *Swift*/BAT sample, Group 1 and Group 2 GRBs, identified within the T_1 interval, mostly consist of bursts from the classical short-duration ($T_{90} < 2$ s) and long-duration ($T_{90} > 2$ s) samples, respectively. However, there are some 'strays', as shown in Table 3 and discussed in Section 5.

The counts measured in Band 3 (50–100 keV) of the first second of *Swift*/BAT 4 ms light curves are summed and normalised by the number of light curves, to obtain an average cumulative counts measure for bursts in each group in Table 3 (Figure 11). The cumulative counts of Group 1 and Group 2 bursts track those of short and long GRBs, respectively, during the first second.

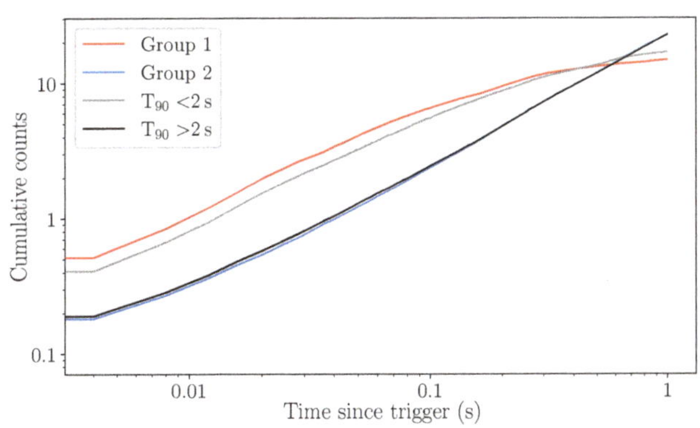

Figure 11. Normalised cumulative counts of Band 3 (50–100 keV) *Swift*/BAT 4 ms light curves. Short ($T_{90} < 2$ s) and long ($T_{90} > 2$ s) duration bursts, and those within the Groups 1 and 2 identified from the first 1 s of prompt emission, are shown.

The results of this analysis suggest that the behaviour of GRB pulses in the first second carries essential information, which is needed to classify GRBs in the vast majority of cases, independent of their duration. The characteristics of the 'long' Group 1 and 'short' Group 2 bursts suggests that they have not been misclassified, but are duration outliers of their identified class. Group 1 and Group 2 bursts evolve in a similar way to the traditional short and long classes, respectively.

Previous studies have interpreted the cumulative GRB light-curves slope as a measure of the cumulative power output of the central engine [143]. Combined with the association of Group 1 and 2 bursts with kilonovae and supernovae, respectively (Section 6), the cumulative counts behaviour in the first second suggests that Group 1 and 2 represent distinct progenitors, namely, the merger and collapsar populations.

8. Conclusions

Wavelet decomposition, combined with PCA and t-SNE, provides an effective method for extracting the similarities between gamma-ray light curves from BATSE, *Swift*/BAT and *Fermi*/GBM. The features extracted from the T_{100} interval of light curves in four energy bands at 64 ms resolution reveal a separation between two groups of bursts. These groups are labelled Group 1 and Group 2. Two groups have also been identified through feature extraction from high resolution (4 ms) light curves within the first seconds of prompt emission. The shortest timescale at which this separation is clear is one second (T_1 interval).

The separation between groups is clearest for *Swift*/BAT and is less distinct for the BATSE and *Fermi*/GBM samples of bursts, perhaps due to instrumental effects. Despite the different timescales and resolutions that were studied, there is >95% agreement between the groups identified within the T_{100} and T_1 interval for *Swift*/BAT. The T_{100} interval is shown to produce different and more classical classifications for some bursts, especially those with long emission episodes. There is also >95% agreement between the results of the T_1 analysis with the results of the Fourier-based feature extraction of *Swift*/BAT light curves by Jespersen et al. [33]. The separation between *Swift*/BAT groups is clearest when all four energy bands are considered. However, energy-dependent characteristics are not the sole effect that drives the classification, as some separation can only be seen when one energy band is considered. Pulse shape and evolution may be important, and the accumulation of counts within the first second is found to be distinct between groups.

Group 1 mostly consists of short-duration, spectrally hard bursts. Group 2 mostly consists of spectrally soft, long-duration bursts. When segmented at T_{90} = 2 s, the traditional dividing line between long and short GRBs, we found that 99% (97%) of *Swift*/BAT Group 2 bursts have durations >2 s when the T_{100} (T_1) interval is used. A total of 32% of the 107 GRBs with $T_{90} < 2$ s are identified as Group 2 bursts when the T_1 interval is used, consistent (within 1σ) with a model in which the duration distribution of *Swift* bursts is fit with a function representing the merger and collapsar distributions, possibly reflecting the amount of collapsar 'contamination' in the short GRB sample. The observed contamination fraction is significantly lower (16%) when the T_{100} interval is used. Thus, the groups can be associated with distinct progenitors, namely, mergers and collapsars. GRBs with associated supernovae are within Group 2, while GRBs with suspected kilonovae lie in Group 1.

Previous studies found that the pulse and spectral properties of the early seconds of long GRBs are similar to those of short GRBs. In this analysis, no significant differences can be identified in pulse or spectral properties to account for Group 1 and Group 2 GRBs being distinguishable in the T_1 interval. Differences in minimum variability timescale, identifiable only when the 4 ms resolution data are used, may account for some of the observed behaviour. However, the two groups in subsequent 1 s intervals should also be evident, which is not the case. The observed different slopes in the first second between the two groups in the combined cumulative counts may point towards differences in the central engine.

The presented results indicate that the nature of a burst may be inferred from the earliest prompt emission, without considering the full burst duration. Prompt classification will be helpful in the era of 'big data' in time-domain astronomy. Gravitational wave detectors will detect mergers at increased rates in the near- and longer-term [144]. State-of-the-art optical surveys such as the Vera Rubin Observatory will deliver an increased number of transient targets in the crowded optical sky [145]. While many optical transients are false positives, the rare gamma-ray transients can pinpoint the unambiguous target of interest. The early detection and classification of these gamma-ray transients will help to prioritise counterpart follow-up for optical telescopes and spectroscopic observations. Classification schemes and triggering algorithms could incorporate a wavelet-based analysis, such as that presented here, to prioritise targets for follow-up observations.

Supplementary Materials: The following are available online at https://www.mdpi.com/article/10.3390/galaxies10040078/s1. The electronic version of this article showcases Figure 4 as three mp4 animations of the t-SNE plots for BATSE, *Swift* and *Fermi*. We provide the full version of Table 2 which includes the classification of *Swift*/BAT GRBs using the first second of prompt emission.

Author Contributions: Conceptualization, L.S., L.H. and A.M.-C.; methodology, L.S., L.H. and A.M.-C.; software, L.S.; validation, L.S., L.H. and A.M.-C.; formal analysis, L.S.; investigation, L.S.; resources, L.S., L.H. and A.M.-C.; data curation, L.S.; writing—original draft preparation, L.S.; writing—review and editing, L.H. and A.M.-C.; visualization, L.S.; supervision, L.H. and A.M.-C.; project administration, L.H. and A.M.-C.; funding acquisition, L.H. and A.M.-C. All authors have read and agreed to the published version of the manuscript.

Funding: L.S. acknowledges the Irish Research Council Postgraduate Scholarship No GOIPG/2017/1525. LH acknowledges support from Science Foundation Ireland (Grant number 19/FFP/6777) and the EU H2020 (Grant agreement 871158).

Institutional Review Board Statement: Not applicable.

Informed Consent Statement: Not applicable.

Data Availability Statement: The BATSE Public Data Archive (https://heasarc.gsfc.nasa.gov/FTP/compton/data/batse/trigger/, accessed on 17/ February 2021) hosts the 64 ms light curves in ascii files, and the TTE data which is used to generate the 4 ms BATSE light curves. The *Swift*/BAT light curves analysed in this paper are available as ascii files from the *Swift*/BAT Gamma-Ray Burst Catalogue (https://swift.gsfc.nasa.gov/results/batgrbcat/, accessed on 29 January 2021). The *Fermi*-GBM Data Tools [81] are used to access *Fermi*/GBM TTE files.

Acknowledgments: This research made use of the following Python packages: NumPy [146], Matplotlib [147], pandas [148,149], scikit-learn [150], GPFlow [83] and PyWavelets [89].

Conflicts of Interest: The authors declare no conflict of interest. The funders had no role in the design of the study; in the collection, analyses, or interpretation of data; in the writing of the manuscript, or in the decision to publish the results.

References

1. Kouveliotou, C.; Meegan, C.A.; Fishman, G.J.; Bhat, N.P.; Briggs, M.S.; Koshut, T.M.; Paciesas, W.S.; Pendleton, G.N. Identification of Two Classes of Gamma-Ray Bursts. *Astrophys. J.* **1993**, *413*, L101. [CrossRef]
2. Perley, D.A.; Niino, Y.; Tanvir, N.R.; Vergani, S.D.; Fynbo, J.P.U. Long-Duration Gamma-Ray Burst Host Galaxies in Emission and Absorption. *Space Sci. Rev.* **2016**, *202*, 111–142. [CrossRef]
3. Galama, T.J.; Vreeswijk, P.M.; van Paradijs, J.; Kouveliotou, C.; Augusteijn, T.; Böhnhardt, H.; Brewer, J.P.; Doublier, V.; Gonzalez, J.F.; Leibundgut, B.; et al. An unusual supernova in the error box of the γ-ray burst of 25 April 1998. *Nature* **1998**, *395*, 670–672. [CrossRef]
4. Hjorth, J.; Sollerman, J.; Møller, P.; Fynbo, J.P.U.; Woosley, S.E.; Kouveliotou, C.; Tanvir, N.R.; Greiner, J.; Andersen, M.I.; Castro-Tirado, A.J.; et al. A very energetic supernova associated with the γ-ray burst of 29 March 2003. *Nature* **2003**, *423*, 847–850. [CrossRef]
5. Stanek, K.Z.; Matheson, T.; Garnavich, P.M.; Martini, P.; Berlind, P.; Caldwell, N.; Challis, P.; Brown, W.R.; Schild, R.; Krisciunas, K.; et al. Spectroscopic Discovery of the Supernova 2003dh Associated with GRB 030329. *Astrophys. J.* **2003**, *591*, L17–L20. [CrossRef]
6. Woosley, S.E.; Bloom, J.S. The Supernova Gamma-Ray Burst Connection. *Annu. Rev. Astron. Astrophys.* **2006**, *44*, 507–556. [CrossRef]
7. MacFadyen, A.I.; Woosley, S.E. Collapsars: Gamma-Ray Bursts and Explosions in "Failed Supernovae". *Astrophys. J.* **1999**, *524*, 262–289. [CrossRef]

8. Eichler, D.; Livio, M.; Piran, T.; Schramm, D.N. Nucleosynthesis, neutrino bursts and γ-rays from coalescing neutron stars. *Nature* **1989**, *340*, 126–128. [CrossRef]
9. Narayan, R.; Paczynski, B.; Piran, T. Gamma-Ray Bursts as the Death Throes of Massive Binary Stars. *Astrophys. J.* **1992**, *395*, L83. [CrossRef]
10. Berger, E. Short-Duration Gamma-Ray Bursts. *Annu. Rev. Astron. Astrophys.* **2014**, *52*, 43–105. [CrossRef]
11. Berger, E. A Short Gamma-ray Burst "No-host" Problem? Investigating Large Progenitor Offsets for Short GRBs with Optical Afterglows. *Astrophys. J.* **2010**, *722*, 1946–1961. [CrossRef]
12. Tunnicliffe, R.L.; Levan, A.J.; Tanvir, N.R.; Rowlinson, A.; Perley, D.A.; Bloom, J.S.; Cenko, S.B.; O'Brien, P.T.; Cobb, B.E.; Wiersema, K.; et al. On the nature of the 'hostless' short GRBs. *Mon. Not. R. Astron. Soc.* **2014**, *437*, 1495–1510. [CrossRef]
13. Goldstein, A.; Veres, P.; Burns, E.; Briggs, M.S.; Hamburg, R.; Kocevski, D.; Wilson-Hodge, C.A.; Preece, R.D.; Poolakkil, S.; Roberts, O.J.; et al. An Ordinary Short Gamma-Ray Burst with Extraordinary Implications: Fermi-GBM Detection of GRB 170817A. *Astrophys. J.* **2017**, *848*, L14. [CrossRef]
14. Savchenko, V.; Ferrigno, C.; Kuulkers, E.; Bazzano, A.; Bozzo, E.; Brandt, S.; Chenevez, J.; Courvoisier, T.L.; Diehl, R.; Domingo, A.; et al. INTEGRAL detection of the first prompt gamma-ray signal coincident with the gravitational-wave event GW170817. *Astrophys. J.* **2017**, *848*, L15. [CrossRef]
15. Abbott, B.P.; Abbott, R.; Abbott, T.D.; Acernese, F.; Ackley, K.; Adams, C.; Adams, T.; Addesso, P.; Adhikari, R.X.; Adya, V.B.; et al. Multi-messenger Observations of a Binary Neutron Star Merger. *Astrophys. J.* **2017**, *848*, L12. [CrossRef]
16. Horváth, I. A Third Class of Gamma-Ray Bursts? *Astrophys. J.* **1998**, *508*, 757–759. [CrossRef]
17. Salmon, L.; Martin-Carrillo, A.; Hanlon, L. Two Dimensional Clustering of *Swift*/BAT and *Fermi*/GBM Gamma-Ray Bursts. *Galaxies* 2022, *submitted*. [CrossRef]
18. Norris, J.P.; Bonnell, J.T. Short Gamma-Ray Bursts with Extended Emission. *Astrophys. J.* **2006**, *643*, 266–275. [CrossRef]
19. Barkov, M.V.; Pozanenko, A.S. Model of the extended emission of short gamma-ray bursts. *Mon. Not. R. Astron. Soc.* **2011**, *417*, 2161–2165. [CrossRef]
20. Norris, J.P.; Gehrels, N.; Scargle, J.D. Threshold for Extended Emission in Short Gamma-Ray Bursts. *Astrophys. J.* **2011**, *217*, 411. [CrossRef]
21. D'Avanzo, P. Short gamma-ray bursts: A review. *J. High Energy Astrophys.* **2015**, *7*, 73–80. [CrossRef]
22. Zhang, B.; Zhang, B.B.; Liang, E.W.; Gehrels, N.; Burrows, D.N.; Mészáros, P. Making a Short Gamma-Ray Burst from a Long One: Implications for the Nature of GRB 060614. *Astrophys. J.* **2007**, *655*, L25–L28. [CrossRef]
23. Zhang, B.; Zhang, B.B.; Virgili, F.J.; Liang, E.W.; Kann, D.A.; Wu, X.F.; Proga, D.; Lv, H.J.; Toma, K.; Mészáros, P.; et al. Discerning the Physical Origins of Cosmological Gamma-ray Bursts Based on Multiple Observational Criteria: The Cases of z = 6.7 GRB 080913, z = 8.2 GRB 090423, and Some Short/Hard GRBs. *Astrophys. J.* **2009**, *703*, 1696–1724. [CrossRef]
24. Li, Y.; Zhang, B.; Yuan, Q. A Comparative Study of Long and Short GRBs. II. A Multiwavelength Method to Distinguish Type II (Massive Star) and Type I (Compact Star) GRBs. *Astrophys. J.* **2020**, *897*, 154. [CrossRef]
25. Tarnopolski, M. Distinguishing short and long Fermi gamma-ray bursts. *Mon. Not. R. Astron. Soc.* **2015**, *454*, 1132–1139. [CrossRef]
26. Goldstein, A.; Preece, R.D.; Briggs, M.S. A New Discriminator for Gamma-ray Burst Classification: The E_{peak}-fluence Energy Ratio. *Astrophys. J.* **2010**, *721*, 1329–1332. [CrossRef]
27. Qin, Y.P.; Chen, Z.F. Statistical classification of gamma-ray bursts based on the Amati relation. *Mon. Not. R. Astron. Soc.* **2013**, *430*, 163–173. [CrossRef]
28. Lü, H.J.; Liang, E.W.; Zhang, B.B.; Zhang, B. A New Classification Method for Gamma-ray Bursts. *Astrophys. J.* **2010**, *725*, 1965–1970. [CrossRef]
29. Lü, H.J.; Zhang, B.; Liang, E.W.; Zhang, B.B.; Sakamoto, T. The 'amplitude' parameter of gamma-ray bursts and its implications for GRB classification. *Mon. Not. R. Astron. Soc.* **2014**, *442*, 1922–1929. [CrossRef]
30. Zhang, S.; Shao, L.; Zhang, B.B.; Zou, J.H.; Sun, H.Y.; Yao, Y.J.; Li, L.L. A Tight Three-parameter Correlation and Related Classification on Gamma-Ray Bursts. *arXiv* **2022**, arXiv:2201.10861.
31. Hakkila, J.; Giblin, T.W.; Roiger, R.J.; Haglin, D.J.; Paciesas, W.S.; Meegan, C.A. How Sample Completeness Affects Gamma-Ray Burst Classification. *Astrophys. J.* **2003**, *582*, 320–329. [CrossRef]
32. Bromberg, O.; Nakar, E.; Piran, T.; Sari, R. Short versus Long and Collapsars versus Non-collapsars: A Quantitative Classification of Gamma-Ray Bursts. *Astrophys. J.* **2013**, *764*, 179. [CrossRef]
33. Jespersen, C.K.; Severin, J.B.; Steinhardt, C.L.; Vinther, J.; Fynbo, J.P.U.; Selsing, J.; Watson, D. An Unambiguous Separation of Gamma-Ray Bursts into Two Classes from Prompt Emission Alone. *Astrophys. J.* **2020**, *896*, L20. [CrossRef]
34. Mallat, S.G. A theory for multiresolution signal decomposition—The wavelet representation. *IEEE Trans. Pattern Anal. Mach. Intell.* **1989**, *11*, 674–693. [CrossRef]
35. Meredith, D.C.; Ryan, J.M.; Young, C.A.; Lestrade, J.P. Wavelet Analysis of Gamma Ray Bursts. *AIP Conf. Proc.* **1994**, *307*, 701. [CrossRef]
36. Young, C.A.; Meredith, D.C.; Ryan, J.M. A Compact Representation of Gamma-Ray Burst Time Series. *Astrophys. Space Sci.* **1995**, *231*, 119–122. [CrossRef]
37. Bagoly, Z.; Balázs, L.G.; Horváth, I.; Mészáros, A. Wavelet Analysis of the BATSE 64ms GRB Lightcurves. *ASP Conf. Ser.* **2004**, *312*, 51.

38. Hurley, K.J.; McBreen, B.; Quilligan, F.; Delaney, M.; Hanlon, L. Wavelet analysis and lognormal distributions in GRBs. *AIP Conf. Ser.* **1998**, *428*, 191–195. [CrossRef]
39. Quilligan, F.; Hurley, K.J.; McBreen, B.; Hanlon, L.; Duggan, P. Characteristic properties of peaks in GRBs. *Astron. Astrophys. Suppl. Ser.* **1999**, *138*, 419–420. [CrossRef]
40. Quilligan, F.; McBreen, B.; Hanlon, L.; McBreen, S.; Hurley, K.J.; Watson, D. Temporal properties of gamma ray bursts as signatures of jets from the central engine. *Astron. Astrophys.* **2002**, *385*, 377–398. [CrossRef]
41. Qin, Y.P.; Liang, E.W.; Xie, G.Z.; Su, C.Y. Statistical Properties of the Highest Pulses in Gamma-Ray Bursts. *Chin. J. Astron. Astrophys.* **2003**, *3*, 38–48. [CrossRef]
42. Walker, K.C.; Schaefer, B.E.; Fenimore, E.E. Gamma-Ray Bursts Have Millisecond Variability. *Astrophys. J.* **2000**, *537*, 264–269. [CrossRef]
43. MacLachlan, G.A.; Shenoy, A.; Sonbas, E.; Dhuga, K.S.; Eskandarian, A.; Maximon, L.C.; Parke, W.C. The minimum variability time-scale and its relation to pulse profiles of Fermi GRBs. *Mon. Not. R. Astron. Soc.* **2012**, *425*, L32–L35. [CrossRef]
44. MacLachlan, G.A.; Shenoy, A.; Sonbas, E.; Dhuga, K.S.; Cobb, B.E.; Ukwatta, T.N.; Morris, D.C.; Eskandarian, A.; Maximon, L.C.; Parke, W.C. Minimum variability time-scales of long and short GRBs. *Mon. Not. R. Astron. Soc.* **2013**, *432*, 857–865. [CrossRef]
45. Golkhou, V.Z.; Butler, N.R. Uncovering the Intrinsic Variability of Gamma-Ray Bursts. *Astrophys. J.* **2014**, *787*, 90. [CrossRef]
46. Golkhou, V.Z.; Butler, N.R.; Littlejohns, O.M. The Energy Dependence of GRB Minimum Variability Timescales. *Astrophys. J.* **2015**, *811*, 93. [CrossRef]
47. Varughese, M.M.; von Sachs, R.; Stephanou, M.; Bassett, B.A. Non-parametric transient classification using adaptive wavelets. *Mon. Not. R. Astron. Soc.* **2015**, *453*, 2848–2861. [CrossRef]
48. Lochner, M.; McEwen, J.D.; Peiris, H.V.; Lahav, O.; Winter, M.K. Photometric Supernova Classification with Machine Learning. *Astrophys. J.* **2016**, *225*, 31. [CrossRef]
49. Vargas dos Santos, M.; Quartin, M.; Reis, R.R.R. On the cosmological performance of photometrically classified supernovae with machine learning. *Mon. Not. R. Astron. Soc.* **2020**, *497*, 2974–2991. [CrossRef]
50. Norris, J.P.; Share, G.H.; Messina, D.C.; Dennis, B.R.; Desai, U.D.; Cline, T.L.; Matz, S.M.; Chupp, E.L. Spectral Evolution of Pulse Structures in Gamma-Ray Bursts. *Astrophys. J.* **1986**, *301*, 213. [CrossRef]
51. Golenetskii, S.V.; Mazets, E.P.; Aptekar, R.L.; Ilinskii, V.N. Correlation between luminosity and temperature in γ-ray burst sources. *Nature* **1983**, *306*, 451–453. [CrossRef]
52. Fenimore, E.E.; in 't Zand, J.J.M.; Norris, J.P.; Bonnell, J.T.; Nemiroff, R.J. Gamma-Ray Burst Peak Duration as a Function of Energy. *Astrophys. J.* **1995**, *448*, L101. [CrossRef]
53. Nemiroff, R.J.; Norris, J.P.; Kouveliotou, C.; Fishman, G.J.; Meegan, C.A.; Paciesas, W.S. Gamma-Ray Bursts Are Time-asymmetric. *Astrophys. J.* **1994**, *423*, 432. [CrossRef]
54. Norris, J.P.; Nemiroff, R.J.; Bonnell, J.T.; Scargle, J.D.; Kouveliotou, C.; Paciesas, W.S.; Meegan, C.A.; Fishman, G.J. Attributes of Pulses in Long Bright Gamma-Ray Bursts. *Astrophys. J.* **1996**, *459*, 393. [CrossRef]
55. Hakkila, J.; Preece, R.D. Unification of Pulses in Long and Short Gamma-Ray Bursts: Evidence from Pulse Properties and Their Correlations. *Astrophys. J.* **2011**, *740*, 104. [CrossRef]
56. Hakkila, J.; Horváth, I.; Hofesmann, E.; Lesage, S. Properties of Short Gamma-ray Burst Pulses from a BATSE TTE GRB Pulse Catalog. *Astrophys. J.* **2018**, *855*, 101. [CrossRef]
57. Cheng, L.X.; Ma, Y.Q.; Cheng, K.S.; Lu, T.; Zhou, Y.Y. The time delay of gamma-ray bursts in the soft energy band. *Astron. Astrophys.* **1995**, *300*, 746.
58. Norris, J.P.; Marani, G.F.; Bonnell, J.T. Connection between Energy-dependent Lags and Peak Luminosity in Gamma-Ray Bursts. *Astrophys. J.* **2000**, *534*, 248–257. [CrossRef]
59. Wu, B.; Fenimore, E. Spectral Lags of Gamma-Ray Bursts From Ginga and BATSE. *Astrophys. J.* **2000**, *535*, L29–L32. [CrossRef]
60. Ukwatta, T.N.; Dhuga, K.S.; Stamatikos, M.; Dermer, C.D.; Sakamoto, T.; Sonbas, E.; Parke, W.C.; Maximon, L.C.; Linnemann, J.T.; Bhat, P.N.; et al. The lag-luminosity relation in the GRB source frame: An investigation with Swift BAT bursts. *Mon. Not. R. Astron. Soc.* **2012**, *419*, 614–623. [CrossRef]
61. Ukwatta, T.N.; Stamatikos, M.; Dhuga, K.S.; Sakamoto, T.; Barthelmy, S.D.; Eskandarian, A.; Gehrels, N.; Maximon, L.C.; Norris, J.P.; Parke, W.C. Spectral Lags and the Lag-Luminosity Relation: An Investigation with Swift BAT Gamma-ray Bursts. *Astrophys. J.* **2010**, *711*, 1073–1086. [CrossRef]
62. Hakkila, J.; Giblin, T.W.; Norris, J.P.; Fragile, P.C.; Bonnell, J.T. Correlations between Lag, Luminosity, and Duration in Gamma-Ray Burst Pulses. *Astrophys. J.* **2008**, *677*, L81. [CrossRef]
63. Bernardini, M.G.; Ghirlanda, G.; Campana, S.; Covino, S.; Salvaterra, R.; Atteia, J.L.; Burlon, D.; Calderone, G.; D'Avanzo, P.; D'Elia, V.; et al. Comparing the spectral lag of short and long gamma-ray bursts and its relation with the luminosity. *Mon. Not. R. Astron. Soc.* **2015**, *446*, 1129–1138. [CrossRef]
64. Hakkila, J.; Preece, R.D. Gamma-Ray Burst Pulse Shapes: Evidence for Embedded Shock Signatures? *Astrophys. J.* **2014**, *783*, 88. [CrossRef]
65. Bhat, P.N.; Fishman, G.J.; Meegan, C.A.; Wilson, R.B.; Kouveliotou, C.; Paciesas, W.S.; Pendleton, G.N.; Schaefer, B.E. Spectral Evolution of a Subclass of Gamma-Ray Bursts Observed by BATSE. *Astrophys. J.* **1994**, *426*, 604. [CrossRef]
66. Band, D.L. Gamma-Ray Burst Spectral Evolution through Cross-Correlations of Discriminator Light Curves. *Astrophys. J.* **1997**, *486*, 928–937. [CrossRef]

67. Ford, L.A.; Band, D.L.; Matteson, J.L.; Briggs, M.S.; Pendleton, G.N.; Preece, R.D.; Paciesas, W.S.; Teegarden, B.J.; Palmer, D.M.; Schaefer, B.E.; et al. BATSE Observations of Gamma-Ray Burst Spectra. II. Peak Energy Evolution in Bright, Long Bursts. *Astrophys. J.* **1995**, *439*, 307. [CrossRef]
68. Borgonovo, L.; Ryde, F. On the Hardness-Intensity Correlation in Gamma-Ray Burst Pulses. *Astrophys. J.* **2001**, *548*, 770–786. [CrossRef]
69. Peng, Z.Y.; Ma, L.; Lu, R.J.; Fang, L.M.; Bao, Y.Y.; Yin, Y. Spectral hardness evolution characteristics of tracking gamma-ray burst pulses. *New Astron.* **2009**, *14*, 311–320. [CrossRef]
70. Lu, R.J.; Hou, S.J.; Liang, E.W. The E_p-flux Correlation in the Rising and Decaying Phases of gamma-ray Burst Pulses: Evidence for Viewing Angle Effect? *Astrophys. J.* **2010**, *720*, 1146–1154. [CrossRef]
71. Hakkila, J.; Lien, A.; Sakamoto, T.; Morris, D.; Neff, J.E.; Giblin, T.W. Swift Observations of Gamma-Ray Burst Pulse Shapes: GRB Pulse Spectral Evolution Clarified. *Astrophys. J.* **2015**, *815*, 134. [CrossRef]
72. Ghirlanda, G.; Nava, L.; Ghisellini, G. Spectral-luminosity relation within individual Fermi gamma rays bursts. *Astron. Astrophys.* **2010**, *511*, A43. [CrossRef]
73. Ghirlanda, G.; Ghisellini, G.; Nava, L.; Burlon, D. Spectral evolution of Fermi/GBM short gamma-ray bursts. *Mon. Not. R. Astron. Soc.* **2011**, *410*, L47–L51. [CrossRef]
74. Lu, R.J.; Wei, J.J.; Liang, E.W.; Zhang, B.B.; Lü, H.J.; Lü, L.Z.; Lei, W.H.; Zhang, B. A Comprehensive Analysis of Fermi Gamma-Ray Burst Data. II. E_p Evolution Patterns and Implications for the Observed Spectrum-Luminosity Relations. *Astrophys. J.* **2012**, *756*, 112. [CrossRef]
75. Kocevski, D.; Petrosian, V. On the Lack of Time Dilation Signatures in Gamma-Ray Burst Light Curves. *Astrophys. J.* **2013**, *765*, 116. [CrossRef]
76. Littlejohns, O.M.; Tanvir, N.R.; Willingale, R.; Evans, P.A.; O'Brien, P.T.; Levan, A.J. Are gamma-ray bursts the same at high redshift and low redshift? *Mon. Not. R. Astron. Soc.* **2013**, *436*, 3640–3655. [CrossRef]
77. Fishman, G.; Johnson, W.N. *Proceedings of the Gamma Ray Observatory Science Workshop*; Johnson, W.N., Ed.; NASA/GSFC Greenbelt: Greenbelt, MD, USA, 1989; pp. 2–39.
78. Gehrels, N.; Chincarini, G.; Giommi, P.; Mason, K.O.; Nousek, J.A.; Wells, A.A.; White, N.E.; Barthelmy, S.D.; Burrows, D.N.; Cominsky, L.R.; et al. The Swift Gamma-Ray Burst Mission. *Astrophys. J.* **2004**, *611*, 1005–1020. [CrossRef]
79. Meegan, C.; Lichti, G.; Bhat, P.N.; Bissaldi, E.; Briggs, M.S.; Connaughton, V.; Diehl, R.; Fishman, G.; Greiner, J.; Hoover, A.S.; et al. The Fermi Gamma-ray Burst Monitor. *Astrophys. J.* **2009**, *702*, 791–804. [CrossRef]
80. Lien, A.; Sakamoto, T.; Barthelmy, S.D.; Baumgartner, W.H.; Cannizzo, J.K.; Chen, K.; Collins, N.R.; Cummings, J.R.; Gehrels, N.; Krimm, H.A.; et al. The Third Swift Burst Alert Telescope Gamma-Ray Burst Catalog. *Astrophys. J.* **2016**, *829*, 7. [CrossRef]
81. Goldstein, A.; Cleveland, W.H.; Kocevski, D. Fermi GBM Data Tools: V1.04, 2020. Available online: https://fermi.gsfc.nasa.gov/ssc/data/analysis/gbm (accessed on 26 April 2022).
82. Sacks, J.; Welch, W.J.; Mitchell, T.J.; Wynn, H.P. Design and analysis of computer experiments. *Stat. Sci.* **1989**, *4*, 409–423. [CrossRef]
83. De, G.; Matthews, A.G.; Van Der Wilk, M.; Nickson, T.; Fujii, K.; Boukouvalas, A.; León-Villagrá, P.; Ghahramani, Z.; Hensman, J. GPflow: A Gaussian process library using TensorFlow. *J. Mach. Learn. Res.* **2017**, *18*, 1299–1304.
84. Abadi, M.; Agarwal, A.; Barham, P.; Brevdo, E.; Chen, Z.; Citro, C.; Corrado, G.S.; Davis, A.; Dean, J.; Devin, M.; et al. TensorFlow: Large-Scale Machine Learning on Heterogeneous Systems. 2015. Available online: tensorflow.org (accessed on 15 March 2021).
85. Daubechies, I. *Ten Lectures on Wavelets*; SIAM: Philadelphia, PA, USA, 1992.
86. Holschneider, M.; Kronland-Martinet, R.; Morlet, J.; Tchamitchian, P. A Real-Time Algorithm for Signal Analysis with the Help of the Wavelet Transform. In *Wavelets. Time-Frequency Methods and Phase Space*; Combes, J.M., Grossmann, A., Tchamitchian, P., Eds.; Spinger: Berlin/Heidelberg, Germany, 1989; p. 286.
87. Addison, P.S. *The Illustrated Wavelet Transform Handbook: Introductory Theory and Applications in Science, Engineering, Medicine and Finance*; CRC Press: Boca Raton, FL, USA, 2017.
88. Rhif, M.; Ben Abbes, A.; Farah, I.R.; Martínez, B.; Sang, Y. Wavelet transform application for/in non-stationary time-series analysis: A review. *Appl. Sci.* **2019**, *9*, 1345. [CrossRef]
89. Lee, G.; Gommers, R.; Waselewski, F.; Wohlfahrt, K.; O'Leary, A. PyWavelets: A Python package for wavelet analysis. *J. Open Source Softw.* **2019**, *4*, 1237. [CrossRef]
90. Daubechies, I. Orthonormal bases of compactly supported wavelets. *Commun. Pure Appl. Math.* **1988**, *41*, 909–996. [CrossRef]
91. Pearson, K. LIII. On lines and planes of closest fit to systems of points in space. *Lond. Edinb. Dublin Philos. Mag. J. Sci.* **1901**, *2*, 559–572. [CrossRef]
92. Hotelling, H. Analysis of a complex of statistical variables into principal components. *J. Educ. Psychol.* **1933**, *24*, 417. [CrossRef]
93. Hinton, G.E.; Roweis, S.T. Stochastic neighbor embedding. *Adv. Neural Inf. Process. Syst.* **2003**, *15*, 857–864.
94. Kullback, S.; Leibler, R.A. On Information and Sufficiency. *Ann. Math. Stat.* **1951**, *22*, 79–86. [CrossRef]
95. Maaten, L.v.d.; Hinton, G. Visualizing data using t-SNE. *J. Mach. Learn. Res.* **2008**, *9*, 2579–2605.
96. Fraley, C.; Raftery, A.E. MCLUST: Software for model-based cluster analysis. *J. Classif.* **1999**, *16*, 297–306. [CrossRef]
97. Scrucca, L.; Fop, M.; Murphy, T.B.; Raftery, A.E. mclust 5: Clustering, classification and density estimation using Gaussian finite mixture models. *R J.* **2016**, *8*, 289–317. [CrossRef] [PubMed]
98. Schwarz, G. Estimating the dimension of a model. *Ann. Stat.* **1978**, *6*, 461–464. [CrossRef]

99. Baudry, J.P.; Raftery, A.E.; Celeux, G.; Lo, K.; Gottardo, R. Combining Mixture Components for Clustering. *J. Comput. Graph. Stat.* **2010**, *19*, 332–353. [CrossRef]
100. Gibson, S.L.; Wynn, G.A.; Gompertz, B.P.; O'Brien, P.T. Fallback accretion on to a newborn magnetar: Short GRBs with extended emission. *Mon. Not. R. Astron. Soc.* **2017**, *470*, 4925–4940. [CrossRef]
101. Racusin, J.L.; Oates, S.R.; Schady, P.; Burrows, D.N.; de Pasquale, M.; Donato, D.; Gehrels, N.; Koch, S.; McEnery, J.; Piran, T.; et al. Fermi and Swift Gamma-ray Burst Afterglow Population Studies. *Astrophys. J.* **2011**, *738*, 138. [CrossRef]
102. Burgess, J.M.; Greiner, J.; Bégué, D.; Giannios, D.; Berlato, F.; Lipunov, V.M. Viewing Short Gamma-Ray Bursts From a Different Angle. *Front. Astron. Space Sci.* **2020**, *7*, 40. [CrossRef]
103. Koen, C.; Bere, A. On multiple classes of gamma-ray bursts, as deduced from autocorrelation functions or bivariate duration/hardness ratio distributions. *Mon. Not. R. Astron. Soc.* **2012**, *420*, 405–415. [CrossRef]
104. Tarnopolski, M. Analysis of gamma-ray burst duration distribution using mixtures of skewed distributions. *Mon. Not. R. Astron. Soc.* **2016**. [CrossRef]
105. Tarnopolski, M. Analysis of the Duration-Hardness Ratio Plane of Gamma-Ray Bursts Using Skewed Distributions. *Astrophys. J.* **2019**, *870*, 105. [CrossRef]
106. Bromberg, O.; Nakar, E.; Piran, T.; Sari, R. An Observational Imprint of the Collapsar Model of Long Gamma-Ray Bursts. *Astrophys. J.* **2012**, *749*, 110. [CrossRef]
107. Cano, Z.; Wang, S.Q.; Dai, Z.G.; Wu, X.F. The Observer's Guide to the Gamma-Ray Burst Supernova Connection. *Adv. Astron.* **2017**, *2017*, 8929054. [CrossRef]
108. Ashall, C.; Mazzali, P.A.; Pian, E.; Woosley, S.E.; Palazzi, E.; Prentice, S.J.; Kobayashi, S.; Holmbo, S.; Levan, A.; Perley, D.; et al. GRB 161219B/SN 2016jca: A powerful stellar collapse. *Mon. Not. R. Astron. Soc.* **2019**, *487*, 5824–5839. [CrossRef]
109. Melandri, A.; Malesani, D.B.; Izzo, L.; Japelj, J.; Vergani, S.D.; Schady, P.; Sagués Carracedo, A.; de Ugarte Postigo, A.; Anderson, J.P.; Barbarino, C.; et al. GRB 171010A/SN 2017htp: A GRB-SN at z = 0.33. *Mon. Not. R. Astron. Soc.* **2019**, *490*, 5366–5374. [CrossRef]
110. Izzo, L.; Rossi, A.; Malesani, D.; Heintz, K.; Selsing, J.; Schady, P.; Starling, R.; Sollerman, J.; Leloudas, G.; Cano, Z.; et al. GRB 180728A: Discovery of the associated supernova. *GRB Coord. Netw.* **2018**, *23142*, 1.
111. Selsing, J.; Izzo, L.; Rossi, A.; Malesani, D.; Heintz, K.; Schady, P.; Starling, R.; Sollerman, J.; Leloudas, G.; Cano, Z.; et al. GRB 180728A: Classification of the associated SN 2018fip. *GRB Coord. Netw.* **2018**, *23181*, 1.
112. Hu, Y.D.; Castro-Tirado, A.J.; Kumar, A.; Gupta, R.; Valeev, A.F.; Pandey, S.B.; Kann, D.A.; Castellón, A.; Agudo, I.; Aryan, A.; et al. 10.4 m GTC observations of the nearby VHE-detected GRB 190829A/SN 2019oyw. *Astron. Astrophys.* **2021**, *646*, A50. [CrossRef]
113. Ahumada, T.; Singer, L.P.; Anand, S.; Coughlin, M.W.; Kasliwal, M.M.; Ryan, G.; Andreoni, I.; Cenko, S.B.; Fremling, C.; Kumar, H.; et al. Discovery and confirmation of the shortest gamma ray burst from a collapsar. *arXiv* **2021**, arXiv:2105.05067.
114. Rossi, A.; Rothberg, B.; Palazzi, E.; Kann, D.A.; D'Avanzo, P.; Klose, S.; Perego, A.; Pian, E.; Savaglio, S.; Stratta, G.; et al. The peculiar short-duration GRB 200826A and its supernova. *arXiv* **2021**, arXiv:2105.03829.
115. Zhang, B.B.; Liu, Z.K.; Peng, Z.K.; Li, Y.; Lü, H.J.; Yang, J.; Yang, Y.S.; Yang, Y.H.; Meng, Y.Z.; Zou, J.H.; et al. A Peculiarly Short-duration Gamma-Ray Burst from Massive Star Core Collapse. *arXiv* **2021**, arXiv:2105.05021.
116. Berger, E.; Fong, W.; Chornock, R. An r-process Kilonova Associated with the Short-hard GRB 130603B. *Astrophys. J.* **2013**, *774*, L23. [CrossRef]
117. Tanvir, N.R.; Levan, A.J.; Fruchter, A.S.; Hjorth, J.; Hounsell, R.A.; Wiersema, K.; Tunnicliffe, R.L. A 'kilonova' associated with the short-duration γ-ray burst GRB 130603B. *Nature* **2013**, *500*, 547–549. [CrossRef] [PubMed]
118. Jin, Z.P.; Fan, Y.Z.; Wei, D.M. An r- process macronova/kilonova in GRB 060614: Evidence for the merger of a neutron star-black hole binary. In Proceedings of the 13th International Symposium on Origin of Matter and Evolution of Galaxies (OMEG2015), Beijing, China, 24–27 June 2015; EDP Sciences: Les Ulis, France, 2016; Volume 109, p. 08002.
119. Yang, B.; Jin, Z.P.; Li, X.; Covino, S.; Zheng, X.Z.; Hotokezaka, K.; Fan, Y.Z.; Piran, T.; Wei, D.M. A possible macronova in the late afterglow of the long-short burst GRB 060614. *Nat. Commun.* **2015**, *6*, 7323. [CrossRef] [PubMed]
120. Perley, D.A.; Metzger, B.D.; Granot, J.; Butler, N.R.; Sakamoto, T.; Ramirez-Ruiz, E.; Levan, A.J.; Bloom, J.S.; Miller, A.A.; Bunker, A.; et al. GRB 080503: Implications of a Naked Short Gamma-Ray Burst Dominated by Extended Emission. *Astrophys. J.* **2009**, *696*, 1871–1885. [CrossRef]
121. Jin, Z.P.; Hotokezaka, K.; Li, X.; Tanaka, M.; D'Avanzo, P.; Fan, Y.Z.; Covino, S.; Wei, D.M.; Piran, T. The Macronova in GRB 050709 and the GRB-macronova connection. *Nat. Commun.* **2016**, *7*, 12898. [CrossRef]
122. Troja, E.; Ryan, G.; Piro, L.; van Eerten, H.; Cenko, S.B.; Yoon, Y.; Lee, S.K.; Im, M.; Sakamoto, T.; Gatkine, P.; et al. A luminous blue kilonova and an off-axis jet from a compact binary merger at z = 0.1341. *Nat. Commun.* **2018**, *9*, 4089. [CrossRef]
123. Kasliwal, M.M.; Korobkin, O.; Lau, R.M.; Wollaeger, R.; Fryer, C.L. Infrared Emission from Kilonovae: The Case of the Nearby Short Hard Burst GRB 160821B. *Astrophys. J.* **2017**, *843*, L34. [CrossRef]
124. Jin, Z.P.; Li, X.; Wang, H.; Wang, Y.Z.; He, H.N.; Yuan, Q.; Zhang, F.W.; Zou, Y.C.; Fan, Y.Z.; Wei, D.M. Short GRBs: Opening Angles, Local Neutron Star Merger Rate, and Off-axis Events for GRB/GW Association. *Astrophys. J.* **2018**, *857*, 128. [CrossRef]
125. Lamb, G.P.; Tanvir, N.R.; Levan, A.J.; de Ugarte Postigo, A.; Kawaguchi, K.; Corsi, A.; Evans, P.A.; Gompertz, B.; Malesani, D.B.; Page, K.L.; et al. Short GRB 160821B: A Reverse Shock, a Refreshed Shock, and a Well-sampled Kilonova. *Astrophys. J.* **2019**, *883*, 48. [CrossRef]

126. Troja, E.; Castro-Tirado, A.J.; Becerra González, J.; Hu, Y.; Ryan, G.S.; Cenko, S.B.; Ricci, R.; Novara, G.; Sánchez-Rámirez, R.; Acosta-Pulido, J.A.; et al. The afterglow and kilonova of the short GRB 160821B. *Mon. Not. R. Astron. Soc.* **2019**, *489*, 2104–2116. [CrossRef]
127. Fong, W.; Laskar, T.; Rastinejad, J.; Escorial, A.R.; Schroeder, G.; Barnes, J.; Kilpatrick, C.D.; Paterson, K.; Berger, E.; Metzger, B.D.; et al. The Broadband Counterpart of the Short GRB 200522A at z = 0.5536: A Luminous Kilonova or a Collimated Outflow with a Reverse Shock? *Astrophys. J.* **2021**, *906*, 127. [CrossRef]
128. Xu, D.; Starling, R.L.C.; Fynbo, J.P.U.; Sollerman, J.; Yost, S.; Watson, D.; Foley, S.; O'Brien, P.T.; Hjorth, J. In Search of Progenitors for Supernovaless Gamma-Ray Bursts 060505 and 060614: Re-examination of Their Afterglows. *Astrophys. J.* **2009**, *696*, 971–979. [CrossRef]
129. Price, P.A.; Berger, E.; Fox, D.B. GRB 060614: Redshift. *GRB Coord. Netw.* **2006**, *5275*, 1.
130. Fugazza, D.; Malesani, D.; Romano, P.; Tagliaferri, G.; Covino, S.; Chincarini, G.; Della Valle, M.; Fiore, F.; Stella, L. GRB 060614: Redshift confirmation. *GRB Coord. Netw.* **2006**, *5276*, 1.
131. Della Valle, M.; Chincarini, G.; Panagia, N.; Tagliaferri, G.; Malesani, D.; Testa, V.; Fugazza, D.; Campana, S.; Covino, S.; Mangano, V.; et al. An enigmatic long-lasting γ-ray burst not accompanied by a bright supernova. *Nature* **2006**, *444*, 1050–1052. [CrossRef] [PubMed]
132. Fynbo, J.P.U.; Watson, D.; Thöne, C.C.; Sollerman, J.; Bloom, J.S.; Davis, T.M.; Hjorth, J.; Jakobsson, P.; Jørgensen, U.G.; Graham, J.F.; et al. No supernovae associated with two long-duration γ-ray bursts. *Nature* **2006**, *444*, 1047–1049. [CrossRef]
133. Gal-Yam, A.; Fox, D.B.; Price, P.A.; Ofek, E.O.; Davis, M.R.; Leonard, D.C.; Soderberg, A.M.; Schmidt, B.P.; Lewis, K.M.; Peterson, B.A.; et al. A novel explosive process is required for the γ-ray burst GRB 060614. *Nature* **2006**, *444*, 1053–1055. [CrossRef]
134. Gehrels, N.; Norris, J.P.; Barthelmy, S.D.; Granot, J.; Kaneko, Y.; Kouveliotou, C.; Markwardt, C.B.; Mészáros, P.; Nakar, E.; Nousek, J.A.; et al. A new γ-ray burst classification scheme from GRB060614. *Nature* **2006**, *444*, 1044–1046. [CrossRef]
135. Lu, Y.; Huang, Y.F.; Zhang, S.N. A Tidal Disruption Model for the Gamma-Ray Burst of GRB 060614. *Astrophys. J.* **2008**, *684*, 1330–1335. [CrossRef]
136. Meng, Y.Z.; Geng, J.J.; Zhang, B.B.; Wei, J.J.; Xiao, D.; Liu, L.D.; Gao, H.; Wu, X.F.; Liang, E.W.; Huang, Y.F.; et al. The Origin of the Prompt Emission for Short GRB 170817A: Photosphere Emission or Synchrotron Emission? *Astrophys. J.* **2018**, *860*, 72. [CrossRef]
137. Li, L.; Ryde, F.; Pe'er, A.; Yu, H.F.; Acuner, Z. Bayesian Time-Resolved Spectroscopy of Multi-Pulsed GRBs: Variations of Emission Properties amongst Pulses. *arXiv* **2020**, arXiv:2012.03038.
138. Dereli-Bégué, H.; Pe'er, A.; Ryde, F. Classification of Photospheric Emission in Short GRBs. *Astrophys. J.* **2020**, *897*, 145. [CrossRef]
139. Li, L. Multipulse Fermi Gamma-Ray Bursts. I. Evidence of the Transition from Fireball to Poynting-flux-dominated Outflow. *Astrophys. J.* **2019**, *242*, 16. [CrossRef]
140. Acuner, Z.; Ryde, F.; Pe'er, A.; Mortlock, D.; Ahlgren, B. The Fraction of Gamma-Ray Bursts with an Observed Photospheric Emission Episode. *Astrophys. J.* **2020**, *893*, 128. [CrossRef]
141. Acuner, Z.; Ryde, F. Clustering of gamma-ray burst types in the Fermi GBM catalogue: Indications of photosphere and synchrotron emissions during the prompt phase. *Mon. Not. R. Astron. Soc.* **2018**, *475*, 1708–1724. [CrossRef]
142. Ryde, F. The Cooling Behavior of Thermal Pulses in Gamma-Ray Bursts. *Astrophys. J.* **2004**, *614*, 827–846. [CrossRef]
143. McBreen, S.; McBreen, B.; Hanlon, L.; Quilligan, F. Cumulative light curves of gamma-ray bursts and relaxation systems. *Astron. Astrophys.* **2002**, *393*, L29–L32. [CrossRef]
144. Abbott, B.P.; Abbott, R.; Abbott, T.D.; Abraham, S.; Acernese, F.; Ackley, K.; Adams, C.; Adya, V.B.; Affeldt, C.; Agathos, M.; et al. Prospects for observing and localizing gravitational-wave transients with Advanced LIGO, Advanced Virgo and KAGRA. *Living Rev. Relativ.* **2020**, *23*, 3. [CrossRef]
145. Ivezić, Ž.; Kahn, S.M.; Tyson, J.A.; Abel, B.; Acosta, E.; Allsman, R.; Alonso, D.; AlSayyad, Y.; Anderson, S.F.; Andrew, J.; et al. LSST: From Science Drivers to Reference Design and Anticipated Data Products. *Astrophys. J.* **2019**, *873*, 111. [CrossRef]
146. Harris, C.R.; Millman, K.J.; van der Walt, S.J.; Gommers, R.; Virtanen, P.; Cournapeau, D.; Wieser, E.; Taylor, J.; Berg, S.; Smith, N.J.; et al. Array programming with NumPy. *Nature* **2020**, *585*, 357–362. [CrossRef]
147. Hunter, J.D. Matplotlib: A 2D graphics environment. *Comput. Sci. Eng.* **2007**, *9*, 90–95. [CrossRef]
148. Pandas Development Team. Pandas-dev/Pandas: Pandas. 2020. Available online: https://doi.org/10.5281/zenodo.3509134 (accessed on 23 June 2021).
149. McKinney, W. Data Structures for Statistical Computing in Python. In Proceedings of the 9th Python in Science Conference, Austin, TX, USA, 28 June–3 July 2010; pp. 56–61. [CrossRef]
150. Pedregosa, F.; Varoquaux, G.; Gramfort, A.; Michel, V.; Thirion, B.; Grisel, O.; Blondel, M.; Prettenhofer, P.; Weiss, R.; Dubourg, V.; et al. Scikit-learn: Machine Learning in Python. *J. Mach. Learn. Res.* **2011**, *12*, 2825–2830.

galaxies

MDPI

Review

Electromagnetic Precursors of Short Gamma-Ray Bursts as Counterparts of Gravitational Waves

Jieshuang Wang [1,2,*] and Liangduan Liu [3,*]

1 Tsung-Dao Lee Institute, Shanghai Jiao Tong University, Shanghai 200240, China
2 Max-Planck-Institut für Kernphysik, Saupfercheckweg 1, D-69117 Heidelberg, Germany
3 Institute of Astrophysics, Central China Normal University, Wuhan 430079, China
* Correspondence: jiesh.wang@gmail.com (J.W.); liuld@mail.ccnu.edu.cn (L.L.)

Abstract: Precursor emissions are found in some short gamma-ray bursts (SGRBs). In this paper, we review the theories and observations of the SGRB precursor and discuss its prospect as an electromagnetic counterpart of the gravitational wave event produced by neutron star (NS) mergers. The observed luminosity, spectrum, and duration of precursors are explained by the magnetospheric interaction model during the inspiral or the cocoon/jet shock breakout model during the jet propagation. In general, these two models predict that the precursor will be weaker than the main GRB, but will be of a larger opening angle, which makes it an advantageous gamma-ray counterpart for NS mergers in the local Universe, especially for NS - black hole mergers with very low mass ratios, in which the main GRBs are not expected. The joint observation of the precursor, SGRB, and gravitational wave will help to reveal the jet launch mechanism and post-merger remnant.

Keywords: gamma-ray burst; gravitational wave; neutron stars; magnetosphere; shock breakout

Citation: Wang, J.; Liu, L. Electromagnetic Precursors of Short Gamma-Ray Bursts as Counterparts of Gravitational Waves. *Galaxies* **2021**, *9*, 104. https://doi.org/10.3390/galaxies9040104

Academic Editors: Elena Moretti and Francesco Longo

Received: 27 September 2021
Accepted: 13 November 2021
Published: 15 November 2021

1. Introduction

On 17 August 2017, the *Fermi*/gamma-ray burst monitor (GBM) was triggered by a short gamma-ray burst (SGRB)-GRB 170817A [1–3]. Independently, the gravitational-wave (GW) event GW170817 produced by the double neutron star (NS) merger was detected by the advanced LIGO and Virgo detectors [4,5]. The joint detection of GW170817/GRB 170817A confirms that at least some SGRBs originate from NS mergers, and herald the multi-messenger astronomy [1–5]. It also enables better localization, which benefited the multi-wavelength follow-up observations. The detection of the associated kilonova, AT 2017gfo, led to the discovery of the host-galaxy NGC 4993 at a distance of $\sim$40 Mpc, which shed light on the physics of nucleosynthesis [1,6–11].

Joint detection can provide abundant information to study some fundamental physics. Using the GW data alone, constraints on the NS equation of state can be obtained (e.g., [4]). Combining with the electromagnetic (EM) observations, (1) the GW event can be treated as a standard siren to study cosmology [12]; (2) one can also constrain the difference between the speed of gravity and the light speed, test the violation of Lorentz invariance and the equivalence principle [5,13]. It can also be used to study the launching mechanism, structure, composition, and radiation mechanism of GRB jets (e.g., [3,14–23]).

Since the discovery of GW170817/GRB 170817A, many efforts have been put into the follow-up observations of GW events to search for their EM counterparts. No new confident joint detection is observed, except for a sub-threshold event: the sub-threshold GRB (GBM-190816) [20,24] was found to be possibly associated with a sub-threshold NS merger event GW190425 [25]. But in the archived *Fermi*/GBM data, a small sample of GRB 170817A-like events has been found [26]. In theory, many EM signals are expected for the NS merger. The observation and theory of SGRBs, afterglows, and kilonova were summarized in many reviews [27–32]. However, little attention has been placed on the pre-merger EM radiation. As a complement, we focused on the precursor emissions of SGRB in this review.

Precursors were initially identified as weak signals in long GRBs (e.g., [33–39]). Later precursors of SGRBs were found in the *Swift*/Burst Alert Telescope (BAT) data [40]. Within the standard fireball scenario, precursors are suggested to be associated with the transition of the fireball from optically thick to optically thin, leading to photospheric blackbody emissions [29,41–44]. This applies to both long and short GRBs. It is also suggested that a precursor can be generated by the shock breakout (SBO) of a jet or a cocoon. For long GRBs, this links to the SBO from the stellar surface [32,45–50], some research proposed that breakout of a radiation mediated shocks train can naturally generate a band-like spectrum [51,52]. For SGRB, this relates to the SBO from the ejecta produced during the NS merger [32,50,53,54]. Besides, there are two more scenarios proposed only for SGRB precursors. During the inspiral phase of the NS–NS/black hole (BH) binary, the magnetospheric interaction of the binary [55–63], or the crust crack of the NS [64–66] may also generate gamma-ray emissions. As such, precursors of SGRBs may shed light on the physical processes right before or shortly after the merger.

Moreover, the magnetospheric interaction model [62] and the SBO model [32,53,54] predict the precursor, although fainter than the main GRB, would have a much larger opening angle, as the radiation is generated by a mild relativistic component. In this case, the precursor can serve as an independent EM counterpart for GWs, even though the prompt GRB points away from the line of sight. It has been suggested that GRB 170817A can be such a case [32,53,54]. This feature would be greatly appreciated for follow-up observations. Thus, research on precursors is important for multi-messenger astronomy. This review aims to summarize the current understanding of SGRB precursors and discuss the possibility for future observations. In the next section, we review the feasible precursor models. Observations are summarized in Section 3. In Section 4, the discussion and prospects are presented.

2. Precursor Models

Various research studies have shown that a gamma-ray precursor event can be produced prior to the main GRB event. Here, we divide the precursor models into two categories based on their relative time to the merger: pre-merger models and post-merger models. More specifically, in the pre-merger phase, magnetospheric interaction in the NS binary and the crustal failure triggered by tidal interactions could lead to precursor emissions. While during the post-merger phase, it is suggested that the photospheric emission from the fireball and the SBO can also result in precursors.

We summarize the luminosity, spectrum, duration, and opening angle of these precursor models below, which relate to their detectability. To make sure the precursor is detectable at an extra-galactic distance, its luminosity should satisfy $L > 4\pi D^2 S$, where S is the sensitivity of the detector, and D is the distance. Recently, researchers have searched for SGRBs in the local Universe in the *Swift* catalog and found that the four closest SGRBs could locate at $D \sim 100$–200 Mpc [67]. Thus we adopt $D > 100$ Mpc. For sensitivity, we use the gravitational wave high-energy electromagnetic counterpart all-sky monitor (GECAM) as an example, which has $S \approx 2 \times 10^{-8}$ erg cm^{-2} s^{-1} in $8 - 2000$ keV [68]. The corresponding lower limit on luminosity is then

$$L \gtrsim 2.4 \times 10^{46} \text{ erg s}^{-1}. \qquad (1)$$

2.1. Pre-Merger Models

2.1.1. Magnetospheric Interactions of NS–NS/BH Binaries

It has long been suggested that the magnetospheric interactions in compact star binaries can lead to energy dissipation (e.g., [55]). Following [62,69], three cases are considered in this review as shown in Figure 1: case 0 with $B_c < \mu_* a^{-3}$, case 1 with $\mu_* \sim -\mu_c$, and case 2 with $\mu_* \sim \mu_c$, where the subscripts ($*$, c) represent the NS and its companion, respectively. The magnetic dipole moment is $\mu = BR^3$, where B is the surface magnetic field and R is the radius. We consider the binary to be of a separation a, a mass ratio $q = M_c/M_*$, and an orbital angular velocity $\Omega = [GM_*(1+q)/a^3]^{1/2}$. Within this framework, the energy dissipation rate of the NS binary system can be well formulated.

Case 0 can be well understood within the unipolar induction directcurrent (DC) circuit model, i.e., the weakly magnetized NS or non-magnetic BH is moving across the magnetic field lines inside the magnetosphere of the NS. This generates an electromotive force (EMF) $\mathcal{E} \simeq 2R_c|\mathbf{E}|$ on the two poles of the companion, where $\mathbf{E} = \mathbf{v} \times \mathbf{B}_c/c$, $\mathbf{v} = (\mathbf{\Omega} - \mathbf{\Omega}_*) \times a$, and $\mathbf{\Omega}_*$ is the spin of the NS. This EMF can drive currents along the magnetic field lines, which makes a closed DC circuit. Note this DC circuit may not always be stable [56]. The resistance of the magnetosphere is $\mathcal{R} = 4\pi/c$ [70]. The luminosity can then be estimated as [56],

$$L_{\rm UI} \approx 1.2 \times 10^{42} M_*(1+q)\mu_{*,30}^2 (R_c/10\,{\rm km})^2 (a/30\,{\rm km})^{-7}\,{\rm erg/s}, \quad (2)$$

where the mass M_* is in units of solar mass $M_\odot$. Note that, here and below, we adopt the approximation of $\Omega \gg \Omega_*$, this is appropriate, as we are considering the last few seconds before the merger. Simulations of inspiraling NS–NS/BH binaries indicate that the main features are well captured by this model (e.g., [58,59,61,63,71,72]).

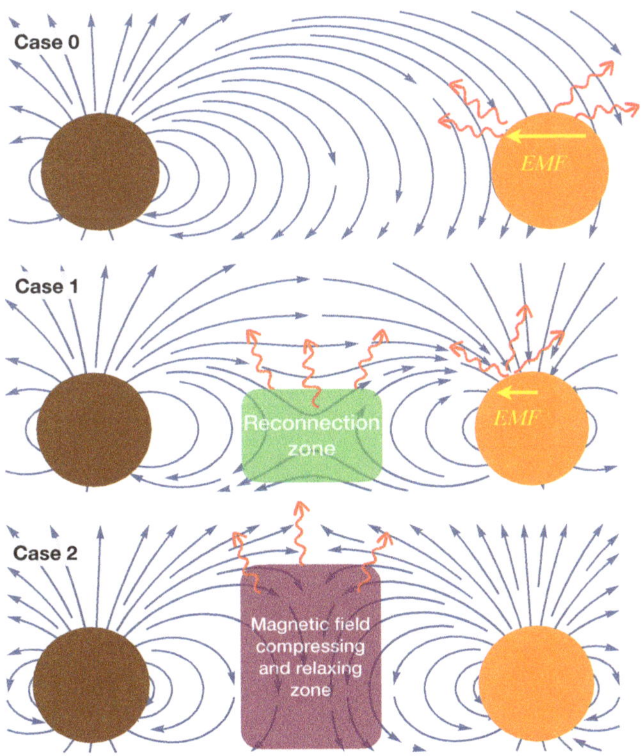

Figure 1. We show the schematics of three typical magnetic field configurations in inspiraling NS binaries: case 0 with $B_c < \mu_* a^{-3}$, case 1 with $\mu_* \sim -\mu_c$, and case 2 with $\mu_* \sim \mu_c$. The red winding arrows represent the emitted photons.

The other cases are more complicated. During the shrinking of the orbit, the magnetospheres of NSs would interact with each other, dissipating the orbital kinetic energy. The location of the interaction is around $r_i = a/(1+\epsilon^{1/3})$, where $\mu_* r_i^{-3} = \mu_c(a - r_i)^{-3}$

and $\epsilon = \mu_c/\mu_*$. If the magnetic field lines from two stars are anti-parallel with each other (Case 1), magnetic reconnection is expected. After the reconnection, the magnetic field lines would connect the two stars directly, leading to the formation of a DC circuit, such as in Case 0 driven by the EMF with an electric field $E \approx a\Omega B_c c^{-1}$. We find that the energy dissipation rate from the DC circuit is generally larger than that in the magnetic reconnection in the last few seconds before the merger, and it can be formulated as [62],

$$L_{a,\mathrm{UI}} \approx 3.8 \times 10^{44} (R_c/10\,\mathrm{km})^{-3} (\epsilon^{5/3} + \epsilon^2) \mu_{*,30}^2 (a/30\,\mathrm{km})^{-2}\,\mathrm{erg/s}\,. \quad (3)$$

If the magnetic field lines are parallel with each other (Case 2), the field lines would experience compressing at r_i, and the compression location would rotate around the main star at an angular speed $\Omega - \Omega_*$. When the compression location moves away, the compressed field lines will relax. This alternate compression and relaxation would lead to an electric field $E \sim \mu_* r_i^{-2} \Omega c^{-1}$ and an energy dissipation rate [62],

$$L_p \approx 1.8 \times 10^{43} (0.19/\eta - 0.08)(1 + \epsilon^{1/3})^3 (1 + \epsilon) \mu_{*,30}^2 (a/30\,\mathrm{km})^{-9/2}\,\mathrm{erg/s}\,, \quad (4)$$

where ηr_i is the width of the compression region.

It can be found that the energy dissipation rate increases non-linearly with time. Comparing with Equation (1), we found that it would be detectable only in the last few milliseconds to seconds depending on the magnetic field and distance to the observer. In general, Case 1 would have the highest energy dissipation rate, while Case 0 would have the lowest. We noticed that, for real cases, the magnetic axis may have an inclination angle with respect to the orbital axis, and in these cases, the energy dissipation rate would lie in between the above scenarios. As the poloidal field is the dominant component, we ignored the contribution of the toroidal magnetic field, which are caused by the revolution of the binary system and are observed in many simulations (e.g., [58,59,61,63,71,72]).

The opening angle of the radiation depends on the actual magnetic configuration and the orbital phase. In all cases, the generated acceleration electric field is not parallel to the B field. The radiation process would then be dominated by the synchrotron radiation [62]. Based on the threshold (Equation (1)), the magnetic field should be $B > 10^{13}$ G. While in such a high magnetic field, the high-energy photon will be absorbed, leading to a synchrotron-pair cascade. Using Monte-Carlo simulations, we find the spectral energy distribution (SED) can be well described by a cutoff-power law, with a photon index around $-2/3$ and peak energy at <MeV [62,73]. This could be understood as the synchrotron radiation by the mildly relativistic electrons with $\gamma \lesssim 10$, as high-energy photons emitted by higher-energy electrons will be absorbed to produce pairs. Therefore, the radiation cone will be of half opening-angle $\sim 1/\gamma = \mathcal{O}(0.1)$, and this radiation cone is rotating with an angular speed at $\Omega - \Omega_*$. Note that the magnetospheric interaction can create more open field lines than the isolated NSs, we would expect the outer gap acceleration to operate at around r_i, and so the curvature radiation may dominate after the electrons/positrons, losing their perpendicular moment. Overall, in these cases, the radiation opening angle will be much larger than that of jetted GRBs [62]. This can also be seen from the Poynting flux direction from magneto-hydrodynamics simulations (e.g., [57,58]).

2.1.2. NS Crust Crack Model

During the inspiral of the NS binary, tidal interactions can distort the NS, inducing ellipticities. Once the ellipticity becomes large enough that after the crystalline structure of the NS crust cannot respond linearly, a crust failure may be induced [40,64]. It has been suggested that the crust breaking strain is around 0.1 [74], which corresponds to a critical ellipticity of $\epsilon_c \approx \delta R/R \approx 4 \times 10^{-6}$, where $R + \delta R$ is the elongated NS radius. This can be easily reached by the tidally-induced f-mode oscillation in the last seconds before the merger [65] (see also Figure 3 of Reference [75]). Recent works show that the g-mode can also lead to the breaking of the NS crust [66].

If the energy is dissipated into heat, the crust can be heated up to $T_c = E_{cc}/C \approx 27.2E_{cc,46}^{1/2}$ keV with $C \approx 10^{29}T_c$ erg/K [76] and $E_{cc} = E_{cc,46}10^{46}$ erg. The corresponding thermal luminosity from the crust surface with $R_* \approx 10^6$ cm is then [69],

$$L_{cc} \approx 4\pi R_*^2 a_S T_c^4 \sim 4.5 \times 10^{42} E_{cc,46}^2 \text{ erg s}^{-1}, \tag{5}$$

where a_S is the Stefan–Boltzmann constant. This is too faint to be observed at an extragalactic distance.

It has also been proposed that, if the NS is highly magnetized, i.e., being a magnetar with $B \gg 10^{13}$ G, the crust failure may trigger a violent reconstruction of magnetic fields, leading to a magnetar-giant-flare-like event (e.g., [40,66]). However, it should be noted that for magnetar giant flares, the crust failure is believed to be caused by the sudden rearrangement of magnetic field [77]. It is unclear whether a crust failure could lead to the amplification and rearrangement of magnetar magnetic field. Nevertheless, in this case, the luminosity may be estimated as

$$L_{\rm Flare} = E_{cc}/t_f = 10^{46} E_{cc,46}/t_f \text{ erg s}^{-1}, \tag{6}$$

where t_f is the duration of the flare. The SED and opening angle in this case would be similar to observed giant flares.

2.2. Post-Merger Models

Within the standard fireball model for GRBs, a photospheric blackbody precursor can be produced when the fireball changes from optically thick regime to optically thin (see Section 7.3.3 of Reference [29], and references therein). The luminosity is determined by the transition radius. However, it has been found that, to fit the observation, the fireball Lorentz factor should be $\sim$30, much smaller than the expected value (>100) [40,69]. Therefore, this is unlikely to be responsible for the precursors, and we mainly focus on the SBO model for post-merger models.

SBO Model

Both NS–NS and NS–BH mergers can launch relativistic jets. As the relativistic jet propagates through the sub-relativistic expanding merger ejecta, a high-pressure bubble would be generated, which engulfs the jet and affects its propagation. This will lead to the formation of a jet-cocoon system, which is a structured relativistic outflow with a wide-angle. A successful SGRB jet is expected to penetrate through this ejecta and produce gamma-rays by the internal dissipation processes within the jet. It has been proposed that when the shock driven by a mildly relativistic cocoon breaks out of the ejecta, gamma-ray emission would also be produced [53,54]. This process differs significantly from what occurs in ordinary SGRBs.

For a low-power and short-duration jet, it may not penetrate through the ejecta, and the jet is choked. Both successful jet and chocked jet can drive an at least mildly relativistic SBO from the expanding merger ejecta [53]. The SBO of the relativistic jet or the mildly relativistic cocoon from the sub-relativistic expanding merger ejecta could release a tiny fraction, e.g., $\sim 10^{-4}$, of the total kinetic energy of the outflow into gamma-ray.

Two key physical parameters, the final Lorentz factor of the breakout layer $\Gamma_{\rm SBO}$ and the radius of the SBO $R_{\rm SBO}$, determine the main properties of the observed emissions. The SBO occurring in a sub-relativistic expanding ejecta is very different from that occurring in a static stellar envelope. The shock velocity in the lab frame would determine the boost of the emission to the observer, and the shock velocity in the ejecta frame would determine the strength of the shock. The SBO from the breakout layer would have an optical depth $\tau \sim c/v'_{\rm sh}$, where $v'_{\rm sh}$ is the shock velocity seen in the ejecta frame. The shock quantities seen in the unshocked merger ejecta frame are marked with a prime. The mass of the breakout layer can be estimated to be

$$M_{\rm SBO} \sim \frac{4\pi R_{\rm SBO}^2}{\kappa \beta'_{\rm SBO}} = 4 \times 10^{-8} M_{\odot} \beta'^{-1}_{\rm SBO} \left(\frac{R_{\rm SBO}}{10^{12}\,{\rm cm}}\right)^2 \left(\frac{\kappa}{0.16\,{\rm cm}^2\,{\rm g}^{-1}}\right)^{-1}, \tag{7}$$

where the opacity $\kappa = 0.16$ cm^2 g^{-1} is expected for fully ionized heavy elements, $\beta'_{\rm SBO} = v'_{\rm sh}/c$.

If the shock is fast enough (e.g., $v'_{\rm sh} > 0.5c$), the radiation temperature behind the shock reaches $\sim$50 keV at which pair production becomes important [49]. In this case, the mean photon energy is in the γ-ray regime. The observed energy from the breakout layer $E_{\rm SBO}$ can be roughly estimated by the internal energy of the shocked breakout layer and boosted to the observer frame as [32,50]

$$E_{\rm SBO} \sim M_{\rm SBO} c^2 \Gamma_{\rm SBO} (\Gamma'_{\rm SBO} - 1) \simeq 7 \times 10^{46}\,{\rm erg} \left(\frac{R_{\rm SBO}}{10^{12}\,{\rm cm}}\right)^2 \frac{\Gamma_{\rm SBO}(\Gamma'_{\rm SBO} - 1)}{\beta'_{\rm SBO}}. \tag{8}$$

In the case of a spherical breakout, the difference between the light travel time of photons emitted along the line of sight determines the duration of the breakout signal

$$\tau_{\rm SBO} \sim \frac{R_{\rm SBO}}{2c\Gamma_{\rm SBO}^2} = 0.67\,{\rm s} \left(\frac{R_{\rm SBO}}{10^{12}\,{\rm cm}}\right) \left(\frac{\Gamma_{\rm SBO}}{5}\right)^{-2}. \tag{9}$$

The bolometric luminosity of an SBO could then be roughly estimated as [69]

$$L_{\rm SBO} \sim \frac{E_{\rm SBO}}{t_{\rm SBO}} = \zeta E_{\rm K,iso} t_{\rm SBO}^{-1}. \tag{10}$$

where $E_{\rm K,iso}$ is the total kinetic energy of the outflow, and ζ is the fraction of the total explosion energy emitted in γ-rays.

In the framework of the SBO scenario, three SBO parameters: the breakout radius $R_{\rm SBO}$, the ejecta Lorentz factor $\gamma_{\rm ej,SBO}$, and the shock Lorentz factor $\gamma_{\rm SBO}$, are related with three main observables: the total observed isotropic equivalent energy $E_{\rm SBO}$, the duration $\tau_{\rm SBO}$, and the breakout temperature $T_{\rm SBO}$. The SBO temperature $T_{\rm SBO}$ is roughly the immediate downstream temperature of the breakout layer, as observed in the observer frame. The rest-frame temperature at the time that the photons are released is $\sim$50 keV, the observed temperature of SBO can be estimated as

$$T_{\rm SBO} \sim 50 \Gamma_{\rm SBO}\,{\rm keV}. \tag{11}$$

The three breakout observed quantities, $\tau_{\rm SBO}$, $E_{\rm SBO}$, and $T_{\rm SBO}$, satisfy a closure relation [32,49]

$$\tau_{\rm SBO} \sim 20\,{\rm s} \left(\frac{E_{\rm SBO}}{10^{46}\,{\rm erg}}\right)^{1/2} \left(\frac{T_{\rm SBO}}{50\,{\rm keV}}\right)^{-\frac{9+\sqrt{3}}{4}}, \tag{12}$$

This closure relation can be used to see if the detected γ-ray flash is consistent with a relativistic SBO origin. It is worth noting that this relation is strongly dependent on the breakout temperature $T_{\rm SBO}$, which is difficult to determine precisely because the SBO spectrum could deviate from a blackbody spectrum.

There are three generic properties of a relativistic SBO from a moving ejecta: (1) the light curve is smooth; (2) only a tiny fraction of the total energy would be emitted at the SBO stage; (3) the spectrum shows a hard to soft evolution [49,53]. Thus for precursors produced by SBO, the observed energies could be orders of magnitude lower, but depend on the viewing angle of the jet. Note, interestingly, all of these properties are observed in GRB 170817A. Therefore, a mildly relativistic cocoon shock breaking out from the merger ejecta provides a natural explanation of the observational properties of GRB 170817A.

3. Observational Results

Searching SGRB precursors has been performed in various space observatories, such as INTEGRAL [78], *Swift* [40,79], and *Fermi*/GBM [69,79]. Detecting weak signals before SGRBs will be subject to instrumental biases (the energy range and sensitivity). For example, although *Fermi*/GBM covers a broad energy range ($\sim$8 keV–40 MeV), *Swift*/BAT has a higher sensitive in the 15–150 keV band. Thus *Swift* will be stronger to detect soft weak precursors, consistent with observations.

In the pioneer research, precursors are searched by visual inspection in binned light curves with a certain fixed bin width (e.g., [33]). Later, wavelet analysis is introduced to analyze such binned light curves [34,40]. Recently, the Bayesian block algorithm [80] has been widely applied in both the binned light curve and the time-tagged event data [69,79,81]. Yet in the Bayesian block algorithm, the false alarm probability is adopted; thus, additional analysis is required to obtain the significance of the precursor [69,79]. By applying these methods, precursors are found in both long GRBs and SGRBs, and the detection rate of precursor is higher in long GRBs (e.g., [35,81]).

For SGRBs, the fraction of precursor events is less than 0.4% for INTEGRAL [78], and is $\sim$8–10% for *Swift*/BAT [40]. For the combined *Swift* and *Fermi*/GBM sample, the fraction is found to be 2.7% [79]. Applying the Bayesian block algorithm in the *Fermi*/GBM sample alone, we found a fraction of 3.0% SGRBs are associated with precursor activities [69], while only a fraction of 1.2% is found in Reference [81]. The major difference in the detection fraction may arise from the selection criteria for the precursors. The precursor events provided in Reference [69] are found of the signal-to-noise ratio (SNR) $\gtrsim 4.5\sigma$, where the SNR is obtained in the optimized energy range for the precursor. Thus, we adopt this precursor sample for *Fermi*/GBM in the following analysis. In Table 1, we list the SGRBs with precursors detected by *Fermi*/GBM [69] and/or *Swift*/BAT [40,79], where we show the duration of the precursor ($\tau_{\rm pre}$), waiting time ($\tau_{\rm wt}$), and the main SGRB ($\tau_{\rm GRB}$). For the *Fermi*/GBM sample [69], the duration is provided following the common definition of T_{90}, during which 90% of the total counts have been detected. However, for the *Swift*/BAT sample, the duration is directly provided by the wavelet analysis or Bayesian block analysis following References [40,79].

Precursors of SGRBs are usually too faint to perform spectral analysis. Therefore, the hardness ratio is often used to indicate the spectral properties [35,40,69,79]. Previous research found that there is no significant spectral difference between the precursors and the main GRBs for *Swift*/BAT events [35,40], while for *Fermi*/GBM events, a slight difference is found [79]. There could be two possibilities for precursor events having similar spectra to the main GRBs. On the observational side, this might be caused by the narrow bandpass of *Swift*/BAT and the lack of photon counts [40]. On the theoretical side, there is a possibility that the precursor and the main GRB are mimicked by two episodes of activities produced by collapsars with only the "tip-of-the-iceberg" of the light curve being observed, which makes them have similar spectral shapes [36,82–85]. However, the latter possibility is disfavored by the f-factor analysis [85] for most of the events in Table 1; thus, we focus here on the scenario that precursors have different origins from the main GRBs.

For the *Fermi*/GBM events, there are several events that have enough photon counts to do spectral analyses as shown in Table 1 of Reference [69]. The precursors of GRB111117A and GRB160804180 are found to be in favor of non-thermal spectra and can be well explained by the magnetospheric interaction model; the precursors of GRB081216 and GRB141102A favor thermal spectra and can be explained by the SBO model [69]. For the magnetospheric interaction model, the precursor duration relates to the chirp signal time of GW radiation. For the SBO model, this relates to the radius and Lorentz factor of the shock. Note that for GRB090510, there are two precursors, and the second one may be described by the thermal SBO model [69], while the first one could then originate from magnetospheric interactions.

The GRB duration is usually described as $\tau_{\rm GRB} \approx R_{\rm GRB}(1+z)/(2\Gamma^2 c)$, where $R_{\rm GRB}$ is the jet dissipation radius, Γ is the bulk Lorentz factor at $R_{\rm GRB}$, and z is the redshift of the source. For the magnetospheric interaction model, the waiting time consists of the jet launch time ($\Delta t_{\rm jet}$) and the jet propagation time ($\Delta t_{\rm GRB} \sim \tau_{\rm GRB}$). For the SBO model, the waiting time relates to the jet propagation from the SBO radius to the jet dissipation radius, i.e., $\tau_{\rm wt} = (R_{\rm GRB} - R_{\rm SBO})(1+z)/(2\Gamma^2 c)$. For the cases with $\Delta t_{\rm jet} \ll \tau_{\rm GRB}$ and $R_{\rm SBO} \ll R_{\rm GRB}$, we would expect $\tau_{\rm wt} \sim \tau_{\rm GRB}$. Yet there is an exception case for magnetospheric interaction model, in which the NS merger remnant is a stable NS (SNS) formed after the spin-down of the initially-formed uniform-rotation-supported supramassive NS with $\Delta t_{\rm jet} > \tau_{\rm GRB}$ [19,86]. Note that for the SBO model, one can constrain the ratio of the radii $R_{\rm SBO}/R_{\rm GRB} \approx 1 - \tau_{\rm wt}/\tau_{\rm GRB}$ from observations.

Previous results based on *Fermi*/GBM events have indicated that $\tau_{\rm wt} \sim \tau_{\rm GRB}$ can be generally satisfied [69]; here, in Figure 2, we show the updated $\tau_{\rm wt} - \tau_{\rm GRB}$ diagram, which includes both *Fermi*/GBM and *Swift*/BAT events. The fitting of the data (red line) shows that $\tau_{\rm wt} \approx 1.9\tau_{\rm GRB}$, largely consistent with theories and previous results [69]. However, note here that the fitting errors are not provided, as the errors of the data points are not available for the *Swift* events [40,79]. However, there are two outliers, GRB090510 (the first precursor event) and GRB191221802, with $\tau_{\rm wt} \gg \tau_{\rm GRB}$. This might suggest that SNSs are formed in these two events.

Table 1. The durations of the precursor ($\tau_{\rm pre}$), waiting time ($\tau_{\rm wt}$), and the main SGRB ($\tau_{\rm GRB}$) are taken from [69] for *Fermi*/GBM detected bursts, and from [40,79] for *Swift* detected bursts (marked with '+'). [a] For the events only detected by *Fermi*/GBM, their names are provided following the Fermi GBM Burst Catalog. The redshift is 0.287 for GRB060502B, and GRB090510 for 0.903.

Name [a]	$\tau_{\rm pre}$ (s)	$\tau_{\rm wt}$ (s)	$\tau_{\rm GRB}$ (s)
GRB060502B [+,a]	0.09	0.32	0.24
GRB071112B [+]	0.01	0.59	0.27
GRB080702A [+]	0.31	0.13	0.64
GRB100213A [+]	0.44	0.68	1.04
GRB081024A [+]	0.06	0.91	0.94
GRB081216	$0.15^{+0.05}_{-0.03}$	$0.53^{+0.04}_{-0.05}$	$0.24^{+0.02}_{-0.02}$
GRB090510 [+,a]	0.4	12.9	-
GRB090510 [a]	$0.05^{+0.07}_{-0.03}$	$0.52^{+0.04}_{-0.08}$	$0.30^{+0.01}_{-0.01}$
GRB100223110	$0.02^{+0.03}_{-0.01}$	$0.08^{+0.02}_{-0.03}$	$0.12^{+0.01}_{-0.01}$
GRB100827455	$0.11^{+0.05}_{-0.04}$	$0.34^{+0.06}_{-0.06}$	$0.09^{+0.02}_{-0.01}$
GRB101208498	$0.17^{+0.12}_{-0.08}$	$1.17^{+0.10}_{-0.14}$	$1.03^{+0.03}_{-0.04}$
GRB111117A	$0.18^{+0.05}_{-0.03}$	$0.22^{+0.03}_{-0.06}$	$0.09^{+0.01}_{-0.01}$
GRB140209A	$0.61^{+0.08}_{-0.08}$	$1.10^{+0.08}_{-0.08}$	$1.03^{+0.04}_{-0.06}$
GRB141102A	$0.06^{+0.10}_{-0.06}$	$1.26^{+0.11}_{-0.15}$	$0.48^{+0.04}_{-0.04}$
GRB150604434	$0.17^{+0.25}_{-0.01}$	$0.64^{+0.02}_{-0.29}$	$0.21^{+0.03}_{-0.02}$
GRB150922A	$0.05^{+0.01}_{-0.01}$	$0.03^{+0.01}_{-0.01}$	$0.08^{+0.01}_{-0.01}$
GRB160804180	$0.16^{+0.02}_{-0.02}$	$0.17^{+0.02}_{-0.02}$	$0.26^{+0.02}_{-0.02}$
GRB170709334	$0.46^{+0.01}_{-0.27}$	$0.17^{+0.30}_{-0.07}$	$0.15^{+0.07}_{-0.04}$
GRB170802638	$0.15^{+0.17}_{-0.11}$	$1.85^{+0.14}_{-0.21}$	$0.33^{+0.04}_{-0.04}$
GRB180511437	$2.80^{+1.38}_{-1.69}$	$12.72^{+1.80}_{-1.57}$	$3.33^{+0.18}_{-0.24}$
GRB181126A	$0.72^{+0.18}_{-0.27}$	$0.85^{+0.40}_{-0.29}$	$0.46^{+0.11}_{-0.13}$
GRB191221802	$0.03^{+0.59}_{-0.03}$	$19.36^{+1.24}_{-3.19}$	$0.37^{+0.26}_{-0.13}$

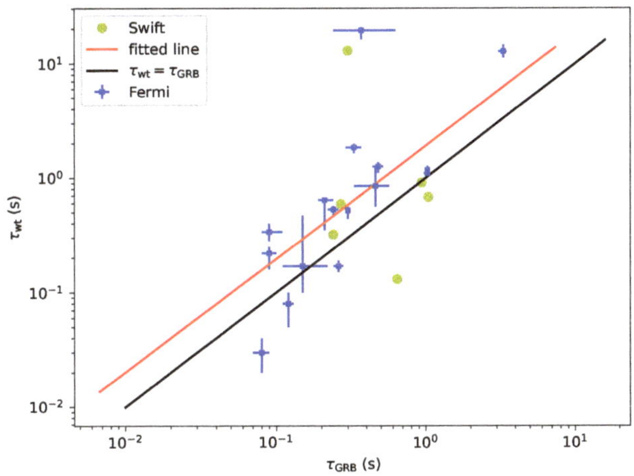

Figure 2. The waiting time and GRB duration are taken from [69] for *Fermi* detected bursts, and from [40,79] for *Swift* detected bursts. The black is the $\tau_{wt} = \tau_{GRB}$ line, and the red line is the fitting of the data.

4. Discussion and Prospects

Precursors have been detected for a small fraction of SGRBs. Here, we briefly review the models for precursors, mainly focusing on the magnetospheric interaction model and the SBO model, while the crust crack model and the fireball photospheric radiation model are found to be less likely based on current observations [69]. We focused on the explanation of the major physical processes in these models. To directly compare with observations, we estimated the luminosity, spectrum, duration, and opening angle for these models.

For the magnetospheric interaction model, the precursor will be produced simultaneously with GWs. A cutoff-power-law spectrum is expected with a photon index $\sim -2/3$ and a cutoff at MeV. Moreover, fast radio bursts (FRBs) are suggested to be produced during the magnetospheric interaction (e.g., [60,63]). It should be noted that for NS–BH binaries with mass ratio <0.2, the NS would be swallowed by the BH without producing a GRB and, thus, only the precursor is available.

While for the SBO model, the precursor is produced after the merger, but before the main GRB. Although GRB 170817A was classified as an SGRB, with a duration $\approx$0.5 s, it was fainter than the faintest SGRB previously detected by roughly three orders of magnitude, with the isotropic equivalent energy of $E_{\gamma,\rm iso} = 3 \times 10^{46}$ erg. The delay time between the GW signal and the γ-rays, $\tau_{{\rm GW},\gamma} = 1.7$ s. The peak energy of the observed integrated spectrum is $E_p = 185 \pm 62$ eV [2,3]. The breakout layer parameter that could produce the observables of GRB 170817A are $R_{\rm SBO} \approx 6 \times 10^{11}$ cm and $\Gamma_{\rm SBO} \approx 4$. SBO emissions from the jet-cocoon system seems to provide a natural explanation for this observed event because of the low radiation efficiency and the wide emission angle. However, the event rate for cocoon-SBO-induced GRBs should be very small, considering that most GRBs are observed at cosmological distances [87].

Compared with the main GRBs, we found that precursors are usually much weaker, but with a larger opening angle. Thus, for the NS mergers that occurred within several hundred Mpc, the detection of precursors is very likely. This will greatly benefit the search for gamma-ray counterparts of GW events and FRBs, which can be well tested by the current and near-future observatories, e.g., *Fermi*/GBM and *Swift*/BAT, GECAM [68], and the space-based multi-band astronomical variable objects monitor (SVOM) [88]. Furthermore, the time delay between precursors and GRBs or GW can be used to constrain the jet launch mechanism and post-merger remnant [19,69]. For the magnetospheric interaction model, photon splitting could be important, and it might significantly change the polarization state of emitted photons [89,90]. This can also be tested by the future gamma-ray polarimeter detector POLAR-2 [91].

Funding: J.W. acknowledges the sponsorship of the Alexander von Humboldt Foundation.

Data Availability Statement: No new data were created or analyzed in this study.

Acknowledgments: The authors thank the reviewers for the valuable comments and suggestions.

Conflicts of Interest: The authors declare no conflict of interest.

References

1. Abbott, B.P.; Abbott, R.; Abbott, T.D.; Acernese, F.; Ackley, K.; Adams, C.; Adams, T.; Addesso, P.; Adhikari, R.X.; Adya, V.B.; et al. Multi-messenger Observations of a Binary Neutron Star Merger. *Astrophys. J. Lett.* **2017**, *848*, L12. [CrossRef]
2. Goldstein, A.; Veres, P.; Burns, E.; Briggs, M.S.; Hamburg, R.; Kocevski, D.; Wilson-Hodge, C.A.; Preece, R.D.; Poolakkil, S.; Roberts, O.J.; et al. An Ordinary Short Gamma-Ray Burst with Extraordinary Implications: Fermi-GBM Detection of GRB 170817A. *Astrophys. J. Lett.* **2017**, *848*, L14. [CrossRef]
3. Zhang, B.B.; Zhang, B.; Sun, H.; Lei, W.H.; Gao, H.; Li, Y.; Shao, L.; Zhao, Y.; Hu, Y.D.; Lü, H.J.; et al. A peculiar low-luminosity short gamma-ray burst from a double neutron star merger progenitor. *Nat. Commun.* **2018**, *9*, 447. [CrossRef] [PubMed]
4. Abbott, B.P.; Abbott, R.; Abbott, T.D.; Acernese, F.; Ackley, K.; Adams, C.; Adams, T.; Addesso, P.; Adhikari, R.X.; Adya, V.B.; et al. GW170817: Observation of Gravitational Waves from a Binary Neutron Star Inspiral. *Phys. Rev. Lett.* **2017**, *119*, 161101. [CrossRef] [PubMed]
5. Abbott, B.P.; Abbott, R.; Abbott, T.D.; Acernese, F.; Ackley, K.; Adams, C.; Adams, T.; Addesso, P.; Adhikari, R.X.; Adya, V.B.; et al. Gravitational Waves and Gamma-Rays from a Binary Neutron Star Merger: GW170817 and GRB 170817A. *Astrophys. J. Lett.* **2017**, *848*, L13. [CrossRef]
6. Coulter, D.A.; Foley, R.J.; Kilpatrick, C.D.; Drout, M.R.; Piro, A.L.; Shappee, B.J.; Siebert, M.R.; Simon, J.D.; Ulloa, N.; Kasen, D.; et al. Swope Supernova Survey 2017a (SSS17a), the Optical Counterpart to a Gravitational Wave Source. *arXiv* **2017**, arXiv:astro-ph.HE/1710.05452.
7. Soares-Santos, M.; Holz, D.E.; Annis, J.; Chornock, R.; Herner, K.; Berger, E.; Brout, D.; Chen, H.Y.; Kessler, R.; Sako, M.; et al. The Electromagnetic Counterpart of the Binary Neutron Star Merger LIGO/Virgo GW170817. I. Discovery of the Optical Counterpart Using the Dark Energy Camera. *Astrophys. J. Lett.* **2017**, *848*, L16. [CrossRef]
8. Valenti, S.; Sand, D.J.; Yang, S.; Cappellaro, E.; Tartaglia, L.; Corsi, A.; Jha, S.W.; Reichart, D.E.; Haislip, J.; Kouprianov, V. The Discovery of the Electromagnetic Counterpart of GW170817: Kilonova AT 2017gfo/DLT17ck. *Astrophys. J. Lett.* **2017**, *848*, L24. [CrossRef]
9. Arcavi, I.; Hosseinzadeh, G.; Howell, D.A.; McCully, C.; Poznanski, D.; Kasen, D.; Barnes, J.; Zaltzman, M.; Vasylyev, S.; Maoz, D.; et al. Optical emission from a kilonova following a gravitational-wave-detected neutron-star merger. *Nature* **2017**, *551*, 64–66. [CrossRef]
10. Tanvir, N.R.; Levan, A.J.; González-Fernández, C.; Korobkin, O.; Mandel, I.; Rosswog, S.; Hjorth, J.; D'Avanzo, P.; Fruchter, A.S.; Fryer, C.L.; et al. The Emergence of a Lanthanide-rich Kilonova Following the Merger of Two Neutron Stars. *Astrophys. J. Lett.* **2017**, *848*, L27. [CrossRef]
11. Lipunov, V.M.; Gorbovskoy, E.; Kornilov, V.G.; Tyurina, N.; Balanutsa, P.; Kuznetsov, A.; Vlasenko, D.; Kuvshinov, D.; Gorbunov, I.; Buckley, D.A.H.; et al. MASTER Optical Detection of the First LIGO/Virgo Neutron Star Binary Merger GW170817. *Astrophys. J. Lett.* **2017**, *850*, L1. [CrossRef]
12. Abbott, B.P.; Abbott, R.; Abbott, T.D.; Acernese, F.; Ackley, K.; Adams, C.; Adams, T.; Addesso, P.; Adhikari, R.X.; Adya, V.B.; et al. A gravitational-wave standard siren measurement of the Hubble constant. *Nature* **2017**, *551*, 85–88. [CrossRef]
13. Wei, J.J.; Zhang, B.B.; Wu, X.F.; Gao, H.; Mészáros, P.; Zhang, B.; Dai, Z.G.; Zhang, S.N.; Zhu, Z.H. Multimessenger tests of the weak equivalence principle from GW170817 and its electromagnetic counterparts. *J. Cosmol. Astropart. Phys.* **2017**, *2017*, 035. [CrossRef]
14. Troja, E.; Piro, L.; van Eerten, H.; Wollaeger, R.T.; Im, M.; Fox, O.D.; Butler, N.R.; Cenko, S.B.; Sakamoto, T.; Fryer, C.L.; et al. The X-ray counterpart to the gravitational-wave event GW170817. *Nature* **2017**, *551*, 71–74. [CrossRef]
15. Hallinan, G.; Corsi, A.; Mooley, K.P.; Hotokezaka, K.; Nakar, E.; Kasliwal, M.M.; Kaplan, D.L.; Frail, D.A.; Myers, S.T.; Murphy, T.; et al. A radio counterpart to a neutron star merger. *Science* **2017**, *358*, 1579–1583. [CrossRef]
16. Mooley, K.P.; Nakar, E.; Hotokezaka, K.; Hallinan, G.; Corsi, A.; Frail, D.A.; Horesh, A.; Murphy, T.; Lenc, E.; Kaplan, D.L.; et al. A mildly relativistic wide-angle outflow in the neutron-star merger event GW170817. *Nature* **2018**, *554*, 207–210. [CrossRef]
17. Mooley, K.P.; Deller, A.T.; Gottlieb, O.; Nakar, E.; Hallinan, G.; Bourke, S.; Frail, D.A.; Horesh, A.; Corsi, A.; Hotokezaka, K. Superluminal motion of a relativistic jet in the neutron-star merger GW170817. *Nature* **2018**, *561*, 355–359. [CrossRef] [PubMed]
18. Gill, R.; Nathanail, A.; Rezzolla, L. When Did the Remnant of GW170817 Collapse to a Black Hole? *Astrophys. J.* **2019**, *876*, 139. [CrossRef]
19. Zhang, B. The delay time of gravitational wave—Gamma-ray burst associations. *Front. Phys.* **2019**, *14*, 64402. [CrossRef]

20. Yang, Y.S.; Zhong, S.Q.; Zhang, B.B.; Wu, S.; Zhang, B.; Yang, Y.H.; Cao, Z.; Gao, H.; Zou, J.H.; Wang, J.S.; et al. Physical Implications of the Subthreshold GRB GBM-190816 and Its Associated Subthreshold Gravitational-wave Event. *Astrophys. J.* **2020**, *899*, 60. [CrossRef]
21. Troja, E.; van Eerten, H.; Ryan, G.; Ricci, R.; Burgess, J.M.; Wieringa, M.H.; Piro, L.; Cenko, S.B.; Sakamoto, T. A year in the life of GW 170817: The rise and fall of a structured jet from a binary neutron star merger. *Mon. Notices Royal Astron. Soc.* **2019**, *489*, 1919–1926. [CrossRef]
22. Ryan, G.; van Eerten, H.; Piro, L.; Troja, E. Gamma-Ray Burst Afterglows in the Multimessenger Era: Numerical Models and Closure Relations. *Astrophys. J.* **2020**, *896*, 166. [CrossRef]
23. Balasubramanian, A.; Corsi, A.; Mooley, K.P.; Brightman, M.; Hallinan, G.; Hotokezaka, K.; Kaplan, D.L.; Lazzati, D.; Murphy, E.J. Continued Radio Observations of GW170817 3.5 yr Post-merger. *Astrophys. J. Lett.* **2021**, *914*, L20. [CrossRef]
24. Goldstein, A.; Hamburg, R.; Wood, J.; Hui, C.M.; Cleveland, W.H.; Kocevski, D.; Littenberg, T.; Burns, E.; Dal Canton, T.; Veres, P.; et al. Updates to the Fermi GBM Targeted Sub-threshold Search in Preparation for the Third Observing Run of LIGO/Virgo. *arXiv* **2019**, arXiv:astro-ph.HE/1903.12597.
25. Abbott, B.P.; Abbott, R.; Abbott, T.D.; Abraham, S.; Acernese, F.; Ackley, K.; Adams, C.; Adhikari, R.X.; Adya, V.B.; Affeldt, C.; et al. GW190425: Observation of a Compact Binary Coalescence with Total Mass $\sim$3.4 $M_{\odot}$. *Astrophys. J. Lett.* **2020**, *892*, L3. [CrossRef]
26. von Kienlin, A.; Veres, P.; Roberts, O.J.; Hamburg, R.; Bissaldi, E.; Briggs, M.S.; Burns, E.; Goldstein, A.; Kocevski, D.; Preece, R.D.; et al. Fermi-GBM GRBs with Characteristics Similar to GRB 170817A. *Astrophys. J.* **2019**, *876*, 89. [CrossRef]
27. Berger, E. Short-Duration Gamma-Ray Bursts. *Annu. Rev. Astron. Astrophys.* **2014**, *52*, 43–105. [CrossRef]
28. Fernández, R.; Metzger, B.D. Electromagnetic Signatures of Neutron Star Mergers in the Advanced LIGO Era. *Annu. Rev. Nucl. Part. Sci.* **2016**, *66*, 23–45. [CrossRef]
29. Zhang, B. *The Physics of Gamma-Ray Bursts*; Cambridge University Press: Cambridge, UK, 2018. [CrossRef]
30. Metzger, B.D. Kilonovae. *Living Rev. Relativ.* **2019**, *23*, 1. [CrossRef] [PubMed]
31. Burns, E. Neutron star mergers and how to study them. *Living Rev. Relativ.* **2020**, *23*, 4. [CrossRef]
32. Nakar, E. The electromagnetic counterparts of compact binary mergers. *Phys. Rep.* **2020**, *886*, 1–84. [CrossRef]
33. Koshut, T.M.; Kouveliotou, C.; Paciesas, W.S.; van Paradijs, J.; Pendleton, G.N.; Briggs, M.S.; Fishman, G.J.; Meegan, C.A. Gamma-Ray Burst Precursor Activity as Observed with BATSE. *Astrophys. J.* **1995**, *452*, 145. [CrossRef]
34. Lazzati, D. Precursor activity in bright, long BATSE gamma-ray bursts. *Mon. Notices Royal Astron. Soc.* **2005**, *357*, 722–731. [CrossRef]
35. Burlon, D.; Ghirlanda, G.; Ghisellini, G.; Lazzati, D.; Nava, L.; Nardini, M.; Celotti, A. Precursors in Swift Gamma Ray Bursts with Redshift. *Astrophys. J. Lett.* **2008**, *685*, L19. [CrossRef]
36. Burlon, D.; Ghirlanda, G.; Ghisellini, G.; Greiner, J.; Celotti, A. Time resolved spectral behavior of bright BATSE precursors. *Astron. Astrophys.* **2009**, *505*, 569–575. [CrossRef]
37. Hu, Y.D.; Liang, E.W.; Xi, S.Q.; Peng, F.K.; Lu, R.J.; Lü, L.Z.; Zhang, B. Internal Energy Dissipation of Gamma-Ray Bursts Observed with Swift: Precursors, Prompt Gamma-Rays, Extended Emission, and Late X-ray Flares. *Astrophys. J.* **2014**, *789*, 145. [CrossRef]
38. Lan, L.; Lü, H.J.; Zhong, S.Q.; Zhang, H.M.; Rice, J.; Cheng, J.G.; Du, S.S.; Li, L.; Lin, J.; Lu, R.J.; et al. Characteristics of Two-episode Emission Patterns in Fermi Long Gamma-Ray Bursts. *Astrophys. J.* **2018**, *862*, 155. [CrossRef]
39. Zhang, B.B.; Zhang, B.; Castro-Tirado, A.J.; Dai, Z.G.; Tam, P.H.T.; Wang, X.Y.; Hu, Y.D.; Karpov, S.; Pozanenko, A.; Zhang, F.W.; et al. Transition from fireball to Poynting-flux-dominated outflow in the three-episode GRB 160625B. *Nat. Astron.* **2018**, *2*, 69–75. [CrossRef]
40. Troja, E.; Rosswog, S.; Gehrels, N. Precursors of Short Gamma-Ray Bursts. *Astrophys. J.* **2010**, *723*, 1711–1717. [CrossRef]
41. Paczynski, B. Gamma-Ray bursters at cosmological distances. *Astrophys. J. Lett.* **1986**, *308*, L43–L46. [CrossRef]
42. Mészáros, P.; Rees, M.J. Steep Slopes and Preferred Breaks in Gamma-Ray Burst Spectra: The Role of Photospheres and Comptonization. *Astrophys. J.* **2000**, *530*, 292–298. [CrossRef]
43. Daigne, F.; Mochkovitch, R. The expected thermal precursors of gamma-ray bursts in the internal shock model. *Mon. Notices Royal Astron. Soc.* **2002**, *336*, 1271–1280. [CrossRef]
44. Ruffini, R.; Aksenov, A.G.; Bernardini, M.G.; Bianco, C.L.; Caito, L.; Dainotti, M.G.; de Barros, G.; Guida, R.; Vereshchagin, G.V.; Xue, S.S. The canonical Gamma-ray Bursts and their "precursors". In Proceedings of the 2008 Nanjing Gamma-ray Burst Conference, Nanjing, China, 11 November 2008; Huang, Y.F., Dai, Z.G., Zhang, B., Eds.; American Institute of Physics Conference Series; Volume 1065, pp. 219–222. [CrossRef]
45. Ramirez-Ruiz, E.; MacFadyen, A.I.; Lazzati, D. Precursors and $e^{+/-}$ pair loading from erupting fireballs. *Mon. Notices Royal Astron. Soc.* **2002**, *331*, 197–202. [CrossRef]
46. Waxman, E.; Mészáros, P. Collapsar Uncorking and Jet Eruption in Gamma-Ray Bursts. *Astrophys. J.* **2003**, *584*, 390–398. [CrossRef]
47. Zhang, W.; Woosley, S.E.; MacFadyen, A.I. Relativistic Jets in Collapsars. *Astrophys. J.* **2003**, *586*, 356–371. [CrossRef]
48. Lazzati, D.; Begelman, M.C. Universal GRB Jets from Jet-Cocoon Interaction in Massive Stars. *Astrophys. J.* **2005**, *629*, 903–907. [CrossRef]
49. Nakar, E.; Sari, R. Relativistic Shock Breakouts—A Variety of Gamma-Ray Flares: From Low-luminosity Gamma-Ray Bursts to Type Ia Supernovae. *Astrophys. J.* **2012**, *747*, 88. [CrossRef]

50. Levinson, A.; Nakar, E. Physics of radiation mediated shocks and its applications to GRBs, supernovae, and neutron star mergers. *Phys. Rep.* **2020**, *866*, 1–46. [CrossRef]
51. Keren, S.; Levinson, A. Sub-photospheric, Radiation-mediated Shocks in Gamma-Ray Bursts: Multiple Shock Emission and the Band Spectrum. *Astrophys. J.* **2014**, *789*, 128. [CrossRef]
52. Ito, H.; Matsumoto, J.; Nagataki, S.; Warren, D.C.; Barkov, M.V.; Yonetoku, D. The photospheric origin of the Yonetoku relation in gamma-ray bursts. *Nat. Commun.* **2019**, *10*, 1504. [CrossRef]
53. Gottlieb, O.; Nakar, E.; Piran, T.; Hotokezaka, K. A cocoon shock breakout as the origin of the γ-ray emission in GW170817. *Mon. Notices Royal Astron. Soc.* **2018**, *479*, 588–600. [CrossRef]
54. Bromberg, O.; Tchekhovskoy, A.; Gottlieb, O.; Nakar, E.; Piran, T. The γ-rays that accompanied GW170817 and the observational signature of a magnetic jet breaking out of NS merger ejecta. *Mon. Notices Royal Astron. Soc.* **2018**, *475*, 2971–2977. [CrossRef]
55. Hansen, B.M.S.; Lyutikov, M. Radio and X-ray signatures of merging neutron stars. *Mon. Notices Royal Astron. Soc.* **2001**, *322*, 695–701. [CrossRef]
56. Lai, D. DC Circuit Powered by Orbital Motion: Magnetic Interactions in Compact Object Binaries and Exoplanetary Systems. *Astrophys. J. Lett.* **2012**, *757*, L3. [CrossRef]
57. Palenzuela, C.; Lehner, L.; Ponce, M.; Liebling, S.L.; Anderson, M.; Neilsen, D.; Motl, P. Electromagnetic and Gravitational Outputs from Binary-Neutron-Star Coalescence. *Phys. Rev. Lett.* **2013**, *111*, 061105. [CrossRef] [PubMed]
58. Palenzuela, C.; Lehner, L.; Liebling, S.L.; Ponce, M.; Anderson, M.; Neilsen, D.; Motl, P. Linking electromagnetic and gravitational radiation in coalescing binary neutron stars. *Phys. Rev. D* **2013**, *88*, 043011. [CrossRef]
59. Paschalidis, V.; Etienne, Z.B.; Shapiro, S.L. General-relativistic simulations of binary black hole-neutron stars: Precursor electromagnetic signals. *Phys. Rev. D* **2013**, *88*, 021504. [CrossRef]
60. Wang, J.S.; Yang, Y.P.; Wu, X.F.; Dai, Z.G.; Wang, F.Y. Fast Radio Bursts from the Inspiral of Double Neutron Stars. *Astrophys. J. Lett.* **2016**, *822*, L7. [CrossRef]
61. Paschalidis, V. General relativistic simulations of compact binary mergers as engines for short gamma-ray bursts. *Class. Quantum Gravity* **2017**, *34*, 084002. [CrossRef]
62. Wang, J.S.; Peng, F.K.; Wu, K.; Dai, Z.G. Pre-merger Electromagnetic Counterparts of Binary Compact Stars. *Astrophys. J.* **2018**, *868*, 19. [CrossRef]
63. Most, E.R.; Philippov, A.A. Electromagnetic Precursors to Gravitational-wave Events: Numerical Simulations of Flaring in Pre-merger Binary Neutron Star Magnetospheres. *Astrophys. J. Lett.* **2020**, *893*, L6. [CrossRef]
64. Tsang, D.; Read, J.S.; Hinderer, T.; Piro, A.L.; Bondarescu, R. Resonant Shattering of Neutron Star Crusts. *Phys. Rev. Lett.* **2012**, *108*, 011102. [CrossRef] [PubMed]
65. Suvorov, A.G.; Kokkotas, K.D. Precursor flares of short gamma-ray bursts from crust yielding due to tidal resonances in coalescing binaries of rotating, magnetized neutron stars. *Phys. Rev. D* **2020**, *101*, 083002. [CrossRef]
66. Kuan, H.J.; Suvorov, A.G.; Kokkotas, K.D. General-relativistic treatment of tidal g-mode resonances in coalescing binaries of neutron stars—II. As triggers for precursor flares of short gamma-ray bursts. *Mon. Notices Royal Astron. Soc.* **2021**, *508*, 1732–1744. [CrossRef]
67. Dichiara, S.; Troja, E.; O'Connor, B.; Marshall, F.E.; Beniamini, P.; Cannizzo, J.K.; Lien, A.Y.; Sakamoto, T. Short gamma-ray bursts within 200 Mpc. *Mon. Notices Royal Astron. Soc.* **2020**, *492*, 5011–5022. [CrossRef]
68. Zhang, D.; Li, X.; Xiong, S.; Li, Y.; Sun, X.; An, Z.; Xu, Y.; Zhu, Y.; Peng, W.; Wang, H.; et al. Energy response of GECAM gamma-ray detector based on $LaBr_3$:Ce and SiPM array. *Nucl. Instruments Methods Phys. Res. A* **2019**, *921*, 8–13. [CrossRef]
69. Wang, J.S.; Peng, Z.K.; Zou, J.H.; Zhang, B.B.; Zhang, B. Stringent Search for Precursor Emission in Short GRBs from Fermi/GBM Data and Physical Implications. *Astrophys. J. Lett.* **2020**, *902*, L42. [CrossRef]
70. Mottez, F.; Heyvaerts, J. Magnetic coupling of planets and small bodies with a pulsar wind. *Astron. Astrophys.* **2011**, *532*, A21. [CrossRef]
71. Ponce, M.; Palenzuela, C.; Lehner, L.; Liebling, S.L. Interaction of misaligned magnetospheres in the coalescence of binary neutron stars. *Phys. Rev. D* **2014**, *90*, 044007. [CrossRef]
72. Carrasco, F.; Shibata, M.; Reula, O. Magnetospheres of black hole-neutron star binaries. *arXiv* **2021**, arXiv:2106.09081.
73. Wang, J.S.; Liu, R.Y.; Aharonian, F.; Dai, Z.G. Analytical treatment for the development of electromagnetic cascades in intense magnetic fields. *Phys. Rev. D* **2018**, *97*, 103016. [CrossRef]
74. Horowitz, C.J.; Kadau, K. Breaking Strain of Neutron Star Crust and Gravitational Waves. *Phys. Rev. Lett.* **2009**, *102*, 191102. [CrossRef] [PubMed]
75. Wang, J.S.; Lai, D. Evolution of inspiralling neutron star binaries: Effects of tidal interactions and orbital eccentricities. *Phys. Rev. D* **2020**, *102*, 083005. [CrossRef]
76. Yakovlev, D.G.; Levenfish, K.P.; Shibanov, Y.A. Cooling of neutron stars and superfluidity in their cores. *Phys. Usp.* **1999**, *42*, 737. [CrossRef]
77. Thompson, C.; Duncan, R.C. The Giant Flare of 1998 August 27 from SGR 1900+14. II. Radiative Mechanism and Physical Constraints on the Source. *Astrophys. J.* **2001**, *561*, 980–1005. [CrossRef]
78. Minaev, P.; Pozanenko, A.; Molkov, S. Precursors of short gamma-ray bursts detected by the INTEGRAL observatory. *Int. J. Mod. Phys. D* **2018**, *27*, 1844013. [CrossRef]

79. Zhong, S.Q.; Dai, Z.G.; Cheng, J.G.; Lan, L.; Zhang, H.M. Precursors in Short Gamma-Ray Bursts as a Possible Probe of Progenitors. *Astrophys. J.* **2019**, *884*, 25. [CrossRef]
80. Scargle, J.D.; Norris, J.P.; Jackson, B.; Chiang, J. Studies in Astronomical Time Series Analysis. VI. Bayesian Block Representations. *Astrophys. J.* **2013**, *764*, 167. [CrossRef]
81. Coppin, P.; de Vries, K.D.; van Eijndhoven, N. Identification of gamma-ray burst precursors in Fermi-GBM bursts. *Phys. Rev. D* **2020**, *102*, 103014. [CrossRef]
82. Zhang, B.; Zhang, B.B.; Virgili, F.J.; Liang, E.W.; Kann, D.A.; Wu, X.F.; Proga, D.; Lv, H.J.; Toma, K.; Mészáros, P.; et al. Discerning the Physical Origins of Cosmological Gamma-ray Bursts Based on Multiple Observational Criteria: The Cases of z = 6.7 GRB 080913, z = 8.2 GRB 090423, and Some Short/Hard GRBs. *Astrophys. J.* **2009**, *703*, 1696–1724. [CrossRef]
83. Virgili, F.J.; Zhang, B.; O'Brien, P.; Troja, E. Are All Short-hard Gamma-ray Bursts Produced from Mergers of Compact Stellar Objects? *Astrophys. J.* **2011**, *727*, 109. [CrossRef]
84. Bromberg, O.; Nakar, E.; Piran, T.; Sari, R. Short versus Long and Collapsars versus Non-collapsars: A Quantitative Classification of Gamma-Ray Bursts. *Astrophys. J.* **2013**, *764*, 179. [CrossRef]
85. Lü, H.J.; Zhang, B.; Liang, E.W.; Zhang, B.B.; Sakamoto, T. The 'amplitude' parameter of gamma-ray bursts and its implications for GRB classification. *Mon. Notices Royal Astron. Soc.* **2014**, *442*, 1922–1929. [CrossRef]
86. Metzger, B.D.; Giannios, D.; Thompson, T.A.; Bucciantini, N.; Quataert, E. The protomagnetar model for gamma-ray bursts. *Mon. Notices Royal Astron. Soc.* **2011**, *413*, 2031–2056. [CrossRef]
87. Matsumoto, T.; Piran, T. On short GRBs similar to GRB 170817A detected by Fermi-GBM. *Mon. Notices Royal Astron. Soc.* **2020**, *492*, 4283–4290. [CrossRef]
88. Yu, S.; Gonzalez, F.; Wei, J.; Zhang, S.; Cordier, B. SVOM: A Joint Gamma-ray Burst Detection Mission. *Chin. Astron. Astrophys.* **2020**, *44*, 269–282. [CrossRef]
89. Adler, S.L. Photon splitting and photon dispersion in a strong magnetic field. *Ann. Phys.* **1971**, *67*, 599–647. [CrossRef]
90. Baring, M.G.; Harding, A.K. Photon Splitting and Pair Creation in Highly Magnetized Pulsars. *Astrophys. J.* **2001**, *547*, 929–948. [CrossRef]
91. De Angelis, N.; Burgess, J.M.; Cadoux, F.; Greiner, J.; Hulsman, J.; Kole, M.; Li, H.C.; Mianowski, S.; Pollo, A.; Produit, N.; et al. Development and science perspectives of the POLAR-2 instrument: A large scale GRB polarimeter. *arXiv* **2021**, arXiv:2109.02978.

galaxies

MDPI

Article

Progenitors of Long-Duration Gamma-ray Bursts

Arpita Roy

Cosmology Research Group, Scuola Normale Superiore, Piazza dei Cavalieri 7, 56126 Pisa, Italy; arpita.roy@sns.it

Abstract: We review the current scenario of long-duration Gamma-ray burst (LGRB) progenitors, and in addition, present models of massive stars for a mass range of 10–150 $M_{\odot}$ with $\Delta M = 10\,M_{\odot}$ and rotation rate $v/v_{\rm crit} = 0$ to 0.6 with a velocity resolution $\Delta v/v_{\rm crit} = 0.1$. We further discuss possible metallicity and rotation rate distribution from our models that might be preferable for the creation of successful LGRB candidates given the observed LGRB rates and their metallicity evolution. In the current understanding, LGRBs are associated with Type-Ic supernovae (SNe). To establish LGRB-SN correlation, we discuss three observational paths: (i) space-time coincidence, (ii) evidence from photometric light curves of LGRB afterglows and SN Type-Ic, (iii) spectroscopic study of both LGRB afterglow and SN. Superluminous SNe are also believed to have the same origin as LGRBs. Therefore, we discuss constraints on the progenitor parameters that can possibly dissociate these two events from a theoretical perspective. We further discuss the scenario of single star versus binary star as a more probable pathway to create LGRBs. Given the limited parameter space in the mass, mass ratio and separation between the two components in a binary, binary channel is less likely to create LGRBs to match the observed LGRB rate. Despite effectively-single massive stars are fewer in number compared to interacting binaries, their chemically homogeneous evolution (CHE) might be the major channel for LGRB production.

Keywords: Gamma-ray bursts; massive stars; supernova; spectroscopy; binary stars; metallicity; rotation rate

Citation: Roy, A. Progenitors of Long-Duration Gamma-ray Bursts. *Galaxies* **2021**, *9*, 79. https://doi.org/10.3390/galaxies9040079

Academic Editors: Elena Moretti and Francesco Longo

Received: 16 July 2021
Accepted: 1 October 2021
Published: 19 October 2021

1. Introduction

Gamma-ray bursts (GRBs) are intense flashes of high energy electromagnetic radiation of ∼a few 100 keV with a very brief duration of 1 s to a few minutes reaching Earth isotropically from unpredictable directions. GRBs are observationally classified in two groups: short-duration hard (∼1 s, 350 keV) and long-duration soft (∼a few 10 s of seconds to ∼1 min, ∼200 keV). These two classes are assumed to have different physical origins. In general, a small amount of matter is required to be accelerated to ultra-relativistic speeds and beamed at a small solid angle to produce GRBs. After their discovery in 1973 [1], understanding the origin of GRBs has been of utmost importance to comprehend the cosmic evolution of the early universe. Given the association of GRBs with the death of massive stars and them being observable in high redshift, in principle as high as $z \sim 20$ [2], GRBs are often considered to probe the star formation histories over cosmic time [3].

Massive stars ($\gtrsim$10 $M_{\odot}$) enter the Wolf-Rayet (WR) phase towards the end of or after the main-sequence [4]. Depending on the spectroscopic identification of heavy elements, such as helium, nitrogen, carbon, oxygen [4,5], and their excitation states, WR stars are primarily classified in three categories: WN, WC, and WO (see Table 4 of [5]). Theoretically, WN and WC stars are believed to be progenitors of Type Ib and Ic core-collapse supernovae (SNe), respectively, given their pre-SN He-/CO- core masses and the absence of surface H and H/He [6]. However, there is no direct evidence that suggests single WR stars as Type Ib/Ic SNe progenitors. Given the WR lifetimes of a few 10^5 years, one would need to observe $\sim 10^4$ single WR stars to draw a firm connection between them and core-collapse SNe (CCSNe), in a timescale of a few 10s of years. This considerably high number means these are not field stars; rather they are in clusters. Given that more than 70% of massive stars are in binaries, the lower mass interacting binaries might be alternative progenitors of SN Type Ib/Ic [7].

Energetically, long-duration GRB (LGRB) and CCSNe both fall under the same category. In many long-duration GRBs (LGRB), the beaming-corrected total Gamma-ray energy is estimated to be $\sim 10^{51}$ erg. In addition, the total kinetic energy of the core collapse SN is $\sim 10^{51}$ erg, comparable to that of GRB jets. This naturally makes one consider the possibility of finding the connection between these two extreme phenomena. To confirm this hypothesis, there are three primary approaches available. Firstly, one can find the causal connection—whether both SN and GRB are coincident in space and time. Secondly, one can study photometry to obtain any overlap in the SN and GRB afterglow light curves. Finally, one can study spectroscopy to establish the SN-GRB connection.

Another class of SNe, Type I superluminous supernovae (henceforth SLSNe), is also believed to have a similar origin as LGRBs. SLSNe are characterized by luminosities 10–100 times larger than "typical" SNe [8–10], and their ^{56}Ni mass of $\sim$20–30 $M_{\odot}$ is much higher compared to 1 $M_{\odot}$ for "typical" SNe ([11] and references therein). Their spectra show the absence of H and He, same as Type-Ic SN, and they are bare carbon and oxygen cores [12]. This indicates that SLSNe progenitors have gone through intense mass-loss and/or mixing of chemical elements that made their envelopes depleted of H and He. However, the nature of SLSNe progenitors is yet unknown. Although the most commonly believed theory for their progenitors, based on the observed properties of SLSNe, is the magnetar model—where the newly formed millisecond magnetar, i.e., rapidly rotating, highly magnetized neutron stars (NS), deposits continuous energy to the SN ejecta. LGRBs, on the other hand, are formed in the framework of collapsar model [13]—where the rapidly rotating stellar core of a massive star collapses into a black hole (BH). Determining the final fate of a massive star that forms a NS or a BH is a complex and poorly understood astrophysical problem. Several recent theoretical models invoke a few diagnostic parameters of the progenitors at the pre-SN or pre-collapse phase that determine the final fates [14–17]. For example, one such diagnostic parameter is the "so-called" core compactness parameter, ξ_M [14]. We discuss these criteria later in this paper.

This paper aims to review the possible progenitors of LGRBs, find the connection between CCSNe and GRB, and dissociate progenitors of LGRB and SLSN based on several theoretical constraints on the pre-SN cores. Finally, we present a set of models of massive stars for varying mass, rotation rate and metallicity to narrow down a range of values that are favoured by the observed LGRB rate and its metallicity evolution. This paper is organized as follows: in Section 2, we discuss the existing observations that associate SNe with LGRBs; in Section 3, we present the leading models for the progenitors of SLSNe and LGRBs; in Section 4, we illustrate whether single or binary stars are more suitable candidates for LGRB progenitors; in Section 5, we describe the properties of WR stars that are required to form LGRB progenitors, and in Section 6, we summarise the salient points of this review article and pose the open questions that will be the focus of research for LGRB astrophysics in the coming years.

2. SN-GRB Connection

Space-time coincidence: Despite the striking similarity in kinetic energy of SNe and Gamma-ray in GRBs, astronomers did not consider any relation between them [18,19] for decades, only because it was difficult to find out the exact location and thus luminosity of GRBs until the late 1990s. After the confirmation of cosmological length scales and the localization of long-wavelength counterparts, it became more evident that GRBs are associated with young star-forming regions of distant galaxies [20,21], rather than being part of the old galaxies as was previously hypothesized by merger theories ([6] and references therein). The strong connection of SN-GRB was first established after the discovery of GRB980425 in conjunction with the most unusually bright SN 1998bw [22], both SN and GRB were coincident in space and time. Another SN 2003dh was also found to be correlated with GRB 030329. Unfortunately, it is difficult to associate each GRB with active star-forming regions in high-redshiftgiven the current instrumental limitations of not resolving $\lesssim$100 pc at $\sim$100 Mpc, although the statistical studies reveal a strong correlation between GRBs and blue active star-forming regions of galaxies ([6] and references therein).

Evidence from photometric light curves: Observation of GRB 980326 [23] at redshift unity showed red-emission bump in the optical afterglow [24,25]. This bump in optical was hypothesized to be caused by a consecutive SN followed by the GRB event. The data of red-emission were also consistent with dust re-radiation [26] from the surrounding material of GRB980326, which again supports the hypothesis of a consecutive SN event. Also, a reanalysis of the GRB970228 afterglow showed the signature of "bump" rising at a similar time as GRB 980326 [27–29]. It was difficult to confirm the absolute magnitude of the peak and the type of SN without the spectroscopic redshift of GRB and multi-band photometry of the bump. Future multi-epoch ground- and space- based observations of several GRBs confirmed that the red-emission bump is indeed associated with SNe [30,31] followed by a GRB event.

The brightness of the GRB 980326 "bump" matched with typical Type Ic supernova SN 1998bw. Most of the long-soft GRBs are accompanied by Type Ic SNe [6]. Studies of these SNe show high velocity ejecta causing broad emission lines, and therefore these SNe are classified as "Type Ic-BL" [1]. However, there are a few exceptions. For example, GRB 060614 is a LGRB with a duration of 102 s, but it has no SN counterpart [32–34]. Instead, the re-analysis of its optical afterglow was identified as "so-called" kilonova emission associated with the compact object merger origin [35,36]. Despite the few exceptions, in most cases, LGRBs are found to be associated with Type Ic-BL SNe. Nonetheless, even though most LGRBs are associated with Type Ic SNe, not all Type Ic SNe are correlated with LGRBs. The reason for different stars to follow different paths is rotation, mass, and metallicity. GRBs are produced by rapidly rotating massive stars that end up with sufficiently rotating pre-SN cores that can create an accretion disk while collapsing into BHs. To retain enough rotation until the pre-SN phase, these stars must not have strong mass loss, and that can easily be achieved at low-metallicity environments. Therefore, metal-poor stars are favoured for the creation of LGRBs. On the other hand, $\sim$66% core-collapse SNe originate from massive stars irrespective of their rotation rates and therefore can happen at both low and high metallicity [6].

Spectroscopy: GRB host galaxies have higher star-formation rate $\sim$a few 10 s of $M_\odot$ yr^{-1}, larger than typical field galaxies, as determined by sub-mm observations [37] and by [Ne III] to [OII] line ratio [38]. At high redshift (z), GRBs track the global star-formation rate [39–41]. The observations of GRBs on a large scale (high-z) confirms predictions of star-formation on small scales. Hence, GRBs, in general, are considered to be good tracers of active star-forming regions. Moreover, the absorption-line spectroscopies give the metallicity estimates of the GRB host galaxies, or in general, the regions through which GRB afterglows are viewed.

Even though there were several GRBs observed whose "bumps" showed characteristics of Type Ic SN, the solid SN-GRB connection was made after the discovery of the low redshift, $z \sim 0.169$, GRB 030329 [42] and the accompanied SN 2003dh. Detailed spectroscopy of this GRB afterglow [43,44] showed the deviation from a pure power-law and also a broad SN spectral feature. SN 2003dh spectroscopy was studied in detail as the afterglow faded, and it showed striking similarity with SN 1998bw. Broad spectral lines indicating high velocities $\sim$25,000 km s^{-1} were observed along with the absence of hydrogen, helium and Si II 6355 absorption lines confirming this SN as Type Ic-BL. There were a few other spectroscopic SN-GRB association: for example, at redshift $z = 0.1055$ GRB 031203 [45] accompanied with SN 2003lw [46], SN associated with GRB021211 at $z = 1.006$ [47], SN with Swift burst GRB050525a [48].

3. Models for LGRB and SLSNe

Several observations and theoretical predictions lead to similar origins for both LGRB and SLSNe, given the similarities in their environments, spectra, and energetics [11]. For LGRB, one needs a central engine that can drive a collimated relativistic jet that produces beamed emission [49–51] of energetic γ-rays. The jet typically has a power of $\sim 10^{50}$ erg s^{-1} within a narrow opening angle of 0.1 radian. Therefore, models should provide $\gtrsim 10^{52}$ ergs of energy in a wider angle of ~ 1 radian for LGRB to be accompanied with SN like SN 1998bw and SN 2003dh [52] along with the jet emission. This energy is $\gtrsim 10$ times the "typical" SN energy. The jet head generally travels at subrelativistic speed with power $\sim 3 \times 10^{48}$ erg s^{-1} inside the star, and it takes 8–25 s to reach the surface with power $\sim 3 \times 10^{50}$ erg s^{-1} [53]. If the jet is interrupted (there are several interruption scenarios, for example, [54]) or its direction changes in that timespan, then the flow will remain subrelativistic, and therefore will not make a LGRB. Hence, theoretical models suitable for LGRB production need to provide $\gtrsim 10^{50}$ erg s^{-1} of relativistic, beamed power for $\gtrsim 10$ s. Considering these constraints, the most acceptable model for LGRB production is the "collapsar" model [13][2], where a rapidly rotating stellar core collapses to a BH. The suitable progenitor for LGRB is the metal-poor, rapidly rotating either single massive star [6,57] that undergoes quasi chemically homogeneous evolution due to rotational mixing, or, a massive star in a closely interacting binary [58,59].

SLSNe progenitors are also not yet understood completely, there are several leading theories: (i) continuous energy injection to the SN ejecta by the spin-down of the newly-formed central millisecond magnetar [60–62], (ii) accretion of surrounding ejecta onto the central compact object [63], (iii) SN ejecta-circumstellar medium interaction [64], (iv) radioactive decay of large amount of ^{56}Ni (20–30 $M_\odot$) produced by pair-instability explosion in very massive stars [12]. Amongst them, the widely accepted theory for SLSNe is the "magnetar" model (for details, see [11] and references therein). In the spirit of the magnetar model, a large number of SLSNe light curves have been analyzed to obtain the distribution of ejecta mass, magnetar spin period and the strength of the magnetic field to reproduce the observables [62,65–67]. The ejecta masses of SLSNe are estimated to be 3.6–40 $M_\odot$ that is significantly different from Type Ib/Ic ejecta masses, i.e., strictly $>10\,M_\odot$ [11]. Estimated magnetar spin period and magnetic fields are 1 to 8 ms and 0.3 to 10×10^{14} G, respectively [11]. Ref. [11] found that even the SLSNe progenitors are rapidly rotating, metal-poor massive stars.

Therefore, it is difficult to theoretically predict which stars will explode as SLSNe with NS remnant and which ones will collapse as BH. There are five commonly used parameters to determine a star's explodability criteria. Stellar core compactness is one such parameter, as mentioned in Section 1, is defined as,

$$\xi_M = \frac{M/M_\odot}{R(M_{\rm baryon} = M)/1000\,{\rm km}}, \tag{1}$$

where $R(M_{\rm baryon} = M)$ is the radius where the progenitor's core baryon mass $M_{\rm baryon} = M$. ξ_M is the collapse indicator in non-rotating star. [15] found that ξ_M is well determined at the Lagrangian mass coordinate of 2.5 $M_\odot$ at the core collapse, where the infall velocity in the core reaches 1000 km/s. Thus ξ_M is typically denoted as $\xi_{2.5}$. Stellar cores with $\xi_{2.5} \lesssim 3.0$–4.5 explode as SLSNe (neutrino winds being the cause of the explosion) leaving behind NS and higher values $of \xi_{2.5}$ produce BHs [11,16,68]. The other four parameters [11,68] are:

$$M_4 = m(s = 4)/M_\odot, \tag{2}$$

m is the Lagrangian mass at specific entropy (in the units of k_B) $s = 4$;

$$\mu_4 = \left.\frac{dm/M_\odot}{dr/1000\,{\rm km}}\right|_{s=4}, \tag{3}$$

where dm is calculated at M_4, in practice, it is set as $dm = 0.3\,M_\odot$, and dr is the change in radius between M_4 and $M_4 + dm$; the dynamo-generated magnetic field strength averaged within the innermost $1.5\,M_\odot$,

$$\langle B_\phi \rangle = \frac{\int_0^{1.5\,M_\odot} B_\phi(m)dm}{\int_0^{1.5\,M_\odot} dm}; \tag{4}$$

the mass averaged specific angular momentum within the innermost mass M,

$$\bar{j}_M = \frac{\int_0^M j_M dm}{\int_0^M dm}. \tag{5}$$

Typical values of $\langle B_\phi \rangle$ are $\sim 10^{14}$–10^{15} G for NS and an order of magnitude higher for collapsars [11]. The average specific angular momentum are $\bar{j}_{1.5\,M_\odot} \sim 10^{15}$ (within the innermost $1.5\,M_\odot$) and $\bar{j}_{5\,M_\odot} \sim 10^{16}$ cm^2 s^{-1} (within the innermost $5\,M_\odot$) for NS and BH progenitors, respectively [11].

4. Single Stars versus Binary Stars as LGRB Progenitors

Having discussed the connection between LGRB and Type Ic BL SNe, and the physical models that differentiate the formation of SLSNe and LGRBs, in this section, we discuss scenarios where a single and/or a binary system is a likely progenitor of LGRB. For the LGRB association with Type-Ic SN, the star needs to lose the hydrogen and helium envelopes[3] and have a large production of ^{56}Ni. The mass loss of a massive star is dependent on its mass, metallicity and rotation rate. The higher the rotation rate, the larger the mass loss [4,5,69]. With mass loss, the angular momentum transported from the core to the surface eventually gets lost to the interstellar medium. This loss in angular momentum reduces the massive star's rotation rate. Although the star will still have to be sufficiently rotating even after the mass loss to produce the accretion disk around the central BH, which is required for the GRB production. Therefore, the right combination of metallicity and rotation rate needs to be satisfied to create LGRB progenitors. For a detailed discussion of the effect of both rotation rate and metallicity, see Section 5.

Single stars: Both single and binary channels are possible for LGRB production. For counting as a single star, it might be born "single" or can be a part of a wide binary with no/minimal interaction between the binary components. In the non-rotating massive stars, heavier elements produced by nucleosynthesis in the core are quasi- chemically homogeneously mixed within the inner convective regions. There are outer layers of lighter elements with gradual decline in atomic masses in the radiative outer zones. In most non-rotating stars, the surface abundance does not evolve much from its initial abundance during the MS [4]. In rotating stars, chemical mixing due to several rotational instabilities (see Section 5 for details) dredges up the elements from the inner convective core to the surface, crossing the radiative barrier. Therefore, a rapidly rotating massive star can be quasi- chemically homogeneously mixed [4–6,57,70] without a clear chemical boundary between the inner convective core and the radiative envelope. The chemically homogeneous evolution (CHE) supplies hydrogen to the core for a longer time producing a larger He-rich core compared to the non-rotating star, leaving little or no hydrogen [57,71]. Therefore, rotating stars form relatively massive cores from modest initial masses [57]. Chemically homogeneous massive stars are observed in the Milky Way [71], even though smaller in number compared to the Magellanic clouds [72]. The major problem with CHE is that these stars lose their angular momentum while shedding off the outer H- and He-layers via line-driven wind mass loss. This channel helps to create Type Ic SNe at the expense of reduced rotation rate that might make it difficult to produce a rotating BH at the centre. Rapid rotators with $v/v_{crit} \gtrsim 0.4$ might satisfy the criteria for both Type-Ic SN and centrally rotating BHs that can produce an accretion disk around it. Thus, we conclude that a sufficiently rotating massive single star can be a legitimate channel for the production of LGRB progenitor.

Binary stars: Binary channel is important for both short and long GRBs [71]. Binary stars are abundant both at solar and subsolar metallicities. For example, ~60–80% MS stars in our Milky Way are in binaries [73–77]. Given the bias towards metal-poor environments for the production of GRBs, it is of particular interest to study the Large Magellanic Cloud (LMC) stellar populations. Observations of compact star-clusters NGC 1818, NGC 1805, NGC 1831 and NGC 1868 in the LMC show that almost 55–100% stars are in binaries [71]. Given the majority of the massive stars residing in binaries, it is worth investigating the physical mechanisms for gaining or retaining angular momentum in binary systems.

Similar to single stars, binary routes also transfer angular momentum to the stellar core during the MS. The star can be spun up via mass transfer from the companion star, or it can also be spun down if the star itself loses mass. It is observed that an older star in the binary gets rejuvenated by the mass transfer from its much younger companion [78,79]. In the context of the binary mass transfer, if the more massive companion loses mass, then it again becomes challenging to produce a sufficiently rotating central BH, similar to the issues faced by single stars. If the mass ratio of the primary and secondary is significantly high, then the massive component (secondary) transfers mass on to the less massive one while on the MS and core He burning phase, and in this scenario, a further little amount of mass transfer from the secondary (after its core He exhaustion) to the CO core of the primary makes the less massive primary companion to spin up and explode energetically to eject the common envelope [71,80].

An alternative channel to obtain the necessary angular momentum is the merger of two companions in a binary. The proposed mechanisms for this channel are either the merger of two He cores or a NS/BH with a He core. The latter channel is thought to be the possible cause of the Christmas-day burst, GRB 101225A [81]. In a merger event, the orbital angular momentum of the two components of the binary is combined in a single merged object. In the case of He core-He core merging, there is a little time lapse between the merging event and SN. In the other scenario of BH-He core merger, BH, in principle, can immediately produce the LGRB via mass accretion from the He-core [71].

Although there are several channels to produce LGRBs via binary evolution, it is yet not certain if they can produce LGRBs at the necessary rate to match the observed LGRB populations. In each of these binary channels, obtaining the required angular momentum is a major issue, and there is a limited parameter space in initial mass, mass ratio and separation between the two components that can produce the necessary angular momentum (as discussed above) in order to create LGRBs, and therefore, each of these routes contributes to only moderate LGRB-production rates. Hence, even though the majority of the massive stars are in binaries, the binary route of LGRB production is less favoured compared to CHE of single stars to match the observed LGRB production rate, especially at sub-solar metallicity [71].

5. Properties of WR Stars Required for LGRB Candidates

Following the discussion in Sections 2 and 3, it is clear that we need sufficiently rotating massive stars for simultaneous production of LGRBs and Type-Ic SNe. The star should rotate at a speed that helps the stellar core retain enough angular momentum to create an accretion disk when collapsing into a BH, and also produce H- and He- depleted SN. Wind mass loss plays a vital role in determining the end phase angular momentum. Metallicity, being the significant determinant of wind mass loss, along with rotation, decides the final fate of a massive star. Therefore, in this section, we discuss the importance of rotation and metallicity in evolving a massive star into a WN star that eventually might become WC star later in its evolution. The eventual transition to the WC phase is essential because WC stars are believed to be the progenitors of Type-Ic SNe [7,82] from the theoretical perspective. In this section, we do not directly study the properties of LGRB progenitors and their dependence on mass, rotation rate, and metallicity. We rather follow an indirect approach—we study the mass, rotation rate, and metallicity dependence of massive stars that evolve into WN, and subsequently to WC stars which are SN Type-Ic progenitors. We follow this indirect method because Type-Ic SNe are proxies for LGRB association given their observed correlation.

Rotation rate: In this section, we investigate the minimum rotation rate required for an O star to evolve into WN phase. To do that, we study the evolution of surface helium and nitrogen mass fractions for varying mass, metallicity and rotation rate because a certain surface enhancement in He and N makes the transition from O to WN stars. Requirements of He- and N- surface enrichments that determine the WN Late-type (L) phase is given in Table 1.

Table 1. Taken from [5]. Criteria to classify stars as O, WNL, WNE, WC, and WO based on the surface enhancement of various elements and core burning status. Surface state and core state refer to conditions at the stellar surface and the centre of the star, respectively; X_Q is the mass fraction of element Q. See [5] for detail.

Classification	Surface State			Central State	
O stars	$X_{He} < 0.4$			$X_H > 10^{-4}$	
WNL	$X_{He} = 0.4$–0.9	$X_C/X_N < 10$		-	
WNE	$X_{He} > 0.9$	$X_C/X_N < 10$		-	
WC	$0.1 < X_C < 0.6$	$X_O < 0.1$	$X_C/X_N > 10$	$X_H < 10^{-4}$	$X_{He} > 0.1$
WO	$X_C > 0.1$	$X_O > 0.1$	$(X_C + X_O)/X_{He} > 1$	$X_H < 10^{-4}$	$X_{He} < 0.1$

Figure 1 shows the surface He mass fraction contours as a function of time and v/v_{crit} for a range of mass and metallicity. We notice that all masses show WNL features with surface ^{4}He mass fraction $\gtrsim$40% quite early on the MS, $t \lesssim 0.5 \times t_{MS}$, if they are moderately or rapidly rotating with $v/v_{crit} \gtrsim 0.4$ irrespective of metallicity. This value of v/v_{crit} is also supported by the theory of massive star formation [83]. Therefore, we use $v/v_{crit} = 0.4$ as our fiducial case. At solar metallicity, stars show surface enhancement of He when they have lost $\lesssim$50% of their initial masses on the MS irrespective of rotation rate. However, at low metallicities, [Fe/H] $\lesssim -1.0$, we see the surface He enrichment only for $v/v_{crit} \gtrsim 0.4$. In these metal-poor moderately or rapidly rotating stars, the rotational mixing of chemical elements dredges up the heavy nucleosynthetic by-products from the core to the surface, even though weak mass loss does not strip off much of their outer layers.

We show the evolution of surface nitrogen mass fraction in Figure 2, similar to Figure 1. The surface N enrichment is consistent with the values inferred for WNL stars. For metal rich stars with [Fe/H] = 0.0, the surface enrichment is ~17 times the initial N abundance (6.73×10^{-3}). This results in surface N mass fraction of 0.011, a factor of 2 less than that is observed in WNL stars in Arches cluster [84]. For metal-poor stars, the surface enrichment is a factor to 28–30 compared to their initial abundances of 6.97×10^{-4}, and 7.024×10^{-5} for [Fe/H] = −1.0 and −2.0, respectively.

Metallicity: Having discussed the required rotation rate that is favourable for WN and eventual WC production, in this section, we discuss the metallicity range that might be optimum for the production of Type-Ic SN. To study this, we run our models for a mass grid of $10\,M_\odot$ to $150\,M_\odot$ with a mass resolution $\Delta M = 5\,M_\odot$ until the end of core ^{12}C exhaustion (t_C) for our fiducial rotation rate, $v/v_{crit} = 0.4$, for three metallicities, [Fe/H] = 0, −1.0, −2.0, similar to Figure 1 of [5]. We use the 1-D stellar evolution code MESA [85–87] Isochrone Stellar Tracks-II (MIST-II [4,5], Dotter et al., 2021, in prep.). For details of mass-grid and the simulation setup, see [4,5].

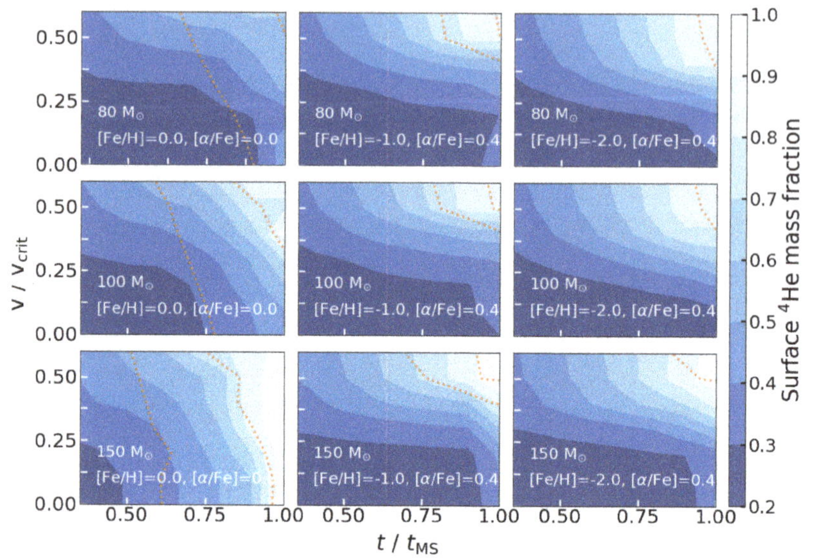

Figure 1. ^{4}He surface mass fraction, as denoted by the colorbar, as a function of time and rotation rate, $v/v_{\rm crit}$ for three metallicities, [Fe/H] = 0.0 (leftmost panels), −1.0 (middle panels), and −2.0 (rightmost panels), and for three masses, 80 M$_\odot$ (top panels), 100 M$_\odot$ (middle panels), 150 M$_\odot$ (bottom panels). We normalize the time by the main-sequence (MS) lifetime, $t_{\rm MS}$. The orange dotted lines show the points where 20% and 50% of the initial mass are lost. For details, see the discussion of Figure 1 of [4].

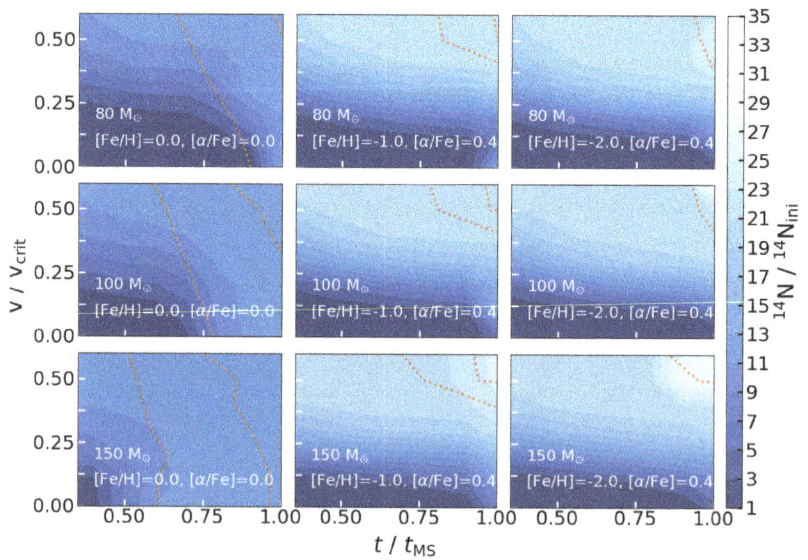

Figure 2. Same as Figure 1, except for the N color contour denoting the ^{14}N mass-fraction compared to the initial ^{14}N (^{14}N$_{\rm ini}$) abundance. ^{14}N$_{\rm ini}$ are 6.73×10^{-3}, 6.97×10^{-4}, and 7.024×10^{-5}, respectively, for [Fe/H] = 0.0, −1.0, −2.0, respectively. For details, see the discussion of Figure 2 of [4].

Definitions of massive star classifications, such as O, WNL, WN Early-type (E), WC, WO, based on surface elemental mass fraction and central burning state, are given in Table 1. For a detailed discussion of our classifications, see [5]. Here, we briefly summarize them:

- **O stars.** Stars that are core H burning with mass fraction $X_{\mathrm{H,core}} > 10^{-4}$. Nucleosynthetic products do not yet contaminate surfaces of these stars, and therefore, the surface He mass fraction is low, $X_{\mathrm{He,surf}} < 0.4$.
- **WNL and WNE stars.** Stars whose surfaces are enriched with He and N due to the mixing of chemical elements driven by various rotational instabilities, and loss of outer envelopes by line-driven wind mass-loss during and after the MS. WNL and WNE stars are defined based on surface abundances of He, N, and C, such as $X_{\mathrm{He,surf}} = 0.4 - 0.9$ & $X_{\mathrm{C,surf}}/X_{\mathrm{N,surf}} < 10$ and $X_{\mathrm{He,surf}} > 0.9$ & $X_{\mathrm{C,surf}}/X_{\mathrm{N,surf}} < 10$ for WNL and WNE, respectively.
- **WC and WO stars.** Massive stars for which surfaces are contaminated by significant amount of C, later in their evolutions during the late core He burning phase, due to both rotational dredge-up and loss of outer envelopes. These stars might eventually evolve into WO stars given their surfaces have significant amount of O as well. Requirements for surface elemental abundances for the classification of WC and WO are $0.1 < X_{\mathrm{C,surf}} < 0.6$ & $X_{\mathrm{O,surf}} < 0.1$ & $X_{\mathrm{C,surf}}/X_{\mathrm{N,surf}} > 10$, and $X_{\mathrm{C,surf}} > 0.1$ & $X_{\mathrm{O,surf}} > 0.1$ & $(X_{\mathrm{C,surf}} + X_{\mathrm{O,surf}})/X_{\mathrm{He,surf}} > 1$, respectively. The core H-, He- mass fractions for both these classes are $X_{\mathrm{H,core}} < 10^{-4}$, $X_{\mathrm{He,core}} < 0.1$.

Having discussed definitions of different classifications of massive stars, we show the fraction of time a star of a given initial mass spends in these individual phases in Figure 3. We find that stars $<45\,\mathrm{M}_\odot$ spend their entire lives as O stars, and beyond this mass, they spend a significant fraction of their lifetimes in the WNL phase for solar metallicity. They enter WNL phase for even lower mass stars $\sim 20\,\mathrm{M}_\odot$–$25\,\mathrm{M}_\odot$ for subsolar metallicities, [Fe/H] $\lesssim -1.0$, independent of metallicity, as can be seen also in Figures 1 and 2. Nonetheless, there is a peculiar metallicity evolution of the stars that show WC and WO features. At solar metallicity, WC and WO features appear for masses $\gtrsim 55\,\mathrm{M}_\odot$. At subsolar metallicities; however, the minimum mass (M_{min}) for showing up WC and WO features takes an anomalous turn. It shifts to a larger mass $\gtrsim 115\,\mathrm{M}_\odot$ for [Fe/H] $= -1.0$, and then again comes down to a lower mass $\sim 50\,\mathrm{M}_\odot$–$55\,\mathrm{M}_\odot$ for [Fe/H] $= -2.0$. Metal-rich stars undergo strong mass loss, and the mass loss rate decreases with decreasing metallicities. Along with the mass-loss, stars lose angular momentum and therefore slow down, causing less rotational dredge-up at later epochs of their evolutions. Therefore, on the one hand, metal-rich stars, such as solar metallicity stars in our model grid, show heavy metals, such as C and O, on their surfaces when significant mass-loss exposes the metal-rich inner cores [4] for $M_{\mathrm{min}} \sim 55\,\mathrm{M}_\odot$. On the other hand, significantly metal-poor stars, for example [Fe/H] $= -2.0$ in our models, can retain sufficient angular momentum due to weaker mass loss, and therefore might have surfaces enriched with C and O for $M_{\mathrm{min}} \sim 50\,\mathrm{M}_\odot$ because of rotational dredge-up, even though they do not expose their metal-rich inner cores. At intermediate metallicity, however, mass loss rates are too weak to expose the metal-rich inner cores but strong enough to lose the angular momentum and therefore to inhibit the chemical mixing driven by rotational dredge-up. This might be the reason why only the most massive stars $\gtrsim 115\,\mathrm{M}_\odot$ show WC- WO- features at intermediate metallicity [Fe/H] $= -1.0$. This, however, needs to be speculated in detail, and we plan to address this issue in our future paper Roy et al., 2021, in prep. Note that this result may depend strongly on the adopted mass-loss schemes and their metallicity dependance, and we will explore that in our follow-up paper Roy et al., 2021, in prep.

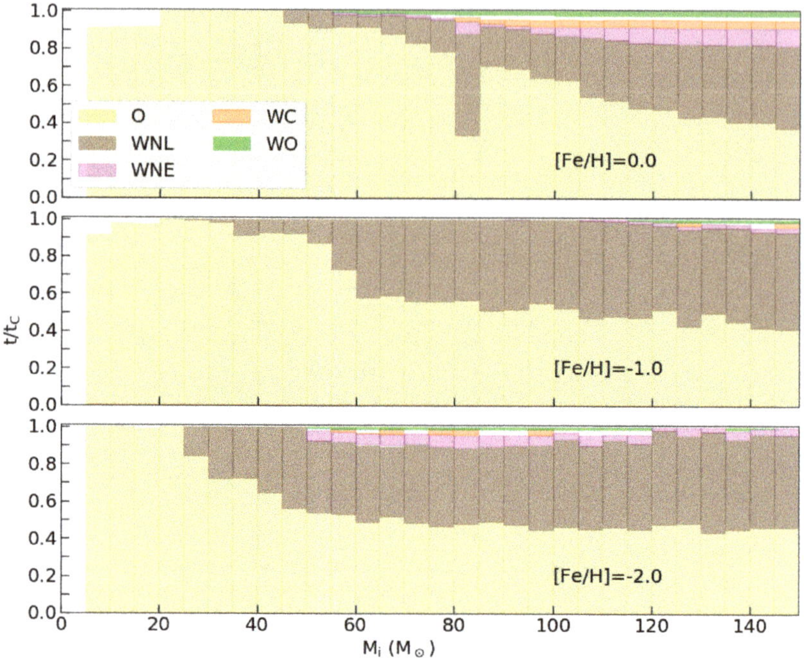

Figure 3. The fraction of time each star with a specific intial mass (horizontal-axis) spends in a particular phase classified as O, WNL, WNE, WC, WO, until we stop our simulations at the end of core ^{12}C exhaustion (t_C), similar to Figure 1 of [5]. These different phases of massive stars are shown by different colors indicated in the figure legend. The definitions for these claasifications of different phases are given in Table 1. Models shown here are for our fiducial rotation rate, $v/v_{\rm crit} = 0.4$, and the three panels refer to three metallicities, as indicated.

Having discussed the metallicity evolution of $M_{\rm min}$, we expect to have a larger number of WC and WO stars at solar metallicity, and the number to decrease at [Fe/H] $= -1.0$ and to increase again at [Fe/H] $= -2.0$. Theoretically, WC and WO stars should be progenitors of Type-Ic SN as they lose their H and He envelopes [7,82]. Therefore, we expect to see a larger number of Type-Ic SNe at solar metallicity decreasing at intermediate subsolar metallicity and increasing again at significantly lower metallicity of [Fe/H]~ -2.0. In addition, there is a one-to-one correlation between LGRB and Type-Ic SNe. We conclude that LGRB numbers starting from solar metallicity might reduce initially with decreasing metallicity and then increase again for significantly metal-poor stars at [Fe/H] $= -2.0$. In metal-poor environments, the LGRB production rate is also favourable because at low metallicities, stellar cores can retain enough angular momentum to form an accretion disk around the collapsing core that creates the central BH. Observations of LGRB also agree with this theoretical prediction that the LGRB rate increases with decreasing metallicities, however, not at excessively low metallicity (e.g., [88–90]). In fact, most observations show that there is a rapid drop-off in the LGRB rate somewhere between solar [90,91] and 1/3 solar [92], i.e., between the Milky Way (almost solar) and the Small Magellanic Cloud ($\approx$half solar). However, to draw a tight constraint on the metallicity evolution of LGRB rate from both theoretical and observational perspectives, one needs to have observations at metallicities much lower than SMC, and also models with finer metallicity resolutions at these low metallicities. We plan to study this in detail in a follow-up paper Roy et al., 2021, in prep.

Even though most LGRBs are observed to be associated with SN Type-Ic, there are a few exceptions as discussed in Section 2. Therefore, for a conclusive study of LGRB progenitors and their possible connection to SN Type-Ic, we need to investigate the properties of stellar cores and their explodability criteria as given in Equations (1)–(5), and also their dependence on mass, rotation rate, and metallicity in detail. We plan to study all these detailed theoretical aspects in a follow-up paper Roy et al., 2021, in prep.

6. Summary and Discussion

It is almost five decades after the first GRB was discovered in the 1960s, and since then, there are observations of $\sim$ 1000 GRBs at different redshifts. These observations have made progress over the years in detecting GRB luminosity and the light curve of the afterglow more accurately, and in advanced spectra aiming to decisively determine the nature of the GRB progenitor and progenitor-circumstellar medium (CSM) interaction. This progenitor-CSM interaction provides hints to the ISM compositions of the host galaxies, and thereby has enriched our understanding of the high-redshift universe. Moreover, the precision to pinpoint the GRB location has also enhanced our knowledge of the host galaxies. A strong correlation between LGRB and Type-Ic SN has broadened our understanding of LGRB progenitors as well.

Even though all these advances have made a clearer picture of GRB production, there are still several open questions yet to be answered:

- What fraction of O stars really produce GRBs? What does it say about the evolution of their progenitors?
- How does the rate of GRB production vary with metallicity and redshift?
- What is the binary fraction that produces LGRBs? How does this fraction vary with metallicity?
- How do we obtain a stronger constraints differentiating LGRB and SLSNe production pathways?

Funding: This research received no external funding.

Data Availability Statement: Data underlying this article will be shared on reasonable request to the author.

Acknowledgments: We thank anonymous referees for their insightful comments to enrich the manuscript. AR is grateful to Mark Krumholz and Alexander Heger for their valuable inputs and suggestions. AR gratefully acknowledges the support from Andrea Ferrara's Italian funding scheme "The quest for the first stars" (Cod. 2017T4ARJ5_001). AR also gratefully acknowledges support from Mark Krumholz's Australian Research Council's *Discovery Projects* and *Future Fellowship* funding scheme, awards DP190101258 and FT180100375. Parts of this research were conducted by the Australian Research Council Centre of Excellence for All Sky Astrophysics in 3 Dimensions (ASTRO 3D), through project number CE170100013. This research/project was undertaken with the assistance of resources and services from the National Computational Infrastructure (NCI)'s supercomputer Gadi, which is supported by the Australian Government, and the Australian National University's Research School of Astronomy & Astrophysics's cluster Avatar.

Conflicts of Interest: The author declares no conflict of interest.

Notes

1 'BL' stands for broad line. The subclassification of "Type Ic-BL" is purely based on observations, there is no model to predict its progenitor [6].

2 Contrary to the "collapsar" model, [55,56] proposed a different origin for GRBs. They argued that when a massive star explodes as SN and the stellar core collapses as a neutron star (NS), the NS acquires high velocity ($\sim$500 km/s) due to a substantial "kick" at birth, and as a result, a recoiling ultra-relativistic outflow can be launched in the opposite direction. This outflow can be energetic enough ($\sim 10^{52}$ erg) to produce a long GRB.

3 However, in Type-Ic SN, some He may still be present, but just not visible.

References

1. Klebesadel, R.W.; Strong, I.B.; Olson, R.A. Observations of Gamma-ray Bursts of Cosmic Origin. *Astrophys. J. Lett.* **1973**, *182*, L85. [CrossRef]
2. Wijers, R.A.M.J.; Bloom, J.S.; Bagla, J.S.; Natarajan, P. Gamma-ray bursts from stellar remnants: Probing the Universe at high redshift. *Mon. Not. R. Astron. Soc.* **1998**, *294*, L13–L17. [CrossRef]
3. Bromm, V.; Larson, R.B. The First Stars. *Annu. Rev. Astron. Astrophys.* **2004**, *42*, 79–118. [CrossRef]
4. Roy, A.; Sutherland, R.S.; Krumholz, M.R.; Heger, A.; Dopita, M.A. Helium and nitrogen enrichment in massive main-sequence stars: Mechanisms and implications for the origin of WNL stars. *Mon. Not. R. Astron. Soc.* **2020**, *494*, 3861–3879. [CrossRef]
5. Roy, A.; Dopita, M.A.; Krumholz, M.R.; Kewley, L.J.; Sutherland, R.S.; Heger, A. On the origin of nitrogen at low metallicity. *Mon. Not. R. Astron. Soc.* **2021**, *502*, 4359–4376. [CrossRef]
6. Woosley, S.E.; Bloom, J.S. The Supernova Gamma-ray Burst Connection. *Annu. Rev. Astron. Astrophys.* **2006**, *44*, 507–556. [CrossRef]
7. Crowther, P.A. Physical Properties of Wolf-Rayet Stars. *Annu. Rev. Astron. Astrophys.* **2007**, *45*, 177–219. [CrossRef]
8. Gal-Yam, A. Luminous Supernovae. *Science* **2012**, *337*, 927. [CrossRef]
9. Howell, D.A. Superluminous Supernovae. In *Handbook of Supernovae*; Alsabti, A.W., Murdin, P., Eds.; Springer: Berlin/Heidelberg, Germany, 2017; p. 431. [CrossRef]
10. Inserra, C. Observational properties of extreme supernovae. *Nat. Astron.* **2019**, *3*, 697–705. [CrossRef]
11. Aguilera-Dena, D.R.; Langer, N.; Antoniadis, J.; Müller, B. Precollapse Properties of Superluminous Supernovae and Long Gamma-ray Burst Progenitor Models. *Astrophys. J.* **2020**, *901*, 114. [CrossRef]
12. Gal-Yam, A. A Simple Analysis of Type I Superluminous Supernova Peak Spectra: Composition, Expansion Velocities, and Dynamics. *Astrophys. J.* **2019**, *882*, 102. [CrossRef]
13. Woosley, S.E. Gamma-ray Bursts from Stellar Mass Accretion Disks around Black Holes. *Astrophys. J.* **1993**, *405*, 273. [CrossRef]
14. O'Connor, E.; Ott, C.D. Black Hole Formation in Failing Core-Collapse Supernovae. *Astrophys. J.* **2011**, *730*, 70. [CrossRef]
15. Sukhbold, T.; Woosley, S.E. The Compactness of Presupernova Stellar Cores. *Astrophys. J.* **2014**, *783*, 10. [CrossRef]
16. Müller, B.; Heger, A.; Liptai, D.; Cameron, J.B. A simple approach to the supernova progenitor-explosion connection. *Mon. Not. R. Astron. Soc.* **2016**, *460*, 742–764. [CrossRef]
17. Sukhbold, T.; Woosley, S.E.; Heger, A. A High-resolution Study of Presupernova Core Structure. *Astrophys. J.* **2018**, *860*, 93. [CrossRef]
18. Colgate, S.A. Prompt Gamma rays and X-rays from supernovae. *Can. J. Phys. Suppl.* **1968**, *46*, 476. [CrossRef]
19. Paczynski, B. Gamma-ray bursters at cosmological distances. *Astrophys. J. Lett.* **1986**, *308*, L43–L46. [CrossRef]
20. Costa, E.; Frontera, F.; Heise, J.; Feroci, M.; in't Zand, J.; Fiore, F.; Cinti, M.N.; Dal Fiume, D.; Nicastro, L.; Orlandini, M.; et al. Discovery of an X-ray afterglow associated with the γ-ray burst of 28 February 1997. *Nature* **1997**, *387*, 783–785. [CrossRef]
21. van Paradijs, J.; Groot, P.J.; Galama, T.; Kouveliotou, C.; Strom, R.G.; Telting, J.; Rutten, R.G.M.; Fishman, G.J.; Meegan, C.A.; Pettini, M.; et al. Transient optical emission from the error box of the γ-ray burst of 28 February 1997. *Nature* **1997**, *386*, 686–689. [CrossRef]
22. Galama, T.J.; Vreeswijk, P.M.; van Paradijs, J.; Kouveliotou, C.; Augusteijn, T.; Böhnhardt, H.; Brewer, J.P.; Doublier, V.; Gonzalez, J.F.; Leibundgut, B.; et al. An unusual supernova in the error box of the γ-ray burst of 25 April 1998. *Nature* **1998**, *395*, 670–672. [CrossRef]
23. Bloom, J.S.; Kulkarni, S.R.; Harrison, F.; Prince, T.; Phinney, E.S.; Frail, D.A. Expected Characteristics of the Subclass of Supernova Gamma-ray Bursts. *Astrophys. J. Lett.* **1998**, *506*, L105–L108. [CrossRef]
24. Bloom, J.S.; Kulkarni, S.R.; Djorgovski, S.G.; Eichelberger, A.C.; Côté, P.; Blakeslee, J.P.; Odewahn, S.C.; Harrison, F.A.; Frail, D.A.; Filippenko, A.V.; et al. The unusual afterglow of the γ-ray burst of 26 March 1998 as evidence for a supernova connection. *Nature* **1999**, *401*, 453–456. [CrossRef]
25. Castro-Tirado, A.J.; Gorosabel, J. Optical observations of GRB afterglows: GRB 970508 and GRB 980326 revisited. *Astron. Astrophys. Suppl. Ser.* **1999**, *138*, 449–450. [CrossRef]
26. Waxman, E.; Draine, B.T. Dust Sublimation by Gamma-ray Bursts and Its Implications. *Astrophys. J.* **2000**, *537*, 796–802. [CrossRef]
27. Reichart, D.E. GRB 970228 Revisited: Evidence for a Supernova in the Light Curve and Late Spectral Energy Distribution of the Afterglow. *Astrophys. J. Lett.* **1999**, *521*, L111–L115. [CrossRef]
28. Reichart, D.E. Light Curves and Spectra of Dust Echoes from Gamma-ray Bursts and Their Afterglows: Continued Evidence That GRB 970228 Is Associated with a Supernova. *Astrophys. J.* **2001**, *554*, 643–659. [CrossRef]
29. Galama, T.J.; Tanvir, N.; Vreeswijk, P.M.; Wijers, R.A.M.J.; Groot, P.J.; Rol, E.; van Paradijs, J.; Kouveliotou, C.; Fruchter, A.S.; Masetti, N.; et al. Evidence for a Supernova in Reanalyzed Optical and Near-Infrared Images of GRB 970228. *Astrophys. J.* **2000**, *536*, 185–194. [CrossRef]
30. Price, P.A.; Kulkarni, S.R.; Berger, E.; Fox, D.W.; Bloom, J.S.; Djorgovski, S.G.; Frail, D.A.; Galama, T.J.; Harrison, F.A.; McCarthy, P.; et al. Discovery of GRB 020405 and Its Late Red Bump. *Astrophys. J.* **2003**, *589*, 838–843. [CrossRef]
31. Stanek, K.Z.; Garnavich, P.M.; Nutzman, P.A.; Hartman, J.D.; Garg, A.; Adelberger, K.; Berlind, P.; Bonanos, A.Z.; Calkins, M.L.; Challis, P.; et al. Deep Photometry of GRB 041006 Afterglow: Hypernova Bump at Redshift z = 0.716. *Astrophys. J. Lett.* **2005**, *626*, L5–L9. [CrossRef]

32. Fynbo, J.P.U.; Watson, D.; Thöne, C.C.; Sollerman, J.; Bloom, J.S.; Davis, T.M.; Hjorth, J.; Jakobsson, P.; Jørgensen, U.G.; Graham, J.F.; et al. No supernovae associated with two long-duration γ-ray bursts. *Nature* **2006**, *444*, 1047–1049. [CrossRef]
33. Della Valle, M.; Chincarini, G.; Panagia, N.; Tagliaferri, G.; Malesani, D.; Testa, V.; Fugazza, D.; Campana, S.; Covino, S.; Mangano, V.; et al. An enigmatic long-lasting γ-ray burst not accompanied by a bright supernova. *Nature* **2006**, *444*, 1050–1052. [CrossRef] [PubMed]
34. Gal-Yam, A.; Fox, D.B.; Price, P.A.; Ofek, E.O.; Davis, M.R.; Leonard, D.C.; Soderberg, A.M.; Schmidt, B.P.; Lewis, K.M.; Peterson, B.A.; et al. A novel explosive process is required for the γ-ray burst GRB 060614. *Nature* **2006**, *444*, 1053–1055. [CrossRef] [PubMed]
35. Yang, B.; Jin, Z.P.; Li, X.; Covino, S.; Zheng, X.Z.; Hotokezaka, K.; Fan, Y.Z.; Piran, T.; Wei, D.M. A possible macronova in the late afterglow of the long-short burst GRB 060614. *Nat. Commun.* **2015**, *6*, 7323. [CrossRef]
36. Jin, Z.P.; Li, X.; Cano, Z.; Covino, S.; Fan, Y.Z.; Wei, D.M. The Light Curve of the Macronova Associated with the Long-Short Burst GRB 060614. *Astrophys. J. Lett.* **2015**, *811*, L22. [CrossRef]
37. Berger, E.; Cowie, L.L.; Kulkarni, S.R.; Frail, D.A.; Aussel, H.; Barger, A.J. A Submillimeter and Radio Survey of Gamma-ray Burst Host Galaxies: A Glimpse into the Future of Star Formation Studies. *Astrophys. J.* **2003**, *588*, 99–112. [CrossRef]
38. Bloom, J.S.; Djorgovski, S.G.; Kulkarni, S.R. The Redshift and the Ordinary Host Galaxy of GRB 970228. *Astrophys. J.* **2001**, *554*, 678–683. [CrossRef]
39. Loredo, T.J.; Wasserman, I.M. Inferring the Spatial and Energy Distribution of Gamma-ray Burst Sources. II. Isotropic Models. *Astrophys. J.* **1998**, *502*, 75–107. [CrossRef]
40. Bloom, J.S.; Fox, D.; van Dokkum, P.G.; Kulkarni, S.R.; Berger, E.; Djorgovski, S.G.; Frail, D.A. The First Two Host Galaxies of X-Ray Flashes: XRF 011030 and XRF 020427. *Astrophys. J.* **2003**, *599*, 957–963. [CrossRef]
41. Jakobsson, P.; Björnsson, G.; Fynbo, J.P.U.; Jóhannesson, G.; Hjorth, J.; Thomsen, B.; Møller, P.; Watson, D.; Jensen, B.L.; Östlin, G.; et al. Ly-α and ultraviolet emission from high-redshift Gamma-ray burst hosts: To what extent do Gamma-ray bursts trace star formation? *Mon. Not. R. Astron. Soc.* **2005**, *362*, 245–251. [CrossRef]
42. Greiner, J.; Klose, S.; Reinsch, K.; Martin Schmid, H.; Sari, R.; Hartmann, D.H.; Kouveliotou, C.; Rau, A.; Palazzi, E.; Straubmeier, C.; et al. Evolution of the polarization of the optical afterglow of the γ-ray burst GRB030329. *Nature* **2003**, *426*, 157–159. [CrossRef]
43. Stanek, K.Z.; Matheson, T.; Garnavich, P.M.; Martini, P.; Berlind, P.; Caldwell, N.; Challis, P.; Brown, W.R.; Schild, R.; Krisciunas, K.; et al. Spectroscopic Discovery of the Supernova 2003dh Associated with GRB 030329. *Astrophys. J. Lett.* **2003**, *591*, L17–L20. [CrossRef]
44. Hjorth, J.; Sollerman, J.; Møller, P.; Fynbo, J.P.U.; Woosley, S.E.; Kouveliotou, C.; Tanvir, N.R.; Greiner, J.; Andersen, M.I.; Castro-Tirado, A.J.; et al. A very energetic supernova associated with the γ-ray burst of 29 March 2003. *Nature* **2003**, *423*, 847–850. [CrossRef]
45. Prochaska, J.X.; Bloom, J.S.; Chen, H.W.; Hurley, K.C.; Melbourne, J.; Dressler, A.; Graham, J.R.; Osip, D.J.; Vacca, W.D. The Host Galaxy of GRB 031203: Implications of Its Low Metallicity, Low Redshift, and Starburst Nature. *Astrophys. J.* **2004**, *611*, 200–207. [CrossRef]
46. Cobb, B.E.; Bailyn, C.D.; van Dokkum, P.G.; Buxton, M.M.; Bloom, J.S. The Supernova Associated with GRB 031203: SMARTS Optical-Infrared Light Curves from 0.2 to 92 Days. *Astrophys. J. Lett.* **2004**, *608*, L93–L96. [CrossRef]
47. Della Valle, M.; Malesani, D.; Benetti, S.; Testa, V.; Hamuy, M.; Antonelli, L.A.; Chincarini, G.; Cocozza, G.; Covino, S.; D'Avanzo, P.; et al. Evidence for supernova signatures in the spectrum of the late-time bump of the optical afterglow of GRB 021211. *Astron. Astrophys.* **2003**, *406*, L33–L37. [CrossRef]
48. Della Valle, M.; Malesani, D.; Bloom, J.S.; Benetti, S.; Chincarini, G.; D'Avanzo, P.; Foley, R.J.; Covino, S.; Melandri, A.; Piranomonte, S.; et al. Hypernova Signatures in the Late Rebrightening of GRB 050525A. *Astrophys. J. Lett.* **2006**, *642*, L103–L106. [CrossRef]
49. Rhoads, J.E. How to Tell a Jet from a Balloon: A Proposed Test for Beaming in Gamma-ray Bursts. *Astrophys. J. Lett.* **1997**, *487*, L1–L4. [CrossRef]
50. Rhoads, J.E. The Dynamics and Light Curves of Beamed Gamma-ray Burst Afterglows. *Astrophys. J.* **1999**, *525*, 737–749. [CrossRef]
51. Frail, D.A.; Kulkarni, S.R.; Sari, R.; Djorgovski, S.G.; Bloom, J.S.; Galama, T.J.; Reichart, D.E.; Berger, E.; Harrison, F.A.; Price, P.A.; et al. Beaming in Gamma-ray Bursts: Evidence for a Standard Energy Reservoir. *Astrophys. J. Lett.* **2001**, *562*, L55–L58. [CrossRef]
52. Woosley, S.E.; Heger, A. The Progenitor Stars of Gamma-ray Bursts. *Astrophys. J.* **2006**, *637*, 914–921. [CrossRef]
53. Zhang, W.; Woosley, S.E.; Heger, A. The Propagation and Eruption of Relativistic Jets from the Stellar Progenitors of Gamma-ray Bursts. *Astrophys. J.* **2004**, *608*, 365–377. [CrossRef]
54. Zou, Z.C.; Zhang, B.B.; Huang, Y.F.; Zhao, X.H. Gamma-ray Burst in a Binary System. *arXiv* **2021**, arXiv:2106.12838.
55. Dar, A.; Plaga, R. Galactic gamma-ray bursters—An alternative source of cosmic rays at all energies. *Astron. Astrophys.* **1999**, *349*, 259–266.
56. Huang, Y.F.; Dai, Z.G.; Lu, T.; Cheng, K.S.; Wu, X.F. Gamma-ray Bursts from Neutron Star Kicks. *Astrophys. J.* **2003**, *594*, 919–923. [CrossRef]

57. Yoon, S.C.; Langer, N. Evolution of rapidly rotating metal-poor massive stars towards Gamma-ray bursts. *Astron. Astrophys.* **2005**, *443*, 643–648. [CrossRef]
58. Cantiello, M.; Yoon, S.C.; Langer, N.; Livio, M. Binary star progenitors of long Gamma-ray bursts. *Astron. Astrophys.* **2007**, *465*, L29–L33. [CrossRef]
59. Detmers, R.G.; Langer, N.; Podsiadlowski, P.; Izzard, R.G. Gamma-ray bursts from tidally spun-up Wolf-Rayet stars? *Astron. Astrophys.* **2008**, *484*, 831–839. [CrossRef]
60. Kasen, D.; Bildsten, L. Supernova Light Curves Powered by Young Magnetars. *Astrophys. J.* **2010**, *717*, 245–249. [CrossRef]
61. Metzger, B.D.; Margalit, B.; Kasen, D.; Quataert, E. The diversity of transients from magnetar birth in core collapse supernovae. *Mon. Not. R. Astron. Soc.* **2015**, *454*, 3311–3316. [CrossRef]
62. Nicholl, M.; Guillochon, J.; Berger, E. The Magnetar Model for Type I Superluminous Supernovae. I. Bayesian Analysis of the Full Multicolor Light-curve Sample with MOSFiT. *Astrophys. J.* **2017**, *850*, 55. [CrossRef]
63. Moriya, T.J.; Nicholl, M.; Guillochon, J. Systematic Investigation of the Fallback Accretion-powered Model for Hydrogen-poor Superluminous Supernovae. *Astrophys. J.* **2018**, *867*, 113. [CrossRef]
64. Chatzopoulos, E.; Wheeler, J.C.; Vinko, J. Generalized Semi-analytical Models of Supernova Light Curves. *Astrophys. J.* **2012**, *746*, 121. [CrossRef]
65. Blanchard, P.K.; Nicholl, M.; Berger, E.; Chornock, R.; Margutti, R.; Milisavljevic, D.; Fong, W.; MacLeod, C.; Bhirombhakdi, K. The Type I Superluminous Supernova PS16aqv: Lightcurve Complexity and Deep Limits on Radioactive Ejecta in a Fast Event. *Astrophys. J.* **2018**, *865*, 9. [CrossRef]
66. Blanchard, P.K.; Nicholl, M.; Berger, E.; Chornock, R.; Milisavljevic, D.; Margutti, R.; Gomez, S. A Hydrogen-poor Superluminous Supernova with Enhanced Iron-group Absorption: A New Link between SLSNe and Broad-lined Type Ic SNe. *Astrophys. J.* **2019**, *872*, 90. [CrossRef]
67. Blanchard, P.K.; Berger, E.; Nicholl, M.; Villar, V.A. The Pre-explosion Mass Distribution of Hydrogen-poor Superluminous Supernova Progenitors and New Evidence for a Mass-Spin Correlation. *Astrophys. J.* **2020**, *897*, 114. [CrossRef]
68. Ertl, T.; Janka, H.T.; Woosley, S.E.; Sukhbold, T.; Ugliano, M. A Two-parameter Criterion for Classifying the Explodability of Massive Stars by the Neutrino-driven Mechanism. *Astrophys. J.* **2016**, *818*, 124. [CrossRef]
69. Choi, J.; Dotter, A.; Conroy, C.; Cantiello, M.; Paxton, B.; Johnson, B.D. Mesa Isochrones and Stellar Tracks (MIST). I. Solar-scaled Models. *Astrophys. J.* **2016**, *823*, 102. [CrossRef]
70. Yoon, S.C.; Dierks, A.; Langer, N. Evolution of massive Population III stars with rotation and magnetic fields. *Astron. Astrophys.* **2012**, *542*, A113. [CrossRef]
71. Levan, A.; Crowther, P.; de Grijs, R.; Langer, N.; Xu, D.; Yoon, S.C. Gamma-ray Burst Progenitors. *Space Sci. Rev.* **2016**, *202*, 33–78. [CrossRef]
72. Martins, F.; Hillier, D.J.; Bouret, J.C.; Depagne, E.; Foellmi, C.; Marchenko, S.; Moffat, A.F. Properties of WNh stars in the Small Magellanic Cloud: Evidence for homogeneous evolution. *Astron. Astrophys.* **2009**, *495*, 257–270. [CrossRef]
73. Mason, B.D.; Henry, T.J.; Hartkopf, W.I.; ten Brummelaar, T.; Soderblom, D.R. A Multiplicity Survey of Chromospherically Active and Inactive Stars. *Astron. J.* **1998**, *116*, 2975–2983. [CrossRef]
74. Kouwenhoven, M.B.N.; Brown, A.G.A.; Zinnecker, H.; Kaper, L.; Portegies Zwart, S.F. The primordial binary population. I. A near-infrared adaptive optics search for close visual companions to A star members of Scorpius OB2. *Astron. Astrophys.* **2005**, *430*, 137–154. [CrossRef]
75. Raghavan, D.; McAlister, H.A.; Henry, T.J.; Latham, D.W.; Marcy, G.W.; Mason, B.D.; Gies, D.R.; White, R.J.; ten Brummelaar, T.A. A Survey of Stellar Families: Multiplicity of Solar-type Stars. *Astrophys. J. Suppl.* **2010**, *190*, 1–42. [CrossRef]
76. Sana, H.; de Mink, S.E.; de Koter, A.; Langer, N.; Evans, C.J.; Gieles, M.; Gosset, E.; Izzard, R.G.; Le Bouquin, J.B.; Schneider, F.R.N. Binary Interaction Dominates the Evolution of Massive Stars. *Science* **2012**, *337*, 444. [CrossRef]
77. Kobulnicky, H.A.; Kiminki, D.C.; Lundquist, M.J.; Burke, J.; Chapman, J.; Keller, E.; Lester, K.; Rolen, E.K.; Topel, E.; Bhattacharjee, A.; et al. Toward Complete Statistics of Massive Binary Stars: Penultimate Results from the Cygnus OB2 Radial Velocity Survey. *Astrophys. J. Suppl.* **2014**, *213*, 34. [CrossRef]
78. Eldridge, J.J.; Stanway, E.R. Spectral population synthesis including massive binaries. *Mon. Not. R. Astron. Soc.* **2009**, *400*, 1019–1028. [CrossRef]
79. Stanway, E.R.; Eldridge, J.J.; Becker, G.D. Stellar population effects on the inferred photon density at reionization. *Mon. Not. R. Astron. Soc.* **2016**, *456*, 485–499. [CrossRef]
80. Podsiadlowski, P.; Ivanova, N.; Justham, S.; Rappaport, S. Explosive common-envelope ejection: Implications for Gamma-ray bursts and low-mass black-hole binaries. *Mon. Not. R. Astron. Soc.* **2010**, *406*, 840–847. [CrossRef]
81. Thöne, C.C.; de Ugarte Postigo, A.; Fryer, C.L.; Page, K.L.; Gorosabel, J.; Aloy, M.A.; Perley, D.A.; Kouveliotou, C.; Janka, H.T.; Mimica, P.; et al. The unusual γ-ray burst GRB 101225A from a helium star/neutron star merger at redshift 0.33. *Nature* **2011**, *480*, 72–74. [CrossRef] [PubMed]
82. Heger, A.; Fryer, C.L.; Woosley, S.E.; Langer, N.; Hartmann, D.H. How Massive Single Stars End Their Life. *Astrophys. J.* **2003**, *591*, 288–300. [CrossRef]
83. Rosen, A.L.; Krumholz, M.R.; Ramirez-Ruiz, E. What Sets the Initial Rotation Rates of Massive Stars? *Astrophys. J.* **2012**, *748*, 97. [CrossRef]

84. Figer, D.F.; Najarro, F.; Gilmore, D.; Morris, M.; Kim, S.S.; Serabyn, E.; McLean, I.S.; Gilbert, A.M.; Graham, J.R.; Larkin, J.E.; et al. Massive Stars in the Arches Cluster. *Astrophys. J.* **2002**, *581*, 258–275. [CrossRef]
85. Paxton, B.; Bildsten, L.; Dotter, A.; Herwig, F.; Lesaffre, P.; Timmes, F. Modules for Experiments in Stellar Astrophysics (MESA). *Astrophys. J. Suppl.* **2011**, *192*, 3. [CrossRef]
86. Paxton, B.; Cantiello, M.; Arras, P.; Bildsten, L.; Brown, E.F.; Dotter, A.; Mankovich, C.; Montgomery, M.H.; Stello, D.; Timmes, F.X. Modules for Experiments in Stellar Astrophysics (MESA): Planets, Oscillations, Rotation, and Massive Stars. *Astrophys. J. Suppl.* **2013**, *208*, 4. [CrossRef]
87. Paxton, B.; Marchant, P.; Schwab, J.; Bauer, E.B.; Bildsten, L.; Cantiello, M.; Dessart, L.; Farmer, R.; Hu, H.; Langer, N. Modules for Experiments in Stellar Astrophysics (MESA): Binaries, Pulsations, and Explosions. *Astrophys. J. Suppl.* **2015**, *220*, 15. [CrossRef]
88. Fruchter, A.S.; Levan, A.J.; Strolger, L.; Vreeswijk, P.M.; Thorsett, S.E.; Bersier, D.; Burud, I.; Castro Cerón, J.M.; Castro-Tirado, A.J.; Conselice, C.; et al. Long γ-ray bursts and core-collapse supernovae have different environments. *Nature* **2006**, *441*, 463–468. [CrossRef]
89. Graham, J.F.; Fruchter, A.S. The Metal Aversion of Long-duration Gamma-ray Bursts. *Astrophys. J.* **2013**, *774*, 119. [CrossRef]
90. Krühler, T.; Malesani, D.; Fynbo, J.P.U.; Hartoog, O.E.; Hjorth, J.; Jakobsson, P.; Perley, D.A.; Rossi, A.; Schady, P.; Schulze, S.; et al. GRB hosts through cosmic time. VLT/X-Shooter emission-line spectroscopy of 96 γ-ray-burst-selected galaxies at $0.1 < z < 3.6$. *Astron. Astrophys.* **2015**, *581*, A125. [CrossRef]
91. Perley, D.A.; Perley, R.A.; Hjorth, J.; Michałowski, M.J.; Cenko, S.B.; Jakobsson, P.; Krühler, T.; Levan, A.J.; Malesani, D.; Tanvir, N.R. Connecting GRBs and ULIRGs: A Sensitive, Unbiased Survey for Radio Emission from Gamma-ray Burst Host Galaxies at $0 < z < 2.5$. *Astrophys. J.* **2015**, *801*, 102. [CrossRef]
92. Graham, J.F.; Schady, P. The Absolute Rate of LGRB Formation. *Astrophys. J.* **2016**, *823*, 154. [CrossRef]

galaxies

MDPI

Article

A Roadmap to Gamma-Ray Bursts: New Developments and Applications to Cosmology

Orlando Luongo [1,2,3,*,†] and Marco Muccino [3,4,†]

1 Scuola di Scienze e Tecnologie, Università di Camerino, Via Madonna delle Carceri, 62032 Camerino, Italy
2 Dipartimento di Matematica, Università di Pisa, Largo B. Pontecorvo 5, 56127 Pisa, Italy
3 National Nanotechnology Laboratory of Open Type (NNLOT), Al-Farabi Kazakh National University, Al-Farabi av. 71, Almaty 050040, Kazakhstan
4 Laboratori Nazionali di Frascati, Istituto Nazionale di Fisica Nucleare (INFN), 00044 Frascati, Italy; Marco.Muccino@lnf.infn.it
* Correspondence: Orlando.Luongo@unicam.it
† These authors contributed equally to this work.

Abstract: Gamma-ray bursts are the most powerful explosions in the universe and are mainly placed at very large redshifts, up to $z \simeq 9$. In this short review, we first discuss gamma-ray burst classification and morphological properties. We then report the likely relations between gamma-ray bursts and other astronomical objects, such as black holes, supernovae, neutron stars, etc., discussing in detail gamma-ray burst progenitors. We classify long and short gamma-ray bursts, working out their timescales, and introduce the standard fireball model. Afterwards, we focus on direct applications of gamma-ray bursts to cosmology and underline under which conditions such sources would act as perfect standard candles if correlations between photometric and spectroscopic properties were not jeopardized by the *circularity problem*. In this respect, we underline how the shortage of low-z gamma-ray bursts prevents anchor gamma-ray bursts with primary distance indicators. Moreover, we analyze in detail the most adopted gamma-ray burst correlations, highlighting their main differences. We therefore show calibration techniques, comparing such treatments with non-calibration scenarios. For completeness, we discuss the physical properties of the correlation scatters and systematics occurring during experimental computations. Finally, we develop the most recent statistical methods, star formation rate, and high-redshift gamma-ray burst excess and show the most recent constraints obtained from experimental analyses.

Keywords: gamma ray bursts; fireball model; circularity problem; standard candles; calibration; dark energy; dark matter; cosmography; cosmological parameters

Citation: Luongo, O.; Muccino, M. A Roadmap to Gamma-Ray Bursts: New Developments and Applications to Cosmology. *Galaxies* **2021**, *9*, 77. https://doi.org/10.3390/galaxies9040077

Academic Editors: Elena Moretti and Francesco Longo

Received: 30 June 2021
Accepted: 7 October 2021
Published: 12 October 2021

1. Introduction

Gamma-ray bursts (GRBs) represent powerful extra-galactic transient that emit in γ-rays [1,2]. They are commonly associated with the death of massive stars or with binary compact object mergers. As expected, due to their enormous luminosity, after the aforementioned processes, there would be the presence a newborn stellar mass black hole (BH) that provides particle accelerations and emits a relativistic collimated outflow, in the form of jets. At the same time, this new system furnishes non-thermal emissions at almost all wavelengths. The above picture lies on the standard model describing GRBs and requires isotropic energies in the range 10^{44}–10^{47} J, or 10^{51}–10^{54} erg, mostly larger than the brightest supernova (SN) emission, lying on 10^{42} J, or 10^{49} erg [3,4]. Thereby, the need of singling out GRB progenitors is essential to disclose their fundamental properties as well as the physical conditions that permit relativistic jets to form and accelerate. Even though a clear landscape for GRB progenitor is still unclear, in view of their duration, it is plausible to classify GRBs into long and short ones.

Clearly, in our *Precision Cosmology*, epoch GRBs could open new windows[1] toward the universe description at intermediate redshifts[2] [5–7], i.e., much larger than SN ones [8]. Thus, several new observations have been developed, with always better accuracy, trying to standardize GRBs and to handle their emissions in analogy to SNe. In general, the most tricky challenge for cosmology is measuring distances and arguing luminosity in the cosmic scenario, understanding from astronomical emission at which distance the emitter is placed [9].

Unfortunately, this is not exactly the case of GRBs that are not standard candles, i.e., they do not provide the above requirement on distance and luminosity [10,11]. In fact, their highly variable γ-ray emission, mostly evident during the prompt phase, is thought to be associated with jet internal energy dissipation. However, the jet kinematics, among all its speed, collimation, energy, magnetization, etc., are all properties not well clarified, as well as energy dissipation mechanisms and/or shock acceleration efficiency. Hence, it is hard to relate luminosity to GRB distances as their microphysics is not well understood. Although the above caveats plague the overall GRB scenario, both short and long GRBs have relativistic outflows and share analogous properties[3] and many attempts have been spent to standardize GRBs for both clarifying their nature, internal structure, and origin together with employing these objects for cosmological purposes [12,13].

In this review, we first introduce the concept of GRB and their main observable quantities. As stated above, according to time duration, we introduce the role of the t_{90} duration to classify GRBs following the standard guidelines and underline the issues related to such a classification, e.g., ultra-long GRBs and X-ray flashes. To this end, we introduce the concepts of GRB progenitor, showing quantitatively the physical reasons that limit GRBs to be fully considered as genuine standard candles. However, we also emphasize how using luminosity correlations found in prompt and afterglow phases would be useful to characterize some sort of standardization technique. In this respect, we portray the main observable quantities coming from GRBs and deeply introduce the standard picture of GRB formation and evolution, dubbed the *fireball model*.

From all the above aspects, we expect GRBs to able to reconcile the cosmic expansion history at small and intermediate redshifts, connecting de facto late with early times, trying to open new windows toward the comprehension of cosmology. We therefore explain how GRBs serve as complementary probes to frame DE and cosmic expansion throughout the universe evolution, together with other standard candles, e.g., type Ia SN (SNeIa), baryonic acoustic oscillation (BAO), cosmic microwave background (CMB), Hubble differential data, etc. We show how to combine such data sets with GRBs and write the main features of experimental analysis for cosmological purposes. Great emphasis will be devoted to the circularity problem that essentially plagues cosmology with GRBs. Once introduced, we also underline strategies that do not take into account its role for fitting cosmological models with GRBs.

Hence, we provide how to challenge the standard cosmological model, namely the ΛCDM paradigm, with GRBs. To do so, we provide the main and evident features of cosmology with GRBs by showing how to perform error analyses, Bayesian treatments, and how to handle systematics for several GRB correlations. We therefore develop model dependent and independent techniques of calibrations and report a few numerical outcomes related to GRBs, showing the most recent cosmological bounds, found with distinct procedures.

The review is split as follows. After this short introduction, in Section 2, we classify GRBs and we report the most interesting properties, among all the classification, the progenitors, and the main observable quantities coming from GRBs. In Section 3, we work out the standard GRB model, namely the fireball paradigm. Here, we also discuss about particle and radiative processes, giving emphasis to the possible emissions coming from GRBs. In Section 4, we start introducing the concept of cosmology with GRBs. We thus highlight distance indicators and the concept of standard candles. In Section 5, we explain in detail the experimental tools useful for getting Bayesian analysis with GRBs. Finally, in Section 6, we provocatively report the concept of standardizing GRBs to permit those objects to be used in the same manner as other probes. Several issues have been raised in Section 7, although likely the most serious one, the circularity problem, is described in detail in Section 8, where we also stress the opposite view in which one can also avoid calibration. Last but not least, we report the most recent developments of cosmology with GRBs in Section 9, while we conclude our journey in Section 10 with our final outlooks and perspectives of this work.

2. GRB Classification and Properties

To achieve a recognized GRB classification, the strategy is to take into account the most relevant astronomical properties of such objects. Thus, as the most prominent GRB component is represented by the prompt γ-ray emission, it is straightforward to use it to define GRB classes based on similarity criteria.

The prompt γ-ray emission is characterized by highly-variable and multi-peaked light curves composed of either overlapping or distinct pulses with variable duration. The duration of these pulses spreads within a wide time range. Since the duration is not fixed a priori, it is natural to wonder whether one can arbitrarily define a time in which the above measures can be obtained. Hence, it is a consolidated convention to take the total burst duration in a time interval, dubbed t_{90}, evaluated in the observer frame over which the 90%, from 5% to 95%, of the total background-subtracted counts are experimentally detectable.

In view of such a property, one gets a plausible classification, as we report below.

2.1. Classification: Short and Long GRBs

The light curve analysis of the first BATSE GRB catalogue showed a clear bimodal distribution of the t_{90} duration, separated at roughly 2 s, and in the hardness ratio (HR), namely the ratio of the total counts of the hard 100–300 keV energy band over the softer 50–100 keV band [3,14,15].

This leads to the widely-adopted classification into

- *short–hard* ($t_{90} \lesssim 2$ s) GRBs, hereafter SGRBs,
- *long–soft* ($t_{90} \gtrsim 2$ s) GRBs, hereafter LGRBs.

The significance of such a classification scheme has been strengthened with the full 2704 GRBs detected by BATSE and later GRB missions, providing strong evidence for two GRB progenitor channels (see, e.g., Figure 1).

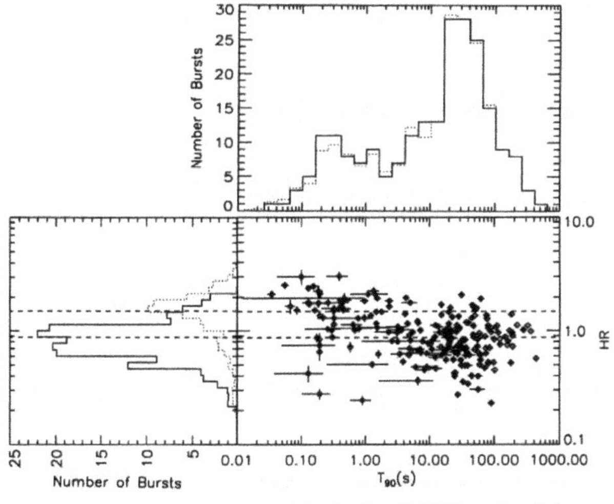

Figure 1. GRB distribution provided by the first BATSE catalog, lying on the t_{90}–HR plane. The solid HR histogram shows LGRBs, whereas the dotted one is for SGRBs. The dashed horizontal lines mark mean HRs for both classes. The solid t_{90} histogram represents the raw data whilst the dotted one shows the error-convolved data, credit from Ref. [3].

However, a significant overlap in the distributions of SGRBs and LGRBs suggests that *a more robust classification scheme based on physical properties is still missing*.

2.2. Intermediate GRBs?

We ended the previous subsection with asserting the need of a more robust classification order. This scheme is veritably challenged by the existence of an intermediate class of SGRBs with extended emission (SGRBEEs), characterized by an initial short duration and spectrally-hard γ-ray pulse, followed by a softer emission lasting up to tens of seconds [15,16]. Depending on the sensitivity and energy range of the GRB alert instrument and, based on the above classification scheme, a SGRBEE could be classified as short or long. A GRB detector with low sensitivity at low-energy ranges in γ-ray could detect only the initial hard part of the burst (resulting in an SGRB), whereas a GRB detector with a higher sensitivity extending down to lower energies could also detect the softer extended emission (falling in the LGRB class).

A possible explanation to the origin of this extended emission involves a highly magnetized neutron star (NS) dipole spin-down emission (see Ref. [17] and Sections 2.4.2 and 3.1.3).

2.3. Ultra-Long GRBs and X-ray Flashes

Furthermore, the detection of rare events characterized by extremely long-lived prompt emissions lasting $\gtrsim 10^3$ s, named ultra-long GRBs (ULGRBs), represents an additional classification threat, since it is still unclear whether ULGRBs represent a distinct class of LGRBs [18], or whether they are the high-end tail of the t_{90} distribution of LGRBs [19].

Finally, it has been reported the existence of extragalactic transient X-ray sources, dubbed X-ray flashes (XRFs), with spatial distribution, spectral, and temporal characteristics similar to LGRBs [20,21]. The distinguishing properties of XRFs are

(a) their observed prompt emission spectrum that peaks at energies which are an order of magnitude lower than those of standard LGRBs;
(b) their time integrated flux in the 2–30 keV X-ray band greater than that in the 30–400 keV γ-ray band.

In view of these hazy results, classifying GRBs through t_{90} and HR criteria only turns out to be puzzling, since the measured t_{90} varies with energy range. Thus, the definition of a novel GRB classification scheme requires multi-wavelength criteria to better understand the physical properties behind the GRB emission.

In this respect, attempts to recategorize GRBs from the popular long/short classes have been made in Ref. [22], introducing alternative classes of *Type I* and *Type II GRBs*. According to this scheme, the Type I class includes short/hard GRBs and SGRBEEs with no SN association, typically found in regions of their host galaxy with low star formation, and very likely originating in compact star mergers (see details in the next Section 2.4). On the other hand, the Type II class includes long and relatively soft GRBs with SN association, usually found in star forming regions within irregular host galaxies, and thus associated with young stellar populations and likely originating in the core-collapses of massive stars (again, see details in the next Section 2.4).

Though the above scheme seems to be promising, further research on this issue is still ongoing. Therefore, for historical reasons in Section 2.4, we pursue the description of the progenitor systems keeping the bimodal classification in LGRBs and SGRBs.

2.4. Progenitors and Open Questions

Beside the above discussion, the working definition of LGRBs and SGRBs suggests the existence of two different progenitor channels. In summary,

I: LGRBs could arise from the core-collapse of a massive star or *collapsar* [23],
II: SGRBs could originate from the binary neutron star-black hole (NS-BH) or NS-NS mergers [24].

The huge observed isotropic equivalent energy release of $\sim 10^{49}$–10^{55} erg implies that: for LGRBs, up to $\sim 10\ M_{\odot}$ are converted into radiation during the prompt emission duration of ~ 100 s, whereas SGRBs up to $\sim 1\ M_{\odot}$ are converted into radiation within ~ 1 s [25]. The energy reservoir and the efficiency of the involved physical processes in producing the emitted energy represent a stringent requirement, especially for LGRBs[4].

The commonly called *jets* substantially alleviate this issue by reducing the GRB energy release by jet's correction factor $f = 1 - \cos\theta$. Jets can be thought of, in an oversimplified picture, as outflows of relativistic matter ejected into a double-cone structure of opening angle θ. In general, the jet correction is poorly constrained because it requires very challenging measurements of θ and the observer viewing angle relative to the jet axis. This makes it troublesome to distinguish between geometric and dynamical effects. Indeed, very soft GRBs could be bursts viewed off-axis, whereas low luminosity GRBs may be the result of large jet opening angles [15].

Measurements of θ can be obtained by the predicted signature of the achromatic *jet breaks*, observable in the afterglow light curve at all frequencies. This feature can be explained by the dynamics of the GRB ejecta as follows. At the beginning, at high values of the bulk Lorentz Γ factor[5], the ejecta is narrowly beamed into the jets while its Lorentz factor is $\Gamma^{-1} < \theta$ and, regardless of the hydrodynamic evolution, a GRB is observed only from a small fraction of the ejecta [15]. As the ejecta decelerates, Γ decreases below θ^{-1}, the beaming angle becomes larger, and a larger portion of the ejecta becomes observable. Continuous deceleration leads to the point that the entire surface of the jet is observable and the jet begins to spread sideways, producing a break in the light curve across the entire afterglow spectrum [27,28]. The sharpness of this break and the change in the afterglow decay rate depend on how long the jet remains collimated and on the jet radial density profile and energy distribution [29,30]. The time of the jet break is related to the jet opening angle, the bulk Lorentz factor, and the density of the circumburst medium (CBM). The above description has two effects:

- an *"on-axis"* observer detects the prompt emission and then, as the jet decelerates, the afterglow emission and finally a *jet-break* due to the faster spreading of the emitted radiation;
- an *"off-axis"* observer cannot detect the prompt emission but detects an *orphan afterglow*, namely an afterglow without a preceding GRB.

In the pre-*Swift* era, simultaneous breaks in the optical and near-infrared (NIR) afterglow light curves were frequently interpreted as jet breaks. Nevertheless, the improved temporal and spectral coverage of GRB afterglows, especially in X-rays by *Swift*, have revealed within the first few hours after the prompt emission a complex structure made of flares, plateaus, and chromatic breaks [31–34]. The detected achromatic breaks are observed in a few cases. The absence of jet-break signatures in most GRB afterglows has been interpreted as due to the over-simplified assumption homogenous jets with sharp edges, whereas more complex models now include structured jets that produce several chromatic jet-breaks, or much smoother breaks, or jets that can keep their structure for longer than previously thought making difficult to detect breaks without a wide temporal coverage [30].

Besides the jet modeling issue, any GRB model has to deal with features like very luminous X-ray flares occurring up to a few 10^4 s after the GRB trigger and with shape and spectra similar to those flares observed during prompt emission and extended plateau phases that last for a few hours during the early afterglow evolution [33,34]. Both features imply an extended central-engine activity with a continuous source of energy injection lasting the above 10^4 s. In the standard picture, such long-lived energy injection requires the accretion of a significant mass onto the central BH via very large ($\sim 1\ M_{\odot}$) and low-viscosity ($\alpha < 10^{-2}$) accretion disk formed at the core collapse time, or via fall-back material continuously replenishing the accretion disk [35].

2.4.1. The LGRB-Supernova Connection

The possible connection between LGRB and massive progenitor stars has been speculated long before the first afterglow detection [23,36]. The first observational evidence came with the association between the broad line Type Ic (Ic-BL) SN 1998bw and the low-luminous LGRB 980425 at $z = 0.0085$ and with lack of an optical afterglow [37]. Later on, this association was also confirmed between the Type Ic-BL SN 2003dh, temporally and spatially coincident with the standard more luminous long GRB 030329 at $z = 0.1685$, with an optical afterglow light curve comparable with other cosmological GRBs [38].

The launch of *Swift* has increased the sample of GRB-SN pairs, both spectroscopic, at $z \lesssim 0.5$ and most of them with isotropic-equivalent γ-ray energies $E_{\rm iso} < 10^{49}$ erg, and photometric, in the form of SN bumps appearing in the optical afterglows 10–30 days[6] after the GRB, at $z \gtrsim 0.5$ and $E_{\rm iso} \approx 10^{51} - 10^{52}$ erg [36]. Most of the GRB-SN pairs belong to this second kind, very likely at the hand of a selection effect: the more common low-luminosity LGRBs per unit volume are not detectable at high redshift, whereas luminous LGRBs, with higher detectability at high redshift, are observed from a larger volumetric area [39].

SNe Ic associated with some long GRBs are characterized by no hydrogen (H) and no weak helium (He) lines [36]. Their occurrence close to star-forming regions offers very strong evidence that long GRBs could be associated with massive star death [36]. In this regard, the best progenitor candidates are the Wolf–Rayet stars, very massive stars with a hydrogen envelope largely depleted, endowed with a fast rotation [23,40]. Within the collapsar model, very massive stars are able to fuse material in their centers all the way to iron (Fe). At this point, they cannot continue to generate energy through fusion and collapse mechanisms forming a BH. Matter from the star around the core rains down towards the center and swirls into a high-density accretion disk. In this picture, the core carries high angular momentum to form a pair of relativistic jetsout along the rotational axis where the matter density is much lower than in the accretion disk. Jets propagate through the stellar envelope at velocities approaching the speed of light, creating a relativistic shock wave at the front [15,41]. If the star is not surrounded by a thick, diffuse H envelope, the leading shock accelerates as the stellar matter density decreases. Thus, by the time it reaches the star surface, $\Gamma \geq 100$ is attained and the energy is released in the form of γ-ray photons [15,41].

The collapsar model attempts to explain the time structure of GRBs' prompt emission, through the modulation of the jets by their interaction with the surrounding medium, which could produce the variable Lorentz factor needful for internal shock occurrence [23]. As the relativistic jet propagation through the stellar envelope of a collapsing star proceeds, its collimation was shown to occur analytically and numerically [15]. Another prediction of this model is the prolonged activity of the central engine which can potentially contribute to the GRB afterglow [23,40,41]. This occurs because the jet and the disk are inefficient at ejecting all the matter in the equatorial plane of the pre-collapse star and some continues to fall back and accrete [23,40,41].

The SNe associated with LGRBs appear to belong to the bright tail of type Ic SNe and can be considered as a "subclass" of SNe Ic, alternatively addressed as *hypernovae*, in order to emphasize the extremely high energy involved in these explosions. Remarkably, the SNe associated with both low- and high-luminous (XRFs and normal LGRBs, respectively) share very similar spectra and their peak luminosities span only two orders of magnitude, whereas the associated GRBs isotropic luminosities span six orders of magnitude [36]. Another distinctive feature of the GRB-SN pairs is the high photospheric expansion velocity, up to $0.1c$ [36]. In this scenario, one has to also fit the class of ULGRBs. The spectroscopic detection of the SN 2011kl coincident with the ULGRB 111209A [42] favors a common core-collapse origin for LGRBs and ULGRBs. This SN exhibited a peculiar, very blue and featureless spectral shape, which was unlike other SNe Ic associated with LGRBs, but more like the newly discovered class of superluminous SNe [43]. Other ULGRBs have either been too far or too dust-extinct to secure any detection of an underlying SN, whereas other cases proved the indicative flattening from a rising SN in their optical and NIR light curves at 10–20 days after the GRB trigger [36].

In this picture, however, exceptions to the LGRB-SN association have been found from deep optical observations in two nearby bursts, GRB 060505 and GRB 060614, for which the hypothetical accompanying SN would have been a hundred times fainter than SN 1998bw [44–46].

To conclude, ULGRBs and SNless LGRBs give proof for the existence of further progenitor channels for LGRBs.

2.4.2. SGRBs, Macronovae, and Gravitational Waves

The *Swift* satellite has enabled rapid and precise localizations and an increase in the number of X-ray and optical afterglow detection of both LGRBs and SGRBs. However, SGRBs have less luminous afterglows than those of LGRBs and this fact makes difficult to obtain optical spectra and a precise burst location to plan optical follow-up to search for host galaxy associations. The lack of any associated core-collapse SNe, the typically large offsets of the GRB position with respect to galaxy center, and the frequent association with galaxies with no ongoing star formation, provide evidence in support of a compact binary merger progenitor scenario [47].

The proposed progenitors for SGRBs are NS–NS and/or NS–BH binary mergers [24,48–50]. These mergers take place as binary orbits decay due to gravitational radiation emission [51]. A merger releases 5×10^{53} erg, but most of this energy is due to low energy neutrinos and gravitational waves. Thus, there is enough energy available to produce a GRB, notwithstanding how a merger generates the relativistic wind required to power a burst is still the object of speculations and not well understood. It has been argued that about one out of thousand of these neutrinos annihilates and produces pairs that in turn produces γ-rays via $\nu\bar{\nu} \rightarrow e^+e^- \rightarrow \gamma\gamma$, but it has been pointed out that a large fraction of the neutrinos would be swallowed by the newly-born BH [15].

A further confirmation to the binary merger scenario consists of the detection of the so-called *macronova* (MN). The MN emission originates in NS-BH or NS-NS mergers from a fast-moving, rapidly-cooling ejected debris of neutron-rich radioactive species that decay to form transient emission and create atomic nuclei heavier than iron through neutron capture process, named the *r-process* [52]. The opacities of these produced heavy elements lead to a dim MN emission, requiring deep follow-up observations down to NIR bands. The first indication of a MN, in the form of a re-brightening, detected approximately nine days after the GRB trigger, has been obtained by extensive follow-up of the SGRB 130603B, one of the nearest and brightest SGRBs ever detected [53]. An MN emission accompanies also the nearest SGRB ever detected, SGRB 160821B [54,55], and the recently detected SGRB 200522A [56]. For a list of other MN emissions, see Ref. [57].

In the binary merger scenario, SGRBs are expected to be significant sources of gravitational waves (GWs). The smoking gun occurred on 17 August 2017, when the Advanced LIGO and Virgo detectors observed the event GW 170817, unambiguously detected in spatial and temporal coincident with the SGRB 170817A independently measured by the *Fermi* Gamma-ray Burst Monitor, and the Anti-Coincidence Shield for the Spectrometer for the International Gamma-Ray Astrophysics Laboratory [58][7].

As a further confirmation on the nature of the progenitor system of SGRB 170817A, an intense observing campaign from radio to X-ray wavelengths over the following days and weeks after the trigger led to the spectroscopic identification of a MN emission, dubbed AT 2017gfo [59].

The observation of SGRB, GW, and MN emission has improved our understanding of the physical properties related to the binary merger, such as the mass of the compact object, the ejected mass, and the details of the CBM surrounding the merger site.

2.5. Observable Quantities from GRBs

Understanding GRB physics passes through the experimental evidence of the energy that can be collected from detectors. In particular, we can start discussing about GRB prompt emission. It is typically observed in the hard-X (above ~ 5 keV) and γ-ray energy domain.

The operative duration of the prompt emission is due to the previously defined t_{90}. Within this time interval, and also within any sub-interval with enough photons to perform a significant analysis[8], the observed spectral energy distribution (SED) of GRBs is non-thermal, and it is best fitted by a phenomenological model composed of a smoothly joined broken power-law called *Band* model [60] (see Figure 2). Its functional form is

$$N_{\rm E}(E) = K \begin{cases} \left(\dfrac{E}{100}\right)^{\alpha} \exp\left[\dfrac{(2+\alpha)E}{E_{\rm p}^{\rm obs}}\right] & , \ E \leq \left(\dfrac{\alpha-\beta}{2+\alpha}\right) E_{\rm p}^{\rm obs} \\ \left(\dfrac{E}{100}\right)^{\beta} \exp(\beta-\alpha) \left[\dfrac{(\alpha-\beta)E_{\rm p}^{\rm obs}}{(2+\alpha)}\right]^{\alpha-\beta} & , \ E > \left(\dfrac{\alpha-\beta}{2+\alpha}\right) E_{\rm p}^{\rm obs} \end{cases} \quad (1)$$

where typical power-law index values are $-1.5 \lesssim \alpha \lesssim 0$ (with an average $\langle\alpha\rangle \simeq -1$) and $-2.5 \leq \beta \leq -1.5$ (with an average $\langle\beta\rangle \simeq -2$), while the peak energy at the maximum of the of the $E^2 N_{\rm E}$ (or $EF_{\rm E}$) spectrum lies within 100 keV $\leq E_{\rm p}^{\rm obs} \lesssim$ few MeV (with an average of $\langle E_{\rm p}^{\rm obs}\rangle \simeq 200$ keV). Finally, K is the normalization constant with units of photons $\rm cm^{-2} s^{-1} keV^{-1}$. In some cases, the SED is also best fitted by a power-law model[9] composed of a power-law plus an exponential cutoff. However, these models are purely mathematical, i.e., not yet physically linked to GRB intrinsic properties. Hence, fitting data with them do not provide any insight about the emission physical origin but may be useful for the classification scheme of GRBs and for comparing the fitted results with the predictions of different theoretical models.

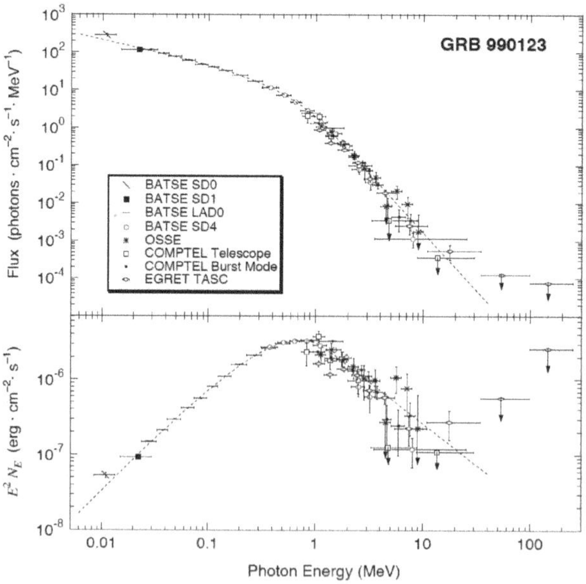

Figure 2. Band spectral model applied to the data of GRB 990123. In the upper panel, the photon spectrum is shown; in the lower panel, the $E^2 N_{\rm E}$ (or $EF_{\rm E}$) spectrum is shown. Courtesy from Ref. [60].

In the recent years, with a much broader spectral coverage enabled by detectors such as Fermi, evidence for more complicated broad band spectra fitted by a combined Band+thermal model has been found in an increasing number of bursts [61–64], where the peak of the thermal component is always observed below $E_{\rm p}^{\rm obs}$.

However, the search of the best-fit model in describing GRB prompt emission spectra depends on the analysis method. Typically, a significant spectral analysis is performed when enough photons are collected. For weak bursts, only time-integrated spectral analyses can be done, and this implies that important time-dependent features may be lost or averaged, leading to a wrong theoretical interpretation. Another issue is that the chosen spectral model is convolved with the detector response and, because of the nonlinearity of the detector response matrix, this procedure cannot be inverted. Therefore, two different models can equally provide a similar minimal difference between the model and the detected counts' spectrum and lead to different theoretical interpretations.

From the fit of the time-integrated prompt emission spectrum, one can get the flux F (in units of erg cm^{-2}s^{-1}) on a detector energy bandpass $E_{\rm min}$–$E_{\rm max}$ as

$$F = \kappa \int_{E_{\rm min}}^{E_{\rm max}} E N_{\rm E}(E) dE\,, \tag{2}$$

where κ is a constant, commonly used to convert the energy, expressed in keV, to erg.

To compute the total energy emitted by a GRB in all wavelengths, a bolometric spectrum is needed. However, the GRB prompt emission triggers γ-rays detectors in a given energy bandpass; therefore, a limited part of the spectrum is available, instead of a bolometric one. Moreover, GRBs are cosmological sources spread over a wide redshift range, so, for GRBs observed by the same detector, the measured energy range corresponds to different energy bands in the cosmological rest frame of the sources.

To standardize all GRBs, fluxes are computed in the fixed rest-frame band 1–10^4 keV, which is a range larger than that of most of the γ-ray detectors. The "bolometric" time-integrated flux is then given by

$$F_{\rm bolo} = F \times \frac{\int_{1/(1+z)}^{10^4/(1+z)} E N_{\rm E}(E) dE}{\int_{E_{\rm min}}^{E_{\rm max}} E N_{\rm E}(E) dE}\,, \tag{3}$$

and the total isotropically-emitted energy and luminosity are, respectively,

$$E_{\rm iso} = 4\pi d_L^2 F_{\rm bolo} t_{90} (1+z)^{-1} \tag{4}$$
$$L_{\rm iso} = 4\pi d_L^2 F_{\rm bolo} \tag{5}$$

where the factor $(1+z)^{-1}$ corrects the t_{90} duration from the observer frame to the GRB cosmological rest-frame. In a similar way, the peak luminosity $L_{\rm p}$, computed from the observed peak flux $F_{\rm p}$ within the time interval of 1 s around the most intense peak of the burst light curve and in the rest frame 30–10^4 keV energy band[10], is given by

$$L_{\rm p} = 4\pi d_{\rm L}^2 F_{\rm p}\,. \tag{6}$$

The luminosity distance $d_{\rm L}$ depends upon the cosmological models adopted as backgrounds and can be related to the continuity equation recast as

$$\frac{d\rho}{dz} = 3\left(\frac{P+\rho}{1+z}\right), \tag{7}$$

which relates the total energy density ρ and pressure P to the barotropic factor $\omega(z) \equiv P/\rho$ of a given cosmological model. For a two component flat background cosmology composed of standard pressure-less matter with $\omega = 0$ and a generic DE component with $\omega(z)$ (dubbed generically ωCDM), the luminosity distance is then given by[11]

$$d_L(z) = (1+z)\frac{c}{H_0}\int_0^z \frac{dz'}{\sqrt{\Omega_m(1+z')^3 + \Omega_x f_x(z')}}, \tag{8}$$

where H_0 is the Hubble constant, Ω_m and Ω_x are the cosmological density parameters of matter and DE, respectively, and $f_x(z)$ is given by

$$f_x(z') = \exp\left[3\int_0^{z'} \frac{1+w(\bar{z})}{1+\bar{z}} d\bar{z}\right]. \tag{9}$$

For the concordance paradigm, namely the ΛCDM model, the DE equation of state is $w(z) \equiv -1$ corresponding to a cosmological constant Λ. Thus, $f_x \equiv 1$ and $\Omega_x \equiv \Omega_\Lambda$. In the following, the choice $w(z) \equiv -1$ is adopted, unless otherwise specified.

The above isotropic energy output can be corrected for the beaming (see Section 3), once the jet opening angle θ is known, leading to beam corrected energy

$$E_\gamma = (1 - \cos\theta)E_{iso}. \tag{10}$$

It is important to stress that the prompt emission is not limited to the γ-rays and that, differently from the afterglow emission starting $\sim$ 100 s after the GRB trigger, current information in other energy bands is extremely difficult to observe without fast triggering. Observations at lower energies (optical and X-rays) have been enabled only for GRBs with a precursor or a very long prompt emission duration, which gave the possibility of performing fast pointing to the source during the prompt phase [66].

Regarding the GeV energy domain, a delayed (with respect to the trigger), long lived emission ($\gtrsim 10^2$ s), and separate lightcurve [67] with a decaying luminosity as a power law in time, $L_{GeV} \propto t^{-1.2}$ has been observed [67]. These distinctive features point towards a separate origin of the GeV with respect to the lower energy photons.

After $\sim$ 100 s since the trigger, the prompt emission starts to decay in flux and, in many cases, this feature is caught by X-ray detectors *Swift*-XRT within the 0.3–10 keV energy band. In general, X-ray afterglow light curves show complex behaviors [15] consisting of (see Figure 3):

(1) an early steep decay, interpreted as the tail of the prompt emission at large angles, followed by a very shallow decay, called the *plateau*, usually accompanied by spectral parameter variations, and
(2) a final decay, less steep than the first one.

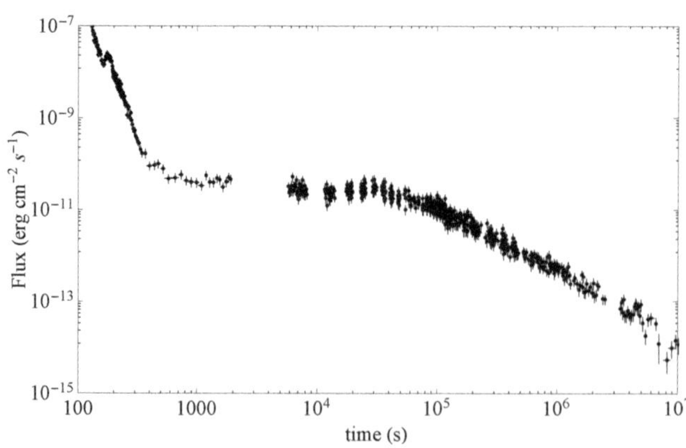

Figure 3. The X-ray afterglow of GRB 060729 with all the three power-law segments and an initial flare clearly shown.

The X-ray afterglow is also characterized by the presence of a flaring activity [15]. The observed behavior of these flares, the rapid rise, and exponential decay together with a fluence comparable in some cases to the prompt emission, points out that the same mechanism for the prompt emission is responsible for the flaring activity [15]. Concluding, as already stressed above, these X-ray afterglow features are important to understand the nature of GRB progenitors.

Timescales and Characteristic Energy as Observable Signature of GRBs

There are other GRB observable quantities often employed in the literature, e.g., to construct GRB correlations (see Section 6.1). They span from characteristic energies to timescales measured in several wavelengths. More specifically, a selection of them is summarized below:

- $t_{\rm b}$, the time at which the late X-ray afterglow power-law decline suddenly steepens due to the slowing down of the jet until the relativistic beaming angle roughly equals the jet opening angle θ.
- $\tau_{\rm lag}$, the time lag is computed as the difference of arrival times to the observer of the high energy photons ($100-300$ keV) and low energy photons (25–50 keV)[12].
- $t_{\rm X}$, the rest-frame time, defined by a broken power-law fit of the X-ray afterglow light curve, at which a late power-law decay after the plateau phase is established.
- τ, the rest-frame time marking the end of the plateau phase, defined from a fit of the X-ray afterglow with a smooth function given in Ref. [68].
- $F_{\rm X}$ and F_0 are the observed X-ray fluxes respective to $t_{\rm X}$ and τ, whereas the corresponding rest-frame 0.3–10 keV luminosities $L_{\rm X}$ and L_0 are computed as follows:[13]

$$L_{\rm X/0} = 4\pi d_{\rm L}^2 F_{\rm X/0} \frac{\int_{0.3/(1+z)}^{10/(1+z)} E N_{\rm E}^{\rm X/0}(E) dE}{\int_{0.3}^{10} E N_{\rm E}^{\rm X/0}(E) dE} = 4\pi d_{\rm L}^2 F_{\rm X/0} (1+z)^{\gamma-2} \tag{11}$$

  where we used the fact that X-ray data are observed by the *Swift*-XRT in the 0.3–10 keV energy band and the SED is in general a power-law spectrum with $N_{\rm E}^{\rm X/0}(E) \propto E^{-\gamma}$ and power-law index $\gamma > 0$.
- V, the variability of the GRB light curve. It is computed by taking the difference between the observed light curve and its smoothed version, squaring this difference, summing these squared differences over time intervals, and appropriately normalizing the resulting sum.

3. Theory of GRB Progenitors

GRBs require progenitor systems able to guarantee enough energy for their powerful explosions to occur and emission mechanisms that can explain the above discussed spectral features. Although it is essential to better understand the physics of GRBs, neither clear evidence for consolidated classes of suitable progenitors nor a definitive GRB model have been yet established, as stressed above. However, observations, in the form of GRB spectra and light curves (see Section 2.5) and correlations between observable quantities (see Sections 6.1 and 7), enhanced our comprehension of these phenomena and led to a general agreement on a few aspects listed below [69]:

- GRB progenitors harbor a BH[14] which acts as a central engine powering the GRB emission.
- The burst energy must be gravitational, and it is released in a very short time and from a compact region.
- Substantial part of this energy is converted into kinetic energy and a relativistic jetted outflow is formed.
- The acceleration process and the role played by magnetic fields are still unclear.
- The dissipation of part of the kinetic energy produces the observed prompt emission.
- The thermal emission of the prompt emission may be the relic of the photons emitted during the initial explosion, whose energy has not been converted into kinetic form.

- Afterward, relativistic jets interact with the CBM, gradual energy conversion occurs, and the afterglow (from X-ray down to radio) is produced.

The observed spectra have a considerable amount of γ-ray photons. Photons with high energy E_1 annihilate with those at a low energy E_2 and produce e^+e^- pairs if $\sqrt{E_1 E_2} \gtrsim m_e c^2$ (up to an angular factor), where m_e is the electron mass. If GRBs were not relativistic sources, the observed light curve variability time scale of $\delta t \approx 10$ ms would imply that their emission would originate from a very compact region not larger than $R = c\delta t \approx 3000$ km. For typical values of the luminosity distance $d_L \approx 3\text{Gpc} \approx 10^{22}$ cm and fluence $S \approx 10^{-7}\text{erg}\,\text{cm}^{-2}$ (energy at the detector per unit area) of GRBs, the opacity for pair creation is enormous, and it is given by [69]

$$\tau_{\gamma\gamma} = f_{e^\pm} \frac{\sigma_T d_l^2 S}{m_e c^2 (c\delta t)^2} \approx 10^{14} f_{e^\pm} \tag{12}$$

where $f_{e^\pm}$ is the fraction of photons with energies sufficient to produce pairs and σ_T is the Thomson cross-section. Such a large optical depth would imply that that the source must be optically thick leading to a thermal spectrum. On the contrary, observations indicate that GRB spectra are typically non-thermal, pointing to the opposite conclusion that their source must be optically thin. This issue is called the *compactness problem* [69].

However, the problem is only apparent, once relativistic effects are taken into account. In fact, the causality limit of a source moving relativistically with bulk Lorentz factor $\Gamma \gg 1$ towards the observer is $R \leq \Gamma^2 c\delta t$. Consequently, the observed photons are blue-shifted and their energy at the source is lower by a factor $\approx 1/\Gamma$, which may be insufficient for pair production. This leads to a decrease in the opacity, by a factor $\Gamma^{-2(\beta+1)}$, where the β is the high-energy power-law index of a photon spectrum of the burst. For $\Gamma \gtrsim 100$, one obtains the optically thin condition of the source. Ultra-relativistic expansion of GRBs is unprecedented in astrophysics. There are indications that relativistic jets in active galactic nuclei have $\Gamma \sim 2$–10, but some GRBs have $\Gamma \gtrsim 100$. These large expansion velocities in GRB outflows find confirmations from the radio scintillation observed in their afterglows, and also from the apparent observation of self-absorption in the radio spectrum of the afterglow, where it is possible to obtain independent estimates of the dimensions of the afterglow relic [15].

3.1. The Fireball Model

The GRB standard model considers a homogeneous *fireball* [69]. For a pure radiation fireball, a large fraction of the initial energy released by the newly-formed BH is converted directly into photons. Close to the BH, at a radius r_0 larger than the Schwarzschild radius, $R_S = 2GM/c^2$, the photon temperature is

$$T_0 = \left(\frac{L}{4\pi\, a\, c\, r_0^2}\right)^{1/4} = 1.2\, L_{52}^{1/4} r_{0,7}^{-1/2}\,\text{MeV} \tag{13}$$

where a is the radiation constant, and the luminosity L and the radius r_0 are expressed, respectively, as $L_{52} = L/10^{52}$ erg/s and $r_{0,7} = r_0/10^7$ cm. In the following, to understand the order of magnitude of the key physical parameters characterizing GRBs, we use the notation $Q_x = Q/10^x$, where the quantity Q is given in cgs units. The temperature T_0 is above the threshold for pair production, hence a large number of $e^\pm$ pairs are created via photon–photon interactions, leading to a fully thermalized pairs-photons plasma with the opacity in Equation (12)[15].

GRB luminosities are many orders of magnitude above the Eddington luminosity, $L_E = 1.25 \times 10^{38} (M/M_\odot)$ erg s^{-1}; therefore, the radiation pressure is much larger than self gravity and the fireball expands under its own pressure up to $\Gamma \approx 10^2$–10^3 [48,49]. Since the final kinetic energy cannot exceed the initial explosion energy E_{tot}, the maximum attainable Lorentz factor is defined as $\Gamma_{\max} = E_{\text{tot}}/Mc^2$ and depends upon the amount of baryons (baryon load) of rest mass M within the fireball [69].

3.1.1. Photon-Dominated Scenario

The simplest scenario considers a photon-dominated expanding shell of width $\delta r'$ "instantaneously" releasing its energy. From here on, prime symbols indicate quantities measured in the comoving frame of the shell, namely from an observer within it. On the other hand, r is the radial coordinate of the laboratory frame, a frame outside the shell where the observer is sitting on the central engine.

Enforcing energy and entropy conservation laws, the shell keeps accelerating up to $\Gamma_{\max} \simeq \eta$, which is attained at at the so-called dissipation radius $r_s \sim \eta r_0$; beyond it, most of the internal energy of the shell has been converted into the kinetic one, so the flow no longer accelerates and it coasts. Thus, the fireball obeys the following scaling laws of the shell comoving temperature, Lorentz factor, and comoving volume, respectively

$$\begin{cases} T'(r) \propto r^{-1} & , \quad \Gamma(r) \propto r \quad , \quad V'(r) \propto r^3 \quad , \quad r < r_s \\ T'(r) \propto r^{-2/3} & , \quad \Gamma(r) = \eta \quad , \quad V'(r) \propto r^2 \quad , \quad r \gtrsim r_s \end{cases} . \tag{14}$$

from which it follows that, as the shell accelerates (as Γ increases with r), its internal energy drops (as T' decreases with r). Finally, the evolution of the observed temperature is given by

$$T^{ob}(r) = \Gamma(r) T'(r) = \begin{cases} T_0 & , \quad r < r_s \\ T_0 (r/r_s)^{-2/3} & , \quad r \gtrsim r_s \end{cases} . \tag{15}$$

3.1.2. Internal Shock Scenario and Photospheric Emission

In the case of LGRBs, the progenitor continuously emits energy at a rate L, over a longer duration $t \gg r_0/c$, and ejects mass at a rate $\dot{M} = L/(\eta c^2)$. In this case, the scaling laws for the instantaneous release are still valid, provided that E is replaced by L and M by $\dot{M}$, and a further equation for the mass conservation of the baryons (within the spherical symmetry assumption) is required [72]

$$n'_p(r) = \frac{\dot{M}}{4\pi r^2 m_p c \Gamma(r)} = \frac{L}{4\pi r^2 m_p c^3 \eta \Gamma(r)} \tag{16}$$

where $n'_p(r)$ is the comoving number density of baryons and m_p is the proton mass.

For longer activity of the inner engine, fluctuations in the energy emission rate would result in the propagation of independent shells, each of them with analogous thickness r_0 and dynamics. For two consecutive shells with a difference in their Lorentz factors $\delta\Gamma \sim \eta$ or velocities $\delta v \approx c/(2\eta^2)$, collisions become possible after a typical time $t_{col} = r_0/\delta v$ and an observer frame radius [73]

$$r_{col} = v t_{col} \simeq c t_{col} \simeq 2\eta^2 r_0 . \tag{17}$$

Above r_{col}, which is a factor η larger than r_s, collisions occur, dissipate the kinetic energy, and convert it into the observed radiation [74,75]. The advantages of the internal shock scenario are listed below:

1. **Light curve variability.** The time delay between the photons produced by the collisions and a photon emitted from the central engine towards the observer, i.e., $\delta t^{ob} \simeq r_{col}/(2\eta^2 c) \sim r_0/c$, is similar to the central engine variability and can explain the observed variability ($\gtrsim 1$ ms).
2. **Particle acceleration.** Shell collisions generate internal shock waves, which can accelerate particles to high energies via Fermi mechanism and produce γ-rays.
3. **Thermal radiation.** Equation (13) states the fireball is optically thick [48,49,76]. For $r > r_s$, an effective photosphere radius $r_{ph} \simeq L\sigma_T/(8\pi m_p c^3 \Gamma \eta^2) \simeq 2 \times 10^{11}\, L_{52}\, \eta_{2.5}^{-3}$ cm can defined by requesting $\tau(r_{ph}) = 1$ [77]. Internal shocks take place at $r_{col} \sim \eta r_s > r_{ph}$. In a more realistic picture, photons decouple the plasma on "photospheric surface" [78] and the emerging emission is a convolution of different Doppler boosts

and different adiabatic energy losses of photons [62,78]. This emission explains the thermal-like emission embedded in the non-thermal spectra of some GRBs [62,79,80].

However, the internal shock scenario manifests some drawbacks.

1. **Efficiency.** From the energy and momentum conservation, the kinetic energy dissipation is highly efficient only if two shells have masses $m_1 \simeq m_2$ and Lorentz factors $\Gamma_1/\Gamma_2 \gg 1$. The average over several collisions leads to a low global efficiency of 1-10% [81,82], which contrasts with the much higher efficiency $\sim 50\%$ inferred from afterglow measurements [31,33]. Higher efficiency up to the $\sim 60\%$ can be attained by considering larger contrasts $\Gamma_1/\Gamma_2 \gg 10$ [82]. However, these Lorentz factor contrasts unlikely occur within the traditional collapsar or the merger scenarios.
2. **Observed spectra.** This model does not explain the observed spectra and needs further assumptions on how the dissipated energy produces photons (i.e., involving standard radiative processes such as synchrotron emission or Compton scattering).

3.1.3. Magnetized (or Poynting-Flux Dominated) Outflows

The Poynting-flux dominated model speculates that the gravitational energy produces very strong magnetic fields, which may be crucial in the jet formation of GRBs, similarly to the Active Galactic Nuclei (AGN), where magnetic energy is converted into particle acceleration via Blandford–Znajek [83] or Blandford–Payne [84] mechanisms. The idea behind this model is that the collapse of a white dwarf (WD) induced by accretion from a massive star, the core collapse of a massive star, or NS merger does not immediately form a BH, but rather a rapidly-spinning (with a period of ~ 1 ms) and highly-magnetized NS (with a magnetic field $B \gtrsim 10^{15}$ G) NS, known as *magnetar* [70]. The maximum amount of magnetic energy that can be stored is $\sim 2 \times 10^{52}$ erg, and it can be extracted in a short timescale of ~ 10 s and drives a jet along the polar axis of the NS powering the prompt emission [71]. The decay of rotational or magnetic energy may continue to power late time flaring or afterglow emission. The dipole radiation naturally produces a plateau phase up to the dipole spin-down time scale [15].

In this model, the magnetic field is essentially toroidal (i.e., $\vec{B} \perp \vec{\beta}$) and its polarity in the flow changes on small scale defined by the light cylinder in the central engine. The total luminosity is given by $L = L_k + L_M$, where $L_k = \Gamma \dot{M} c^2$ is the kinetic part and $L_M = 4\pi r^2 c[B^2/(4\pi)]$ is the magnetic part [85,86]. The key parameter is the magnetization $\sigma \equiv L_M/L_k = B^2/(4\pi\Gamma^2 n m_p c^2)$, which plays a similar role to the baryon load in the classical model and defines the maximum attainable Lorentz factor $\Gamma_{\max} \approx \sigma^{3/2}$, whereas, during the acceleration phase, one gets $\Gamma(r) \propto r^{1/3}$ [85,86].

In this model, the rapid variability observed in GRBs and the low efficiency in dissipating the kinetic energy via shock waves in highly magnetized plasmas are still open issues. Recent recipes suggest that central engine variability leads to the ejection of magnetized plasma shells which expand due to internal magnetic pressure gradient and collide at a distance r_{col}. The ordered magnetic field lines of the ejecta get distorted and fast reconnection occurs. The induced relativistic turbulence may be able to overcome the low efficiency difficulty of the classical internal shock scenario [87].

3.1.4. Particle Acceleration

To produce the non-thermal GRB spectra, part of the kinetic energy needs to be dissipated and used to increase the random motion of the outflow particles and/or accelerates some fraction of them to a non-thermal distribution. Once accelerated, these high-energy particles emit non-thermal photons.

The most widely proposed particle acceleration mechanism within the internal shock scenario is the Fermi mechanism [88]. In this process, the accelerated particles cross the shock fronts, and, during each crossing, their energy increases at a constant rate $\Delta E/E \sim 1$. The accelerated particles have a power-law energy distribution $N(E) \propto E^{-\delta}$ with index $\delta \approx 2.0$–2.4 [89].

Dissipation mechanisms in magnetized outflows have been discussed at length (see Ref. [90] and references therein). Furthermore, particles may also be accelerated via Fermi mechanism in shock waves, but it has been pointed out that, in highly magnetized plasma, this process may be inefficient [91].

3.1.5. Radiative Processes

After kinetic energy dissipation and particle acceleration, energy conversion is needed to produce the non-thermal spectra observed in GRBs. The most discussed radiative model in the literature is the optically thin synchrotron emission [74,92–95], accompanied by synchrotron-self Compton (SSC) at high energies [96–98].

Recently, with mounting evidence of thermal components in GRB spectra [63,99], the photospheric model has acquired growing relevance [61,100,101]. This model is not in contrast with and has to be considered complementary to the synchrotron+SSC emission, which originates from a different region of the outflow.

- *Synchrotron Emission*

The synchrotron emission has been extensively studied for first interpreting the non-thermal emission in AGNs and then the GRB afterglow emission [102]. Regarding the explanation of the GRB prompt emission spectra [74,93–95,103], the synchrotron emission model has several advantages:

(1) requires energetic particles and strong magnetic fields, both expected in shock waves;
(2) has a broad-band spectrum with characteristic peak, associated with the observed peak energy;
(3) for typical parameters, energetic electrons radiate nearly 100% of their energy.

A source at redshift z, expanding with $\Gamma = \sqrt{1-\beta^2}$ and at an angle θ with respect to the observer, emits photons which are seen with a Doppler boost $\mathcal{D} = [\Gamma(1-\beta\cos\theta)]^{-1}$. In the comoving frame, electrons move in a magnetic field B and thus have random Lorentz factor γ_e. Their typical energy is [102]

$$\varepsilon_{ob} = \frac{3q\hbar B\gamma_e^2}{2m_e c}\frac{\mathcal{D}}{(1+z)} = 1.75\times10^{-19}B\gamma_e^2\frac{\mathcal{D}}{(1+z)}\,\mathrm{erg}\,. \tag{18}$$

Typical GRB peak energies $\varepsilon_{ob} \approx 200$ keV require strong magnetic fields and very energetic electrons, both feasible for Poynting flux-dominated outflows or photon-dominated outflows where strong magnetic fields may be generated via Weibel instabilities [104][16].

On the other hand, strong magnetic fields imply the comoving cooling time of the electrons to be $t'_{cool} \lesssim t'_d \sim R/(\Gamma c)$. Thus, the expected synchrotron spectrum below the peak energy would be $F_\nu \propto \nu^{-1/2}$ (or $N_E \propto E^{-3/2}$) [106,107], which is inconsistent with the average low energy spectral slope $\langle\alpha\rangle = -1$ (see Figure 2) and, hence, the value $\alpha = -3/2$ is called "synchrotron line of death". To overcome this problem, electrons must cool slowly, leading to a spectrum below the peak given by $F_\nu \propto \nu^{1/3}$ (or $N_E \propto E^{-2/3}$), which is roughly consistent with the observations. However, the condition $t'_{cool} \gtrsim t'_d$ leads to high values of γ_e, whereas B would be very low and, in order to explain the observed flux, the electron energy would be several orders of magnitude higher than that stored in the magnetic field [108]. To overcome this, the inverse Compton contribution has to be significant, producing $\sim$ TeV emission. To avoid a substantial increase of the total energy budget, the emission radius should be $R \gtrsim 10^{17}$ cm but cannot explain the rapid variability observed [108].

Suggested modifications (and drawbacks) to the synchrotron scenario can be found in the literature [25,98,109–113].

- *Photospheric Emission*

For $r_{ph} \gg r_s$, a large fraction of the kinetic energy is dissipated below the photosphere [114]. The produced non-thermal photons cannot directly escape and are advected with the flow until the transparency. Within the flow, multiple Compton scatterings occur and modify the synchrotron spectrum of the heated electrons, which rapidly cool, mainly by IC scattering. The electron distribution becomes quasi-Maxwellian, with a temperature determined by the balance between heating (external and by direct Compton scattering of energetic photons), and cooling (adiabatic and radiative) [115]. Finally, the photon field is modified by the scattering from the quasi-Maxwellian electron distribution [114].

Furthermore, the thermal photons of the fireball contribute as seed for IC scattering, hence the non-thermal electrons, heated by energy dissipation below the photosphere, rapidly cool, and reach a quasi-steady state distribution [101]. The result is a *two temperature plasma*, with electron temperature $T_e > T_{ph}$. If dissipation processes occur at intermediate optical depth τ ~few–few tens, the resulting spectrum is:

(1) similar to the Rayleigh–Jeans part of the thermal spectrum, for $T < T_{ph}$;
(2) $F_\nu \propto \nu^{-1}$ (or $N_E \propto E^{-2}$) because of multiple Compton scattering, for $T_{ph} < T < T_e$;
(3) an exponential cutoff, for for $T > T_e$.

The spectral slope obtained in the above 2) is similar to the high energy spectral slope in GRB spectra, $\langle\beta\rangle \sim -2$; thus, it could be concluded that E_p is associated with T_{ph}. However, recently, Fermi data have shown thermal peaks at lower energies than E_p, which points rather to the more natural interpretation that the thermal peak is associated with T_{ph} and suggests that E_p may be associated with T_e or with the synchrotron emission. Moreover, if dissipation occurs at $\tau \gtrsim 10^2$, the resulting spectra is thermal-like. On the other hand, for $\tau \lesssim$, a few more complex spectrum forms, with the main contribution coming from synchrotron photons (emitted by the electrons) below the thermal peak and above it from multiple IC scatterings (leading to a nearly flat energy spectrum) [115]. All of the above discussions are viable for dissipation processes from highly magnetized plasma as well [86,116].

However, the above model also suffers two major drawbacks, since it cannot explain

(1) low energy spectral slopes less steep than the Rayleigh—Jeans part of a Planck spectrum;
(2) the observed GeV emission, which may originate from some dissipation above the photosphere.

4. Reconciling Cosmological Indicators to GRBs

After the first part of this review, in which we faced the main properties of GRBs, their possible theoretical background and progenitors, we are now in a condition to relate our understanding directly to cosmology. In fact, for cosmological purposes, it is essential to get the distance of astronomical objects and thus the use of GRBs would help in computing such distances up to very large redshifts. In particular, source physical parameters mostly depend on luminosity and size and then cosmic bounds can be inferred if there exists a relation between distances and redshifts. This prerogative is intimately related to two distinct concepts, i.e., distance indicators[17] and standard candles[18].

Below, we elucidate the main properties of such objects and the most important consequences they have in observational cosmology.

4.1. Distance Indicators

At the beginning of our review, we emphasized how distances in cosmology are relevant to compute GRB luminosity/energy. A further step consists of noticing the distance measurements are classifiable by

- **Absolute measures**, as they are computed through previously known information, e.g., trigonometric parallax.
- **Relative measures**, as they involve empirical relations based on indirect or direct probes, e.g., Cepheids period–luminosity relation, for which the distance measures are calibrated against an absolute method to enable those measurements to be somehow *anchored*.

Standard cosmology shows how to relate the redshift to metric distances in both of the above cases. The machinery of dynamical distance indicators involves tightly packing all the ingredients of cosmological physics. We thus require the cosmological principle to hold in an expanding universe in the context of general relativity. Despite it being obvious, there is no direct analogy to classical dynamical distance indicators, as the laboratory in which measurements are obtained is moving as well. Precision cosmology would enrich data during the incoming years, as future surveys will provide resources of data to constrain and refine our understanding about distances and cosmological parameters.

Using current data catalogs, it appears evident that GRBs can be significantly investigated once the calibration of the correlation functions are deduced from absolute confidence. Recently, techniques of non-calibration have been more often used, overcoming the problem of standardizing GRBs that are, as known, not perfect standard candles for cosmological distance tests. Later on, we confront the calibration and non-calibration procedures, emphasizing how to single out the most promising treatment to handle GRBs in cosmology.

4.2. Standard Candles

Above, we stated astrophysical distances are crucial for picturing the current universe. Though essential, estimating cosmic distances mainly remains a complicated prerogative. In view of the above classification, the distance estimation passes through the use of *standard candles*. These objects hold the fundamental property of relating the intrinsic luminosity, namely L, to some known property, enabling one to get constraints over it. Once the luminosity is known, the distance can be computed accordingly.

A standard procedure is to get measures of the energy emitted from astrophysical objects. The energy bounds are obtained in a precise time interval, say Δt and by virtue of $E = L \cdot \Delta t$, i.e., the relation between luminosity and energy, it is possible to get distances from the energy itself, through a well-consolidated strategy, reported below.

Detectors are able to catch fractions E_d of the emitted energy E, which is proportional to the ratio between the detector area A and the spherical shell $4\pi d_L^2$ in which one defines the cosmic distance d_L, i.e.,

$$E_{\rm d} = \frac{E\,A}{4\pi d_L^2}\,. \tag{19}$$

A general relation for $d_L(z)$ is written as

$$d_L(z,\theta) = c\,(1+z)\int_0^z \frac{{\rm d}z'}{H(z',\theta)} \tag{20}$$

where the set of free parameters to constrain is indicated by θ. Exploring a given cosmological model is equivalent to obtaining θ.

Thereby, combining the aforementioned quantities, we obtain the energy per unit detector area A and per unitary time Δt, which defines the flux expressed by

$$F = \frac{E_{\rm d}}{A\,\Delta t} = \frac{L}{4\pi d_L^2}\,. \tag{21}$$

As we highlighted, the luminosity L is known for standard candles, thus one can measure F in order to get a given astrophysical object distance.

4.3. Classifying Standard Candles

We above stressed that physical laws underlying a particular astronomical object permit one to know the luminosity of standard candles. Clearly, such rules are essentially based on thermodynamic or chemical processes of a given astrophysical object. Consequently, one can classify standard candles by means of these physical laws and, according to the simplest classification scheme, we can handle at least two kinds of standard candles summarized below [117].

- *Standard candles as primary distance indicators*, which can be calibrated within the Milky Way galaxy.
- *Standard candles as secondary distance indicators*, which can be observed at larger distances than Milky Way scales. However, they require calibration, typically performed using known primary distance indicators within distant galaxies.

4.3.1. Primary Distance Indicators

The above first typology mainly includes *Variable stars*, i.e., among which, Cepheids, RR Lyrae, and Mira. Here, the variable star type is based on the possible correlation between their period of variation, steadily measured, and their luminosity. Even though this set of stars mainly constitutes the primary indicators, further typologies are main sequence and red clump stars. Here, using the luminosity-temperature relations from the standard Hertzsprung–Russell diagram, one deduces stellar luminosity within a fairly narrow range. Last but not least, eclipsing binaries are also primary distance indicators, since their luminosity is computed by the Stefan–Boltzmann law through a direct estimate of their radius, by means of a Doppler measurement of orbital velocities combined with the light–curve data, together with the temperature, deduced from the spectrum.

4.3.2. Secondary Distance Indicators

On the other hand, the second class of standard candles is essentially based on very different indicators with respect to the first case. For instance, the prototypes of such indicators are the *properties of galaxies*, among all, the Tully–Fisher relation. This law matches spiral galaxy rotation speed and stellar luminosity. In particular, to argue the spiral galaxy rotation speed, one can consider, for example, the spectral line width. Another relation, widely adopted as an underlying second type of indicator, is the Faber–Jackson relation. Here, it is possible to infer elliptical galaxy random stellar velocities using the total luminosity. Again, the way to get these velocities consists of the use of spectral line widths. Another quite relevant relation is the fundamental plane law, i.e., a treatment that extends the Faber–Jackson one by including surface brightness as an additional observable parameter.

Besides galaxy properties, another second typology of standard candles is represented by *SNe Ia*, i.e., probably the most used cosmological standard candles to accredit the late time cosmic speed up. The scenario in which they form is due to thermonuclear explosions of WDs that exceed the Chandrasekhar's limit, namely $\sim 1.4M_\odot$. For such objects, we see a correlation between the time scale of the explosion and the peak luminosity. The corresponding light curves follow given shapes, in agreement with the so-called Phillips curve [118]. As stated, SNe Ia are the most fruitful standard candles. For each event, even if the luminosity is clearly different for every SN, the Phillips curve relates the B magnitude peak to the luminous decay after 15 days with an overall set of SNe distributed in the range $z = 0$–2.5. These redshifts span between decelerating and accelerating phases of universe's evolution, corresponding to the matter and DE dominated epochs[19]. Last but not least, these indicators are present in all galaxies, except in the arms of spiral galaxies, but their physical internal processes are still the object of investigations as they are not fully-interpreted.

5. Going Ahead with Standard Indicators: The χ^2 Analysis

Using standard candles, it is possible to establish data catalogs that can be used and matched with GRB data. Hence, to experimentally fit a given model with a given set of free parameters, one requires the definition of a merit function that quantifies the overall agreement between the working model with the aforementioned cosmic data. Equivalently, it is of utmost importance to get best fit parameters and corresponding estimates of error bars, together with a method to possibly measure the goodness of fit. The parameter fitting treatment commonly makes use of least-squares analyses, based on the combination among data points, say D_i, a model for these data, namely the $y(x, \vec{\theta})$, function of θ. Naively, the simplest approach to least squares for uncorrelated data becomes

$$\chi^2 = \sum_i w_i [D_i - y(x_i|\vec{\theta})]^2 \tag{22}$$

where the weights w_i reach the maximum variance in case $w_i = 1/\sigma_i^2$, with σ_i the data point errors. For correlated data, we have

$$\chi^2 = \sum_{ij} (D_i - y(x_i|\theta)) Q_{ij}^{-1} (D_j - y(x_j|\theta)) \tag{23}$$

in which the inverse of covariance matrix, Q, has been introduced describing the degree of correlations among data. Minimizing the χ^2 is equivalent to getting suitable sets of findings that represent the best fit for our procedure. Different χ^2 values lead to probability distribution around the minimum.

5.1. Probability Distribution

Analyzing the probability distribution, once the above treatment is worked out, becomes essential. In particular, probabilities p that the observed χ^2 exceeds by chance a value $\hat{\chi}$ for the correct model is clearly calculable and, in fact, Q provides a measure of the goodness of fit, as one infers it at the minimum of χ^2. Two limiting cases, unfortunately, are possible, Q is too small or too large. The first occurrence leads to the fact that the model is either wrong or errors are underestimated and/or they do not distribute Gaussian. The second occurrence happens when either errors are overestimated or data are correlated while rarely it could also happen that the distribution is non-Gaussian.

In general, the statistical procedure suggests that χ^2 is roughly comparable with the data number. Consequently, using the reduced chi square, as the ratio between the chi square and the number of degrees of freedom, could be a useful trick to handle experimental workarounds.

5.2. The SNe Ia Measurements

SNe are widely-adopted in astrophysics as standard candles. Thereby, several SN catalogs are often updated, furnishing today a large number of data points that combined with other data sets enable one to fix tighter constraints over the universe expansion history in terms of its constituents. In particular, SNe Ia are likely the most used objects that constrain DE at late times. The standard procedure makes use of the luminosity distance $d_L(z)$ and of apparent magnitude. A general relation for $d_L(z)$ has been previously written, with θ the set of free parameters of a given model. Then, we can notice that exploring a given cosmological model is equivalent to getting the whole set of parameters, θ.

In particular, when one adopts a given cosmological model, then an indirect requirement naturally holds: *the underlying cosmological model is the most suitable one*. This is clearly a limitation because this hypothesis does not always coincide with the most feasible statistical model. Thus, more than one scenario can lead to subtle bounds, indicating a degeneracy problem among different models. This justifies the need of analyzing different cosmological paradigms working out data set hierarchy, i.e., combining more than one data catalog. In addition, statistical criteria are also crucial to check the goodness of a given paradigm.

For SNe Ia, by virtue of Equation (20), it is possible to relate the brightness to fluxes to get the distance modulus

$$\mu(z) = 25 + 5 \log\left(\frac{d_{\rm L}}{\rm Mpc}\right). \tag{24}$$

Neglecting error bars on z, we underline errors on μ, namely σ_μ, whereas the best fit is determined by the standard maximization of the underlying likelihood function, or simply minimizing the χ^2, provided by

$$\chi^2(\theta_{min}) = \sum_{i=1}^{N_s} \left[\frac{\mu_i(z_i, \theta) - \mu_{obs,i}(z_i)}{\sigma_{\mu,i}} \right]^2 \tag{25}$$

where the subscript *min* refers to the set of values that minimize the chi square function, as requested above. Theoretical models can be therefore tested by χ^2 statistics, leading to probing DE by inferring d_L in units of megaparsecs and using it by means of the apparent magnitude.

Again, intertwining more than one data set with other surveys is quite essential to determine the whole set of parameters, with refined accuracy. For instance, SNe alone, as well as GRBs[20], H_0 cannot be arguable. In fact, expanding up to the first order the luminosity distance, valid up to $z \lesssim 0.001$, one gets

$$d_{\rm L}(z, H_0) \simeq \frac{cz}{H_0} \tag{26}$$

that clearly vanishes at $z = 0$, implying that H_0 cannot be constrained with SNe Ia alone. In addition, a multiplicative degeneracy between H_0 and the other free parameters occurs.

Once the chi square statistic is computed, the confidence regions are planes with fixed χ^2. For example, one can get $\Omega_M - \theta_i$ planes by marginalizing the likelihood functions over H_0. This procedure consists of integrating the probability density $p \propto \exp(-\chi^2/2)$ for all values of H_0. Marginalization is a generic technique, clearly not limited to H_0. In fact, one who desires to simultaneously constrain a few parameters and in the meantime wants to get the corresponding probability distribution regardless of the values of a given parameter, say $\theta^\star$, can proceed with marginalizing. Let us call $\theta^\star$ the parameter we do not care about; the marginalized probability density, computed for example for Ω_m, is given by $p(\Omega_m) = \int d\theta^\star\, p(\Omega_m, \theta^\star)$.

5.3. BAO Measurements

The BAO measurements are due to overdensity of baryonic matter due to acoustic waves. These waves propagate in the early universe [119,120] and represent the *standard ruler* for cosmological length scale. This signature, in the large-scale clustering of galaxies, constrains cosmological parameters by detection of a peak in the correlation function [121], by defining the A parameter as follows:

$$A = \frac{\sqrt{\Omega_m}}{z_1} \left[\frac{z_1}{E(\mathbf{x}, z_1)} \frac{1}{|\Omega_k|} \mathrm{sinn}^2 \left(\sqrt{|\Omega_k|} \int_0^{z_1} \frac{dz}{E(\mathbf{x}, z)} \right) \right]^{\frac{1}{3}} \tag{27}$$

where $\mathbf{x}$ is the set of cosmological density parameters, $E(\mathbf{x}, z) = H(\mathbf{x}, z)/H_0$, and $\mathrm{sinn}(x) = \sinh(x)$ for the curvature parameter $\Omega_k > 0$, $\mathrm{sinn}(x) = x$ for $\Omega_k = 0$, and $\mathrm{sinn}(x) = \sin(x)$ for $\Omega_k < 0$. The A parameter has been measured from the SDSS data and reads to be $A = 0.469(0.95/0.98)^{-0.35} \pm 0.017$, with $z_1 = 0.35$, so the χ^2 in terms of A reads $\chi^2_{\rm BAO} = (A - 0.469)^2/0.017^2$. The BAO corresponding angular distance measures can be defined by means of

$$d_z(\mathbf{x}, z) \equiv r_s(z_d) \left[\frac{cz}{H(\mathbf{x}, z)} \right]^{-1/3} \left[\frac{d_{\rm L}(\mathbf{x}, z)}{1+z} \right]^{-2/3}. \tag{28}$$

The corresponding χ^2 is given by

$$\chi^2_{\rm BAO} = \sum_{i=1}^{N_{\rm BAO}} \left[\frac{d_{\rm z}^{\rm th}(\mathbf{x}, z_i) - d_{\rm z,i}^{\rm obs}}{\sigma_{d_{\rm z,i}}} \right]^2. \tag{29}$$

It is clear that BAO measures are slightly model-dependent as they depend on the comoving sound horizon $r_s(z_d)$. In particular, in Equation (28), the sound horizon depends upon the baryon drag redshift z_d. This quantity requires calibration that typically is performed with CMB data, adopting a given background model that commonly is the ΛCDM scenario. Very often, the best expected values are given by $z_d = 1059.62 \pm 0.31$ and $r_s(z_d) = 147.41 \pm 0.30$ [122].

5.4. Differential Age and Hubble Measurements

Another intriguing treatment, widely used in observational cosmology and also for calibrating GRB correlations, has been firstly proposed in Ref. [123]. The idea is to measure the Hubble rate by using galaxies, in a quite model-independent way. In the context of GRBs, the Hubble catalog has been widely explored. For example, in Ref. [123], the core idea is to match the observational Hubble rate data (OHD) with model independent expansion of H made by Bézier polynomials. At a first glance, this *differential age method* (see, e.g., Refs. [124,125]) does not require any assumption over the form of H, although spatial curvature can affect the overall treatment if it varies with time, instead of being fixed[21].

To better introduce the method, we notice that it is well known that spectroscopic measurements of the age difference Δt and redshift difference Δz of couples of passively evolving galaxies lead to$\Delta z/\Delta t \equiv dz/dt$ and so, if galaxies formed at the same time (redshift z), the Hubble rate can be approximated by

$$H(z) = -(1+z)^{-1}\Delta z/\Delta t\,. \tag{30}$$

Consequently, model-independent estimates may come from cosmic chronometers based on the assumption that observable Hubble rates are given by the exact formula

$$H_{obs} = -\frac{1}{(1+z)}\left(\frac{dt}{dz}\right)^{-1} \tag{31}$$

if approximated as in Equation (30). The χ^2 from the current 31 OHD measurements reads

$$\chi^2_{OHD} = \sum_{i=1}^{31}\left[\frac{H_{th}(\mathrm{x}, z_i) - H_{obs}(z_i)}{\sigma_{H,i}}\right]^2. \tag{32}$$

This procedure has the great advantage of directly considering H without passing through any cosmic distance.

5.5. The $\mathcal{R}$ Parameter

The CMB represents a cosmic recombination epoch remnant and contains abundant early universe information. Consequently, the acoustic peak positions [120,126] can be used to characterize a given cosmological model by means of the shift parameter, defined as [127]

$$\mathcal{R} = \frac{\sqrt{\Omega_M}}{\sqrt{|\Omega_k|}}\mathrm{sinn}\left(\sqrt{|\Omega_k|}\int_0^{z_{\rm ls}}\frac{dz}{E(\mathbf{x},z)}\right) = 1.70 \pm 0.03\,. \tag{33}$$

The last scattering redshift, namely $z_{\rm ls}$, is fixed to

$$z_{\rm ls} = 1048[1 + 0.00124(\Omega_b h^2)^{-0.738}][1 + g_1(\Omega_M h^2)^{g_2}], \tag{34}$$

where $g_1 = 0.078(\Omega_b h^2)^{-0.238}[1 + 39.5(\Omega_b h^2)^{0.763}]^{-1}$ and $g_2 = 0.56[1 + 21.1(\Omega_b h^2)^{1.81}]^{-1}$ [128] and the χ^2 reads

$$\chi^2_{\rm CMB} = \frac{(\mathcal{R} - 1.70)^2}{0.03^2}. \tag{35}$$

In analogy with BAO measures, the shift parameter is not fully model-independent.

5.6. Confidence Levels and Uncertainties

As one performs fits combining GRBs with other observable quantities, meaningful information on the best-fit parameters is achieved by computing their confidence limits or contour plots, which define the allowed parameter phase-space. These are essentially regions constructed around a set of best fit parameters obtained from computation. One does not mind about the number of dimensional parameter space, namely m, corresponding de facto to the number of parameters, since, to make those regions compact, one holds constant χ^2 boundaries, fixing the chi squared values to specific numbers. Thus, one takes m to be the number of parameters, n the number of data, and p to be the confidence limit that one desires to reach. Assuming to shift by solving $Q[n-m, min(\chi^2)+\Delta\chi^2] = p$, and to find the parameter region where $\chi^2 \leq min(\chi^2)+\Delta\chi^2$, immediately one gets the requested confidence region. Once the regions have been computed, it is necessary to obtain uncertainties. To do so, expanding the log likelihood in Taylor series $\ln\mathcal{L} = \ln\mathcal{L}(\theta_0) + \frac{1}{2}\sum_{ij}(\theta_i - \theta_{i,0})\left.\frac{\partial^2\ln\mathcal{L}}{\partial\theta_i\partial\theta_j}\right|_{\theta_0}(\theta_j - \theta_{j0}) + \ldots$, we define the Hessian matrix by

$$\mathcal{H}_{ij} = -\frac{\partial^2\ln\mathcal{L}}{\partial\theta_i\partial\theta_j}. \tag{36}$$

Since its non diagonal terms indicate correlated parameters, one can assume the errors on a given i parameter to be $1/\sqrt{\mathcal{H}_{ii}}$. This naive representation of errors is a coarse-grained approach, dubbed conditional error, not frequently adopted in the literature. On the other hand, one can compute the Fisher information matrix, as a forecast expression for error bars

$$F_{ij} = \langle\mathcal{H}\rangle = -\left\langle\frac{\partial^2\ln\mathcal{L}}{\partial\theta_i\partial\theta_j}\right\rangle, \tag{37}$$

with the ensemble average over observational data. In analogy to conditional errors, we write $\sigma_{ij}^2 \geq (F^{-1})_{ij}$, while the marginalized errors become $\sigma_{\theta_i} \geq (F^{-1})_{ii}^{1/2}$.

We underlined above that the Fisher matrix is somehow related to error bars. In this respect, we mean that the Fisher Information matrix enables to estimate the parameters errors before the experiment is performed. Hence, it permits to explore different experimental set ups that could optimize the experiment itself. For these reasons, the Fisher matrix is largely adopted in the literature.

5.7. Binning Procedure

In several cases, it is useful to get constraints directly on the universe equation of state. Thus, fitting it for the late-time universe constituents is extremely important to understand the dark energy evolution. In particular, pointing out a possible variation of the equation of state of dark energy is essential to *disentangle* the standard model predictions from possible theoretical extensions and, in this respect, GRBs can be seen as intermediate redshift probes to disclose such an evolution.

To do that, an intriguing strategy consists of binning the dark energy equation of state, say w, in short intervals of z and then fit w in each bin, assuming it is constant in each bin. Indicating with a generic function $f(z)$ the dark energy evolution, we have

$$f(z_{n-1} < z \leq z_n) = (1+z)^{3(1+w_n)}\prod_{i=0}^{n-1}(1+z_i)^{3(w_i - w_{i+1})}, \tag{38}$$

where w_i is the barotropic factor within the ith redshift bin. The bin is built up by an upper boundary at z_i, whereas the zeroth bin is defined as $z_0 = 0$.

Therefore, uncorrelated sub-equations of state in every bin can be experimentally refined adding data points and, in particular, GRBs, being calibrated as we will discuss later. Several indications have shown good agreement with the standard paradigm, up to

$z \simeq 9$, albeit relevant deviations have been found, indicating that the situation is not still clear.

6. Standardizing GRBs

Being successful in standardizing GRB data is of utmost importance to characterize new data catalogs up to high redshifts. In particular, getting redshifts, or more generally spectroscopic observations, is essential for GRB-related science, as we summarized below:

(1) Computing the luminosity function for GRBs, constructing it from the prompt emission as well as afterglows. This treatment is analogous to what we do for SNe Ia.
(2) Computing the redshift distribution of GRBs. This enables one to use GRBs as tracers for the cosmic star-formation history. Consequently, spotting very high redshift GRBs will shed light on their distribution at intermediate epochs of the universe evolution.
(3) Studying the host galaxies, in particular those faint, high-redshift galaxies that are unlikely to be found and studied with other methods, characterizing the dust extinction curves of high-z galaxies.
(4) Studying GRB-selected absorption line systems and probing cosmic chemical evolution with GRBs.
(5) Studying if and how much GRBs can be used for determining the cosmological parameters of dark energy models and/or to rule out a few models. Analogously, the use of GRBs can be tested in view of determining cosmographic parameters, i.e., getting model independent bounds over the cosmic evolution.

6.1. GRB Correlations and Related Issues

Since the first discovery of GRBs independent groups has found different correlations that represent a key to using GRBs for cosmological purposes, the basic idea is to intertwine different quantities of such objects among them. The observable quantities of interest are in relation with the cosmological model that lies on the background. This fact permits GRBs to be distance indicators at a first glance but limits their use because it requires postulating the underlying cosmological model, providing a circularity in the process itself, which is known as the circularity problem.

The widest majority of GRB correlations prompts the same requirement: the GRB standardization in terms of cosmological tools. Attempts for new correlations have been severely investigated, relating different observable quantities with each other. The way in which this is realized provides the theoretical interpretation behind the relation itself. In other words, evidence for a given correlation leads to interpreting particular physical processes. Thus, achieving the goal of standardizing GRBs brings the certainty of getting feasible bounds on cosmological parameters. Intriguingly, a narrow set of correlations enables one to also estimate GRB redshifts. Even though this is still under speculation, in general, a wide number of correlations could provide information about GRB progenitors.

More precisely, standardizing GRBs for cosmological purposes aims at reaching further hints toward progenitors of *different groups of GRBs*. Multi-wavelength instruments of recently-adopted satellites have significantly increased the number of GRBs that could be observed to check the validity of a given relation. Thus, it is even possible that a few correlations may be derived from experimental evidence, instead of theoretically. Unfortunately, this could open further issues related to data processing whose outputs can be biased in the overall computations.

Going ahead, it is certainly possible to constantly observe new hints undertaking novel correlations to allow free theoretical speculations that deeply probe into new physics beyond the standard comprehension of GRBs.

6.2. Prompt Emission GRB Correlations

The correlations that make use of prompt emission quantities are listed below with the corresponding properties of each of them. For details on the involved quantities, see Section 2.5.

6.2.1. $L_{\rm iso}$–$\tau_{\rm lag}$ Correlation

This correlation holds for LGRBs and indicates that the more luminous bursts possess shorter time lags, i.e., $L_{\rm iso} \propto \tau_{\rm lag}^{-1.25}$ [129]. It has been used as a GRB redshift indicator and to constrain cosmological parameters. However, the existence of this correlation is challenged by recent studies. For details, see Ref. [130].

6.2.2. $L_{\rm iso}$–V Correlation

This correlation holds for LGRBs and indicates that the more luminous bursts have the more variable light curves [131]. However, the intrinsic scatter is very large, and the index is still not completely settled, also in view of the fact that the variability V is different for various instruments. For details, see Ref. [130].

6.2.3. Amati or $E_{\rm p}$–$E_{\rm iso}$ Correlation

This correlation is of the form $E_{\rm p} \propto E_{\rm iso}^{0.52}$ and shows that $E_{\rm iso}$ is correlated with the rest-frame spectral peak energy, namely $E_{\rm p} = E_{\rm p}^{\rm obs}(1+z)$ [132]. Observations by *Swift* and *Fermi* detectors confirmed this correlation for LGRBs. An analogous $E_{\rm p}$–$E_{\rm iso}$ correlation with a slope similar to that of LGRBs but a larger value of the normalization holds also for SGRBs, though with a much smaller data set [133]. Moreover, the $E_{\rm p} - E_{\rm iso}$ correlation also holds within individual GRBs using time-resolved spectra, and the slopes are consistent with the correlation from time-integrated spectra. For details, see Ref. [130].

6.2.4. Yonetoku or $L_{\rm p}$–$E_{\rm p}$ Correlation

This correlation reads $L_{\rm p} \propto E_{\rm p}^2$ [65] and holds for both LGRBs and SGRBs [133]. Similarly to the $E_{\rm p}$–$E_{\rm iso}$ correlation, the $L_{\rm p}$–$E_{\rm p}$ correlation holds also within individual GRBs using time-resolved spectra. For details, see Ref. [130].

6.2.5. Ghirlanda or $E_{\rm p}$–E_γ Correlation

This represents a tight, less scattered, correlation between $E_{\rm p}$ and E_γ, valid for LGRBs [134]. One of the major drawbacks is the lack of achromatic breaks in the *Swift* afterglow light curves of most of the GRBs. This fact limits the increase in the correlation sample. For details, see Ref. [130].

6.3. Prompt and Afterglow Emission Correlations

The following correlations involve prompt and afterglow emission observables. Below, we report the most common correlations. For details on the involved quantities, see Section 2.5.

6.3.1. Liang–Zhang or $E_{\rm p}$–$E_{\rm iso}$–$t_{\rm b}$ Correlation

It is a correlation valid for LGRBs among $E_{\rm iso}$, $E_{\rm p}$ and the rest-frame break time in the optical band $t_{\rm b}$, i.e., $E_{\rm p} \propto E_{\rm iso}^{0.52} t_{\rm b}^{0.64}$ [135]. If we take the optical break time as the jet break time, this correlation is similar to the Ghirlanda one. However, the inclusion of additional GRBs made this correlation more scattered. For details, see Ref. [130].

6.3.2. Dainotti or $L_{\rm X}$–$t_{\rm X}$ Correlation

This correlation links the X-ray luminosity $L_{\rm X}$ and rest-frame time $t_{\rm X}$, i.e., $L_{\rm X} \propto t_{\rm X}^{-1}$, at the time when the X-ray afterglow light curve establishes a power-law decay after the plateau phase [136]. This correlation holds for LGRBs and SGRBEEs. By adding a third parameter, $E_{\rm iso}$, the new correlation of the form $L_{\rm X} \propto t_{\rm X}^{-0.87} E_{\rm iso}^{0.88}$ [137] has been found. However, both relations are quite scattered and seem to be a selection effect due to the flux detection limit of *Swift*-XRT instrument. For details, see Ref. [130].

6.3.3. $E_{\rm iso}^{\rm X}$–$E_{\rm iso}$–$E_{\rm p}$ Correlation

This is a universal correlation for both LGRBs and SGRBs which links $E_{\rm iso}$ and $E_{\rm p}$ to the isotropic energy of the X-ray afterglow $E_{\rm iso}^{\rm X}$ computed in the rest-frame energy band 0.3–

30 keV, i.e., $E_{iso}^{X} \propto E_{iso}^{1.00} E_{p}^{0.60}$ [138]. However, due to the fact that this correlation depends upon two cosmology-dependent quantities, it is unsuitable to constrain cosmological parameters. For details, see Ref. [139].

6.3.4. Combo Correlation

The correlation represents the combination of the Amati correlation and the E_{iso}^{X}–E_{iso}–E_{p} correlation and, like the Amati correlation, it holds for LGRBs only [140]. It relates the prompt emission E_p and the X-ray afterglow luminosity L_0 and rest-frame duration τ of the plateau phase, and the late power-law decay index α, i.e., $L_0 \propto E_{p}^{0.90} \tau^{-1} |1+\alpha|$. The main drawback is that it is much more complicated than other correlations, as it depends upon four parameters. For details, see Ref. [141].

6.3.5. *L*–*T*–*E* Correlation

This correlation connects the rest-frame end time t_X and luminosity L_X of the X-ray afterglow plateau phase with E_{iso}, i.e., $L_X \propto t_X^{-1.01} E_{iso}^{0.84}$ [142]. This correlation is very similar to the Combo correlation, but, unlike it, the *L*–*T*–*E* one holds for a few SGRBs and requires being corrected for the redshift evolution effects. For more details, see Ref. [142].

7. Further Issues Related to Constructing GRB Correlations

Despite having so many correlations reported in the literature, all of them suffer from several issues related to constructing the correlations themselves. Besides the circularity problem, we face several issues to be addressed. In the following, we list the most common ones.

7.1. *Evolution Effects*

GRBs are observed from a large redshift range and in principle correlation parameters may evolve with the redshift. In some cases, to estimate the cosmological parameters, a correction of the kind of $(1+z)^{-d}$ to the energy or luminosity parameters is needed, introducing a further parameter to fit. However, though tested for subsamples of GRBs and several correlations, this issue is still ongoing [143].

7.2. *Instrumental Selection Effects*

A still open issue is the instrumental selection effects that may affect the observed GRB energy or luminosity correlations. There are at least two kinds of issues due to (a) the trigger threshold, i.e., the minimum photon peak flux that a burst must have in order to be detected by a given instrument, and (b) the spectral analysis threshold, i.e., the minimum fluence to perform a reliable spectral analysis and determine the SED parameters.

Several analyses have been performed in the literature, using different samples and correlations of GRBs and searching for outliers and inconsistent GRBs. Several conclusions were drawn: (a) some correlations may exist, though due to selection effects; (b) other correlations may exist when accounting for the intrinsic scatter; and (c) some correlations may have statistical significance, though affected by the thresholds of GRB detectors, etc. See Ref. [144], for more details. Interestingly, using the time-resolved spectra, similar correlations were found in individual bursts, strongly supporting the fact that the correlations may be physical [64].

7.3. *Systematic Errors*

Sources of systematic errors for GRBs are the sensitivity of the detectors, the differences in the estimated spectral parameters depending on detectors and/or fitting models, the lack of unknown parameters, etc. All of these might dominate over the intrinsic dispersions of GRBs.

In general, due to the vast number of systematic errors, a technique to consider them in GRB fitting procedures consists of deriving those errors requiring the chi-square to be comparable to the fitted degrees of freedom ν, namely $\chi^2 \simeq \nu$, and then summing them in quadrature with the statistical errors [145].

However, studies on different GRB correlations suggest that the systematic uncertainties of correlation parameters are not sensitive to the assumptions about cosmological parameters [146].

7.4. Issues and Interpretation of Prompt Emission GRB Correlations

In the following, attention is given to the above described GRB prompt emission correlations, focusing on possible issues and their physical interpretation. Particular emphasis is given to the functional form of $E_{\rm p}$–$E_{\rm iso}$ (or Amati), Ghirlanda and Yonetoku correlations, the most popular and most quoted correlations for prompt emission observables.

7.4.1. $L_{\rm iso}$–$\tau_{\rm lag}$ Correlation

As established in Ref. [147], this correlation holds also for X-ray flares (in the rest frame energy band 0.3–10 keV) and proposed that their underlying mechanism is similar. However, this correlation is affected by evolution effects, as discussed in Ref. [130].

One of the latest proposed explanations of the $L_{\rm iso}$–$\tau_{\rm lag}$ correlation involves only kinematic effects [148], as the observed time-lag is $\tau_{\rm lag} \propto \mathcal{D}^{-1}$ and the luminosity is $L_{\rm iso} \propto \mathcal{D}$, where $\mathcal{D}$ is the Doppler boost defined in Section 3.1.5.

As discussed in Ref. [149], this correlation was constructed from a small sample of heterogeneously collected GRBs and is severely affected by sample incompleteness.

7.4.2. $L_{\rm iso}$–V Correlation

This correlation has a non-negligible scatter; thus, it is the least reliable one among all GRB prompt emission correlations [130]. The physical origin of the $L_{\rm iso}$–V correlation is still unclear. Within the internal shock scenario (see Section 3.1.2), it seems to be related to the activity of the central engine through the values of Γ and the jet-opening angles [150] and, based on this interpretation, $L_{\rm iso}$ (V) is proportional to a high (low) power of Γ, hence high-luminosity pulses imply high variability prompt light curves [145].

As discussed above for the $L_{\rm iso}$–$\tau_{\rm lag}$ correlation, the luminosity–variability correlation as well is severely plagued by sample incompleteness [149].

7.4.3. Amati or $E_{\rm p}$–$E_{\rm iso}$ Correlation

Likely, the most used and investigated relation is represented by the so-called $E_{\rm p}$–$E_{\rm iso}$ or Amati correlation [151] that can be recast here by

$$\log\left(\frac{E_{\rm p}}{\rm keV}\right) = a_0 + a_1\left[\log\left(\frac{E_{\rm iso}}{\rm erg}\right) - 52\right]. \tag{39}$$

Here, we have two free constants, namely a_0 and a_1, that represent the calibration constants to determine once the relation is somehow calibrated. A possible limitation of the $E_{\rm p}$–$E_{\rm iso}$ correlation is due to the extra source of variability $\sigma_{\rm a}$. This is thought as a direct consequence of hidden variables that contributes to the overall calibration, albeit we cannot directly observe them [152].

A possible explanation for the $E_{\rm p}$–$E_{\rm iso}$ correlation considers the thermal radiation emitted when the GRB jet drills through the core of the progenitor star (see Section 2.4.1), responsible for the thermal peak in the spectrum, and the Compton scattering of this radiation by relativistic electrons outside the photosphere (see Section 3.1.5 and Ref. [153], for details).

There are claims that the Amati correlation is caused by some selection effect of observations, rather than being an intrinsic property of GRBs [149,154]. However, there is a general consensus on the fact that the correlation is real [155–157], though detector sensitivity affects the correlations and a weak fluence dependence may be larger than the statistical uncertainty and contributes to the dispersion of the correlation [158,159].

7.4.4. Ghirlanda or $E_{\rm p}$–E_γ Correlation

From theoretical and observational arguments in favor of the jetted nature of GRBs [36], the radiated GRB energy can be corrected by means of the collimation factor $f = 1 - \cos\theta$, leading to $E_\gamma = f\,E_{\rm iso}$. In particular, the jet opening angle θ is evaluated at the characteristic time $t_{\rm b}$ for specific assumptions on the circumburst medium that can be assumed to be homogeneous [28]. The functional form adopted here for the Ghirlanda relation reads

$$\log\left(\frac{E_{\rm p}}{\rm keV}\right) = b_0 + b_1\left[\log\left(\frac{E_\gamma}{\rm erg}\right) - 50\right] \tag{40}$$

in which, as usual, b_0 and b_1 are the two free constants, fixed by means of calibration. The extra scatter, $\sigma_{\rm b}$, behooves us to better constrain the relation itself.

The Ghirlanda correlation shares with the $E_{\rm p}$–$E_{\rm iso}$ one a similar physical interpretation (see Section 3.1.5 and Ref. [153], for details). This correlation also takes into account the jet correction in the computation of the GRB energy output (see Section 2.4).

The Ghirlanda correlation is linked to the Amati one and, thus, criticisms/analyses against/in favor of being an intrinsic property of GRBs. In Ref. [160], it was shown that as many as 33% of the BATSE bursts would not be consistent with the Ghirlanda relation, but these results depended upon the assumed distribution for the jet's correction factor f [161]. This fact limits the increase in the correlation sample. For details, see Ref. [130]. Likewise for the Amati correlation, the Ghirlanda one is statistically real but strongly affected by the thresholds of GRB detectors [159].

7.4.5. Yonetoku or $L_{\rm p}$–$E_{\rm p}$ Correlation

The Yonetoku or $L_{\rm p}$–$E_{\rm p,}$ [65] correlation functional form here adopted reads

$$\log\left(\frac{L_{\rm p}}{\rm erg/s}\right) - 52 = m_0 + m_1\log\left(\frac{E_{\rm p}}{\rm keV}\right). \tag{41}$$

Here, the free terms are m_0 and m_1 and require to be calibrated. Again, $\sigma_{\rm m}$ is the extra scatter term.

Detailed hydrodynamical simulations suggest that Yonetoku correlation may be due to the emission of photons from the photosphere of a relativistic jet, where the outflow becomes optically thin, whereas most of it is still optically thick (see Section 3.1.5). Quasi-thermal radiation is thus expected and the expected spectral shapes are obtained [162].

Like previous correlations, for the Yonetoku correlation there are also ongoing discussions whether it is a by-product of some selection effect or not [149,154–157]. For this correlation, however, a weak redshift dependence has been confirmed, which may contribute to the dispersion of the correlation [158].

7.5. Issues and Interpretation of Prompt and Afterglow Emission GRB Correlations

Analogous with the prompt emission correlations, we here focus on possible issues and physical interpretation of prompt and afterglow emission correlations. The emphasis on the functional form is given to the Combo correlation, one of the less scattered correlations for both prompt and X-ray observables without evolution effects.

7.5.1. Liang–Zhang or $E_{\rm p}$–$E_{\rm iso}$–$t_{\rm b}$ Correlation

The $E_{\rm p}$–$E_{\rm iso}$–$t_{\rm b}$ correlation [135] has been proposed by purely considering phenomenological considerations, thus avoiding any theoretical assumption, unlike that made for the Ghirlanda correlation. However, as stated above, the $E_{\rm p}$–$E_{\rm iso}$–$t_{\rm b}$ somehow shares similar implications and drawbacks to the Ghirlanda one, as well as analogous physical interpretation (see Sections 2.4 and 3.1.5 and Ref. [153], for details).

As proposed in Ref. [163], this correlation, like the Ghirlanda one, appears to be affected by a selection effect on $E_{\rm p}$ (whereas, for the Ghirlanda correlation, E_γ is affected as well) and suffers sample incompleteness.

7.5.2. Dainotti or L_X–t_X Correlation

The Dainotti correlation, akin to the $L_{\rm iso}$–$\tau_{\rm lag}$ correlation, may be retrieved from kinematic effects (see Section 3.1.5 and Ref. [148], for details), pointing out to a common origin between the two correlations.

As already discussed above, this correlation is quite scattered and seems to be a selection effect due to the flux detection limit of *Swift*-XRT instrument [130]. Moreover, it is also affected by evolution effects with the redshift [164].

7.5.3. $E^{\rm X}_{\rm iso}$–$E_{\rm iso}$–$E_{\rm p}$ Correlation

This correlation depends upon two cosmology-dependent quantities, although it is unsuitable to constrain cosmological parameters. Since it holds for both SGRBs and LGRBs, with different progenitor and surrounding medium properties (see Sections 2.4.1 and 2.4.2), its physical interpretation has not been yet established. There is a speculation that it may be connected with the Γ of the outflow, which might regulate the efficiency of conversion from γ-rays to X-rays [138].

The $E^{\rm X}_{\rm iso}$–$E_{\rm iso}$–$E_{\rm p}$ correlation utilizes the prompt emission observables $E_{\rm iso}$ and $E_{\rm p}$ on which the Amati correlation is based. For this reason, it is straightforward to deduce that the biases and selection effects, at work for the Amati correlation, partially affect this hybrid correlation. Moreover, unlike pure afterglow correlations such as the Dainotti one, this correlation is also plagued by double truncation in the flux limit, both in the prompt and X-ray afterglow emissions, making the correction for any selection effect difficult and the use as redshift estimators and cosmological tool (see discussions in Ref. [164]).

7.5.4. Combo Correlation

The Combo correlation is a hybrid correlation linking the prompt emission $E_{\rm p}$ and the observable quantities determined from the X-ray afterglow light curve, i.e., among all the rest-frame 0.3–10 keV plateau luminosity L_0, its rest-frame duration τ, and the late power-law decay index α [140]. For each GRB, L_0, τ, and α can be obtained by fitting the rest-frame 0.3–10 keV flare-filtered afterglow luminosity light curves with the function $L(t) = (1 + t/\tau)^{\alpha}$.[22] The general expression is much more complicated than previous ones and reads

$$\log\left(\frac{L_0}{\rm erg/s}\right) = k_0 + k_1 \log\left(\frac{E_{\rm p}}{\rm keV}\right) - \log\left(\frac{\tau/{\rm s}}{|1+\alpha|}\right). \tag{42}$$

Here, the constants k_0 and k_1 need to be determined by means of the calibration procedure. Again, the correlation is characterized by an extra scatter $\sigma_{\rm k}$.

The Combo correlation can be explained by the external shock scenario (see Section 3). The correlation is the result of the synchrotron emission from the electrons accelerated in a relativistic shock (see Section 3.1.5). The shock propagates through the external CBM and interacts with the magnetic field of the the turbulent plasma. Hence, the relationship among $E_{\rm p}$, L_0, and τ and the corresponding comoving quantities scale with the initial Lorentz factor of the bulk motion Γ_0, whereas the intrinsic scatter is due to the uncertainties on the source spectral energy distribution [165].

Keeping in mind the hybrid nature and the Combo correlation, which is a combination of the Amati and the $E^{\rm X}_{\rm iso}$–$E_{\rm iso}$–$E_{\rm p}$ correlations, the same biases and selection effects at work in the $E^{\rm X}_{\rm iso}$–$E_{\rm iso}$–$E_{\rm p}$ affect the Combo correlation as well.

7.5.5. *L*–*T*–*E* Correlation

As already stated, this correlation shares similarities with the Combo correlation but, unlike it, needs to be corrected for the redshift evolution effects [142].

The *L*–*T*–*E* correlation, as well as the Combo correlation, may be explained within the magnetar scenario, which justifies the plateau phase observed in the X-ray afterglow light curves as due to the continuous energy supply from a supra-massive NS (see Section 3.1.3 and Ref. [71], for details).

Like the Combo correlation, which shares similar features, the same discussion on the biases and selection effects holds for the L–T–E correlation as well. In addition, in Ref. [142], it is shown that GRBs of the sample at high redshifts usually have relatively larger L_X and E_{iso}. This is very likely a selection effect due to the fact that very distant GRBs with small L_X and E_{iso} cannot be observed in view of the limited threshold of current detectors.

8. Circularity or Not Circularity?

Even though GRBs could be thought of as indicators toward the determination of the universe's expansion history, it is remarkable to stress that small redshift GRB data are still missing. Consequently, if one requires to calibrate GRBs with small redshift data, there is the strict need of other data sets, whose data points lie around $z \simeq 0$ that permit performing the calibration procedure. The calibration procedure is essentially a consequence of the correlation functional forms that, by virtue of the above considerations, are commonly written as $y = a\,x + b$, with a, b real constants that depend upon the relation itself. Bearing this in mind, we focus below on different calibration strategies. In this respect, we highlight whether it is absolutely necessary or not to calibrate our correlations in order to confront cosmological models with GRB data. Below, we start with such considerations, critically discussed.

8.1. Calibration Versus Non-Calibration

Constraining cosmological parameters using GRBs is plagued by several conceptual and practical issues.

First, all GRB correlations by definition are built assuming an *a priori* background cosmology (see Sections 2.5 and 6.1 for details) and, consequently, introducing a circularity problem [166]. Second, the majority of GRB correlations holds for LGRBs (see Section 6.1 for details), whose observational rate falls off rapidly at low-z and, in some cases, such a nearby LGRB seems to be intrinsically different from the *cosmological* ones [37,167]. Third, unlike SNe Ia, which are calibrated with a selected sub-sample at a very low redshift[23] by anchoring them to primary distance indicators as Cepheids, the shortage of low-z GRBs prevents anchoring them to primary distance indicators.

Focusing on the above circularity problem, this is essentially an epistemological issue due to the lack of very low-z GRBs and arising from the need of a background cosmology to compute the above-defined E_{iso}, E_γ, and L_{iso} entering GRB correlations [166]. For example, calibrating GRBs through the standard ΛCDM model, the estimate of cosmological parameters of any dark energy framework inevitably returns an overall agreement with the concordance model. Debates toward its use in cosmology seem to indicate that this effect could be minor, albeit it plagues cosmological constraints obtained from GRBs.

Possible ways out known in the literature involve calibration techniques based on the use of SNe Ia distance moduli, cosmographic series, cosmic chronometers, etc. All of these procedures represent plausible solutions, but at the same time introduce possible issues that we are going to describe below.

The above calibration procedures have to compare with an alternative method which completely by-passes the calibration procedure [10]. This uncalibrated procedure consists of a simultaneous fit of correlation parameters together with the cosmological model parameters. This uncalibrated procedure consists of a simultaneous fit of correlation and cosmological model parameters. This procedure and the related issues are also discussed in the following.

8.2. Fitting Procedures with Calibration

8.2.1. SN Calibration

A widely-used method to calibrate GRB correlations is through the use of SNe Ia that span within $z \lesssim 2.3$. In such a way, assuming this could work for any LGRBs, the GRB data points are mixed with SNe in order to build up a whole, quite large, Hubble diagram, where in the small redshift domain one has the majority of SNe, while, at large z, GRBs are the most. Here, the simplest error bars on distance modulus are [168,169]

$$\sigma_\mu = \left([(z_{i+1} - z)/(z_{i+1} - z_i)]^2 \epsilon_{\mu,i}^2 + [(z - z_i)/(z_{i+1} - z_i)]^2 \epsilon_{\mu,i+1}^2\right)^{1/2}, \tag{43}$$

where $\epsilon_{\mu,i}$ and $\epsilon_{\mu,i+1}$ and μ_i and μ_{i+1} are the errors and distance moduli of the SNe Ia at z_i and z_{i+1}, respectively.

For each SN catalog, we could find different interpolating functions to model the SN distribution. Thus, calibrating GRBs with SNe would seriously depend on the choice of these expressions for each catalog. Hence, GRB calibrations may turn out to be extremely sensitive to SNe Ia and the approach should be carefully handled since GRB luminosity correlations may no longer be fully independent from SN data points.

8.2.2. Model Dependent Calibration

The model dependent procedure fixes the background cosmology with a given cosmological model, where typically the dark energy evolution is assumed a priori. Since the background cosmology[24] enters the correlation functions, generically in the form $y = ax + b$; this means that one has to (1) assume a background cosmology, (2) fix the most suitable numerical bounds over the free coefficients of the background cosmology, and (3) calibrate the correlation.

As it appears evident, this strategy consists of determining an accredited cosmological model with particular choices of the free parameters, determined elsewhere.

This procedure is obviously strongly plagued by the circularity problem. It fixes the cosmological evolution with a given model and does not permit constraining suitably another cosmological paradigm. In fact, if one calibrates with a generic model, say $H^{(1)}(z)$, any other statistical expectations on a different model, say $H^{(2)}(z)$, would favor the model that better matches $H^{(1)}(z)$. In other words, calibrating with $H^{(1)}(z)$ implies that the best fits are statistically argued for $H^{(2)}(z) \simeq H^{(1)}(z)$.

Another dramatic fact is that one has to constrain the free parameters of the background scenario by means of additional fits, with different data sets. This implies that an overall analysis would be plagued by error propagation between different catalogs of data and limits severely the analysis itself. To avoid other fits, one can assume exact versions of the cosmological models that should be used as backgrounds. In such a way, the corresponding error propagation reduces, albeit one does not take a real tested cosmological scenario, but rather a simplified version of it.

8.2.3. Model Independent Calibration

Calibrating correlations via model independent treatments permits using GRBs as distance indicators, although the calibration is made by means of other standard candles. The idea of model-independent calibrations, however, enables getting the luminosity distance $d_{\rm L}$ without a priori postulating the background cosmology, healing de facto the circularity problem.

A nice possibility consists of relating distances with model-independent quantities written in terms of a Taylor series expansion of the scale factor. Thus, we first notice

$$H(z) = \left\{ \frac{d}{dz} \left[\frac{d_{\rm L}(z)}{1+z} \right] \right\}^{-1} \tag{44}$$

and then we consider the following expansions:

$$d_L(z) = \frac{z}{H_0} \sum_{n=0}^{N} \frac{\alpha_n}{n!} z^n \tag{45}$$

$$H(z) = \sum_{m=0}^{M} \frac{H_m}{m!} z^m \tag{46}$$

where α_n are the coefficients of the luminosity distance expansion and H_m are the coefficients of the Hubble rate expansion. Thus, baptizing the cosmographic set, q_0, j_0, s_0, as the present values of the following quantities:

$$q(t) = -\frac{1}{a}\frac{d^2a}{dt^2}\left[\frac{1}{a}\frac{da}{dt}\right]^{-2}, \quad j(t) = +\frac{1}{a}\frac{d^3a}{dt^3}\left[\frac{1}{a}\frac{da}{dt}\right]^{-3}, \quad s(t) = +\frac{1}{a},\frac{d^4a}{dt^4}\left[\frac{1}{a}\frac{da}{dt}\right]^{-4} \tag{47}$$

where the scale factor a has been considered, with the requirement $a \equiv (1+z)^{-1}$. The above quantities are named *deceleration*, *jerk*, and *snap* parameters, respectively, we formally have

$$d_L^{(4)} \simeq \frac{z}{H_0}\left(\alpha_0 + \alpha_1 z + \alpha_2\frac{z^2}{2} + \alpha_3\frac{z^3}{6}\right). \tag{48}$$

The coefficients in Equations (45) and (46), say $\alpha_i \equiv \alpha_i(q_0, j_0, s_0)$ and $H_i \equiv H_i(q_0, j_0, s_0)$, can be determined directly with data, without considering a cosmological model a priori. This treatment is known with the name of *cosmography* or *cosmokinetics*, i.e., the part of cosmology that reconstructs the universe's kinematics model-independently. Thus, at $z = 0$, we have

$$\alpha_1 = \frac{1}{2}(1 - q_0) \tag{49}$$

$$\alpha_2 = -\frac{1}{6}(1 - q_0 - 3q_0^2 + j_0) \tag{50}$$

$$\alpha_3 = \frac{1}{24}(2 - 2q_0 - 15q_0^2 - 15q_0^3 + 5j_0 + 10q_0j_0 + s_0) \tag{51}$$

and

$$H_1 = 1 + q_0 \tag{52}$$

$$H_2 = \frac{1}{2}(j_0 - q_0^2) \tag{53}$$

$$H_3 = \frac{1}{6}\left[-3q_0^2 - 3q_0^3 + j_0(3 + 4q_0) + s_0\right]. \tag{54}$$

Although powerful, the above formalism suffers from shortcomings due to the convergence at higher redshifts[25], i.e., the high GRB redshifts are very far from $z = 0$. In other words, the standard cosmographic approach fails to be predictive if one employs data at higher redshift domains, which is exactly the case of GRBs.

Healing the convergence problem leads to a high-redshift cosmography. In this respect, several strategies have been suggested. For instance, one could (1) extend the limited convergence radii of Taylor series by changing variables of expansion, using the so-called *auxiliary variables*, or (2) changing the mathematical technique in which the expansions are performed, i.e., involving expansions different from Taylor ones, etc.

In the case of auxiliary variables, one employs a tricky method in which the expansion variable is reformulated as a function of the redshift itself, but with particular convergence properties. In other words, we cosmic quantities are rewritten in a more complicated function of the redshift z, namely y. Changing the redshift variable from z to y modifies accordingly the convergence radius. Formally speaking, we write $y \equiv \mathcal{F}(z)$ [9], where we assume $\mathcal{F}(z)$ a generic function of the redshift. The function $\mathcal{F}(z)$ is properly chosen from physical prime principles. All $\mathcal{F}(z)$ prototypes, however, might fulfill a few mathematical conditions:

1. $\mathcal{F}(z)\Big|_{z=0} = 0$
2. $\mathcal{F}(z)\Big|_{z=0} < \infty$.

The first guarantees that at $z = 0$, our time, even y is zero. The second that the auxiliary variable does not diverge; otherwise, the convergence problem would still persist. In addition, a further requirement is helpful in constructing $\mathcal{F}(z)$:

3. $\mathcal{F}(z)\big|_{z=-1} < \infty$.

The former condition enables y to converge at a future time, as well as z. Using these hints toward the formulation of $\mathcal{F}(z)$, we can suggest a couple of well-consolidated examples of $\mathcal{F}(z)$:

$$y_1 = \frac{z}{1+z} \tag{55}$$

$$y_2 = \arctan z\,. \tag{56}$$

The second possibility is offered by rational approximations, where the expansion is thought to be a rational function, instead of a polynomial. This guarantees to optimize the Taylor series with rational approximants that better approach large z than the Taylor series, guaranteeing mathematical stability of the new series if data points exceed $z = 0$. Among all the possible choices, here the attention is given to the Padé polynomials, firstly introduced in Ref. [170]. This technique of approximations turns out to be a bookkeeping device to keep the calculations manageable for the cosmography convergence issue. Thus, provided we have Taylor expansions of $f(z)$ under the form $f(z) = \sum_{i=0}^{\infty} c_i z^i$, with $c_i = f^{(i)}(0)/i!$, it is possible to obtain the (n,m) Padé approximant by

$$P_{n,m}(z) = \left(\sum_{i=0}^{n} a_i z^i\right)\left(1 + \sum_{j=1}^{m} b_j z^j\right)^{-1} \tag{57}$$

and requires that $b_0 = 1$. Furthermore, it is important that $f(z) - P_{n,m}(z) = \mathcal{O}(z^{n+m+1})$ and the coefficients b_i come from solving the homogeneous system of linear equations $\sum_{j=1}^{m} b_j\, c_{n+k+j} = -b_0\, c_{n+k}$, valid for $k = 1, \dots, m$. Once b_i are known, a_i can be obtained using the formula $a_i = \sum_{k=0}^{i} b_{i-k}\, c_k$. Just for an example, we report the $(2,1)$ Padé polynomial as

$$P_{2,1}(z) = \frac{z}{H_0}\left\{\frac{6(q_0-1) + [q_0(8+3q_0) - 5 - 2j_0]z}{2q_0(3+z+3q_0 z) - 2(3+z+j_0 z)}\right\}. \tag{58}$$

8.3. The Use of Bézier Polynomials

The left term of Equation (30) can be approximated by means of particular choices, such as using model-independent Bézier parametric curves. They are constructed to be stable at the lower degrees of control points. They can be rotated and translated by performing the operations on the points and assuming a degree n. They formally are defined as

$$H_n(z) = \sum_{d=0}^{n} \beta_d h_n^d(z) \quad , \quad h_n^d(z) \equiv \frac{n!(z/z_{\rm m})^d}{d!(n-d)!}\left(1 - \frac{z}{z_{\rm m}}\right)^{n-d} \tag{59}$$

where we notice the linear combination of Bernstein basis polynomials $h_n^d(z)$. Assuming the coefficients β_d to be positive in the range of $0 \leq z/z_{\rm m} \leq 1$, where $z_{\rm m}$ is the maximum z of OHD, we soon can classify those polynomials by means of the exponent n.

In particular, besides the constant case, $n = 0$, both linear growth that happens for $n = 1$ and oscillatory regimes, say $n > 2$, work well. This implies that a suitable choice is $n = 2$. In this case, we have

$$H_2(z) = \beta_0\left(1 - \frac{z}{z_{\rm m}}\right)^2 + 2\beta_1\left(\frac{z}{z_{\rm m}}\right)\left(1 - \frac{z}{z_{\rm m}}\right) + \beta_2\left(\frac{z}{z_{\rm m}}\right)^2. \tag{60}$$

The comparison between $H_2(z)$ and the OHD data points give $\beta_0 = 67.76 \pm 3.68$, $\beta_1 = 103.3 \pm 11.1$, and $\beta_2 = 208.4 \pm 14.3$, all in units of km s^{-1} Mpc^{-1}.

After having approximated $H(z)$ with Equation (60), for spatially flat cosmology, $\Omega_k = 0$, the calibrating luminosity distance becomes

$$d_{\rm cal}(z) \simeq (1+z)\int_0^z \frac{dz'}{H_2(z')}\,. \tag{61}$$

Once the luminosity distance is written, it is possible to calibrate $E_{\rm iso}$, E_γ, $L_{\rm p}$, and L_0 for the correlations that we intend to test. For $E_{\rm p} - E_{\rm iso}$, Ghirlanda, Yonetoku, and Combo correlations, we report in Table 1 the corresponding numerical outcomes related to the calibration process.

Table 1. For brevity, we report in this table only a few calibrated correlations. In particular, in the columns, we prompt four correlations, i.e., Amati, Ghirlanda, Yonetoku, and Combo, with the data set number points and the corresponding last update year. On the right, we display the calibrated best fit parameters. The statistical method behind these calibrations is reported in Section 9.2.

Correlation	N	Update		Parameters	
Amati	193	2015	$a_0 = 2.06 \pm 0.03$	$a_1 = 0.50 \pm 0.02$	$\sigma_{\rm a} = 0.20 \pm 0.01$
Ghirlanda	27	2007	$b_0 = 2.09 \pm 0.04$	$b_1 = 0.63 \pm 0.04$	$\sigma_{\rm b} = 0.10 \pm 0.02$
Yonetoku	101	2018	$m_0 = -3.43 \pm 0.21$	$m_1 = 1.51 \pm 0.08$	$\sigma_{\rm m} = 0.35 \pm 0.03$
Combo	60	2015	$q_0 = 50.04 \pm 0.27$	$q_1 = 0.71 \pm 0.11$	$\sigma_{\rm q} = 0.35 \pm 0.04$

Once calibrated, the corresponding distance moduli from Equation (24) are computed for each correlation.

8.3.1. Simultaneous Fits

Another relevant strategy is based on the idea to constrain the cosmological parameters *together with* the luminosity correlation [171,172]. In particular, the real distance modulus can be computed as

$$\mu_{\rm fit} = \frac{\sum_i \mu_i/\sigma^2_{\mu_i}}{\sum_i \sigma^{-2}_{\mu_i}}, \tag{62}$$

where the sum is over a given number of different correlations. In particular, μ_i is the best estimated distance modulus and the subscript i-th refers to the correlation, with σ_{μ_i} the error bars. The uncertainty of the distance modulus for each burst is $\sigma_{\mu_{\rm fit}} = \left(\sum_i \sigma^{-2}_{\mu_i}\right)^{-1/2}$.

A great advantage is that, as one computes bounds on cosmological parameters, the normalization functions and slopes of each correlations are marginalized. Consequently, we write down the χ^2 as

$$\chi^2_{\rm GRB} = \sum_{i=1}^{N} \frac{[\mu_i(z_i, H_0, \Omega_M, \Omega_{DE}) - \mu_{{\rm fit},i}]^2}{\sigma^2_{\mu_{{\rm fit},i}}}, \tag{63}$$

where $\mu_{{\rm fit},i}$ and $\sigma_{\mu_{{\rm fit},i}}$ are the fitted distance modulus and its error, respectively.

8.3.2. Narrow Calibration

Another intriguing technique consists of calibrating standard candles using GRBs in a narrow redshift range, hereafter δz. This short interval is placed near a fiducial redshift [173,174] with the great advantage that, in some cases, see e.g., Ref. [173], no low-redshift GRB sample is necessary.

8.4. Fitting Procedures without Calibration

Constraints on the cosmological parameters can be obtained with an alternative method which completely by-passes the calibration procedure. It consists of taking all best data sets of any GRB correlations, introducing the lowest intrinsic dispersion. The method is described below.

Any correlation between a generic energy/luminosity quantity $\mathcal{Y}$ and a GRB observable $\mathcal{X}$ has the form

$$\log \mathcal{Y}^{\rm obs} = a \log \mathcal{X} + b\,. \quad (64)$$

The energy/luminosity quantity in general contains the information on the cosmological parameters $\Omega_{\rm i}$ through $d_{\rm L}(z, \Omega_{\rm i})$, defined by the theoretical model describing the background cosmology

$$\mathcal{Y}^{\rm th} = 4\pi d_{\rm L}^2(z, \Omega_{\rm i})\mathcal{F}_{\rm bolo} \quad (65)$$

where $\mathcal{F}_{\rm bolo}$ may be the rest-frame bolometric fluence $S_{\rm bolo}(1+z)^{-1}$ for Amati-like correlations, or the bolometric observed flux $F_{\rm bolo}$ for Yonetoku-like correlations. Please notice that we only focus on these two relations just for giving an example. The same can be reformulated for other correlations, although, for brevity, we do not report other treatments here.

The best cosmological and correlation parameters are then obtained by maximizing the log-likelihood function [152]

$$\ln \mathcal{L} = -\frac{1}{2}\sum_{i=1}^{N}\left[\frac{\left(\log \mathcal{Y}_{\rm i}^{\rm obs} - \log \mathcal{Y}_{\rm i}^{\rm th}\right)^2}{\sigma_i^2} + \ln(2\pi\sigma_i^2)\right], \quad (66)$$

where $\sigma_i^2 = \sigma^2_{\log \mathcal{Y}_{\rm i}^{\rm obs}} + a^2\sigma^2_{\log \mathcal{X}_{\rm i}} + \sigma^2_{\rm ext}$. Here, $\sigma_{\log \mathcal{Y}_{\rm i}^{\rm obs}}$ is the error in the measured value of $\log \mathcal{Y}_{\rm i}^{\rm obs}$, $\sigma_{\log \mathcal{X}_{\rm i}}$ is the error in $\log \mathcal{X}_{\rm i}$, and $\sigma_{\rm ext}$ is the intrinsic dispersion of the correlation.

The above treatment avoids calibrating GRB data. This procedure has been applied to an uncalibrated $E_{\rm p}$–$E_{\rm iso}$ correlation in Ref. [175] and to an uncalibrated Combo correlation in Ref. [10]. In both cases, the correlations have been built up from samples of GRBs that have lower intrinsic dispersion. As byproducts, the resulting GRB correlations are close for different cosmological models, which can be interpreted with the fact that this procedure is model-independent. However, the application of this method seems to indicate that current GRB data are not able to put stringent constraints on cosmological parameters, though consistent with those resulting from better-established cosmological probes.

However, as hinted by the above results, this method may introduce a possible bias, namely that the GRB correlation may adjust itself to the cosmology that maximizes Equation (66), rather than allowing the derivation of $\Omega_{\rm i}$ from a cosmology independent calibrated correlation. Moreover, it may be possible that a more exotic cosmological model would lead to best-fit GRB correlations significantly different from simpler models, thus failing in providing a model-independent procedure. Therefore, this method is still the subject of ongoing studies.

9. Recent Developments of Cosmology with Gamma-Ray Bursts

9.1. Numerical Results Using Correlations

Observational data indicate that the cosmological expansion is currently accelerating. They also indicate that, in the recent past, the expansion was decelerated. The standard spatially-flat ΛCDM model [176–178] is the simplest model consistent with these observations [122,179–181]. Here, a cosmological constant Λ dominates the current energy budget and fuels the currently-accelerating cosmological expansion. In this model, above a redshift $z \approx 0.75$, non-relativistic cold dark matter and baryons dominate over Λ and powered the then-decelerating cosmological expansion. While the observations are consistent with dark energy being time- and space-independent, they do not rule out slowly-evolving and weakly spatially-inhomogeneous dynamical dark energy or spatial flatness.

Significant constraints on cosmological parameters come from the CMB anisotropy data—that primarily probe the $z \sim 1100$ part of redshift space—as well as from BAO observations—the highest of which reach to $z \sim 2.3$—and other lower-redshift SNeIa and OHD measurements. Observational data in the intermediate redshift range, between $z \sim 2.3$ and ~ 1100, are not as constraining as the lower and higher redshift data, but hold significant promise. Intermediate redshift observations include those of HII starburst galaxies that reach $z \sim 2.4$ [182–188], quasar angular sizes that reach $z \sim 2.7$ [186,189–193], quasar X-ray, and UV fluxes that reach $z \sim 7.5$ [193–201], as well as GRBs that have now been detected to $z = 9.4$ [202]. Observed correlations between GRB photometric and spectroscopic properties that can be related to an intrinsic burst physical property would allow GRBs to be used as valuable standard candles that reach high z and probe a largely unexplored region of cosmological redshift space (see, e.g., [136,140,142,145,151,203–206], and references therein), similar to how SNeIa are used as standard candles [118] at $z < 2.3$. However, as stressed many times in this review, this is still a challenge for GRBs.

After it was established that GRBs were at cosmological distances, many attempts have been made to use burst correlations to constrain cosmological parameters. The first GRB Hubble diagram of a small sample of 9 bursts, obtained in Ref. [207] from the L_{iso}–V correlation [131], led to a current non-relativistic matter energy density parameter limit of $\Omega_{m0} < 0.35$ at the 1σ confidence level (for the flat ΛCDM model). Soon after, using the Ghirlanda correlation, in Ref. [208], with a sample of 12 bursts, it has been found $\Omega_{m0} = 0.35 \pm 0.15$ for the flat ΛCDM model, and in Ref. [171], with 14 GRBs as well as SNeIa, it has been inferred that $\Omega_{m0} = 0.37 \pm 0.10$ and a cosmological constant energy density parameter $\Omega_{\Lambda} = 0.87 \pm 0.23$, in the non-flat ΛCDM model, and $\Omega_{m0} = 0.29 \pm 0.04$ in the flat model. Similar constraints were obtained in Ref. [135], using the E_{p}–E_{iso}–t_{b} correlation: $0.13 < \Omega_{m0} < 0.49$ and $0.50 < \Omega_{\Lambda} < 0.85$ at 1σ confidence level in the flat ΛCDM model.

More recently, many contrasting results have been reported in the literature. In Ref. [209], cosmological parameter constraints, within the ΛCDM model and dynamical dark energy models from two different GRB data sets, were found to be different from the two data sets and also relatively broad. Similarly, in Ref. [210], it has been shown that at that time GRB data could not significantly constrain cosmological parameters. In addition, in Refs. [211,212], it has been shown that most GRB correlations have large scatter and/or their parameters differ somewhat significantly between low- and high-z GRB data sets. From the calibration of the Ghirlanda correlation, by using a SNeIa distance-redshift relation—through the $(3,2)$ Padé approximant—in Ref. [211], it has been obtained $\Omega_{m0} = 0.302 \pm 0.142$ within the flat ΛCDM model.[26] Based on a cosmographic approach, an updated E_{p}–E_{iso} correlation with 162 GRBs has been used to get cosmological constraints. In Ref. [12], GRBs were calibrated with SNeIa, resulting in $\Omega_{m0} = 0.25^{+0.29}_{-0.12}$ within the flat ΛCDM model, whereas, in Ref. [214], a cosmographic expansion, up to the fifth order, involving SNeIa is used to calibrate the E_{p}–E_{iso} correlation for GRBs, which are then used in conjunction with OHD and BAO measurements to constrain cosmographic parameters, resulting in a 1σ deviation from the ΛCDM cosmological model.

Other recent works (involving GRB data only or in conjuction with other probes) also report inconsistencies with the ΛCDM model. In Ref. [215], the E_{p}–E_{iso} correlation has been used, including also modeling the potential evolution of GRB observables, to conclude that calibrated GRB, SNeIa, and OHD data favor a dynamical dark energy model described by a scalar field with an exponential potential energy density. In Ref. [216], Amati, Ghirlanda, Yonetoku, and Combo correlations have been calibrated in a model-independent way via OHD and jointly analyzed with SNeIa and BAO by using cosmographic methods, such as Taylor expansions, auxiliary variables, and Padé approximations, to conclude that GRB do not favor the flat ΛCDM model but instead favor a mildly evolving dark energy density model. Similarly, in Ref. [217], the E_{p}–E_{iso} and Combo correlations have been calibrated via OHD actual and machine-learned data, and again, based on a joint analysis with SNeIa and BAO, indications against a genuine cosmological constant have been found. Analogously, in Ref. [218], different combinations of SNe Ia, quasar, and GRB data sets have been used for testing the ΛCDM model and dynamical dark energy parametrizations. It was found that GRB and quasar data sets were inconsistent with the flat ΛCDM model, in agreement with Ref. [219] for similar data. In Ref. [220], strong gravitational lensing data in conjunction with SNe Ia and GRBs have been considered, and it has been found that the best-fit value of the spatial curvature parameter favored a closed universe, although a flat universe can be accommodated at the 68% confidence level.

On the other hand, some recent efforts have shown that the $E_{\rm p}$–$E_{\rm iso}$ and Combo correlations calibrated using better-established cosmological data—such as SNe Ia or OHD measurements—provide cosmological constraints that are consistent with the flat ΛCDM model (see Table 2). In Ref. [123], an updated $E_{\rm p}$–$E_{\rm iso}$ correlation with 193 GRBs and a calibration based on an interpolation of the OHD data set have been considered, leading to $\Omega_{m0} = 0.397^{+0.040}_{-0.039}$ in a flat ΛCDM cosmology, though the value of the mass density is higher than the one established by Ref. [122]. In Ref. [221], the $E_{\rm p}$–$E_{\rm iso}$ correlation, calibrated with the latest OHD data set, has been jointly fit with CMB, BAO, and SNe Ia data in a search for cosmological parameter constraints within the standard cosmological model, as well as in dynamical dark energy parametrizations, finding no evidence in favor of the alternatives to the ΛCDM model. Finally, by using the Combo correlation with 174 GRBs calibrated in a semi-model independent way, in Ref. [141], it has been found: a) for a flat ΛCDM model $\Omega_{m0} = 0.32^{+0.05}_{-0.05}$ and $\Omega_{m0} = 0.22^{+0.04}_{-0.03}$ for the two values of the Hubble constant H_0 of Ref. [122] and Ref. [222], respectively, and b) for a non-flat ΛCDM model $\Omega_{m0} = 0.34^{+0.08}_{-0.07}$ and $\Omega_\Lambda = 0.91^{+0.22}_{-0.35}$ for the H_0 of Ref. [122], and $\Omega_{m0} = 0.24^{+0.06}_{-0.05}$ and $\Omega_\Lambda = 1.01^{+0.15}_{-0.25}$ for the H_0 of Ref. [222].

Table 2. Summary of some recent cosmological constraints obtained by using Amati and Combo correlations, with or without other well established cosmological probes, within the flat and non-flat ΛCDM models. The numbers near the correlations name indicate the size of the GRB sample. For details on the names of the other probes, see the text.

Correlation	Sample	H_0 (km s^{-1}Mpc^{-1})	$\Omega_{\rm m0}$	Ω_Λ	$\Omega_{\rm k0}$	Reference
Amati (193)	GRB + SNIa	67.76 ± 3.68 [a]	$0.397^{+0.040}_{-0.039}$	$0.603^{+0.040}_{-0.039}$	0	[123]
Amati (74)	GRB + SNIa + BAO + CMB	70.81 ± 3.68 [b]	0.3180 ± 0.0006	0.6820 ± 0.0006	0	[221]
Combo (174)	GRB	74.03 ± 1.42 [c]	$0.22^{+0.04}_{-0.03}$	$0.78^{+0.03}_{-0.04}$	0	[141]
			$0.24^{+0.06}_{-0.05}$	$0.68^{+0.05}_{-0.05}$	$-0.24^{+0.16}_{-0.25}$	
		67.4 ± 0.5 [d]	$0.32^{+0.05}_{-0.05}$	$0.68^{+0.05}_{-0.05}$	0	
			$0.34^{+0.08}_{-0.07}$	$0.91^{+0.22}_{-0.35}$	$-0.24^{+0.24}_{-0.35}$	
Amati (118)	GRB + $H(z)$ + BAO	$68.544^{+0.871}_{-0.862}$	0.316 ± 0.016	0.684 ± 0.016	0	[175]
		$67.499^{+2.281}_{-2.279}$	0.310 ± 0.016	$0.639^{+0.072}_{-0.078}$	$0.051^{+0.094}_{-0.088}$	
Amati (118)	GRB + $H(z)$ + BAO + QSO + HIIG	69.3 ± 1.2	0.313 ± 0.013	0.687 ± 0.013	0	[186]

[a] Inferred from the interpolation of the OHD data by using Bézier polynomials. [b] Inferred from the interpolation of the OHD data with additional systematic errors [221] by using Bézier polynomials. [c] Value from Ref. [222]. [d] Value from Ref. [122].

Again, by examining an uncalibrated $E_{\rm p}$–$E_{\rm iso}$ correlation built up from a sample of bright *Fermi*-LAT GRBs [223] and another GRB sample with lower average fluence GRBs [224], in Ref. [175], cosmological parameter constraints have been obtained in a number of cosmological models, concluding that current GRB data are not able to restrictively constrain cosmological parameters, and that cosmological parameter constraints from the more-reliable GRBs are consistent with those resulting from better-established cosmological probes. In Ref. [187], a joint $H(z)$+BAO+quasar (QSO)+HII starburst galaxy (HIIG)+GRB fit determined $\Omega_{m0} = 0.313 \pm 0.013$ in the flat ΛCDM model, consistency with a cosmological constant and zero spatial curvature, though mild dark energy dynamics or a little spatial curvature are not ruled out at all.

Mixing all together the cosmological results summarized above, obtained through GRB data, seem to be mutually inconsistent. This reflects all the efforts made so far to employ GRB as distance indicators are still affected by a certain number of issues, as we outlined previously.

First of all, we recall that GRB correlations involve a number of observable quantities affected by the so-called circularity problem [166], caused by having to compute the GRB correlations in an a priori assumed background cosmological model, being not fully model-independent [12,136,138,140,151,206,209,214]. However, even uncalibrated GRB correlations, in principle free from the circularity issue, are not able to put stringent constraints on the cosmological parameters, though consistent with those resulting from better-established cosmological probes. In addition, we recall that all GRB correlations are characterized by large intrinsic dispersions, conceivably caused by unknown large systematic errors[27] [64,143–145] in comparison to the case of better-established probes, such as BAO, OHD, and SNeIa, where many error sources have been better modeled. On the other hand, the influence of possible selection bias and evolution effects are currently debated [154–158]. One may therefore conclude that the large intrinsic dispersions of GRB correlations could be a consequence of yet undiscovered GRB intrinsic properties and/or a yet unidentified sub-class within the population of GRBs, analogously to SN populations.

9.2. Applications of Statistical Analysis with GRBs

In this section, we describe a few applications of statistical analysis using GRBs. Clearly, we focus on a particular choice and, in principle, it is possible to work out different fits and/or experimental procedures. In particular, we here propose a calibration at the very beginning, adopting the most consolidated route to handle GRBs. Above, we also described the non-calibration procedure that, for brevity, we do not report here.

A widely consolidated approach is based on sampling the original catalog by means of a Monte Carlo technique, essentially built up using Markov Chain Monte Carlo simulations that are sampled within the widest possible parameter space. Commonly, the most adopted algorithm is the *Metropolis–Hastings* and the standard approach to get limits uses the minimization of the total χ^2 function. As we stated above, in this review, the idea of combining more than one sample is essential in order to refine cosmological bounds on the model parameters. Hereafter, we denote with $\mathbf{x}$ the set of parameters and include in our analysis SNe Ia and BAO data sets together with the calibrated GRB data. The former data have been obtained through calibrating the correlations. For the sake of brevity, as well as above, we only consider Amati, Ghirlanda, Yonetoku, and Combo correlations. In the specific case of our three samples, i.e., GRBs, SNeIa, and BAO, we combine the chi square functions by

$$\chi^2_{\rm tot} = \chi^2_{\rm GRB} + \chi^2_{\rm SN} + \chi^2_{\rm BAO} \tag{67}$$

with the following recipe:

- **GRB** χ^2. Here, we define

$$\chi^2_{\rm GRB} = \sum_{i=1}^{N_{\rm GRB}} \left[\frac{\mu^{\rm obs}_{\rm GRB,i} - \mu^{\rm th}_{\rm GRB}(\mathbf{x}, z_i)}{\sigma_{\mu_{\rm GRB,i}}} \right]^2 \tag{68}$$

  where $N_{\rm GRB}$ and $\mu^{\rm th}_{\rm GRB}$ are the experimental and theoretical GRB distance moduli.
- **SN** χ^2. Here, by virtue of the above discussion concerning SN statistical analysis, we rewrite the chi square function in Equation (25) by

$$\chi^2_{\rm SN} = (\Delta\mu_{\rm SN} - \mathcal{M}\mathbf{1})^{\rm T}\mathbf{C}^{-1}(\Delta\mu_{\rm SN} - \mathcal{M}\mathbf{1}) \tag{69}$$

  where $\Delta\mu_{\rm SN} \equiv \mu_{\rm SN} - \mu^{\rm th}_{\rm SN}(\mathbf{x}, z_i)$ is the module of the vector of residuals, and $\mathbf{C}$ the covariance matrix.
  In particular, we prompt the distance modulus for the most recent SN catalog, named *Pantheon Sample*. This represents the current largest SN sample consisting of 1048 SNe Ia lying on $0.01 < z < 2.3$ [180]. The corresponding magnitudes read

$$\mu_{\rm SN} = m_{\rm B} - (\mathcal{M} - \alpha\mathcal{X}_1 + \beta\mathcal{C} - \Delta_{\rm M} - \Delta_{\rm B})\,. \tag{70}$$

  Here, $\mathcal{M}$ and $m_{\rm B}$ are the B-band absolute and apparent magnitudes, respectively. The above distance moduli also depend upon other quantities required to standardize/correct the light curves of SNe Ia. The quantities $\mathcal{X}_1$ and $\mathcal{C}$ are the light curve shape and color parameters, respectively, whereas α and β are the coefficients of the

luminosity–stretch and luminosity–color relationships, respectively. Δ_M is a distance correction determined on host galaxy mass of SNe, while Δ_B a distance correction that is built up from predicted biases determined by means of simulations.
By marginalizing over $\mathcal{M}$ through a flat prior, it is possible to demonstrate that SN uncertainties do not depend on $\mathcal{M}$, and this permits one to simplify the chi square function through

$$\chi^2_{SN,\mathcal{M}} = a + \log\frac{e}{2\pi} - \frac{b^2}{e} \tag{71}$$

where $a \equiv \vec{\Delta}^T_{SN}\mathbf{C}^{-1}\vec{\Delta}_{SN}$, $b \equiv \vec{\Delta}^T_{SN}\mathbf{C}^{-1}\vec{\mathbf{1}}$, $e \equiv \vec{\mathbf{1}}^T\mathbf{C}^{-1}\vec{\mathbf{1}}$.
- **BAO** χ^2. The chi square function for BAO data is given in Equations (27)–(29).

Below, we summarize a couple of statistical methods applied to GRB data to extract cosmological constraints.

9.2.1. Bézier Polynomials and Cosmographic Series

The first method utilizes GRB data, calibrated through the above Bézier polynomials, to extract cosmological constraints by means of cosmographic model-independent series and heal de facto the circularity problem without postulating the model a priori [216]. This method has been implemented to Amati, Ghirlanda, Yonetoku, and Combo GRB correlations in conjunction with SNe Ia and BAO data sets to get more stable and narrow constraints. We considered the most recent approaches to cosmography, comparing among them Taylor expansions with z and y_2 series, and Padé polynomials. Two hierarchies have been considered: *hierarchy 1*, up to j_0, and *hierarchy 2*, up to s_0.

Reasonable results have been found for both hierarchies through several MCMC fits showing possible matching with the standard paradigm (see Tables 3–6). Moreover, we only partially alleviated the tension on local H_0 measurements as hierarchy 2 is considered. Taylor outcomes are quite stable within each hierarchy, as portrayed by the results in Table 3, and work well with Amati, Ghirlanda, and Yonetoku correlations in the sense that the corresponding numerical outcomes are consistent within 1–σ with previous findings. Again, this suggests a spatially flat ΛCDM paradigm as a statistically favored model, with mass density parameter $\Omega_m = 2(1+q_0)/3 \sim 0.3$ for Combo correlation, whereas the other correlations seem to indicate smaller values.

The auxiliary y_2 variable is not stable enough compared to Taylor expansions. It significantly enlarges h_0, see, e.g., Table 4, and the overall results are however quite non-predictive at the level of hierarchy 1. Moreover, Padé fits seem to improve the quality of Taylor expansion hierarchy 1, as expected by construction (see Table 5). This is particularly evident for Combo and Yonetoku correlations, while, for Amati and Ghirlanda correlations, it is not. It is worth noticing that, to go further, jerk term implies $\geq$(3,1), leading to higher orders than $P_{3,1}$, quite unconstrained at higher redshift domains.

Quite surprisingly, our findings summarized in Tables 3–6 show that the ΛCDM model is not fully confirmed using GRBs. Although this can be an indication that more refined analyses are necessary, as GRBs are involved, simple indications seem to be against a genuine cosmological constant [122] and may be interpreted either with a barotropic dark energy contribution or with the need of non-zero spatial curvature [216]. Nevertheless, at this stage, our findings are in line with recent claims on tensions with the ΛCDM model [195,219,225].

Table 3. Cosmographic best fits and 1–σ (2–σ) errors from Taylor expansions labeled as *hierarchy 1* (h_0, q_0, j_0) and *hierarchy 2* (h_0, q_0, j_0, s_0). Letters A, G, Y, and C indicate Amati, Ghirlanda, Yonetoku, and Combo correlations, respectively.

Taylor Fits							
	Hierarchy 1			*Hierarchy* 2			
	h_0	q_0	j_0	h_0	q_0	j_0	s_0
A	$0.740^{+0.005\,(+0.010)}_{-0.006\,(-0.013)}$	$-0.68^{+0.03\,(+0.06)}_{-0.02\,(-0.04)}$	$0.77^{+0.08\,(+0.16)}_{-0.10\,(-0.20)}$	$0.700^{+0.007\,(+0.014)}_{-0.008\,(-0.015)}$	$-0.51^{+0.02\,(+0.03)}_{-0.01\,(-0.02)}$	$0.71^{+0.06\,(+0.12)}_{-0.05\,(-0.10)}$	$-0.36^{+0.05\,(+0.13)}_{-0.10\,(-0.20)}$
G	$0.716^{+0.006\,(+0.013)}_{-0.006\,(-0.014)}$	$-0.63^{+0.03\,(+0.06)}_{-0.03\,(-0.05)}$	$0.76^{+0.09\,(+0.17)}_{-0.09\,(-0.18)}$	$0.691^{+0.008\,(+0.016)}_{-0.007\,(-0.015)}$	$-0.50^{+0.02\,(+0.05)}_{-0.02\,(-0.05)}$	$0.64^{+0.06\,(+0.15)}_{-0.10\,(-0.19)}$	$-0.42^{+0.10\,(+0.17)}_{-0.08\,(-0.16)}$
Y	$0.737^{+0.008\,(+0.014)}_{-0.008\,(-0.015)}$	$-0.73^{+0.03\,(+0.06)}_{-0.01\,(-0.04)}$	$0.88^{+0.02\,(+0.13)}_{-0.13\,(-0.23)}$	$0.695^{+0.007\,(+0.014)}_{-0.008\,(-0.015)}$	$-0.54^{+0.02\,(+0.04)}_{-0.01\,(-0.03)}$	$0.70^{+0.07\,(+0.13)}_{-0.05\,(-0.11)}$	$-0.36^{+0.08\,(+0.16)}_{-0.09\,(-0.18)}$
C	$0.706^{+0.007\,(+0.013)}_{-0.007\,(-0.013)}$	$-0.59^{+0.03\,(+0.07)}_{-0.03\,(-0.06)}$	$0.72^{+0.09\,(+0.18)}_{-0.10\,(-0.18)}$	$0.693^{+0.006\,(+0.014)}_{-0.009\,(-0.015)}$	$-0.52^{+0.02\,(+0.05)}_{-0.01\,(-0.03)}$	$0.73^{+0.06\,(+0.13)}_{-0.09\,(-0.15)}$	$-0.38^{+0.06\,(+0.14)}_{-0.10\,(-0.19)}$

Table 4. Cosmographic best fits and 1–σ (2–σ) errors from expansions with y_2 labeled as *hierarchy 1* (h_0, q_0, j_0) and *hierarchy 2* (h_0, q_0, j_0, s_0). Letters A, G, Y, and C indicate Amati, Ghirlanda, Yonetoku, and Combo correlations, respectively.

y_2 Fits							
	Hierarchy 1			*Hierarchy* 2			
	h_0	q_0	j_0	h_0	q_0	j_0	s_0
A	$0.76^{+0.01\,(+0.02)}_{-0.01\,(-0.02)}$	$-1.35^{+0.05\,(+0.09)}_{-0.04\,(-0.08)}$	$3.85^{+0.25\,(+0.49}_{-0.28\,(-0.50)}$	$0.78^{+0.01\,(+0.02)}_{-0.01\,(-0.02)}$	$-0.53^{+0.06\,(+0.11)}_{-0.04\,(-0.10)}$	$-2.52^{+0.31\,(+0.71)}_{-0.42\,(-0.78)}$	$-4.41^{+1.00\,(+1.82)}_{-0.58\,(-1.29)}$
G	$0.75^{+0.01\,(+0.02)}_{-0.01\,(-0.02)}$	$-1.04^{+0.05\,(+0.10)}_{-0.05\,(-0.10)}$	$2.40^{+0.24\,(+0.47)}_{-0.23\,(-0.47)}$	$0.74^{+0.01\,(+0.02)}_{-0.01\,(-0.02)}$	$-0.45^{+0.10\,(+0.17)}_{-0.06\,(-0.13)}$	$-2.17^{+0.42\,(+0.96)}_{-0.61\,(-1.14)}$	$-3.08^{+1.41\,(+2.68)}_{-0.56\,(-1.28)}$
Y	$0.75^{+0.01\,(+0.02)}_{-0.01\,(-0.02)}$	$-1.05^{+0.05\,(+0.10)}_{-0.04\,(-0.09)}$	$2.47^{+0.22\,(0.50)}_{-0.23\,(-0.49)}$	$0.74^{+0.01\,(+0.02)}_{-0.01\,(-0.02)}$	$-0.43^{+0.03\,(+0.08)}_{-0.10\,(-0.21)}$	$-2.19^{+0.62\,(+1.62)}_{-0.32\,(-0.63)}$	$-2.70^{+0.37\,(+0.86)}_{-1.04\,(-1.51)}$
C	$0.75^{+0.01\,(+0.02)}_{-0.01\,(-0.02)}$	$-1.01^{+0.04\,(+0.9)}_{-0.05\,(-0.09)}$	$2.29^{+0.23\,(+0.44)}_{-0.20\,(-0.40)}$	$0.74^{+0.01\,(+0.02)}_{-0.01\,(-0.02)}$	$-0.43^{+0.06\,(+0.16)}_{-0.09\,(-0.17)}$	$-2.19^{+0.64\,(+1.19)}_{-0.38\,(-1.03)}$	$-2.79^{+0.90\,(+2.59)}_{-0.82\,(-1.50)}$

Table 5. Cosmographic best fits and 1–σ (2–σ) errors from Padé expansions labeled as *hierarchy 1*. Letters A, G, Y, and C indicate Amati, Ghirlanda, Yonetoku, and Combo correlations, respectively.

Padé Fits			
	Hierarchy 1		
	h_0	q_0	j_0
A	$0.70^{+0.01\,(+0.03)}_{-0.02\,(-0.03)}$	$-0.33^{+0.05\,(+0.09)}_{-0.03\,(-0.08)}$	$0.240^{+0.010\,(+0.020)}_{-0.010\,(-0.020)}$
G	$0.70^{+0.02\,(+0.03)}_{-0.01\,(-0.02)}$	$-0.31^{+0.02\,(+0.06)}_{-0.05\,(-0.09)}$	$0.235^{+0.013\,(+0.027)}_{-0.002\,(-0.006)}$
Y	$0.68^{+0.01\,(+0.02)}_{-0.01\,(-0.02)}$	$-0.32^{+0.02\,(+0.06)}_{-0.04\,(-0.07)}$	$0.240^{+0.010\,(+0.021)}_{-0.005\,(-0.010)}$
C	$0.68^{+0.01\,(+0.02)}_{-0.01\,(-0.02)}$	$-0.33^{+0.03\,(+0.06)}_{-0.03\,(-0.06)}$	$0.244^{+0.009\,(+0.019)}_{-0.006\,(-0.012)}$

Table 6. χ^2 values of the cosmographic fits performed over the considered approximants. For each GRB correlation, the number of degrees of freedom (DoF) and the considered hierarchy are reported. Correlations are sorted for increasing values of the ratio χ^2/DoF with respect to the Taylor hierarchy 1 expansion.

Sample	DoF	Hierarchy	Approximant χ^2		
			Taylor	Function y_2	Padé $P_{2,1}$
Combo	1113	1	1116.84	1230.71	1113.77
	1112	2	1089.25	1160.04	
Ghirlanda	1080	1	1120.19	1271.92	2203.16
	1079	2	1075.01	1184.42	
Yonetoku	1154	1	1235.08	1350.27	1178.07
	1153	2	1147.72	1227.25	
Amati	1246	1	2334.35	2818.25	2202.75
	1245	2	2174.13	2539.98	

9.2.2. Bézier Polynomials and ΛCDM and ωCDM Cosmological Models

In a second method summarized here, the circularity problem affecting GRBs is again overcome by using Bézier polynomials to calibrate, in this case, the Amati correlation alone [123]. Unlike the previous method, GRB data are utilized now in conjuction with the SNe Ia JLA data set alone [226] and employed to explicitly constrain two different cosmological scenarios: the concordance ΛCDM model and the ωCDM model, with the dark energy equation of the state parameter free to vary [123].

In the Monte Carlo integration, through the Metropolis–Hastings algorithm, H_0 has been fixed to the best-fit value obtained from the model-independent analysis over OHD data, i.e., $H_0 = 67.76$ km s^{-1} Mpc^{-1}. The results for Ω_m and w (see Table 7) agree with previous findings making use of GRBs. The statistical performance of the models under study has been evaluated through the Akaike information criterion (AIC) criterion [227]

$$\mathrm{AIC} \equiv 2p - 2\ln\mathcal{L}_{max}$$

where p is the number of free parameters in the model and $\mathcal{L}_{max}$ is the maximum probability function calculated at the best-fit point, and the deviance information criterion (DIC) criterion [228]

$$\mathrm{DIC} \equiv 2p_{eff} - 2\ln\mathcal{L}_{max}\,,$$

where $p_{eff} = \langle -2\ln\mathcal{L}\rangle + 2\ln\mathcal{L}_{max}$ is the number of parameters that a data set can effectively constrain[28]. The best model is the one that minimizes the AIC and DIC values. Unlike the AIC criterion, the DIC statistics do not penalize for the total number of free parameters of the model, but only for those which are constrained by the data [229]. Differently from the previous approach, we found that the ΛCDM model is preferred with respect to the minimal ωCDM extension (see Table 7) and then conclude that no modifications of the standard paradigm are expected as intermediate redshifts are involved. However, future efforts dedicated to the use of our new technique to fix refined constraints over dynamical dark energy models are encouraged in order to fix the apparent dichotomy in the results of the two described methods.

Table 7. 95% confidence level results of the MCMC analysis for the SN+GRB data. The AIC and DIC differences are intended with respect to the ΛCDM model.

Model	ω	Ω_m	ΔAIC	ΔDIC
ΛCDM	-1	$0.397^{+0.040}_{-0.039}$	0	0
wCDM	$-0.86^{+0.36}_{-0.38}$	$0.34^{+0.13}_{-0.15}$	1.44	1.24

9.3. *The Role of Spatial Curvature*

An updated sample of GRBs has been developed in 2020, in which the Combo relation extracts bounds on the spatial curvature with no other probes [141], differently from previous attempts that commonly assume a spatially flat background, with the inclusion of SNe Ia and BAO.

The way in which the Combo relation is calibrated is without an OHD data set, but rather invokes two step methods. In this picture, we assume [140]

I. the terms k_1 and $\sigma_{\rm k}$ are obtained from small GRB sub-samples with almost the same redshift;
II. k_0 is determined from the use of SNe Ia limited to the lowest redshift of the GRBs of the Combo data set, in which the calibration of SNe Ia is negligible [140].

GRB sub-samples with the same z are chosen among those ones providing well constrained best-fit parameters.

Considering that in each sub-sample the GRB luminosity distances $d_{\rm L}$ are quite the same, we employ the rest-frame 0.3–10 keV energy flux F_0. This improves the dependence on the model and enables one to render our procedure cosmology-independent. In the specific example, reported here, seven sub-samples have been suggested and the best fits reported in Table 8, showing no evident trends with z within the errors. This technique is used for Combo relation since previous results suggest its advantages in the fitting strategies described above[29]. One can perform a simultaneous fit of the above sub-samples with the same k_1 and σ_k implying $k_1 = 0.90 \pm 0.13$ and $\sigma_k = 0.28 \pm 0.03$.

Table 8. Best-fit parameters of the seven sub-samples at average redshift $\langle z \rangle$: the slope $q_{1,z}$, the normalization $\log F_{0,z}$, and the extrascatter $\sigma_{q,z}$ are shown.

$\langle z \rangle$	$k_{1,z}$	$\log[F_{0,z}/(\text{erg cm}^{-2}\text{s}^{-1})]$	$\sigma_{k,z}$
0.54 ± 0.01	0.81 ± 0.49	-7.38 ± 1.08	0.29 ± 0.10
1.18 ± 0.07	0.83 ± 0.31	-7.93 ± 0.77	0.26 ± 0.08
1.46 ± 0.05	0.80 ± 0.32	-8.29 ± 0.79	0.26 ± 0.08
1.70 ± 0.06	0.94 ± 0.32	-8.92 ± 0.84	0.19 ± 0.08
2.05 ± 0.05	1.05 ± 0.32	-9.44 ± 0.90	0.34 ± 0.08
2.27 ± 0.07	0.78 ± 0.20	-8.57 ± 0.54	0.16 ± 0.06
2.69 ± 0.08	1.02 ± 0.34	-9.38 ± 0.87	0.34 ± 0.10

The calibration of k_0 is performed by means of the nearest couple of GRBs of the employed sample with the same redshift. In particular, it is possible to take GRB 130702A at $z = 0.145$ and GRB 161219B at $z = 0.1475$. Then, μ_C^{obs} can be replaced via its average distance modulus $\langle \mu_{SNIa} \rangle = 39.21 \pm 0.24$, determined by SNe Ia with the same z as the above two GRBs, considering the bound over k_1 and the values of F_0, E_p, τ, and α for the two GRBs adopted throughout the computation. The computed value is $k_0 = 49.54 \pm 0.21$.

At this stage, comparing between GRB distance moduli μ_C^{obs}, with uncertainties $\sigma\mu_C^{obs}$, with theoretical expectations, it is possible to get constraints over background cosmologies. In particular, for a non-flat ΛCDM model, i.e., the simplest scenario to work with, we write

$$d_L = \frac{c}{H_0}\frac{(1+z)}{\sqrt{|\Omega_k|}}\text{sinn}\left(\int_0^z \frac{\sqrt{|\Omega_k|}dz'}{\sqrt{\Omega_m(1+z)^3+\Omega_\Lambda+\Omega_k(1+z)^2}}\right), \tag{72}$$

and we compute numerical bounds once H_0 is marginalized[30] as in Table 9.

In particular, slightly larger estimations on matter density are obtained from GRBs, i.e., $\Omega_m = 0.32^{+0.05}_{-0.05}$ with H_0 of Ref. [122] and the opposite, i.e., $\Omega_m = 0.22^{+0.04}_{-0.03}$, for the H_0 of Ref. [222]. Analogous results, i.e., compatible with the flat case, are computed using the non-flat ΛCDM model.

Table 9. Best-fit parameters with 1–σ uncertainties for the various cosmological cases discussed in this work. The last column lists the values of the χ^2. H_0 is fixed to the values given by Ref. [122,222], respectively the Planck and Riess expectation values.

H_0	Ω_m	Ω_Λ	Ω_k	χ^2
		Flat ΛCDM (DOF = 173)		
P18	$0.32^{+0.05}_{-0.05}$	$0.68^{+0.05}_{-0.05}$	0	165.54
R19	$0.22^{+0.04}_{-0.03}$	$0.78^{+0.04}_{-0.03}$	0	171.32
		ΛCDM (DOF = 172)		
P18	$0.34^{+0.08}_{-0.07}$	$0.91^{+0.22}_{-0.35}$	$-0.24^{+0.24}_{-0.35}$	164.38
R19	$0.24^{+0.06}_{-0.05}$	$1.01^{+0.15}_{-0.25}$	$-0.24^{+0.16}_{-0.25}$	169.23

10. Further Application of GRBs as Probes of the High-Redshift Universe

Above, we faced some of the standard statistical methods associated with GRBs, and we proposed a few applications with the corresponding experimental bounds. Now we shortly review other phenomenological applications of GRBs that can shed light on the high-redshift Universe.

10.1. Star Formation Rate from GRBs

LGRBs are likely associated with core-collapse SNe [38,230]. By virtue of this fact, GRB-SN associations may provide a complementary technique that measures a high-redshift star formation rate [231–235]. Again, a problem related to calibration occurs, i.e., how to calibrate a GRB event rate with a star formation rate. Moreover, the luminosity function is not known a priori, and its reconstruction depends on the particular model selected for the analysis [236–239]. The typical functional structures for the luminosity functions are (i) a broken power law [240] and (ii) a single power law with an exponential cut-off at low luminosities [237]. Reconstructed SFRs from GRBs are typically larger than those from other observations [241]. The reason behind this apparent inconsistency may reside in the fact that usually SFR at high-z is obtained from the observations of the brightest galaxies, whereas GRBs, in view of their high luminosities, may help in detecting faint galaxies at high-z otherwise unobserved [241,242].

10.2. High-Redshift GRB Rate Excess

Although appealing, the above developments do not show why GRBs do not follow the star formation history being enhanced by hidden high-redshift mechanisms [242–247]. In particular, the star formation rate at high-redshift, namely $z > 6$, appears too large if confronted with the star formation rate obtained from high-redshift galaxy surveys [248].

A natural origin of the high-redshift GRB rate excess can be found in the metallicity evolution, as LGRBs seem to prefer a low-metallicity environment, as supported by recent studies that favor such a requirement[31]. Typical mass bounds on stars suggest $> 30M_{\odot}$, being responsible for BH remnants.

10.3. Gravitationally-Lensed GRBs

Gravitationally lensed GRBs (GLGRBs) have been proposed in Ref. [49], where it was speculated that such a phenomenon would result in multiple light curves detected at different times, as due to the different light paths of the produced multiple GRB images.

Quests for GLGRBs were mostly based on strong lensing[32] and similarities among GRB light curves with identical spectra and close locations in the sky, as primary search criteria [251–255]. However, such searches led to null results, possibly due to Poisson noise uncertainties, affecting GRB light curves especially at low signal-to-noise ratios, which may have introduced large differences between the lensed GRB images [254]. On the other hand, some GLGRB could exhibit time delays shorter than (or comparable to) the burst duration, hence leading to unresolved (or locally separated) peaks separated by the time delay [256–258].

Several searches have been performed in the literature, resulting in a few or null candidates, based on different techniques and lens models, such as globular cluster with a mass of $\approx 10^5$–$10^7\ M_{\odot}$ [256], Population III stars with a mass range of 10^2–$10^3\ M_{\odot}$ [257], diverse objects with a mass range of 10^2–$10^7\ M_{\odot}$ [259], or a supermassive BH with a mass in the range of $\approx 10^5$–$10^7\ M_{\odot}$ [260].

Considering models where the lens is a supermassive BH [259,260], GLGRB candidates can provide an opportunity to estimate the number density of massive compact objects at cosmological distances by calculating the rate of GRB lensing. Assuming such supermassive BH lenses with mass $\approx 10^6\ M_{\odot}$ as dark matter compact objects [261], the density parameter of BHs is $\Omega_{\rm BH} = 0.007 \pm 0.004$, or equivalently $\Omega_{BH}/\Omega_M = 0.027 \pm 0.016$ [260]. In this respect, finding more GLGRBs candidates from supermassive BH may enhance our understanding of the matter content of the Universe.

11. Conclusions

In this review, we outlined some of the most recent developments toward GRB physics, properties, and their applications to cosmology. In particular, the review is structured into two main parts.

In the first part, we discussed the basic demands of GRBs, including their main observable quantities, their classification scheme into LGRBs and SGRBs and the corresponding issues, emphasizing new possible ways of classification, still the object of speculation. Afterwards, we put emphasis on GRB progenitors and on their fundamental microphysics, in view of the experimental evidence characterizing prompt and afterglow emissions, etc. LGRB connections with SNe have been explored as well along with SGRB matching with KNe and GWs. Great emphasis has been devoted to portray the standard GRB formation, working with the well consolidated *fireball model*. Particle acceleration and radiative processes, predicted in such a picture, have been largely reported, with particular concern for observable signatures and the standard model frontiers. Even in this part, we illustrated that GRBs cannot be contemplated as genuine standard candles, since there is no consensus toward their internal processes that we depicted throughout the manuscript. Accordingly, their luminosity cannot be easily related to their redshifts as, for instance, one does for SNe.

For these aspects and for the overall limitations described above, we developed in the second part considerable cosmological applications of GRB physics. We tried to standardize GRBs by means of the most recent techniques and accentuated GRBs are essential to reconcile small with intermediate redshift domains, opening new scenarios toward our universe comprehension. In this respect, we featured how GRBs could be used as complementary and outstanding probes to trace dark energy's evolution in support of other indicators, e.g., SNeIa, BAO, CMB, Hubble differential data, etc. Thereby, we have shown a few statistical treatments related to Bayesian analysis in cosmology, able to combine GRBs with other catalogs of data, reporting the most recent cosmological constraints on dark energy models. To do so, we expounded the bristly circularity problem, burdening GRBs in cosmological set ups. In particular, we also changed perspective, showing how to avoid calibration, i.e., how not to employ the circularity. We confronted the two methods and checked which departures could be expected from the standard cosmological model through the use of GRBs in both of the cases. Details on error propagation and GRB systematics have been discussed for several cosmic GRB correlations. Model dependent and independent techniques of calibrations have been likewise portrayed.

Perspectives about GRB developments will be based on clarifying the overall issues raised in this review. In particular, it is of utmost importance to shed light on how to standardize GRBs, in view of a likely self-consistent evolutionary paradigm, so far missing. With this recipe, we expect in the incoming years to improve GRB use in cosmology and get rid of circularity and greatly reduce the systematics and all the other issues that affect GRB data and challenge their use in cosmology. In particular, some models akin to those characterizing other cosmic indicators will spell out how to describe in toto GRB physics and evolution.

Author Contributions: Conceptualization, O.L. and M.M.; methodology, O.L. and M.M.; software, O.L. and M.M.; validation, O.L. and M.M.; formal analysis, O.L. and M.M.; investigation, O.L. and M.M.; resources, O.L. and M.M.; data curation, O.L. and M.M.; writing—original draft preparation, O.L. and M.M.; writing—review and editing, O.L. and M.M.; visualization, O.L. and M.M.; supervision, O.L. and M.M.; project administration, O.L. and M.M.; funding acquisition, O.L. and M.M. All authors have read and agreed to the published version of the manuscript.

Funding: O.L. and M.M. acknowledge partial support of the Ministry of Education and Science of the Republic of Kazakhstan, Grant No. IRN AP08052311. M. M. is supported by INFN as part of the MoonLIGHT-2 experiment in the framework of the research activities of CSN2.

Data Availability Statement: Data are taken from the papers quoted in the text.

Acknowledgments: We acknowledge the financial support provided by the Ministry of Education and Science of the Republic of Kazakhstan, Grant No. IRN AP08052311.

Conflicts of Interest: The authors declare no conflict of interest.

Notes

1 New data come from the newly born gravitational wave, neutrino, and BH astronomy. We remark that precision cosmology is essential to shed light on the mysteries that jeopardize the standard cosmological puzzle. In this respect, GRBs could play a significant role since they represent outstanding explosions whose nature can trace the dark energy (DE) and BH natures.

2 GRBs approximately span in the range $z \simeq 2$–10.

3 Like the long-lived GeV emission, which is consistent with the afterglow emission of a blast wave in adiabatic expansion

4 For a different perspective, see, e.g., [26]

5 Hereafter the bulk Lorentz factor is indicated with Γ to avoid confusion with the power-law photon index γ, describing simple power-law GRB spectra.

6 In the observer frame.

7 The GW signal, originating from the shell elliptical galaxy NGC 4993, had a duration of $\sim$ 100 s. By the characteristics in intensity and frequency, GW 170817 has been unambiguously associated with the inspiraling of a binary NS-SN merger of total mass $2.82^{+0.09}_{-0.47}$ $M_{\odot}$, which is consistent with the masses of all known binary NS systems.

8 Typically dubbed time-integrated and time-resolved analyses, respectively.

9 However, in this case the spectral break is very likely below or above the detector bandpass.

10 This energy band is the one established in the original work by Ref. [65].

11 Additional details on $d_L(z)$ will be summarized later in this review. Here, we stress that this definition has been written for spatially flat DE models.

12 The time lag is historically computed in these energy bands which are the BATSE energy channels 3 and 1, respectively.

13 As a convention, the X-ray luminosities are computed in a rest-frame energy band with similar extrema with respect to the observed one; with this prescription, their expressions are simple, as portrayed in Equation (11).

14 An alternative scenario proposes that an NS remnant could be left after a GRB emission, though this issue is still under debate. For details, see Refs. [15,70,71] and Section 3.1.3.

15 Astrophysical fireballs include also some baryons from the surrounding medium, remnant of the progenitor system.

16 However, both B and γ_e are much higher than the ones inferred from the fit with the synchrotron model of the GRB afterglow, whose microphysics of particle acceleration and magnetic field generation should be similar to that of the prompt emission environment [105].

17 Examples are main sequence fitting method, variable stars, Tully–Fisher and Faber–Jackson relations, etc.

18 Examples are SNe Ia, Cosmic Microwave Background (CMB) measurements, Baryonic Acoustic Oscillations (BAO), etc.

19 They refer to "Type Ia" for the absence of hydrogen and the presence of once ionized silicon (SiII) in their early-time spectra.

20 Weak similarities between GRBs and SNe Ia may occur at the level of formal computation, although the GRB nature is absolutely different from SNe. The core idea is to write a GRB luminosity distance as well and proceed analogously.

21 We here focus on vanishing spatial curvature, i.e., $\Omega_k = 0$ [122].

22 The flare-filtered luminosity light curves are iteratively fitted with the above function: at every iteration, data points with the largest positive residual are discarded, until a final fit with a p-value > 0.3 is obtained.

23 Where their luminosity distance is essentially independent from the choice of the cosmological model

24 The cosmological model under exam, or background cosmology, is intimately determined by knowing the functional form and evolution of $H(z)$.

25 For the sake of completeness, this problem is not related to GRB redshifts only, but it remains an open issue of cosmography.

26 Similar conclusions have been reached in Ref. [213], who confirmed that only the Ghirlanda correlation has no redshift dependence, and determined $\Omega_{m0} = 0.307^{+0.065}_{-0.073}$ in the flat ΛCDM model from SNeIa calibrated GRB data.

27 Possibly including those associated with detector sensitivity, and the differences in estimated spectral parameters determined from measurements taken with different detectors or from different models.

28 Here, the brackets indicate the average over the posterior distribution.

29 By construction, these sub-samples exhibit the same correlation with the same q_1 and σ_q but involve different normalizations.

30 Constraints come from $H_0 = (67.4 \pm 0.5)$ km s^{-1}Mpc^{-1} [122] and $H_0 = (74.03 \pm 1.42)$ km s^{-1}Mpc^{-1} [222].

31 There is uncertainty on measuring GRB metallicity at high-redshifts, due to chemical inhomogeneity, for example [249,250]. Thus, this approach cannot be seen as definitive.

32 In the case of strong lensing, the time delay between the images is larger than the duration of the burst.

References

1. Piron, F. Gamma-ray bursts at high and very high energies. *Comptes Rendus Phys.* **2016**, *17*, 617–631. [CrossRef]
2. Nakar, E. Short-hard gamma-ray bursts. *Phys. Rep.* **2007**, *442*, 166–236. [CrossRef]
3. Kouveliotou, C.; Meegan, C.A.; Fishman, G.J.; Bhat, N.P.; Briggs, M.S.; Koshut, T.M.; Paciesas, W.S.; Pendleton, G.N. Identification of two classes of gamma-ray bursts. *Astrophys. J. Lett.* **1993**, *413*, L101. [CrossRef]
4. von Kienlin, A.; Meegan, C.A.; Paciesas, W.S.; Bhat, P.N.; Bissaldi, E.; Briggs, M.S.; Burgess, J.M.; Byrne, D.; Chaplin, V.; Cleveland, W.; et al. The Second Fermi GBM Gamma-Ray Burst Catalog: The First Four Years. *Astrophys. J. Suppl.* **2014**, *211*, 13. [CrossRef]
5. D'Avanzo, P. The link between short gamma-ray bursts and gravitational waves: Perspectives for the THESEUS mission. *Mem. Soc. Ast. It.* **2018**, *89*, 213.
6. Ghirlanda, G.; Salafia, O.S.; Pescalli, A.; Ghisellini, G.; Salvaterra, R.; Chassande-Mottin, E.; Colpi, M.; Nappo, F.; D'Avanzo, P.; Melandri, A.; et al. Short gamma-ray bursts at the dawn of the gravitational wave era. *Astron. Astrophys.* **2016**, *594*, A84. [CrossRef]
7. Berger, E.; Kulkarni, S.R.; Fox, D.B.; Soderberg, A.M.; Harrison, F.A.; Nakar, E.; Kelson, D.D.; Gladders, M.D.; Mulchaey, J.S.; Oemler, A.; et al. Afterglows, Redshifts, and Properties of Swift Gamma-Ray Bursts. *Astrophys. J.* **2005**, *634*, 501. [CrossRef]
8. Rodney, S.A.; Riess, A.G.; Scolnic, D.M.; Jones, D.O.; Hemmati, S.; Molino, A.; McCully, C.; Mobasher, B.; Strolger, L.G.; Graur, O.; et al. Two SNe Ia at Redshift ~2: Improved Classification and Redshift Determination with Medium-band Infrared Imaging. *Astron. J.* **2015**, *150*, 156. [CrossRef]
9. Capozziello, S.; D'Agostino, R.; Luongo, O. Extended gravity cosmography. *Int. J. Mod. Phys. D* **2019**, *28*, 1930016. [CrossRef]
10. Khadka, N.; Luongo, O.; Muccino, M.; Ratra, B. Do gamma-ray burst measurements provide a useful test of cosmological models? *arXiv* **2021**, arXiv:2105.12692.
11. Lin, H.N.; Li, X.; Wang, S.; Chang, Z. Are long gamma-ray bursts standard candles? *Mon. Not. R. Astron. Soc.* **2015**, *453*, 128–132. [CrossRef]
12. Demianski, M.; Piedipalumbo, E.; Sawant, D.; Amati, L. Cosmology with gamma-ray bursts. I. The Hubble diagram through the calibrated $E_{p,I}$-E_{iso} correlation. *Astron. Astrophys.* **2017**, *598*, A112. [CrossRef]
13. Demianski, M.; Lusso, E.; Paolillo, M.; Piedipalumbo, E.; Risaliti, G. Investigating dark energy equation of state with high redshift Hubble diagram. *Front. Astron. Space Sci.* **2020**, *7*, 69. [CrossRef]
14. Mazets, E.P.; Golenetskii, S.V.; Ilinskii, V.N.; Panov, V.N.; Aptekar, R.L.; Gurian, I.A.; Proskura, M.P.; Sokolov, I.A.; Sokolova, Z.I.; Kharitonova, T.V. Catalog of cosmic gamma-ray bursts from the KONUS experiment data. I. *Astrophys. Space Sci.* **1981**, *80*, 3–83. [CrossRef]
15. Kouveliotou, C.; Wijers, R.A.M.J.; Woosley, S. *Gamma-Ray Bursts*; Cambridge University Press: Cambridge, UK, 2012.
16. Norris, J.P.; Bonnell, J.T. Short Gamma-Ray Bursts with Extended Emission. *Astrophys. J.* **2006**, *643*, 266. [CrossRef]
17. Lü, H.J.; Zhang, B.; Lei, W.H.; Li, Y.; Lasky, P.D. The Millisecond Magnetar Central Engine in Short GRBs. *Astrophys. J.* **2015**, *805*, 89. [CrossRef]
18. Levan, A.J.; Tanvir, N.R.; Starling, R.L.C.; Wiersema, K.; Page, K.L.; Perley, D.A.; Schulze, S.; Wynn, G.A.; Chornock, R.; Hjorth, J.; et al. A New Population of Ultra-long Duration Gamma-Ray Bursts. *Astrophys. J.* **2014**, *781*, 13. [CrossRef]
19. Zhang, B.B.; Zhang, B.; Murase, K.; Connaughton, V.; Briggs, M.S. How Long does a Burst Burst? *Astrophys. J.* **2014**, *787*, 66. [CrossRef]
20. Heise, J.; in't Zand, J.; Kippen, R.M.; Woods, P.M. X-ray Flashes and X-ray Rich Gamma Ray Bursts. In *Gamma-ray Bursts in the Afterglow Era*; Costa, E., Frontera, F., Hjorth, J., Eds.; Springer: Berlin, Heidelberg, 2001; p. 16.
21. Piran, T.; Bromberg, O.; Nakar, E.; Sari, R. The long, the short and the weak: The origin of gamma-ray bursts. *R. Soc. Lond. Philos. Trans. Ser. A* **2013**, *371*, 20273. [CrossRef] [PubMed]
22. Zhang, B.; Zhang, B.B.; Liang, E.W.; Gehrels, N.; Burrows, D.N.; Mészáros, P. Making a Short Gamma-Ray Burst from a Long One: Implications for the Nature of GRB 060614. *Astrophys. J. Lett.* **2007**, *655*, L25–L28. [CrossRef]
23. Woosley, S.E.; Bloom, J.S. The Supernova Gamma-Ray Burst Connection. *Annu. Rev. Astron. Astrophys.* **2006**, *44*, 507–556. [CrossRef]
24. Berger, E. Short-Duration Gamma-Ray Bursts. *Annu. Rev. Astron. Astrophys.* **2014**, *52*, 43–105. [CrossRef]
25. Piran, T.; Sari, R.; Zou, Y.C. Observational limits on inverse Compton processes in gamma-ray bursts. *Mon. Not. R. Astron. Soc.* **2009**, *393*, 1107. [CrossRef]
26. Ruffini, R.; Vereshchagin, G.; Xue, S.S. Electron-positron pairs in physics and astrophysics: From heavy nuclei to black holes. *Phys. Rep.* **2010**, *487*, 1. [CrossRef]
27. Panaitescu, A.; Mészáros, P. Dynamical Evolution, Light Curves, and Spectra of Spherical and Collimated Gamma-Ray Burst Remnants. *Astrophys. J.* **1999**, *526*, 707. [CrossRef]
28. Sari, R.; Piran, T.; Halpern, J.P. Jets in Gamma-Ray Bursts. *Astrophys. J. Lett.* **1999**, *519*, L17. [CrossRef]
29. Racusin, J.L.; Liang, E.W.; Burrows, D.N.; Falcone, A.; Sakamoto, T.; Zhang, B.B.; Zhang, B.; Evans, P.; Osborne, J. Jet Breaks and Energetics of Swift Gamma-Ray Burst X-Ray Afterglows. *Astrophys. J.* **2009**, *698*, 43–74. [CrossRef]
30. Lamb, G.P.; Kann, D.A.; Fernández, J.J.; Mandel, I.; Levan, A.J.; Tanvir, N.R. GRB jet structure and the jet break. *Mon. Not. R. Astron. Soc.* **2021**, *506*, 4163–4174. [CrossRef]

31. Nousek, J.A.; Kouveliotou, C.; Grupe, D.; Page, K.L.; Granot, J.; Ramirez-Ruiz, E.; Patel, S.K.; Burrows, D.N.; Mangano, V.; Barthelmy, S.; et al. Evidence for a Canonical Gamma-Ray Burst Afterglow Light Curve in the Swift XRT Data. *Astrophys. J.* **2006**, *642*, 389. [CrossRef]
32. Panaitescu, A.; Mészáros, P.; Burrows, D.; Nousek, J.; Gehrels, N.; O'Brien, P.; Willingale, R. Evidence for chromatic X-ray light-curve breaks in Swift gamma-ray burst afterglows and their theoretical implications. *Mon. Not. R. Astron. Soc.* **2006**, *369*, 2059. [CrossRef]
33. Zhang, B.; Fan, Y.Z.; Dyks, J.; Kobayashi, S.; Mészáros, P.; Burrows, D.N.; Nousek, J.A.; Gehrels, N. Physical Processes Shaping Gamma-Ray Burst X-Ray Afterglow Light Curves: Theoretical Implications from the Swift X-Ray Telescope Observations. *Astrophys. J.* **2006**, *642*, 354. [CrossRef]
34. Chincarini, G.; Moretti, A.; Romano, P.; Falcone, A.D.; Morris, D.; Racusin, J.; Campana, S.; Covino, S.; Guidorzi, C.; Tagliaferri, G.; et al. The First Survey of X-Ray Flares from Gamma-Ray Bursts Observed by Swift: Temporal Properties and Morphology. *Astrophys. J.* **2007**, *671*, 1903. [CrossRef]
35. Kumar, P.; Narayan, R.; Johnson, J.L. Mass fall-back and accretion in the central engine of gamma-ray bursts. *Mon. Not. R. Astron. Soc.* **2008**, *388*, 1729–1742. [CrossRef]
36. Hjorth, J. The supernova-gamma-ray burst-jet connection. *Philos. Trans. R. Soc. Lond. Ser. A* **2013**, *371*, 20120275. [CrossRef]
37. Galama, T.J.; Vreeswijk, P.M.; van Paradijs, J.; Kouveliotou, C.; Augusteijn, T.; Böhnhardt, H.; Brewer, J.P.; Doublier, V.; Gonzalez, J.; Leibundgut, B.; et al. An unusual supernova in the error box of the γ-ray burst of 25 April 1998. *Nature* **1998**, *395*, 670. [CrossRef]
38. Hjorth, J.; Sollerman, J.; Møller, P.; Fynbo, J.P.U.; Woosley, S.E.; Kouveliotou, C.; Tanvir, N.R.; Greiner, J.; Andersen, M.I.; Castro-Tirado, A.J.; et al. A very energetic supernova associated with the γ-ray burst of 29 March 2003. *Nature* **2003**, *423*, 847. [CrossRef] [PubMed]
39. Soderberg, A.M.; Kulkarni, S.R.; Nakar, E.; Berger, E.; Cameron, P.B.; Fox, D.B.; Frail, D.; Gal-Yam, A.; Sari, R.; Cenko, S.B.; et al. Relativistic ejecta from X-ray flash XRF 060218 and the rate of cosmic explosions. *Nature* **2006**, *442*, 1014–1017. [CrossRef]
40. MacFadyen, A.I.; Woosley, S.E.; Heger, A. Supernovae, Jets, and Collapsars. *Astrophys. J.* **2001**, *550*, 410–425. [CrossRef]
41. Zhang, W.; Woosley, S.E.; MacFadyen, A.I. Relativistic Jets in Collapsars. *Astrophys. J.* **2003**, *586*, 356–371. [CrossRef]
42. Greiner, J.; Mazzali, P.A.; Kann, D.A.; Krühler, T.; Pian, E.; Prentice, S.; Olivares E., F.; Rossi, A.; Klose, S.; Taubenberger, S.; et al. A very luminous magnetar-powered supernova associated with an ultra-long γ-ray burst. *Nature* **2015**, *523*, 189–192. [CrossRef] [PubMed]
43. Gal-Yam, A. Luminous Supernovae. *Science* **2012**, *337*, 927. [CrossRef]
44. Fynbo, J.P.U.; Watson, D.; Thöne, C.C.; Sollerman, J.; Bloom, J.S.; Davis, T.M.; Hjorth, J.; Jakobsson, P.; Jørgensen, U.G.; Graham, J.F.; et al. No supernovae associated with two long-duration γ-ray bursts. *Nature* **2006**, *444*, 1047–1049. [CrossRef]
45. Della Valle, M.; Chincarini, G.; Panagia, N.; Tagliaferri, G.; Malesani, D.; Testa, V.; Fugazza, D.; Campana, S.; Covino, S.; Mangano, V.; et al. An enigmatic long-lasting γ-ray burst not accompanied by a bright supernova. *Nature* **2006**, *444*, 1050. [CrossRef]
46. Gal-Yam, A.; Fox, D.B.; Price, P.A.; Ofek, E.O.; Davis, M.R.; Leonard, D.C.; Soderberg, A.M.; Schmidt, B.P.; Lewis, K.M.; Peterson, B.A.; et al. A novel explosive process is required for the γ-ray burst GRB 060614. *Nature* **2006**, *444*, 1053–1055. [CrossRef]
47. Gehrels, N.; Sarazin, C.L.; O'Brien, P.T.; Zhang, B.; Barbier, L.; Barthelmy, S.D.; Blustin, A.; Burrows, D.N.; Cannizzo, J.; Cummings, J.R.; et al. A short γ-ray burst apparently associated with an elliptical galaxy at redshift z = 0.225. *Nature* **2005**, *437*, 851. [CrossRef]
48. Goodman, J. Are gamma-ray bursts optically thick? *Astrophys. J. Lett.* **1986**, *308*, L47. [CrossRef]
49. Paczynski, B. Gamma-ray bursters at cosmological distances. *Astrophys. J. Lett.* **1986**, *308*, L43. [CrossRef]
50. Eichler, D.; Livio, M.; Piran, T.; Schramm, D.N. Nucleosynthesis, neutrino bursts and gamma-rays from coalescing neutron stars. *Nature* **1989**, *340*, 126–128. [CrossRef]
51. Taylor, J.H.; Weisberg, J.M. Further experimental tests of relativistic gravity using the binary pulsar PSR 1913 + 16. *Astrophys. J.* **1989**, *345*, 434. [CrossRef]
52. Li, L.X.; Paczyński, B. Transient Events from Neutron Star Mergers. *Astrophys. J. Lett.* **1998**, *507*, L59–L62. [CrossRef]
53. Tanvir, N.R.; Levan, A.J.; Fruchter, A.S.; Hjorth, J.; Hounsell, R.A.; Wiersema, K.; Tunnicliffe, R.L. A 'kilonova' associated with the short-duration γ-ray burst GRB 130603B. *Nature* **2013**, *500*, 547–549. [CrossRef]
54. Lamb, G.P.; Tanvir, N.R.; Levan, A.J.; de Ugarte Postigo, A.; Kawaguchi, K.; Corsi, A.; Evans, P.A.; Gompertz, B.; Malesani, D.B.; Page, K.L.; et al. Short GRB 160821B: A Reverse Shock, a Refreshed Shock, and a Well-sampled Kilonova. *Astrophys. J.* **2019**, *883*, 48. [CrossRef]
55. Troja, E.; Castro-Tirado, A.J.; Becerra González, J.; Hu, Y.; Ryan, G.S.; Cenko, S.B.; Ricci, R.; Novara, G.; Sánchez-Rámirez, R.; Acosta-Pulido, J.A.; et al. The afterglow and kilonova of the short GRB 160821B. *Mon. Not. R. Astron. Soc.* **2019**, *489*, 2104–2116. [CrossRef]
56. Fong, W.; Laskar, T.; Rastinejad, J.; Escorial, A.R.; Schroeder, G.; Barnes, J.; Kilpatrick, C.D.; Paterson, K.; Berger, E.; Metzger, B.D.; et al. The Broadband Counterpart of the Short GRB 200522A at z = 0.5536: A Luminous Kilonova or a Collimated Outflow with a Reverse Shock? *Astrophys. J.* **2021**, *906*, 127. [CrossRef]
57. Gompertz, B.P.; Levan, A.J.; Tanvir, N.R.; Hjorth, J.; Covino, S.; Evans, P.A.; Fruchter, A.S.; González-Fernández, C.; Jin, Z.P.; Lyman, J.D.; et al. The Diversity of Kilonova Emission in Short Gamma-Ray Bursts. *Astrophys. J.* **2018**, *860*, 62. [CrossRef]

58. LIGO Scientific Collaboration; Virgo Collaboration; Fermi Gamma-ray Burst Monitor; INTEGRAL. Gravitational Waves and Gamma-Rays from a Binary Neutron Star Merger: GW170817 and GRB 170817A. *Astrophys. J. Lett.* **2017**, *848*, L13. [CrossRef]
59. Troja, E.; Ryan, G.; Piro, L.; van Eerten, H.; Cenko, S.B.; Yoon, Y.; Lee, S.K.; Im, M.; Sakamoto, T.; Gatkine, P.; et al. A luminous blue kilonova and an off-axis jet from a compact binary merger at z = 0.1341. *Nat. Commun.* **2018**, *9*, 4089. [CrossRef]
60. Band, D.; Matteson, J.; Ford, L.; Schaefer, B.; Palmer, D.; Teegarden, B.; Cline, T.; Briggs, M.; Paciesas, W.; Pendleton, G.; et al. BATSE observations of gamma-ray burst spectra. I—Spectral diversity. *Astrophys. J.* **1993**, *413*, 281. [CrossRef]
61. Ryde, F.; Axelsson, M.; Zhang, B.B.; McGlynn, S.; Pe'er, A.; Lundman, C.; Larsson, S.; Battelino, M.; Zhang, B.; Bissaldi, E.; et al. Identification and Properties of the Photospheric Emission in GRB090902B. *Astrophys. J. Lett.* **2010**, *709*, L172. [CrossRef]
62. Ryde, F.; Pe'er, A.; Nymark, T.; Axelsson, M.; Moretti, E.; Lundman, C.; Battelino, M.; Bissaldi, E.; Chiang, J.; Jackson, M.S.; et al. Observational evidence of dissipative photospheres in gamma-ray bursts. *Mon. Not. R. Astron. Soc.* **2011**, *415*, 3693–3705. [CrossRef]
63. Guiriec, S.; Connaughton, V.; Briggs, M.S.; Burgess, M.; Ryde, F.; Daigne, F.; Mészáros, P.; Goldstein, A.; McEnery, J.; Omodei, N.; et al. Detection of a Thermal Spectral Component in the Prompt Emission of GRB 100724B. *Astrophys. J. Lett.* **2011**, *727*, L33. [CrossRef]
64. Ghirlanda, G.; Ghisellini, G.; Nava, L. Short and long gamma-ray bursts: Same emission mechanism? *Mon. Not. R. Astron. Soc.* **2011**, *418*, L109–L113. [CrossRef]
65. Yonetoku, D.; Murakami, T.; Nakamura, T.; Yamazaki, R.; Inoue, A.K.; Ioka, K. Gamma-Ray Burst Formation Rate Inferred from the Spectral Peak Energy-Peak Luminosity Relation. *Astrophys. J.* **2004**, *609*, 935–951. [CrossRef]
66. Burlon, D.; Ghirlanda, G.; Ghisellini, G.; Lazzati, D.; Nava, L.; Nardini, M.; Celotti, A. Precursors in Swift Gamma Ray Bursts with Redshift. *Astrophys. J. Lett.* **2008**, *685*, L19. [CrossRef]
67. Ajello, M.; Arimoto, M.; Axelsson, M.; Baldini, L.; Barbiellini, G.; Bastieri, D.; Bellazzini, R.; Bhat, P.N.; Bissaldi, E.; Blandford, R.D.; et al. A Decade of Gamma-Ray Bursts Observed by Fermi-LAT: The Second GRB Catalog. *Astrophys. J.* **2019**, *878*, 52. [CrossRef]
68. Ruffini, R.; Muccino, M.; Bianco, C.L.; Enderli, M.; Izzo, L.; Kovacevic, M.; Penacchioni, A.V.; Pisani, G.B.; Rueda, J.A.; Wang, Y. On binary-driven hypernovae and their nested late X-ray emission. *Astron. Astrophys.* **2014**, *565*, L10. [CrossRef]
69. Piran, T. The physics of gamma-ray bursts. *Rev. Mod. Phys.* **2005**, *76*, 1143. [CrossRef]
70. Usov, V.V. Millisecond pulsars with extremely strong magnetic fields as a cosmological source of gamma-ray bursts. *Nature* **1992**, *357*, 472–474.
71. Bucciantini, N.; Quataert, E.; Arons, J.; Metzger, B.D.; Thompson, T.A. Relativistic jets and long-duration gamma-ray bursts from the birth of magnetars. *Mon. Not. R. Astron. Soc.* **2008**, *383*, L25–L29. [CrossRef]
72. Pe'er, A.; Zhang, B.B.; Ryde, F.; McGlynn, S.; Zhang, B.; Preece, R.D.; Kouveliotou, C. The Connection Between Thermal and Non-Thermal Emission in Gamma-ray Bursts: General Considerations and GRB090902B as a Case Study. *Mon. Not. R. Astron. Soc.* **2012**, *420*, 468–482. [CrossRef]
73. Meszaros, P.; Laguna, P.; Rees, M.J. Gasdynamics of relativistically expanding gamma-ray burst sources - Kinematics, energetics, magnetic fields, and efficiency. *Astrophys. J.* **1993**, *415*, 181. [CrossRef]
74. Rees, M.J.; Meszaros, P. Relativistic fireballs - Energy conversion and time-scales. *Mon. Not. R. Astron. Soc.* **1992**, *258*, 41. [CrossRef]
75. Sari, R.; Piran, T. Variability in Gamma-Ray Bursts: A Clue. *Astrophys. J.* **1997**, *485*, 270. [CrossRef]
76. Shemi, A.; Piran, T. The appearance of cosmic fireballs. *Astrophys. J. Lett.* **1990**, *365*, L55. [CrossRef]
77. Paczynski, B. Cosmological gamma-ray bursts. *Acta Astron.* **1991**, *41*, 257–267.
78. Pe'er, A. Temporal Evolution of Thermal Emission from Relativistically Expanding Plasma. *Astrophys. J.* **2008**, *682*, 463. [CrossRef]
79. Ryde, F. The Cooling Behavior of Thermal Pulses in Gamma-Ray Bursts. *Astrophys. J.* **2004**, *614*, 827. [CrossRef]
80. Ryde, F.; Pe'er, A. Quasi-blackbody Component and Radiative Efficiency of the Prompt Emission of Gamma-ray Bursts. *Astrophys. J.* **2009**, *702*, 1211. [CrossRef]
81. Kobayashi, S.; Piran, T.; Sari, R. Can Internal Shocks Produce the Variability in Gamma-Ray Bursts? *Astrophys. J.* **1997**, *490*, 92. [CrossRef]
82. Kobayashi, S.; Sari, R. Ultraefficient Internal Shocks. *Astrophys. J.* **2001**, *551*, 934. [CrossRef]
83. Blandford, R.D.; Znajek, R.L. Electromagnetic extraction of energy from Kerr black holes. *Mon. Not. R. Astron. Soc.* **1977**, *179*, 433–456. [CrossRef]
84. Blandford, R.D.; Payne, D.G. Hydromagnetic flows from accretion disks and the production of radio jets. *Mon. Not. R. Astron. Soc.* **1982**, *199*, 883–903. [CrossRef]
85. Giannios, D.; Spruit, H.C. Spectra of Poynting-flux powered GRB outflows. *Astron. Astrophys.* **2005**, *430*, 1. [CrossRef]
86. Giannios, D. Prompt emission spectra from the photosphere of a GRB. *Astron. Astrophys.* **2006**, *457*, 763. [CrossRef]
87. Zhang, B.; Yan, H. The Internal-collision-induced Magnetic Reconnection and Turbulence (ICMART) Model of Gamma-ray Bursts. *Astrophys. J.* **2011**, *726*, 90. [CrossRef]
88. Fermi, E. On the Origin of the Cosmic Radiation. *Phys. Rev.* **1949**, *75*, 1169–1174. [CrossRef]
89. Ellison, D.C.; Double, G.P. Diffusive shock acceleration in unmodified relativistic, oblique shocks. *Astropart. Phys.* **2004**, *22*, 323–338. [CrossRef]
90. Pe'er, A. Physics of Gamma-Ray Bursts Prompt Emission. *Adv. Astron.* **2015**, *2015*, 907321. [CrossRef]

91. Sironi, L.; Spitkovsky, A. Particle Acceleration in Relativistic Magnetized Collisionless Electron-Ion Shocks. *Astrophys. J.* **2011**, *726*, 75. [CrossRef]
92. Meszaros, P.; Rees, M.J. Relativistic fireballs and their impact on external matter—Models for cosmological gamma-ray bursts. *Astrophys. J.* **1993**, *405*, 278. [CrossRef]
93. Tavani, M. A Shock Emission Model for Gamma-Ray Bursts. II. Spectral Properties. *Astrophys. J.* **1996**, *466*, 768. [CrossRef]
94. Sari, R. Hydrodynamics of Gamma-Ray Burst Afterglow. *Astrophys. J. Lett.* **1997**, *489*, L37. [CrossRef]
95. Daigne, F.; Mochkovitch, R. Gamma-ray bursts from internal shocks in a relativistic wind: Temporal and spectral properties. *Mon. Not. R. Astron. Soc.* **1998**, *296*, 275. [CrossRef]
96. Chiang, J.; Dermer, C.D. Synchrotron and Synchrotron Self-Compton Emission and the Blast-Wave Model of Gamma-Ray Bursts. *Astrophys. J.* **1999**, *512*, 699. [CrossRef]
97. Stern, B.E.; Poutanen, J. Gamma-ray bursts from synchrotron self-Compton emission. *Mon. Not. R. Astron. Soc.* **2004**, *352*, L35–L39. [CrossRef]
98. Nakar, E.; Ando, S.; Sari, R. Klein-Nishina Effects on Optically Thin Synchrotron and Synchrotron Self-Compton Spectrum. *Astrophys. J.* **2009**, *703*, 675–691. [CrossRef]
99. Ryde, F. Is Thermal Emission in Gamma-Ray Bursts Ubiquitous? *Astrophys. J. Lett.* **2005**, *625*, L95. [CrossRef]
100. Mészáros, P.; Rees, M.J. Steep Slopes and Preferred Breaks in Gamma-Ray Burst Spectra: The Role of Photospheres and Comptonization. *Astrophys. J.* **2000**, *530*, 292. [CrossRef]
101. Pe'er, A.; Mészáros, P.; Rees, M.J. The Observable Effects of a Photospheric Component on GRB and XRF Prompt Emission Spectrum. *Astrophys. J.* **2006**, *642*, 995. [CrossRef]
102. Rybicki, G.B.; Lightman, A.P. *Radiative Processes in Astrophysics*; A Wiley-Interscience Publication: New York, NY, USA, 1979.
103. Meszaros, P.; Rees, M.J. Gamma-Ray Bursts: Multiwaveband Spectral Predictions for Blast Wave Models. *Astrophys. J. Lett.* **1993**, *418*, L59. [CrossRef]
104. Spitkovsky, A. On the Structure of Relativistic Collisionless Shocks in Electron-Ion Plasmas. *Astrophys. J. Lett.* **2008**, *673*, L39. [CrossRef]
105. Kumar, P.; Panaitescu, A. Afterglow Emission from Naked Gamma-Ray Bursts. *Astrophys. J. Lett.* **2000**, *541*, L51. [CrossRef]
106. Sari, R. The Observed Size and Shape of Gamma-Ray Burst Afterglow. *Astrophys. J. Lett.* **1998**, *494*, L49. [CrossRef]
107. Sari, R.; Piran, T.; Narayan, R. Spectra and Light Curves of Gamma-Ray Burst Afterglows. *Astrophys. J. Lett.* **1998**, *497*, L17. [CrossRef]
108. Kumar, P.; McMahon, E. A general scheme for modelling γ-ray burst prompt emission. *Mon. Not. R. Astron. Soc.* **2008**, *384*, 33. [CrossRef]
109. Granot, J.; Piran, T.; Sari, R. The Synchrotron Spectrum of Fast Cooling Electrons Revisited. *Astrophys. J. Lett.* **2000**, *534*, L163. [CrossRef]
110. Panaitescu, A.; Mészáros, P. Gamma-Ray Bursts from Upscattered Self-absorbed Synchrotron Emission. *Astrophys. J. Lett.* **2000**, *544*, L17. [CrossRef]
111. Pe'er, A.; Zhang, B. Synchrotron Emission in Small-Scale Magnetic Fields as a Possible Explanation for Prompt Emission Spectra of Gamma-Ray Bursts. *Astrophys. J.* **2006**, *653*, 454. [CrossRef]
112. Gupta, N.; Zhang, B. Diagnosing the site of gamma-ray burst prompt emission with spectral cut-off energy. *Mon. Not. R. Astron. Soc.* **2008**, *384*, L11–L15. [CrossRef]
113. Asano, K.; Mészáros, P. Delayed Onset of High-energy Emissions in Leptonic and Hadronic Models of Gamma-Ray Bursts. *Astrophys. J.* **2012**, *757*, 115. [CrossRef]
114. Pe'er, A.; Waxman, E. Prompt Gamma-Ray Burst Spectra: Detailed Calculations and the Effect of Pair Production. *Astrophys. J.* **2004**, *613*, 448–459. [CrossRef]
115. Pe'er, A.; Mészáros, P.; Rees, M.J. Peak Energy Clustering and Efficiency in Compact Objects. *Astrophys. J.* **2005**, *635*, 476. [CrossRef]
116. Gao, H.; Zhang, B. Photosphere Emission from a Hybrid Relativistic Outflow with Arbitrary Dimensionless Entropy and Magnetization in GRBs. *Astrophys. J.* **2015**, *801*, 103. [CrossRef]
117. Weinberg, S. *Cosmology*; Oxford University Press: Oxford, UK, 2008.
118. Phillips, M.M. The absolute magnitudes of Type IA supernovae. *Astrophys. J. Lett.* **1993**, *413*, L105–L108. [CrossRef]
119. Silk, J. Cosmic Black-Body Radiation and Galaxy Formation. *Astrophys. J.* **1968**, *151*, 459. [CrossRef]
120. Peebles, P.J.E.; Yu, J.T. Primeval Adiabatic Perturbation in an Expanding Universe. *Astrophys. J.* **1970**, *162*, 815. [CrossRef]
121. Eisenstein, D.J.; Zehavi, I.; Hogg, D.W.; Scoccimarro, R.; Blanton, M.R.; Nichol, R.C.; Scranton, R.; Seo, H.J.; Tegmark, M.; Zheng, Z.; et al. Detection of the Baryon Acoustic Peak in the Large-Scale Correlation Function of SDSS Luminous Red Galaxies. *Astrophys. J.* **2005**, *633*, 560–574. [CrossRef]
122. Planck Collaboration; Aghanim, N.; Akrami, Y.; Ashdown, M.; Aumont, J.; Baccigalupi, C.; Ballardini, M.; Banday, A.J.; Barreiro, R.B.; Bartolo, N.; et al. Planck 2018 results. VI. Cosmological parameters. *Astron. Astrophys.* **2020**, *641*, A6.
123. Amati, L.; D'Agostino, R.; Luongo, O.; Muccino, M.; Tantalo, M. Addressing the circularity problem in the E_p-E_{iso} correlation of gamma-ray bursts. *Mon. Not. R. Astron. Soc.* **2019**, *486*, L46–L51. [CrossRef]
124. Jimenez, R.; Loeb, A. Constraining Cosmological Parameters Based on Relative Galaxy Ages. *Astrophys. J.* **2002**, *573*, 37–42. [CrossRef]

125. Moresco, M.; Pozzetti, L.; Cimatti, A.; Jimenez, R.; Maraston, C.; Verde, L.; Thomas, D.; Citro, A.; Tojeiro, R.; Wilkinson, D. A 6% measurement of the Hubble parameter at z~0.45: Direct evidence of the epoch of cosmic re-acceleration. *J. Cosmology Astropart. Phys.* **2016**, *2016*, 014. [CrossRef]
126. Bond, J.R.; Efstathiou, G. Cosmic background radiation anisotropies in universes dominated by nonbaryonic dark matter. *Astrophys. J. Lett.* **1984**, *285*, L45–L48. [CrossRef]
127. Bond, J.R.; Efstathiou, G.; Tegmark, M. Forecasting cosmic parameter errors from microwave background anisotropy experiments. *Mon. Not. R. Astron. Soc.* **1997**, *291*, L33–L41.
128. Hu, W.; Sugiyama, N. Small-Scale Cosmological Perturbations: An Analytic Approach. *Astrophys. J.* **1996**, *471*, 542. [CrossRef]
129. Norris, J.P.; Scargle, J.D.; Bonnell, J.T. Pulses, spectral lags, durations, and hardness ratios in long GRBs. In *Gamma 2001: Gamma-Ray Astrophysics*; Ritz, S., Gehrels, N., Shrader, C.R., Eds.; American Institute of Physics Conference Series; American Institute of Physics: College Park, MD, USA, 2001; Volume 587, pp. 176–180.
130. Dainotti, M.G.; Del Vecchio, R.; Tarnopolski, M. Gamma-Ray Burst Prompt Correlations. *Adv. Astron.* **2018**, *2018*, 4969503. [CrossRef]
131. Fenimore, E.E.; Ramirez-Ruiz, E. Redshifts For 220 BATSE Gamma-Ray Bursts Determined by Variability and the Cosmological Consequences. *arXiv* **2000**, arXiv:astroph:0004176.
132. Amati, L.; Frontera, F.; Tavani, M.; in't Zand, J.J.M.; Antonelli, A.; Costa, E.; Feroci, M.; Guidorzi, C.; Heise, J.; Masetti, N.; et al. Intrinsic spectra and energetics of BeppoSAX Gamma-Ray Bursts with known redshifts. *Astron. Astrophys.* **2002**, *390*, 81. [CrossRef]
133. Calderone, G.; Ghirlanda, G.; Ghisellini, G.; Bernardini, M.G.; Campana, S.; Covino, S.; D'Elia, V.; Melandri, A.; Salvaterra, R.; Sbarufatti, B.; et al. There is a short gamma-ray burst prompt phase at the beginning of each long one. *Mon. Not. R. Astron. Soc.* **2015**, *448*, 403–416. [CrossRef]
134. Ghirlanda, G.; Ghisellini, G.; Lazzati, D. The Collimation-corrected Gamma-Ray Burst Energies Correlate with the Peak Energy of Their νF_ν Spectrum. *Astrophys. J.* **2004**, *616*, 331. [CrossRef]
135. Liang, E.; Zhang, B. Model-independent Multivariable Gamma-Ray Burst Luminosity Indicator and Its Possible Cosmological Implications. *Astrophys. J.* **2005**, *633*, 611–623. [CrossRef]
136. Dainotti, M.G.; Cardone, V.F.; Capozziello, S. A time-luminosity correlation for γ-ray bursts in the X-rays. *Mon. Not. R. Astron. Soc.* **2008**, *391*, L79–L83. [CrossRef]
137. Dainotti, M.G.; Ostrowski, M.; Willingale, R. Towards a standard gamma-ray burst: Tight correlations between the prompt and the afterglow plateau phase emission. *Mon. Not. R. Astron. Soc.* **2011**, *418*, 2202–2206. [CrossRef]
138. Bernardini, M.G.; Margutti, R.; Zaninoni, E.; Chincarini, G. A universal scaling for short and long gamma-ray bursts: $E_{X,iso}$-$E_{?,iso}$-E_{pk}. *Mon. Not. R. Astron. Soc.* **2012**, *425*, 1199–1204. [CrossRef]
139. Margutti, R.; Zaninoni, E.; Bernardini, M.G.; Chincarini, G.; Pasotti, F.; Guidorzi, C.; Angelini, L.; Burrows, D.N.; Capalbi, M.; Evans, P.A.; et al. The prompt-afterglow connection in gamma-ray bursts: A comprehensive statistical analysis of Swift X-ray light curves. *Mon. Not. R. Astron. Soc.* **2013**, *428*, 729–742. [CrossRef]
140. Izzo, L.; Muccino, M.; Zaninoni, E.; Amati, L.; Della Valle, M. New measurements of Ω_m from gamma-ray bursts. *Astron. Astrophys.* **2015**, *582*, A115. [CrossRef]
141. Muccino, M.; Izzo, L.; Luongo, O.; Boshkayev, K.; Amati, L.; Della Valle, M.; Pisani, G.B.; Zaninoni, E. Tracing Dark Energy History with Gamma-Ray Bursts. *Astrophys. J.* **2021**, *908*, 181. [CrossRef]
142. Tang, C.H.; Huang, Y.F.; Geng, J.J.; Zhang, Z.B. Statistical Study of Gamma-Ray Bursts with a Plateau Phase in the X-Ray Afterglow. *Astrophys. J. Suppl.* **2019**, *245*, 1. [CrossRef]
143. Basilakos, S.; Perivolaropoulos, L. Testing gamma-ray bursts as standard candles. *Mon. Not. R. Astron. Soc.* **2008**, *391*, 411–419. [CrossRef]
144. Nava, L.; Ghirlanda, G.; Ghisellini, G. Selection effects on GRB spectral-energy correlations. In *Gamma-Ray Burst: Sixth Huntsville Symposium*; Meegan, C., Kouveliotou, C., Gehrels, N., Eds.; American Institute of Physics Conference Series; American Institute of Physics: College Park, MD, USA, 2009; Volume 1133, pp. 350–355.
145. Schaefer, B.E. The Hubble Diagram to Redshift > 6 from 69 Gamma-Ray Bursts. *Astrophys. J.* **2007**, *660*, 16. [CrossRef]
146. Wang, Y. Model-independent distance measurements from gamma-ray bursts and constraints on dark energy. *Phys. Rev. D* **2008**, *78*, 123532. [CrossRef]
147. Margutti, R.; Guidorzi, C.; Chincarini, G.; Bernardini, M.G.; Genet, F.; Mao, J.; Pasotti, F. Lag-luminosity relation in γ-ray burst X-ray flares: A direct link to the prompt emission. *Mon. Not. R. Astron. Soc.* **2010**, *406*, 2149–2167. [CrossRef]
148. Sultana, J.; Kazanas, D.; Fukumura, K. Luminosity Correlations for Gamma-Ray Bursts and Implications for Their Prompt and Afterglow Emission Mechanisms. *Astrophys. J.* **2012**, *758*, 32. [CrossRef]
149. Shahmoradi, A.; Nemiroff, R.J. A Catalog of Redshift Estimates for 1366 BATSE Long-Duration Gamma-Ray Bursts: Evidence for Strong Selection Effects on the Phenomenological Prompt Gamma-Ray Correlations. *arXiv* **2019**, arXiv:1903.06989.
150. Salmonson, J.D.; Galama, T.J. Discovery of a Tight Correlation between Pulse Lag/Luminosity and Jet-Break Times: A Connection between Gamma-Ray Bursts and Afterglow Properties. *Astrophys. J.* **2002**, *569*, 682–688. [CrossRef]
151. Amati, L.; Della Valle, M. Measuring Cosmological Parameters with Gamma Ray Bursts. *Int. J. Mod. Phys. D* **2013**, *22*, 1330028. [CrossRef]

152. D'Agostini, G. Fits, and especially linear fits, with errors on both axes, extra variance of the data points and other complications. *arXiv* **2005**, arXiv:physics/0511182.
153. Thompson, C.; Mészáros, P.; Rees, M.J. Thermalization in Relativistic Outflows and the Correlation between Spectral Hardness and Apparent Luminosity in Gamma-Ray Bursts. *Astrophys. J.* **2007**, *666*, 1012–1023. [CrossRef]
154. Butler, N.R.; Kocevski, D.; Bloom, J.S.; Curtis, J.L. A Complete Catalog ofSwiftGamma-Ray Burst Spectra and Durations: Demise of a Physical Origin for Pre-Swift High-Energy Correlations. *Astrophys. J.* **2007**, *671*, 656–677. [CrossRef]
155. Ghirlanda, G.; Nava, L.; Ghisellini, G.; Firmani, C.; Cabrera, J.I. The E_{peak}-E_{iso} plane of long gamma-ray bursts and selection effects. *Mon. Not. R. Astron. Soc.* **2008**, *387*, 319–330. [CrossRef]
156. Nava, L.; Ghirlanda, G.; Ghisellini, G.; Firmani, C. Peak energy of the prompt emission of long gamma-ray bursts versus their fluence and peak flux. *Mon. Not. R. Astron. Soc.* **2008**, *391*, 639–652. [CrossRef]
157. Amati, L.; Frontera, F.; Guidorzi, C. Extremely energetic Fermi gamma-ray bursts obey spectral energy correlations. *Astron. Astrophys.* **2009**, *508*, 173–180. [CrossRef]
158. Yonetoku, D.; Murakami, T.; Tsutsui, R.; Nakamura, T.; Morihara, Y.; Takahashi, K. Possible Origins of Dispersion of the Peak Energy-Brightness Correlations of Gamma-Ray Bursts. *Publ. Astron. Soc. Jpn.* **2010**, *62*, 1495. [CrossRef]
159. Shahmoradi, A.; Nemiroff, R.J. The possible impact of gamma-ray burst detector thresholds on cosmological standard candles. *Mon. Not. R. Astron. Soc.* **2011**, *411*, 1843–1856. [CrossRef]
160. Band, D.L.; Preece, R.D. Testing the Gamma-Ray Burst Energy Relationships. *Astrophys. J.* **2005**, *627*, 319–323. [CrossRef]
161. Friedman, A.S.; Bloom, J.S. Toward a More Standardized Candle Using Gamma-Ray Burst Energetics and Spectra. *Astrophys. J.* **2005**, *627*, 1–25. [CrossRef]
162. Ito, H.; Matsumoto, J.; Nagataki, S.; Warren, D.C.; Barkov, M.V.; Yonetoku, D. The photospheric origin of the Yonetoku relation in gamma-ray bursts. *Nat. Commun.* **2019**, *10*, 1504. [CrossRef] [PubMed]
163. Lu, R.J.; Wei, J.J.; Qin, S.F.; Liang, E.W. Selection Effects on the Observed Redshift Dependence of Gamma-Ray Burst Jet Opening Angles. *Astrophys. J.* **2012**, *745*, 168. [CrossRef]
164. Dainotti, M.G.; Del Vecchio, R. Gamma Ray Burst afterglow and prompt-afterglow relations: An overview. *New Astron. Rev.* **2017**, *77*, 23–61. [CrossRef]
165. Muccino, M.; Boshkayev, K. Physical insight into the Combo-relation. *Mon. Not. R. Astron. Soc.* **2017**, *468*, 570–576. [CrossRef]
166. Kodama, Y.; Yonetoku, D.; Murakami, T.; Tanabe, S.; Tsutsui, R.; Nakamura, T. Gamma-ray bursts in $1.8<z<5.6$ suggest that the time variation of the dark energy is small. *Mon. Not. R. Astron. Soc.* **2008**, *391*, L1–L4.
167. Soderberg, A.M.; Berger, E.; Page, K.L.; Schady, P.; Parrent, J.; Pooley, D.; Wang, X.Y.; Ofek, E.O.; Cucchiara, A.; Rau, A.; et al. An extremely luminous X-ray outburst at the birth of a supernova. *Nature* **2008**, *453*, 469–474. [CrossRef] [PubMed]
168. Liang, N.; Xiao, W.K.; Liu, Y.; Zhang, S.N. A Cosmology-Independent Calibration of Gamma-Ray Burst Luminosity Relations and the Hubble Diagram. *Astrophys. J.* **2008**, *685*, 354–360. [CrossRef]
169. Wei, H. Observational constraints on cosmological models with the updated long gamma-ray bursts. *J. Cosmology Astropart. Phys.* **2010**, *2010*, 020. [CrossRef]
170. Gruber, C.; Luongo, O. Cosmographic analysis of the equation of state of the universe through Padé approximations. *Phys. Rev. D* **2014**, *89*, 103506. [CrossRef]
171. Ghirlanda, G.; Ghisellini, G.; Lazzati, D.; Firmani, C. Gamma-Ray Bursts: New Rulers to Measure the Universe. *Astrophys. J. Lett.* **2004**, *613*, L13–L16. [CrossRef]
172. Wang, F.Y.; Dai, Z.G.; Zhu, Z.H. Measuring Dark Energy with Gamma-Ray Bursts and Other Cosmological Probes. *Astrophys. J.* **2007**, *667*, 1–10. [CrossRef]
173. Liang, E.; Zhang, B. Calibration of gamma-ray burst luminosity indicators. *Mon. Not. R. Astron. Soc.* **2006**, *369*, L37–L41. [CrossRef]
174. Ghirlanda, G.; Ghisellini, G.; Firmani, C.; Nava, L.; Tavecchio, F.; Lazzati, D. Cosmological constraints with GRBs: Homogeneous medium vs. wind density profile. *Astron. Astrophys.* **2006**, *452*, 839–844. [CrossRef]
175. Khadka, N.; Ratra, B. Constraints on cosmological parameters from gamma-ray burst peak photon energy and bolometric fluence measurements and other data. *Mon. Not. R. Astron. Soc.* **2020**, *499*, 391–403. [CrossRef]
176. Peebles, P.J.E. Tests of cosmological models constrained by inflation. *Astrophys. J.* **1984**, *284*, 439–444. [CrossRef]
177. Di Valentino, E.; Mena, O.; Pan, S.; Visinelli, L.; Yang, W.; Melchiorri, A.; Mota, D.F.; Riess, A.G.; Silk, J. In the Realm of the Hubble tension—A Review of Solutions. *arXiv* **2021**, arXiv:2103.01183.
178. Perivolaropoulos, L.; Skara, F. Challenges for ΛCDM: An update. *arXiv* **2021**, arXiv:2105.05208.
179. Farooq, O.; Ranjeet Madiyar, F.; Crandall, S.; Ratra, B. Hubble Parameter Measurement Constraints on the Redshift of the Deceleration-Acceleration Transition, Dynamical Dark Energy, and Space Curvature. *Astrophys. J.* **2017**, *835*, 26. [CrossRef]
180. Scolnic, D.M.; Jones, D.O.; Rest, A.; Pan, Y.C.; Chornock, R.; Foley, R.J.; Huber, M.E.; Kessler, R.; Narayan, G.; Riess, A.G.; et al. The Complete Light-curve Sample of Spectroscopically Confirmed SNe Ia from Pan-STARRS1 and Cosmological Constraints from the Combined Pantheon Sample. *Astrophys. J.* **2018**, *859*, 101. [CrossRef]
181. eBOSS Collaboration. The Completed SDSS-IV extended Baryon Oscillation Spectroscopic Survey: Cosmological Implications from two Decades of Spectroscopic Surveys at the Apache Point observatory. *arXiv* **2020**, arXiv:2007.08991.
182. Mania, D.; Ratra, B. Constraints on dark energy from H II starburst galaxy apparent magnitude versus redshift data. *Phys. Lett. B* **2012**, *715*, 9–14. [CrossRef]

183. Chávez, R.; Terlevich, R.; Terlevich, E.; Bresolin, F.; Melnick, J.; Plionis, M.; Basilakos, S. The L-σ relation for massive bursts of star formation. *Mon. Not. R. Astron. Soc.* **2014**, *442*, 3565–3597. [CrossRef]
184. González-Morán, A.L.; Chávez, R.; Terlevich, R.; Terlevich, E.; Bresolin, F.; Fernández-Arenas, D.; Plionis, M.; Basilakos, S.; Melnick, J.; Telles, E. Independent cosmological constraints from high-z H II galaxies. *Mon. Not. R. Astron. Soc.* **2019**, *487*, 4669–4694. [CrossRef]
185. González-Morán, A.L.; Chávez, R.; Terlevich, E.; Terlevich, R.; Fernández-Arenas, D.; Bresolin, F.; Plionis, M.; Melnick, J.; Basilakos, S.; Telles, E. Independent cosmological constraints from high-z HII galaxies: New results from VLT-KMOS data. *arXiv* **2021**, arXiv:2105.04025.
186. Cao, S.; Ryan, J.; Ratra, B. Cosmological constraints from H II starburst galaxy apparent magnitude and other cosmological measurements. *Mon. Not. R. Astron. Soc.* **2020**, *497*, 3191–3203. [CrossRef]
187. Cao, S.; Ryan, J.; Khadka, N.; Ratra, B. Cosmological constraints from higher redshift gamma-ray burst, H II starburst galaxy, and quasar (and other) data. *Mon. Not. R. Astron. Soc.* **2021**, *501*, 1520–1538. [CrossRef]
188. Johnson, J.P.; Sangwan, A.; Shankaranarayanan, S. Cosmological perturbations in the interacting dark sector: Observational constraints and predictions. *arXiv* **2021**, arXiv:2102.12367.
189. Cao, S.; Zheng, X.; Biesiada, M.; Qi, J.; Chen, Y.; Zhu, Z.H. Ultra-compact structure in intermediate-luminosity radio quasars: building a sample of standard cosmological rulers and improving the dark energy constraints up to z 3. *Astron. Astrophys.* **2017**, *606*, A15. [CrossRef]
190. Ryan, J.; Chen, Y.; Ratra, B. Baryon acoustic oscillation, Hubble parameter, and angular size measurement constraints on the Hubble constant, dark energy dynamics, and spatial curvature. *Mon. Not. R. Astron. Soc.* **2019**, *488*, 3844–3856. [CrossRef]
191. Cao, S.; Ryan, J.; Ratra, B. Using Pantheon and DES supernova, baryon acoustic oscillation, and Hubble parameter data to constrain the Hubble constant, dark energy dynamics, and spatial curvature. *Mon. Not. R. Astron. Soc.* **2021**, *504*, 300–310. [CrossRef]
192. Zheng, X.; Cao, S.; Biesiada, M.; Li, X.; Liu, T.; Liu, Y. Multiple measurements of quasars acting as standard probes: Model independent calibration and exploring the dark energy equation of states. *Sci. China Physics, Mech. Astron.* **2021**, *64*, 259511. [CrossRef]
193. Lian, Y.; Cao, S.; Biesiada, M.; Chen, Y.; Zhang, Y.; Guo, W. Probing modified gravity theories with multiple measurements of high-redshift quasars. *arXiv* **2021**, arXiv:2105.04992.
194. Risaliti, G.; Lusso, E. A Hubble Diagram for Quasars. *Astrophys. J.* **2015**, *815*, 33. [CrossRef]
195. Risaliti, G.; Lusso, E. Cosmological Constraints from the Hubble Diagram of Quasars at High Redshifts. *Nat. Astron.* **2019**, *3*, 272–277. [CrossRef]
196. Khadka, N.; Ratra, B. Quasar X-ray and UV flux, baryon acoustic oscillation, and Hubble parameter measurement constraints on cosmological model parameters. *Mon. Not. R. Astron. Soc.* **2020**, *492*, 4456–4468. [CrossRef]
197. Khadka, N.; Ratra, B. Using quasar X-ray and UV flux measurements to constrain cosmological model parameters. *Mon. Not. R. Astron. Soc.* **2020**, *497*, 263–278. [CrossRef]
198. Khadka, N.; Ratra, B. Determining the range of validity of quasar X-ray and UV flux measurements for constraining cosmological model parameters. *Mon. Not. R. Astron. Soc.* **2021**, *502*, 6140–6156. [CrossRef]
199. Yang, T.; Banerjee, A.; Ó Colgáin, E. Cosmography and flat Λ CDM tensions at high redshift. *Phys. Rev. D* **2020**, *102*, 123532. [CrossRef]
200. Lusso, E.; Risaliti, G.; Nardini, E.; Bargiacchi, G.; Benetti, M.; Bisogni, S.; Capozziello, S.; Civano, F.; Eggleston, L.; Elvis, M.; et al. Quasars as standard candles. III. Validation of a new sample for cosmological studies. *Astron. Astrophys.* **2020**, *642*, A150. [CrossRef]
201. Li, X.; Keeley, R.E.; Shafieloo, A.; Zheng, X.; Cao, S.; Biesiada, M.; Zhu, Z.H. Hubble diagram at higher redshifts: Model independent calibration of quasars. *arXiv* **2021**, arXiv:2103.16032.
202. Cucchiara, A.; Levan, A.J.; Fox, D.B.; Tanvir, N.R.; Ukwatta, T.N.; Berger, E.; Krühler, T.; Küpcü Yoldaş, A.; Wu, X.F.; Toma, K.; et al. A Photometric Redshift of z ~ 9.4 for GRB 090429B. *Astrophys. J.* **2011**, *736*, 7. [CrossRef]
203. Amati, L.; Guidorzi, C.; Frontera, F.; Della Valle, M.; Finelli, F.; Landi, R.; Montanari, E. Measuring the cosmological parameters with the $E_{p,i}$–E_{iso} correlation of gamma-ray bursts. *Mon. Not. R. Astron. Soc.* **2008**, *391*, 577. [CrossRef]
204. Capozziello, S.; Izzo, L. Cosmography by gamma ray bursts. *Astron. Astrophys.* **2008**, *490*, 31–36. [CrossRef]
205. Izzo, L.; Capozziello, S.; Covone, G.; Capaccioli, M. Extending the Hubble diagram by gamma ray bursts. *Astron. Astrophys.* **2009**, *508*, 63–67. [CrossRef]
206. Wei, J.J.; Wu, X.F.; Melia, F.; Wei, D.M.; Feng, L.L. Cosmological tests using gamma-ray bursts, the star formation rate and possible abundance evolution. *Mon. Not. R. Astron. Soc.* **2014**, *439*, 3329–3341. [CrossRef]
207. Schaefer, B.E. Gamma-Ray Burst Hubble Diagram to z=4.5. *Astrophys. J. Lett.* **2003**, *583*, L67–L70. [CrossRef]
208. Dai, Z.G.; Liang, E.W.; Xu, D. Constraining Ω_M and Dark Energy with Gamma-Ray Bursts. *Astrophys. J. Lett.* **2004**, *612*, L101–L104. [CrossRef]
209. Samushia, L.; Ratra, B. Constraining Dark Energy with Gamma-ray Bursts. *Astrophys. J.* **2010**, *714*, 1347–1354. [CrossRef]
210. Liu, J.; Wei, H. Cosmological models and gamma-ray bursts calibrated by using Padé method. *Gen. Relativ. Gravit.* **2015**, *47*, 141. [CrossRef]

211. Lin, H.N.; Li, X.; Chang, Z. Model-independent distance calibration of high-redshift gamma-ray bursts and constrain on the ΛCDM model. *Mon. Not. R. Astron. Soc.* **2016**, *455*, 2131–2138. [CrossRef]
212. Huang, L.; Huang, Z.; Luo, X.; He, X.; Fang, Y. Revisiting the Redshift Evolution of GRB Luminosity Correlation with PAge Approximation. *arXiv* **2020**, arXiv:2012.02474.
213. Tang, L.; Li, X.; Lin, H.N.; Liu, L. Model-independently Calibrating the Luminosity Correlations of Gamma-Ray Bursts Using Deep Learning. *Astrophys. J.* **2021**, *907*, 121. [CrossRef]
214. Demianski, M.; Piedipalumbo, E.; Sawant, D.; Amati, L. Cosmology with gamma-ray bursts. II. Cosmography challenges and cosmological scenarios for the accelerated Universe. *Astron. Astrophys.* **2017**, *598*, A113. [CrossRef]
215. Demianski, M.; Piedipalumbo, E.; Sawant, D.; Amati, L. High redshift constraints on dark energy models and tension with the flat LambdaCDM model. *arXiv* **2019**, arXiv:1911.08228.
216. Luongo, O.; Muccino, M. Kinematic constraints beyond $z \simeq 0$ using calibrated GRB correlations. *Astron. Astrophys.* **2020**, *641*, 15. [CrossRef]
217. Luongo, O.; Muccino, M. Model independent calibrations of gamma ray bursts using machine learning. *arXiv* **2020**, arXiv:2011.13590.
218. Rezaei, M.; Pour-Ojaghi, S.; Malekjani, M. A Cosmography Approach to Dark Energy Cosmologies: New Constraints Using the Hubble Diagrams of Supernovae, Quasars, and Gamma-Ray Bursts. *Astrophys. J.* **2020**, *900*, 70. [CrossRef]
219. Lusso, E.; Piedipalumbo, E.; Risaliti, G.; Paolillo, M.; Bisogni, S.; Nardini, E.; Amati, L. Tension with the flat ΛCDM model from a high-redshift Hubble diagram of supernovae, quasars, and gamma-ray bursts. *Astron. Astrophys.* **2019**, *628*, L4. [CrossRef]
220. Kumar, D.; Jain, D.; Mahajan, S.; Mukherjee, A.; Rani, N. Constraining cosmological and galaxy parameters using strong gravitational lensing systems. *Phys. Rev. D* **2021**, *103*, 063511. [CrossRef]
221. Montiel, A.; Cabrera, J.I.; Hidalgo, J.C. Improving sampling and calibration of gamma-ray bursts as distance indicators. *Mon. Not. R. Astron. Soc.* **2021**, *501*, 3515–3526.
222. Riess, A.G.; Casertano, S.; Yuan, W.; Macri, L.M.; Scolnic, D. Large Magellanic Cloud Cepheid Standards Provide a 1% Foundation for the Determination of the Hubble Constant and Stronger Evidence for Physics beyond ΛCDM. *Astrophys. J.* **2019**, *876*, 85. [CrossRef]
223. Fana Dirirsa, F.; Razzaque, S.; Piron, F.; Arimoto, M.; Axelsson, M.; Kocevski, D.; Longo, F.; Ohno, M.; Zhu, S. Spectral Analysis of Fermi-LAT Gamma-Ray Bursts with Known Redshift and their Potential Use as Cosmological Standard Candles. *Astrophys. J.* **2019**, *887*, 13. [CrossRef]
224. Wang, J.S.; Wang, F.Y.; Cheng, K.S.; Dai, Z.G. Measuring dark energy with the E_{iso} - E_p correlation of gamma-ray bursts using model-independent methods. *Astron. Astrophys.* **2016**, *585*, A68. [CrossRef]
225. Yang, T.; Banerjee, A.; Colgáin, E.Ó. On cosmography and flat ΛCDM tensions at high redshift. *arXiv* **2019**, arXiv:1911.01681.
226. Betoule, M.; Kessler, R.; Guy, J.; Mosher, J.; Hardin, D.; Biswas, R.; Astier, P.; El-Hage, P.; Konig, M.; Kuhlmann, S.; et al. Improved cosmological constraints from a joint analysis of the SDSS-II and SNLS supernova samples. *Astron. Astrophys.* **2014**, *568*, A22. [CrossRef]
227. Akaike, H. A New Look at the Statistical Model Identification. *IEEE Trans. Autom. Control* **1974**, *19*, 716–723. [CrossRef]
228. Kunz, M.; Trotta, R.; Parkinson, D. Measuring the effective complexity of cosmological models. *Phys. Rev.* **2006**, *D74*, 023503. [CrossRef]
229. Liddle, A.R. Information criteria for astrophysical model selection. *Mon. Not. Roy. Astron. Soc.* **2007**, *377*, L74–L78. [CrossRef]
230. Stanek, K.Z.; Matheson, T.; Garnavich, P.M.; Martini, P.; Berlind, P.; Caldwell, N.; Challis, P.; Brown, W.R.; Schild, R.; Krisciunas, K.; et al. Spectroscopic Discovery of the Supernova 2003dh Associated with GRB 030329. *Astrophys. J. Lett.* **2003**, *591*, L17. [CrossRef]
231. Totani, T. Cosmological Gamma-Ray Bursts and Evolution of Galaxies. *Astrophys. J. Lett.* **1997**, *486*, L71–L74. [CrossRef]
232. Wijers, R.A.M.J.; Bloom, J.S.; Bagla, J.S.; Natarajan, P. Gamma-ray bursts from stellar remnants: Probing the Universe at high redshift. *Mon. Not. R. Astron. Soc.* **1998**, *294*, L13–L17. [CrossRef]
233. Lamb, D.Q.; Reichart, D.E. Gamma-Ray Bursts as a Probe of the Very High Redshift Universe. *Astrophys. J.* **2000**, *536*, 1. [CrossRef]
234. Porciani, C.; Madau, P. On the Association of Gamma-Ray Bursts with Massive Stars: Implications for Number Counts and Lensing Statistics. *Astrophys. J.* **2001**, *548*, 522–531. [CrossRef]
235. Bromm, V.; Loeb, A. The Expected Redshift Distribution of Gamma-Ray Bursts. *Astrophys. J.* **2002**, *575*, 111–116. [CrossRef]
236. Wanderman, D.; Piran, T. The luminosity function and the rate of Swift's gamma-ray bursts. *Mon. Not. R. Astron. Soc.* **2010**, *406*, 1944–1958. [CrossRef]
237. Cao, X.F.; Yu, Y.W.; Cheng, K.S.; Zheng, X.P. The luminosity function of Swift long gamma-ray bursts. *Mon. Not. R. Astron. Soc.* **2011**, *416*, 2174–2181. [CrossRef]
238. Tan, W.W.; Cao, X.F.; Yu, Y.W. Determining the Luminosity Function of Swift Long Gamma-Ray Bursts with Pseudo-redshifts. *Astrophys. J. Lett.* **2013**, *772*, L8. [CrossRef]
239. Pescalli, A.; Ghirlanda, G.; Salvaterra, R.; Ghisellini, G.; Vergani, S.D.; Nappo, F.; Salafia, O.S.; Melandri, A.; Covino, S.; Götz, D. The rate and luminosity function of long gamma ray bursts. *Astron. Astrophys.* **2016**, *587*, A40. [CrossRef]
240. Virgili, F.J.; Zhang, B.; O'Brien, P.; Troja, E. Are All Short-hard Gamma-ray Bursts Produced from Mergers of Compact Stellar Objects? *Astrophys. J.* **2011**, *727*, 109. [CrossRef]
241. Wang, F.Y.; Dai, Z.G.; Liang, E.W. Gamma-ray burst cosmology. *New Astron. Rev.* **2015**, *67*, 1–17. [CrossRef]

242. Wang, F.Y. The high-redshift star formation rate derived from gamma-ray bursts: Possible origin and cosmic reionization. *Astron. Astrophys.* **2013**, *556*, A90. [CrossRef]
243. Le, T.; Dermer, C.D. On the Redshift Distribution of Gamma-Ray Bursts in the Swift Era. *Astrophys. J.* **2007**, *661*, 394–415. [CrossRef]
244. Salvaterra, R.; Chincarini, G. The Gamma-Ray Burst Luminosity Function in the Light of the Swift 2 Year Data. *Astrophys. J. Lett.* **2007**, *656*, L49–L52. [CrossRef]
245. Kistler, M.D.; Yüksel, H.; Beacom, J.F.; Stanek, K.Z. An Unexpectedly Swift Rise in the Gamma-Ray Burst Rate. *Astrophys. J. Lett.* **2008**, *673*, L119. [CrossRef]
246. Wang, F.Y.; Dai, Z.G.; Qi, S. Constraints on generalized Chaplygin gas model including gamma-ray bursts. *Res. Astron. Astrophys.* **2009**, *9*, 547–557. [CrossRef]
247. Robertson, B.E.; Ellis, R.S. Connecting the Gamma Ray Burst Rate and the Cosmic Star Formation History: Implications for Reionization and Galaxy Evolution. *Astrophys. J.* **2012**, *744*, 95. [CrossRef]
248. Bouwens, R.J.; Illingworth, G.D.; Bradley, L.D.; Ford, H.; Franx, M.; Zheng, W.; Broadhurst, T.; Coe, D.; Jee, M.J. z ˜ 7-10 Galaxies Behind Lensing Clusters: Contrast with Field Search Results. *Astrophys. J.* **2009**, *690*, 1764. [CrossRef]
249. Levesque, E.M.; Kewley, L.J.; Berger, E.; Zahid, H.J. The Host Galaxies of Gamma-ray Bursts. II. A Mass-metallicity Relation for Long-duration Gamma-ray Burst Host Galaxies. *Astron. J.* **2010**, *140*, 1557–1566. [CrossRef]
250. Niino, Y. Revisiting the metallicity of long-duration gamma-ray burst host galaxies: The role of chemical inhomogeneity within galaxies. *Mon. Not. R. Astron. Soc.* **2011**, *417*, 567–572. [CrossRef]
251. Marani, G.F.; Nemiroff, R.J.; Norris, J.P.; Bonnell, J.T. A new gravitational lens search for gamma ray bursts. In *Gamma-Ray Bursts: 3rd Huntsville Symposium*; Kouveliotou, C., Briggs, M.F., Fishman, G.J., Eds.; American Institute of Physics Conference Series; American Institute of Physics: College Park, MD, USA, 1996; Volume 384, pp. 487–491.
252. Davidson, R.; Bhat, P.N.; Li, G. Are there Gravitationally Lensed Gamma-ray Bursts detected by GBM? In *Gamma Ray Bursts 2010*; McEnery, J.E., Racusin, J.L., Gehrels, N., Eds.; American Institute of Physics Conference Series; American Institute of Physics: College Park, MD, USA, 2011; Volume 1358, pp. 17–20.
253. Li, C.; Li, L. Search for strong gravitational lensing effect in the current GRB data of BATSE. *Sci. China Physics, Mech. Astron.* **2014**, *57*, 1592–1599. [CrossRef]
254. Hurley, K.; Tsvetkova, A.E.; Svinkin, D.S.; Aptekar, R.L.; Frederiks, D.D.; Golenetskii, S.V.; Kokomov, A.A.; Kozlova, A.V.; Lysenko, A.L.; Ulanov, M.V.; et al. A Search for Gravitationally Lensed Gamma-Ray Bursts in the Data of the Interplanetary Network and Konus-Wind. *Astrophys. J.* **2019**, *871*, 121. [CrossRef]
255. Ahlgren, B.; Larsson, J. A Search for Lensed Gamma-Ray Bursts in 11 yr of Observations by Fermi GBM. *Astrophys. J.* **2020**, *897*, 178. [CrossRef]
256. Ougolnikov, O.S. A Search for Possible Mesolensing of Cosmic Gamma-Ray Bursts: II. Double and Triple Bursts in the BATSE Catalog. *Cosm. Res.* **2003**, *41*, 141–146. [CrossRef]
257. Hirose, Y.; Umemura, M.; Yonehara, A.; Sato, J. Imprint of Gravitational Lensing by Population III Stars in Gamma-Ray Burst Light Curves. *Astrophys. J.* **2006**, *650*, 252–260. [CrossRef]
258. Ji, L.; Kovetz, E.D.; Kamionkowski, M. Strong lensing of gamma ray bursts as a probe of compact dark matter. *Phys. Rev. D* **2018**, *98*, 123523. [CrossRef]
259. Paynter, J.; Webster, R.; Thrane, E. Evidence for an intermediate-mass black hole from a gravitationally lensed gamma-ray burst. *Nat. Astron.* **2021**, *5*, 560–568. [CrossRef]
260. Kalantari, Z.; Ibrahim, A.; Rahimi Tabar, M.R.; Rahvar, S. Imprints of Gravitational Microlensing on the Light Curve of GRBs. *arXiv* **2021**, arXiv:2105.00585.
261. Marani, G.F.; Nemiroff, R.J.; Norris, J.P.; Hurley, K.; Bonnell, J.T. Gravitationally Lensed Gamma-Ray Bursts as Probes of Dark Compact Objects. *Astrophys. J. Lett.* **1999**, *512*, L13–L16. [CrossRef]

galaxies

MDPI

Article

Accounting for Selection Bias and Redshift Evolution in GRB Radio Afterglow Data

Maria Dainotti [1,2,3,*,†], Delina Levine [4,†], Nissim Fraija [5] and Poonam Chandra [6]

1 National Astronomical Observatory of Japan, 2-21-1 Osawa, Tokyo 181-8588, Japan
2 Space Science Institute, Boulder, CO 80301, USA
3 Department of Astronomical Science, School of Physical Sciences, The Graduate University for Advanced Studies, SOKENDAI, Shonankokusaimura, Miura District, Hayama 240-0193, Japan
4 Department of Astronomy, University of Maryland, College Park, MD 20742, USA; dmlevine@umd.edu
5 Instituto de Astronomia, Universidad Nacional Autónoma de México, Mexico City 04510, Mexico; nifraija@astro.unam.mx
6 National Centre for Radio Astrophysics, Tata Institute of Fundamental Research, Pune 411007, India; poonam@ncra.tifr.res.in
* Correspondence: maria.dainotti@nao.ac.jp
† These authors contributed equally to this work.

Abstract: Gamma-ray Bursts (GRBs) are highly energetic events that can be observed at extremely high redshift. However, inherent bias in GRB data due to selection effects and redshift evolution can significantly skew any subsequent analysis. We correct for important variables related to the GRB emission, such as the burst duration, T^*_{90}, the prompt isotropic energy, $E_{\rm iso}$, the rest-frame end time of the plateau emission, $T^*_{\rm a,radio}$, and its correspondent luminosity $L_{\rm a,radio}$, for radio afterglow. In particular, we use the Efron–Petrosian method presented in 1992 for the correction of our variables of interest. Specifically, we correct $E_{\rm iso}$ and T^*_{90} for 80 GRBs, and $L_{\rm a,radio}$ and $T^*_{\rm a,radio}$ for a subsample of 18 GRBs that present a plateau-like flattening in their light curve. Upon application of this method, we find strong evolution with redshift in most variables, particularly in $L_{\rm a,radio}$, with values similar to those found in past and current literature in radio, X-ray and optical wavelengths, indicating that these variables are susceptible to observational bias. This analysis emphasizes the necessity of correcting observational data for evolutionary effects to obtain the intrinsic behavior of correlations to use them as discriminators among the most plausible theoretical models and as reliable cosmological tools.

Keywords: GRB; radio; redshift evolution

Citation: Dainotti, M.; Levine, D.; Fraija, N.; Chandra, P. Accounting for Selection Bias and Redshift Evolution in GRB Radio Afterglow Data. *Galaxies* **2021**, *9*, 95. https://doi.org/10.3390/galaxies9040095

Academic Editors: Elena Moretti and Francesco Longo

Received: 30 September 2021
Accepted: 4 November 2021
Published: 7 November 2021

1. Introduction

Gamma-ray bursts (GRBs) can be observed at extremely high redshift, making them a valuable tool for studies of the early Universe. The ability to observe these highly energetic objects at redshifts out to $z = 9.4$ [1–11] has created significant interest in using them as standardizable candles, similar to Type Ia supernovae. However, observations of GRBs have shown a very diverse population with few common characteristics.

Phenomenologically, GRBs are characterized by the main event, called the prompt emission, which is usually observed in gamma-rays, hard X-rays and sometimes in optical, while the afterglow is the counterpart in soft X-rays ($\approx$66% of observed GRBs), in optical ($\approx$38% of observed GRBs) and sometimes in radio ($\approx$6.6% of observed GRBs). GRB radio afterglows are very difficult to observe, indeed, similar to the X-ray observations which are characterized by the detector limits, and additional difficulties rise due to the limited allocated time for the follow-up observations in the radio band after the GRB trigger. Bursts are classified following the duration of the prompt episode (T_{90}). The population of short GRBs (sGRBs) usually has harder spectra and a duration of less than 2 s. In contrast, the population of long GRBs (lGRBs) has softer spectra and a duration larger than 2 s [12].

However, this classification is still in debate, and so in some bursts, it is not clear if GRBs with intermediate features belong to short or long GRBs [13,14]. LGRBs are associated with the core collapses of dying massive stars [15,16] and sGRBs by merging two compact objects; a black hole (BH) with a neutron star (NS) and two NSs [17–19].

A crucial breakthrough in the analysis of GRB features is the discovery of the plateau emission by the Neils Gehrels Swift Observatory [20]. The plateau emission is a flat part of the lightcurves which follows the decay phase of the prompt emission [21–24]. In the current paper we focus on properties resulting from this plateau emission, as well as both the prompt and afterglow emission. In general, attempts have been made to find standardizable properties, such as a plateau of GRBs or through prompt and afterglow correlation studies. We here mention a few of them: Yonetoku et al. [25], Dai and Wang [26], Ghirlanda [27], Dainotti et al. [28], Amati et al. [29], Dainotti et al. [30–34], Dainotti [35].

However, it is clear from past and current studies [36–39] that observations of GRBs are further susceptible to selection bias and evolutionary effects, which may change the results of any subsequent analysis and can substantially impact the results related to cosmological application of GRB relations [40]. In GRB studies, it is, therefore, crucial to know whether the studied correlations are intrinsic or artificially created as a result of observational biases and redshift evolution. "Redshift evolution" is the dependence of the variable of interest on redshift, and thus "independent of redshift" indicates the absence of such evolution.

In the study of GRB correlations, all variables must be computed in the rest-frame, as we are comparing objects at different epochs. This introduces another source of redshift dependence included in the definition of luminosity distance:

$$D_L = (1+z)\frac{c}{H_0}\int_0^z \frac{dz'}{\sqrt{\Omega_M(1+z')^3+\Omega_\Lambda}}, \tag{1}$$

where H_0 is the Hubble constant at the present day and Ω_M is the matter density in a flat Universe assuming the equation of state parameter w = -1. Indeed, usually one of the variables in the correlation is either a luminosity or energy which, by definition, depends on the luminosity distance. Ideally, all correlations we use must be corrected for redshift evolution, if any, requiring the removal of any existing redshift dependence.

There do exist statistical techniques that are capable of correcting for these effects, as well as correcting for data truncation from detector limits [41–43]. Among the methods to remove evolution, we consider here the Efron–Petrosian (EP) [41] method. The EP method is a well-established example of these kinds of techniques, and has been used to recover intrinsic relationships in many correlations in the past [7,40,44–48].

Lloyd et al. [44] discuss the correlation between E_p, or the peak of the νF_ν spectrum, with flux and fluence in GRBs, later investigated in the rest-frame and known as the $E_p - E_{\rm iso}$ relation [49]. A further modification of this relation is the one discovered by Yonetoku et al. [25] in which E_p is correlated with the prompt isotropic luminosity, $L_{\rm iso}$. The EP method provides an explanation on how to perform analysis on truncated data, and in Yonetoku et al. [25], Amati et al. [49] it is illustrated that the method is capable of recovering the correlation present in the original "parent" sample with the truncated data.

This technique has been further explored regarding the luminosity function and formation rate of sGRBs in a recent study by Dainotti et al. [48]. They look at the intrinsic distributions of these variables using the EP method, and introduce a method of accounting for incompleteness of redshift data with the Kolmogorov–Smirnov (KS) test (this is described in more detail in Section 2.2). They find a strong evolution of luminosity with redshift, emphasizing the necessity of this correction. The analysis presented in Dainotti et al. [48] is also relevant, as it emphasizes that both sGRBs and lGRBs undergo strong redshift evolution.

It should also be noted that though this method is mainly applied to GRB correlation studies, it has been also successfully applied in studies of Active Galactic Nuclei as well [45].

Among GRB correlations in particular we focus our attention to the rest-frame time at the end of the plateau emission, $T^*_{\rm a,radio}$, and its correspondent luminosity $L_{\rm a,radio}$, this is an extension in radio of the so-called 2D Dainotti relation in X-rays [29–31] and optical [50]. For the very recent analysis on the 2D Dainotti relation in radio see Levine et al. (2021) in preparation. For a review of the subject of GRB correlations and selection biases in the prompt and afterglow see Dai and Wang [26], Ghirlanda [27], Dainotti [35], Dainotti and Del Vecchio [51], Dainotti et al. [52], Dainotti and Amati [53].

One of the main problems in the application of GRB relationships as theoretical model discriminators and as cosmological tools is the fact that correlations must be intrinsic to the physics and not induced by biases. There are several examples of how the correlations are used to interpret theoretical models both in the prompt and afterglow emission. The photospheric emission and the Comptonization models [54–62] are the two main models used to test the E_{peak}–$E_{\rm iso}$ and the Yonetoku et al. [25] correlations, the latter between E_{peak} and the isotropic energy in the prompt emission. Otherwise, the parameter space pinpointed by those correlations can be the effect of selection biases and not of the true underlying physics. To this end, it is necessary to apply these correlations as model discriminators only after correction for such biases. Indeed, for example the plateau emission in X-rays and optical, which reconciles with the existence of the 2D Dainotti relation, can be derived through the equations of a fast rotating NS, the so-called magnetar model [17–19,63–65]. In Rowlinson et al. [66], Rea et al. [67], Stratta et al. [68] the derivation of the parameter space of the magnetic field and spin period have been computed accounting for selection biases and redshift evolution. The current status in the literature is that only a few correlations have been corrected for selection biases and evolutionary effects through the EP method, such as Dainotti et al. [7,48,69,70,71].

Specifically, Dainotti et al. [7] examine this correlation in X-ray for a sample of 101 GRBs that present a plateau, or flattening, in their light curves. After correction for evolutionary effects using the EP method, they conclude that the observed correlation is intrinsic at the 12 σ level. In mimicking the evolution of each variable with redshift, they tested both a simple and more complex model, finding similar results in both cases. Dainotti et al. [40] further examine the importance of these corrections when studying the cosmological properties of GRBs, applying the EP method to a simulated correlation between luminosity and time at the end of the plateau emission for 101 GRBs and testing whether a 5 σ deviation from the intrinsic values strongly changes the cosmological results. They demonstrated that their results change with this deviation by 13% regarding the values of Ω_M, emphasizing the necessity of applying such corrections. The problem of evolution of the variables and their correction is not only important for GRB-cosmology studies, but also for more general cosmological studies. Indeed, in Dainotti et al. [72] it has been shown that there is indication of a possible evolution even of the Hubble constant. If this is not due to selection biases, a new physical cosmological model which relies on alternative theories, such as the modified theory of gravity, must be accounted for.

Regarding the correlations in GRB afterglows, Lloyd-Ronning et al. [46,73] discussed the correlation of not only luminosity with redshift, but also isotropic energy, $E_{\rm iso}$, T^*_{90}, and the jet opening angle, θ_j, for a sample of 376 GRBs. They emphasize the difficulty of obtaining intrinsic values for these quantities due to inherent biases in observation methods, and additional truncation from detector limits that can introduce false correlations in the data [74]. They find strong evolution with redshift for each of these variables, indicating that achromatic properties of GRBs are also susceptible to selection bias. A further study by Lloyd-Ronning et al. [47] discusses the evolution of θ_j with redshift in greater detail, using the EP method to recover the intrinsic behavior of the jet opening angle.

In this study, we seek to determine whether the strong evolution of $E_{\rm iso}$ and T^*_{90} vs. redshift initially found by Lloyd-Ronning et al. [46,73] is still the same for GRBs with observed radio afterglow. In addition to the isotropic energy, we apply the EP method to the luminosity, and break time in radio wavelengths to determine if these variables are strongly affected by inherent bias and evolutionary effects.

We here point out that we are aware that the plateau sample is a subsample of a more extended population of plateaus that we cannot see. We have fixed the issue of the biases related to the redshift evolution and due to the selection threshold with the Efron–Petrosian method; however, we cannot account for the missing population of GRBs for which the follow-up has not been tackled. Nonetheless, it is crucial to discuss the $L_{a,radio}$ versus the redshift, since this correlation has been studied in X-rays and optical extensively and it is important to investigate if this correlation holds true in radio with comparable slopes to optical and X-ray. The first step to investigate the correlation is to determine if the variables involved are subjected to redshift evolution and selection biases.

In summary, the main point of the paper is the study of the redshift evolution and the removal of selection biases through the EP method. The analysis of the true correlations can be done only if we first determine the evolution among the variables. The plateau emission has been extensively investigated in X-rays and optical, but so far there has not been a statistical analysis of the existence of the plateau in radio. The radio observations of the plateau emission can cast a light on whether or not the end time of the plateau is indeed a jet break. This point can be revealed only through such a study. The evaluation of the jet break allows one to better understand the evolution of the GRB and its physics in relation to the standard fireball model [75,76] or other scenarios.

This paper is organized as follows: in Section 2, we discuss the selection of our sample, as well as the formulation of the EP method and its application to our sample. In Section 3, we present the results of this analysis, and in Section 4, we discuss the implications of our study, as well as a comparison to previous studies, and present our conclusions.

2. Methods

2.1. Variables of Interest

The EP method uses a modified version of Kendall's τ statistics to test for independence of variables in a truncated data set. Here, τ is defined as:

$$\tau = \frac{\sum_i (\mathcal{R}_i - \mathcal{E}_i)}{\sqrt{\sum_i \mathcal{V}_i}} \tag{2}$$

with R_i defined as the rank $\mathcal{E}_i = (1/2)(i+1)$ defined as the expectation value, and $\mathcal{V}_i = (1/12)(i^2+1)$ defined as the variance. For a more complete discussion of the method and the algebra involved, see Dainotti et al. [7,34], Efron and Petrosian [41], Singal et al. [45], Dainotti et al. [77], Petrosian et al. [78], Lloyd-Ronning and Petrosian [79]. Here, we use the EP method to determine the impact of redshift evolution and selection bias on four variables: T^*_{90}, where the star denotes the rest-frame, $E_{\rm iso}$, the radio light curve break time $T^*_{\rm a,radio}$, and the radio luminosity at the time of break $L_{\rm a,radio}$. These variables are considered to be they are pertinent to the correlations analyzed in Levine et al. (2021 in preparation). Throughout our analysis, we consider these variables in logarithmic scale for convenience.

We look at the $\log E_{\rm iso}$ and $\log T^*_{90}$ for a sample of 80 GRBs with observed radio afterglow published in the literature [80–98]. Values of $\log E_{\rm iso}$ are taken from the literature. If no $E_{\rm iso}$ value could be found, the $\log E_{\rm iso}$, in units of ergs, is calculated using the equation:

$$E_{\rm iso} = 4\pi D_L^2(z) S K, \tag{3}$$

where S is the fluence in units of erg cm^{-2}, $D_L^2(z)$ is the luminosity distance assuming a flat ΛCDM model with $\Omega_M = 0.3$ and $H_0 = 70$ km s^{-1} Mpc^{-1} (see Equation (1)), and K is the correction for cosmic expansion [99]:

$$K = \frac{1}{(1+z)^{1-\beta}}, \tag{4}$$

with β as the spectral index of the GRB. Fluence and β values are taken from the literature.

To analyze the impact of selection bias and redshift evolution for $\log L_{a,radio}$ and $\log T^*_{a,radio}$, we first fit each of the 80 GRBs with a broken power law (BPL) according to the formulation:

$$F(t) = \begin{cases} F_a(\frac{t}{T^*_a})^{-\alpha_1} & t < T^*_a \\ F_a(\frac{t}{T^*_a})^{-\alpha_2} & t \geq T^*_a \end{cases} \tag{5}$$

where F_a refers to the flux at the break, T^*_a refers to the rest-frame time of break, α_1 refers to the slope before the break, and α_2 refers to the slope after the break. We can only obtain values of $\log L_{a,radio}$ and $\log T^*_{a,radio}$ for light curves that show a "plateau" feature, or a flattening of the light curve before a clear break. In our analysis, we consider a light curve to display a plateau if $|\alpha_1| < 0.5$. Therefore, we discard fits to light curves with scattered observations, unreliable error bars, or shapes incompatible with a BPL and plateau. In our subsequent analysis we include those light curves whose $\Delta\chi^2$ analysis of the BPL best-fit parameters are suitable following the Avni [100] methodology. After the rejection process, we are left with 18 GRBs that present a plateau and clear break in the light curve.

The luminosity $\log L_{a,radio}$ in units of erg s^{-1} is computed at time $\log T^*_{a,radio}$ using the equation:

$$L_a = 4\pi D^2_L(z) F_a(T_a) K, \tag{6}$$

where F_a is the observed flux at $T_{a,radio}$, $D^2_L(z)$ is defined as in Equation (3), and K is the k-correction:

$$K = \frac{1}{(1+z)^{\alpha_1 - \beta}}, \tag{7}$$

with β as the radio spectral index and α_1 as the fitted BPL temporal index before the break. β values are taken from Chandra and Frail [80] or other literature—if no β value could be found, the average of published spectral indices, $\beta = 0.902 \pm 0.17$, was assigned. We show the distribution of spectral indices for the plateau sample in Figure 1.

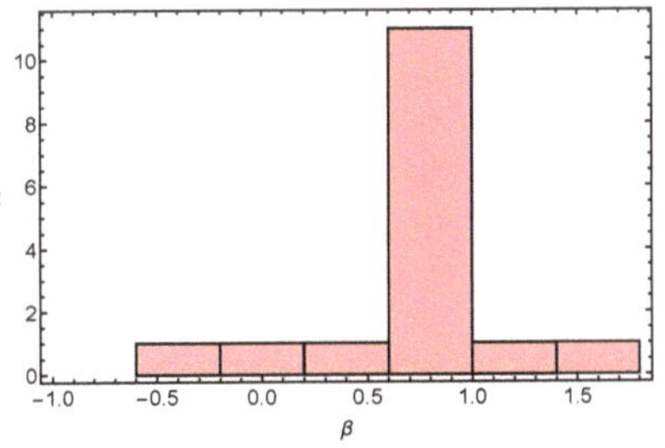

Figure 1. Distribution of spectral indices (β) for sample of 18 GRBs that display a plateau in their light curve.

2.2. Limiting Fluxes, Fluences, and Times

The EP method can overcome selection bias for a particular variable of interest, but it must first be determined if the variable is dependent or independent of redshift.

It is then necessary to define limiting values for each of the variables. In Dainotti et al. [7], it was demonstrated that a good choice of limiting times and luminosities retains at least 90% of the total sample. For time variables, $\log T^*_{90}$ and $\log T^*_{a,radio}$ (in units of seconds), we define a general form for the limiting values as $\frac{T_{min}}{(1+z)}$, where T_{min} is the minimum observed time. We need to choose a compromise between a limit which is representative of the population of data points, but still retains most of the sample size. It has been shown in Monte Carlo simulations in [7] that such a strategy with limiting values is accurate. For $\log T^*_{90}$, we find the best limiting duration to be $\log T^*_{90_{min,obs}} = -0.54$ s, with a limiting boundary defined as $\frac{-0.30}{(1+z)}$ s, which excludes 5/80 (<10%) GRBs. The limiting line for $\log T^*_{90}$ is shifted at a higher value to be allow the sample data to be representative of the whole population. For $\log T^*_{a,radio}$, we find the limiting time to be the observed minimum $\log T^*_{\text{a,radio}_{\text{min,obs}}} = 4.94$ s, thus defining the boundary as $\log T^*_{\text{a,radio}} = \frac{4.94}{(1+z)}$ s, which does not exclude any data points.

For the isotropic energy, we instead define the limiting energy according to the methodology of Dainotti et al. [7], in which the limiting fluence should be representative of the population while including at least 90% of the sample. We use the following formula:

$$E_{\text{iso,lim}} = 4\pi D_L^2(z) S_{\text{lim}}, \tag{8}$$

where S_{lim} is the fluence limit. For our sample, we define S_{lim} as 6.3×10^{-8} erg cm^{-2}. Applying this limit excludes 8/80 GRBs, which is 10% of our sample. In all the method described here we use GRBs that have $\log E_{\text{iso}} > \log E_{\text{iso,lim}}$, $\log T^*_{90} > \log T^*_{90}$, $\log T_{\text{a,radio}} > \log T_{\text{a,radio,lim}}$, and $\log L_{\text{a,radio}} > \log L_{\text{a,radio,lim}}$. For the luminosity, however, a caveat should be posed when we consider the total distribution of the parent population of GRBs with and without redshift (see [48]).

Using the method presented in Dainotti et al. [48], we compare the parent sample of all GRBs with observed radio afterglow and known peak flux to a smaller "subsample" of GRBs with known peak flux and known redshift. We then apply cuts to the data by defining limiting fluxes at regular intervals. Considering only the data with values above the limiting fluxes, fluences and time, we conduct a two-sample Kolmogorov–Smirnov (KS) test between the data of the total sample and the data for which the limiting cuts have been applied to determine the distribution of the probability that the subsample was drawn from the parent sample, as well as the geometric distance between the two samples as determined by the KS test. We take the limiting flux to be the value of f_{lim} where the probability as a function of limiting flux reaches a plateau in which the probability that two samples are drawn by the same population is 100%. In our sample, we find this limit to be $\log f_{\text{lim}} = -17.2$. We define the flux throughout our analysis in units of erg cm^{-2} s.

We show the distribution of the parent sample and subsample in the left panel of Figure 2, with the limiting line shown in red. We plot this probability as a function of flux limit (blue), as well as the distance between the distributions (orange), in the right panel of Figure 2.

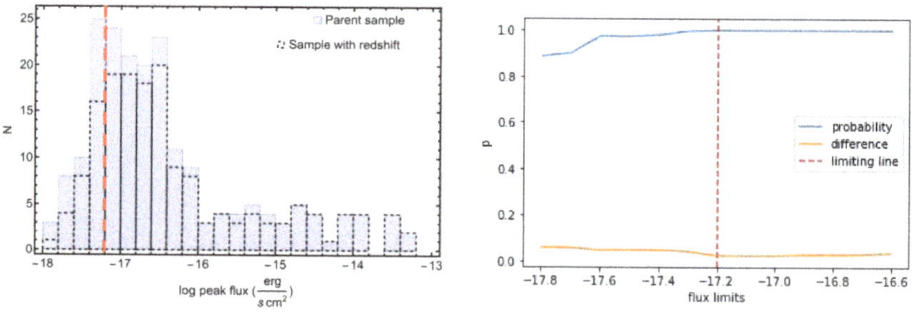

Figure 2. (**Left**): peak flux distribution for "parent" sample and subsample with known redshift. Limiting flux shown in red. (**Right**): plot of probability (blue) and distance between samples as given by the KS test (orange) as a function of flux limit. Limiting line f_{lim} shown in red.

2.3. Removing Selection Bias and Redshift Evolution

After defining the limiting lines, we can then mimic the evolution of this variable with redshift using a simpler function of redshift, $f(z) = \frac{1}{(1+z)^\delta}$. We here adopt the choice of a simple function, but in principle it is possible to use a more complex function as already shown in [77] and obtain compatible results. Using a modified version of Kendall's τ statistics, we can compute an evolutionary function $f(z)$ where the slope of the function is defined by δ. The values for δ where $\tau = 0$ represent the removal of evolutionary effects. We define errors on δ out to 1 σ, corresponding to the δ values where $\tau = \pm 1$.

3. Results

We here clarify that the purpose of our analysis is to show how similar our results are, compared to other ones in the literature given that our sample size differs from other studies for $E_{\rm iso}$ and T^*_{90} (our sample has 80 GRBs), while this is the first time in the literature that we compare the results for the $L_{\rm a,radio}$ and $T^*_{\rm a,radio}$ (our sample has 18 GRBs) with the previous results in the literature performed in X-rays and optical. This is an essential comparison to allow the determination of the intrinsic nature of the $L_{\rm a,radio}$ and $T^*_{\rm a,radio}$ correlation and to check for the universality of the results related to the evolutionary functions for these variables with the EP method.

Using the procedure outlined above, we correct $\log E_{\rm iso}$, $\log T^*_{90}$ using the formulation $\log E'_{\rm iso} = \log E_{\rm iso} - \log((1+z)^{\delta_{E_{\rm iso}}})$ and $\log T^{*'}_{90} = \log T^*_{90} - \log((1+z)^{\delta_{T^*_{90}}})$, where all quantities that have $'$ symbol are the variables for which the evolution has been removed, thus they are no longer dependent on the redshift. We find $\delta_{T^*_{90}} = -0.65 \pm 0.27$, with the 1 σ errors defined as the average of the values of $\tau = 1$ and $\tau = -1$, and $\delta_{E_{\rm iso}} = 0.39 \pm 0.88$. Figure 3 shows these results—the top left panel shows $\log T^*_{90}$ for the sample of 80 GRBs as a function of redshift, with the limiting value shown in red. The top right panel highlights the evolutionary function for $\log T^*_{90}$, with dashed lines at $\tau = 0, \pm 1$. The same plots for $\log E_{\rm iso}$ are shown in the bottom panels.

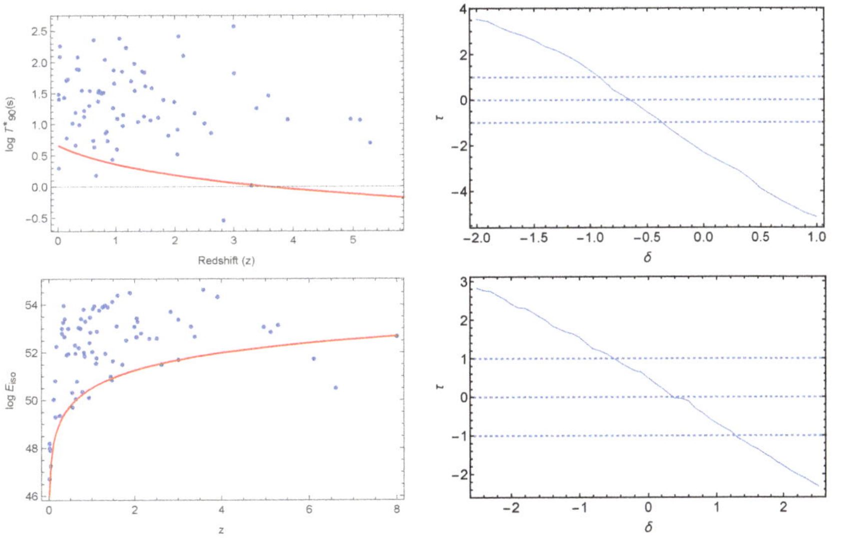

Figure 3. The (**upper left**) panel shows the values of $\log T^*_{90}$ vs. redshift in blue and the limiting $\log T^*_{90}$ in the rest-frame in red. The (**upper right**) panel shows the Kendall τ vs. the slope of the evolutionary function with 1 σ errors shown with dashed blue lines. As with the upper panels, the (**lower left**) panel shows values of $\log E_{\rm iso}$ vs. redshift in blue with the limiting line in red, and the (**right**) panel shows the Kendall τ vs. slope of the evolutionary function with 1 σ errors as dashed blue lines.

For the plateau sample of 18 GRBs, we use the same formulation to obtain $\log L'_{\rm a,radio} = \log L_{\rm a,radio} - \log((1+z)^{\delta_{L_{\rm a,radio}}})$ and $\log T^{*'}_{\rm a,radio} = \log T^*_{\rm a,radio} - \log((1+z)^{\delta_{T^*_{\rm a,radio}}})$. We find the values of δ for $\log L_{\rm a,radio}$ and $\log T^*_{\rm a,radio}$ as $\delta_{T^*_{\rm a,radio}} = -1.94 \pm 0.86$ and $\delta_{L_{\rm a,radio}} = 3.15 \pm 1.65$. These results are shown in Figure 4—$\log T^*_{\rm a,radio}$ as a function of redshift is shown in the top left panel, with the limiting values in red. The evolutionary function is shown in the top right panel, with dashed blue lines at $\tau = 0, \pm 1$. The bottom two panels display the same plots for $\log L_{\rm a,radio}$.

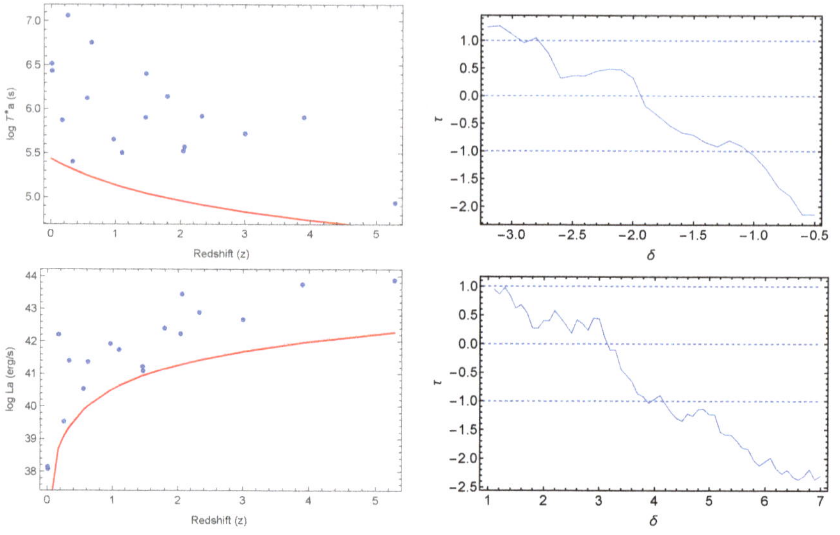

Figure 4. The (**upper left**) panel shows the values of $\log T^*_{\mathrm{a,radio}}$ vs. redshift in blue and the limiting $\log T^*_{\mathrm{a,radio}}$ in the rest-frame in red. The (**upper right**) panel shows the Kendall τ vs. the slope of the evolutionary function with 1 σ errors shown with dashed blue lines. As with the upper panels, the (**lower left**) panel shows values of $\log L_{\mathrm{a,radio}}$ vs. redshift in blue with the limiting line in red, and the (**right**) panel shows the Kendall τ vs. slope of the evolutionary function with 1 σ errors as dashed blue lines.

4. Discussion and Conclusions

Using a sample of 80 GRBs with observed radio afterglow, we test the use of the EP method on $\log E_{\mathrm{iso}}$ and $\log T^*_{90}$ for a subsample 80 GRBs, and on $\log L_{\mathrm{a,radio}}$ and $\log T^*_{\mathrm{a,radio}}$ for a subsample of 18 GRBs. We find that when considering $\log E_{\mathrm{iso}}$ and $\log T^*_{90}$, we obtain indices for the parameter of redshift evolution, δ, of $\delta_{T^*_{90}} = -0.65 \pm 0.27$ and $\delta_{E_{\mathrm{iso}}} = 0.39 \pm 0.88$, while the values of δ for $\log L_{\mathrm{a,radio}}$ and $\log T^*_{\mathrm{a,radio}}$ are $\log T^*_{\mathrm{a,radio}}$ as $\delta_{T^*_{\mathrm{a,radio}}} = -1.94 \pm 0.86$ and $\delta_{L_{\mathrm{a,radio}}} = 3.15 \pm 1.65$.

For $\log T^*_{90}$, for $\log T_{\mathrm{a,radio}}$, and $\log L_{\mathrm{a,radio}}$, we find relatively strong evolution of each variable with redshift. The luminosity presents the strongest correlation with redshift, emphasizing the necessity of correction for these effects before using data in correlation analysis. E_{iso}, by contrast, appears to be the most independent from redshift, with the smallest value for $|\delta|$.

We find that our values are comparable to values obtained in previous studies. A study by Lloyd-Ronning et al. [46] of the cosmological evolution of isotropic energy E_{iso}, burst duration T^*_{90}, jet opening angle θ_j, and luminosity L_j reports δ values compatible with our findings T^*_{90} and $L_{\mathrm{a,radio}}$ within 1 σ, and values compatible with our $\delta_{E_{\mathrm{iso}}}$ within approximately 2 σ. Specifically, they find a value of $\delta_{E_{\mathrm{iso}}} = 2.3 \pm 0.5$, which agrees with our value of $\delta_{E_{\mathrm{iso}}} = 0.39 \pm 0.88$ within 2.17 σ. We also see agreement with $\delta_{T^*_{90}} = -0.8 \pm 0.3$, with a 0.55 σ difference from our value of $\delta_{T^*_{90}} = -0.65 \pm 0.27$, and in the luminosity with $\delta_{L_j} = 3.5 \pm 0.5$, a 0.2 σ difference from our value of $\delta_{L_{\mathrm{a,radio}}} = 3.15 \pm 1.65$.

These results also agree with previous values of δ for L_{a} and T^*_{a} in X-ray and optical wavelengths. Dainotti et al. [7] conduct a similar analysis of the luminosity and rest-frame end time of the plateau emission using X-ray data. Their value for correction for $\delta_{T^*_{\mathrm{a}}}$, reported as $\delta_{T^*_{\mathrm{a,X}}} = -0.85 \pm 0.3$, is compatible with our value of $\delta_{T^*_{\mathrm{a,radio}}} = -1.94 \pm 0.86$ within 1.23 σ. However, they find a very slow evolution in luminosity, with a value of $\delta_{L_{\mathrm{a,X}}} = -0.05 \pm 0.35$, which is a 1.94 σ difference from our value of $\delta_{L_{\mathrm{a,radio}}} = 3.15 \pm 1.65$. This discrepancy is likely due in part to the small sample size, which may exaggerate the extent of the evolution present in our sample, but may also be due to differences in the behavior of the X-ray and radio emission.

We have corrected the luminosity and time in X-rays with 222 GRB lightcurves with a given redshift, presenting plateaus according the Willingale et al. [23] model and in optical with 181 GRBs with plateaus taken from Dainotti et al. [50], but with the additional analysis of 80 GRBs found in the literature. For tackling this analysis, we followed the same procedure described in the current paper. The results of this analysis reports σ, and $\delta_{L_{\mathrm{a,X}}} = 2.42 \pm 0.58$, which is a 0.44 σ difference from our value. They also report values in optical of $\delta_{T^*_{\mathrm{a,opt}}} = -2.11 \pm 0.49$ and $\delta_{L_{\mathrm{a,opt}}} = 3.96 \pm 0.58$, which both agree with our result within 1 σ.

In general, it can be seen that a larger sample size is preferred when applying the EP method. In our case, for the sample pertinent to E_{iso}, and T^*_{90} we choose the limiting values while excluding $\leq$ 10% of the overall sample. However, for the smaller sample of 18 plateau GRBs, the limiting values are chosen so that we do not exclude any data points due to the small sample size. In addition, this conservative choice would allow us to have smaller error bars on the slope of the evolutionary functions. However, the δ values obtained for $L_{\mathrm{a,radio}}$ and $T^*_{\mathrm{a,radio}}$ are similar to values found in previous studies of larger sample sizes, thus indicating that the EP method is still successful even with a small dataset.

GRB correlations in radio afterglows related to the plateau emission are crucial to understand if the jet break is coincident with the end of the plateau emission. To investigate this point, a multiwavelength analysis not only in optical and X-rays must be performed together with the radio data. Since the evolution of the variables with redshift can change the time at which the break happens, it is crucially important to correct for the redshift evolution. This study is also the preliminary step to the investigation of the intrinsic nature of the plateau emission correlations.

After our analysis on the radio observations both for the prompt emission in relation to the variables of E_{iso} and T^*_{90} and for the afterglow emission in relation to $L_{\mathrm{a,radio}}$, $T^*_{\mathrm{a,radio}}$, we can conclude:

1. After testing intrinsic properties of a GRB, such as E_{iso} and T^*_{90}, as well as properties such as $L_{\mathrm{a,radio}}$ and $T^*_{\mathrm{a,radio}}$, we see T^*_{90}, $L_{\mathrm{a,radio}}$ and $T^*_{\mathrm{a,radio}}$ present strong correlation with redshift, thus indicating that they are susceptible to redshift evolution.
2. The δ values obtained in this work agree with those of previous studies, indicating that this trend of strong correlation with redshift still holds true in radio wavelengths.
3. The study of these evolutionary functions is the first necessary step to determine the true intrinsic nature of the correlations, object of our study.

Author Contributions: Conceptualization: M.D.; Methodology: M.D.; Software: M.D.; Validation: M.D. and D.L.; Formal Analysis: D.L.; Investigation, D.L. and M.D.; Data Curation: P.C.; Writing—Original Draft Preparation: D.L.; Writing—Review and Editing Draft: M.D., N.F. and D.L.; Supervision: M.D. and N.F. All authors have read and agreed to the published version of the manuscript.

Funding: This research was funded by the National Astronomical Observatory of Japan (NAOJ).

Institutional Review Board Statement: Not applicable.

Informed Consent Statement: Not applicable.

Data Availability Statement: The data are taken from Chandra and Frail [80], Zauderer et al. [81], Laskar et al. [82,83], Cucchiara et al. [84], Bright et al. [85], Anderson et al. [86], Singer et al. [87], Laskar et al. [88], Kangas et al. [89], Laskar et al. [90], Chen et al. [91], Bhalerao et al. [92], Hallinan et al. [93], Mooley et al. [94], Resmi et al. [95], Margutti et al. [96], Maity and Chandra [97] and Rhodes et al. [98].

Acknowledgments: This work was made possible in part by the United States Department of Energy, Office of Science, Office of Workforce Development for Teachers and Scientists (WDTS) under the Science Undergraduate Laboratory Internships (SULI) program. We thank Cuellar for managing the SULI program at Stanford National Accelerator Laboratory. We also acknowledge the National Astronomical Observatory of Japan for their support in making this research possible, as well as Kevin J. Zvonarek for his help in preparing the dataset in this work. NF acknowledges financial support from UNAM-DGAPA-PAPIIT through grant IN106521. D. Levine acknowledges support from NAOJ division of Science.

Conflicts of Interest: The authors declare no conflict of interest. The founding sponsors had no role in the design of the study; in the collection, analyses, or interpretation of data; in the writing of the manuscript, and in the decision to publish the results.

References

1. Cucchiara, A.; Levan, A.J.; Fox, D.B.; Tanvir, N.R.; Ukwatta, T.N.; Berger, E.; Krühler, T.; Küpcü Yoldaş, A.; Wu, X.F.; Toma, K.; et al. A Photometric Redshift of z~9.4 for GRB 090429B. *Astrophys. J.* **2011**, *736*, 7. [CrossRef]
2. Cardone, V.F.; Capozziello, S.; Dainotti, M.G. An updated gamma-ray bursts Hubble diagram. *Mon. Not. R. Astron. Soc.* **2009**, *400*, 775–790. [CrossRef]
3. Salvaterra, R.; Della Valle, M.; Campana, S.; Chincarini, G.; Covino, S.; D'Avanzo, P.; Fernández-Soto, A.; Guidorzi, C.; Mannucci, F.; Margutti, R.; et al. GRB090423 at a redshift of z~8.1. *Nature* **2009**, *461*, 1258–1260. [CrossRef] [PubMed]
4. Tanvir, N.R.; Fox, D.B.; Levan, A.J.; Berger, E.; Wiersema, K.; Fynbo, J.P.U.; Cucchiara, A.; Krühler, T.; Gehrels, N.; Bloom, J.S.; et al. A γ-ray burst at a redshift of z~8.2. *Nature* **2009**, *461*, 1254–1257. [CrossRef]
5. Cardone, V.F.; Dainotti, M.G.; Capozziello, S.; Willingale, R. Constraining cosmological parameters by gamma-ray burst X-ray afterglow light curves. *Mon. Not. R. Astron. Soc.* **2010**, *408*, 1181–1186. [CrossRef]
6. Postnikov, S.; Dainotti, M.G.; Hernandez, X.; Capozziello, S. Nonparametric Study of the Evolution of the Cosmological Equation of State with SNeIa, BAO, and High-redshift GRBs. *Astrophys. J.* **2014**, *783*, 126. [CrossRef]
7. Dainotti, M.G.; Petrosian, V.; Singal, J.; Ostrowski, M. Determination of the Intrinsic Luminosity Time Correlation in the X-ray Afterglows of Gamma-ray Bursts. *Astrophys. J.* **2013**, *774*, 157. [CrossRef]
8. Littlejohns, O.M.; Tanvir, N.R.; Willingale, R.; Evans, P.A.; O'Brien, P.T.; Levan, A.J. Are gamma-ray bursts the same at high redshift and low redshift? *Mon. Not. R. Astron. Soc.* **2013**, *436*, 3640–3655. [CrossRef]
9. Laskar, T.; Berger, E.; Tanvir, N.; Zauderer, B.A.; Margutti, R.; Levan, A.; Perley, D.; Fong, W.F.; Wiersema, K.; Menten, K.; et al. GRB 120521C at z~6 and the Properties of High-redshift γ-Ray Bursts. *Astrophys. J.* **2014**, *781*, 1. [CrossRef]
10. McGuire, J.T.W.; Tanvir, N.R.; Levan, A.J.; Trenti, M.; Stanway, E.R.; Shull, J.M.; Wiersema, K.; Perley, D.A.; Starling, R.L.C.; Bremer, M.; et al. Detection of Three Gamma-ray Burst Host Galaxies at z~6. *Astrophys. J.* **2016**, *825*, 135. [CrossRef]
11. Tanvir, N.R.; Laskar, T.; Levan, A.J.; Perley, D.A.; Zabl, J.; Fynbo, J.P.U.; Rhoads, J.; Cenko, S.B.; Greiner, J.; Wiersema, K.; et al. The Properties of GRB 120923A at a Spectroscopic Redshift of $z \approx 7.8$. *Astrophys. J.* **2018**, *865*, 107. [CrossRef]
12. Gehrels, N.; Razzaque, S. Gamma-ray bursts in the swift-Fermi era. *Front. Phys.* **2013**, *8*, 661–678. [CrossRef]
13. Norris, J.P.; Bonnell, J.T. Short Gamma-ray Bursts with Extended Emission. *Astrophys. J.* **2006**, *643*, 266–275. [CrossRef]
14. Norris, J.P.; Gehrels, N.; Scargle, J.D. Heterogeneity in Short Gamma-ray Bursts. *Astrophys. J.* **2011**, *735*, 23. [CrossRef]
15. Woosley, S.E. Gamma-ray Bursts from Stellar Mass Accretion Disks around Black Holes. *Astrophys. J.* **1993**, *405*, 273. [CrossRef]
16. Paczyński, B. Are Gamma-ray Bursts in Star-Forming Regions? *Astrophys. J.* **1998**, *494*, L45–L48. [CrossRef]
17. Usov, V.V. Millisecond pulsars with extremely strong magnetic fields as a cosmological source of gamma-ray bursts. *Nature* **1992**, *357*, 472–474. [CrossRef]
18. Duncan, R.C.; Thompson, C. Formation of Very Strongly Magnetized Neutron Stars: Implications for Gamma-ray Bursts. *Astrophys. J.* **1992**, *392*, L9. [CrossRef]
19. Metzger, B.D.; Giannios, D.; Thompson, T.A.; Bucciantini, N.; Quataert, E. The protomagnetar model for gamma-ray bursts. *Mon. Not. R. Astron. Soc.* **2011**, *413*, 2031–2056. [CrossRef]
20. Gehrels, N.; Chincarini, G.; Giommi, P.; Mason, K.O.; Nousek, J.A.; Wells, A.A.; White, N.E.; Barthelmy, S.D.; Burrows, D.N.; Cominsky, L.R.; et al. The Swift Gamma-ray Burst Mission. *Astrophys. J.* **2004**, *611*, 1005–1020. [CrossRef]
21. Sakamoto, T.; Hill, J.E.; Yamazaki, R.; Angelini, L.; Krimm, H.A.; Sato, G.; Swindell, S.; Takami, K.; Osborne, J.P. Evidence of Exponential Decay Emission in the Swift Gamma-ray Bursts. *Astrophys. J.* **2007**, *669*, 1115–1129. [CrossRef]
22. O'Brien, P.T.; Willingale, R.; Osborne, J.; Goad, M.R.; Page, K.L.; Vaughan, S.; Rol, E.; Beardmore, A.; Godet, O.; Hurkett, C.P.; et al. The Early X-ray Emission from GRBs. *Astrophys. J.* **2006**, *647*, 1213–1237. [CrossRef]
23. Willingale, R.; O'Brien, P.T.; Osborne, J.P.; Godet, O.; Page, K.L.; Goad, M.R.; Burrows, D.N.; Zhang, B.; Rol, E.; Gehrels, N.; et al. Testing the Standard Fireball Model of Gamma-ray Bursts Using Late X-ray Afterglows Measured by Swift. *Astrophys. J.* **2007**, *662*, 1093–1110. [CrossRef]
24. Pe'er, A. Physics of Gamma-ray Bursts Prompt Emission. *Adv. Astron.* **2015**, *2015*, 907321. [CrossRef]
25. Yonetoku, D.; Murakami, T.; Nakamura, T.; Yamazaki, R.; Inoue, A.K.; Ioka, K. Gamma-ray Burst Formation Rate Inferred from the Spectral Peak Energy-Peak Luminosity Relation. *Astrophys. J.* **2004**, *609*, 935–951. [CrossRef]
26. Dai, Z.G.; Wang, F.Y. Recent progress on GRB cosmology. *Adv. Space Res.* **2007**, *40*, 1244–1249. [CrossRef]
27. Ghirlanda, G. Gamma-ray bursts spectral correlations and their cosmological use. *Philos. Trans. R. Soc. Lond. Ser. A* **2007**, *365*, 1385–1394. [CrossRef] [PubMed]

28. Dainotti, M.G.; Cardone, V.F.; Capozziello, S. A time-luminosity correlation for γ-ray bursts in the X-rays. *Mon. Not. R. Astron. Soc.* **2008**, *391*, L79–L83. [CrossRef]
29. Amati, L.; Guidorzi, C.; Frontera, F.; Della Valle, M.; Finelli, F.; Landi, R.; Montanari, E. Measuring the cosmological parameters with the $E_{p,i}$-E_{iso} correlation of gamma-ray bursts. *Mon. Not. R. Astron. Soc.* **2008**, *391*, 577–584. [CrossRef]
30. Dainotti, M.G.; Willingale, R.; Capozziello, S.; Fabrizio Cardone, V.; Ostrowski, M. Discovery of a Tight Correlation for Gamma-ray Burst Afterglows with "Canonical" Light Curves. *Astrophys. J.* **2010**, *722*, L215–L219. [CrossRef]
31. Dainotti, M.G.; Fabrizio Cardone, V.; Capozziello, S.; Ostrowski, M.; Willingale, R. Study of Possible Systematics in the L^*_X-T^*_a Correlation of Gamma-ray Bursts. *Astrophys. J.* **2011**, *730*, 135. [CrossRef]
32. Dainotti, M.G.; Ostrowski, M.; Willingale, R. Towards a standard gamma-ray burst: Tight correlations between the prompt and the afterglow plateau phase emission. *Mon. Not. R. Astron. Soc.* **2011**, *418*, 2202–2206. [CrossRef]
33. Dainotti, M.G.; Postnikov, S.; Hernandez, X.; Ostrowski, M. A Fundamental Plane for Long Gamma-ray Bursts with X-ray Plateaus. *Astrophys. J.* **2016**, *825*, L20. [CrossRef]
34. Dainotti, M.G.; Nagataki, S.; Maeda, K.; Postnikov, S.; Pian, E. A study of gamma ray bursts with afterglow plateau phases associated with supernovae. *Astron. Astrophys.* **2017**, *600*, A98. [CrossRef]
35. Dainotti, M. *Gamma-ray Burst Correlations: Current Status and Open Questions*; IOP Publishing Ltd.: Bristol, UK, 2019. [CrossRef]
36. Kocevski, D.; Butler, N.; Bloom, J.S. Pulse Width Evolution of Late-Time X-ray Flares in Gamma-ray Bursts. *Astrophys. J.* **2007**, *667*, 1024–1032. [CrossRef]
37. Shahmoradi, A.; Nemiroff, R.J. The possible impact of gamma-ray burst detector thresholds on cosmological standard candles. *Mon. Not. R. Astron. Soc.* **2011**, *411*, 1843–1856. [CrossRef]
38. Shahmoradi, A.; Nemiroff, R.J. Short versus long gamma-ray bursts: A comprehensive study of energetics and prompt gamma-ray correlations. *Mon. Not. R. Astron. Soc.* **2015**, *451*, 126–143. [CrossRef]
39. Coward, D.M.; Howell, E.J.; Wan, L.; MacPherson, D. Selection biases in the gamma-ray burst Eiso-Lopt, X correlation. *Mon. Not. R. Astron. Soc.* **2015**, *449*, L6–L10. [CrossRef]
40. Dainotti, M.G.; Cardone, V.F.; Piedipalumbo, E.; Capozziello, S. Slope evolution of GRB correlations and cosmology. *Mon. Not. R. Astron. Soc.* **2013**, *436*, 82–88. [CrossRef]
41. Efron, B.; Petrosian, V. A Simple Test of Independence for Truncated Data with Applications to Redshift Surveys. *Astrophys. J.* **1992**, *399*, 345. [CrossRef]
42. Lynden-Bell, D. A method of allowing for known observational selection in small samples applied to 3CR quasars. *Mon. Not. R. Astron. Soc.* **1971**, *155*, 95. [CrossRef]
43. Efron, B.; Petrosian, V. Nonparametric Methods for Doubly Truncated Data. *arXiv* **1998**, arXiv:astro-ph/9808334.
44. Lloyd, N.M.; Petrosian, V.; Mallozzi, R.S. Cosmological versus Intrinsic: The Correlation between Intensity and the Peak of the νF_ν Spectrum of Gamma-ray Bursts. *Astrophys. J.* **2000**, *534*, 227–238. [CrossRef]
45. Singal, J.; Petrosian, V.; Lawrence, A.; Stawarz, Ł. On the Radio and Optical Luminosity Evolution of Quasars. *Astrophys. J.* **2011**, *743*, 104. [CrossRef]
46. Lloyd-Ronning, N.M.; Aykutalp, A.; Johnson, J.L. On the cosmological evolution of long gamma-ray burst properties. *Mon. Not. R. Astron. Soc.* **2019**, *488*, 5823–5832.10.1093/mnras/stz2155. [CrossRef]
47. Lloyd-Ronning, N.; Hurtado, V.U.; Aykutalp, A.; Johnson, J.; Ceccobello, C. The evolution of gamma-ray burst jet opening angle through cosmic time. *Mon. Not. R. Astron. Soc.* **2020**, *494*, 4371–4381. [CrossRef]
48. Dainotti, M.G.; Petrosian, V.; Bowden, L. Cosmological Evolution of the Formation Rate of Short Gamma-ray Bursts with and without Extended Emission. *Astrophys. J.* **2021**, *914*, L40. [CrossRef]
49. Amati, L.; Frontera, F.; Tavani, M.; In't Zand, J.J.M.; Antonelli, A.; Costa, E.; Feroci, M.; Guidorzi, C.; Heise, J.; Masetti, N.; et al. Intrinsic spectra and energetics of BeppoSAX Gamma-ray Bursts with known redshifts. *Astron. Astrophys.* **2002**, *390*, 81–89. [CrossRef]
50. Dainotti, M.G.; Livermore, S.; Kann, D.A.; Li, L.; Oates, S.; Yi, S.; Zhang, B.; Gendre, B.; Cenko, B.; Fraija, N. The Optical Luminosity-Time Correlation for More than 100 Gamma-ray Burst Afterglows. *Astrophys. J.* **2020**, *905*, L26. [CrossRef]
51. Dainotti, M.G.; Del Vecchio, R. Gamma ray Burst afterglow and prompt-afterglow relations: An overview. *nar* **2017**, *77*, 23–61. [CrossRef]
52. Dainotti, M.G.; Del Vecchio, R.; Tarnopolski, M. Gamma-ray Burst Prompt Correlations. *Adv. Astron.* **2018**, *2018*, 4969503. [CrossRef]
53. Dainotti, M.G.; Amati, L. Gamma-ray Burst Prompt Correlations: Selection and Instrumental Effects. *Adv. Astron.* **2018**, *130*, 051001. [CrossRef]
54. Titarchuk, L.; Farinelli, R.; Frontera, F.; Amati, L. An Upscattering Spectral Formation Model for the Prompt Emission of Gamma-ray Bursts. *Astrophys. J.* **2012**, *752*, 116. [CrossRef]
55. Beloborodov, A.M.; Mészáros, P. Photospheric Emission of Gamma-ray Bursts. *Space Sci. Rev.* **2017**, *207*, 87–110. [CrossRef]
56. Dereli-Bégué, H.; Pe'er, A.; Ryde, F. Classification of Photospheric Emission in Short GRBs. *Astrophys. J.* **2020**, *897*, 145. [CrossRef]
57. Toma, K.; Wu, X.F.; Mészáros, P. Photosphere-internal shock model of gamma-ray bursts: Case studies of Fermi/LAT bursts. *Mon. Not. R. Astron. Soc.* **2011**, *415*, 1663–1680. [CrossRef]
58. Zheng, Z.; Lu, Y.; Zhao, Y.H. A Comptonization Model for the Prompt Optical and Infrared Emission of GRB 041219A. *Astrophys. J.* **2006**, *646*, L25–L28. [CrossRef]

59. Ito, H.; Matsumoto, J.; Nagataki, S.; Warren, D.C.; Barkov, M.V.; Yonetoku, D. The photospheric origin of the Yonetoku relation in gamma-ray bursts. *Nat. Commun.* **2019**, *10*, 1504. [CrossRef]
60. Ahlgren, B.; Larsson, J.; Nymark, T.; Ryde, F.; Pe'er, A. Confronting GRB prompt emission with a model for subphotospheric dissipation. *Mon. Not. R. Astron. Soc.* **2015**, *454*, L31–L35. [CrossRef]
61. Ghirlanda, G.; Pescalli, A.; Ghisellini, G. Photospheric emission throughout GRB 100507 detected by Fermi. *Mon. Not. R. Astron. Soc.* **2013**, *432*, 3237–3244. [CrossRef]
62. Ito, H.; Nagataki, S.; Matsumoto, J.; Lee, S.H.; Tolstov, A.; Mao, J.; Dainotti, M.; Mizuta, A. Spectral and Polarization Properties of Photospheric Emission from Stratified Jets. *Astrophys. J.* **2014**, *789*, 159. [CrossRef]
63. Dall'Osso, S.; Stratta, G.; Guetta, D.; Covino, S.; De Cesare, G.; Stella, L. Gamma-ray bursts afterglows with energy injection from a spinning down neutron star. *Astron. Astrophys.* **2011**, *526*, A121. [CrossRef]
64. Rowlinson, A.; O'Brien, P.T.; Metzger, B.D.; Tanvir, N.R.; Levan, A.J. Signatures of magnetar central engines in short GRB light curves. *Mon. Not. R. Astron. Soc.* **2013**, *430*, 1061–1087. [CrossRef]
65. Fraija, N.; Veres, P.; Beniamini, P.; Galvan-Gamez, A.; Metzger, B.D.; Barniol Duran, R.; Becerra, R.L. On the Origin of the Multi-GeV Photons from the Closest Burst with Intermediate Luminosity: GRB 190829A. *Astrophys. J.* **2021**, *918*, 12. [CrossRef]
66. Rowlinson, A.; Gompertz, B.P.; Dainotti, M.; O'Brien, P.T.; Wijers, R.A.M.J.; van der Horst, A.J. Constraining properties of GRB magnetar central engines using the observed plateau luminosity and duration correlation. *Mon. Not. R. Astron. Soc.* **2014**, *443*, 1779–1787. [CrossRef]
67. Rea, N.; Gullón, M.; Pons, J.A.; Perna, R.; Dainotti, M.G.; Miralles, J.A.; Torres, D.F. Constraining the GRB-Magnetar Model by Means of the Galactic Pulsar Population. *Astrophys. J.* **2015**, *813*, 92. [CrossRef]
68. Stratta, G.; Dainotti, M.G.; Dall'Osso, S.; Hernandez, X.; De Cesare, G. On the Magnetar Origin of the GRBs Presenting X-ray Afterglow Plateaus. *Astrophys. J.* **2018**, *869*, 155. [CrossRef]
69. Dainotti, M.G.; Del Vecchio, R.; Shigehiro, N.; Capozziello, S. Selection Effects in Gamma-ray Burst Correlations: Consequences on the Ratio between Gamma-ray Burst and Star Formation Rates. *Astrophys. J.* **2015**, *800*, 31. [CrossRef]
70. Dainotti, M.G.; Hernandez, X.; Postnikov, S.; Nagataki, S.; O'brien, P.; Willingale, R.; Striegel, S. A Study of the Gamma-ray Burst Fundamental Plane. *Astrophys. J.* **2017**, *848*, 88. [CrossRef]
71. Dainotti, M.G.; Lenart, A.Ł.; Sarracino, G.; Nagataki, S.; Capozziello, S.; Fraija, N. The X-ray Fundamental Plane of the Platinum Sample, the Kilonovae, and the SNe Ib/c Associated with GRBs. *Astrophys. J.* **2020**, *904*, 97. [CrossRef]
72. Dainotti, M.G.; De Simone, B.; Schiavone, T.; Montani, G.; Rinaldi, E.; Lambiase, G. On the Hubble Constant Tension in the SNe Ia Pantheon Sample. *Astrophys. J.* **2021**, *912*, 150. [CrossRef]
73. Lloyd-Ronning, N.M.; Gompertz, B.; Pe'er, A.; Dainotti, M.; Fruchter, A. A Comparison between Radio Loud and Quiet Gamma-ray Bursts, and Evidence for a Potential Correlation between Intrinsic Duration and Redshift in the Radio Loud Population. *Astrophys. J.* **2019**, *871*, 118. [CrossRef]
74. Bryant, C.M.; Osborne, J.A.; Shahmoradi, A. How unbiased statistical methods lead to biased scientific discoveries: A case study of the Efron-Petrosian statistic applied to the luminosity-redshift evolution of gamma-ray bursts. *Mon. Not. R. Astron. Soc.* **2021**, *504*, 4192–4203. [CrossRef]
75. Piran, T. Gamma-ray bursts and the fireball model. *Phys. Rep.* **1999**, *314*, 575–667. [CrossRef]
76. Sari, R.; Piran, T. GRB 990123: The Optical Flash and the Fireball Model. *Astrophys. J.* **1999**, *517*, L109–L112. [CrossRef]
77. Dainotti, M.; Petrosian, V.; Willingale, R.; O'Brien, P.; Ostrowski, M.; Nagataki, S. Luminosity-time and luminosity-luminosity correlations for GRB prompt and afterglow plateau emissions. *Mon. Not. R. Astron. Soc.* **2015**, *451*, 3898–3908. [CrossRef]
78. Petrosian, V.; Kitanidis, E.; Kocevski, D. Cosmological Evolution of Long Gamma-ray Bursts and the Star Formation Rate. *Astrophys. J.* **2015**, *806*, 44. [CrossRef]
79. Lloyd-Ronning, N.M.; Petrosian, V. Interpreting the Behavior of Time-resolved Gamma-ray Burst Spectra. *Astrophys. J.* **2002**, *565*, 182–194. [CrossRef]
80. Chandra, P.; Frail, D.A. A Radio-selected Sample of Gamma-ray Burst Afterglows. *Astrophys. J.* **2012**, *746*, 156. [CrossRef]
81. Zauderer, B.A.; Berger, E.; Margutti, R.; Levan, A.J.; Olivares, E.F.; Perley, D.A.; Fong, W.; Horesh, A.; Updike, A.C.; Greiner, J.; et al. Illuminating the Darkest Gamma-ray Bursts with Radio Observations. *Astrophys. J.* **2013**, *767*, 161. [CrossRef]
82. Laskar, T.; Berger, E.; Margutti, R.; Perley, D.; Zauderer, B.A.; Sari, R.; Fong, W.F. Energy Injection in Gamma-ray Burst Afterglows. *Astrophys. J.* **2015**, *814*, 1. [CrossRef]
83. Laskar, T.; van Eerten, H.; Schady, P.; Mundell, C.G.; Alexander, K.D.; Barniol Duran, R.; Berger, E.; Bolmer, J.; Chornock, R.; Coppejans, D.L.; et al. A Reverse Shock in GRB 181201A. *Astrophys. J.* **2019**, *884*, 121. [CrossRef]
84. Cucchiara, A.; Veres, P.; Corsi, A.; Cenko, S.B.; Perley, D.A.; Lien, A.; Marshall, F.E.; Pagani, C.; Toy, V.L.; Capone, J.I.; et al. Happy Birthday Swift: Ultra-long GRB 141121A and Its Broadband Afterglow. *Astrophys. J.* **2015**, *812*, 122. [CrossRef]
85. Bright, J.S.; Horesh, A.; van der Horst, A.J.; Fender, R.; Anderson, G.E.; Motta, S.E.; Cenko, S.B.; Green, D.A.; Perrott, Y.; Titterington, D. A detailed radio study of the energetic, nearby, and puzzling GRB 171010A. *Mon. Not. R. Astron. Soc.* **2019**, *486*, 2721–2729. [CrossRef]
86. Anderson, G.E.; van der Horst, A.J.; Staley, T.D.; Fender, R.P.; Wijers, R.A.M.J.; Scaife, A.M.M.; Rumsey, C.; Titterington, D.J.; Rowlinson, A.; Saunders, R.D.E. Probing the bright radio flare and afterglow of GRB 130427A with the Arcminute Microkelvin Imager. *Mon. Not. R. Astron. Soc.* **2014**, *440*, 2059–2065. [CrossRef]

87. Singer, L.P.; Kasliwal, M.M.; Cenko, S.B.; Perley, D.A.; Anderson, G.E.; Anupama, G.C.; Arcavi, I.; Bhalerao, V.; Bue, B.D.; Cao, Y.; et al. The Needle in the 100 deg^2 Haystack: Uncovering Afterglows of Fermi GRBs with the Palomar Transient Factory. *Astrophys. J.* **2015**, *806*, 52. [CrossRef]
88. Laskar, T.; Berger, E.; Margutti, R.; Zauderer, B.A.; Williams, P.K.G.; Fong, W.F.; Sari, R.; Alexander, K.D.; Kamble, A. A VLA Study of High-redshift GRBs. II. The Complex Radio Afterglow of GRB 140304A: Shell Collisions and Two Reverse Shocks. *Astrophys. J.* **2018**, *859*, 134. [CrossRef]
89. Kangas, T.; Fruchter, A.S.; Cenko, S.B.; Corsi, A.; de Ugarte Postigo, A.; Pe'er, A.; Vogel, S.N.; Cucchiara, A.; Gompertz, B.; Graham, J.; et al. The Late-time Afterglow Evolution of Long Gamma-ray Bursts GRB 160625B and GRB 160509A. *Astrophys. J.* **2020**, *894*, 43. [CrossRef]
90. Laskar, T.; Alexander, K.D.; Berger, E.; Fong, W.F.; Margutti, R.; Shivvers, I.; Williams, P.K.G.; Kopač, D.; Kobayashi, S.; Mundell, C.; et al. A Reverse Shock in GRB 160509A. *Astrophys. J.* **2016**, *833*, 88. [CrossRef]
91. Chen, W.J.; Urata, Y.; Huang, K.; Takahashi, S.; Petitpas, G.; Asada, K. Two-component Jets of GRB 160623A as Shocked Jet Cocoon Afterglow. *Astrophys. J.* **2020**, *891*, L15. [CrossRef]
92. Bhalerao, V.; Kasliwal, M.M.; Bhattacharya, D.; Corsi, A.; Aarthy, E.; Adams, S.M.; Blagorodnova, N.; Cantwell, T.; Cenko, S.B.; Fender, R.; et al. A Tale of Two Transients: GW 170104 and GRB 170105A. *Astrophys. J.* **2017**, *845*, 152. [CrossRef]
93. Hallinan, G.; Corsi, A.; Mooley, K.P.; Hotokezaka, K.; Nakar, E.; Kasliwal, M.M.; Kaplan, D.L.; Frail, D.A.; Myers, S.T.; Murphy, T.; et al. A radio counterpart to a neutron star merger. *Science* **2017**, *358*, 1579–1583. [CrossRef] [PubMed]
94. Mooley, K.P.; Nakar, E.; Hotokezaka, K.; Hallinan, G.; Corsi, A.; Frail, D.A.; Horesh, A.; Murphy, T.; Lenc, E.; Kaplan, D.L.; et al. A mildly relativistic wide-angle outflow in the neutron-star merger event GW170817. *Nature* **2018**, *554*, 207–210. [CrossRef] [PubMed]
95. Resmi, L.; Schulze, S.; Ishwara-Chandra, C.H.; Misra, K.; Buchner, J.; De Pasquale, M.; Sánchez-Ramírez, R.; Klose, S.; Kim, S.; Tanvir, N.R.; et al. Low-frequency View of GW170817/GRB 170817A with the Giant Metrewave Radio Telescope. *Astrophys. J.* **2018**, *867*, 57. [CrossRef]
96. Margutti, R.; Alexander, K.D.; Xie, X.; Sironi, L.; Metzger, B.D.; Kathirgamaraju, A.; Fong, W.; Blanchard, P.K.; Berger, E.; MacFadyen, A.; et al. The Binary Neutron Star Event LIGO/Virgo GW170817 160 Days after Merger: Synchrotron Emission across the Electromagnetic Spectrum. *Astrophys. J.* **2018**, *856*, L18. [CrossRef]
97. Maity, B.; Chandra, P. 1000 Days of the Lowest-frequency Emission from the Low-luminosity GRB 171205A. *Astrophys. J.* **2021**, *907*, 60. [CrossRef]
98. Rhodes, L.; van der Horst, A.J.; Fender, R.; Monageng, I.M.; Anderson, G.E.; Antoniadis, J.; Bietenholz, M.F.; Böttcher, M.; Bright, J.S.; Green, D.A.; et al. Radio afterglows of very high-energy gamma-ray bursts 190829A and 180720B. *Mon. Not. R. Astron. Soc.* **2020**, *496*, 3326–3335. [CrossRef]
99. Bloom, J.S.; Frail, D.A.; Sari, R. The Prompt Energy Release of Gamma-ray Bursts using a Cosmological [ITAL]k[/ITAL]-Correction. *Astron. J.* **2001**, *121*, 2879–2888. [CrossRef]
100. Avni, Y. Parameter estimation for X-ray sources: Positions. *Astron. Astrophys.* **1978**, *66*, 307–309.

galaxies 2022, 10, 24

Article

On the Evolution of the Hubble Constant with the SNe Ia Pantheon Sample and Baryon Acoustic Oscillations: A Feasibility Study for GRB-Cosmology in 2030

Maria Giovanna Dainotti [1,2,3,*], **Biagio De Simone** [4,5], **Tiziano Schiavone** [6,7], **Giovanni Montani** [8,9], **Enrico Rinaldi** [10,11,12], **Gaetano Lambiase** [4,5], **Malgorzata Bogdan** [13,14] and **Sahil Ugale** [15]

1 National Astronomical Observatory of Japan, 2 Chome-21-1 Osawa, Mitaka, Tokyo 181-8588, Japan
2 School of Physical Sciences, The Graduate University for Advanced Studies, Shonankokusaimura, Hayama, Miura District, Kanagawa 240-0193, Japan
3 Space Science Institute, Boulder, CO 80301, USA
4 Department of Physics "E.R. Caianiello", University of Salerno, Via Giovanni Paolo II, 132, Fisciano, I-84084 Salerno, Italy; bdesimone@unisa.it (B.D.S.); glambiase@unisa.it (G.L.)
5 INFN Gruppo Collegato di Salerno—Sezione di Napoli—c/o Dipartimento di Fisica "E.R. Caianiello", Ed. F, Università di Salerno—Via Giovanni Paolo II, 132, Fisciano, I-84084 Salerno, Italy
6 Department of Physics "E. Fermi", University of Pisa, Polo Fibonacci, Largo B. Pontecorvo 3, I-56127 Pisa, Italy; tiziano.schiavone@phd.unipi.it
7 INFN, Istituto Nazionale di Fisica Nucleare, Sezione di Pisa, Polo Fibonacci, Largo B. Pontecorvo 3, I-56127 Pisa, Italy
8 ENEA, Fusion and Nuclear Safety Department, C.R. Frascati, Via E. Fermi 45, Frascati, I-00044 Rome, Italy; giovanni.montani@enea.it
9 Physics Department, "Sapienza" University of Rome, P.le Aldo Moro 5, I-00185 Rome, Italy
10 Physics Department, University of Michigan, Ann Arbor, MI 48109, USA; erinaldi.work@gmail.com
11 Theoretical Quantum Physics Laboratory, Center for Pioneering Research, RIKEN, 2-1 Hirosawa, Wako, Saitama 351-0198, Japan
12 Interdisciplinary Theoretical & Mathematical Science Program, RIKEN (iTHEMS), 2-1 Hirosawa, Wako, Saitama 351-0198, Japan
13 Department of Mathematics, University of Wroclaw, plac Uniwersytecki 1, 50-137 Wrocław, Poland; malgorzata.bogdan@uwr.edu.pl
14 Department of Statistics, Lund University, P.O. Box 117, SE-221 00 Lund, Sweden
15 Department of Physics, Mithibai College, Mumbai 400056, India; sahil.ugale@svkmmumbai.onmicrosoft.com
* Correspondence: maria.dainotti@nao.ac.jp

Citation: Dainotti, M.G.; De Simone, B.; Schiavone, T.; Montani, G.; Rinaldi, E.; Lambiase, G.; Bogdan, M.; Ugale, S. On the Evolution of the Hubble Constant with the SNe Ia Pantheon Sample and Baryon Acoustic Oscillations: A Feasibility Study for GRB-Cosmology in 2030. *Galaxies* **2022**, *10*, 24. https://doi.org/10.3390/galaxies10010024

Academic Editor: Elena Moretti and Francesco Longo

Received: 5 November 2021
Accepted: 25 January 2022
Published: 29 January 2022

Abstract: The difference from 4 to 6 σ in the Hubble constant (H_0) between the values observed with the local (Cepheids and Supernovae Ia, SNe Ia) and the high-z probes (Cosmic Microwave Background obtained by the Planck data) still challenges the astrophysics and cosmology community. Previous analysis has shown that there is an evolution in the Hubble constant that scales as $f(z) = \mathcal{H}_0/(1+z)^\eta$, where $\mathcal{H}_0$ is $H_0(z=0)$ and η is the evolutionary parameter. Here, we investigate if this evolution still holds by using the SNe Ia gathered in the Pantheon sample and the Baryon Acoustic Oscillations. We assume $H_0 = 70\ \mathrm{km\,s^{-1}\,Mpc^{-1}}$ as the local value and divide the Pantheon into three bins ordered in increasing values of redshift. Similar to our previous analysis but varying two cosmological parameters contemporaneously (H_0, Ω_{0m} in the ΛCDM model and H_0, w_a in the w_0w_aCDM model), for each bin we implement a Markov-Chain Monte Carlo analysis (MCMC) obtaining the value of H_0 assuming Gaussian priors to restrict the parameters spaces to values we expect from our prior knowledge of the current cosmological models and to avoid phantom Dark Energy models with $w < -1$. Subsequently, the values of H_0 are fitted with the model $f(z)$. Our results show that a decreasing trend with $\eta \sim 10^{-2}$ is still visible in this sample. The η coefficient reaches zero in 2.0 σ for the ΛCDM model up to 5.8 σ for w_0w_aCDM model. This trend, if not due to statistical fluctuations, could be explained through a hidden astrophysical bias, such as the effect of stretch evolution, or it requires new theoretical models, a possible proposition is the modified gravity theories, $f(R)$. This analysis is meant to further cast light on the evolution of H_0 and it does not specifically focus on constraining the other parameters. This work is also a preparatory to understand how the combined probes still show an evolution of the H_0 by redshift and what is the current status of simulations on GRB cosmology to obtain the uncertainties on the Ω_{0m} comparable with the ones achieved through SNe Ia.

Keywords: supernovae; Ia; cosmology; Hubble; tension; ΛCDM; evolution; modified; gravity; theories

---

1. Introduction

The ΛCDM model is one of the most accredited models, which implies an accelerated expansion phase [1,2]. Although it represents the favored paradigm, it is affected by great challenges: the fine-tuning, the coincidence [3,4], and the Dark Energy nature's problems.

More importantly, the H_0 tension represents a big challenge for modern cosmology. Indeed, the 4.4 up to 6.2 σ discrepancy, depending on the sample used [5–7], between the local value of H_0 obtained with Cepheids observations and SNe Ia, $H_0 = 74.03 \pm 1.42\,\mathrm{km\,s^{-1}\,Mpc^{-1}}$ [8,9], and the Planck data of Cosmic Microwave background radiation (CMB), $H_0 = 67.4 \pm 0.5\,\mathrm{km\,s^{-1}\,Mpc^{-1}}$ from the Planck Collaboration [10] requires further investigation. From now on, H_0 will be in the units $\mathrm{km\,s^{-1}\,Mpc^{-1}}$.

We stress that other probes report values of $H_0 \approx 72 \pm 2$, similar to the value obtained with the SNe Ia. Surely, to solve the Hubble tension it is necessary to use probes that are standard candles. SNe Ia, considered one of the best standard candles, are observed only up to a low redshift range: the farthest so far discovered is at $z = 2.26$ [11].

It is important for studying the evolution of the cosmological parameters to investigate probes at high redshift. One of the best candidates in this regard is represented by the Gamma-ray Bursts (GRBs).

GRBs are observed up to cosmological redshifts (the actual record is of $z = 9.4$ [12]) and surpassed even the quasars (the most distant being at $z = 7.64$ [13]). Due to their detectability at high redshift, GRBs allow extending the current Hubble diagram to new redshift ranges [14–18].

Indeed, it is important to stress that once we have established if the Hubble constant undergoes redshift evolution, the Pantheon sample can safely be combined with other probes. Surely the advantage of the use of the SNe Ia is that their emission mechanism is pretty clear, namely they originate from the thermonuclear explosion of carbon–oxygen white dwarfs (C/O WDs).

For GRBs, more investigation about their progenitor mechanism is needed. We here stress that this work can be also preparatory to the work of future application of GRBs as cosmological tools together with SNe Ia and Baryon Acoustic Oscillations (BAOs) through well-established correlations among the prompt variables, such as: the Amati relation [19], which connects the peak in the νF_ν spectrum to the isotropic energy in the prompt emission (E_{iso}), the Yonetoku relation [20,21] between E_{peak} and the peak luminosity of the prompt emission, L_{peak}, the Liang and Zhang relation [22] between E_{iso}, the rest-frame break time of the GRB t'_b and the peak energy spectrum in the rest frame E'_p, the Ghirlanda relation ($E_{peak} - E_{jet} = E_{iso} \times (1 - cos\theta)$) [23], and the prompt-afterglow relations for the GRBs with the plateau emission investigated in [24–38], which have as common emission mechanism most likely the magnetar model, where a neutron star with an intense magnetic field undergoes a fast-spinning down [39–43].

A feasibility study shows that GRBs can give relevant constraints on the cosmological parameters [17,44]. We here give a list of examples of other probes used for measuring the Hubble constant tension. One of them is the use of data from time-delay measurements and strong lens systems [45,46]. On the contrary, additional probes carry similar values of H_0 to the ones of Planck, based on the Cosmic Chronometers (CC) ($H_0 = 67.06 \pm 1.68$) in [8]. Besides, there is a series of independent probes, such as quasars [47], the Tip of the Red-Giant Branch (TRGB) calibration through SNe Ia [48], and also GRBs [14,15,17,18,49,50], which bring estimates of H_0 ranging between the values obtained with local measurements (SNe Ia and Cepheids) and Baryon Acoustic Oscillations (BAO)+CMB. Ref. [51] discuss possible reasons behind the H_0 tension in the Pantheon sample: selection biases of parameters of the SNe Ia, unknown systematics, internal inconsistencies in the Planck data, or alternative theoretical interpretations compared to the standard cosmological model. Furthermore, the use of type 1 Active Galactic Nuclei (AGN) represents another promising cosmological probe given the peculiarity of their spectral emission [52].

To date, a wide range of different solutions to the Hubble constant tension has been provided by several groups [53–86]. Concerning the observational solutions, we here detail a series of proposals [87–130]. In [131], the simulations of data taken from the anomalously fast-colliding El Gordo galaxy clusters allow discussing the probability of observing such a scenario in a ΛCDM framework. Ref. [132] perform a re-calibration of Cepheids in NGC 5584, thus obtaining a relation between the periods of Cepheids and their amplitude ratios (tighter than the one obtained in SH0ES [9]) which could be useful to better estimate the value of H_0. In [133], the UV and X-ray data coming from quasars are used to constrain H_0 in the Finslerian cosmology. Ref. [123] demonstrate that the Planetary Nebula Luminosity Function (PNLF) can be extended beyond the Cepheid distances, thus promoting it to be an additive probe for constraining H_0. In [134], the analysis of Pan-STARRS telescope SNe Ia data provides a value of H_0 which lies between the SH0ES and Planck values.

Ref. [135] investigated how the H_0 measurements can depend on the choice of different probes (SNe, BAO, Cepheid, CC, etc.), showing also that through the set of filters on cosmological models, such as fiducial values for cosmological parameters ($w = -1$, with w parameter for the equation of state, or $\Omega_k = 0$, namely the curvature parameter set to zero), the tension can be alleviated.

Ref. [136] extended a Hubble diagram up to redshift $z \sim 8$ combining galaxies and high-redshift quasars to test the late-time cosmic expansion history, giving a constraint on the upper-value of H_0 which is only marginally consistent with the results obtained by the Cepheids.

Ref. [137] further tests the wCDM (with varying parameters of the equation of state), and oCDM models (with varying curvature) through the merging of BAOs, SNe Ia, CC, GRBs, and quasars data, after the analysis of the standard ΛCDM model.

In [138], the combination of strongly lensed quasars and SNe Ia led the authors to conclude that the solution to the tension should be found outside of the Friedmann–Lemaitre–Robertson–Walker metric.

Refs. [125,139] detect in the Stochastic Gravitational Wave Background a new method to alleviate the tension, while [140,141] focuses on the gravitational-wave signals from compact star mergers as probes that can give constraints on the H_0 value.

Ref. [142] combine the SNe Ia and the VLT-KMOS HII galaxies data to put new constraints on the cosmokinetic parameters. The proposed solutions deal also with models that are alternative to the standard ΛCDM or, in other cases, that can extend it.

Ref. [143] constrain the Brans–Dicke (BD) theory through CMB and BAOs. The TRGB method, combined with SNe Ia, gives a value of H_0 compatible with the one from CMB [144].

Ref. [145] obtain an 8%-precise value of H_0 through the Fast Radio Bursts (FRB).

In [146], the Cepheids calibration parameters are allowed to vary, thus leading to an estimated value of H_0 which is compatible with the CMB one. The possibility that the Solar system's proper motion may induce a bias in the measurement of H_0 has been subject to study in [147], finding out that there is no degeneracy between the cosmological parameters and the parameters of the Solar system motion.

Ref. [148] measure H_0 through the galaxies parallax having as reference the CMB rest-frame, being this parallax caused by the peculiar motions.

Ref. [149] verified through the measurements on GRBs and quasars that the Hubble constant has a bigger value in the sky directions aligned with the CMB dipole polarization, suggesting that a detachment from the FLRW should be considered.

Refs. [150–153] investigate how the dark sirens producing gravitational waves could help to probe H_0. Despite being a promising method, the incompleteness of galaxy catalogs may hinder the outcome of this method, thus [154] proposes a pixelated-sky approach to overcome the issue of event redshifts which are missing but may be retrieved through the galaxies present on the line of sight.

A review of the most promising emerging probes to measure the Hubble constant can be found in [155].

Recent results on the measurements of the Hubble parameter and constant through the Third LIGO-Virgo-KAGRA Gravitational-Wave Transient Catalog (GWTC-3) can be found in [156]. An evolving trend for H_0 may be naturally predicted in Teleparallelism [157–161], as well as in modified gravity theories [162–167]. Refs. [168–170] study the $f(Q,T)$ models in Teleparallel Gravity through CC and SNe Ia, thus obtaining a value of H_0 compatible in 1 σ with the SH0ES result. The linear theory of perturbation for the $f(Q,T)$ theory is investigated in [171], allowing the future tests of this model through CMB data.

In [172], the Unimodular Gravity model is constrained with Planck 2018 [10], SH0ES, SNe Ia, and H0LiCOW collaboration [7]. Furthermore, the Axi–Higgs model is tested with CMB, BAO, Weak Lensing data (WL), and SNe [173]: in another paper, it is shown how this model relaxes the Hubble tension [173]. Ref. [174] describe the modified inflationary models considering constant-roll inflation. Ref. [175] give boundaries on the Hubble constant value with the gravitino mass conjecture.

Refs. [176–180] show the role of cosmological second-order perturbations of the flat ΛCDM model in the H_0 tension. Ref. [181] discuss how Dark Energy may be generated by quantum fluctuations of an inflating field and how the Hubble tension may be reduced by the spatial correlations induced by this effect. The Dark Energy itself may be subject to evolution, as pointed out in [182]. Ref. [183] show how a modification of the Friedmann equation may naturally explain the inconsistency between the local and the cosmological measurements of the Hubble constant.

Ref. [184] explain how the search for low-frequency gravitational waves (GWs) justifies the Hubble tension's solution through the assumption of neutrino-dark sector interactions. Ref. [185] show how the $R_K^{(*)}$ anomalies (namely, the discrepancy between the theoretical ratio of the fractions $B \to K^*\mu^+\mu^- / B \to K^*e^+e^-$ for the dilepton invariant mass bins from the Standard Model and the observed one, see [186]) and the H_0 tension can be solved by Dirac neutrinos in a two-Higgs-doublet theory.

The introduction of models where the cosmological axio-dilation is present may lead to a solution of the Hubble tension [187].

Refs. [188–191] discuss how the Early Dark Energy models (EDE) can be used to alleviate the H_0 tension. Ref. [192] analyze how the phantom Dark Energy models can give a limited reduction of the H_0 tension, while [193] explore how the Kaniadakis holographic Dark Energy model alleviates the H_0 tension.

In [194], the Viscous Generalized Chaplygin Gas (VGCG) model is used to diminish the Hubble tension. The holographic Dark Energy models are pointed as a possible solution through the study of unparticle cosmology [195].

Ref. [196] test seven cosmological models through the constraints of SNe Ia, BAO, CMB, Planck lensing, and Cosmic Chronometers with the outcome that in the ΛCDM scenario a flat universe is favored.

Ref. [197] discuss how the new physical scenarios before the recombination epoch imply the shift of cosmological parameters and how these shifts are related to the discrepancy between the local and non-local values of H_0.

Ref. [198] proposes that the H_0 tension may be solved if the speed of light is treated as a function of the scale factor (as in [199]), and applies this scenario to SNe Ia data.

Refs. [200,201] discuss the implementation of the alternative Phenomenologically Emergent Dark Energy model (PEDE), which can be also extended to a Generalized Emergent Dark Energy model (GEDE) with the addition of an extra free parameter. This shows the possibility of obtaining the PEDE or the ΛCDM cosmology as sub-cases of the GEDE scenario.

Ref. [202] consider a scenario of modified gravity predicting the increase of the expansion rate in the late-universe, thus proving that in this scenario the Hubble tension reduces significantly.

Ref. [203] study the ΛCDM model constrained, at the early-time universe, by the presence of the early Integrated Sachs–Wolfe (eISW) effect, proving that the early-time models aimed at attenuating the Hubble tension should be able to reproduce the same eISW effect just like the ΛCDM does. The observations of a locally higher value for H_0 led to the discussion of local measurements, constraints, and modeling. In this regards, the assumption of a local void [204–206] may produce locally an increased value for H_0.

The Universe appears locally inhomogeneous below a scale of roughly 100 Mpc. The question some cosmologists are attempting to solve is whether local inhomogeneities have impact on cosmological measurements and the Hubble diagram. Many observables are related to photons paths, which may be directly affected by the matter distribution. Many theoretical attempts were made during the last few decades to develop the necessary average prescription to evaluate the photon propagation on the observer's past light cone based on covariant and gauge-invariant observables [207–210]. Local inhomogeneities and cosmic structure cause scattering and bias effects in the Hubble diagram, which are due to peculiar velocities, selection effects, and gravitational lensing, but also to non-linear relativistic corrections [210–212]. This question was addressed in [213] utilising the N-body simulation of cosmic structure formation through the numerical code *gevolution*. This non-perturbative approach pointed out discrepancies in the luminosity distance between a homogeneous and inhomogeneous scenario, showing, in particular, the presence of non-Gaussian effects at higher redshifts. These studies related to distance indicators will become even more significant considering the large number of the forthcoming surveys designed to the observations on the Large Scale Structure of the Universe in the next decade (for instance, the Euclid survey [214,215] and the Vera C. Rubin Observatory's LSST [216]). The effect of local structures in an inhomogeneous universe should be considered in the locally measured value of H_0 [217,218]. The local under-density interpretation was also studied in Milgromian dynamics [131,219,220], but in [221] it is shown how this interpretation does not solve the tension. Ref. [219] study the KBC local void which is in contrast with the ΛCDM, thus proposing the Milgromian dynamics as an alternative to standard cosmology. Milgromian dynamics are studied also in [222] where, through the galactic structures and clusters, it is shown how this model can be consistent at different scales and alleviate the Hubble tension.

Ref. [223] describe the late time approaches and their effect on the Hubble parameter. The bulk viscosity of the universe is also considered the link between the early and late universe values of H_0 [224].

Ref. [225] explain how the local measurements over-constrain the cosmological models and propose the graphical analysis of the impact that these constraints have on the H_0 estimation through ad hoc triangular plots.

Refs. [220,226,227] describes the effects of inhomogeneities at small scales in the baryon density. Ref. [228] find out that the late time modifications can solve the tension between the H_0 SH0ES and CMB values through a parametrization of the comoving distance.

Ref. [229] propose to alleviate the Hubble tension considering an abrupt modification of the effective gravitational constant at redshift $z \approx 0.01$.
Other proposals are focused on the existence of different approaches.

Refs. [230,231] show how H_0 evolves with redshift at local scales.

Ref. [232] discuss how the breakdown of Friedmann–Lemaitre–Robertson–Walker (FLRW [233]) may be a plausible assumption to alleviate the Hubble tension. Ref. [234] investigate the binary neutron stars mergers and, with the analysis of simulated catalogs, show their potential to help to alleviate the H_0 tension.

Ref. [235] explain how Gaussian process (GP) and locally weighted scatter plot smoothing are used in conjunction with simulation and extrapolation (LOESS-Simex) methods can reproduce different sets of data with a high level of precision, thus giving new perspectives on the Hubble tension through the simulation of Cosmic Chronometers, SNe Ia, and BAOs data sets.

Ref. [236] focus on the GP and state the necessity of lower and upper bounds on the hyperparameters to obtain a reliable estimation of H_0.

On the other hand, Ref. [237] suggested a novel approach to measure H_0 based on the distance duality relation, namely a method that connects the luminosity distance of a source to its angular diameter. In this case, data do not require a calibration phase and the relative constraints are not dependent on the underlying cosmological model.

Ref. [238] showed how the tension can be solved with a modified weak-field General Relativity theory, thus defining a local H_0 and a global H_0 value.

Ref. [239] investigated how a specific Dark Energy model in the generalized Proca theory can alleviate the tension.

In [240], the Horndeski model can describe with significantly good precision the late expansion of the universe thanks to the Hubble parameter data. The same model is considered promising for the solution of the H_0 tension in [241].

Ref. [242] described how the transition observed in Tully–Fisher data could imply an evolving gravitational strength and explain the tension.

Ref. [197] explain how the physical models of the pre-recombination era could cause the observed H_0 values discrepancy and suggest that if the local H_0 measurements are consistent then a scale-invariant Harrison–Zeldovich spectrum should be considered to solve the H_0 issue. The Dynamical Dark Energy (DDE) models are the object of study in [243,244]: in the former, the DDE is proposed as an alternative to ΛCDM, while in the latter it is shown how the Chevallier–Polarski–Linder (CPL) parametrization [245,246] is insensitive to Dark Energy at low redshift scales.

Refs. [247,248] propose Dark Radiation as a new surrogate of the Standard Model.

In [249], the scalar field cosmological model is used, together with the parametrization of the equation of state, to obtain H_0 and investigate the nature of Dark Energy. The possibility of a scalar field non minimally coupled to gravity as a probable solution to the H_0 tension is investigated in [250].

Ref. [251] highlight the advantage of the braneworld models to predict the local higher values of H_0 and, contemporaneously, respect the CMB constraints. Another approach is to solve the H_0 tension by allowing variations in the fundamental constants [252].

Ref. [253] propose a non-singular Einstein–Cartan cosmological model with a simple parametrization of spacetime torsion to alleviate the tension, while [254] propose a model where the Dark Matter is annihilated to produce Dark Radiation.

Ref. [255] introduce a hidden sector of atomic Dark Matter in a realistic model that avoids the fine-tuning problem. The observed weak effect of primordial magnetic fields can create clustering at small scales for baryons and this could explain the H_0 tension [256].

Ref. [257] test the General Relativity at galactic scales through Strong Gravitational Lensing. The Strong Lensing is a promising probe for obtaining new constraints on H_0, thanks to the next generation DECIGO and B-DECIGO space interferometers [258].

In [259], the cosmological constant Λ is considered a dynamical quantity in the context of the running vacuum models and this assumption could tackle the H_0 tension. Ref. [260] show the singlet Majoron model to explain the acceleration of the expansion at later times and prove that this is consistent with large-scale data: this model has been subsequently discussed in other works [261]. The vacuum energy density value is affected by the Hubble tension as well and its measurement may cast more light on this topic [262].

Ref. [263] discuss the outcomes of the Oscillatory Tracker Model with an H_0 value that agrees with the CMB measurements. In [264], it is explained how the Generalized Uncertainty Principle and the Extended Uncertainty Principle can modify the Hubble parameter. [265] explore the implication of the Mirror Twin Higgs model and the need for future measurements to alleviate the tension.

The artificial neural networks can be applied to reconstruct the behavior of large scale structure cosmological parameters [266].

Another alternative is given by the gravitational transitions at low redshift which can solve the H_0 tension better than the late-time $H(z)$ smooth deformations [84,229].

Another comparison between the late-time gravitational transition models and other models which predict a smooth deformation of the Hubble parameter can be found in [267].

Ref. [268], as modifications to the ΛCDM model, consider as plausible scenarios or a Dark Matter component with negative pressure or the decay of Dark Energy into Dark Matter.

Ref. [269] does not observe the H_0 tension through the Effective Field Theory of Large Scale Structure and the Baryon Oscillation Spectroscopic Survey (BOSS) Correlation Function.

Considering the Dark Matter particles with two new charges, Ref. [270] reproduce a repulsive force which has similar effects to the Λ cosmological constant. Furthermore, the models where interaction between Dark Matter and Dark Energy is present are promising for a solution of the Hubble constant tensions, see [271].

In [272], it is shown how two independent sets of cosmological parameters, the background (geometrical) and the matter density (growth) component parameters, respectively, give consistent results and how the preference for high values of H_0 is less significant in their analysis.

Ref. [273] introduce a global parametrization based on the cosmic age which rules out the early-time and the late-time frameworks.

Ref. [274] point out, through the use of non-parametric methods, how the cosmological models may induce biases in the cosmological parameters. In the same way, the statistical analysis of galaxies' redshift value and distance estimations may be affected by biases which could, in turn, affect the estimation of H_0.

Ref. [275]. This consideration holds also for the quadruply lensed quasars which are another method to measure H_0 [276].

Ref. [277] use the machine learning techniques to measure time delays in lensed SNe Ia, these being an independent method to measure H_0.

Additionally, in [231] it is explained how an evolution of H_0 with the redshift is to be expected. If a statistical approach on the different H_0 values is used instead, together with the assumption of an alternative cosmology, another solution to the tension could be naturally implied [278].

Ref. [279] use data to reconstruct the $f(T)$ gravity function without assuming any cosmological model: this $f(T)$ could in turn represent a solution to the H_0 tension.

Ref. [124] discuss how the addition of scalar fields with particle physics motivation to the cosmological model which predicts Dark Matter can retrieve the observed abundances of the Big Bang Nucleosynthesis.

In [280], a Dark Matter production mechanism is proposed to alleviate the H_0 tension. A general review of the perspectives and proposals concerning the H_0 tension can be found in [281–283].

SNe Ia represents a very good example of standard candles. Here we consider also the contribution of geometrical probes, the so-called *standard rulers*: while standard candles show a constant intrinsic luminosity (or obey an intrinsic relation between their luminosity and other physical parameters independent of luminosity), standard rulers are characterized by a typical scale dimension. This property allows estimating their distance according to the apparent angular size. Among the possible standard rulers, the BAOs assume great importance for cosmological purposes.

We here investigate the H_0 tension in the Pantheon sample (hereafter PS) from [284] and we add the contribution of BAOs to the cosmological computations to check if the trend of H_0 found in [36] is present also with the addition of other probes. We here point out that the current analysis is not meant to constrain Ω_{0m} or any other cosmological parameters, but it is focused to study the reliability of the trend of H_0 as a function of the redshift.

We here point out that this analysis is not meant to constrain Ω_{0m} or any other cosmological parameters, but it is focused to study the reliability of the trend of H_0 as a function of the redshift. The range of redshift in the PS goes from $z = 0.01$ to $z = 2.26$. We tackle the problem with a redshift binning approach of H_0, the same used in [51], but here we adopt a starting value of $H_0 = 70$ instead of 73.5: if a trend with redshift exists, it should be independent on the initial value for H_0. The systematic contributions for the PS are calibrated through a reference cosmological model, where H_0 is 70.0 [284]. In the current paper, the aforementioned systematic uncertainties are considered for the analysis. Our approach has a two-fold advantage: on the one hand, it is relatively simple and on the other hand, it avoids the re-estimation of the SNe Ia uncertainties and may be able to highlight a residual dependence on the SNe Ia parameters with redshift.

While a slow varying Einstein constant with the redshift, as it emerges in a modified $f(R)$ gravity, appears as the most natural explanation for a trend $H_0(z)$, the analysis of Section 7 seems to indicate that such effect is not necessarily related with the Dark Energy contribution of the late universe. Since the Hu–Sawicki gravity lacks of reproducing the correct profile $H_0(z)$ shows that a Dark Energy model in the late Universe may not be enough to explain the observed effect since the scalar mode dynamics can not easily conciliate the Dark Energy contribution with the decreasing trend of $H_0(z)$. Thus, it may be necessary a modified gravity scenario more general than a Dark Energy model in the late Universe.

The current paper is composed as expressed in the following: in Section 2 the ΛCDM and $w_0 w_a$CDM models are briefly introduced together with SNe Ia properties; Section 3 describes the use of BAOs as cosmological rulers; Section 4 contains our binned analysis results, after slicing the PS in 3 redshift bins for the aforementioned models, and assuming locally $H_0 = 70$; in Section 5, we investigate, through simulated events, how the GRBs will be contributing to cosmological investigations by 2030; in Section 6 we discuss the results; in Section 7 we test the Hu–Sawicki model through a binning approach; in Section 8 we report an overview on the requirements that a suitable $f(R)$ model should have to properly describe the observed trend of H_0 and in Section 9 our conclusions are reported.

2. SNe Ia Cosmology

SNe Ia are characterized by an intrinsic luminosity that is almost uniform. Because of this, SNe Ia are considered reliable *standard candles*. We compare the theoretical distance moduli μ_{th} with the observed distance moduli μ_{obs} of SNe Ia belonging to the PS. The theoretical distance moduli are defined through the luminosity distance $d_L(z)$ which we need to define based on the cosmological model of interest. We here show the CPL parametrization which describes the w parameter as a function of redshift ($w(z) = w_0 + w_a \times z/(1+z)$) in the $w_0 w_a$CDM model. In the usual assumptions $w_0 \sim -1$ and $w_a \sim 0$, and $d_L(z)$ is defined as the following [285]:

$$d_L(z, H_0 ...) = \frac{c(1+z)}{H_0} \int_0^z \frac{dz^*}{\sqrt{\Omega_{0m}\,(1+z^*)^3 + \Omega_{0\Lambda}\,(1+z^*)^{3\,(w_0+w_a+1)}\,e^{-3 w_a \frac{z^*}{1+z^*}}}}, \quad (1)$$

where $\Omega_{0\Lambda}$ is the Dark Energy component, c is the speed of light, and z is the redshift. We stress that in this context the relativistic components are ignored. Moreover, since in the present universe the radiation density parameter $\Omega_{0r} \approx 10^{-5}$, this contribution can be neglected. If we substitute $w_a = 0$, $w_0 = -1$ in Equation (1) the luminosity distance expression for ΛCDM model is automatically retrieved. According to the distance luminosity expression, the theoretical distance modulus can be written in the following form:

$$\mu_{th} = 5\, log_{10}\, d_L(z, H_0, \ldots) + 25, \quad (2)$$

which is usually expressed in Megaparsec (Mpc). The observed distance modulus, $\mu_{obs} = m'_B - M$, taken from PS contains the apparent magnitude in the B-band corrected for statistical and systematic effects (m'_B) and the absolute in the B-band for a fiducial SN Ia with a null value of stretch and color corrections (M). Considering the color and stretch population models for SNe Ia, in our approach we average the distance moduli given by the [286] (G2010) and [287] (C2011) models. We here remind the reader that H_0 and M are degenerate parameters: in the PS release, $M = -19.35$ such that $H_0 = 70.0$.

Ref. [51] obtain information on H_0 by comparing μ_{obs} in [284][1] with μ_{th} for each SN. Moreover, they fix Ω_{0m} to a fiducial value to better constrain the H_0 parameter. Furthermore, according to [288], we consider the correction of the luminosity distance keeping into account the peculiar velocities of the host galaxies which contain the SNe Ia. To perform our analysis, we define the χ^2 for SNe:

$$\chi^2_{SN} = \Delta\mu^T \cdot \mathcal{C}^{-1} \cdot \Delta\mu. \quad (3)$$

Here $\Delta\mu = \mu_{obs} - \mu_{th}$, and $\mathcal{C}$ denotes the 1048 × 1048 *covariance matrix*, given by [284]. As for the μ_{obs} values of G2010 and C2011, the systematic uncertainty matrices of the two models have been averaged. After building the $\mathcal{C}$ total matrix from Equation (16) in [51], we slice the PS in redshift bins, and then we divide $\mathcal{C}$ into submatrices considering the order in redshift. More in detail, starting from the 1048 SNe Ia redshift-ordering, we divide the SNe Ia into 3 equally populated bins made up of ≈349 SNe Ia. Concerning only D_{stat}, it is trivial to build its submatrices considering that the statistical matrix is diagonal. Hence, a single matrix element is related to a given SN of the PS. On the other hand, if the non-diagonal matrix C_{sys} is included, a customized code will be used[2] to build the submatrices. Our code was developed to select only the total covariance matrix elements related to SNe Ia having redshift within the considered bin.

The choice of three bins is justified by the high number of SNe Ia (around hundreds of SNe per bin) that can still constitute statistically illustrative subsamples of the PS and that can properly consider the contribution of systematic uncertainties. Subsequently to the bins division, we focus on the optimal values of H_0 to minimize the χ^2 in Equation (3). H_0 is regarded as a nuisance parameter, which is free to vary, to better analyze a possible redshift function of H_0. We follow the assumptions on the fiducial value of $M = -19.35$: while in [51] M was estimated assuming a local ($z = 0$) value of $H_0 = 73.5$, we here consider the conventional H_0 value of the PS release, namely $H_0 = 70.0$ for three bins. Our choice of a starting value of $H_0 = 70$ is dictated by the presence in the current literature of more than 50 papers that are using the PS in combination with other probes to estimate the value of H_0, see [172,198,204,205,288–307,307–339]. Thus, if an evolutionary effect is present, it is necessary to investigate to which extent this can affect current and future results largely based on the PS sample. Conversely, we fix $\Omega_{0m} = 0.298 \pm 0.022$ according to [284] for a standard flat ΛCDM model. More specifically, after the minimization of χ^2, we extract the H_0 value in each redshift bin, via the *Cobaya* code [340]. To this end, we execute an MCMC using the D'Agostini method to obtain the confidence intervals for H_0 at the 68% and 95% levels, in three bins.

3. The Contribution of BAOs

The environment of relativistic plasma in the early universe was crossed by the sound waves that were generated by cosmological perturbations. At redshift $z_d \sim 1059.3$, which marks the ending of the drag period [341], the recombination of electrons and protons into a neutral gas interrupted the propagation of the sound waves while the photons were able to propagate further [342]. In the period between the formation of the perturbations and the recombination, the different modes produced a sequence of peaks and minima in the anisotropy power spectrum. Given the huge fraction of baryons in the universe, it is expected by cosmological models that the oscillations may affect also the distribution of baryons in the late universe. As a consequence, the BAOs manifest as a local maximum in the correlation function of the galaxies distribution in correspondence of the comoving sound horizon scale at the given redshift z_d, namely $r_s(z_d)$: this is associated with the stopping of the propagation of the acoustic waves.

To use the BAOs data for cosmology, we first need to define the following variables:

$$D_V(z) = \left[\frac{czd_L^2(z)}{(1+z)^2 H(z)}\right]^{1/3}, \qquad d_z(z) = \frac{r_s(z_d)}{D_V(z)}. \tag{4}$$

The value of the redshift z_d, which corresponds to the drag era ending and marks the decoupling of the photons, allows estimating the sound horizon scale:

$$(r_d \cdot h)_{fid} = 104.57\,\text{Mpc}, \qquad r_s(z_d) = \frac{(r_d \cdot h)_{fid}}{h}, \tag{5}$$

where we use the adimensional ratio $h = H_0/100(\text{km s}^{-1}\ \text{Mpc}^{-1})$. To estimate r_s, the following approximated formula [343] can be applied:

$$r_s \approx \frac{55.154 \cdot e^{-72.3(\omega_\nu + 0.0006)^2}}{\omega_{0m}^{0.25351}\omega_b^{0.12807}}\,\text{Mpc}, \tag{6}$$

where $\omega_i = \Omega_i \cdot h^2$, and $i = m, \nu, b$ represent matter, neutrino and baryons. We here assume $\omega_\nu = 0.00064$ [344] and $\omega_b = 0.02237$ [10]. Given these quantities, we define the χ^2 for BAOs as follows:

$$\chi^2_{BAO} = \Delta d^T \cdot \mathcal{M}^{-1} \cdot \Delta d, \tag{7}$$

where $\Delta d = d_z^{obs}(z_i) - d_z^{theo}(z_i)$ and $\mathcal{M}$ is the covariance matrix for the BAO $d_z^{obs}(z_i)$ values. In this binned analysis, a subset of the 26 BAO observations set available in [341] will be employed.

4. Multidimensional Binned Analysis with SNe Ia and BAOs

To investigate the H_0 tension through the SNe Ia and BAOs data, we combine the χ^2 Equations (3) and (7) to obtain the total χ^2

$$\chi^2 = \frac{1}{2}\chi^2_{SN} + \frac{1}{2}\chi^2_{BAO}, \tag{8}$$

In our work, we combine each SNe bin with only 1 BAO data point which has a redshift value within the SNe bin: this approach of using one BAO comes from [18]. In this way we do not have the problem of a different number of BAOs in different bins. Through Equation (8), we investigate if a redshift evolution of $H_0(z)$ is present, obtaining it from the binning of SNe Ia+BAOs considering three bins with the ΛCDM and w_0w_aCDM models. A feasibility study done in [51] performed with different bins selections has highlighted how the maximum number of bins in which the PS should be divided is 3, otherwise the statistical fluctuations would dominate on a multi-dimensional analysis, leading to relatively large uncertainties which would mask any evolving trend, if present. Furthermore, for the same reason, it is not advisable to leave free to vary more than two parameters at the same time, thus in the current section, we will analyze the behavior of H_0 in three bins when it is varied together with a second cosmological parameter. The same considerations make necessary the choice of more tight priors since we are basing the current analysis on the prior knowledge, avoiding the degeneracies among the parameter space, and letting the priors have more weight in the process of posteriors estimation. Differently from [51], for the ΛCDM model, we will let the parameters H_0 and Ω_{0m} vary simultaneously, while in the w_0w_aCDM model the varying parameters are H_0 and w_a. We decided to leave w_a free to vary since, according to the CPL parametrization, w_a gives direct information about the evolution of the $w(z)$ while w_0 is considered a constant in the same model. Concerning the fiducial values and the priors assignment for the MCMC computations, we apply Gaussian priors with mean equal to the central values of $\Omega_{0m} = 0.298 \pm 0.022$ and $H_0 = 70.393 \pm 1.079$ for Ω_{0m} and H_0, respectively, and with $1\ \sigma = 2 * 0.022$ and $1\ \sigma = 2 * 1.079$ for Ω_{0m} and H_0, respectively. In summary, to draw the Gaussian priors, we consider the mean value of the parameters as the expected one of the Gaussian distribution and we double the σ value which is then considered the new standard deviation for the distribution. Concerning the $w_0 w_a$CDM model, we fix $w_0 = -0.905$ and we consider the priors on w_a with the mean = -0.129 taken from Table 13 of [284], while 1 σ = is the 20% of its central value. Such an assumption with small prior is needed since we need to assume that $w(z) > -1.168$ as the value tabulated in [284]. Besides, since we are here dealing with standard cosmologies, with this constraint we are avoiding some of the phantom Dark Energy models.

After the χ^2 minimization for each bin, we perform a MCMC simulation to draw the mean value of H_0 and its uncertainty. Once H_0 is obtained for each bin, we perform a fit of H_0 using a simple function largely employed to characterize the evolution of many astrophysical objects, such as GRBs and quasars [17,29,31,35,345–349]. More specifically, the fitting of H_0 is given by

$$f(z) = H_0(z) = \frac{\mathcal{H}_0}{(1+z)^{\eta}}, \tag{9}$$

in which $\mathcal{H}_0$ and η are the fitting parameters. The former $\mathcal{H}_0 \equiv H_0$ at $z = 0$, while the latter η coefficient describes a possible evolutionary trend of H_0. We consider the 68% confidence interval at, namely 1 σ uncertainty.

In the current treatment, we consider the calibration of the PS with $H_0 = 70$ as provided by [284]. Results are presented in the panels of Table 1. We here stress that the fiducial magnitude value is assumed to be $M = -19.35$ for each SNe bin, thus it will not be mentioned in the same Table. All the uncertainties in the tables in this paper are in 1 σ. As reported in the upper half of Table 1, namely with the ΛCDM model, if we do not include the BAOs then the η coefficient is compatible with 0 in 2.0 σ for the three bins case. When we introduce the BAOs within the ΛCDM model, we observe again a reduction of the η/σ_η ratio for three bins down to 1.2. Concerning the lower half of Table 1 with the $w_0 w_a$CDM model, when BAOs are not included we have η non compatible with 0 in 5.7 σ and, including the BAOs, the compatibility with 0 is given in 5.8 σ. The increasing of the ratio η/σ_η is observed when BAOs are added in the case of $w_0 w_a$CDM model in three bins. The results can be visualized in Figure 1. Comparing the η/σ_η ratios with the ones reported in [51] (Table 1) we have that for the ΛCDM model the current η values are compatible in 1 σ with the α reported in [51], while the η estimated in the $w_0 w_a$CDM model are compatible in 3 σ with the α values in the same reference paper.

Table 1. Upper half. Fit parameters of $H_0(z)$ for three bins (flat ΛCDM model, varying H_0 and Ω_{0m}) in the cases with SNe only and with the SNe + BAOs contribution. The columns are: (1) the number of bins; (2) $\mathcal{H}_0$, (3) η; (4) how many σs the evolutionary parameter η is compatible with zero (namely, η/σ_η). **Lower half.** Similarly to the upper half, the lower half shows the fit parameters of $H_0(z)$ (flat $w_0 w_a$CDM model, varying H_0 and w_a) without and with the BAOs.

Flat ΛCDM model, without BAOs, varying H_0 and Ω_{0m}			
Bins	$\mathcal{H}_0$	η	$\frac{\eta}{\sigma_\eta}$
3	70.093 ± 0.102	0.009 ± 0.004	2.0
Flat ΛCDM model, including BAOs, varying H_0 and Ω_{0m}			
Bins	$\mathcal{H}_0$	η	$\frac{\eta}{\sigma_\eta}$
3	70.084 ± 0.148	0.008 ± 0.006	1.2

Table 1. *Cont.*

Flat w_0w_aCDM model, without BAOs, varying H_0 and w_a			
Bins	$\mathcal{H}_0$	η	$\frac{\eta}{\sigma_\eta}$
3	69.847 ± 0.119	0.034 ± 0.006	5.7
Flat w_0w_aCDM model, including BAOs, varying H_0 and w_a			
Bins	$\mathcal{H}_0$	η	$\frac{\eta}{\sigma_\eta}$
3	69.821 ± 0.126	0.033 ± 0.005	5.8

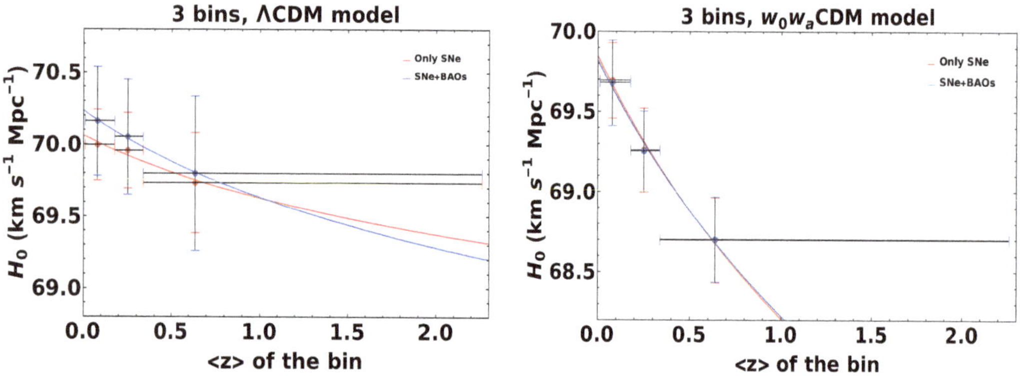

Figure 1. Left panel The $H_0(z)$ vs. z wit varying also $\Omega_0 M$. The red color indicates the case with only SNe Ia as probes, while the blue refers to the case of SNe + 1 BAO per bin. This color-coded will be applied also in the right panel. **Right.** The same plot for the w_0w_aCDM model, considering the local fiducial value $H_0 = 70$, where both H_0 and w_a are left free to vary.

5. Perspective of the Future Contribution of GRB-Cosmology in 2030

The discussion of GRBs as possible cosmological tools has been going on for more than two decades [350,351]. The best bet is yet to come since we need first to identify the tightest correlation possible with a solid physical grounding. Among the many correlations proposed [19–23] we here choose to apply the *fundamental plane* (or Dainotti relation) [30,352–354], namely the three-dimensional relation between the end of the plateau emission's luminosity, L_a, its time in the rest-frame, T^*_a, and the peak luminosity of the GRB, L_{peak}: it is possible to estimate how many GRBs are needed to obtain constraints for the cosmological parameters that are comparable with the ones obtained from the other probes, such as SNe Ia and BAOs. After a selection of the best fundamental plane sample through the trimming of GRBs, a simulation of a sample of 1500 and 2000 GRBs according to the properties of the fundamental plane relation has been performed. The fundamental plane relation can be expressed as the following:

$$log_{10} L_a = a \times log_{10} T^*_a + b \times log_{10} L_{peak} + c, \quad (10)$$

where a, b are the parameters of the plane and c is the normalization constant. It is important to stress that here the variables L_a, T^*_a, and L_{peak} have been corrected for evolutionary effects with redshift applying the Efron and Petrosian method [355]. Based on Equation (10), we perform the maximization of the following log-likelihood for the simulated sample of GRBs:

$$
\begin{aligned}
ln\mathcal{L}_{GRB} = &-\frac{1}{2}\left(ln(\sigma^2 + (a*\delta_{log_{10}T_a^*})^2 + (b*\delta_{log_{10}L_{peak}})^2 + \delta^2_{log_{10}L_a})\right) \\
&-\frac{1}{2}\left(\frac{(log_{10}L_{a,th} - log_{10}L_a)^2}{\sigma^2 + (a*\delta_{log_{10}T_a^*})^2 + (b*\delta_{log_{10}L_{peak}})^2 + \delta^2_{log_{10}L_a}}\right),
\end{aligned} \tag{11}
$$

where $L_{a,th}$ is the theoretical luminosity computed through the fundamental plane in Equation (10), σ is the intrinsic scatter of the plane and $\delta_{log_{10}T_a^*}$, $\delta_{log_{10}L_{peak}}$, and $\delta_{log_{10}L_a}$ are the errors on the rest-frame time at the end of the plateau emission, the peak luminosity and the luminosity at the end of the plateau, respectively.

After performing an MCMC analysis using the D'Agostini method [356] and letting vary the parameters $a, b, c, \sigma, \Omega_{0m}$, the results are shown in Figure 2. Through the simulations of 1000 GRBs, with 9500 steps and keeping the same errors (errors undivided) as the ones observed in the fundamental plane ($n = 1$, see the upper left panel of Figure 2) we obtain a value of $\Omega_{0m} = 0.310$ with a symmetrized uncertainty of $\sigma_{\Omega_{0m}} = 0.078$. In the case of 2000 GRBs with 13,000 steps and $n = 1$ (see the upper right panel of Figure 2) instead, we have $\Omega_{0m} = 0.300$, $\sigma_{\Omega_{0m}} = 0.052$. If we consider the division of the errors on the variables of the fundamental plane by a factor 2 (halved errors, $n = 2$) we obtain, in the case of 1500 GRBs with 11,100 steps, $\Omega_{0m} = 0.300$, $\sigma_{\Omega_{0m}} = 0.037$ (see the lower-left panel of Figure 2), while through 2000 simulated GRBs in 14,600 steps (still with $n = 2$, see the lower right panel of Figure 2) we have $\Omega_{0m} = 0.310$, $\sigma_{\Omega_{0m}} = 0.034$. The idea of considering halved errors comes from the prospects for improvement in the fitting procedures of GRB light curves. Through this approach, the GRBs have provided constraints on the value of Ω_{0m} that are compatible with the ones of previous samples of SNe Ia: in the $n = 1$ cases, the values of the uncertainties are comparable with the ones from [357], while for the $n = 2$ cases the values are close to the ones found in [358] with 2000 GRBs. Furthermore, the GRBs have proven to be promising standardizable candles and, given the bigger redshift span they can cover if compared with SNe Ia, GRBs will provide more complete information about the structure and the evolution of the early universe after the Big Bang, together with quasars [359,360]. After discussing the potentiality of GRBs as future standard candles, we estimate the frequency of GRBs with a plateau emission over the total number of GRBs observed to date. We can expect that by 2030 we will have reached several GRB observations such that these—as standalone probes that respect the properties of the GRB platinum sample [35]—will give constraints as precise as the ones from [357] in the case of not halved errors. In case of halved errors, we can reach the level of precision of [357] even now. In addition, if we consider a machine learning analysis [361,362] for which we can double the size of the sample we are able to reach the precision of [357] now with the case of $n = 1$. If we consider the case of reconstructing the light curves and thus we have a sample which has the 47% of cases with halved errors we can reach the limit of [357] in 2022 if $n = 1$ and now if $n = 2$.

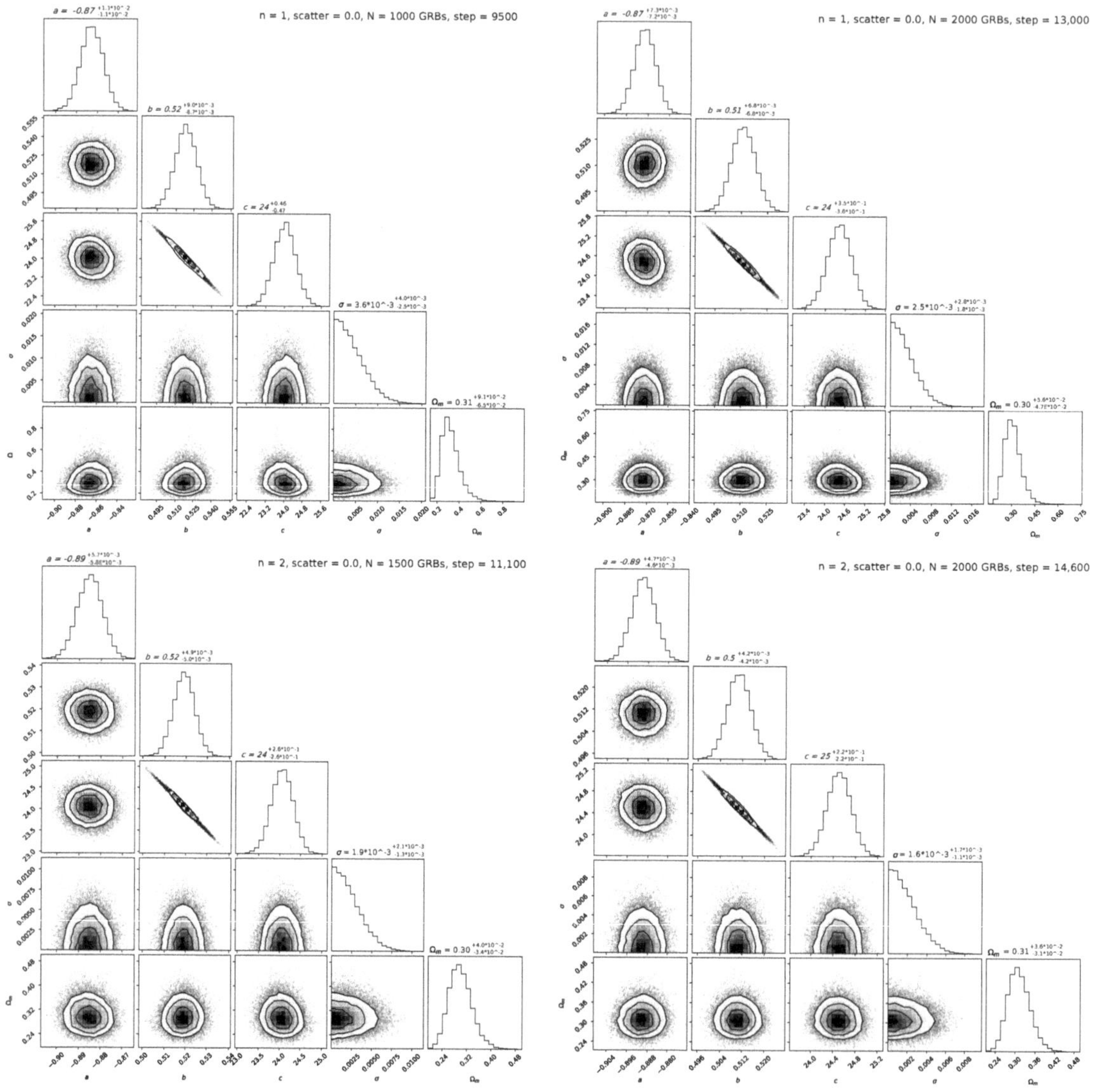

Figure 2. Upper left. An example of 1000 simulated GRBs with the posterior distribution of the fundamental plane parameters a, b, c, and its intrinsic scatter σ together with the total matter density parameter Ω_{0m}. In this case, the steps of the simulation are 9500 and the errors on the variables of the fundamental plane have not been divided by any factor ($n = 1$). **Upper right.** The same case of the upper left panel, but considering 2000 GRBs and a number of steps of 13,000. **Lower left.** The results of 1500 simulated GRBs, dividing by two the errors on the fundamental plane variables (halved errors, $n = 2$): here the steps are 11,100. **Lower right.** The same result of the lower-left panel, considering 2000 GRBs and 14,600 steps instead.

6. Discussions on the Results

Our results can be interpreted because of astrophysical selection biases or theoretical models alternatives to the standard cosmological models.

6.1. Astrophysical Effects

The main effect that has a stake in the SNe Ia luminosity variation is the presence of metallicity and the difference in stellar ages. Indeed, ref. [284] correct the PS with a mass-step contribution (ΔM). Despite this term improving the results, other effects need to be accounted for. Considering the stretch and the color, ref. [363] claim that the Hubble residuals, after being properly corrected according to the stretch and color observations, for SNe Ia in low mass and high mass host galaxies show a difference of 0.077 ± 0.014 mag, compatibly with the result of [284]. SNe Ia age metallicity and age are believed to be responsible for the observed behavior: those can replicate the Hubble residual trends consistent with the ones of [363]. In the PS, to account for the evolutions of stretch (α) and color (β), the parametrization utilized is the following: $\alpha(z) = \alpha_0 + (\alpha_1 \times z)$, $\beta(z) = \beta_0 + (\beta_1 \times z)$. According to [284], there is no clear dependence on the redshift for $\alpha(z)$ and $\beta(z)$, thus α_1 and β_1 are set to zero. Only the selection effect for color is noteworthy and [284] consider the uncertainty on β_1 as a statistical contribution. Concerning the stretch evolution in the PS calibration, it appears to be negligible and is not included at any level.

Conversely, ref. [364] recently studied the SALT2.4 lightcurve stretch and showed that the SN stretch parameter is redshift-dependent. According to their analysis, the asymmetric Gaussian model assumed by [284] for describing the populations of SNe Ia does not take into consideration the redshift drift of the PS, thus leaving a residual evolutionary trend that manifests at higher redshifts. Indeed, the simulations performed by [284] for studying the systematics calibration reach redshifts up to $z = 0.7$: this threshold is present in the third bin of our analysis. The effect from $0.7 \leq z \leq 2.26$ needs additional investigations. It is worth noting that this decreasing trend of H_0 (with a given value of η) found in [51] is consistent in 1 σ for the ΛCDM both in the cases of SNe Ia only and SNe+BAOs. When we consider the w_0w_aCDM, the η values are compatible in 3 σ with the ones with SNe Ia only and SNe + BAOs. We here have two cosmological parameters varying at the same time, differently from [51]. Therefore, one of the possible astrophysical reasons behind the observed trend is the residual stretch evolution with redshift. If so, in our work the effect is simply switched from stretch to H_0. The forthcoming release of the Pantheon+ data [107,365–369] will give the chance to test if these evolutionary effects may be still visible, but this analysis goes far beyond the scope of the current paper. The astrophysical interpretation seems to be favored, but also many theoretical explanations may be possible to describe the outcome of these results.

6.2. Theoretical Interpretations

We now investigate possible theoretical explanations for our results, focusing particular attention on modified gravity models. We first discuss a general scalar-tensor formulation and, then, we concentrate our attention on the so-called metric $f(R)$ gravity.

6.2.1. The Scalar Tensor Theory of Gravity

The action of the scalar tensor theories (STTs) of gravity is given by $S = S^{JF} + S_m$ [370–374] with the Jordan Frame (JF) action

$$S^{JF} = \frac{1}{16\pi} \int d^4x \sqrt{-\tilde{g}} \left[\Phi^2 \tilde{R} + 4\,\omega(\Phi) \tilde{g}^{\mu\nu} \partial_\mu \Phi \partial_\nu \Phi - 4\tilde{V}(\Phi) \right], \tag{12}$$

where $\tilde{R}$ is the Ricci scalar obtained with the physical metric $\tilde{g}_{\mu\nu}$, while the matter fields Ψ_m couple to the metric tensor $\tilde{g}_{\mu\nu}$ and not to Φ, i.e., $S_m = S_m[\Psi_m, \tilde{g}_{\mu\nu}]$.

In this Section we adopt natural units such that $c = 1$ and $G = 1$. Different STTs follow with the appropriate choice of the two functions $\omega(\Phi)$ and $\tilde{V}(\Phi)$: e.g., the Brans–Dicke (BD) theory [375–378] can be obtained for $\omega(\Phi) = \omega$ (const.) and $\tilde{V}(\Phi) = 0$, while the metric $f(R)$ gravity, discussed in the next subsection, would correspond to $\omega \equiv 0$.

The action S^{JF} can be rewritten in the Einstein Frame (EF), where one defines $\tilde{g}_{\mu\nu} \equiv A^2(\varphi)g_{\mu\nu}$, $\Phi^2 \equiv 8\pi M_*^2 A^{-2}(\varphi)$, $V(\varphi) \equiv \frac{A^4(\varphi)}{4\pi}\tilde{V}(\Phi)$, $\gamma(\varphi) \equiv \frac{d\log A(\varphi)}{d\varphi}$, and $\gamma^2(\varphi) = \frac{1}{4\omega(\Phi)+6}$, to get

$$S^{EF} = \frac{M_*^2}{2}\int d^4x\sqrt{-g}\left[R + g^{\mu\nu}\partial_\mu\varphi\partial_\nu\varphi - \frac{2}{M_*^2}V(\varphi)\right]. \tag{13}$$

Matter is coupled to φ only through a purely metric coupling, $S_m = S_m[\Psi_m, A^2(\varphi)g_{\mu\nu}]$ and M_* is the Planck mass.

The physical quantities in the Jordan and Einstein frame are related by $d\tilde{\tau} = A(\varphi)d\tau$, $\tilde{a} = A(\varphi)a$, $\tilde{\rho} = A(\varphi)^{-4}\rho$, $\tilde{p} = A(\varphi)^{-4}p$, where τ is the synchronous time variable. Defining $N \equiv \log\frac{a}{a_0}$, $\lambda \equiv \frac{V(\varphi)}{\rho}$, $w \equiv \frac{p}{\rho}$, and $\varphi' = \frac{d\varphi}{dN} = a\frac{d\varphi}{da}$, the combination of cosmological equations allows to write the equation for φ in the form (for a flat Friedmann–Robertson–Walker geometry) [379]

$$\frac{2}{3}\,\frac{1+\lambda}{1-\varphi'^2/6}\,\varphi'' + [(1-w)+2\lambda]\varphi' = -\sqrt{2}\,\gamma(\varphi)\,(1-3w) - 2\,\lambda\,\frac{V_\varphi(\varphi)}{V}, \tag{14}$$

Moreover, the Jordan- and Einstein-frame Hubble parameters, $\tilde{H} \equiv d\log\tilde{a}/d\tilde{\tau}$ and $H \equiv d\log a/d\tau$, respectively, are related as

$$\tilde{H} = \frac{1+\gamma(\varphi)\,\varphi'}{A(\varphi)}\,H. \tag{15}$$

For our purpose, we consider $A(\varphi) = A_0 e^{c_1\varphi + c_2\varphi^2/2}$, which implies $\gamma(\varphi) = c_1 + c_2\varphi$, where $c_{1,2}$ are constants. Under the following conditions $\varphi''/\varphi \ll 1$, $\varphi'^2/\varphi^2 \ll 1$, and $\frac{V_\varphi(\varphi)}{\varphi V\rho} \ll 1$, the solution of Equation (14) is $\varphi(z) = C(1+z)^K - \frac{c_1}{c_2}$, where $K = \frac{1-3w}{1+w}\sqrt{2}\,c_2$, and C is an integration constant. We are looking for solutions such that $H = f(\varphi)\tilde{H}_0$, so that $\tilde{H} = \frac{\tilde{H}_0}{(1+z)^\eta}$, where $\tilde{H}_0$ is constant. These relations and (15) allow to derive $f(\varphi)$ (the expression of $f(\varphi)$ is quite involved, and in the case in which $c_{1,2} \ll 1$, it is a polynomial in φ). The scalar field Φ in the (physical) JF can be cast in the form $\Phi(z) = \Phi_0(1+z)^{\tilde{K}}$, where $\Phi_0 \equiv \frac{\sqrt{8\pi}M_*}{A_0}\left[1 - C\left(c_1 - \frac{Cc_2}{c_1}\right)\right]$, $\tilde{K} = -\frac{KC(c_1+Cc_2)}{1-C\left(c_1+\frac{Cc_2}{2}\right)}$, and $z < 1$ has been used (note: $\tilde{K}$ is positive for c_1 or c_2 negative). The scalar field Φ reduces to ϕ for $\Phi_0 \to 1$ and $\tilde{K} \to 2\eta$. From the Friedmann Equation [379]

$$\left(\frac{\dot{a}}{a}\right)^2 = \frac{1}{3M_*^2}\left[\rho + \frac{M_*^2}{2}\dot{\varphi}^2 + V(\varphi)\right], \tag{16}$$

with ρ given by matter ($\rho = \rho_{0m}/a^3 = \rho_{0m}(1+z)^3$), and $c_{1,2} \ll 1$, one infers the effective potential

$$\frac{\tilde{V}}{3m^2} = \frac{4\pi M_*^2}{A_0^2}\left[f_0^2 - \frac{1}{\Omega_{0m}}\left(\frac{\Phi}{\Phi_0}\right)^{\frac{3}{2\eta}} - \frac{C^2K^2\varphi_0^2}{6\Omega_{0m}}\left(\frac{\Phi}{\Phi_0}\right)^{\frac{K-\eta}{\eta}}\right], \tag{17}$$

where we recall that $\Omega_{0m} = \rho_{0m}/\rho_{cr}$, $\rho_{cr} = 3M_*^2\tilde{H}_0^2$, $f_0 = f(\varphi = 0)$, and $m^2 = \Omega_{0m}\tilde{H}_0^2$. For redshift $0 \leqslant z < 0.3$, to which we are interested, the scalar field varies slowly with z, $\Phi \sim \Phi_0$, so that the effective potential behaves like a cosmological constant. We see how the proposed scalar-tensor formulation has the right degrees of freedom to reproduce, in the JF, the required behavior of the (physical) trend of $H_0(z)$. In the next subsection, we analyze a sub-case of the general paradigm discussed above, which leads to the well-known $f(R)$ gravity, which is among the most popular modified gravity formulations.

6.2.2. Metric f(R) Gravity in the Jordan Frame

The observed decaying behavior of the Hubble constant H_0 with the redshift draws significant attention for an explanation and, if it is not due to selection effects or systematics in the sample data, we need to interpret our results from a physical point of view. As already argued in [51,380], the simplest way to account for this unexpected behavior of $H_0(z)$ is that the Einstein constant $\chi = 8\pi G$ (where G denotes the gravitational constant), mediating the gravity-matter interaction, is subjected itself to a slow decaying profile with the redshift. In this Section, we consider $c = 1$ for the speed of light. More specifically, since the critical energy density $\rho_{c0} = 3\,H_0^2/\chi$ today must be a constant, we need an evolution for $\chi \sim (1+z)^{-2\eta}$, considering the function $H_0(z)$ given by Equation (9). The evolution of $\chi(z)$ is not expected within the cosmological Einsteinian gravity, therefore we are led to think of it as a pure dynamical effect, associated with a modified Lagrangian for the gravitational field beyond the ΛCDM cosmological model. Ref. [143] obtained cosmological constraints within the Brans–Dicke theory considering how the evolution of the gravitational constant G, contained in χ, affects the SNe Ia peak luminosity. The most natural extended framework is the $f(R)$-gravity proposal [162,163,167,381] which contains only an additional scalar degree of freedom. For instance, ref. [323] try to alleviate the H_0 tension considering exponential and power-law $f(R)$ models.

The formulation of the $f(R)$ theories in an equivalent scalar-tensor paradigm turns out to be particularly intriguing for our purposes: the function $f(R)$ is restated as a real scalar field ϕ, which is non-minimally coupled to the metric in the JF. The information about the function f turns into the expression of the scalar field potential $V(\phi)$. The relevance of modified gravity models relies on the possibility that this revised scenario for the gravitational field can account for the physics of the so-called "dark universe"component without the need for a cosmological constant. Indeed, the observed cosmic acceleration in the late universe via the SNe Ia data is a pure dynamical effect, i.e., associated with a modification of the Einsteinian gravity at very large scales (in the order of the present Hubble length).

According to the standard literature on this field (which includes a large number of proposals), three specific $f(R)$ models, i.e., the Hu–Sawicki [382], the Starobinsky [383], and Tsujikawa models [384,385], successfully describe the Dark Energy component (say an effective parameter for the Dark Energy $w = w(z) < -1/3$) and overcome all local constraints. The difference in the form of the Lagrangian densities associated with $f(R)$ models is reflected in the morphology of the potential term governing the dynamics of the scalar field. For instance, the scalar field potential related to the Hu–Sawicki $f(R)$ proposal, with the power index $n = 1$, in the JF is given by

$$V(\phi) = \frac{m^2}{c_2}\left[c_1 + 1 - \phi - 2\sqrt{c_1(1-\phi)}\right], \tag{18}$$

where we have two free parameters c_1 and c_2, while $m^2 = \chi\,\rho_{0m}/3$. The scalar-tensor dynamics in the JF for a flat FLRW metric with a matter component is summarized by

$$H^2 = \frac{\chi\,\rho}{3\,\phi} - H\,\frac{\dot{\phi}}{\phi} + \frac{V(\phi)}{6\,\phi} \tag{19}$$

$$\frac{\ddot{a}}{a} = -\frac{\chi\,\rho}{3\,\phi} - \frac{V(\phi)}{6\,\phi} + \frac{1}{6}\frac{dV}{d\phi} + \frac{\dot{a}\,\dot{\phi}}{a\,\phi} \tag{20}$$

$$3\ddot{\phi} - 2\,V(\phi) + \phi\,\frac{dV}{d\phi} + 9\,H\,\dot{\phi} = \chi\,\rho, \tag{21}$$

which are the generalized Friedmann equation, the generalized cosmic acceleration equation and the scalar field equation, respectively [167]. We recall that $\phi = \phi(t)$ is a function of the time (or the redshift z) only for an isotropic universe. Considering the first term on the right-hand side of Equation (19), it is possible to recognize that ϕ mediates the gravity-matter coupling, and therefore it mimics a space-time varying Einstein constant. Hence, to account for our observed decay of $H_0(z)$, we have to require that the scalar field assumes a specific behavior with the redshift, i.e.,

$$\phi(z) = (1+z)^{2\eta}. \tag{22}$$

Moreover, the remaining terms contained in the gravitational field equations must be negligible. This situation is naturally reached when the potential term is sufficiently slow-varying in a given time interval. We see that the hypothesis of a near-frozen scalar field evolution is a possible assumption, as far as the potential term should provide a dynamical impact, sufficiently close to a cosmological constant term. These simple considerations lead us to claim that this scenario is worth to be investigated for the behavior of $H_0(z)$ here observed.

The specific cosmological models affect the expression of the luminosity distance and this should be the starting point of a careful test of a $f(R)$ theory versus the comprehension of the H_0 tension. A new binned analysis of the PS, using the corrected luminosity distance obtained through a reliable $f(R)$, may in principle shed new light on the observed decaying trend of $H_0(z)$, testing also new physics. This analysis is performed in the next Section.

As a preliminary approach, we try to understand which profile we could expect for the scalar field potential, inferred from the behavior of $H_0(z)$. This is quite different from a standard analysis of $f(R)$ models. Generally, a specific $f(R)$ function is defined a priori, and then the dynamical equations are studied to obtain constraints on the free parameters. Here, instead, starting from the observed decreasing trend of $H_0(z)$ and assuming $\phi(z)$ from Equation (22), we wonder what the scalar field potential would be in a scalar-tensor dynamics. Eventually, we should have a scalar field in near-frozen dynamics, i.e., a slow-roll of the scalar field potential, mimicking a cosmological constant term ($\phi \to 1$). To this end, we rewrite the generalized Friedmann Equation (19) and calculate $V(\phi)$:

$$V(\phi) = 6(1-2\eta)\left(\frac{dz}{dt}\right)^2 \phi^{1-1/\eta} - 6m^2\,\phi^{3/\,2\eta}\,, \tag{23}$$

where we have used the standard definition of redshift and the relation (22) for $\phi(z)$. Moreover, we recall that for a matter component $\rho \sim (1+z)^3$. As a final step, we need to calculate the term $\frac{dz}{dt}$. Starting again from the redshift definition, it is well known that

$$\frac{dz}{dt} = -(1+z)\,H(z). \tag{24}$$

In principle, we would need to compute the Hubble parameter $H(z)$ from the field equations, and then replace $H(z)$ in the term $\frac{dz}{dt}$. However, this procedure is not viable, since we need to fix a well-defined $V(\phi)$ to solve the field equations. Moreover, $H(z)$ appears also in the right-hand-side of Equation (19), because of the non-minimal coupling with the scalar field. Therefore, we can not calculate exactly $\frac{dz}{dt}$ to get $V(\phi)$ in the JF.

Then, to obtain $V(\phi)$ inferred from the trend of $H_0(z)$, we require that the Hubble function provides the same physical mechanism suggested from our binned analysis in Section 4, i.e., simply replacing H_0 with $H_0(z)$ given by Equation (9) in the standard Friedmann equation in the ΛCDM model. With this new definition of $\mathcal{H}_0$, we write the following condition on the Hubble function:

$$H(z) = \frac{\mathcal{H}_0}{(1+z)^\eta}\sqrt{\Omega_{0m}\,(1+z)^3 + 1 - \Omega_{0m}}\,. \tag{25}$$

In doing so, using Equations (23)–(25), we determine the form of the scalar field potential

$$\frac{V(\phi)}{m^2} = 6\,(1-2\eta)\left(\frac{1-\Omega_{0m}}{\Omega_{0m}}\right) - 12\eta\,\phi^{3/2\eta} \tag{26}$$

inferred from the decreasing trend of $H_0(z)$. In other words, the potential Equation (26) might provide an effective Hubble constant that evolves with redshift. In the computation, we have used the expression $\Omega_{0m} = m^2/\mathcal{H}_0^2$.

In Figure 3, we plotted this potential profile, observing that, as expected, a flat region consistently appears, validating our guess on the feasibility of $f(R)$-gravity in the JF to account for the observed behavior of $H_0(z)$. We set $\eta = 0.009$ in Figure 3, according to our binned analysis results for three bins (see Table 1). We stress that the flatness of the potential does not emerge throughout the Pantheon sample redshift range, $0 < z < 2.3$, but it appears only in a narrow region for $0 < z \lesssim z^*$, where $z^* = 0.3$ is the redshift at the Dark Energy and Matter components equivalence of the universe. This form of $V(\phi)$ is reasonable since the Dark Energy contribution, provided by the scalar field in the JF gravity, dominates the matter component only for $0 < z \ll z^*$. It is the weak dependence of H_0 on z that ensures the existence of a flat region of the potential, according to the theoretical scenario argued above.

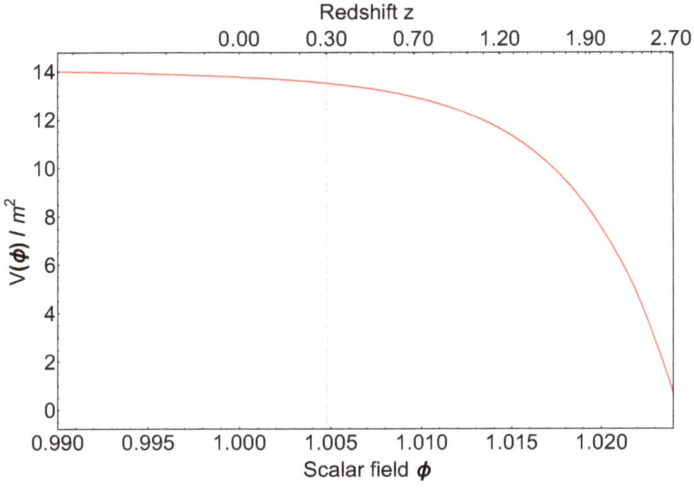

Figure 3. Profile of the scalar field potential $V(\phi)$ in the JF equivalent scalar-tensor formalism of the f(R) modified gravity. The form of $V(\phi)$) is inferred from the behavior of $H_0(z)$ (Equation (9)). Note that $V(\phi)/m^2$ is a dimensionless quantity. A flat profile of $V(\phi)$ occurs only at low redshifts, for $0 < z \lesssim 0.3$ or equivalently $\phi \lesssim 1.005$. Note, also, the non-linearity of the scale for the redshift axis on top, considering the relation (22) between ϕ and z. In this plot, $\eta = 0.009$.

Finally, we can calculate the form of the $f(R)$ function associated with the potential profile. Recalling the following general relations in the JF [167]:

$$R = \frac{dV}{d\phi}, \tag{27}$$

$$f(R) = R\,\phi(R) - V(\phi(R)), \tag{28}$$

we can obtain:

$$f(R) = -6\,m^2\left[(1-2\eta)\frac{1-\Omega_{0m}}{\Omega_{0m}} + (3-2\eta)\left(-\frac{R}{18\,m^2}\right)^{\frac{3}{3-2\eta}}\right]. \tag{29}$$

Note that the formula above provides a generalization of the Einstein theory of gravity, as it should be in the context of a $f(R)$ model. Indeed, if $\eta = 0$, then $f(R) \equiv R$ reproduces exactly the Einstein–Hilbert Lagrangian density in GR with a cosmological constant Λ, as soon as you recognize that $\Lambda = 3m^2(1 - \Omega_{0m}) \,/\, \Omega_{0m}$ for a flat geometry, using $m^2 = \mathcal{H}_0^2\,\Omega_{0m}$. In particular, expanding the function (29) for $\eta \sim 0$, we can see explicitly the deviation from the Einstein–Hilbert term:

$$f(R) \approx \left(R - 6\,m^2\,\frac{1-\Omega_{0m}}{\Omega_{0m}}\right) + \frac{2}{3}\eta\,\left[R\,\ln\left(-\frac{R}{m^2}\right) - (1+\ln 18)\,R + 18m^2\,\frac{1-\Omega_{0m}}{\Omega_{0m}}\right] + O\left(\eta^2\right). \tag{30}$$

The first term at the zero-th order in η is exactly the Einstein–Hilbert Lagrangian density, while the linear term in η provides the correction to GR. Therefore, η, in addition to being the physical parameter that describes the evolution of $H_0(z)$, also denotes the deviation from GR and the standard cosmological model. It is worthwhile to remark that the expression above may not be the final form of the underlying modified theory of gravity, associated with the global universe dynamics, but only its asymptotic form in the late Universe, i.e., as the scalar of curvature approaches the value corresponding to the cosmological constant in the ΛCDM model. In all these computations we do not consider relativistic or radiation components at very high redshifts, but it may be interesting to test this model with other local probes in the late Universe.

In this discussion, we infer that the dependence of H_0 on the redshift points out the necessity of new physics in the description of the universe dynamics and that such a new framework may be identified in the modified gravity, related to metric theories.

7. The Binned Analysis with Modified f(R) Gravity

To try to explain the observed trend of $H_0(z)$, we focus on $f(R)$ theories of gravity, and then we perform the same binned analysis, using the correction for the distance luminosity according to the modified gravity. We start from the gravitational field action [167]:

$$S_g = \frac{1}{2\chi}\int d^4x\sqrt{-g}f(R), \tag{31}$$

where $f(R)$, as a function of the Ricci scalar R, is an extra degree of freedom compared to General Relativity. We rewrite $f(R) = R + F(R)$ to highlight the deviation from the standard gravity. Varying the total action with respect to the metric, we obtain the flat FLRW metric field equations:

$$H^2(1+F_R) = \frac{\chi\rho}{3} + \left[\frac{R\,F_R - F}{6} - F^{RR}H\dot{R}\right], \tag{32}$$

where $F_R \equiv \frac{dF(R)}{dR}$. The Ricci scalar R can be cast in the form

$$R = 12H^2 + 6HH', \tag{33}$$

where the Hubble parameter H is expressed as a function of $\gamma \equiv ln(a)$, and the prime indicates the derivative with respect to γ.
Now, we introduce two dimensionless variables [382]

$$y_H = \frac{H^2}{m^2} - \frac{1}{a^3}, \qquad y_R = \frac{R}{m^2} - \frac{3}{a^3}, \tag{34}$$

which denote the deviation of H^2 and R with respect to the matter contribution when compared to the ΛCDM model. We rewrite the modified Friedmann Equation (32) and the

Ricci scalar relation (33) in terms of y_H and y_R. Then, we have a set of coupled ordinary differential equations:

$$y'_H = \frac{1}{3}y_R - 4y_H \tag{35}$$

$$y'_R = \frac{9}{a^3} - \frac{1}{y_H + a^{-3}}\frac{1}{m^2 F_{RR}}\left[y_H - F_R\left(\frac{1}{6}y_R - y_H - \frac{a^{-3}}{2}\right) + \frac{1}{6}\frac{F}{m^2}\right]. \tag{36}$$

The solution of this coupled first-order differential equations system above can not be obtained analytically, but can be numerically calculated. We need initial conditions such that this scenario mimics the ΛCDM model in the matter dominated universe at initial redshift $z_i \gg z^*$. Hence, we impose the following conditions for y_H and y_R at the redshift z_i:

$$y_H(z_i) = \frac{\Omega_{0\Lambda}}{\Omega_{0m}} \tag{37}$$

$$y_R(z_i) = 12\frac{\Omega_{0\Lambda}}{\Omega_{0m}}. \tag{38}$$

The standard ΛCDM model is reached for $z = z_i$ or asymptotically, and we consider a flat geometry, such that $\Omega_{0\Lambda} = 1 - \Omega_{0m}$. Finally, the luminosity distance can be written as

$$d_L(z) = \frac{(1+z)}{H_0}\int_0^z \frac{dz'}{\sqrt{\Omega_{0m}\left(y_H(z') + (1+z')^3\right)}}, \tag{39}$$

including the solution $y_H(z)$ from Equation (34) [386].

Hu–Sawicki Model

We focus on the Hu–Sawicki model with $n = 1$, considering a late-time gravity modification, described by the following function [382]:

$$f(R) = R + F(R) = R - m^2\frac{c_1(R/m^2)^n}{c_2(R/m^2)^n + 1}, \tag{40}$$

corresponding to the potential $V(\phi)$ in Equation (18). The parameters c_1 and c_2 are fixed by the following conditions [382]

$$\frac{c_1}{c_2} \approx 6\frac{\Omega_{0\Lambda}}{\Omega_{0m}} \tag{41}$$

$$F_{R0} \approx -\frac{c_1}{c_2^2}\left(\frac{12}{\Omega_{0m}} - 9\right)^{-2}, \tag{42}$$

where F_{R0} is the value of the field $F_R \equiv dF/dR$ at the present time, and $F(R)$ is the deviation from the Einstein–Hilbert Lagrangian density. Cosmological constraints provide $|F_{R0}| \leq 10^{-7}$ from gravitational lensing and $|F_{R0}| \leq 10^{-3}$ from Solar system [387,388]. We explore several choices of F_{R0}.

To simplify the numerical integration of the modified luminosity distance (39), we approximate the numerical solution y_H, obtained from the system (35), by a polynomial of order 8. This function is an accurate representation of y_H when we restrict the solution to the range of PS (see Figure 4).

As a consequence, we obtain constraints on c_1 and c_2, according to Equations (41) and (42). Then, we perform the same binned analysis of Section 4 using the Hu–Sawicki model and the modified luminosity distance (39).

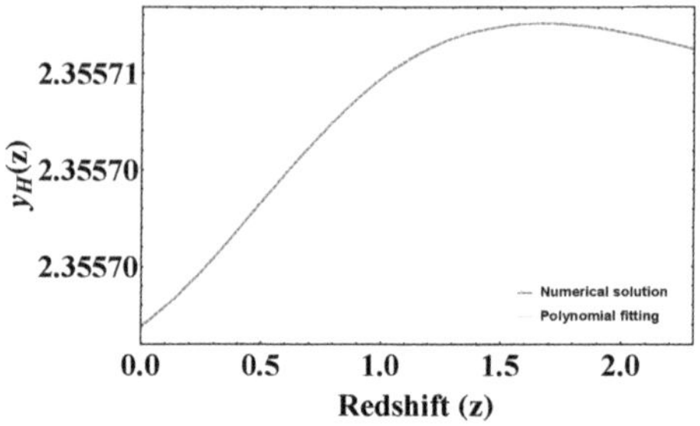

Figure 4. The numerical solution for Equation (35) (blue dashed curve) plotted together with its polynomial fitting (green continuous curve) in the case of $F_{R0} = -10^{-7}$. The assumption of a function of redshift in the form of a order-8 polynomial allows an accurate fit for the numerical values. The same fitting procedure has been used in the $F_{R0} = -10^{-4}$ case.

We here run the analysis both for the case of Ω_{0m} fixed to a fiducial value of 0.298 and for several values of $F_{R0} = -10^{-7}, -10^{-6}, -10^{-5}, -10^{-4}$ (see Table 2 and Figure 5) or we let Ω_{0m} vary with the two values of $F_{R0} = -10^{-7}, -10^{-4}$ (see Table 3 and Figure 6) for the SNe alone and with SNe +BAOs. Note also that the η parameters are all consistent for the several values of F_{R0} in 1 σ, as you can see in Table 2, for both SNe Ia and SNe Ia + BAOs. Moreover, the values of η are consistent in 1 σ with the ones obtained from the analysis of the ΛCDM model (see also Table 1). We consider the cases $F_{R0} = -10^{-7}$, $F_{R0} = -10^{-4}$ and, to study how these results may vary according to the different values of Ω_{0m} chosen, we tested the model with four values of $\Omega_{0m} = (0.301, 0.303, 0.305)$ taken from the 1 σ from a Gaussian distribution centred around the most probable value of 0.298, see [368].

We show in Figure 6 the comparison between the different applications of the Hu–Sawicki model: in the left panels (upper and lower), we consider SNe Ia only, while in the right panels (upper and lower) we combine SNe Ia+BAOs. We here remind that the assumed values for $|F_{R0}|$ of 10^{-4} and 10^{-7} are well constrained by the $f(T)$ theories [167,382].

Table 2. Fitting parameters of $H_0(z)$ for three bins within the Hu–Sawicki model, with SNe only and SNe + BAOs with a fixed value of $\Omega_{0m} = 0.298$ and with several values of F_{R0} : $-10^{-4}, -10^{-5}, -10^{-6}, -10^{-7}$. The columns contains: (1) the number of bins; (2) $\mathcal{H}_0$, (3) is η, according to Equation (9); (4) how many σs η is compatible with zero (namely, the ratio η/σ_η); (5) F_{R0} values; (6) the sample used.

Hu–Sawicki Model, Results of the Redshift Binned Analysis					
Bins	$\mathcal{H}_0$	η	$\frac{\eta}{\sigma_\eta}$	F_{R0}	**Sample**
3	70.089 ± 0.144	0.008 ± 0.006	1.2	-10^{-4}	SNe
3	70.127 ± 0.128	0.008 ± 0.006	1.4	-10^{-4}	SNe + BAOs
3	70.045 ± 0.052	0.007 ± 0.002	3.0	-10^{-5}	SNe
3	70.062 ± 0.132	0.007 ± 0.005	1.3	-10^{-5}	SNe + BAOs
3	70.125 ± 0.046	0.010 ± 0.002	5.4	-10^{-6}	SNe
3	70.115 ± 0.153	0.008 ± 0.007	12.1	-10^{-6}	SNe + BAOs
3	70.118 ± 0.131	0.011 ± 0.006	1.9	-10^{-7}	SNe
3	70.053 ± 0.150	0.007 ± 0.007	1.1	-10^{-7}	SNe + BAOs

Thus, the existence of this trend is, once again, confirmed, and it remains unexplained also in the modified gravity scenario. Indeed, a suitable modified gravity model which would be able to predict the observed trend of H_0, would allow observing a flat profile of $H_0(z)$ after a binned analysis. Further analysis must be carried out with other Dark Energy models or other modified gravity theories to investigate this issue in the future, for instance focusing on the proposed model in Section 6.2.2.

Table 3. Fitting parameters of $H_0(z)$ for three bins within the Hu–Sawicki model, with SNe and SNe + BAOs by fixing several values of $\Omega_M = 0.298, 0.303, 0.301, 0.305$ and values of $F_{R0} = -10^{-4}$ and $F_{R0} = -10^{-7}$. The columns are as follows: (1) the Ω_{0m} value; (2) $\mathcal{H}_0$, (3) η, according to Equation (9); (4) how many σs the evolutionary parameter η is compatible with zero (namely, η/σ_η); (5) F_R0; (6) the sample used.

Hu–Sawicki Model, Results of the 3 Bins Analysis					
Ω_{0m}	$\mathcal{H}_0$	η	$\frac{\eta}{\sigma_\eta}$	F_{R0}	Sample
0.298	70.140 ± 0.045	0.011 ± 0.002	5.1	-10^{-7}	SNe
0.298	70.050 ± 0.126	0.007 ± 0.006	1.2	-10^{-7}	SNe + BAOs
0.303	70.088 ± 0.075	0.012 ± 0.004	3.0	-10^{-7}	SNe
0.303	70.004 ± 0.139	0.009 ± 0.007	1.3	-10^{-7}	SNe + BAOs
0.301	70.054 ± 0.056	0.009 ± 0.003	3.0	-10^{-7}	SNe
0.301	70.072 ± 0.170	0.010 ± 0.008	1.2	-10^{-7}	SNe + BAOs
0.305	70.048 ± 0.034	0.012 ± 0.002	6.0	-10^{-7}	SNe
0.305	70.004 ± 0.140	0.010 ± 0.007	1.4	-10^{-7}	SNe + BAOs
0.298	70.135 ± 0.080	0.009 ± 0.004	2.2	-10^{-4}	SNe
0.298	70.087 ± 0.155	0.009 ± 0.007	1.2	-10^{-4}	SNe + BAOs
0.303	70.096 ± 0.146	0.012 ± 0.007	1.7	-10^{-4}	SNe
0.303	70.044 ± 0.129	0.009 ± 0.006	1.5	-10^{-4}	SNe + BAOs
0.301	70.111 ± 0.158	0.012 ± 0.008	1.5	-10^{-4}	SNe
0.301	70.038 ± 0.170	0.009 ± 0.008	1.1	-10^{-4}	SNe + BAOs
0.305	70.074 ± 0.026	0.016 ± 0.001	16.0	-10^{-4}	SNe
0.305	70.028 ± 0.090	0.011 ± 0.004	2.4	-10^{-4}	SNe + BAOs

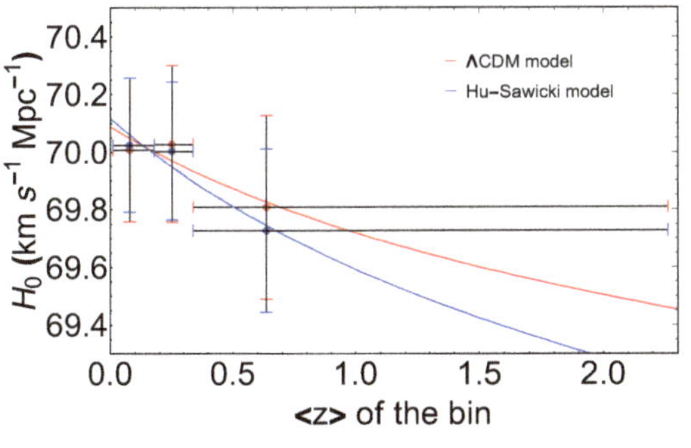

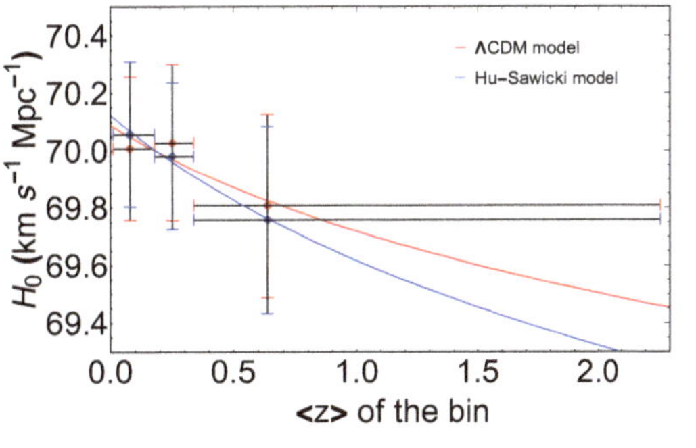

Figure 5. *Cont.*

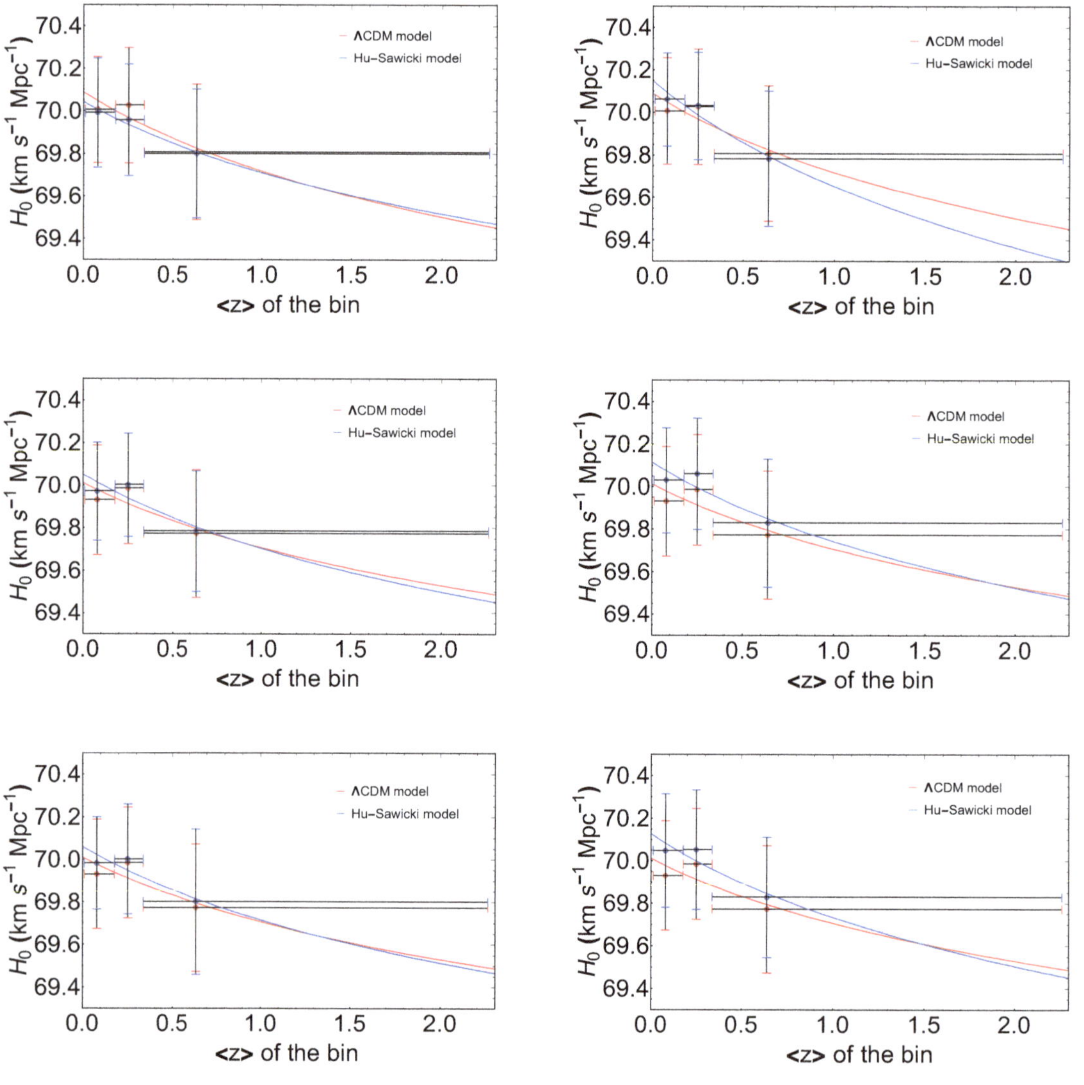

Figure 5. The first four panels deal with H_0 vs. z for SNe, the four bottom panels include BAO measurements for the H-S model. The upper 4 panels show from the left to the right $F_{R0} = -10^{-7}, -10^{-6}, -10^{-5}, -10^{-4}$, respectively. The standard ΛCDM cosmology is shown in red and the Hu-Sawicki model in blue. Analogously, the bottom panels have the same notation about the values of F_{R0}.

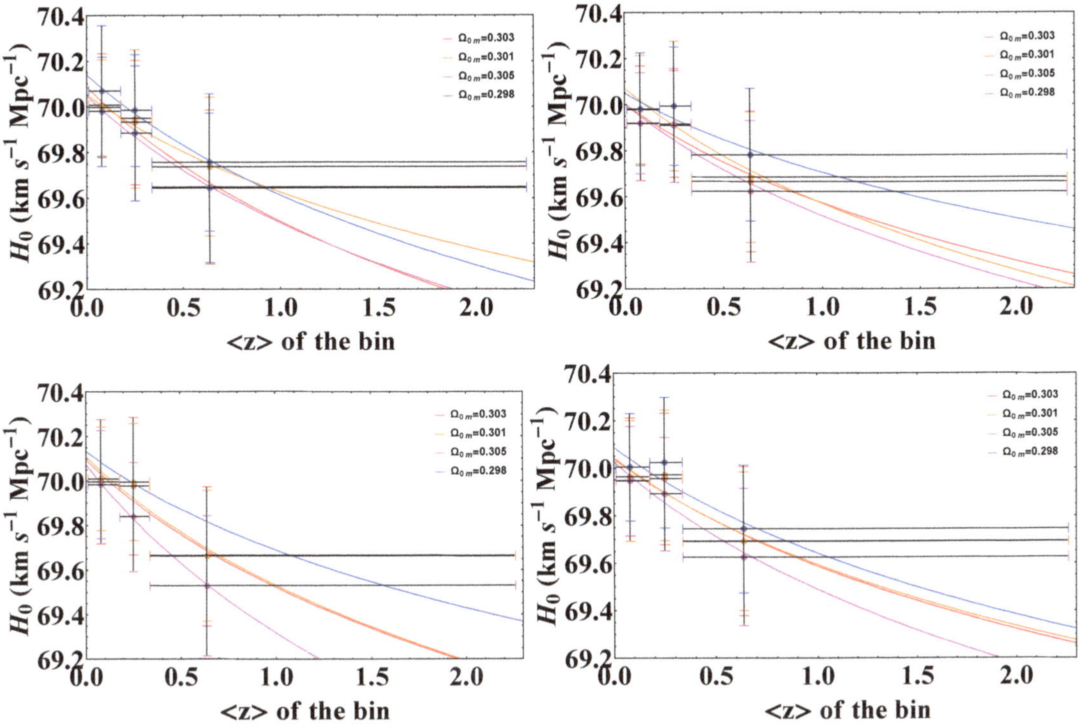

Figure 6. The Hubble constant versus redshift plots for the three bins of SNe Ia only, considering the Hu–Sawicki model. **Upper left panel.** The condition of $F_{R0} = -10^{-7}$ is applied to the case of SNe only, with the different values of $\Omega_{0m} = 0.301, 0.303, 0.305$. **Upper right panel.** The same of the upper left, but with the contribution of BAOs. **Lower left panel.** The SNe only case with the $F_{R0} = -10^{-4}$ condition, considering the different values of $\Omega_{0m} = 0.301, 0.303, 0.305$. **Lower right panel.** The same as the lower left, but with the contribution of BAOs. The orange color refers to $\Omega_{0m} = 0.301$, the red to $\Omega_{0m} = 0.303$, the magenta to $\Omega_{0m} = 0.305$, and the blue to $\Omega_{0m} = 0.298$.

8. Requirements for a Suitable f(R) Model

Since the Hu–Sawicki model seems to be inadequate to account for the observed phenomenon of the decaying $H_0(z)$, in what follows, we provide some general properties that an $f(R)$ model in the JF must possess to induce the necessary scenario of a slowly varying Einstein constant. Now, we consider again the dynamical impact of the scalar field ϕ, related to the $f(R)$ function. Let us observe that the following relation holds in the following way:

$$\frac{d\phi}{dz} = -\frac{1}{1+z}\frac{\dot{\phi}}{H}. \tag{43}$$

In order to get the desired behavior $\phi \simeq (1+z)^{2\eta}$, we must deal with a dynamical regime where the following request is satisfied:

$$\frac{\dot{\phi}}{H} = -2\eta\phi. \tag{44}$$

We consider a slow-rolling evolution of the scalar field ϕ in the late universe, near enough to $\phi \simeq 1$. Then, we consider in Equation (19) $\rho \sim 0$, because we are in the Dark Energy dominated phase, and we consider $H_0\,\dot{\phi}$ small with respect to the potential term $V(\phi \simeq 1)$. We neglect, also, the term $\ddot{\phi}$. Under these conditions, Equations (19) and (21) become

$$H^2 = \frac{V}{6\phi} \tag{45}$$

and

$$\frac{\dot{\phi}}{H} = \frac{1}{9H^2}\left(2V - \phi\frac{dV}{d\phi}\right), \tag{46}$$

respectively.

Referring to Equation (45) at $z \sim 0$, we make the identification $H_0^2 \equiv V(\phi \simeq 1)/6$. Hence, in order to reproduce Equation (44), we must require that for $\phi \to 1$, the following relation holds:

$$\eta = \frac{1}{3V}\left(\phi\frac{dV}{d\phi} - 2V\right). \tag{47}$$

The analysis above states the general features that a $f(R)$ model in the JF has to exhibit to provide a viable candidate to reproduce the observed decay behavior of $H_0(z)$ (Equation (36)). We conclude by observing that the picture depicted above relies on the concept of a slow-rolling phase of the scalar field, when it approaches the value $\phi \simeq 1$ and, in this respect, the potential term should have for such value a limiting dynamics, which remains there confined for a sufficiently long phase. It is just in such a limit that we are reproducing a ΛCDM model, but with the additional feature of a slowly varying Einstein constant. As far as the value of z increases, the deviation of the considered model from General Relativity becomes more important, but this effect is observed mainly in the gravity-matter coupling. In other words, the motion of the photon, as observed in the gravitational lensing, is not directly affected by the considered deviation, since the geodesic trajectories in the space-time do not directly feel the Einstein constant value. This consideration could allow for a large deviation of ϕ from the unity that is expected in studies of the photons' propagation.

8.1. An Example for Low Redshifts

As a viable example for the Dark Energy dominated Universe (slightly different from the traced above), we consider a potential term (and the associated slow-rolling phase) similar to the one adopted in the so-called chaotic inflation [389,390], i.e.,:

$$V(\phi) = \delta + 6H_0^2\phi^2, \tag{48}$$

where δ is a positive constant, such that $\delta \ll 6H_0^2$. From Equation (45), we immediately get

$$H^2 \simeq H_0^2\phi \sim H_0^2, \tag{49}$$

where, we recall that we are considering the slow-rolling phase near $\phi \to 1$. Analogously, from Equation (47), we immediately get:

$$\eta \sim -\frac{\delta}{9H_0^2}. \tag{50}$$

The negative value of η is coherent with the behavior $H^2 \propto \phi$. Hence, we can reproduce the requested behavior of $\phi(z)$ by properly fixing the value of δ to get η as it comes out from the data analysis of Section 4. Specifically, we get $\delta \sim 10^{-3}H_0^2$ to have $\eta \sim 10^{-2}$. Furthermore, it is easy to check that, for $\phi \to 1$, Equations (44) and (49), we find the relation

$$\ddot{\phi} \sim \mid \eta \mid H_0\dot{\phi} \ll 3H_0\dot{\phi}, \tag{51}$$

which ensures that we are dealing with a real slow-rolling phase.

Finally, we compute the $f(R)$ function corresponding to the potential in Equation (48), recalling the relation (28):

$$f(R) = \frac{R^2}{24H_0^2} - \delta \tag{52}$$

We conclude by observing that this specific model is reliable only as far the universe matter component is negligible, $z < 0.3$. The Einstein constant in front of the matter-energy density ρ would run as $(1+z)^{2\eta}$. The example above confirms that the $f(R)$ gravity in the JF is a possible candidate to account for the observed effect of $H_0(z)$, but the accomplishment of a satisfactory model for the whole ΛCDM phase requires a significant effort in further investigation, especially accounting for the constraints that observations in the local universe provided for modified gravity.

8.2. Discussion

Let us now try to summarize the physical insight that we can get from the analysis above, about the possible theoretical nature of the observed $H_0(z)$ behavior. We can keep as a reliably good starting point the idea that the origin of a modified scaling of the function $H(z)$ with respect to the standard ΛCDM model can be identified in a slowly varying Einstein constant with the redshift. Furthermore, it is a comparably good assumption to search, in the framework of a scalar-tensor formulation of gravity, the natural explanation for such a varying Einstein constant. As shown in Section 6.2, a scalar-tensor formulation can reproduce the required scaling of the function $H(z)$, which we observe as an $H_0(z)$ behavior in the standard ΛCDM model. Hence, we naturally explored one of the most interesting and well-motivated formulations of a scalar-tensor theory, namely the $f(R)$ gravity in the JF. In this respect, in Section 6.2.2, we first evaluated the form of the scalar field potential inferred from the observed decreasing trend of $H_0(z)$, and our data analysis suggested a model described in Equations (26) and (29). Then, we investigated if, one of the most reliable models for reproducing the Dark Energy effect with modified gravity, i.e., the Hu–Sawicki proposal, was able to induce the requested luminosity distance to somehow remove the observed effect, thus accounting for its physical nature. The non-positive result of this investigation leads us to explore theoretically the question of reproducing simultaneously the Dark Energy contribution and the observed $H_0(z)$ effect, by a single $f(R)$ model of gravity in the JF. In Section 8, it has been addressed this theoretical question, by establishing the conditions that a modified gravity model has to satisfy to reach the simultaneous aims mentioned above. Finally, we considered a specific model for the late universe, based on a slow-rolling picture for the scalar field near its today value $\phi \simeq 1$. This model was successful in explaining the Dark Energy contribution and the necessary variation of the Einstein constant, but it seems hard to be reconciled with the earlier Universe behavior, when the role of the matter contribution becomes relevant. Thus, based on this systematic analysis, we can conclude that the explanation for $H_0(z)$ is probably to be attributed to modified gravity dynamics, but it appears more natural to separate its effect from the existence of a Dark Energy contribution. In other words, we are led to believe that what we discovered about the SNe Ia+BAOs binned analysis must be regarded as a modified gravity physics of the scalar-tensor type, but leaving on a standard Universe, well represented by a ΛCDM model a priori.

9. Conclusions

We analyzed the PS together with the BAOs in three bins in both the ΛCDM and w_0w_aCDM models to investigate if an evolutionary trend of H_0 persists also with the contribution of BAOs and by varying two parameters contemporaneously with H_0 (Ω_{0m} and w_a for the ΛCDM and w_0w_aCDM, respectively). The persistence of the trend of H_0 as a function of redshift is also shown in the case of the Hu–Sawicki model. We here stress that the main goal of the current analysis is to highlight the reliability of the trend of $H_0(z)$ and not to further constrain Ω_{0m} or any other cosmological parameters. With the subsequent fitting of H_0 values through the model $g(z) = \mathcal{H}_0/(1+z)^{\eta}$, we obtain

$\eta \sim 10^{-2}$, as in the previous work [51]: those are compatible with zero from 1.2 to 5.8 σ (see Table 1). The multidimensional results could reveal a dependence on the redshift of H_0, assuming that it is observable at any redshift scale. If this evolution is not caused by statistical effects and other selection biases or hidden evolution of SNe Ia parameters [364], we show how $H_0(z)$ could modify the luminosity distance definition within the modified theory of gravity. If we consider a theoretical interpretation for the observed trend, new cosmological scenarios may explain an evolving Hubble constant with the redshift. For instance, we test in Sections 6.2 and 7 a simple class of modified gravity theories given by the $f(R)$ models in the equivalent scalar-tensor formalism. In principle, this could be due to an effective varying Einstein constant governed by a slow evolution of a scalar field which mediates the gravity-matter interaction. However, the slow decreasing trend of H_0 has proven to be independent of the Hu–Sawicki model application. Indeed, if this theory had worked we would have observed the trend of the η parameter to be flattened out and be compatible with 0 in 1 σ at any redshift bin. This is not the case, thus new scenarios must be explored within the modified theories of gravity or slightly alternative approaches (see Section 8.2). We can state that this evolving trend of H_0 is independent of the starting values of the fitting for H_0 (we here have considered $H_0 = 70$) and, thus, on the fiducial M and on the redshift bins and even when we consider two cosmological parameters changing contemporaneously (Ω_{0m} and w_a in ΛCDM and $w_0 w_a$CDM models, respectively). Thus, we need to further investigate the nature of this trend. In addition, the implementation of GRBs as cosmological probes together with SNe Ia and BAOs has proven to be not only possible in a near future but also necessary since the redshift range that GRBs cover is much larger than the one typical of SNe Ia. This last characteristic will surely allow GRBs to give further information on the nature of the early universe and pose new constraints in the future measurements of H_0.

Author Contributions: M.G.D. performed the conceptualization of all project, data curation, formal analysis, methodology, writing original draft, validation, supervision, software. B.D.S. performed data curation, visualization, formal analysis, methodology, writing original draft, software. T.S. performed formal analysis, visualization, methodology, writing original draft on f(R) and revised it, G.M. performed a partial conceptualization limited to the theoretical part of the f(R) gravity theory; E.R. edited and review the analysis on f(R) and participated in the general discussion and conceptualization of the paper. G.L. wrote Section 6.2.1 and gave suggestions on the cosmological constraints on w. M.B. performed the conceptualization on the priors to answer the referee report. S.U. performed a formal analysis on changing the parameters together with H_0. All authors have read and agreed to the published version of the manuscript.

Funding: This research received no external funding.

Institutional Review Board Statement: Not applicable.

Informed Consent Statement: Not applicable.

Data Availability Statement: Not applicable.

Acknowledgments: This work made use of Pantheon sample data [284], which can be found in the GitHub repository: https://github.com/dscolnic/Pantheon (accessed on 21 December 2020). This work made use of data supplied by the UK Swift Science Data Centre at the University of Leicester. We are thankful to V. Nielson, A. Lenart, G. Sarracino, and D. Jyoti for their support on cosmological computations. T. Schiavone is supported in part by INFN under the program TAsP (Theoretical Astroparticle Physics). G.L. and B.D.S. acknowledge the support of INFN. T.S. acknowledges the support of the Department of Physics of the University of PISA. M.G.D. acknowledges the support from NAOJ and NAOJ—Division of Science.

Conflicts of Interest: The authors declare no conflict of interest.

Notes

1 https://github.com/dscolnic/Pantheon (accessed on 21 December 2020).

2 The code is available upon request.

References

1. Riess, A.G.; Filippenko, A.V.; Challis, P.; Clocchiatti, A.; Diercks, A.; Garnavich, P.M.; Gilliland, R.L.; Hogan, C.J.; Jha, S.; Kirshner, R.P.; et al. Observational Evidence from Supernovae for an Accelerating Universe and a Cosmological Constant. *Astron. J.* **1998**, *116*, 1009–1038. [CrossRef]
2. Perlmutter, S.; Aldering, G.; Goldhaber, G.; Knop, R.A.; Nugent, P.; Castro, P.G.; Deustua, S.; Fabbro, S.; Goobar, A.; Groom, D.E.; et al. Measurements of Ω and Λ from 42 High-Redshift Supernovae. *Astrophys. J.* **1999**, *517*, 565–586. [CrossRef]
3. Weinberg, S. The cosmological constant problem. *Rev. Mod. Phys.* **1989**, *61*, 1–23. [CrossRef]
4. Peebles, P.J.; Ratra, B. The cosmological constant and dark energy. *Rev. Mod. Phys.* **2003**, *75*, 559–606. [CrossRef]
5. Riess, A.G.; Casertano, S.; Yuan, W.; Macri, L.M.; Scolnic, D. Large Magellanic Cloud Cepheid Standards Provide a 1% Foundation for the Determination of the Hubble Constant and Stronger Evidence for Physics beyond ΛCDM. *Astrophys. J.* **2019**, *876*, 85. [CrossRef]
6. Camarena, D.; Marra, V. Local determination of the Hubble constant and the deceleration parameter. *Phys. Rev. Res.* **2020**, *2*, 013028. [CrossRef]
7. Wong, K.C.; Suyu, S.H.; Chen, G.C.F.; Rusu, C.E.; Millon, M.; Sluse, D.; Bonvin, V.; Fassnacht, C.D.; Taubenberger, S.; Auger, M.W.; et al. H0LiCOW–XIII. A 2.4 per cent measurement of H_0 from lensed quasars: 5.3σ tension between early- and late-Universe probes. *Mon. Not. R. Astron. Soc.* **2020**, *498*, 1420–1439. [CrossRef]
8. Gómez-Valent, A.; Amendola, L. H_0 from cosmic chronometers and Type Ia supernovae, with Gaussian Processes and the novel Weighted Polynomial Regression method. *J. Cosmol. Astropart. Phys.* **2018**, *2018*, 051. [CrossRef]
9. Reid, M.J.; Pesce, D.W.; Riess, A.G. An Improved Distance to NGC 4258 and Its Implications for the Hubble Constant. *Astrophys. J. Lett.* **2019**, *886*, L27. [CrossRef]
10. Collaboration, P.; Akrami, Y.; Arroja, F.; Ashdown, M.; Aumont, J.; Baccigalupi, C.; Ballardini, M.; Banday, A.J.; Barreiro, R.B.; Bartolo, N.; et al. Planck 2018 results. I. Overview and the cosmological legacy of Planck. *Astron. Astrophys.* **2020**, *641*, A6. [CrossRef]
11. Rodney, S.A.; Riess, A.G.; Scolnic, D.M.; Jones, D.O.; Hemmati, S.; Molino, A.; McCully, C.; Mobasher, B.; Strolger, L.G.; Graur, O.; et al. Erratum: "Two SNe Ia at redshift ~2: Improved Classification and redshift determination with medium-band infrared imaging" (2015, AJ, 150, 156). *Astron. J.* **2016**, *151*, 47. [CrossRef]
12. Cucchiara, A.; Levan, A.J.; Fox, D.B.; Tanvir, N.R.; Ukwatta, T.N.; Berger, E.; Krühler, T.; Yoldaş, A.K.; Wu, X.F.; Toma, K.; et al. A photometric redshift OFz~ 9.4 for grb 090429B. *Astrophys. J.* **2011**, *736*, 7. [CrossRef]
13. Wang, F.; Yang, J.; Fan, X.; Hennawi, J.F.; Barth, A.J.; Banados, E.; Bian, F.; Boutsia, K.; Connor, T.; Davies, F.B.; et al. A Luminous Quasar at Redshift 7.642. *Astrophys. J. Lett.* **2021**, *907*, L1. [CrossRef]
14. Cardone, V.F.; Capozziello, S.; Dainotti, M.G. An updated gamma-ray bursts Hubble diagram. *Mon. Not. R. Astron. Soc.* **2009**, *400*, 775–790. [CrossRef]
15. Cardone, V.F.; Dainotti, M.G.; Capozziello, S.; Willingale, R. Constraining cosmological parameters by gamma-ray burst X-ray afterglow light curves. *Mon. Not. R. Astron. Soc.* **2010**, *408*, 1181–1186. [CrossRef]
16. Cardone, V.F.; Perillo, M.; Capozziello, S. Systematics in the gamma-ray burst Hubble diagram. *Mon. Not. R. Astron. Soc.* **2011**, *417*, 1672–1683. [CrossRef]
17. Dainotti, M.G.; Cardone, V.F.; Piedipalumbo, E.; Capozziello, S. Slope evolution of GRB correlations and cosmology. *Mon. Not. R. Astron. Soc.* **2013**, *436*, 82–88. [CrossRef]
18. Postnikov, S.; Dainotti, M.G.; Hernandez, X.; Capozziello, S. Nonparametric Study of the Evolution of the Cosmological Equation of State with Sneia, Bao, and High-Redshift Grbs. *Astrophys. J.* **2014**, *783*, 126. [CrossRef]
19. Amati, L.; Frontera, F.; Tavani, M.; Antonelli, A.; Costa, E.; Feroci, M.; Guidorzi, C.; Heise, J.; Masetti, N.; Montanari, E.; et al. Intrinsic spectra and energetics of BeppoSAX Gamma-Ray Bursts with known redshifts. *Astron. Astrophys.* **2002**, *390*, 81–89. [CrossRef]
20. Yonetoku, D.; Murakami, T.; Nakamura, T.; Yamazaki, R.; Inoue, A.K.; Ioka, K. Gamma-Ray Burst Formation Rate Inferred from the Spectral Peak Energy–Peak Luminosity Relation. *Astrophys. J.* **2004**, *609*, 935–951. [CrossRef]
21. Ito, H.; Matsumoto, J.; Nagataki, S.; Warren, D.C.; Barkov, M.V.; Yonetoku, D. The photospheric origin of the Yonetoku relation in gamma-ray bursts. *Nat. Commun.* **2019**, *10*. [CrossRef] [PubMed]
22. Liang, E.; Zhang, B. Model-independent Multivariable Gamma-Ray Burst Luminosity Indicator and Its Possible Cosmological Implications. *Astrophys. J.* **2005**, *633*, 611–623. [CrossRef]
23. Ghirlanda, G.; Nava, L.; Ghisellini, G. Spectral-luminosity relation within individual Fermi gamma rays bursts. *Astron. Astrophys.* **2010**, *511*, A43. [CrossRef]
24. Dainotti, M.G.; Cardone, V.F.; Capozziello, S. A time-luminosity correlation for γ-ray bursts in the X-rays. *Mon. Not. R. Astron. Soc.: Letters* **2008**, *391*, L79–L83. [CrossRef]
25. Dainotti, M.G.; Willingale, R.; Capozziello, S.; Cardone, V.F.; Ostrowski, M. Discovery of a Tight Correlation for Gamma-Ray Burst Afterglows with "Canonical" Light Curves. *Astrophys. J. Lett.* **2010**, *722*, L215–L219. [CrossRef]
26. Dainotti, M.G.; Cardone, V.F.; Capozziello, S.; Ostrowski, M.; Willingale, R. Study of possible systematics in the L*X - Ta* correlation of Gamma Ray Bursts. *Astrophys. J.* **2011**, *730*, 135. [CrossRef]
27. Dainotti, M.G.; Petrosian, V.; Singal, J.; Ostrowski, M. Determination of the intrinsic luminosity time correlation in the X-ray afterglows of gamma-ray bursts. *Astrophys. J.* **2013**, *774*, 157. [CrossRef]

28. Dainotti, M.G.; Vecchio, R.D.; Shigehiro, N.; Capozziello, S. Selection effects in Gamma-Ray Burst Correlations: Consequences on the Ratio Between Gamma-Ray Burst And Star Formation rates. *Astrophys. J.* **2015**, *800*, 31. [CrossRef]
29. Dainotti, M.; Petrosian, V.; Willingale, R.; O'Brien, P.; Ostrowski, M.; Nagataki, S. Luminosity–time and luminosity–luminosity correlations for GRB prompt and afterglow plateau emissions. *Mon. Not. R. Astron. Soc.* **2015**, *451*, 3898–3908. [CrossRef]
30. Dainotti, M.G.; Postnikov, S.; Hernandez, X.; Ostrowski, M. A fundamental Plane for Long Gamma-Ray Bursts with X-Ray plateaus. *Astrophys. J. Lett.* **2016**, *825*, L20. [CrossRef]
31. Dainotti, M.G.; Nagataki, S.; Maeda, K.; Postnikov, S.; Pian, E. A study of gamma ray bursts with afterglow plateau phases associated with supernovae. *Astron. Astrophys.* **2017**, *600*, A98. [CrossRef]
32. Dainotti, M.G.; Hernandez, X.; Postnikov, S.; Nagataki, S.; O'brien, P.; Willingale, R.; Striegel, S. A Study of the Gamma-Ray Burst Fundamental Plane. *Astrophys. J.* **2017**, *848*, 88. [CrossRef]
33. Dainotti, M.G.; Amati, L. Gamma-ray Burst Prompt Correlations: Selection and Instrumental Effects. *Astrophys. J.* **2018**, *130*, 051001. [CrossRef]
34. Dainotti, M.G.; Del Vecchio, R.; Tarnopolski, M. Gamma-Ray Burst Prompt Correlations. *Adv. Astron.* **2018**, *2018*, 1–31. [CrossRef]
35. Dainotti, M.G.; Lenart, A.Ł.; Sarracino, G.; Nagataki, S.; Capozziello, S.; Fraija, N. The X-ray Fundamental Plane of the Platinum Sample, the Kilonovae, and the SNe Ib/c Associated with GRBs. *Astrophys. J.* **2020**, *904*, 97. [CrossRef]
36. Dainotti, M.G.; Omodei, N.; Srinivasaragavan, G.P.; Vianello, G.; Willingale, R.; O'Brien, P.; Nagataki, S.; Petrosian, V.; Nuygen, Z.; Hernandez, X.; et al. On the Existence of the Plateau Emission in High-energy Gamma-Ray Burst Light Curves Observed by Fermi-LAT. *Astrophys. J. Suppl. Ser.* **2021**, *255*, 13. [CrossRef]
37. Dainotti, M.; Levine, D.; Fraija, N.; Chandra, P. Accounting for Selection Bias and Redshift Evolution in GRB Radio Afterglow Data. *Galaxies* **2021**, *9*, 95. [CrossRef]
38. Vecchio, R.D.; Dainotti, M.G.; Ostrowski, M. Study of Grb Light-Curve Decay Indices in the Afterglow phase. *Astrophys. J.* **2016**, *828*, 36. [CrossRef]
39. Duncan, R.C. Gamma-ray bursts from extragalactic Magnetar Flares. *AIP Conf. Proc.* **2001**, 586, 495–500. [CrossRef]
40. Dall'Osso, S.; Granot, J.; Piran, T. Magnetic field decay in neutron stars: From soft gamma repeaters to "weak-field magnetars". *Mon. Not. R. Astron. Soc.* **2012**, *422*, 2878–2903. [CrossRef]
41. Rowlinson, A.; Gompertz, B.P.; Dainotti, M.; O'Brien, P.T.; Wijers, R.A.M.J.; van der Horst, A.J. Constraining properties of GRB magnetar central engines using the observed plateau luminosity and duration correlation. *Mon. Not. R. Astron. Soc.* **2014**, *443*, 1779–1787. [CrossRef]
42. Rea, N.; Gullón, M.; Pons, J.A.; Perna, R.; Dainotti, M.G.; Miralles, J.A.; Torres, D.F. Constraining The Grb-Magnetar Model By Means Of The Galactic Pulsar PopulatioN. *Astrophys. J.* **2015**, *813*, 92. [CrossRef]
43. Stratta, G.; Dainotti, M.G.; Dall'Osso, S.; Hernandez, X.; Cesare, G.D. On the Magnetar Origin of the GRBs Presenting X-Ray Afterglow Plateaus. *Astrophys. J.* **2018**, *869*, 155. [CrossRef]
44. Amati, L.; D'Agostino, R.; Luongo, O.; Muccino, M.; Tantalo, M. Addressing the circularity problem in the E_p-E_{iso} correlation of gamma-ray bursts. *Mon. Not. R. Astron. Soc. Lett.* **2019**, *486*, L46–L51. [CrossRef]
45. Liao, K.; Shafieloo, A.; Keeley, R.E.; Linder, E.V. A Model-independent Determination of the Hubble Constant from Lensed Quasars and Supernovae Using Gaussian Process Regression. *Astrophys. J. Lett.* **2019**, *886*, L23. [CrossRef]
46. Keeley, R.E.; Shafieloo, A.; Hazra, D.K.; Souradeep, T. Inflation wars: A new hope. *J. Cosmol. Astropart. Phys.* **2020**, *2020*, 055. [CrossRef]
47. Risaliti, G.; Lusso, E. Cosmological Constraints from the Hubble Diagram of Quasars at High Redshifts. *Nat. Astron.* **2019**, *3*, 272–277. [CrossRef]
48. Freedman, W.L.; Madore, B.F.; Hatt, D.; Hoyt, T.J.; Jang, I.S.; Beaton, R.L.; Burns, C.R.; Lee, M.G.; Monson, A.J.; Neeley, J.R.; et al. The Carnegie-Chicago Hubble Program. VIII. An Independent Determination of the Hubble Constant Based on the Tip of the Red Giant Branch. *Astrophys. J.* **2019**, *882*, 34. [CrossRef]
49. Dutta, K.; Roy, A.; Ruchika; Sen, A.A.; Sheikh-Jabbari, M.M. Cosmology with low-redshift observations: No signal for new physics. *Phys. Rev. D* **2019**, *100*, 103501. [CrossRef]
50. Yang, T.; Banerjee, A.; Ó Colgáin, E. Cosmography and flat ΛCDM tensions at high redshift. *Phys. Rev. D* **2020**, *102*, 123532. [CrossRef]
51. Dainotti, M.G.; De Simone, B.; Schiavone, T.; Montani, G.; Rinaldi, E.; Lambiase, G. On the Hubble Constant Tension in the SNe Ia Pantheon Sample. *Astrophys. J.* **2021**, *912*, 150. [CrossRef]
52. Ingram, A.; Mastroserio, G.; van der Klis, M.; Nathan, E.; Connors, R.; Dauser, T.; García, J.A.; Kara, E.; König, O.; Lucchini, M.; et al. On measuring the Hubble constant with X-ray reverberation mapping of active galactic nuclei. *arXiv* **2021**, arXiv:2110.15651.
53. Agrawal, P.; Cyr-Racine, F.Y.; Pinner, D.; Randall, L. Rock 'n' Roll Solutions to the Hubble Tension. *arXiv* **2019**, arXiv:1904.01016.
54. Arjona, R.; Espinosa-Portales, L.; García-Bellido, J.; Nesseris, S. A GREAT model comparison against the cosmological constant. *arXiv* **2021**, arXiv:2111.13083.
55. Fernandez-Martinez, E.; Pierre, M.; Pinsard, E.; Rosauro-Alcaraz, S. Inverse Seesaw, dark matter and the Hubble tension. *Eur. Phys. J. C* **2021**, *81*, 954. [CrossRef]
56. Ghose, S.; Bhadra, A. Is non-particle dark matter equation of state parameter evolving with time? *Eur. Phys. J. C* **2021**, *81*, 683. [CrossRef]

57. Hart, L.; Chluba, J. Varying fundamental constants principal component analysis: Additional hints about the Hubble tension. *Mon. Not. R. Astron. Soc.* **2021**, *510*, 2206–2227. [CrossRef]
58. Beltrán, J.B.; Bettoni, D.; Figueruelo, D.; Pannia, F.A.T.; Tsujikawa, S. Probing elastic interactions in the dark sector and the role of S8. *Phys. Rev. D.* **2021**, *104*, 103503. [CrossRef]
59. Rezaei, M.; Peracaula, J.; Malekjani, M. Cosmographic approach to Running Vacuum dark energy models: New constraints using BAOs and Hubble diagrams at higher redshifts. *Mon. Not. R. Astron. Soc.* **2022**, *509*, 2593–2608. [CrossRef]
60. Shah, P.; Lemos, P.; Lahav, O. A buyer's guide to the Hubble constant. *Astron. Astrophys. Rev.* **2021**, *29*, 9. [CrossRef]
61. Firouzjahi, H. Cosmological constant problem on the horizon. *arXiv* **2022**, arXiv:2201.02016.
62. Banihashemi, A.; Khosravi, N.; Shafieloo, A. Dark energy as a critical phenomenon: A hint from Hubble tension. *J. Cosmol. Astropart. Phys.* **2021**, *2021*, 003. [CrossRef]
63. Ballardini, M.; Finelli, F.; Sapone, D. Cosmological constraints on Newton's gravitational constant. *arXiv* **2021**, arXiv:2111.09168.
64. Corona, M.A.; Murgia, R.; Cadeddu, M.; Archidiacono, M.; Gariazzo, S.; Giunti, C.; Hannestad, S. Pseudoscalar sterile neutrino self-interactions in light of Planck, SPT and ACT data. *arXiv* **2021**, arXiv:astro-ph.CO/2112.00037.
65. Cyr-Racine, F.Y.; Ge, F.; Knox, L. A Symmetry of Cosmological Observables, and a High Hubble Constant as an Indicator of a Mirror World Dark Sector. *arXiv* **2021**, arXiv:astro-ph.CO/2107.13000.
66. Valentino, E.D.; Gariazzo, S.; Giunti, C.; Mena, O.; Pan, S.; Yang, W. Minimal dark energy: Key to sterile neutrino and Hubble constant tensions? *arXiv* **2021**, arXiv:astro-ph.CO/2110.03990.
67. Valentino, E.D.; Melchiorri, A. Neutrino Mass Bounds in the era of Tension Cosmology. *arXiv* **2021**, arXiv:astro-ph.CO/2112.02993.
68. Drees, M.; Zhao, W. $U(1)_{L_\mu - L_\tau}$ for Light Dark Matter, $g_\mu - 2$, the 511 keV excess and the Hubble Tension. *arXiv* **2021**, arXiv:hep-ph/2107.14528.
69. Gu, Y.; Wu, L.; Zhu, B. Axion Dark Radiation and Late Decaying Dark Matter in Neutrino Experiment. *arXiv* **2021**, arXiv:hep-ph/2105.07232.
70. Khalifeh, A.R.; Jimenez, R. Using Neutrino Oscillations to Measure H_0. *arXiv* **2021**, arXiv:astro-ph.CO/2111.15249.
71. Li, J.; Zhou, Y.; Xue, X. Spatial Curvature and Large Scale Lorentz Violation. *arXiv* **2021**, arXiv:gr-qc/2112.02364.
72. Lulli, M.; Marciano, A.; Shan, X. Stochastic Quantization of General Relativity à la Ricci-Flow. *arXiv* **2021**, arXiv:gr-qc/2112.01490.
73. Mawas, E.; Street, L.; Gass, R.; Wijewardhana, L.C.R. Interacting dark energy axions in light of the Hubble tension. *arXiv* **2021**, arXiv:astro-ph.CO/2108.13317.
74. Moreno-Pulido, C.; Peracaula, J.S. Renormalized $\rho_{\rm vac}$ without m^4 terms. *arXiv* **2021**, arXiv:gr-qc/2110.08070.
75. Naidoo, K.; Massara, E.; Lahav, O. Cosmology and neutrino mass with the Minimum Spanning Tree. *arXiv* **2021**, arXiv:astro-ph.CO/2111.12088.
76. Niedermann, F.; Sloth, M.S. Hot New Early Dark Energy: Towards a Unified Dark Sector of Neutrinos, Dark Energy and Dark Matter. *arXiv* **2021**, arXiv:hep-ph/2112.00759.
77. Nilsson, N.A.; Park, M.I. Tests of Standard Cosmology in Horava Gravity. *arXiv* **2021**, arXiv:hep-th/2108.07986.
78. Ray, P.P.; Tarai, S.; Mishra, B.; Tripathy, S.K. Cosmological models with Big rip and Pseudo rip Scenarios in extended theory of gravity. *arXiv* **2021**, arXiv:gr-qc/2107.04413.
79. Schöneberg, N.; Abellán, G.F.; Sánchez, A.P.; Witte, S.J.; Poulin, V.; Lesgourgues, J. The H_0 Olympics: A fair ranking of proposed models. *arXiv* **2021**, arXiv:astro-ph.CO/2107.10291.
80. Trott, E.; Huterer, D. Challenges for the statistical gravitational-wave method to measure the Hubble constant. *arXiv* **2021**, arXiv:astro-ph.CO/2112.00241.
81. Ye, G.; Zhang, J.; Piao, Y.S. Resolving both H_0 and S_8 tensions with AdS early dark energy and ultralight axion. *arXiv* **2021**, arXiv:astro-ph.CO/2107.13391.
82. Zhou, Z.; Liu, G.; Xu, L. Can late dark energy restore the Cosmic concordance? *arXiv* **2021**, arXiv:astro-ph.CO/2105.04258.
83. Zhu, L.G.; Xie, L.H.; Hu, Y.M.; Liu, S.; Li, E.K.; Napolitano, N.R.; Tang, B.T.; dong Zhang, J.; Mei, J. Constraining the Hubble constant to a precision of about 1 multi-band dark standard siren detections. *arXiv* **2021**, arXiv:astro-ph.CO/2110.05224.
84. Alestas, G.; Perivolaropoulos, L.; Tanidis, K. Constraining a late time transition of $G_{\rm eff}$ using low-z galaxy survey data. *arXiv* **2022**, arXiv:astro-ph.CO/2201.05846.
85. Cea, P. The Ellipsoidal Universe and the Hubble tension. *arXiv* **2022**, arXiv:astro-ph.CO/2201.04548.
86. Gurzadyan, V.G.; Fimin, N.N.; Chechetkin, V.M. On the origin of cosmic web. *Eur. Phys. J. Plus* **2022**, *137*. [CrossRef]
87. Rashkovetskyi, M.; Muñoz, J.; Eisenstein, D.; Dvorkin, C. Small-scale clumping at recombination and the Hubble tension. *Phys. Rev. D* **2021**, *104*, 103517. [CrossRef]
88. Das, A. Self-interacting neutrinos as a solution to the Hubble tension? *arXiv* **2021**, arXiv:2109.03263.
89. Karwal, T.; Kamionkowski, M. Dark energy at early times, the Hubble parameter, and the string axiverse. *Phys. Rev. D.* **2016**, *94*, e103523. [CrossRef]
90. Yang, W.; Pan, S.; Di Valentino, E.; Nunes, R.C.; Vagnozzi, S.; Mota, D.F. Tale of stable interacting dark energy, observational signatures, and the H_0 tension. *J. Cosmol. Astropart. Phys.* **2018**, *2018*, 019. [CrossRef]
91. Di Valentino, E.; Linder, E.V.; Melchiorri, A. Vacuum phase transition solves the H_0 tension. *Phys. Rev. D* **2018**, *97*, 043528. [CrossRef]

92. Di Valentino, E.; Ferreira, R.Z.; Visinelli, L.; Danielsson, U. Late time transitions in the quintessence field and the H_0 tension. *Phys. Dark Univ.* **2019**, *26*, 100385. [CrossRef]
93. Pan, S.; Yang, W.; Di Valentino, E.; Shafieloo, A.; Chakraborty, S. Reconciling H_0 tension in a six parameter space? *J. Cosmol. Astropart. Phys.* **2020**, *2020*, 062. [CrossRef]
94. Di Valentino, E.; Anchordoqui, L.A.; Akarsu, O.; Ali-Haimoud, Y.; Amendola, L.; Arendse, N.; Asgari, M.; Ballardini, M.; Basilakos, S.; Battistelli, E.; et al. Snowmass2021 - Letter of interest cosmology intertwined I: Perspectives for the next decade. *Astropart. Phys.* **2021**, *131*, 102606. [CrossRef]
95. Di Valentino, E.; Anchordoqui, L.A.; Akarsu, O.; Ali-Haimoud, Y.; Amendola, L.; Arendse, N.; Asgari, M.; Ballardini, M.; Basilakos, S.; Battistelli, E.; et al. Snowmass2021—Letter of interest cosmology intertwined II: The hubble constant tension. *Astropart. Phys.* **2021**, *131*, 102605. [CrossRef]
96. Di Valentino, E.; Anchordoqui, L.A.; Akarsu, O.; Ali-Haimoud, Y.; Amendola, L.; Arendse, N.; Asgari, M.; Ballardini, M.; Basilakos, S.; Battistelli, E.; et al. Cosmology Intertwined III: $f\sigma 8$ and S8 *Astropart. Phys.* **2021**, *131*, 102604. [CrossRef]
97. Di Valentino, E.; Anchordoqui, L.A.; Akarsu, O.; Ali-Haimoud, Y.; Amendola, L.; Arendse, N.; Asgari, M.; Ballardini, M.; Basilakos, S.; Battistelli, E.; et al. Snowmass2021—Letter of interest cosmology intertwined IV: The age of the universe and its curvature. *Astropart. Phys.* **2021**, *131*, 102607. [CrossRef]
98. Di Valentino, E.; Bœhm, C.; Hivon, E.; Bouchet, F.R. Reducing the H_0 and σ_8 tensions with dark matter-neutrino interactions. *Phys. Rev. D* **2018**, *97*, 043513. [CrossRef]
99. Di Valentino, E.; Pan, S.; Yang, W.; Anchordoqui, L.A. Touch of neutrinos on the vacuum metamorphosis: Is the H_0 solution back? *Phys. Rev. D* **2021**, *103*, 123527. [CrossRef]
100. Anchordoqui, L.A.; Di Valentino, E.; Pan, S.; Yang, W. Dissecting the H_0 and S_8 tensions with Planck + BAO + supernova type Ia in multi-parameter cosmologies. *J. High Energy Astrophys.* **2021**, *32*, 28–64. [CrossRef]
101. Di Valentino, E.; Mukherjee, A.; Sen, A.A. Dark Energy with Phantom Crossing and the H0 Tension. *Entropy* **2021**, *23*, 404. [CrossRef]
102. Di Valentino, E.; Linder, E.V.; Melchiorri, A. H_0 ex machina: Vacuum metamorphosis and beyond H_0. *Phys. Dark Univ.* **2020**, *30*, 100733. [CrossRef]
103. Di Valentino, E.; Melchiorri, A.; Silk, J. Planck evidence for a closed Universe and a possible crisis for cosmology. *Nat. Astron.* **2020**, *4*, 196–203. [CrossRef]
104. Allali, I.J.; Hertzberg, M.P.; Rompineve, F. Dark sector to restore cosmological concordance. *Phys. Rev. D* **2021**, *104*. [CrossRef]
105. Anderson, R.I. Relativistic corrections for measuring Hubble's constant to 1 stellar standard candles. *arXiv* **2021**, arXiv:astro-ph.CO/2108.09067.
106. Asghari, M.; Sheykhi, A. Observational constraints of the modified cosmology through Barrow entropy. *arXiv* **2021**, arXiv:gr-qc/2110.00059.
107. Brownsberger, S.; Brout, D.; Scolnic, D.; Stubbs, C.W.; Riess, A.G. The Pantheon+ Analysis: Dependence of Cosmological Constraints on Photometric-Zeropoint Uncertainties of Supernova Surveys. *arXiv* **2021**, arXiv:astro-ph.CO/2110.03486.
108. Cyr-Racine, F.Y. Cosmic Expansion: A mini review of the Hubble-Lemaitre tension. *arXiv* **2021**, arXiv:astro-ph.CO/2105.09409.
109. Khosravi, N.; Farhang, M. Phenomenological Gravitational Phase Transition: Early and Late Modifications. *arXiv* **2021**, arXiv:astro-ph.CO/2109.10725.
110. Mantz, A.B.; Morris, R.G.; Allen, S.W.; Canning, R.E.A.; Baumont, L.; Benson, B.; Bleem, L.E.; Ehlert, S.R.; Floyd, B.; Herbonnet, R.; et al. Cosmological constraints from gas mass fractions of massive, relaxed galaxy clusters. *Mon. Not. R. Astron. Soc.* **2021**, *510*, 131–145. [CrossRef]
111. Mortsell, E.; Goobar, A.; Johansson, J.; Dhawan, S. The Hubble Tension Bites the Dust: Sensitivity of the Hubble Constant Determination to Cepheid Color Calibration. *arXiv* **2021**, arXiv:astro-ph.CO/2105.11461.
112. Mortsell, E.; Goobar, A.; Johansson, J.; Dhawan, S. The Hubble Tension Revisited: Additional Local Distance Ladder Uncertainties. *arXiv* **2021**, arXiv:astro-ph.CO/2106.09400.
113. Theodoropoulos, A.; Perivolaropoulos, L. The Hubble Tension, the M Crisis of Late Time H(z) Deformation Models and the Reconstruction of Quintessence Lagrangians. *Universe* **2021**, *7*, 300. [CrossRef]
114. Gómez-Valent, A. Measuring the sound horizon and absolute magnitude of SNIa by maximizing the consistency between low-redshift data sets. *arXiv* **2022**, arXiv:astro-ph.CO/2111.15450.
115. Pol, A.R.; Caprini, C.; Neronov, A.; Semikoz, D. The gravitational wave signal from primordial magnetic fields in the Pulsar Timing Array frequency band. *arXiv* **2022**, arXiv:astro-ph.CO/2201.05630.
116. Wong, J.H.W.; Shanks, T.; Metcalfe, N. The Local Hole: a galaxy under-density covering 90% of sky to 200 Mpc. *arXiv* **2022**, arXiv:astro-ph.CO/2107.08505.
117. Romaniello, M.; Riess, A.; Mancino, S.; Anderson, R.I.; Freudling, W.; Kudritzki, R.P.; Macri, L.; Mucciarelli, A.; Yuan, W. The iron and oxygen content of LMC Classical Cepheids and its implications for the Extragalactic Distance Scale and Hubble constant. *arXiv* **2021**, arXiv:astro-ph.CO/2110.08860.
118. Luu, H.N. Axi-Higgs cosmology: Cosmic Microwave Background and cosmological tensions. *arXiv* **2021**, arXiv:astro-ph.CO/2111.01347.
119. Alestas, G.; Kazantzidis, L.; Perivolaropoulos, L. w -M phantom transition at z_t<0.1 as a resolution of the Hubble tension. *Phys. Rev. D* **2021**, *103*, 083517. [CrossRef]

120. Sakr, Z.; Sapone, D. Can varying the gravitational constant alleviate the tensions? *arXiv* **2021**, arXiv:astro-ph.CO/2112.14173.
121. Wang, Y.Y.; Tang, S.P.; Li, X.Y.; Jin, Z.P.; Fan, Y.Z. Prospects of calibrating afterglow modeling of short GRBs with gravitational wave inclination angle measurements and resolving the Hubble constant tension with a GW/GRB association event. *arXiv* **2021**, arXiv:astro-ph.HE/2111.02027.
122. Safari, Z.; Rezazadeh, K.; Malekolkalami, B. Structure Formation in Dark Matter Particle Production Cosmology. *arXiv* **2022**, arXiv:astro-ph.CO/2201.05195.
123. Roth, M.M.; Jacoby, G.H.; Ciardullo, R.; Davis, B.D.; Chase, O.; Weilbacher, P.M. Towards Precision Cosmology With Improved PNLF Distances Using VLT-MUSE I. Methodology and Tests. *Astrophys. J.* **2021**, *916*, 21. [CrossRef]
124. Gutiérrez-Luna, E.; Carvente, B.; Jaramillo, V.; Barranco, J.; Escamilla-Rivera, C.; Espinoza, C.; Mondragón, M.; Núñez, D. Scalar field dark matter with two components: Combined approach from particle physics and cosmology. *arXiv* **2021**, arXiv:astro-ph.CO/2110.10258.
125. Chang, C. Imprint of early dark energy in stochastic gravitational wave background. *Phys. Rev. D* **2022**, *105*, 023508. [CrossRef]
126. Liu, S.; Zhu, L.; Hu, Y.; Zhang, J.; Ji, M. Capability for detection of GW190521-like binary black holes with TianQin. *Phys. Rev. D* **2022**, *105*, 023019. [CrossRef]
127. Farrugia, C.; Sultana, J.; Mifsud, J. Spatial curvature in f(R) gravity. *Phys. Rev. D* **2021**, *104*, 123503. [CrossRef]
128. Lu, W.; Qin, Y. New constraint of the Hubble constant by proper motions of radio components observed in AGN twin-jets. *Res. Astron. Astrophys.* **2021**, *21*, 261. [CrossRef]
129. Greene, K.; Cyr-Racine, F. Hubble distancing: Focusing on distance measurements in cosmology. *arXiv* **2021**, arXiv:2112.11567.
130. Borghi, N.; Moresco, M.; Cimatti, A. Towards a Better Understanding of Cosmic Chronometers: A new measurement of H(z) at $z \sim 0.7$. *arXiv* **2021**, arXiv:2110.04304.
131. Asencio, E.; Banik, I.; Kroupa, P. A massive blow for ΛCDM—The high redshift, mass, and collision velocity of the interacting galaxy cluster El Gordo contradicts concordance cosmology. *Mon. Not. R. Astron. Soc.* **2021**, *500*, 5249–5267. [CrossRef]
132. Javanmardi, B.; Mérand, A.; Kervella, P.; Breuval, L.; Gallenne, A.; Nardetto, N.; Gieren, W.; Pietrzyński, G.; Hocdé, V.; Borgniet, S. Inspecting the Cepheid Distance Ladder: The Hubble Space Telescope Distance to the SN Ia Host Galaxy NGC 5584. *Astrophys. J.* **2021**, *911*, 12. [CrossRef]
133. Zhao, D.; Xia, J.Q. Constraining the anisotropy of the Universe with the X-ray and UV fluxes of quasars. *arXiv* **2021**, arXiv:2105.03965.
134. Thakur, R.K.; Singh, M.; Gupta, S.; Nigam, R. Cosmological Analysis using Panstarrs data: Hubble Constant and Direction Dependence. *arXiv* **2021**, arXiv:2105.14514.
135. Sharov, G.S.; Sinyakov, E.S. Cosmological models, observational data and tension in Hubble constant. *arXiv* **2020**, arXiv:2002.03599.
136. Vagnozzi, S.; Pacucci, F.; Loeb, A. Implications for the Hubble tension from the ages of the oldest astrophysical objects. *arXiv* **2021**, arXiv:2105.10421.
137. Staicova, D. Hints of the $H_0 - r_d$ tension in uncorrelated Baryon Acoustic Oscillations dataset. *arXiv* **2021**, arXiv:astro-ph.CO/2111.07907.
138. Krishnan, C.; Mohayaee, R.; Colgain, E.O.; Sheikh-Jabbari, M.M.; Yin, L. Hints of FLRW Breakdown from Supernovae. *arXiv* **2021**, arXiv:astro-ph.CO/2106.02532.
139. Li, B.; Shapiro, P.R. Precision cosmology and the stiff-amplified gravitational-wave background from inflation: NANOGrav, Advanced LIGO-Virgo and the Hubble tension. *J. Cosmol. Astropart. Phys.* **2021**, *2021*, 024. [CrossRef]
140. Mozzon, S.; Ashton, G.; Nuttall, L.K.; Williamson, A.R. Does non-stationary noise in LIGO and Virgo affect the estimation of H_0? *arXiv* **2021**, arXiv:astro-ph.CO/2110.11731.
141. Abbott, T.C.; Buffaz, E.; Vieira, N.; Cabero, M.; Haggard, D.; Mahabal, A.; McIver, J. GWSkyNet-Multi: A Machine Learning Multi-Class Classifier for LIGO-Virgo Public Alerts. *arXiv* **2021**, arXiv:astro-ph.IM/2111.04015.
142. Mehrabi, A.; Basilakos, S.; Tsiapi, P.; Plionis, M.; Terlevich, R.; Terlevich, E.; Moran, A.L.G.; Chavez, R.; Bresolin, F.; Arenas, D.F.; et al. Using our newest VLT-KMOS HII Galaxies and other cosmic tracers to test the ΛCDM tension. *Mon. Not. R. Astron. Soc.* **2021**, *509*, 224–231. [CrossRef]
143. Li, J.X.; Wu, F.Q.; Li, Y.C.; Gong, Y.; Chen, X.L. Cosmological constraint on Brans–Dicke Model. *Res. Astron. Astrophys.* **2015**, *15*, 2151–2163. [CrossRef]
144. Freedman, W.L. Measurements of the Hubble Constant: Tensions in Perspective. *Astrophys. J.* **2021**, *919*, 16. [CrossRef]
145. Wu, Q.; Zhang, G.Q.; Wang, F.Y. An 8% Determination of the Hubble Constant from localized Fast Radio Bursts. *arXiv* **2021**, arXiv:astro-ph.CO/2108.00581.
146. Perivolaropoulos, L.; Skara, F. Hubble tension or a transition of the Cepheid Sn Ia calibrator parameters? *arXiv* **2021**, arXiv:astro-ph.CO/2109.04406.
147. Horstmann, N.; Pietschke, Y.; Schwarz, D.J. Inference of the cosmic rest-frame from supernovae Ia. *arXiv* **2021**, arXiv:astro-ph.CO/2111.03055.
148. Ferree, N.C.; Bunn, E.F. Constraining H_0 Via Extragalactic Parallax. *arXiv* **2021**, arXiv:astro-ph.CO/2109.07529.
149. Luongo, O.; Muccino, M.; Colgáin, E.O.; Sheikh-Jabbari, M.M.; Yin, L. On Larger H_0 Values in the CMB Dipole Direction. *arXiv* **2021**, arXiv:astro-ph.CO/2108.13228.

150. de Souza, J.M.S.; Sturani, R.; Alcaniz, J. Cosmography with Standard Sirens from the Einstein Telescope. *arXiv* **2021**, arXiv:gr-qc/2110.13316.
151. Fang, Y.; Yang, H. Orbit Tomography of Binary Supermassive Black Holes with Very Long Baseline Interferometry. *arXiv* **2021**, arXiv:gr-qc/2111.00368.
152. Palmese, A.; Bom, C.R.; Mucesh, S.; Hartley, W.G. A standard siren measurement of the Hubble constant using gravitational wave events from the first three LIGO/Virgo observing runs and the DESI Legacy Survey. *arXiv* **2021**, arXiv:astro-ph.CO/2111.06445.
153. Yang, T.; Lee, H.M.; Cai, R.G.; gil Choi, H.; Jung, S. Space-borne Atom Interferometric Gravitational Wave Detections II: Dark Sirens and Finding the One. *arXiv* **2021**, arXiv:gr-qc/2110.09967.
154. Gray, R.; Messenger, C.; Veitch, J. A Pixelated Approach to Galaxy Catalogue Incompleteness: Improving the Dark Siren Measurement of the Hubble Constant. *arXiv* **2021**, arXiv:astro-ph.CO/2111.04629.
155. Moresco, M.; Amati, L.; Amendola, L.; Birrer, S.; Blakeslee, J.P.; Cantiello, M.; Cantiello, A.; Darling, J.; Della Valle, M.; Fishbach, M.; et al. Unveiling the Universe with Emerging Cosmological Probes. *arXiv* **2022**, arXiv:astro-ph.CO/2201.07241.
156. The LIGO Scientific Collaboration; the Virgo Collaboration; the KAGRA Collaboration; Abbott, R.; Abe, H.; Acernese, F.; Ackley, K.; Adhikari, N.; Adhikari, R.X.; Adkins, V.K.; et al. Constraints on the cosmic expansion history from GWTC-3. *arXiv* **2021**, arXiv:2111.03604.
157. Nunes, R.C.; Pan, S.; Saridakis, E.N.; Abreu, E.M. New observational constraints onf(R) gravity from cosmic chronometers. *J. Cosmol. Astropart. Phys.* **2017**, *2017*, 005. [CrossRef]
158. Nunes, R.C. Structure formation in f(T) gravity and a solution for H0 tension. *J. Cosmol. Astropart. Phys.* **2018**, *2018*, 052. [CrossRef]
159. Benetti, M.; Capozziello, S.; Lambiase, G. Updating constraints on f(T) teleparallel cosmology and the consistency with big bang nucleosynthesis. *Mon. Not. R. Astron. Soc.* **2020**, *500*, 1795–1805. [CrossRef]
160. Räsänen, S. Light propagation in statistically homogeneous and isotropic dust universes. *J. Cosmol. Astropart. Phys.* **2009**, *2009*, 011. [CrossRef]
161. Räsänen, S. Light propagation in statistically homogeneous and isotropic universes with general matter content. *J. Cosmol. Astropart. Phys.* **2010**, *2010*, 018. [CrossRef]
162. Sotiriou, T.P. f(R) gravity and scalar tensor theory. *Class. Quantum Gravity* **2006**, *23*, 5117–5128. [CrossRef]
163. Nojiri, S.; Odintsov, S.D. Introduction to modified gravity and gravitational alternative for dark energy. *Int. J. Geom. Methods Mod. Phys.* **2007**, *04*, 115–145. [CrossRef]
164. Koksbang, S.M. Towards statistically homogeneous and isotropic perfect fluid universes with cosmic backreaction. *Class. Quantum Gravity* **2019**, *36*, 185004. [CrossRef]
165. Koksbang, S. Another look at redshift drift and the backreaction conjecture. *J. Cosmol. Astropart. Phys.* **2019**, *2019*, 036. [CrossRef]
166. Koksbang, S.M. Observations in statistically homogeneous, locally inhomogeneous cosmological toy-models without FLRW backgrounds. *arXiv* **2020**, arXiv:astro-ph.CO/2008.07108.
167. Sotiriou, T.P.; Faraoni, V. $f(R)$ theories of gravity. *Rev. Mod. Phys.* **2010**, *82*, 451–497. [CrossRef]
168. Odderskov, I.; Koksbang, S.; Hannestad, S. The local value ofH0in an inhomogeneous universe. *J. Cosmol. Astropart. Phys.* **2016**, *2016*, 001. [CrossRef]
169. Nájera, A.; Fajardo, A. Testing $f(Q,T)$ gravity models that have ΛCDM as a submodel. *arXiv* **2021**, arXiv:2104.14065.
170. Nájera, A.; Fajardo, A. Fitting $f(Q,T)$ gravity models with a ΛCDM limit using $H(z)$ and Pantheon data. *Phys. Dark Univ.* **2021**, *34*, 100889. [CrossRef]
171. Nájera, A.; Fajardo, A. Cosmological Perturbation Theory in $f(Q,T)$ Gravity. *arXiv* **2021**, arXiv:gr-qc/2111.04205.
172. Linares Cedeño, F.X.; Nucamendi, U. Revisiting cosmological diffusion models in Unimodular Gravity and the H_0 tension. *Phys. Dark Univ.* **2021**, *32*, 100807. [CrossRef]
173. Fung, L.W.; Li, L.; Liu, T.; Luu, H.N.; Qiu, Y.C.; Tye, S.H.H. The Hubble Constant in the Axi-Higgs Universe. *arXiv* **2021**, arXiv:astro-ph.CO/2105.01631.
174. Shokri, M.; Sadeghi, J.; Setare, M.R.; Capozziello, S. Nonminimal coupling inflation with constant slow roll. *arXiv* **2021**, arXiv:2104.00596.
175. Castellano, A.; Font, A.; Herraez, A.; Ibáñez, L.E. A Gravitino Distance Conjecture. *arXiv* **2021**, arXiv:2104.10181.
176. Tomita, K. Cosmological renormalization of model parameters in second-order perturbation theory. *Prog. Theor. Exp. Phys.* **2017**, *2017*, 053E01. [CrossRef]
177. Tomita, K. Cosmological models with the energy density of random fluctuations and the Hubble-constant problem. *Prog. Theor. Exp. Phys.* **2017**, *2017*, 083E04. [CrossRef]
178. Tomita, K. Super-horizon second-order perturbations for cosmological random fluctuations and the Hubble-constant problem. *Prog. Theor. Exp. Phys.* **2018**, *2018*, 021E01. [CrossRef]
179. Tomita, K. Hubble constants and luminosity distance in the renormalized cosmological models due to general-relativistic second-order perturbations. *arXiv* **2019**, arXiv:1906.09519.
180. Tomita, K. Cosmological renormalization of model parameters in the second-order perturbation theory. *Prog. Theor. Exp. Phys.* **2020**, *2020*, 019202. [CrossRef]
181. Belgacem, E.; Prokopec, T. Quantum origin of dark energy and the Hubble tension. *arXiv* **2021**, arXiv:astro-ph.CO/2111.04803.

182. Bernardo, R.C.; Grandón, D.; Said, J.L.; Cárdenas, V.H. Parametric and nonparametric methods hint dark energy evolution. *arXiv* **2022**, arXiv:astro-ph.CO/2111.08289.
183. Ambjorn, J.; Watabiki, Y. Easing the Hubble constant tension? *arXiv* **2021**, arXiv:gr-qc/2111.05087.
184. Di Bari, P.; Marfatia, D.; Zhou, Y.L. Gravitational waves from first-order phase transitions in Majoron models of neutrino mass. *arXiv* **2021**, arXiv:2106.00025.
185. Duan, W.F.; Li, S.P.; Li, X.Q.; Yang, Y.D. Linking $R_{K^{(*)}}$ anomalies to H_0 tension via Dirac neutrino. *arXiv* **2021**, arXiv:hep-ph/2111.05178.
186. Ghosh, D. Explaining the R_K and R_{K^*} anomalies. *Eur. Phys. J. C* **2017**, *77*. [CrossRef]
187. Burgess, C.P.; Dineen, D.; Quevedo, F. Yoga Dark Energy: Natural Relaxation and Other Dark Implications of a Supersymmetric Gravity Sector. *arXiv* **2021**, arXiv:hep-th/2111.07286.
188. Jiang, J.Q.; Piao, Y.S. Testing AdS early dark energy with Planck, SPTpol and LSS data. *arXiv* **2021**, arXiv:astro-ph.CO/2107.07128.
189. Karwal, T.; Raveri, M.; Jain, B.; Khoury, J.; Trodden, M. Chameleon Early Dark Energy and the Hubble Tension. *arXiv* **2021**, arXiv:astro-ph.CO/2106.13290.
190. Nojiri, S.; Odintsov, S.D.; Sáez-Chillón Gómez, D.; Sharov, G.S. Modeling and testing the equation of state for (Early) dark energy. *Phys. Dark Univ.* **2021**, *32*, 100837. [CrossRef]
191. Tian, S.X.; Zhu, Z.H. Early dark energy in k-essence. *Phys. Rev. D* **2021**, *103*, 043518. [CrossRef]
192. Linares Cedeño, F.X.; Roy, N.; Ureña-López, L.A. Tracker phantom field and a cosmological constant: Dynamics of a composite dark energy model. *arXiv* **2021**, arXiv:2105.07103.
193. Hernández-Almada, A.; Leon, G.; Magaña, J.; García-Aspeitia, M.A.; Motta, V.; Saridakis, E.N.; Yesmakhanova, K. Kaniadakis holographic dark energy: Observational constraints and global dynamics. *arXiv* **2021**, arXiv:astro-ph.CO/2111.00558.
194. Hernández-Almada, A.; García-Aspeitia, M.A.; Rodríguez-Meza, M.A.; Motta, V. A hybrid model of viscous and Chaplygin gas to tackle the Universe acceleration. *Eur. Phys. J. C* **2021**, *81*, 295. [CrossRef]
195. Abchouyeh, M.A.; van Putten, M.H.P.M. Late-time Universe, H_0-tension, and unparticles. *Phys. Rev. D* **2021**, *104*, 083511. [CrossRef]
196. Wang, D. Dark energy constraints in light of Pantheon SNe Ia, BAO, cosmic chronometers and CMB polarization and lensing data. *Phys. Rev. D* **2018**, *97*, 123507. [CrossRef]
197. Ye, G.; Hu, B.; Piao, Y.S. Implication of the Hubble tension for the primordial Universe in light of recent cosmological data. *Phys. Rev. D* **2021**, *104*, 063510. [CrossRef]
198. Nguyen, H. Analyzing Pantheon SNeIa data in the context of Barrow's variable speed of light. *arXiv* **2020**, arXiv:2010.10292.
199. Barrow, J.D. Cosmologies with varying light speed. *Phys. Rev. D* **1999**, *59*, 043515. [CrossRef]
200. Artymowski, M.; Ben-Dayan, I.; Kumar, U. More on Emergent Dark Energy from Unparticles. *arXiv* **2021**, arXiv:astro-ph.CO/2111.09946.
201. Yang, W.; Di Valentino, E.; Pan, S.; Shafieloo, A.; Li, X. Generalized Emergent Dark Energy Model and the Hubble Constant Tension. *arXiv* **2021**, arXiv:2103.03815.
202. Adil, S.A.; Gangopadhyay, M.R.; Sami, M.; Sharma, M.K. Late time acceleration due to generic modification of gravity and Hubble tension. *arXiv* **2021**, arXiv:2106.03093.
203. Vagnozzi, S. Consistency tests of ΛCDM from the early ISW effect: Implications for early-time new physics and the Hubble tension. *arXiv* **2021**, arXiv:2105.10425..
204. Alestas, G.; Kazantzidis, L.; Perivolaropoulos, L. H_0 tension, phantom dark energy, and cosmological parameter degeneracies. *Phys. Rev. D* **2020**, *101*, 123516. [CrossRef]
205. Kazantzidis, L.; Koo, H.; Nesseris, S.; Perivolaropoulos, L.; Shafieloo, A. Hints for possible low redshift oscillation around the best-fitting ΛCDM model in the expansion history of the Universe. *Mon. Not. R. Astron. Soc.* **2021**, *501*, 3421–3426. [CrossRef]
206. Martín, M.S.; Rubio, C. Hubble tension and matter inhomogeneities: A theoretical perspective. *arXiv* **2021**, arXiv:astro-ph.CO/2107.14377.
207. Buchert, T. On Average Properties of Inhomogeneous Fluids in General Relativity: Dust Cosmologies. *Gen. Relativ. Gravit.* **2000**, *32*, 105–125. [CrossRef]
208. Gasperini, M.; Marozzi, G.; Veneziano, G. Gauge invariant averages for the cosmological backreaction. *J. Cosmol. Astropart. Phys.* **2009**, *2009*, 011. [CrossRef]
209. Gasperini, M.; Marozzi, G.; Nugier, F.; Veneziano, G. Light-cone averaging in cosmology: Formalism and applications. *J. Cosmol. Astropart. Phys.* **2011**, *2011*, 008. [CrossRef]
210. Fanizza, G.; Gasperini, M.; Marozzi, G.; Veneziano, G. Generalized covariant prescriptions for averaging cosmological observables. *J. Cosmol. Astropart. Phys.* **2020**, *2020*, 017. [CrossRef]
211. Ben-Dayan, I.; Gasperini, M.; Marozzi, G.; Nugier, F.; Veneziano, G. Average and dispersion of the luminosity-redshift relation in the concordance model. *J. Cosmol. Astropart. Phys.* **2013**, *2013*, 002. [CrossRef]
212. Fleury, P.; Clarkson, C.; Maartens, R. How does the cosmic large-scale structure bias the Hubble diagram? *J. Cosmol. Astropart. Phys.* **2017**, *2017*, 062. [CrossRef]
213. Adamek, J.; Clarkson, C.; Coates, L.; Durrer, R.; Kunz, M. Bias and scatter in the Hubble diagram from cosmological large-scale structure. *Phys. Rev. D* **2019**, *100*, 021301. [CrossRef]

214. Amendola, L.; Appleby, S.; Avgoustidis, A.; Bacon, D.; Baker, T.; Baldi, M.; Bartolo, N.; Blanchard, A.; Bonvin, C.; Borgani, S.; et al. Cosmology and fundamental physics with the Euclid satellite. *Living Rev. Relativ.* **2018**, *21*, 2. [CrossRef] [PubMed]
215. Fanizza, G. Precision Cosmology and Hubble tension in the era of LSS survey. *arXiv* **2021**, arXiv:astro-ph.CO/2110.15272.
216. Andreoni, I.; Margutti, R.; Salafia, O.S.; Parazin, B.; Villar, V.A.; Coughlin, M.W.; Yoachim, P.; Mortensen, K.; Brethauer, D.; Smartt, S.J.; et al. Target of Opportunity Observations of Gravitational Wave Events with Vera C. Rubin Observatory. *arXiv* **2021**, arXiv:astro-ph.HE/2111.01945.
217. Ben-Dayan, I.; Durrer, R.; Marozzi, G.; Schwarz, D.J. Value of H_0 in the Inhomogeneous Universe. *Phys. Rev. Lett.* **2014**, *112*, 221301. [CrossRef]
218. Fanizza, G.; Fiorini, B.; Marozzi, G. Cosmic variance of H_0 in light of forthcoming high-redshift surveys. *Phys. Rev. D* **2021**, *104*, 083506. [CrossRef]
219. Haslbauer, M.; Banik, I.; Kroupa, P. The KBC void and Hubble tension contradict ΛCDM on a Gpc scale—Milgromian dynamics as a possible solution. *Mon. Not. R. Astron. Soc.* **2020**, *499*, 2845–2883. [CrossRef]
220. Perivolaropoulos, L. Large Scale Cosmological Anomalies and Inhomogeneous Dark Energy. *Galaxies* **2014**, *2*, 22–61.
221. Castello, S.; Högås, M.; Mörtsell, E. A Cosmological Underdensity Does Not Solve the Hubble Tension. *arXiv* **2021**, arXiv:astro-ph.CO/2110.04226.
222. Banik, I.; Zhao, H. From galactic bars to the Hubble tension—Weighing up the astrophysical evidence for Milgromian gravity. *arXiv* **2021**, arXiv:astro-ph.CO/2110.06936.
223. Alestas, G.; Perivolaropoulos, L. Late-time approaches to the Hubble tension deforming H(z), worsen the growth tension. *Mon. Not. R. Astron. Soc.* **2021**, *504*, 3956–3962. [CrossRef]
224. Normann, B.D.; Brevik, I.H. Can the Hubble tension be resolved by bulk viscosity? *Mod. Phys. Lett. A* **2021**, *36*, 2150198. [CrossRef]
225. Bernal, J.L.; Verde, L.; Jimenez, R.; Kamionkowski, M.; Valcin, D.; Wandelt, B.D. Trouble beyond H_0 and the new cosmic triangles. *Phys. Rev. D* **2021**, *103*, 103533. [CrossRef]
226. Thiele, L.; Guan, Y.; Hill, J.C.; Kosowsky, A.; Spergel, D.N. Can small-scale baryon inhomogeneities resolve the Hubble tension? An investigation with ACT DR4. *arXiv* **2021**, arXiv:2105.03003.
227. Grande, J.; Perivolaropoulos, L. Generalized Lemaître-Tolman-Bondi model with inhomogeneous isotropic dark energy: Observational constraints. *Phys. Rev. D* **2011**, *84*, 023514. [CrossRef]
228. Dinda, B.R. Cosmic expansion parametrization: Implication for curvature and H_0 tension. *arXiv* **2021**, arXiv:2106.02963.
229. Marra, V.; Perivolaropoulos, L. Rapid transition of G_{eff} at $z_t \approx 0.01$ as a possible solution of the Hubble and growth tensions. *Phys. Rev. D* **2021**, *104*. [CrossRef]
230. Krishnan, C.; Colgáin, E.Ó.; Sen, A.A.; Sheikh-Jabbari, M.M.; Yang, T. Is there an early Universe solution to Hubble tension? *Phys. Rev. D* **2020**, *102*, 103525. [CrossRef]
231. Krishnan, C.; Colgain, E.O.; Sheikh-Jabbari, M.M.; Yang, T. Running Hubble tension and a H_0 diagnostic. *Phys. Rev. D* **2021**, *103*, 103509. [CrossRef]
232. Krishnan, C.; Mohayaee, R.; Colgáin, E.Ó.; Sheikh-Jabbari, M.M.; Yin, L. Does Hubble Tension Signal a Breakdown in FLRW Cosmology? *arXiv* **2021**, arXiv:2105.09790.
233. Robertson, H.P. Kinematics and World-Structure. *Astrophys. J.* **1935**, *82*, 284. [CrossRef]
234. Gerardi, F.; Feeney, S.M.; Alsing, J. Unbiased likelihood-free inference of the Hubble constant from light standard sirens. *arXiv* **2021**, arXiv:2104.02728.
235. Escamilla-Rivera, C.; Levi Said, J.; Mifsud, J. Performance of Non-Parametric Reconstruction Techniques in the Late-Time Universe. *arXiv* **2021**, arXiv:2105.14332.
236. Sun, W.; Jiao, K.; Zhang, T.J. Influence of the Bounds of the Hyperparameters on the Reconstruction of Hubble Constant with Gaussian Process. *arXiv* **2021**, arXiv:astro-ph.CO/2105.12618.
237. Renzi, F.; Silvestri, A. A look at the Hubble speed from first principles. *arXiv* **2020**, arXiv:astro-ph.CO/2011.10559.
238. Gurzadyan, V.G.; Stepanian, A. Hubble tension vs. two flows. *Eur. Phys. J. Plus* **2021**, *136*. [CrossRef]
239. Geng, C.Q.; Hsu, Y.T.; Lu, J.R.; Yin, L. A Dark Energy model from Generalized Proca Theory. *Phys. Dark Univ.* **2021**, *32*, 100819. [CrossRef]
240. Reyes, M.; Escamilla-Rivera, C. Improving data-driven model-independent reconstructions and updated constraints on dark energy models from Horndeski cosmology. *arXiv* **2021**, arXiv:2104.04484.
241. Petronikolou, M.; Basilakos, S.; Saridakis, E.N. Alleviating H_0 tension in Horndeski gravity. *arXiv* **2021**, arXiv:gr-qc/2110.01338.
242. Alestas, G.; Antoniou, I.; Perivolaropoulos, L. Hints for a Gravitational Transition in Tully–Fisher Data. *Universe* **2021**, *7*, 366. [CrossRef]
243. Benisty, D.; Staicova, D. A preference for Dynamical Dark Energy? *arXiv* **2021**, arXiv:astro-ph.CO/2107.14129.
244. Ó Colgáin, E.; Sheikh-Jabbari, M.; Yin, L. Can dark energy be dynamical? *Phys. Rev. D* **2021**, *104*. [CrossRef]
245. Chevallier, M.; Polarski, D. Accelerating Universes with Scaling Dark Matter. *Int. J. Mod. Phys. D* **2001**, *10*, 213–223. [CrossRef]
246. Linder, E.V. Exploring the Expansion History of the Universe. *Phys. Rev. Lett.* **2003**, *90*, 091301. [CrossRef]
247. Aloni, D.; Berlin, A.; Joseph, M.; Schmaltz, M.; Weiner, N. A Step in Understanding the Hubble Tension. *arXiv* **2021**, arXiv:astro-ph.CO/2111.00014.

248. Ghosh, S.; Kumar, S.; Tsai, Y. Free-streaming and Coupled Dark Radiation Isocurvature Perturbations: Constraints and Application to the Hubble Tension. *arXiv* **2021**, arXiv:astro-ph.CO/2107.09076.
249. Shrivastava, P.; Khan, A.J.; Goswami, G.K.; Yadav, A.K.; Singh, J.K. The simplest parametrization of equation of state parameter in the scalar field Universe. *arXiv* **2021**, arXiv:astro-ph.CO/2107.05044.
250. Pereira, S.H. An unified cosmological model driven by a scalar field nonminimally coupled to gravity. *arXiv* **2021**, arXiv:astro-ph.CO/2111.00029.
251. Bag, S.; Sahni, V.; Shafieloo, A.; Shtanov, Y. Phantom braneworld and the Hubble tension. *arXiv* **2021**, arXiv:astro-ph.CO/2107.03271.
252. Franchino-Viñas, S.A.; Mosquera, M.E. The cosmological lithium problem, varying constants and the H_0 tension. *arXiv* **2021**, arXiv:astro-ph.CO/2107.02243.
253. Palle, D. Einstein–Cartan cosmology and the high-redshift Universe. *arXiv* **2021**, arXiv:physics.gen-ph/2106.08136.
254. Liu, W.; Anchordoqui, L.A.; Valentino, E.D.; Pan, S.; Wu, Y.; Yang, W. Constraints from High-Precision Measurements of the Cosmic Microwave Background: The Case of Disintegrating Dark Matter with Λ or Dynamical Dark Energy. *arXiv* **2021**, arXiv:astro-ph.CO/2108.04188.
255. Blinov, N.; Krnjaic, G.; Li, S.W. Towards a Realistic Model of Dark Atoms to Resolve the Hubble Tension. *arXiv* **2021**, arXiv:hep-ph/2108.11386.
256. Galli, S.; Pogosian, L.; Jedamzik, K.; Balkenhol, L. Consistency of Planck, ACT and SPT constraints on magnetically assisted recombination and forecasts for future experiments. *arXiv* **2021**, arXiv:astro-ph.CO/2109.03816.
257. Liu, X.H.; Li, Z.H.; Qi, J.Z.; Zhang, X. Galaxy-Scale Test of General Relativity with Strong Gravitational Lensing. *arXiv* **2021**, arXiv:astro-ph.CO/2109.02291.
258. Hou, S.; Fan, X.L.; Zhu, Z.H. Constraining cosmological parameters from strong lensing with DECIGO and B-DECIGO sources. *Mon. Not. R. Astron. Soc.* **2021**, *507*, 761–771. [CrossRef]
259. Sola, J. Running vacuum interacting with dark matter or with running gravitational coupling. Phenomenological implications. *arXiv* **2021**, arXiv:gr-qc/2109.12086.
260. Cuesta, A.J.; Gómez, M.E.; Illana, J.I.; Masip, M. Cosmology of an Axion-Like Majoron. *arXiv* **2021**, arXiv:hep-ph/2109.07336.
261. González-López, M. Neutrino Masses and Hubble Tension via a Majoron in MFV. *arXiv* **2021**, arXiv:hep-ph/2110.15698.
262. Prat, J.; Hogan, C.; Chang, C.; Frieman, J. Vacuum Energy Density Measured from Cosmological Data. *arXiv* **2021**, arXiv:astro-ph.CO/2111.08151.
263. Joseph, A.; Saha, R. Dark energy with oscillatory tracking potential: Observational Constraints and Perturbative effects. *arXiv* **2021**, arXiv:astro-ph.CO/2110.00229.
264. Aghababaei, S.; Moradpour, H.; Vagenas, E.C. Hubble tension bounds the GUP and EUP parameters. *Eur. Phys. J. Plus* **2021**, *136*. [CrossRef]
265. Bansal, S.; Kim, J.H.; Kolda, C.; Low, M.; Tsai, Y. Mirror Twin Higgs Cosmology: Constraints and a Possible Resolution to the H_0 and S_8 Tensions. *arXiv* **2021**, arXiv:hep-ph/2110.04317.
266. Dialektopoulos, K.; Said, J.L.; Mifsud, J.; Sultana, J.; Adami, K.Z. Neural Network Reconstruction of Late-Time Cosmology and Null Tests. *arXiv* **2021**, arXiv:astro-ph.CO/2111.11462.
267. Alestas, G.; Camarena, D.; Valentino, E.D.; Kazantzidis, L.; Marra, V.; Nesseris, S.; Perivolaropoulos, L. Late-transition vs. smooth $H(z)$ deformation models for the resolution of the Hubble crisis. *arXiv* **2021**, arXiv:astro-ph.CO/2110.04336.
268. Parnovsky, S. Possible Modification of the Standard Cosmological Model to Resolve a Tension with Hubble Constant Values. *Ukr. J. Phys.* **2021**, *66*, 739. [CrossRef]
269. Zhang, P.; D'Amico, G.; Senatore, L.; Zhao, C.; Cai, Y. BOSS Correlation Function Analysis from the Effective Field Theory of Large-Scale Structure. *arXiv* **2021**, arXiv:astro-ph.CO/2110.07539.
270. Hansen, S.H. Accelerated expansion induced by dark matter with two charges. *Mon. Not. R. Astron. Soc. Lett.* **2021**, *508*, 22–25. [CrossRef]
271. Gariazzo, S.; Valentino, E.D.; Mena, O.; Nunes, R.C. Robustness of non-standard cosmologies solving the Hubble constant tension. *arXiv* **2021**, arXiv:astro-ph.CO/2111.03152.
272. Ruiz-Zapatero, J.; Stölzner, B.; Joachimi, B.; Asgari, M.; Bilicki, M.; Dvornik, A.; Giblin, B.; Heymans, C.; Hildebrandt, H.; Kannawadi, A.; et al. Geometry versus growth—Internal consistency of the flat model with KiDS-1000. *Astron. Astrophys.* **2021**, *655*, A11. [CrossRef]
273. Cai, R.G.; Guo, Z.K.; Wang, S.J.; Yu, W.W.; Zhou, Y. A No-Go guide for the Hubble tension. *arXiv* **2021**, arXiv:astro-ph.CO/2107.13286.
274. Mehrabi, A.; Vazirnia, M. Non-parametric modeling of the cosmological data, base on the χ^2 distribution. *arXiv* **2021**, arXiv:astro-ph.CO/2107.11539.
275. Parnovsky, S.L. Bias of the Hubble constant value caused by errors in galactic distance indicators. *arXiv* **2021**, arXiv:astro-ph.CO/2109.09645.
276. Baldwin, D.; Schechter, P.L. A Malmquist-like bias in the inferred areas of diamond caustics and the resulting bias in inferred time delays for gravitationally lensed quasars. *arXiv* **2021**, arXiv:astro-ph.CO/2110.06378.

277. Huber, S.; Suyu, S.H.; Ghoshdastidar, D.; Taubenberger, S.; Bonvin, V.; Chan, J.H.H.; Kromer, M.; Noebauer, U.M.; Sim, S.A.; Leal-Taixé, L. HOLISMOKES—VII. Time-delay measurement of strongly lensed SNe Ia using machine learning. *arXiv* **2021**, arXiv:astro-ph.CO/2108.02789.
278. Mercier, C. A New Physics Would Explain What Looks Like an Irreconcilable Tension between the Values of Hubble Constants and Allows H0 to Be Calculated Theoretically Several Ways. *J. Mod. Phys.* **2021**, *12*, 1656–1707. [CrossRef]
279. Ren, X.; Wong, T.H.T.; Cai, Y.F.; Saridakis, E.N. Data-driven reconstruction of the late-time cosmic acceleration with $f(T)$ gravity. *Phys. Dark Univ.* **2021**, *32*, 100812. [CrossRef]
280. Hryczuk, A.; Jodłowski, K. Self-interacting dark matter from late decays and the H0 tension. *Phys. Rev. D* **2020**, *102*. [CrossRef]
281. Di Valentino, E.; Mena, O.; Pan, S.; Visinelli, L.; Yang, W.; Melchiorri, A.; Mota, D.F.; Riess, A.G.; Silk, J. In the realm of the Hubble tension—A review of solutions. *Class. Quantum Gravity* **2021**, *38*, 153001. [CrossRef]
282. Perivolaropoulos, L.; Skara, F. Challenges for ΛCDM: An update. *arXiv* **2021**, arXiv:2105.05208.
283. Saridakis, E.N.; Lazkoz, R.; Salzano, V.; Vargas Moniz, P.; Capozziello, S.; Beltrán Jiménez, J.; De Laurentis, M.; Olmo, G.J.; Akrami, Y.; Bahamonde, S.; et al. Modified Gravity and Cosmology: An Update by the CANTATA Network. *arXiv* **2021**, arXiv:2105.12582.
284. Scolnic, D.M.; Jones, D.O.; Rest, A.; Pan, Y.C.; Chornock, R.; Foley, R.J.; Huber, M.E.; Kessler, R.; Narayan, G.; Riess, A.G.; et al. The Complete Light-curve Sample of Spectroscopically Confirmed SNe Ia from Pan-STARRS1 and Cosmological Constraints from the Combined Pantheon Sample. *Astrophys. J.* **2018**, *859*, 101. [CrossRef]
285. Weinberg, S. *Cosmology*; Oxford University Press: Oxford, UK, 2008.
286. Guy, J.; Sullivan, M.; Conley, A.; Regnault, N.; Astier, P.; Balland, C.; Basa, S.; Carlberg, R.G.; Fouchez, D.; Hardin, D.; et al. The Supernova Legacy Survey 3-year sample: Type Ia supernovae photometric distances and cosmological constraints. *Astron. Astrophys.* **2010**, *523*, A7. [CrossRef]
287. Chotard, N.; Gangler, E.; Aldering, G.; Antilogus, P.; Aragon, C.; Bailey, S.; Baltay, C.; Bongard, S.; Buton, C.; Canto, A.; et al. The reddening law of type Ia supernovae: Separating intrinsic variability from dust using equivalent widths. *Astron. Astrophys.* **2011**, *529*, L4. [CrossRef]
288. Kenworthy, W.D.; Scolnic, D.; Riess, A. The Local Perspective on the Hubble Tension: Local Structure Does Not Impact Measurement of the Hubble Constant. *Astrophys. J.* **2019**, *875*, 145. [CrossRef]
289. Deng, H.K.; Wei, H. Null signal for the cosmic anisotropy in the Pantheon supernovae data. *Eur. Phys. J. C* **2018**, *78*, 755. [CrossRef]
290. Akarsu, Ö.; Kumar, S.; Sharma, S.; Tedesco, L. Constraints on a Bianchi type I spacetime extension of the standard Λ CDM model. *Phys. Rev. D* **2019**, *100*, 023532. [CrossRef]
291. Shafieloo, A.; L'Huillier, B.; Starobinsky, A.A. Falsifying ΛCDM : Model-independent tests of the concordance model with eBOSS DR14Q and Pantheon. *Phys. Rev. D* **2018**, *98*, 083526. [CrossRef]
292. Hossienkhani, H.; Azimi, N.; Zarei, Z. Probing the anisotropy effects on the CPL parametrizations from light-curve SNIa, BAO and OHD datasets. *Int. J. Geom. Methods Mod. Phys.* **2019**, *16*, 1950177. [CrossRef]
293. L'Huillier, B.; Shafieloo, A.; Linder, E.V.; Kim, A.G. Model independent expansion history from supernovae: Cosmology versus systematics. *Mon. Not. R. Astron. Soc.* **2019**, *485*, 2783–2790. [CrossRef]
294. Lusso, E.; Piedipalumbo, E.; Risaliti, G.; Paolillo, M.; Bisogni, S.; Nardini, E.; Amati, L. Tension with the flat ΛCDM model from a high-redshift Hubble diagram of supernovae, quasars, and gamma-ray bursts. *Astron. Astrophys.* **2019**, *628*, L4. [CrossRef]
295. Ma, Y.B.; Cao, S.; Zhang, J.; Qi, J.; Liu, T.; Liu, Y.; Geng, S. Testing Cosmic Opacity with the Combination of Strongly Lensed and Unlensed Supernova Ia. *Astrophys. J.* **2019**, *887*, 163. [CrossRef]
296. Sadri, E. Observational constraints on interacting Tsallis holographic dark energy model. *Eur. Phys. J. C* **2019**, *79*, 762. [CrossRef]
297. Sadri, E.; Khurshudyan, M. An interacting new holographic dark energy model: Observational constraints. *Int. J. Mod. Phys. D* **2019**, *28*, 1950152. [CrossRef]
298. Wagner, J.; Meyer, S. Generalized model-independent characterization of strong gravitational lenses V: Reconstructing the lensing distance ratio by supernovae for a general Friedmann universe. *Mon. Not. R. Astron. Soc.* **2019**, *490*, 1913–1927. [CrossRef]
299. Zhai, Z.; Wang, Y. Robust and model-independent cosmological constraints from distance measurements. *J. Cosmol. Astropart. Phys.* **2019**, *2019*, 005. [CrossRef]
300. Zhao, D.; Zhou, Y.; Chang, Z. Anisotropy of the Universe via the Pantheon supernovae sample revisited. *Mon. Not. R. Astron. Soc.* **2019**, *486*, 5679–5689. [CrossRef]
301. Abdullah, M.H.; Klypin, A.; Wilson, G. Cosmological Constraints on Ω_m and σ_8 from Cluster Abundances Using the GalWCat19 Optical-spectroscopic SDSS Catalog. *Astrophys. J.* **2020**, *901*, 90. [CrossRef]
302. Al Mamon, A.; Saha, S. The logotropic dark fluid: Observational and thermodynamic constraints. *Int. J. Mod. Phys. D* **2020**, *29*, 2050097. [CrossRef]
303. Anagnostopoulos, F.K.; Basilakos, S.; Saridakis, E.N. Observational constraints on Barrow holographic dark energy. *Eur. Phys. J. C* **2020**, *80*, 826. [CrossRef]
304. Brout, D.; Hinton, S.R.; Scolnic, D. Binning is Sinning (Supernova Version): The Impact of Self-calibration in Cosmological Analyses with Type Ia Supernovae. *Astrophys. J. Lett.* **2021**, *912*, L26. [CrossRef]
305. Cai, R.G.; Ding, J.F.; Guo, Z.K.; Wang, S.J.; Yu, W.W. Do the observational data favor a local void? *Phys. Rev. D* **2021**, *103*, 123539. [CrossRef]

306. Chang, Z.; Zhao, D.; Zhou, Y. Constraining the anisotropy of the Universe with the Pantheon supernovae sample. *Chin. Phys. C* **2019**, *43*, 125102. [CrossRef]
307. D'Amico, G.; Senatore, L.; Zhang, P. Limits on wCDM from the EFTofLSS with the PyBird code. *J. Cosmol. Astropart. Phys.* **2021**, *2021*, 006. [CrossRef]
308. Di Valentino, E.; Gariazzo, S.; Mena, O.; Vagnozzi, S. Soundness of dark energy properties. *J. Cosmol. Astropart. Phys.* **2020**, *2020*, 045. [CrossRef]
309. Gao, C.; Chen, Y.; Zheng, J. Investigating the relationship between cosmic curvature and dark energy models with the latest supernova sample. *Res. Astron. Astrophys.* **2020**, *20*, 151. [CrossRef]
310. Garcia-Quintero, C.; Ishak, M.; Ning, O. Current constraints on deviations from General Relativity using binning in redshift and scale. *J. Cosmol. Astropart. Phys.* **2020**, *2020*, 018. [CrossRef]
311. Geng, S.; Cao, S.; Liu, T.; Biesiada, M.; Qi, J.; Liu, Y.; Zhu, Z.H. Gravitational-wave Constraints on the Cosmic Opacity at $z \sim 5$: Forecast from Space Gravitational-wave Antenna DECIGO. *Astrophys. J.* **2020**, *905*, 54. [CrossRef]
312. Ghaffari, S.; Sadri, E.; Ziaie, A.H. Tsallis holographic dark energy in fractal universe. *Mod. Phys. Lett. A* **2020**, *35*, 2050107. [CrossRef]
313. Hu, J.P.; Wang, Y.Y.; Wang, F.Y. Testing cosmic anisotropy with Pantheon sample and quasars at high redshifts. *Astron. Astrophys.* **2020**, *643*, A93. [CrossRef]
314. Huang, Z. Supernova Magnitude Evolution and PAge Approximation. *Astrophys. J. Lett.* **2020**, *892*, L28. [CrossRef]
315. Çamlıbel, A.K.; Semiz, İ.; Feyizoğlu, M.A. Pantheon update on a model-independent analysis of cosmological supernova data. *Class. Quantum Gravity* **2020**, *37*, 235001. [CrossRef]
316. Liao, K.; Shafieloo, A.; Keeley, R.E.; Linder, E.V. Determining Model-independent H_0 and Consistency Tests. *Astrophys. J. Lett.* **2020**, *895*, L29. [CrossRef]
317. Koo, H.; Shafieloo, A.; Keeley, R.E.; L'Huillier, B. Model selection and parameter estimation using the iterative smoothing method. *J. Cosmol. Astropart. Phys.* **2021**, *2021*, 034. [CrossRef]
318. Li, E.K.; Du, M.; Xu, L. General cosmography model with spatial curvature. *Mon. Not. R. Astron. Soc.* **2020**, *491*, 4960–4972. [CrossRef]
319. Luković, V.V.; Haridasu, B.S.; Vittorio, N. Exploring the evidence for a large local void with supernovae Ia data. *Mon. Not. R. Astron. Soc.* **2020**, *491*, 2075–2087. [CrossRef]
320. Luongo, O.; Muccino, M. Kinematic constraints beyond $z \simeq 0$ using calibrated GRB correlations. *Astron. Astrophys.* **2020**, *641*, A174. [CrossRef]
321. Micheletti, S.M.R. Quintessence and tachyon dark energy in interaction with dark matter: Observational constraints and model selection. *Int. J. Mod. Phys. D* **2020**, *29*, 2050057. [CrossRef]
322. Mishra, R.K.; Dua, H. Phase transition of cosmological model with statistical techniques. *Astrophys. Space Sci.* **2020**, *365*, 131. [CrossRef]
323. Odintsov, S.D.; Sáez-Chillón Gómez, D.; Sharov, G.S. Analyzing the H_0 tension in $F(R)$ gravity models. *Nucl. Phys. B* **2021**, *966*, 115377. [CrossRef]
324. Prasad, R.; Singh, M.; Yadav, A.K.; Beesham, A. An exact solution of the observable universe in Bianchi V space-time. *Int. J. Mod. Phys. A* **2021**, *36*, 2150044. [CrossRef]
325. Rezaei, M.; Pour-Ojaghi, S.; Malekjani, M. A Cosmography Approach to Dark Energy Cosmologies: New Constraints Using the Hubble Diagrams of Supernovae, Quasars, and Gamma-Ray Bursts. *Astrophys. J.* **2020**, *900*, 70. [CrossRef]
326. Ringermacher, H.I.; Mead, L.R. Reaffirmation of cosmological oscillations in the scale factor from the Pantheon compilation of 1048 Type Ia supernovae. *Mon. Not. R. Astron. Soc.* **2020**, *494*, 2158–2165. [CrossRef]
327. Tang, L.; Li, X.; Lin, H.N.; Liu, L. Model-independently Calibrating the Luminosity Correlations of Gamma-Ray Bursts Using Deep Learning. *Astrophys. J.* **2021**, *907*, 121. [CrossRef]
328. Wang, K.; Chen, L. Constraints on Newton's constant from cosmological observations. *Eur. Phys. J. C* **2020**, *80*, 570. [CrossRef]
329. Wei, J.J.; Melia, F. Cosmology-independent Estimate of the Hubble Constant and Spatial Curvature using Time-delay Lenses and Quasars. *Astrophys. J.* **2020**, *897*, 127. [CrossRef]
330. Zhang, X.; Huang, Q.G. Measuring H_0 from low-z datasets. *Sci. China Phys. Mech. Astron.* **2020**, *63*, 290402. [CrossRef]
331. Baxter, E.J.; Sherwin, B.D. Determining the Hubble constant without the sound horizon scale: measurements from CMB lensing. *Mon. Not. R. Astron. Soc.* **2021**, *501*, 1823–1835. [CrossRef]
332. Camarena, D.; Marra, V. On the use of the local prior on the absolute magnitude of Type Ia supernovae in cosmological inference. *Mon. Not. R. Astron. Soc.* **2021**, *504*, 5164–5171. [CrossRef]
333. Cao, S.; Ryan, J.; Ratra, B. Using Pantheon and DES supernova, baryon acoustic oscillation, and Hubble parameter data to constrain the Hubble constant, dark energy dynamics, and spatial curvature. *Mon. Not. R. Astron. Soc.* **2021**, *504*, 300–310. [CrossRef]
334. Heisenberg, L.; Bartelmann, M.; Brandenberger, R.; Refregier, A. Model independent analysis of supernova data, dark energy, trans-Planckian censorship and the swampland. *Phys. Lett. B* **2021**, *812*, 135990. [CrossRef]
335. Jesus, J.F.; Valentim, R.; Moraes, P.H.R.S.; Malheiro, M. Kinematic constraints on spatial curvature from supernovae Ia and cosmic chronometers. *Mon. Not. R. Astron. Soc.* **2021**, *500*, 2227–2235. [CrossRef]
336. Lee, S. Constraints on the time variation of the speed of light using Pantheon dataset. *arXiv* **2021**, arXiv:2101.09862.

337. Montiel, A.; Cabrera, J.I.; Hidalgo, J.C. Improving sampling and calibration of gamma-ray bursts as distance indicators. *Mon. Not. R. Astron. Soc.* **2021**, *501*, 3515–3526. [CrossRef]
338. Mukherjee, P.; Banerjee, N. Non-parametric reconstruction of the cosmological jerk parameter. *Eur. Phys. J. C* **2021**, *81*, 36. [CrossRef]
339. Wang, G.J.; Ma, X.J.; Xia, J.Q. Machine learning the cosmic curvature in a model-independent way. *Mon. Not. R. Astron. Soc.* **2021**, *501*, 5714–5722. [CrossRef]
340. Torrado, J.; Lewis, A. Cobaya: Code for Bayesian Analysis of hierarchical physical models. *J. Cosmol. Astropart. Phys.* **2021**, *05*, 057. [CrossRef]
341. Sharov, G.S.; Vasiliev, V.O. How predictions of cosmological models depend on Hubble parameter data sets. *Math. Model. Geom.* **2018**, *6*. [CrossRef]
342. Eisenstein, D.J.; Zehavi, I.; Hogg, D.W.; Scoccimarro, R.; Blanton, M.R.; Nichol, R.C.; Scranton, R.; Seo, H.J.; Tegmark, M.; Zheng, Z.; et al. Detection of the Baryon Acoustic Peak in the Large-Scale Correlation Function of SDSS Luminous Red Galaxies. *Astrophys. J.* **2005**, *633*, 560–574. [CrossRef]
343. Cuceu, A.; Farr, J.; Lemos, P.; Font-Ribera, A. Baryon Acoustic Oscillations and the Hubble constant: Past, present and future. *J. Cosmol. Astropart. Phys.* **2019**, *2019*, 044. [CrossRef]
344. Aubourg, É.; Bailey, S.; Bautista, J.E.; Beutler, F.; Bhardwaj, V.; Bizyaev, D.; Blanton, M.; Blomqvist, M.; Bolton, A.S.; Bovy, J.; et al. Cosmological implications of baryon acoustic oscillation measurements. *Phys. Rev. D* **2015**, *92*, 123516. [CrossRef]
345. Singal, J.; Petrosian, V.; Lawrence, A. On the Radio and Optical Luminosity Evolution of Quasars. *Astrophys. J.* **2011**, *743*, 104. [CrossRef]
346. Singal, J.; Petrosian, V.; Lawrence, A. The Radio and Optical Luminosity Evolution of Quasars. II. The SDSS Sample. *Astrophys. J.* **2013**, *764*, 43. [CrossRef]
347. Petrosian, V.; Kitanidis, E.; Kocevski, D. Cosmological Evolution of Long Gamma-Ray Bursts and the Star Formation Rate. *Astrophys. J.* **2015**, *806*, 44. [CrossRef]
348. Lloyd, N.M.; Petrosian, V. Synchrotron Radiation as the Source of Gamma-Ray Burst Spectra. *Astrophys. J.* **2000**, *543*, 722–732. [CrossRef]
349. Dainotti, M.G.; Petrosian, V.; Bowden, L. Cosmological Evolution of the Formation Rate of Short Gamma-Ray Bursts with and without Extended Emission. *Astrophys. J.* **2021**. *914*, L40. [CrossRef]
350. Atteia, J.L. Gamma-ray bursts: Towards a standard candle luminosity. *Astron. Astrophys.* **1997**, *328*, L21–L24,
351. Atteia, J.L. Choosing a measure of GRB brightness that approaches a standard candle. In *Gamma-Ray Bursts, 4th Hunstville Symposium*; Meegan, C.A., Preece, R.D., Koshut, T.M., Eds.; American Institute of Physics Conference Series; American Institute of Physics: College Park, MD, USA, 1998; Volume 428, pp. 92–96. [CrossRef]
352. Simone, B.D.; Nielson, V.; Rinaldi, E.; Dainotti, M.G. A new perspective on cosmology through Supernovae Ia and Gamma Ray Bursts. *arXiv* **2021**, arXiv:astro-ph.CO/2110.11930.
353. Cao, S.; Dainotti, M.; Ratra, B. Standardizing Platinum Dainotti-correlated gamma-ray bursts, and using them with standardized Amati-correlated gamma-ray bursts to constrain cosmological model parameters. *arXiv* **2022**, arXiv:astro-ph.CO/2201.0524.
354. Cao, S.; Khadka, N.; Ratra, B. Standardizing Dainotti-correlated gamma-ray bursts, and using them with standardized Amati-correlated gamma-ray bursts to constrain cosmological model parameters. *Mon. Not. R. Astron. Soc.* **2022**, *510*, 2928–2947. [CrossRef]
355. Efron, B.; Petrosian, V. A simple test of independence for truncated data with applications to redshift surveys. *Astrophys. J.* **1992**, *399*, 345–352. [CrossRef]
356. D'Agostini, G. A multidimensional unfolding method based on Bayes' theorem. *Nucl. Instrum. Methods Phys. Res. A* **1995**, *362*, 487–498. [CrossRef]
357. Conley, A.; Guy, J.; Sullivan, M.; Regnault, N.; Astier, P.; Balland, C.; Basa, S.; Carlberg, R.G.; Fouchez, D.; Hardin, D.; et al. Supernova constraints and systematic uncertainties from the first three years of the supernova legacy survey. *Astrophys. J. Suppl. Ser.* **2010**, *192*, 1. [CrossRef]
358. Betoule, M.; Kessler, R.; Guy, J.; Mosher, J.; Hardin, D..; Biswas, R.; Astier, P.; El-Hage, P.; Konig, M.; Kuhlmann, S..; et al. Improved cosmological constraints from a joint analysis of the SDSS-II and SNLS supernova samples. *Astron. Astrophys.* **2014**, *568*, A22. [CrossRef]
359. Bargiacchi, G.; Benetti, M.; Capozziello, S.; Lusso, E.; Risaliti, G.; Signorini, M. Quasar cosmology: Dark energy evolution and spatial curvature. *arXiv* **2021**, arXiv:2111.02420.
360. Bargiacchi, G.; Risaliti, G.; Benetti, M.; Capozziello, S.; Lusso, E.; Saccardi, A.; Signorini, M. Cosmography by orthogonalized logarithmic polynomials. *Astron. Astrophys.* **2021**, *649* A65. [CrossRef]
361. Dainotti, M.G.; Bogdan, M.; Narendra, A.; Gibson, S.J.; Miasojedow, B.; Liodakis, I.; Pollo, A.; Nelson, T.; Wozniak, K.; Nguyen, Z.; et al. Predicting the Redshift of γ-Ray-loud AGNs Using Supervised Machine Learning. *Astrophys. J.* **2021**, *920*, 118. [CrossRef]
362. Dainotti, M.; Petrosian, V.; Bogdan, M.; Miasojedow, B.; Nagataki, S.; Hastie, T.; Nuyngen, Z.; Gilda, S.; Hernández, X.; Krol, D. Gamma-ray Bursts as distance indicators through a machine learning approach. *arXiv* **2019**, arXiv:1907.05074.
363. Childress, M.; Aldering, G.; Antilogus, P.; Aragon, C.; Bailey, S.; Baltay, C.; Bongard, S.; Buton, C.; Canto, A.; Cellier-Holzem, F.; et al. Host Galaxies of Type Ia Supernovae from the Nearby Supernova Factory. *Astrophys. J.* **2013**, *770*, 107. [CrossRef]

364. Nicolas, N.; Rigault, M.; Copin, Y.; Graziani, R.; Aldering, G.; Briday, M.; Kim, Y.L.; Nordin, J.; Perlmutter, S.; Smith, M. Redshift evolution of the underlying type Ia supernova stretch distribution. *Astron. Astrophys.* **2021**, *649*, A74. [CrossRef]
365. Brout, D.; Taylor, G.; Scolnic, D.; Wood, C.M.; Rose, B.M.; Vincenzi, M.; Dwomoh, A.; Lidman, C.; Riess, A.; Ali, N.; et al. The Pantheon+ Analysis: SuperCal-Fragilistic Cross Calibration, Retrained SALT2 Light Curve Model, and Calibration Systematic Uncertainty. *arXiv* **2021**, arXiv:astro-ph.CO/2112.03864.
366. Carr, A.; Davis, T.M.; Scolnic, D.; Said, K.; Brout, D.; Peterson, E.R.; Kessler, R. The Pantheon+ Analysis: Improving the Redshifts and Peculiar Velocities of Type Ia Supernovae Used in Cosmological Analyses. *arXiv* **2021**, arXiv:astro-ph.CO/2112.01471.
367. Popovic, B.; Brout, D.; Kessler, R.; Scolnic, D. The Pantheon+ Analysis: Forward-Modeling the Dust and Intrinsic Colour Distributions of Type Ia Supernovae, and Quantifying their Impact on Cosmological Inferences. *arXiv* **2021**, arXiv:astro-ph.CO/2112.04456.
368. Scolnic, D.; Brout, D.; Carr, A.; Riess, A.G.; Davis, T.M.; Dwomoh, A.; Jones, D.O.; Ali, N.; Charvu, P.; Chen, R.; et al. The Pantheon+ Type Ia Supernova Sample: The Full Dataset and Light-Curve Release. *arXiv* **2021**, arXiv:astro-ph.CO/2112.03863.
369. Riess, A.G.; Yuan, W.; Macri, L.M.; Scolnic, D.; Brout, D.; Casertano, S.; Jones, D.O.; Murakami, Y.; Breuval, L.; Brink, T.G.; et al. A Comprehensive Measurement of the Local Value of the Hubble Constant with 1 km/s/Mpc Uncertainty from the Hubble Space Telescope and the SH0ES Team. *arXiv* **2022**, arXiv:astro-ph.CO/2112.04510.
370. Damour, T.; Nordtvedt, K. Tensor-scalar cosmological models and their relaxation toward general relativity. *Phys. Rev. D* **1993**, *48*, 3436–3450. [CrossRef]
371. Damour, T.; Polyakov, A.M. The string dilation and a least coupling principle. *Nucl. Phys. B* **1994**, *423*, 532–558. [CrossRef]
372. Damour, T. Gravitation, experiment and cosmology. Les Houches Summer School on Gravitation and Quantizations, Session 57. *arXiv* **1996**, arXiv:gr-qc/gr-qc/9606079.
373. Boisseau, B.; Esposito-Farèse, G.; Polarski, D.; Starobinsky, A.A. Reconstruction of a Scalar-Tensor Theory of Gravity in an Accelerating Universe. *Phys. Rev. Lett.* **2000**, *85*, 2236–2239. [CrossRef] [PubMed]
374. Esposito-Farèse, G.; Polarski, D. Scalar-tensor gravity in an accelerating universe. *Phys. Rev. D* **2001**, *63*, 063504. [CrossRef]
375. Jordan, P. *Schwerkraft und Weltall: Grundlagen der theoretischen Kosmologie*; Die Wissenschaft: Braunschweig, Germany, 1955.
376. Fierz, M. On the physical interpretation of P. Jordan's extended theory of gravitation. *Helv. Phys. Acta* **1956**, *29*, 128–134.
377. Brans, C.; Dicke, R.H. Mach's Principle and a Relativistic Theory of Gravitation. *Phys. Rev.* **1961**, *124*, 925–935. [CrossRef]
378. Singh, C.P.; Peracaula, J.S. Friedmann cosmology with decaying vacuum density in Brans–Dicke theory. *arXiv* **2021**, arXiv:gr-qc/2110.12411.
379. Catena, R.; Fornengo, N.; Masiero, A.; Pietroni, M.; Rosati, F. Dark matter relic abundance and scalar-tensor dark energy. *Phys. Rev. D* **2004**, *70*, 063519. [CrossRef]
380. Kazantzidis, L.; Perivolaropoulos, L. Hints of a local matter underdensity or modified gravity in the low z Pantheon data. *Phys. Rev. D* **2020**, *102*, 023520. [CrossRef]
381. Capozziello, S.; de Laurentis, M. Extended Theories of Gravity. *Phys. Rep.* **2011**, *509*, 167–321. [CrossRef]
382. Hu, W.; Sawicki, I. Models of f(R) cosmic acceleration that evade solar system tests. *Phys. Rev. D* **2007**, *76*, 064004. [CrossRef]
383. Starobinsky, A.A. Disappearing cosmological constant in f(R) gravity. *Sov. J. Exp. Theor. Phys. Lett.* **2007**, *86*, 157–163. [CrossRef]
384. Amendola, L.; Gannouji, R.; Polarski, D.; Tsujikawa, S. Conditions for the cosmological viability of f(R) dark energy models. *Phys. Rev. D* **2007**, *75*, 083504. [CrossRef]
385. Tsujikawa, S. Observational signatures of f(R) dark energy models that satisfy cosmological and local gravity constraints. *Phys. Rev. D* **2008**, *77*, 023507. [CrossRef]
386. Martinelli, M.; Melchiorri, A.; Amendola, L. Cosmological constraints on the Hu-Sawicki modified gravity scenario. *Phys. Rev. D* **2009**, *79*, 123516. [CrossRef]
387. Burrage, C.; Sakstein, J. Tests of chameleon gravity. *Living Rev. Relativ.* **2018**, *21*, 1. [CrossRef] [PubMed]
388. Liu, T.; Zhang, X.; Zhao, W. Constraining f(R) gravity in solar system, cosmology and binary pulsar systems. *Phys. Lett. B* **2018**, *777*, 286–293. [CrossRef]
389. Linde, A.D. Chaotic inflation. *Phys. Lett. B* **1983**, *129*, 177–181. [CrossRef]
390. Riotto, A. Inflation and the Theory of Cosmological Perturbations. *ICTP Lect. Notes Ser.* **2003**, *14*, 317–413.

galaxies

MDPI

Article

Stable Up-Down Quark Matter Nuggets, Quark Star Crusts, and a New Family of White Dwarfs

Lang Wang [1], Jian Hu [1], Cheng-Jun Xia [1,*], Jian-Feng Xu [2,*], Guang-Xiong Peng [3,4,5,*] and Ren-Xin Xu [6,7,*]

1 School of Information Science and Engineering, NingboTech University, Ningbo 315100, China; wanglang@nit.zju.edu.cn (L.W.); 13655826788@139.com (J.H.)
2 School of Physics & Electrical Engineering, AnYang Normal University, AnYang 455000, China
3 School of Nuclear Science and Technology, University of Chinese Academy of Sciences, Beijing 100049, China
4 Theoretical Physics Center for Science Facilities, Institute of High Energy Physics, P.O. Box 918, Beijing 100049, China
5 Synergetic Innovation Center for Quantum Effects and Application, Hunan Normal University, Changsha 410081, China
6 School of Physics, Peking University, Beijing 100871, China
7 Kavli Institute for Astronomy and Astrophysics, Peking University, Beijing 100871, China
* Correspondence: cjxia@nit.zju.edu.cn (C.-J.X.); jfxu@aynu.edu.cn (J.-F.X.); gxpeng@ucas.ac.cn (G.-X.P.); r.x.xu@pku.edu.cn (R.-X.X.)

Abstract: The possible existence of stable up-down quark matter (*ud*QM) was recently proposed, and it was shown that the properties of *ud*QM stars are consistent with various pulsar observations. In this work we investigate the stability of *ud*QM nuggets and found at certain size those objects are more stable than others if a large symmetry energy and a small surface tension were adopted. In such cases, a crust made of *ud*QM nuggets exists in quark stars. A new family of white dwarfs comprised entirely of *ud*QM nuggets and electrons were also obtained, where the maximum mass approaches to the Chandrasekhar limit.

Keywords: quark matter; up-down quark nuggets; quark star crusts; white dwarfs

Citation: Wang, L.; Hu, J.; Xia, C.-J.; Xu, J.-F.; Peng, G.-X.; Xu, R.-X. Stable Up-Down Quark Matter Nuggets, Quark Star Crusts, and a New Family of White Dwarfs. *Galaxies* **2021**, *9*, 70. https://doi.org/10.3390/galaxies9040070

Academic Editors: Elena Moretti and Francesco Longo

Received: 30 July 2021
Accepted: 23 September 2021
Published: 28 September 2021

1. Introduction

At increasing density of baryonic matter, it is expected that a deconfinement phase transition will take place and form quark matter. The properties of quark matter is of particular interest to us since its absolute stability would permit an explanation of dark matter within the framework of the standard model [1]. In the beginning of the 1970s, it was suggested that strange quark matter (SQM) comprised of u, d, and s quarks may be more stable than nuclear matter [1–3], which can exist in various forms, e.g., strangelets [4–7], nuclearites [8,9], meteorlike compact ultradense objects [10], and strange stars [11–13]. Nevertheless, the absolute stability of SQM was challenged by chiral models due to a too large strange quark mass with dynamical chiral symmetry breaking [14,15]. Then SQM only exists in extreme conditions such as the center of compact stars [16–22] and heavy-ion collisions [23,24]. In recent years, an interesting proposition was raised suggesting that quark matter comprised of only u and d quarks (*ud*QM) may be more stable [25], so that *ud*QM nuggets and *ud*QM stars can exist in the Universe. Due to a much smaller surface tension, the ordinary nuclei would not decay into *ud*QM nuggets [25,26]. In fact, it was shown that in a large parameter space the energy per baryon of *ud*QM nuggets is larger than 930 MeV at $A \lesssim 300$ [25]. The properties of nonstrange quark stars and their astrophysical implications are then examined extensively in recent years, e.g., those in References [27–30]. In particular, the merger of binary quark stars would eject *ud*QM nuggets into space. If those objects become supercritically charged, the e^+e^- pair production would inevitably start and release a large amount of energy. The positron emission of the supercritically charged objects are thus expected to play important roles in the short γ-ray burst during the merger of binary quark stars, the 511 keV continuum emission, and the narrow faint emission lines in X-ray spectra from galaxies and galaxy clusters [26].

Based on various investigations, it was shown that the interface effects of quark matter play key roles in the properties of strangelets, *ud*QM nuggets, compact stars, and the processes of quark-hadron transition [26,31,32]. The energy contribution due to the interface effects is often taken into account with a surface tension σ, while its exact value is still veiled in mystery. Adopting the bag model [33], linear sigma model [34–36], NJL model [37,38], three-flavor Polyakov-quark-meson model [39], Dyson-Schwinger equation approach [40], equivparticle model [41], nucleon-meson model [42], and Fermi gas approximations [43,44], recent estimations indicate a small value with $\sigma \lesssim 30$ MeV/fm^2, while larger values were obtained in previous studies [45–47].

In the framework of the bag model, it was shown that a small strangelet can be destabilized substantially if $\sigma^{1/3} \approx B^{1/4}$ with B being the bag constant [4], while the minimum baryon number for metastable strangelets $A_{\rm min} \propto \sigma^3$ [5,48]. Depending on the values of surface tension, large strangelets and strange stars will face very different fates. On the one hand, if a moderate value for σ is adopted, larger strangelets are more stable than smaller ones and strange stars' surfaces are likely bare [49]. On the other hand, if σ is smaller than a critical value $\sigma_{\rm crit}$, large strangelets will decay via fission [50] and strange stars' surfaces may fragment into crystalline crusts [51]. Adopting linearization for the charge density, it was shown that the critical surface tension can be obtained with [50]

$$\sigma_{\rm crit} = 0.1325 n_Q^2 \lambda_D / \chi_Q, \tag{1}$$

where n_Q is the charge density, $\lambda_D = 1/\sqrt{4\pi\alpha\chi_Q}$ the Debye screening length, and $\chi_Q = \sum_i q_i \frac{\partial n_Q}{\partial \mu_i}$ the electric charge susceptibility of quark matter at zero electric charge chemical potential $\mu_e = 0$. Assuming noninteracting SQM, Equation (1) suggests $\sigma_{\rm crit} \propto m_s^4$ with m_s being the strange quark mass [51,52]. As we increase m_s, the strangeness per baryon f_s for β-stable SQM decreases and eventually reaches $f_s = 0$, where SQM is converted into *ud*QM. We thus expect that the critical surface tension of *ud*QM is much larger than that of SQM, so it is more likely that there exist *ud*QM nuggets at certain size that are more stable than others. Additionally, varying the symmetry energy of quark matter will alter the values of n_Q, λ_D, χ_Q, and consequently $\sigma_{\rm crit}$ according to Equation (1). Since it was shown that the symmetry energy of quark matter plays an important role on the structures of quark stars [53–57], in this work we investigate its impact on the properties of small objects such as *ud*QM nuggets.

The purpose of our current study is thus twofold, i.e., investigate the properties of *ud*QM nuggets with various symmetry energies and discuss their implications on *ud*QM stars' structures. The paper is organized as follows. In Section 2, we discuss briefly the equivparticle model and present the corresponding Lagrangian density. To investigate the impact of symmetry energy, an isospin dependent term is added to the quark mass scaling. Then the properties of *ud*QM nuggets are investigated adopting the method discussed in our previous publications [58–61], where the stability window for *ud*QM is obtained according to the binding energy of the heaviest β-stable nucleus ^{266}Hs. The properties of *ud*QM stars with and without crusts are then examined in Section 4 according to the stability of *ud*QM nuggets. We draw our conclusion in Section 5.

2. Equivparticle Model

As an example, in this work we adopt the equivparticle model to investigate the properties of quark matter and their nuggets. The strong interactions are included with density-dependent quark masses in the equivparticle model, while quarks are considered as quasi-free particles [53,62–71]. It is thus straightforward to write out the Lagrangian density, i.e.,

$$\mathcal{L} = \sum_{i=u,d,s} \bar{\Psi}_i \left[i\gamma^\mu \partial_\mu - m_i(n_{\rm b}) \right] \Psi_i, \tag{2}$$

where Ψ_i represents the Dirac spinor of quark flavor i, $m_i(n_\mathrm{b})$ the equivalent mass with n_b being the baryon number density. The Coulomb interactions can also be included by adding photon field in the Lagrangian density [41].

The strong interaction among quarks are then reproduced by equivalent quark masses, where many different mass scalings were proposed. For example, for density dependent masses $m_i(n_\mathrm{b}) = m_{i0} + m_\mathrm{I}(n_\mathrm{b})$ with $m_{u0} = 2.2$ MeV and $m_{d0} = 4.7$ MeV being the current masses of u and d quarks [72], the inversely linear scaling $m_\mathrm{I} = B/3n_\mathrm{b}$ was obtained by reproducing bag model results in the limit of vanishing densities [73]. Considering the contributions of linear confinement and adopting linearization for the in-medium chiral condensates, an inversely cubic scaling $m_\mathrm{I} = Dn_\mathrm{b}^{-1/3}$ was derived [74], while the one-gluon-exchange interaction was later included with $m_\mathrm{I} = Dn_\mathrm{b}^{-1/3} - Cn_\mathrm{b}^{1/3}$ [66]. Meanwhile, in the limit of large densities, perturbation theory suggests repulsive interactions among quarks and the quark mass scaling becomes $m_\mathrm{I} = Dn_\mathrm{b}^{-1/3} + Cn_\mathrm{b}^{1/3}$ [68]. An isospin dependent term was also introduced to examine the impacts of quark matter symmetry energy, which was given by $m_\mathrm{I} = Dn_\mathrm{b}^{-1/3} - \tau_i \delta D_I n_\mathrm{b}^{\alpha} e^{-\beta n_\mathrm{b}}$ with τ_i being the third component of isospin for quark flavor i and $\delta = 3(n_d - n_u)/(n_d + n_u)$ the isospin asymmetry [53].

In this work we adopt the following quark mass scaling

$$m_\mathrm{I}(n_\mathrm{b}) = Dn_\mathrm{b}^{-1/3} + Cn_\mathrm{b}^{1/3} + C_I \delta^2 n_\mathrm{b}. \tag{3}$$

The first term corresponds to linear confinement with the confinement parameter D connected to the string tension σ_0, the chiral restoration density ρ^*, and the sum of the vacuum chiral condensates $\sum_q \langle \bar{q}q \rangle_0$ [74]. The second term represents the contribution of one-gluon-exchange interaction for $C < 0$ [66] and the leading-order perturbative interaction for $C > 0$ [68], where in both cases C is connected to the strong coupling constant α_s. Finally, we have added the third term in Equation (3) to account for the quark matter symmetry energy, which is increasing with C_I. In fact, in addition to the kinetic contribution, it was shown that the formation of u-d quark Cooper pairs (2SC phase) could effectively enhance the quark matter symmetry energy [54]. The impacts of quark matter symmetry energy on compact star properties were extensively investigated in the past few years, e.g., those in References [53–57]. Note that in contrast to the mass scaling proposed in Reference [53], here m_I is identical for different quark flavor, i.e., neglecting the isovector-scalar channel. Meanwhile, as will be discussed later, the third term leads to the isovector contributions in the vector potentials of quarks.

Adopt mean-field and no-sea approximations, the energy density E of quark matter at zero temperature is obtained with

$$E = \sum_i \int_0^{\nu_i} \frac{g_i p^2}{2\pi^2} \sqrt{p^2 + m_i{}^2} \mathrm{d}p = \sum_i \frac{g_i m_i{}^4}{16\pi^2} f\left(\frac{\nu_i}{m_i}\right), \tag{4}$$

where $f(x) = \left[x(2x^2+1)\sqrt{x^2+1} - \mathrm{arcsh}(x)\right]$, g_i (= 6 for quarks and 2 for leptons) the degeneracy factor for particle type i, ν_i the Fermi momentum and is linked to the number density

$$n_i = \langle \bar{\psi}_i \gamma^0 \psi_i \rangle = \frac{g_i \nu_i^3}{6\pi^2}. \tag{5}$$

Note that the masses of leptons take constant values with $m_e = 0.511$ MeV and $m_\mu = 105.66$ MeV. The chemical potentials μ_i and pressure P can then be obtained according to the basic relations of standard thermodynamics.

For udQM without leptons, to investigate its saturation properties, we then expand the energy per baryon to the second order [26], i.e.,

$$\varepsilon_{2f}(n_\mathrm{b}, f_Z) = \varepsilon_0 + \frac{K_0}{18}\left(\frac{n_\mathrm{b}}{n_0} - 1\right)^2 + \varepsilon_\mathrm{s}(2f_Z - 1)^2, \tag{6}$$

where $n_b = (n_u + n_d)/3$ is the baryon number density, $f_Z = (2n_u - n_d)/(n_u + n_d)$ the charge fraction, ε_0 the minimum energy per baryon at saturation density n_0 and charge fraction $f_Z = 1/2$, K_0 the incompressibility parameter, and ε_s the symmetry energy. The corresponding values obtained with equivparticle model using various combinations of parameters are then indicated in Table 1, where the symmetry energy is found to be increasing with C_I.

Table 1. The adopted parameter sets for the quark mass scaling in Equation (3) and the corresponding properties of *ud*QM.

C	$\sqrt{D}$ MeV	C_I MeV/fm^3	n_0 fm^{-3}	ε_0 MeV	K_0 MeV	ε_s MeV	$\sigma_{\min}$ MeV/fm^2	σ_{crit} MeV/fm^2
		0				16.0	14.0	5.03
−0.5	176	40	0.275	900.9	2571.5	36.5	13.1	14.5
		80				57.5	12.8	24.2
		0				15.0	12.4	4.24
−0.3	167	40	0.241	902.0	2306.4	34.1	11.7	12.1
		80				53.5	11.4	20.1
		0				12.6	12.1	2.77
0.1	149	40	0.172	897.4	1942.0	27.8	11.6	7.58
		80				43.1	11.4	12.5
		0				10.4	7.40	1.77
0.5	135	40	0.120	904.6	1786.3	22.0	7.03	4.65
		80				33.5	6.87	7.65

3. *ud*QM Nuggets

At fixed surface tension values, the energy per baryon of *ud*QM nuggets can be obtained with a liquid-drop type formula, i.e.,

$$\frac{M}{A} = \varepsilon_{2f}(n_0, f_Z) + \frac{1}{5}\sqrt[3]{36\pi n_0 \alpha f_Z^2} A^{2/3} + \sigma\left(\frac{A n_0^2}{36\pi}\right)^{-1/3}, \tag{7}$$

where the bulk (first) term is fixed by Equation (6). The second term represents the Coulomb energy with α being the fine structure constant and the third term corresponds to the surface energy. By minimizing Equation (7) with respect to f_Z, one obtains the charge fraction for β-stable *ud*QM nuggets, i.e.,

$$f_Z = \frac{10\varepsilon_s}{\sqrt[3]{36\pi n_0 \alpha} A^{2/3} + 20\varepsilon_s}. \tag{8}$$

Then we can constrain the parameters by comparing the binding energy of *ud*QM nuggets with finite nuclei, where *ud*QM nuggets should always be unstable at $A \lesssim 300$ and stable as $A \to \infty$. At this moment, the heaviest β-stable nucleus is ^{266}Hs with its energy per baryon being 931.74 MeV [75–77]. We thus require the energy per baryon obtained with Equation (7) at $A = 266$ should always be larger than that of ^{266}Hs, where a lower limit for the surface tension value $\sigma_{\min}$ is found for various combinations of parameters. The corresponding constraints are then indicated in Figure 1 for $C_I = 0$. The red curve indicates the boundary with the minimum energy per baryon $\varepsilon_0 = 922$ MeV, so that *ud*QM can be more stable than nuclear matter at $\delta = 0$ in the lower left region. If we demand *ud*QM star matter to be more stable than neutron star matter, the stable-unstable boundary is expected to shift slightly to the lower left region. Meanwhile, the surface tension value should be larger than $\sigma_{\min}$, which favors the upper right region at a fixed σ. The combination of both constraints indicate an stability window for *ud*QM at a given surface tension value σ.

Finally, it is worth mentioning that the constraints in Figure 1 are insensitive to C_I, where we have only presented the results at $C_I = 0$.

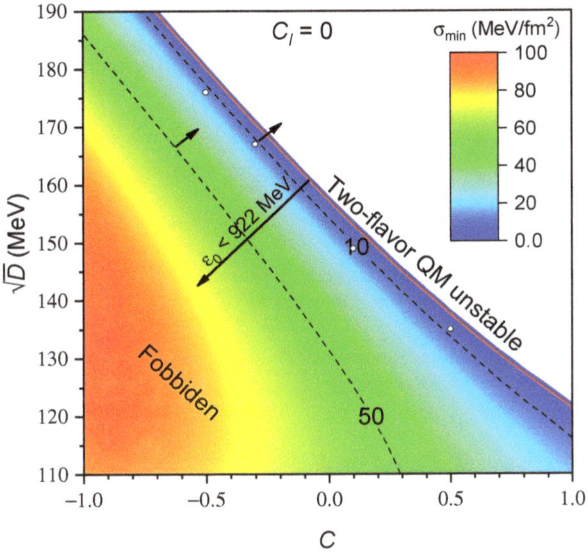

Figure 1. Various constrains for the parameters of the mass scaling in Equation (3) and minimum surface tension value $\sigma_{\rm min}$ for *ud*QM nuggets to be unstable comparing with ^{266}Hs. The open circles correspond to the choices of parameters in Table 1.

The liquid-drop type formula in Equation (7) is valid for small *ud*QM nuggets. If we want to investigate the properties of *ud*QM nuggets at larger baryon numbers, the effects of charge screening and electrons should be considered [78]. In such cases, we adopt Thomas–Fermi approximation as was done in References [26,58–61]. The total energy of the system is obtained with

$$M = \int_0^\infty \left[4\pi r^2 E(r) + \frac{r^2}{2\alpha}\left(\frac{\mathrm{d}\varphi}{\mathrm{d}r}\right)^2\right]\mathrm{d}r + 4\pi R^2\sigma, \tag{9}$$

where the energy density E is given by Equation (4) and the electric potential φ by solving

$$r^2\frac{\mathrm{d}^2\varphi}{\mathrm{d}r^2} + 2r\frac{\mathrm{d}\varphi}{\mathrm{d}r} + 4\pi\alpha r^2\left(\frac{2}{3}n_u - \frac{1}{3}n_d - n_e\right) = 0. \tag{10}$$

The most stable structure of an *ud*QM nugget is fixed by minimizing Equation (9), which follows the constancy of chemical potentials, i.e.,

$$\mu_i(\vec{r}) = \sqrt{\nu_i(\vec{r})^2 + m_i(\vec{r})^2} + V_i(\vec{r}) = \text{constant}. \tag{11}$$

Here the vector potential is given by

$$V_i = \frac{\mathrm{d}m_{\rm I}}{\mathrm{d}n_i}\sum_{i=u,d} n_i^{\rm s} + q_i\varphi, \tag{12}$$

where the scalar density is

$$n_i^{\rm s} = \langle\bar{\psi}_i\psi_i\rangle = \frac{g_i m_i^3}{4\pi^2}g\left(\frac{\nu_i}{m_i}\right) \tag{13}$$

with $g(x) = x\sqrt{x^2+1} - \text{arcsh}(x)$. The quantities n_i^s, n_i, ν_i, m_i and E represent the local properties of udQM and vary with the space coordinates. Note that a density derivative term is added to the vector potential in Equation (12), which is essential in order to maintain the self-consistency of thermodynamics [62,68,79–82]. Meanwhile, since in Equation (3) we have added the third term to account for the symmetry energy of quark matter, in contrast to our previous findings, the density derivative term takes different forms for u and d quarks, i.e.,

$$\frac{\mathrm{d}m_\mathrm{I}}{\mathrm{d}n_u} = \frac{1}{3}\frac{\mathrm{d}m_\mathrm{I}}{\mathrm{d}n_\mathrm{b}} + \frac{\mathrm{d}m_\mathrm{I}}{\mathrm{d}\delta}\frac{3-\delta}{3n_\mathrm{b}}, \tag{14}$$

$$\frac{\mathrm{d}m_\mathrm{I}}{\mathrm{d}n_d} = \frac{1}{3}\frac{\mathrm{d}m_\mathrm{I}}{\mathrm{d}n_\mathrm{b}} - \frac{\mathrm{d}m_\mathrm{I}}{\mathrm{d}\delta}\frac{3+\delta}{3n_\mathrm{b}}. \tag{15}$$

In such cases, even though we have assumed the same equivalent masses for u and d quarks, their vector potentials take different forms, i.e., contributions to the isovector-vector channel.

Beside the constancy of chemical potentials in Equation (11), the dynamic stability of the quark-vacuum interface should also be fulfilled, which determines the radius of the quark core R with

$$P_\mathrm{QM}(R) = \frac{2\sigma}{R}. \tag{16}$$

The obtained energy per baryon for udQM nuggets are then presented in Figure 2, where we have adopted the parameters indicated in Table 1 as well as the smallest possible surface tension value $\sigma = \sigma_\mathrm{min}$. It is found that the energy per baryon for various cases crosses at $A = 266$ and is generally decreasing with A. However, if large symmetry energies with $C_I = 40$ and 80 MeV/fm^3 were considered, large udQM nuggets are destabilized and there may exists udQM nuggets at $A \approx 1000$ that are more stable than others. Particularly, there are even cases where the minimum energy per baryon of udQM star matter is larger than 930 MeV but that of udQM nuggets smaller than 930 MeV, e.g., $C = -0.5$, $\sqrt{D} = 176$ MeV, and $C_I = 40$ MeV/fm^3. Such kind of scenarios occur when the surface tension value is smaller than the critical one σ_crit, which can be roughly estimated with Equation (1) with the corresponding values presented in Table 1. In such cases, the surface of an udQM star will fragment into a lattice of udQM nuggets immersed in a sea of electrons, similar to the cases of strange stars with crusts of strangelets [51,52].

The charge-to-mass ratios of udQM nuggets are indicated in Figure 3, which are decreasing with baryon number due to Coulomb repulsion. Additionally, for the cases with a larger symmetry energy (larger C_I), udQM nuggets are more positively charged, which gives larger Coulomb energy and leads to a barrier of energy per baryon as we increase A for the cases with $\sigma < \sigma_\mathrm{crit}$.

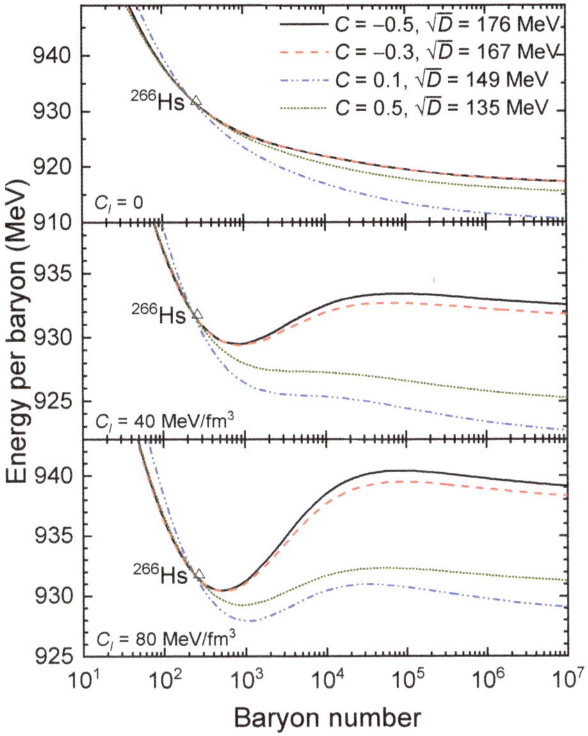

Figure 2. Energy perbaryon for *ud*QM nuggets as functions of the baryon number A. The experimental data for the heaviest β-stable nuclei ^{266}Hs is obtained from the 2016 Atomic Mass Evaluation [75–77].

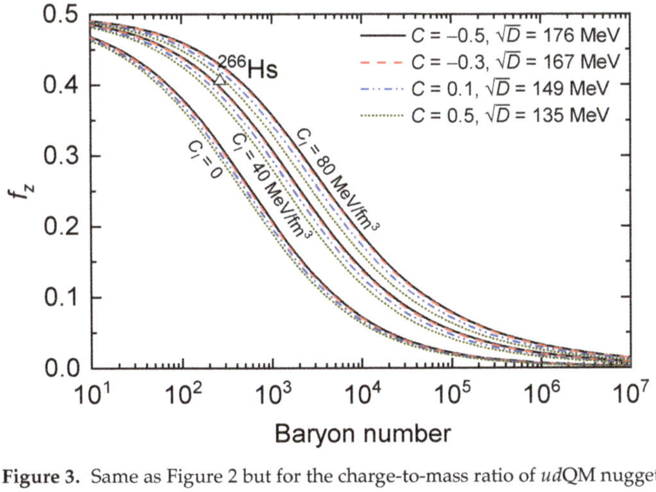

Figure 3. Same as Figure 2 but for the charge-to-mass ratio of *ud*QM nuggets.

4. *ud*QM Star

By fulfilling simultaneously the local charge neutrality condition $\frac{2}{3}n_u - \frac{1}{3}n_d = n_e + n_\mu$ and β-stability condition $\mu_u + \mu_e = \mu_u + \mu_\mu = \mu_d$, the equation of state (EOS) for *ud*QM star matter can be fixed based on the equivparticle model illustrated in Section 2. Then the structure of *ud*QM star can be obtained by solving the TOV equation

$$\frac{\mathrm{d}P}{\mathrm{d}r} = -\frac{GME}{r^2}\frac{(1+P/E)(1+4\pi r^3 P/M)}{1-2GM/r}, \tag{17}$$

$$\frac{\mathrm{d}M}{\mathrm{d}r} = 4\pi E r^2, \tag{18}$$

where the gravity constant is taken as $G = 6.707 \times 10^{-45}$ MeV^{-2}. The tidal deformability can also be estimated with perturbation method, where a special boundary treatment on the surface is required for bare *ud*QM star [83–86].

The energy per baryon of *ud*QM nuggets is decreasing monotonously with baryon number for certain choices of parameters in Figure 2, which excludes a crust of *ud*QM nuggets. The corresponding mass-radius relations of bare *ud*QM stars are then presented in Figure 4. Note that the third term in the quark mass scaling in Equation (3) generally introduces an repulsive interaction among quarks, so that larger masses and radii are obtained for larger symmetry energy (larger C_I). The obtained maximum masses for the parameter sets (C, $\sqrt{D}$ in MeV): (−0.5, 176), (−0.3, 167), (0.1, 149), and (0.5, 135) reach the observational mass ($2.14^{+0.20}_{-0.18} M_\odot$, 95.4% credibility) of PSR J0740+6620 [87], while the radii of the two-solar-mass stars are consistent with that of PSR J0740+6620 ($12.39^{+1.30}_{-0.98}$ km and $2.072^{+0.067}_{-0.066} M_\odot$) only for $C = 0.1$ and $\sqrt{D} = 149$ MeV [88]. Nevertheless, if we examine the tidal deformation of those stars, only the *ud*QM stars obtained with the parameter sets (−0.5, 176) and (−0.3, 167) meet the constraints $70 \leq \Lambda_{1.4} \leq 580$ from the binary neutron star merger event GW170817 [89].

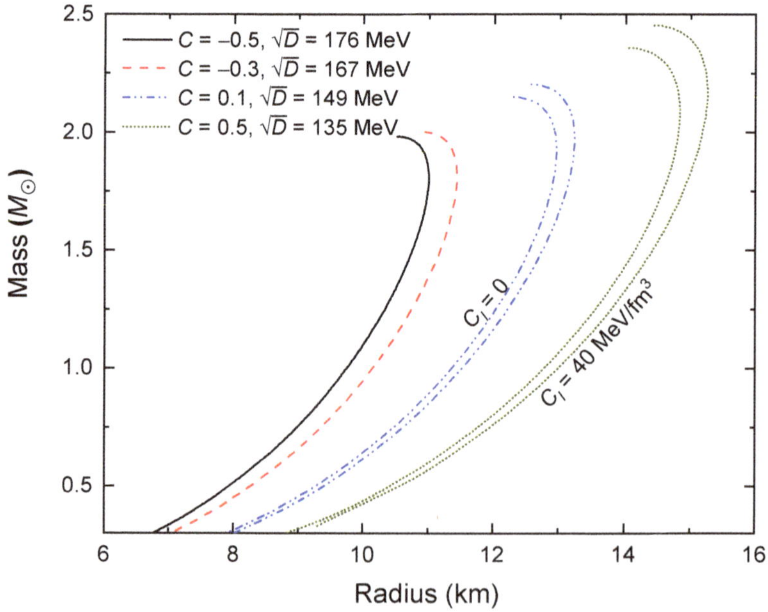

Figure 4. Mass-radius relation of *ud*QM stars without crusts, where $C_I = 0$. For the two cases with larger radii, $C_I = 40$ MeV/fm^3 are adopted.

For the cases where *ud*QM nuggets at certain sizes ($A \approx 1000$) are more stable than others, the surface of *ud*QM star will fragment into a crystalline crust. To obtain the EOSs of the *ud*QM star crust, as an rough estimation, we neglect the effects of charge screening and assume vanishing surface tension, i.e., Gibbs construction. By equating the pressures of the quark phase $[P_{\rm QM}(\mu_{\rm b}, \mu_e) + P_e(\mu_e)]$ and electrons $[P_e(\mu_e)]$, we find that the pressure of pure quark matter should vanish, i.e.,

$$P_{\rm QM}(\mu_{\rm b}, \mu_e) = 0. \tag{19}$$

Then the pressure of the nonuniform phase is exactly the pressure of electrons $P = P_e(\mu_e)$. The volume fraction χ of the quark phase is fixed according to the global charge neutrality condition, i.e.,

$$\chi\left(\frac{2}{3}n_u - \frac{1}{3}n_d\right) = n_e. \tag{20}$$

Combining both the quark phase and electron phase, the energy density is determined by

$$E = \chi E_{\rm QM}(\mu_{\rm b}, \mu_e) + E_e(\mu_e). \tag{21}$$

The obtained EOSs are presented in Figure 5, where the nonuniform phase takes place at $E \lesssim 200$ MeV/fm^3. Note that the nonuniform phase in *ud*QM star is similar to that of the outer crust of neutron stars. Since the surface tension is nonzero and the energy per baryon of the most stable *ud*QM nugget does not reach ε_0, we expect the formation of various geometrical structures in *ud*QM stars. A detailed investigation is thus necessary to obtain the realistic structures and EOSs [50,52], which is intended in our future works. Additionally, we should mention that the possible strong attractive interactions among quark clusters could result in very interesting conclusions [90–92], where the formations of strangeon matter and strangeon stars are expected [93,94].

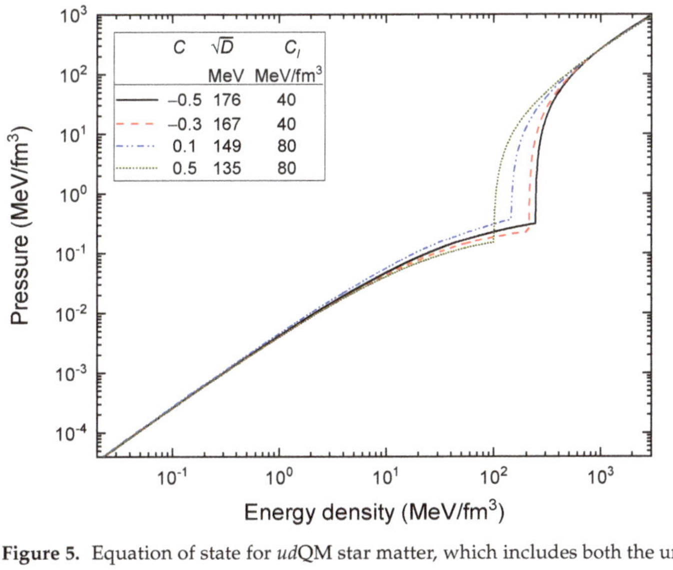

Figure 5. Equation of state for *ud*QM star matter, which includes both the uniform phase at large densities and nonuniform phase at small densities.

Based on the EOSs presented in Figure 5, we solve the TOV Equation (17) and obtain the structures of *ud*QM stars with crusts, where the mass-radius relations are presented in Figure 6. Due to the presence of a crust, the mass-radius relations are similar to those of neutron stars, where the radius is increasing as we decrease the density in the center. This is essentially different from that of hybrid stars with unstable quark matter, where the deconfinement phase transition would reduce the radius and even lead to high-mass twins in case of a strong first-order phase transition [95]. To show this explicitly, the mass-radius relation for hybrid stars is presented in Figure 5, which is obtained with the combination of a density functional PKDD for nuclear matter, *p*QCD with $C_1 = 2.5$ and $\Delta\mu = 770$ MeV for quark matter, and a surface tension value $\sigma = 5\ \mathrm{MeV/fm^2}$ as indicated in Reference [96]. For *ud*QM stars, it is found that the maximum masses for the parameter sets (C, $\sqrt{D}$ in MeV): (−0.3, 167), (0.1, 149), and (0.5, 135) reach the observational mass ($2.14^{+0.20}_{-0.18} M_\odot$, 95.4% credibility) of PSR J0740+6620 [87], while the radii of the two-solar-mass stars coincide with that of PSR J0740+6620 ($12.39^{1.30}_{-0.98}$ km and $2.072^{+0.067}_{-0.066} M_\odot$) only for $C = -0.3$ and $\sqrt{D} = 167$ MeV. Similar as the cases of bare *ud*QM stars, the GW170817 constraint $70 \leq \Lambda_{1.4} \leq 580$ is consistent with the parameter sets (−0.5, 176) and (−0.3, 167) [89]. Combined with all these constraints, only the case of $C = -0.3$ and $\sqrt{D} = 167$ MeV is consistent with various pulsar observations.

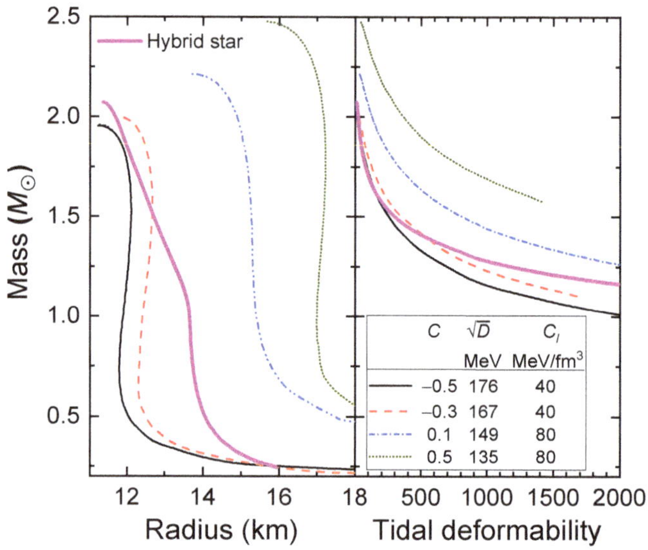

Figure 6. Mass-radius relations (left) and tidal deformabilities (right) of *ud*QM stars with crusts, where the EOSs presented in Figure 5 were adopted. The corresponding values for typical hybrid stars from Reference [96] are presented as well.

In addition to the pulsar-like objects discussed so far, it was realized that for SQM there exists low-mass large-radius strangelet dwarfs if the surface tension is small enough [97]. A strangelet dwarf is comprised of a charge separated phase, which is energetically favorable to form crystalline structures with strangelets and electrons. Adopting the EOSs with $P_{\mathrm{QM}} = 0$ in Figure 5 and solving the TOV Equation (17), we have observed similar objects comprised of only *ud*QM nuggets and electrons as indicated in Figure 7. At sufficiently low temperatures, the *ud*QM nuggets and electrons form crystalline structures inside the star, which is in analogy with white dwarfs comprised of nuclei and electrons. We thus refer to them as *ud*QM dwarfs. Comparing with strangelet dwarfs [97], the masses of *ud*QM dwarfs are much larger and approaching to the Chandrasekhar limit ($\sim 1.4\ M_\odot$), which is mainly due to the large charge-to-mass ratio of *ud*QM nuggets. In such cases, it is likely that some of the observed white dwarfs may in fact be *ud*QM dwarfs. Nevertheless, it is worth mentioning that the results indicated in Figure 7 should be considered as an upper limit, where the emergence of geometrical structures with various surface tension values is expected to play an important role [97]. A detailed investigation is thus necessary and intended for our future study.

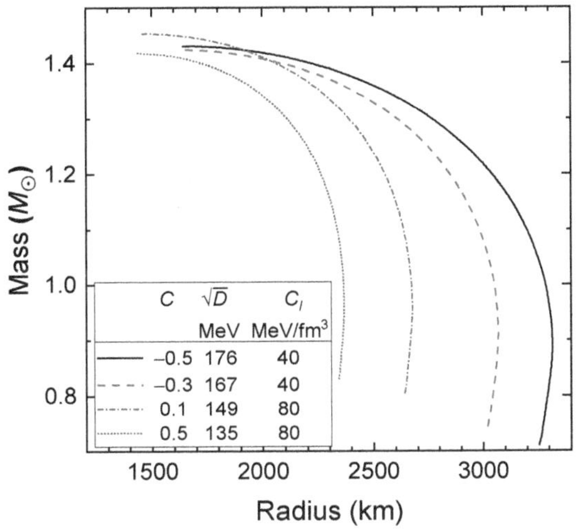

Figure 7. Same as Figure 6 but for the mass-radius relations of *ud*QM dwarfs.

5. Conclusions

In this work we investigate the stability of *ud*QM and the properties of *ud*QM nuggets, *ud*QM stars, and white dwarfs comprised of *ud*QM nuggets. The properties of *ud*QM are obtained based on equivparticle model [62,63,68], where an additional term is introduced to account for the quark matter symmetry energy. By comparing the binding energies of *ud*QM nuggets and the heaviest β-stable nucleus ^{266}Hs, the stability window for *ud*QM is obtained, where there exists a minimum surface tension value below which finite nuclei cannot decay into *ud*QM nuggets. The properties of *ud*QM nuggets are then examined considering the effects of charge screening and electrons [78]. By adopting the minimum possible surface tension values, there are *ud*QM nuggets at certain size that are more stable than others if a large symmetry energy is adopted. In such cases, a crust made of *ud*QM nuggets is expected in *ud*QM stars, where we have obtained the EOSs adopting Gibbs construction. It is found that the mass-radius relation of *ud*QM stars resembles that of traditional neutron stars. Meanwhile, similar to the cases of strange stars [97], a new family of white dwarfs comprised entirely of *ud*QM nuggets and electrons is obtained, where its maximum mass approaches to the Chandrasekhar limit due to the large charge-to-mass ratio of *ud*QM nuggets.

Author Contributions: Conceptualization, C.-J.X., J.-F.X., G.-X.P. and R.-X.X.; Data curation, C.-J.X.; Funding acquisition, C.-J.X.; Investigation, J.-F.X.; Methodology, G.-X.P. and R.-X.X.; Writing—original draft preparation, L.W. and J.H.; Writing—review and editing, J.-F.X., G.-X.P. and R.-X.X. All authors have read and agreed to the published version of the manuscript.

Funding: This research was funded by by National SKA Program of China No. 2020SKA0120300, National Natural Science Foundation of China (Grant Nos. 11705163, 12005005, 11947098, 11875052, and 11673002), Ningbo Natural Science Foundation (Grant Nos. 2019A610066 and 2019A610092), and key research projects of universities in Henan province (Grant No. 20A140003).

Institutional Review Board Statement: Not applicable.

Conflicts of Interest: The authors declare no conflict of interest.

References

1. Witten, E. Cosmic separation of phases. *Phys. Rev.* **1984**, *30*, 272–285. [CrossRef]
2. Bodmer, A.R. Collapsed Nuclei. *Phys. Rev.* **1971**, *4*, 1601–1606. [CrossRef]
3. Terazawa, H. Super-Hypernuclei in the Quark-Shell Model. *J. Phys. Soc. Jpn.* **1989**, *58*, 3555–3563. [CrossRef]
4. Farhi, E.; Jaffe, R.L. Strange matter. *Phys. Rev.* **1984**, *30*, 2379–2390. [CrossRef]

5. Berger, M.S.; Jaffe, R.L. Radioactivity in strange quark matter. *Phys. Rev.* **1987**, *35*, 213–225. [CrossRef]
6. Gilson, E.P.; Jaffe, R.L. Stability of very small strangelets. *Phys. Rev. Lett.* **1993**, *71*, 332–335. [CrossRef]
7. Peng, G.X.; Wen, X.J.; Chen, Y.D. New solutions for the color-flavor locked strangelets. *Phys. Lett.* **2006**, *633*, 314–318. [CrossRef]
8. Rújula, A.D.; Glashow, S.L. Nuclearites–a novel form of cosmic radiation. *Nature* **1984**, *312*, 734. [CrossRef]
9. Lowder, D.M. Experimental searches for strange quark matter in cosmic rays. *Nucl. Phys. (Proc. Suppl.)* **1991**, *24*, 177–183. [CrossRef]
10. Rafelski, J.; Labun, L.; Birrell, J. Compact Ultradense Matter Impactors. *Phys. Rev. Lett.* **2013**, *110*, 111102. [CrossRef]
11. Itoh, N. Hydrostatic Equilibrium of Hypothetical Quark Stars. *Prog. Theor. Phys.* **1970**, *44*, 291–292. [CrossRef]
12. Alcock, C.; Farhi, E.; Olinto, A. Strange stars. *Astrophys. J.* **1986**, *310*, 261–272. [CrossRef]
13. Haensel, P.; Zdunik, J.L.; Schaeffer, R. Strange quark stars. *Astron. Astrophys.* **1986**, *160*, 121–128.
14. Buballa, M.; Oertel, M. Strange quark matter with dynamically generated quark masses. *Phys. Lett.* **1999**, *457*, 261–267. [CrossRef]
15. Klähn, T.; Fischer, T. Vector Interaction Enhanced Bag Model for Astrophysical Applications. *Astrophys. J.* **2015**, *810*, 134. [CrossRef]
16. Weber, F. Strange quark matter and compact stars. *Prog. Part. Nucl. Phys.* **2005**, *54*, 193–288. [CrossRef]
17. Maruyama, T.; Chiba, S.; Schulze, H.J.; Tatsumi, T. Hadron-quark mixed phase in hyperon stars. *Phys. Rev.* **2007**, *76*, 123015. [CrossRef]
18. Peng, G.X.; Li, A.; Lombardo, U. Deconfinement phase transition in hybrid neutron stars from the Brueckner theory with three-body forces and a quark model with chiral mass scaling. *Phys. Rev.* **2008**, *77*, 065807. [CrossRef]
19. Shao, G.Y.; Colonna, M.; Di Toro, M.; Liu, Y.X.; Liu, B. Isoscalar-vector interaction and hybrid quark core in massive neutron stars. *Phys. Rev.* **2013**, *87*, 096012. [CrossRef]
20. Klähn, T.; Łastowiecki, R.; Blaschke, D. Implications of the measurement of pulsars with two solar masses for quark matter in compact stars and heavy-ion collisions: A Nambu-Jona-Lasinio model case study. *Phys. Rev.* **2013**, *88*, 085001. [CrossRef]
21. Zhao, T.; Xu, S.S.; Yan, Y.; Luo, X.L.; Liu, X.J.; Zong, H.S. Studies of two-solar-mass hybrid stars within the framework of Dyson-Schwinger equations. *Phys. Rev.* **2015**, *92*, 054012. [CrossRef]
22. Li, A.; Zuo, W.; Peng, G.X. Massive hybrid stars with a first-order phase transition. *Phys. Rev.* **2015**, *91*, 035803. [CrossRef]
23. Greiner, C.; Koch, P.; Stöcker, H. Separation of strangeness from antistrangeness in the phase transition from quark to hadron matter: Possible formation of strange quark matter in heavy-ion collisions. *Phys. Rev. Lett.* **1987**, *58*, 1825–1828. [CrossRef]
24. Greiner, C.; Stöcker, H. Distillation and survival of strange-quark-matter droplets in ultrarelativistic heavy-ion collisions. *Phys. Rev.* **1991**, *44*, 3517–3529. [CrossRef]
25. Holdom, B.; Ren, J.; Zhang, C. Quark Matter May Not Be Strange. *Phys. Rev. Lett.* **2018**, *120*, 222001. [CrossRef]
26. Xia, C.J.; Xue, S.S.; Xu, R.X.; Zhou, S.G. Supercritically charged objects and electron-positron pair creation. *Phys. Rev.* **2020**, *101*, 103031. [CrossRef]
27. Zhao, T.; Zheng, W.; Wang, F.; Li, C.M.; Yan, Y.; Huang, Y.F.; Zong, H.S. Do current astronomical observations exclude the existence of nonstrange quark stars? *Phys. Rev.* **2019**, *100*, 043018. [CrossRef]
28. Zhang, C. Probing up-down quark matter via gravitational waves. *Phys. Rev.* **2020**, *101*, 043003. [CrossRef]
29. Cao, Z.; Chen, L.W.; Chu, P.C.; Zhou, Y. GW190814: Circumstantial Evidence for Up-Down Quark Star. *arXiv* **2020**, arXiv:2009.00942.
30. Zhang, C.; Mann, R.B. Unified interacting quark matter and its astrophysical implications. *Phys. Rev.* **2021**, *103*, 063018.
31. Lugones, G. From quark drops to quark stars. *Eur. Phys. J. A* **2016**, *52*, 53. [CrossRef]
32. Xia, C.J. Interface effects of strange quark matter. *AIP Conf. Proc.* **2019**, *2127*, 020029.
33. Oertel, M.; Urban, M. Surface effects in color superconducting strange-quark matter. *Phys. Rev.* **2008**, *77*, 074015. [CrossRef]
34. Palhares, L.F.; Fraga, E.S. Droplets in the cold and dense linear sigma model with quarks. *Phys. Rev.* **2010**, *82*, 125018. [CrossRef]
35. Pinto, M.B.; Koch, V.; Randrup, J. Surface tension of quark matter in a geometrical approach. *Phys. Rev.* **2012**, *86*, 025203. [CrossRef]
36. Kroff, D.; Fraga, E.S. Nucleating quark droplets in the core of magnetars. *Phys. Rev.* **2015**, *91*, 025017. [CrossRef]
37. Garcia, A.F.; Pinto, M.B. Surface tension of magnetized quark matter. *Phys. Rev.* **2013**, *88*, 025207. [CrossRef]
38. Ke, W.Y.; Liu, Y.X. Interface tension and interface entropy in the $2+1$ flavor Nambu-Jona-Lasinio model. *Phys. Rev.* **2014**, *89*, 074041.
39. Mintz, B.W.; Stiele, R.; Ramos, R.O.; Schaffner-Bielich, J. Phase diagram and surface tension in the three-flavor Polyakov-quark-meson model. *Phys. Rev.* **2013**, *87*, 036004. [CrossRef]
40. Gao, F.; Liu, Y.X. Interface effect in QCD phase transitions via Dyson-Schwinger equation approach. *Phys. Rev.* **2016**, *94*, 094030. [CrossRef]
41. Xia, C.J.; Peng, G.X.; Sun, T.T.; Guo, W.L.; Lu, D.H.; Jaikumar, P. Interface effects of strange quark matter with density dependent quark masses. *Phys. Rev.* **2018**, *98*, 034031. [CrossRef]
42. Fraga, E.S.; Hippert, M.; Schmitt, A. Surface tension of dense matter at the chiral phase transition. *Phys. Rev.* **2019**, *99*, 014046. [CrossRef]
43. Lugones, G.; Grunfeld, A.G. Surface tension of hot and dense quark matter under strong magnetic fields. *Phys. Rev.* **2019**, *99*, 035804. [CrossRef]

44. Lugones, G.; Grunfeld, A.G. Surface tension of highly magnetized degenerate quark matter. *Phys. Rev.* **2017**, *95*, 015804. [CrossRef]
45. Wen, X.J.; Li, J.Y.; Liang, J.Q.; Peng, G.X. Medium effects on the surface tension of strangelets in the extended quasiparticle model. *Phys. Rev.* **2010**, *82*, 025809. [CrossRef]
46. Lugones, G.; Grunfeld, A.; Ajmi, M.A. Surface tension and curvature energy of quark matter in the Nambu-Jona-Lasinio model. *Phys. Rev.* **2013**, *88*, 045803. [CrossRef]
47. Alford, M.; Rajagopal, K.; Reddy, S.; Wilczek, F. Minimal color-flavor-locked-nuclear interface. *Phys. Rev.* **2001**, *64*, 074017. [CrossRef]
48. Berger, M.S. Surface tension of strange-quark matter in the early Universe. *Phys. Rev.* **1989**, *40*, 2128–2130. [CrossRef] [PubMed]
49. Heiselberg, H. Screening in quark droplets. *Phys. Rev.* **1993**, *48*, 1418–1423. [CrossRef] [PubMed]
50. Alford, M.G.; Rajagopal, K.; Reddy, S.; Steiner, A.W. Stability of strange star crusts and strangelets. *Phys. Rev.* **2006**, *73*, 114016. [CrossRef]
51. Jaikumar, P.; Reddy, S.; Steiner, A.W. Strange Star Surface: A Crust with Nuggets. *Phys. Rev. Lett.* **2006**, *96*, 041101. [CrossRef]
52. Alford, M.; Eby, D. Thickness of the strangelet-crystal crust of a strange star. *Phys. Rev.* **2008**, *78*, 045802. [CrossRef]
53. Chu, P.C.; Chen, L.W. Quark Matter Symmetry Energy and Quark Stars. *Astrophys. J.* **2014**, *780*, 135. [CrossRef]
54. Jeong, K.S.; Lee, S.H. Symmetry energy in cold dense matter. *Nucl. Phys.* **2016**, *945*, 21–41. [CrossRef]
55. Chen, L.W. The symmetry energy in nucleon and quark matter. *Nucl. Phys. Rev.* **2017**, *34*, 20–28.
56. Chu, P.C.; Zhou, Y.; Qi, X.; Li, X.H.; Zhang, Z.; Zhou, Y. Isospin properties in quark matter and quark stars within isospin-dependent quark mass models. *Phys. Rev.* **2019**, *99*, 035802. [CrossRef]
57. Wu, X.; Ohnishi, A.; Shen, H. Quark matter symmetry energy effect on equation of state for neutron stars. *AIP Conf. Proc.* **2019**, *2127*, 020032.
58. Xia, C.J.; Peng, G.X.; Zhao, E.G.; Zhou, S.G. From strangelets to strange stars: A unified description. *Sci. Bull.* **2016**, *61*, 172–177. [CrossRef]
59. Xia, C.J. Strange quark matter: From strangelets to strange stars (in Chinese). *Sci. Sin. Phys. Mech. Astron.* **2016**, *46*, 012021. [CrossRef]
60. Xia, C.J.; Peng, G.X.; Zhao, E.G.; Zhou, S.G. Properties of strange quark matter objects with two types of surface treatments. *Phys. Rev.* **2016**, *93*, 085025. [CrossRef]
61. Xia, C.J.; Zhou, S.G. Stable strange quark matter objects with running masses and coupling constant. *Nucl. Phys. B* **2017**, *916*, 669–687. [CrossRef]
62. Peng, G.X.; Chiang, H.C.; Zou, B.S.; Ning, P.Z.; Luo, S.J. Thermodynamics, strange quark matter, and strange stars. *Phys. Rev.* **2000**, *62*, 025801. [CrossRef]
63. Wen, X.J.; Zhong, X.H.; Peng, G.X.; Shen, P.N.; Ning, P.Z. Thermodynamics with density and temperature dependent particle masses and properties of bulk strange quark matter and strangelets. *Phys. Rev.* **2005**, *72*, 015204. [CrossRef]
64. Wen, X.J.; Peng, G.X.; Chen, Y.D. Charge, strangeness and radius of strangelets. *J. Phys.* **2007**, *34*, 1697. [CrossRef]
65. Xia, C.J.; Chen, S.W.; Peng, G.X. Properties of strangelets in a new quark mass confinement model with one-gluon-exchange interaction. *Sci. China Phys. Mech. Astron.* **2014**, *57*, 1304. [CrossRef]
66. Chen, S.W.; Gao, L.; Peng, G.X. One-gluon-exchange effect on the properties of quark matter and strange stars. *Chin. Phys.* **2012**, *36*, 947. [CrossRef]
67. Chang, Q.; Chen, S.W.; Peng, G.X.; Xu, J.F. Properties of color-flavor locked strange quark matter and strange stars in a new quark mass scaling. *Sci. China Phys. Mech. Astron.* **2013**, *56*, 1730. [CrossRef]
68. Xia, C.J.; Peng, G.X.; Chen, S.W.; Lu, Z.Y.; Xu, J.F. Thermodynamic consistency, quark mass scaling, and properties of strange matter. *Phys. Rev.* **2014**, *89*, 105027. [CrossRef]
69. Hou, J.X.; Peng, G.X.; Xia, C.J.; Xu, J.F. Magnetized strange quark matter in a mass-density-dependent model. *Chin. Phys.* **2015**, *39*, 015101. [CrossRef]
70. Peng, C.; Peng, G.X.; Xia, C.J.; Xu, J.F.; Zhang, S.P. Magnetized strange quark matter in the equivparticle model with both confinement and perturbative interactions. *Nucl. Sci. Tech.* **2016**, *27*, 98. [CrossRef]
71. Chu, P.C.; Chen, L.W. Isovector properties of quark matter and quark stars in an isospin-dependent confining model. *Phys. Rev.* **2017**, *96*, 083019. [CrossRef]
72. Particle Data Group. Review of Particle Physics. *Chin. Phys.* **2016**, *40*, 100001. [CrossRef]
73. Fowler, G.N.; Raha, S.; Weiner, R.M. Confinement and phase transitions. *Z. Phys. C* **1981**, *9*, 271–273. [CrossRef]
74. Peng, G.X.; Chiang, H.C.; Yang, J.J.; Li, L.; Liu, B. Mass formulas and thermodynamic treatment in the mass-density-dependent model of strange quark matter. *Phys. Rev.* **1999**, *61*, 015201. [CrossRef]
75. Audi, G.; Kondev, F.G.; Wang, M.; Huang, W.; Naimi, S. The NUBASE2016 evaluation of nuclear properties. *Chin. Phys.* **2017**, *41*, 030001. [CrossRef]
76. Huang, W.; Audi, G.; Wang, M.; Kondev, F.G.; Naimi, S.; Xu, X. The AME2016 atomic mass evaluation (I). Evaluation of input data; and adjustment procedures. *Chin. Phys.* **2017**, *41*, 030002. [CrossRef]
77. Wang, M.; Audi, G.; Kondev, F.G.; Huang, W.; Naimi, S.; Xu, X. The AME2016 atomic mass evaluation (II). Tables, graphs and references. *Chin. Phys.* **2017**, *41*, 030003. [CrossRef]
78. Xia, C.J. A unified description for strange quark matter objects. *J. Phys: Conf. Ser.* **2017**, *861*, 012022.

79. Brown, G.E.; Rho, M. Scaling effective Lagrangians in a dense medium. *Phys. Rev. Lett.* **1991**, *66*, 2720–2723. [CrossRef]
80. Wang, P. Strange matter in a self-consistent quark mass-density-dependent model. *Phys. Rev.* **2000**, *62*, 015204. [CrossRef]
81. Torres, J.R.; Menezes, D.P. Quark matter equation of state and stellar properties. *Europhys. Lett.* **2013**, *101*, 42003. [CrossRef]
82. Dexheimer, V.; Torres, J.; Menezes, D. Stability windows for proto-quark stars. *Eur. Phys. J. C* **2013**, *73*, 2569. [CrossRef]
83. Damour, T.; Nagar, A. Relativistic tidal properties of neutron stars. *Phys. Rev.* **2009**, *80*, 084035. [CrossRef]
84. Hinderer, T.; Lackey, B.D.; Lang, R.N.; Read, J.S. Tidal deformability of neutron stars with realistic equations of state and their gravitational wave signatures in binary inspiral. *Phys. Rev.* **2010**, *81*, 123016. [CrossRef]
85. Postnikov, S.; Prakash, M.; Lattimer, J.M. Tidal Love numbers of neutron and self-bound quark stars. *Phys. Rev.* **2010**, *82*, 024016. [CrossRef]
86. Zhou, E.P.; Zhou, X.; Li, A. Constraints on interquark interaction parameters with GW170817 in a binary strange star scenario. *Phys. Rev.* **2018**, *97*, 083015. [CrossRef]
87. Cromartie, H.T.; Fonseca, E.; Ransom, S.M.; Demorest, P.B.; Arzoumanian, Z.; Blumer, H.; Brook, P.R.; DeCesar, M.E.; Dolch, T.; Ellis, J.A.; et al. Relativistic Shapiro delay measurements of an extremely massive millisecond pulsar. *Nat. Astron.* **2020**, *4*, 72–76. [CrossRef]
88. Riley, T.E.; Watts, A.L.; Ray, P.S.; Bogdanov, S.; Guillot, S.; Morsink, S.M.; Bilous, A.V.; Arzoumanian, Z.; Choudhury, D.; Deneva, J.S.; et al. A NICER View of the Massive Pulsar PSR J0740+6620 Informed by Radio Timing and XMM-Newton Spectroscopy. *Astrophys. J. Lett.* **2021**, *918*, L27. [CrossRef]
89. LIGO Scientific and Virgo Collaborations. GW170817: Measurements of neutron star radii and equation of state. *Phys. Rev. Lett.* **2018**, *121*, 161101. [CrossRef]
90. Sakai, T.; Mori, J.; Buchmann, A.; Shimizu, K.; Yazaki, K. The interaction between H-dibaryons. *Nucl. Phys.* **1997**, *625*, 192–206. [CrossRef]
91. Glendenning, N.K.; Schaffner-Bielich, J. Neutron star constraints on the H dibaryon. *Phys. Rev.* **1998**, *58*, 1298–1305. [CrossRef]
92. Lai, X.Y.; Gao, C.Y.; Xu, R.X. H-cluster stars. *Mon. Not. R. Astron. Soc.* **2013**, *431*, 3282–3290. [CrossRef]
93. Xu, R.X. Solid Quark Stars? *Astrophys. J.* **2003**, *596*, L59. [CrossRef]
94. Wang, W.; Lu, J.; Tong, H.; Ge, M.; Li, Z.; Men, Y.; Xu, R. The Optical/UV Excess of X-Ray-dim Isolated Neutron Stars. I. Bremsstrahlung Emission from a Strangeon Star Atmosphere. *Astrophys. J.* **2017**, *837*, 81. [CrossRef]
95. Alvarez-Castillo, D.; Blaschke, D.; Typel, S. Mixed phase within the multi-polytrope approach to high-mass twins. *Astron. Nachr.* **2017**, *338*, 1048. [CrossRef]
96. Xia, C.-J.; Maruyama, T.; Yasutake, N.; Tatsumi, T.; Shen, H.; Togashi, H. Systematic study on the quark-hadron mixed phase in compact stars. *Phys. Rev.* **2020**, *102*, 023031. [CrossRef]
97. Alford, M.G.; Han, S.; Reddy, S. Strangelet dwarfs. *J. Phys.* **2012**, *39*, 065201. [CrossRef]

MDPI
St. Alban-Anlage 66
4052 Basel
Switzerland
Tel. +41 61 683 77 34
Fax +41 61 302 89 18
www.mdpi.com

Galaxies Editorial Office
E-mail: galaxies@mdpi.com
www.mdpi.com/journal/galaxies

www.ingramcontent.com/pod-product-compliance
Lightning Source LLC
Chambersburg PA
CBHW061507220926
51850CB00004B/188
9783036568997